The Elements

	Symbol	Atomic Number	Atomic Weight
Actinium	Ac	89	(227)[a]
Aluminum	Al	13	26.98154
Americium	Am	95	(243)[a]
Antimony	Sb	51	121.75
Argon	Ar	18	39.948
Arsenic	As	33	74.9216
Astatine	At	85	(210)[a]
Barium	Ba	56	137.34
Berkelium	Bk	97	(249)[a]
Beryllium	Be	4	9.01218
Bismuth	Bi	83	208.9804
Boron	B	5	10.81
Bromine	Br	35	79.904
Cadmium	Cd	48	112.40
Calcium	Ca	20	40.08
Californium	Cf	98	(251)[a]
Carbon	C	6	12.011
Cerium	Ce	58	140.12
Cesium	Cs	55	132.9054
Chlorine	Cl	17	35.453
Chromium	Cr	24	51.996
Cobalt	Co	27	58.9332
Copper	Cu	29	63.546
Curium	Cm	96	(247)[a]
Dysprosium	Dy	66	162.50
Einsteinium	Es	99	(254)[a]
Erbium	Er	68	167.26
Europium	Eu	63	151.96
Fermium	Fm	100	(253)[a]
Fluorine	F	9	18.99840
Francium	Fr	87	(223)[a]
Gadolinium	Gd	64	157.25
Gallium	Ga	31	69.72
Germanium	Ge	32	72.59
Gold	Au	79	196.9665
Hafnium	Hf	72	178.49
Helium	He	2	4.00260
Holmium	Ho	67	164.9304
Hydrogen	H	1	1.0079
Indium	In	49	114.82
Iodine	I	53	126.9045
Iridium	Ir	77	192.22
Iron	Fe	26	55.847
Krypton	Kr	36	83.80
Lanthanum	La	57	138.9055
Lawrencium	Lr	103	(257)[a]
Lead	Pb	82	207.2
Lithium	Li	3	6.941
Lutetium	Lu	71	174.97
Magnesium	Mg	12	24.305
Manganese	Mn	25	54.9380
Mendelevium	Md	101	(256)[a]
Mercury	Hg	80	200.59
Molybdenum	Mo	42	95.94
Neodymium	Nd	60	144.24
Neon	Ne	10	20.179
Neptunium	Np	93	237.0482[b]
Nickel	Ni	28	58.71
Niobium	Nb	41	92.9064
Nitrogen	N	7	14.0067
Nobelium	No	102	(254)[a]
Osmium	Os	76	190.2
Oxygen	O	8	15.9994
Palladium	Pd	46	106.4
Phosphorus	P	15	30.97376
Platinum	Pt	78	195.09
Plutonium	Pu	94	(242)[a]
Polonium	Po	84	(210)[a]
Potassium	K	19	39.098
Praseodymium	Pr	59	140.9077
Promethium	Pm	61	(145)[a]
Protactinium	Pa	91	231.0359[b]
Radium	Ra	88	226.0254[b]
Radon	Rn	86	(222)[a]
Rhenium	Re	75	186.2
Rhodium	Rh	45	102.9055
Rubidium	Rb	37	85.4678
Ruthenium	Ru	44	101.07
Samarium	Sm	62	150.4
Scandium	Sc	21	44.9559
Selenium	Se	34	78.96
Silicon	Si	14	28.086
Silver	Ag	47	107.868
Sodium	Na	11	22.98977
Strontium	Sr	38	87.62
Sulfur	S	16	32.06
Tantalum	Ta	73	180.9479
Technetium	Tc	43	98.9062[b]
Tellurium	Te	52	127.60
Terbium	Tb	65	158.9254
Thallium	Tl	81	204.37
Thorium	Th	90	232.0381[b]
Thulium	Tm	69	168.9342
Tin	Sn	50	118.69
Titanium	Ti	22	47.90
Tungsten	W	74	183.85
Uranium	U	92	238.029
Vanadium	V	23	50.9414
Xenon	Xe	54	131.30
Ytterbium	Yb	70	173.04
Yttrium	Y	39	88.9059
Zinc	Zn	30	65.38
Zirconium	Zr	40	91.22

[a] Mass number of most stable or best known isotope.

[b] Mass of most commonly available, long-lived isotope.

General Chemistry

Theory and Description

J. Rex Goates   J. Bevan Ott   Eliot A. Butler
Brigham Young University

HBJ
HARCOURT BRACE JOVANOVICH, INC.
New York   San Diego   Chicago   San Francisco   Atlanta
London   Sydney   Toronto

A limited first edition of this book was distributed at Brigham Young University under the title PRINCIPLES OF CHEMISTRY.

Printed in the United States of America

Library of Congress Catalog Card Number: 80-82053

ISBN: 0-15-529535-7

Preface

Our goal in writing this book has been to treat general chemistry as a single, unified subject of exceptional integrity, not as a collection of fascinating but casually related topics. In our years of teaching we have used many general chemistry texts. Each had its strong points but, to our minds, none satisfactorily emphasized the *unity* of the subject. We found it frustrating that chemical equilibrium is commonly treated as if it bears no relationship to thermodynamics, and gas-phase equilibrium as if it were to be solved by one set of tricks while acid-base equilibrium is managed by another. We have seen descriptive chemistry either slighted or treated as if the principles of atomic structure, bonding, and equilibrium were irrelevant. These frustrations, coupled with frequent urgings from our students to "write it out the way you developed it in class," pushed us to this writing.

To achieve our goal of presenting chemistry as a tightly integrated subject, we have carefully constructed the framework of stoichiometry, phase equilibrium, chemical equilibrium, bonding, and descriptive chemistry upon a base of fundamental theory.

Our organization of topics follows a somewhat standard order in that the principles are discussed early in the text and the main body of descriptive chemistry comes later. We wish, however, to explain several important features that we have incorporated into the book.

One is the use of a two-level approach to several subjects, with an elementary discussion presented in the early chapters and then a more penetrating discussion later on. This technique gives the student the advantage of seeing the overall picture before being subjected to all the details and the theoretical interpretation. Examples are the discussions of atomic structure, bonding, and chemical equilibrium that are introduced in Chapters 2 through 4 and then developed more fully in Chapters 9, 10,

13, and 14. This feature also allows one to get quickly into significant laboratory work. Formulas, simple reactions, and periodic properties are introduced in Chapters 1 and 2. Stoichiometry, acid-base definitions, oxidation-reduction, and concentration of solutions are introduced in Chapters 3 and 4.

Since there is the inherent danger of excessive fragmentation in the two-level technique, we have tried to be very selective in the choice of topics discussed in that way. The class testing of preliminary editions of the book with our students has been very helpful in this regard.

You will note that the chapter on thermodynamics comes early in the text. We think that this is important so that thermodynamics can be used as a *tool*, not only in the discussion of equilibrium but in the discussions of solutions, electrochemical cells, and descriptive chemistry as well.

The presentation of the theory part of the course, particularly atomic structure and thermodynamics, is based on a postulate-and-law approach rather than on one that proceeds from historical development. However, once the subject has been presented in what we think is the best pedagogical order, we have occasionally included very brief historical discussions to give perspective. "A Brief Thirty-Year History" at the end of Chapter 13 is an example.

The discussion of properties also starts early. Periodic relations, simple reactions, and nomenclature are introduced in Chapters 1 and 2. An introduction to acid-base and oxidation-reduction reactions is given in Chapter 4. The main body of descriptive chemistry (which comes later in the text) is discussed from three points of view. One approach is in terms of the *type of reaction*—oxidation-reduction (Chapter 11), acid-base (Chapter 16), precipitation (Chapter 17), and complex-ion formation (Chapter 20). The second approach is to correlate the same properties with the position of the elements in the periodic table (Chapters 18, 19, and 21). Finally the large body of descriptive chemistry is tied together in in a review chapter (22). This is the most unusual chapter in the text, and we think that it is one of the most important. The reactions used in this three-pronged approach to descriptive chemistry are carefully *limited* and *selected*, making mastery of this large, important, and often neglected body of information possible. It should also be noted that in our discussion of descriptive chemistry, we refer back continually to the principles discussed in earlier chapters. We have found, too, that the descriptive chapters, with their discussions of detection and separation of elements, provide a superb base for qualitative analysis.

Other special features of the text include the use of *marginal notes*. These consist of interesting facts, historical comments, reminders, and qualifying statements that are helpful for detailed study of the material but that are placed as asides so as not to interrupt the smooth and logical flow of thought. Many *worked examples* are provided, including one for each new principle that is introduced in the text. These are not intended as supplementary material; the text is complete only with these examples.

Included at the end of each chapter is a *summary* of the topics covered and a list of *important terms and concepts* introduced in the chapter. Also included at the end of each chapter are extensive sets of *exercises* and *problems*. The exercises are designed to give the student repetitive practice in mastering the concepts of the chapter. The problems are more thought-provoking and often carry the student into new areas. Of special interest are the "Horatio" problems, in which our fictional hero, Horatio LeChatelier Smith, goes through a series of tongue-in-cheek experiences, encountering some good, solid chemistry problems along the way.

Along with logarithms and tables of data (equilibrium constants, thermodynamic data, standard half-cell potentials, and so on), the *appendixes* include discussions of SI units, reliability of numbers, and chemical nomenclature.

We have attempted to reach a reasonable compromise in the use of units. SI units are used for wavelengths (nanometers) and in most instances for energy (joules). Electron volts are used occasionally in situations requiring the energy per atom, and calories are used in several of the problems only to acquaint the student with the form in which energy appears in all but the most recent literature references. Mass is expressed in grams, and cgs units are used to express velocity and momentum of atomic or molecular systems. We have also retained the angstrom for expressing distance on the atomic level. Instead of using pascals for expressing pressure, we have stayed almost exclusively with the common units of atmosphere and torr.

During the past few years, this book in preliminary form has been used by large numbers of students in our general chemistry courses. The very enthusiastic response we received from these students contributed in large part to our decision to publish the book. Their candor has led to many modifications in approach. They have taught us that while a direct attack on a sound theoretical base may appear more difficult, it proves to be easier and more satisfying to students and instructor alike. It has been especially gratifying to receive this approval of our approach even by students from disciplines outside the physical sciences.

J. Rex Goates
J. Bevan Ott
Eliot A. Butler

Contents

1

Introduction

The subject of chemistry cannot be adequately defined in a few sentences; indeed, this entire textbook might be considered an introductory definition of the subject. Nevertheless, it is helpful to have a tentative answer to the question, "What is chemistry?" To this we offer the time-honored declaration: Chemistry is the study of matter, the changes undergone by matter, and the laws that govern the changes.

In this chapter we look very briefly at the methods used by chemists, at the basis of the laws of science, and at some definitions that are needed in order to begin a careful consideration of the science of chemistry.

1.1 Methods of Science

One hears often of the **scientific method,** which, as it is usually described, is a smooth series of logical steps leading from humble observations to majestic laws. Scientists do have methods—and we will show you some of the methods that they have used—but in chemistry the pathway leading from the views of the early alchemists to the present still-incomplete understanding of modern chemists has been anything but smooth. The history of chemistry is studded with examples of great leaps forward in understanding, but for the most part it is a record of small and difficult steps made by many separate experimenters.

Science, by its nature, attempts to unveil and to simplify, and simplification results from the recognition of the laws that control the behavior of nature. Therefore, we turn our attention first to the *laws* of physical science.

Laws

The laws of chemistry and physics are laws of experience. As the result of observing many times that a certain result follows from a given act, one comes to expect that result. A rock dropped from the hand falls to the ground. If a person tries this simple experiment many times and finds the result to be the same each time, he may say with some conviction that the rock will always fall to the ground. However, the "law" that he expresses is nothing more than a statement of conviction gained through experience. Nothing that he has seen or done rules out the possibility that some rock so dropped may behave differently. Also, even though he and hundreds of millions of other experimenters have repeated the falling-rock experiment and have all observed the same result (thus enhancing the reputation of the law), nothing in all these combined experiences explains *why* the rock falls. It is true that the magnitude of the attraction between earth and rock was expressed by Sir Isaac Newton in his law of universal gravitation, but even that well-established law explains nothing about the reasons for the attraction.

The "law" predicts that the rock will fall; it does not explain why it falls.

It has become common for people to say that "experiments have proved" a scientific law to be true, or that the "law is derived from the facts." But such statements are false. Unfortunately, the aura surrounding the terms "scientific" and "scientific proof" is such that it lends those phrases dignity, but they and the context in which they are often used are meaningless.

Models

Although science does not tell "why" something occurs, it does answer questions about what happens and attempts to provide correlations that can simplify the understanding of many observations. In the discussion of what happens, science relies upon *models* that have been constructed as the result of observations. A good model fits many separate observations and can be used to predict what will happen in a new experiment. The reputation of the model grows as predictions made from it are verified. A poor model is modified or replaced in an effort to obtain one that better fits the observations.

In retrospect there appears to be a smooth, easy path from model to model. You can, however, imagine the pains—both birth and death—during the periods of change.

Consider the various models that have been used by scientists as aids to their understanding of the atom. An early model represented the atom as a hard, incompressible sphere. The model fits many observations. For example, in the solid and liquid states matter is compressed only slightly even under rather large pressures.

The hard-sphere model does not fit certain other observations, however. Experiments made in the nineteenth century demonstrated that chemical change can be caused by passage of an electric current through a solution. Such experiments led investigators to recognize a relationship between electricity and matter. A model was proposed in which the atom was supposed to be a positively charged spherical body in which electrons are embedded like "plums in a pudding."

The plum-pudding model failed to explain experiments reported in 1909 in which positively charged particles (helium nuclei) sent at high velocity toward gold foil passed through the foil in large numbers as if nothing were there. An even more striking observation was that some of the particles were deflected through large angles as they passed through the foil, and a few, indeed, were deflected a full 180° back to their source. A new model was required, and the one developed was a nuclear atom with its positive charge concentrated at the center of the atom (to account for the deflection of alpha particles) and with its negative charge in the form of rapidly moving electrons occupying most of the volume of the atom.

The nuclear-atom model soon required modification to fit observations of the energy emitted by an excited atom. Early forms of the model represented the electrons as orbiting about the nucleus in circular planetlike orbits with the restriction that only certain orbits were available for an electron. A subsequent revision changed the orbits' shape to elliptical so that data were fit more closely.

The revised nuclear-atom model was replaced—again because the model and the new data did not fit—by a mathematical model (a wave equation), which is used today. Earlier models could all be pictured as physical entities. The wave equation, however, serves the same purpose as did the hard-sphere model, but on a more sophisticated level. Much of human knowledge about the atom and its behavior can be expressed in this equation, just as much of human knowledge about the atom a century ago could be expressed by the assumption that the atom was a hard sphere.

The observed trend toward mathematical models should remind the serious student of one's need for competence in mathematics.

The history of the atomic model shows clearly the kinds of changes that occur with a model in science. It tends to be simple and physical initially, but by modification it progresses through increased complexity or refinement toward a mathematical statement. Each successive model is capable of demonstrating that which was shown by its predecessor, and in addition it comes closer to giving complete demonstration of the behavior in nature.

The mere classification or organization of observations or "facts" falls far short of meeting the requirements of a science. In the next chapter we will examine briefly the periodic classification of the elements. It was a momentous day when chemists first were able to arrange the known elements into a simple pattern that placed those having similar properties close to each other. For it was this orderliness that caused people to seek a model that would account for the order and predict additional relationships in the properties and the behavior of elementary substances. However, the brief discussion of the periodic system in Chapter 2 concerns itself only with observations and their tabulations. The *science* of the subject is reserved for Chapter 13, in which the quantum theory is presented. In the intervening chapters, you will encounter a number of observations that will stress the need for a model.

1.2 Terms and Definitions

In starting the study of chemistry we face an immediate need for a language. While it is true that mathematics is the language of the sciences and is used constantly in chemistry, it is also true that people can exchange ideas on the subject only when they agree upon the specific meanings of words and phrases used. Many of the statements that follow are given rather dogmatically, without adequate explanation or derivation, in order that we will have a language available as we proceed. We could have chosen the much slower historical approach in which ideas unfold in the approximate order in which they were first accepted or understood by scientists, but it would be an unusual college freshman who had reached maturity without being qualitatively familiar with most of the terms that are defined below. Therefore, we present some definitions as directly as possible and urge you to make the terms part of your working vocabulary. Explanation is offered subsequently in the text for the statements given here. Keep in mind that these statements are not given as ultimate truths, but, rather, as definitions based on presently accepted models.

The assignment of signs to charges dates back to Benjamin Franklin. Had he made the reverse assignment when he was doing experiments with amber and glass rods, cats' fur, and kites, we would speak of the positively charged electron.

A Few Subatomic Particles

An **electron** is a negatively charged particle of very small mass (9.11×10^{-28} g). The quantity of charge is called the **electronic charge** and is often used as the unit of charge when atomic- or subatomic-sized particles are involved.

The mass of a **proton** is 1836 times that of an electron and has an equal but opposite charge.

A **neutron** has approximately the same mass as a proton but carries no charge.

Elements, Atoms, and Isotopes

The least complex substances are **elements;** these cannot be separated by chemical processes into simpler substances. An **atom** is the ultimate particle of an element; it is the smallest unit of an element that can take part in chemical reactions. An atom consists of a very small nucleus and a much larger electronic cloud. Protons and neutrons are found in the nucleus while electrons make up the negatively charged cloud. The number of protons in the nucleus is referred to as the **atomic number,** and the sum of the numbers of protons and neutrons is called the **mass number.**

All atoms that have the same atomic number are said to be atoms of the same element. Thus, if an atom has six protons in its nucleus, it is a carbon atom regardless of the number of neutrons present. If two carbon atoms differ by having different numbers of neutrons, they are spoken of as **isotopes** of carbon. Isotopes, then, are atoms of the same element that have different mass numbers.

The physical matter of the known universe is made up of a relatively small number of elements. Fewer than 110 elements are known and, of these, nearly a score have not been found in nature. The great complexity of matter that is observed is the result of these relatively few elements intercombining to form compounds.

Molecules and Compounds

A **molecule** consists of two or more atoms bonded together. Each particular kind of molecule has a fixed composition and structure. In a water molecule, for example, the two hydrogen atoms make, within narrow limits, an angle of 105° with respect to each other, and each is a distance of 0.96 Å from the oxygen atom. Two hydrogen atoms and one oxygen atom do not make a water molecule unless they are bonded together in this particular way. If the atoms in a molecule are of at least two kinds, that molecule is the ultimate particle of a **compound.** When two or more identical atoms are bonded together, a molecule, but not a compound, results. Under these circumstances we speak of the element existing as a **diatomic molecule,** as in the case of O_2 or F_2.

Symbols

Symbols are the one or two letters used to designate each element. Thus, H means hydrogen; He, helium; Li, lithium; and so forth. A subscript is sometimes used to indicate the atomic number, as in $_{6}C$. You will notice, however, that this is a redundant statement; if the atomic number is 6, the symbol must be C, since only carbon has atomic number 6. A superscript is used for mass number, as in ^{12}C, which in this example means that the isotope represented has 6 neutrons with its 6 protons, to give mass number 12. Atomic numbers and mass numbers are written as part of the symbol only when interest is centered on nuclear reactions or isotopic differences.

A listing of the known elements and their symbols appears on the inside front cover of this book. All matter found in nature is made of these elements in various combinations.

EXAMPLE 1.1 Complete the following table. Where the symbol of the element does not include both atomic number and mass number, supply those.

SYMBOL OF ISOTOPE	NUMBER OF PROTONS	NUMBER OF NEUTRONS	NUMBER OF ELECTRONS
$^{12}_{6}C$	6	______	______
______	6	8	______
U	______	146	______
Cl	______	20	______

SOLUTION There must be as many electrons as protons in the uncharged atom. Hence there are 6 electrons in the carbon atom. The number of neutrons is given by the difference of the mass number and the atomic number. Therefore the first line should read:

$^{12}_{6}C$ 6 6 6

The element in the second line is clearly carbon, for only carbon has 6 protons. The mass number is 14 (6 protons plus 8 neutrons) and the number of electrons is 6. Thus the second line should read:

$^{14}_{6}C$ 6 8 6

The atomic number of uranium is 92, the mass number for this isotope is 238 (92 + 146), and the number of electrons is 92.

$^{238}_{92}U$ 92 146 92

Similar reasoning leads to the development of the following information for the chlorine isotope:

$^{37}_{17}Cl$ 17 20 17

For compounds, **formulas** are derived from the symbols of the elements they contain. Thus H_2O, H_2SO_4, $KMnO_4$, and CO_2 are formulas for the compounds water, sulfuric acid, potassium permanganate, and carbon dioxide, respectively. The subscript following a symbol tells the number of atoms of that element contained in a molecule of the compound.

Ions

If an atom has a deficiency or an excess of electrons over protons, it is called an **ion.** Thus, a sodium atom with one electron removed is Na^+, the sodium ion. An additional electron on a chlorine atom makes it Cl^-, the chloride ion. Also commonly encountered are ions of greater complexity, such as the sulfate ion, SO_4^{2-}. An uncharged sulfur atom has 16 electrons whereas each oxygen atom has 8. The sulfate ion has two additional electrons, or a total of 50, thus giving a net charge of minus two (−2). A few other very common complex ions are

nitrate ion	NO_3^-
phosphate ion	PO_4^{3-}
carbonate ion	CO_3^{2-}
perchlorate ion	ClO_4^-
ammonium ion	NH_4^+

EXAMPLE 1.2 How many electrons has each of the following: Mg^{2+}, Cl^-, K^+, S^{2-}, Ar, Ca^{2+}, Fe^{3+}?

SOLUTION The neutral magnesium atom has 12 electrons. To become Mg^{2+} this atom had to lose two electrons; therefore, Mg^{2+} has 10 electrons. The neutral chlorine atom has 17 electrons and has to gain one to become Cl^-. Hence, Cl^- has 18 electrons. The other ions have 18, 18, 18, 18, and 23, respectively. (We will see reasons, in the next two chapters, why so many of those species have the same number of electrons.)

EXAMPLE 1.3 How many electrons has each of the following: NO_2^-, HPO_3^{2-}, SO_3^{2-}, ClO_3^-, $Ag(NH_3)_2^+$?

SOLUTION The nitrite ion NO_2^- has 24 electrons. There is one electron (indicated by the negative charge) beyond the electrons of the neutral atoms. The neutral nitrogen atom has 7 electrons and each oxygen atom has 8; hence the total is $7 + 8 + 8 + 1 = 24$ electrons.

The diamine silver ion has 66 electrons. The neutral silver atom has 47 electrons; each nitrogen atom has 7; and each hydrogen atom has 1. The ion has a positive charge, indicating that there is one fewer electron than the neutral atoms would have. Therefore, the total is $47 + 14 + 6 - 1 = 66$ electrons.

Each of the other complex ions has 42 electrons.

Nomenclature of Binary Compounds

The formulas for *binary compounds* (those consisting of just two elements) are written with the symbol for the more metallic element first. A binary compound's name is made up from the names of the two elements it contains, with the second name modified so that "ide" becomes its final syllable. Following are some common examples:

NaCl	sodium chloride
BaO	barium oxide
ZnS	zinc sulfide

If there are multiple atoms of one or both kinds, prefixes may be used to indicate the numbers:

CO_2	carbon dioxide
N_2O_5	dinitrogen pentoxide
Cl_2O_7	dichlorine heptoxide

We will return to the subject of nomenclature in subsequent chapters as we encounter the need. Nomenclature rules are given in Appendix H.

EXAMPLE 1.4 Give names for the following compounds: BaS, AgBr, MgI_2, ZnO, $AlCl_3$.

SOLUTION Barium sulfide, silver bromide, magnesium iodide (magnesium di-iodide would be acceptable also), zinc oxide, aluminum chloride (or aluminum trichloride).

Chemical Equations

A **chemical equation** is written to show in concise form what happens when substances react. For example, the equation

$$CH_4 + 2\ O_2 \rightarrow CO_2 + 2\ H_2O$$

is read, "One molecule of methane reacts with two molecules of oxygen to produce (or yield) one molecule of carbon dioxide and two molecules of water." The equation also can be written to include more information:

$$CH_4(g) + 2\ O_2(g) \rightarrow CO_2(g) + 2\ H_2O(g)$$

This time it explicitly tells us the physical states of the reactants and products. The parenthetical (*g*) is written to show that the reaction occurs between gaseous substances to produce gaseous substances. When substances are in the liquid or solid state, the letters (*l*) or (*s*) are shown in the equation to give this information.

Many reactions can proceed in either direction; indeed, most of the reactions we will be considering have this property. Two arrows or an equal sign are used to denote reactions that can proceed in a reverse direction:†

$$N_2O_4(g) \rightleftarrows 2\ NO_2(g)$$
$$N_2O_4(g) = 2\ NO_2(g)$$

A chemical equation is said to be *balanced* when it meets the two following conditions: There must be as many atoms of each element on the left side as on

† It will be apparent that if a reaction goes almost completely in one direction under a given set of conditions (temperature, pressure, concentration, and so forth), it will not proceed extensively in the opposite direction under identical conditions. Changes in conditions, however, can change the direction of the reaction. In Chapters 7 through 10 you will discover that reversibility is a major concern of chemists.

the right, and there must be the same net charge on both sides. In Chapter 4's discussion of oxidation and reduction equations, we will make use of yet another condition for a balanced equation, but you will find that it is not completely independent from the above statements.

Any number of incorrect but balanced equations can be written if the real world is ignored, so, besides being balanced, a useful equation must be a representation of an actual reaction. One must not fall into the trap of supposing that a balanced equation necessarily represents a reaction that occurs or that could occur. The task of predicting what will happen when substances are combined would be completely overwhelming if one had no alternative to memorization of all the known chemical reactions. But here the scientific models come to our aid and permit us to make predictions based on the periodic relationships of the elements and also on thermodynamic and kinetic relationships. The fact that the models work in so many cases gives substance to the predictions and increases our confidence in their usefulness. The fact that we cannot express all of chemistry and chemical behavior by models reminds us that chemistry is an active field of study—that the aim is not merely to learn and apply what others have discovered, but to aid in improving the models. That is no mean task, but the effort is rewarding.

If you have familiarized yourself with the statements and definitions of the above sections, you are now ready to consider the periodic classification of the elements.

Summary

The scientific method attempts through observation to recognize the laws that control the behavior of nature. A law is never proved but will evolve as observations add to our knowledge of the behavior. Similarly, models in science are constructed from the experience gained from observations. A good model fits many of these observations and thus permits predictions to be made about new experiments. Scientific models also evolve as understanding increases; the evolution is toward more mathematical models.

The study of chemistry must begin with the language of chemistry. Elements, the least complex substances, are made up of atoms, which in turn are composed of electrons, protons, and neutrons. The number of protons in the nucleus is the atomic number; the sum of the numbers of protons and neutrons is the mass number. Isotopes are atoms of the same element that have different mass numbers. An ion is an atom with a deficiency or an excess of electrons over protons. Two or more atoms bond together to form a molecule. If these atoms are of at least two kinds, the molecule is the ultimate particle of a compound. If the atoms are identical, then a diatomic molecule exists.

Each element is represented by a symbol, and these symbols can be combined into the formula of a compound. Chemical equations use these formulas to show concisely what happens when substances react. Though an equation may be balanced, it will only represent an actual reaction if it is written with the aid of certain scientific models.

Important Terms and Concepts

Scientific method
Electron
Electronic charge
Proton
Neutron
Element
Atom
Atomic number
Mass number
Isotope
Molecule
Compound
Diatomic molecule
Symbol
Formula
Ion
Chemical equation

2

Introduction to Atomic Structure and Bonding

The ancient Greeks pictured all matter as composed of just four elements: earth, air, fire, and water. To a university student of the late twentieth century such a notion undoubtedly appears naïve, but keep in mind how unsophisticated your views of nature would be if you had to depend only on your own experience and observations. Today, it is common knowledge that there are elementary substances made up of atoms, and that none of the Greeks' "elements" are really elements. But how is this known? And how is it known that the atom has a very heavy nucleus, and that the mass of the nucleus consists essentially of the mass of the protons plus the mass of the neutrons? This book provides answers to many questions that begin "How do we know . . . ?" You will see that we owe a great debt to many for the knowledge of chemistry that we have. You will also see that chemistry is not merely the study of what is known; it is also the struggle to know more. The learning of known "facts" or even "laws" is not the ultimate goal in the sciences, however important learning these may be. The most important goal is to increase our understanding or, alternatively, to decrease our ignorance.

Students often aptly observe that if one were acquainted with chemistry before one began its study, one could be much more successful in that study. This introduction to atomic structure and bonding is to help provide the acquaintanceship that precedes effective detailed study of the subject.

2.1 Periodic Relationships

The term **periodic** is applied in chemistry to properties of elements that appear repeatedly with increasing atomic number. For example, elements having atomic numbers 3, 11, 19, 37, and 55 are all soft metals that react vigorously with water with the evolution of hydrogen. The similarity of these elements is an example of a periodic property. The recognition more than a century ago of this periodic recurrence of similarities in properties of the elements represented a great step toward the understanding of order in nature.

Experimental Evidence for the Periodic System

The assumption that earth, air, fire, and water were the elements of which matter is composed could not stand the tests of time and experimentation. In Chapter 1, we defined an element as a substance comprising atoms of a single kind; that is, elements are composed of atoms that all have the same atomic number. This definition is acceptable because of present-day techniques for determining mass numbers and atomic numbers. In the nineteenth century, such techniques were not available, and an element was defined as a substance that could not be decomposed into simpler substances. Using this criterion, chemists were able by careful experiments to identify a large fraction of the naturally occurring elements. But before the development of modern spectroscopic methods, there was uncertainty about the nature of a substance; for instance, even if it could not be decomposed into simpler substances, might it not be a very stable compound? Or, again, could it not be a mixture of elements so similar that the methods used for separation were ineffective? These limitations of early technology caused mistakes to be made about the elements. An extreme example of a mixture containing very similar substances is an oxide known as ceria, which was believed by its discoverers to be a pure oxide.

However, when subsequent experiments were performed with more sophisticated methods, ceria was shown to be a mixture of thc oxides of cerium, lanthanum, neodymium, praseodymium, samarium, gadolinium, and europium. Therefore, even though ceria actually contained seven different oxides, they were so similar in their properties that even the best early techniques for separating substances were incapable of detecting the presence of more than one oxide.

In the nineteenth century, several independent investigators observed striking relationships between the atomic weights of elements and their properties. In retrospect, the patterns these chemists observed fit easily into the modern periodic law. At the time, however, their efforts were received in the scientific community with scorn and scoffing, or at best with little notice. It remained for the Russian chemist Dmitri Mendeleev in 1869 to present a periodic classification so strongly supported by experimental evidence and by accurate and detailed predictions that it could not help but be greeted with wide acceptance. Mendeleev left spaces in his table where the properties of known elements did not correspond to the positions where they apparently fell. He then went further and made explicit predictions concerning the yet-undiscovered elements that would fill in those spaces. The entire noble-gas family was unknown to Mendeleev, and his ordering of the elements depended upon atomic weights (or atomic masses[†]) rather than atomic numbers. Otherwise, his century-old table and the modern periodic table (Figure 2.1, and also shown inside the front cover) are surprisingly similar except for the presence in the modern table of new elements, several of which he predicted.

Structure of the Periodic Table

The periodic table consists of horizontal rows, or **periods,** and vertical groups, or **families.** The first period contains only 2 elements, the second and third contain 8 each. The fourth and fifth periods have been enlarged to 18 each, and the sixth period has 32 elements. The seventh period is seen to be incomplete.

Several of the families are given names for easy indentification, particularly when the members of the family are very closely related in their properties. The elements of Group IA are known as **alkali metals,** and those in Group IIA are the **alkaline earth metals,** or the alkaline earths. Group VIIIA elements are called the **noble gases** or, sometimes, the **inert gases,** and the elements in Group VIIA are called the **halogens.** At the bottom of Figure 2.1a are two series of elements (atomic numbers 58 to 71 and 90 to 103) that appear as inserts to the table. These have the names **lanthanum series** and **actinium series.**

Other divisions of the periodic table also are important. The stepped line near the right side approximately separates the **metals** on the left from the **nonmetals** on the right. Some of the elements contiguous to the line are intermediate in character and are called **metalloids.**

An earlier name for the lanthanum series was the **rare earths.** Now that we know more about their abundance, that name seems rather inappropriate.

You will have noticed that most group numbers include the designation A or B. The A groups are known as the **representative elements,** and the B groups are called the **transition elements.** The two 14-membered series at the

[†] Atomic mass is defined at the beginning of Chapter 3 on page 43.

Figure 2.1a
Periodic chart of the elements.

IA	IIA	IIIB	IVB	VB	VIB	VIIB	VIII	VIII	VIII	IB	IIB	IIIA	IVA	VA	VIA	VIIA	VIIIA
1 **H** 1.0079																	2 **He** 4.00260
3 **Li** 6.941	4 **Be** 9.01218											5 **B** 10.81	6 **C** 12.011	7 **N** 14.0067	8 **O** 15.9994	9 **F** 18.99840	10 **Ne** 20.179
11 **Na** 22.98977	12 **Mg** 24.305											13 **Al** 26.98154	14 **Si** 28.086	15 **P** 30.97376	16 **S** 32.06	17 **Cl** 35.453	18 **Ar** 39.948
19 **K** 39.098	20 **Ca** 40.08	21 **Sc** 44.9559	22 **Ti** 47.90	23 **V** 50.9414	24 **Cr** 51.996	25 **Mn** 54.9380	26 **Fe** 55.847	27 **Co** 58.9332	28 **Ni** 58.71	29 **Cu** 63.546	30 **Zn** 65.38	31 **Ga** 69.72	32 **Ge** 72.59	33 **As** 74.9216	34 **Se** 78.96	35 **Br** 79.904	36 **Kr** 83.80
37 **Rb** 85.4678	38 **Sr** 87.62	39 **Y** 88.9059	40 **Zr** 91.22	41 **Nb** 92.9064	42 **Mo** 95.94	43 **Tc** 98.9062[b]	44 **Ru** 101.07	45 **Rh** 102.9055	46 **Pd** 106.4	47 **Ag** 107.868	48 **Cd** 112.40	49 **In** 114.82	50 **Sn** 118.69	51 **Sb** 121.75	52 **Te** 127.60	53 **I** 126.9045	54 **Xe** 131.30
55 **Cs** 132.9054	56 **Ba** 137.34	57† **La** 138.9055	72 **Hf** 178.49	73 **Ta** 180.9479	74 **W** 183.85	75 **Re** 186.2	76 **Os** 190.2	77 **Ir** 192.22	78 **Pt** 195.09	79 **Au** 196.9665	80 **Hg** 200.59	81 **Tl** 204.37	82 **Pb** 207.2	83 **Bi** 208.9804	84 **Po** (210)[a]	85 **At** (210)[a]	86 **Rn** (222)[a]
87 **Fr** (223)[a]	88 **Ra** 226.0254[b]	89‡ **Ac** (227)[a]	104 (260)[a]	105 (260)[a]	106 (263)[a]												

metals ← → nonmetals

†	58 **Ce** 140.12	59 **Pr** 140.9077	60 **Nd** 144.24	61 **Pm** (145)[a]	62 **Sm** 150.4	63 **Eu** 151.96	64 **Gd** 157.25	65 **Tb** 158.9254	66 **Dy** 162.50	67 **Ho** 164.9304	68 **Er** 167.26	69 **Tm** 168.9342	70 **Yb** 173.04	71 **Lu** 174.97
‡	90 **Th** 232.0381[b]	91 **Pa** 231.0359[b]	92 **U** 238.029	93 **Np** 237.0482[b]	94 **Pu** (242)[a]	95 **Am** (243)[a]	96 **Cm** (247)[a]	97 **Bk** (249)[a]	98 **Cf** (251)[a]	99 **Es** (254)[a]	100 **Fm** (253)[a]	101 **Md** (256)[a]	102 **No** (254)[a]	103 **Lr** (257)[a]

[a] Mass number of most stable or best known isotope.
[b] Mass of most commonly available, long-lived isotope.

Figure 2.1b
Schematic outline of periodic chart of the elements.

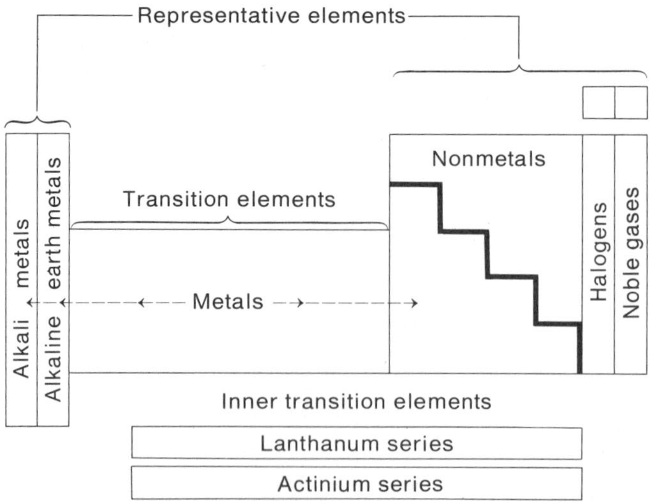

bottom of the table (lanthanum series and actinium series), each to be inserted as indicated, are known as the **inner transition elements.** There are readily observable differences in the relationships of the properties of elements in different divisions. In the representative elements, group similarities are especially pronounced, and the change in properties from group to group is highly distinctive. The transition elements, on the other hand, show less similarity within a group and a greater similarity along a row. The inner transition elements of a given period are strikingly similar in their properties. As a result, chemical separation of the naturally occurring lanthanum series elements from one another is a very difficult undertaking. In Figure 2.1b, an outline of the periodic table is shown in which the named regions are identified.

A name is assigned to a collection of elements only when their properties are very similar. Thus, among the representative elements names are assigned to families, whereas for the inner transition elements names are assigned to portions of horizontal periods.

Now let us briefly examine some of the readily observable reactions and properties of elements that justify the various divisions we have just identified.

Families

The alkali metals (Group IA) are all silver-white soft metals having very high conductivity of heat and electricity. They react vigorously with water, causing hydrogen gas to be evolved and producing a solution that causes litmus to turn blue. They form simple 1:1 compounds with the halogens (Group VIIA), such as NaCl, KBr, and CsF. Nearly all their compounds are soluble in water.

The alkali metals are so soft that they can be cut easily with a knife. They are usually stored under kerosene to prevent reaction with water and oxygen in the air.

The alkaline earths (Group IIA) are most easily described at this point in terms of their differences from the Group IA elements. The alkaline earths are not so soft, not nearly so reactive, and are not such effective conductors of heat and electricity. They form 1:2 compounds with the halogens, such as $MgBr_2$, $BaCl_2$ and CaF_2, and 1:1 compounds with oxygen of general formula MO, where M represents any of the alkaline earths. Many compounds of the alkaline earth metals have low solubility in water.

The elements of Group VIIIA are all **monatomic gases,** are difficult to liquefy, and have resisted nearly all attempts made to cause them to react to form compounds.

In recent years, several compounds involving noble gases have been prepared. These are discussed in Chapter 19.

The halogens (Group VIIA) all exist in elementary form as diatomic molecules: F_2, Cl_2, Br_2, I_2. At room temperature, the first two are gases, bromine is a liquid, and iodine is a solid. Each has a characteristic pungent odor, is poisonous, and can cause serious burns if it contacts the skin. The halogens are very reactive, combining directly with many metals to form such compounds as NaCl, FeI_3, and $BaBr_2$.

Since astatine (At) has no stable isotopes, we do not include it in this first brief mention of the properties of halogens.

The preceding short statement about the four named families of representative elements will subsequently be rounded out with much more information. It is very helpful to have initially a few simple observations clearly in mind and to expand that number little by little. You should become well acquainted with the few properties listed above and then watch in your study for correlations between the structures of these elements and their properties.

Metals and Nonmetals

Referring back to Figure 2.1b, you can see that a preponderance of the elements are metals and that the relatively few nonmetals are grouped toward the upper right-hand corner of the periodic table. The metals are characterized by having luster, high electrical and thermal conductivity, malleability (capability of being beaten or rolled into extended form), and ductility (capability of being drawn through a die to form wire). Metals possess these properties in varying degrees, metalloids possess few of the metallic properties, and nonmetals seldom show any of these properties.

A moment's reflection about a few solid elements that are familiar to everyone will help you recognize the variation in these properties. Copper, silver, gold, tin, chromium, iron, nickel, and aluminum are well-known metals having the identifying properties listed above. Luster, malleability, ductility, and conductivity are all apparent from everyday observations of chromium- and silver-plated objects, aluminum and tin foil, iron and copper wire, and cooking containers made of or clad with aluminum and copper. The contrast with elementary carbon in diamond or graphite or with sulfur is so great that one recognizes clearly the distinction between metals and nonmetals. None of the metalloids is so commonly encountered in the elementary state, and we cannot fall back on everyday experience for recognition of the intermediate properties characteristic of metalloids. A brief statement of the properties of arsenic should help you to see why it is classed as a metalloid. The stable form of elementary arsenic is gray in color, has bright luster, is very brittle, conducts heat, but is a poor conductor of electricity.

2.2 The Electrical Nature of Matter

Two separate series of experiments begun in the mid-nineteenth century led scientists to the conviction that electricity and matter are closely associated. Michael Faraday (1791–1867) performed experiments in which he passed known quantities of electrical charge through various solutions and measured the extent of the chemical change that resulted. In many cases, the chemical change was the deposition of a metal from a metal salt solution onto an electrode. His experiment was analogous to electroplating as carried out today, for example, in the preparation of silver-plated tableware. Faraday was not concerned about the quality of the plating but was interested in the relationship between the quantity of electricity used and the quantity of metal that was deposited on the electrode. He observed a simple numerical relationship between the atomic weight and the weight of the metal deposited by a given quantity of electrical charge.

The other series of experiments began with the use of a Crookes tube, shown in diagram form in Figure 2.2. When the tube contains air or other gas at near-atmospheric pressure, application of a potential difference of about 10,000 volts across the two electrodes causes no effect. If the gas pressure is lowered by means of the vacuum pump, a pressure is reached at which an

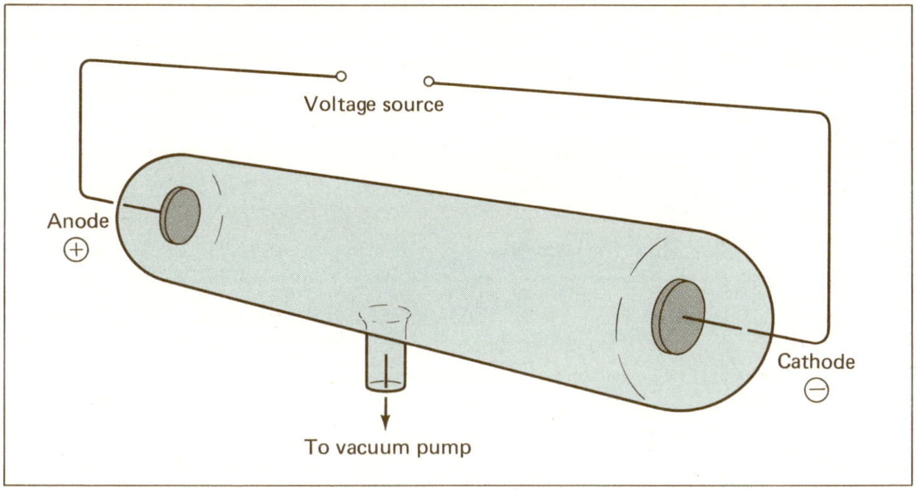

Figure 2.2
The Crookes tube. A potential difference of approximately 10,000 volts is put across the two electrodes. As the tube is evacuated, an electric current flows and the tube glows.

electrical current flows and the tube glows. If the evacuation of the tube is continued, a dark region becomes apparent beginning at one electrode and extending toward the other. At still lower pressures the dark region can be made to extend through the entire interelectrode space. If the experiment is repeated, this time with a small thin shield coated with zinc sulfide placed in the path between the two electrodes, one side of the shield glows brilliantly. When the polarity of the electrodes is reversed, the opposite side of the coated shield glows. Other experiments and their results are indicated in Figure 2.3a, b, and c. In a. the rays, called **cathode rays** since they come always from the cathode (or negative electrode in the Crookes tube) regardless of the composition of the electrodes, are seen to travel in a straight line. In b. the rays cause a pinwheel to turn, demonstrating mass, and in c. the rays are shown to have an electrical charge, for they are deflected by a magnet.

Both series of experiments suggested a close relationship between electricity and matter, and several conclusions were drawn from the observations: (1) Electricity exists in discrete units and these units are associated with atoms; (2) the units of electricity that are emitted from the cathode in the Crookes-tube experiments have both mass and negative charge; (3) these units of electricity (called electrons) travel in straight lines in an evacuated chamber; (4) electrons cause a gas at reduced pressure to glow; and (5) at low enough pressure the collisions between electrons and gas molecules are sufficiently infrequent near the cathode that a dark space is observed.

Thomson's Determination of the Charge-to-Mass Ratio of the Electron

J. J. Thomson (1856–1940) used a modified Crookes tube to make experiments that permitted him to calculate the ratio of the charge of the electron to its mass, e/m. His apparatus is outlined in Figure 2.4. The electron emitted from the cathode passes through an electric field and a magnetic field so oriented that the forces they exert upon the electron are precisely opposed to each other.

Figure 2.3a
The beam of electrons in the Crookes tube travels in a straight line after passing through a slit. This is shown by the glowing line that appears on the zinc sulfide screen.

Figure 2.3b
The beam of electrons in the Crookes tube turns a pinwheel.

Figure 2.3c
The beam of electrons in the Crookes tube is bent by a magnetic field.

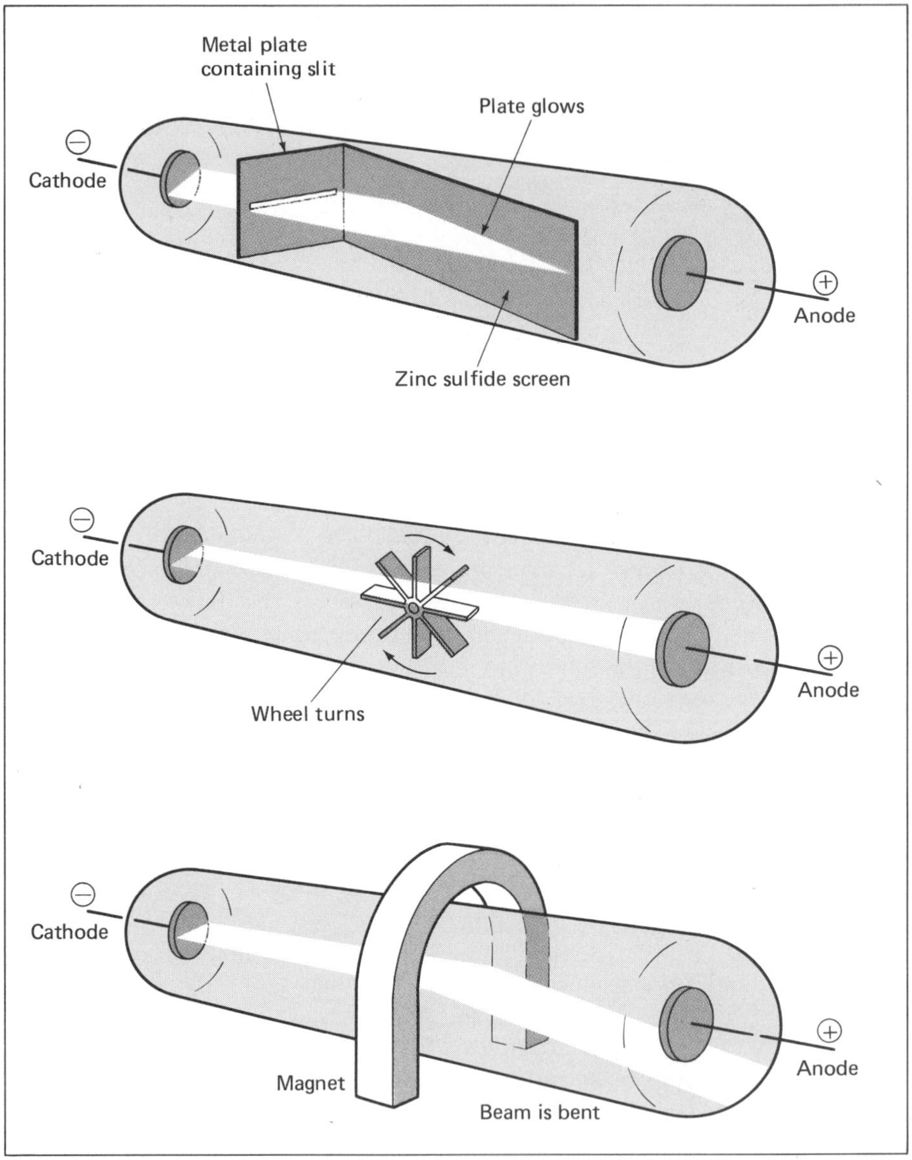

The principle used by Thomson is closely related to that upon which modern **mass spectrometers** are based. These instruments permit accurate measurements of the mass-to-charge ratio of positively charged ions.

With proper adjustment of the variable electric field, Thomson was able to balance the two forces and thus cause the stream of electrons to strike the detecting screen at the same point it struck when subjected to no external force.

From the strengths of the two fields, along with measurement of the angle through which the beam was deflected when one of the opposing fields was removed, Thomson calculated the charge-to-mass ratio. The value obtained from such measurements is $e/m = 1.76 \times 10^8$ coulombs/gram. (See Appendix I for definition of these units.)

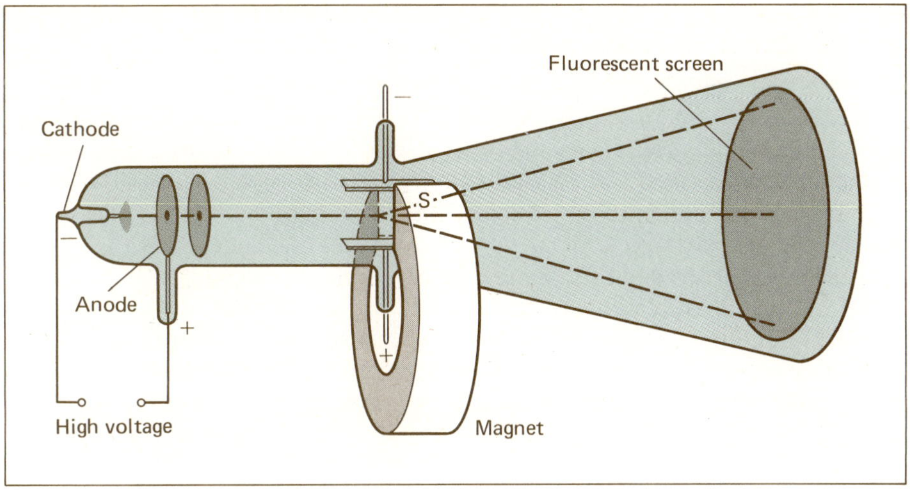

Figure 2.4
The apparatus used by J. J. Thomson to determine e/m. The beam of electrons was bent by the magnetic field between the poles of the magnet. The *e/m* ratio could be calculated from a knowledge of the electric field (between the charged plates) required to restore the beam to an undeflected position.

Millikan's Oil-Drop Experiment

The result of Thomson's experiment was that he could calculate the ratio of charge to mass of the electron. An independent measurement of either value would permit calculation of the other. However, independent measurements of either the electronic mass or charge continued to elude scientists until R. A. Millikan (1868–1953) devised his classic oil-drop experiment. The procedure (see Figure 2.5) consists of making measurements of the rates of fall of a tiny droplet of oil between two oppositely charged plates when (a) the droplet has no electrical charge, (b) it is charged with one electron, and (c) it has larger numbers of electrons. Millikan could not, of course, count the numbers of electrons, but he could deduce the numbers on the oil droplet from his experiments. If the droplet fell at the same (very low) rate with the plates charged as with the plates uncharged, then there was no charge on the droplet. A drop

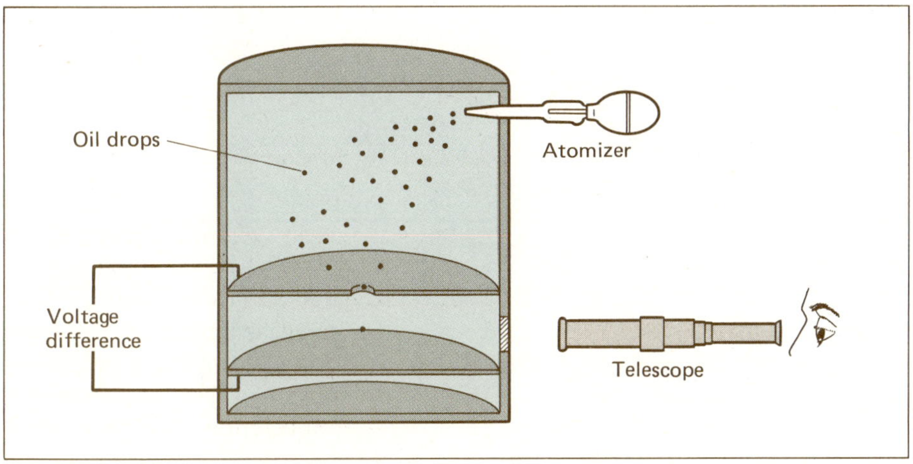

Figure 2.5
Millikan's oil-drop experiment.

that changed its velocity when the field was applied was known to carry an electrical charge. Millikan added charges to the oil drop by passing a beam of x-rays through the space above or below the droplet. The x-ray beam created gaseous ions of appropriate charge, which, impelled by the high electrical field, would stream past the droplet, and one or more would be caught by the drop. Millikan made measurements on many drops. He found that they moved under the influence of the field always at integral multiples of the rate of the slowest of the charged drops. Hence, Millikan concluded that the slowest of the charged drops had the charge carried by one electron.

From his measurements of the rates of rise and fall of oil droplets, from the field strength, and from the viscosity of air, Millikan was able to calculate the charge on the electron. The best measurement of the charge on an electron is 1.60×10^{-19} coulomb. The mass, calculated from the e/m value given above, is 9.1×10^{-28} g.

The Nuclear Atom

One other classic experiment greatly expanded scientists' understanding of the distribution of electrical charge in the atom. Thomson, who measured e/m, had proposed a model in which both mass and positive charge were uniformly distributed throughout the atom. But this view of the atom had to be discarded in 1910 when Ernest Rutherford (1871–1937) demonstrated that both mass and positive charge must be concentrated in a very small fraction of the total atomic volume. His apparatus is outlined in Figure 2.6. A beam of helium ions, He^{2+}, also known as α particles, was directed at a very thin gold foil, beyond which was placed a detecting screen. Small deflections were expected on the basis of the gold atoms having the structure suggested by Thomson; however, the observed behavior was quite different. Most α particles passed through the gold foil virtually undisturbed, showing at most a very small deflection of about one degree. A few α particles were deflected through much

Figure 2.6
The Rutherford experiment.

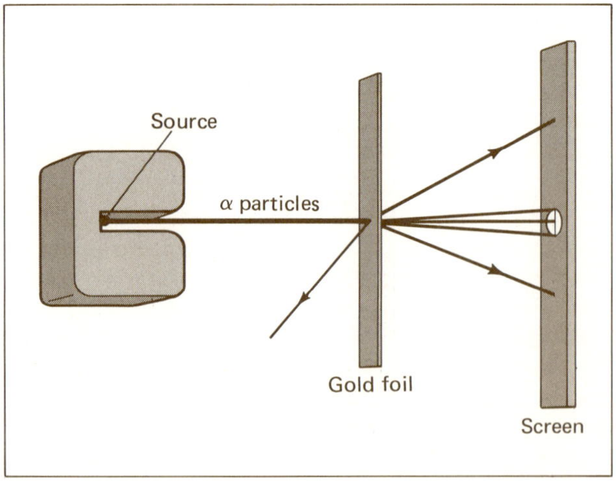

larger angles, and a very small number were deflected 180 degrees! Rutherford described this as being "almost as incredible as if you fired a 15-inch shell at a piece of tissue paper and it came back and hit you."

To account for his observation, Rutherford proposed a nuclear atom in which both positive charge and mass are concentrated in a very small volume at the center of the atom while only the electrons occupy nearly all the atom's volume. This model explains Rutherford's observations in the following way: The electrons are of such low mass that they would have negligible effect on the trajectory of the α particles. The gold nuclei, on the other hand, have large positive charges and large masses. An α particle passing very close to a gold nucleus would suffer a large deflection. Very occasionally, an α particle would approach a gold nucleus directly and would be deflected through an angle approaching 180 degrees. Most particles pass through the gold foil with little effect, being subjected only to small forces from distant nuclei and to negligible forces from electrons.

From his scattering experiments, Rutherford was able to calculate that the radius of the gold *atom* must be about 10^4 times the radius of the gold *nucleus*. It may be a help in grasping the significance of the ratio of sizes to note that the length of a very large football stadium is about 10^4 times the diameter of a ping-pong ball. It is little wonder that most α particles, themselves of nuclear size, passed through the gold foil with no more than small deflection.

We should pause here to reflect upon the striking changes in people's comprehension of the nature of matter that occurred as a result of the experiments briefly discussed in this section. Faraday demonstrated that electricity and matter are intimately related. Thomson made measurements that allowed him to calculate the charge-to-mass ratio of the negatively charged electron, and Millikan directly calculated the charge on the electron from his experiments. Rutherford showed that the negative charges (electrons) and the positive charges were not evenly distributed in the atom, and that the mass and positive charges occupy only a very small part of the atom's volume, with the electrons occupying most of that volume. It is well to recognize that the intensive study and effort of a few great individuals—not a summation of half-hearted curiosity of millions—led to this dramatic increase in knowledge. Phrases such as "now we all know," "people have learned," and "there is an explosion of knowledge" are so commonly used that it is easy to suppose that we are all more intelligent than were members of earlier generations. Our very brief review of some of the results of the work of Faraday, Thomson, Millikan, and Rutherford should make us realize that if we know more, it is the result of hard work by individuals, and if there is a knowledge explosion, it has come about only through disciplined study.

2.3 Electronic Configurations of Atoms

The experiments of Faraday and Thomson gave evidence of electrically charged particles in the atom, and the experiments of Rutherford led to the belief that the negatively charged electrons were in the volume outside a heavy nucleus. However, the energy relationships of the electrons and the structure of the atom came to be understood only after many more careful experiments and much hard work. We will have constant need of some knowledge of this structure, but a detailed study of the basis for our current understanding (the

Figure 2.7
Energy levels available to an electron. The levels are numbered $n = 1, 2, 3, \ldots$. The $n = 1$ level is the lowest energy.

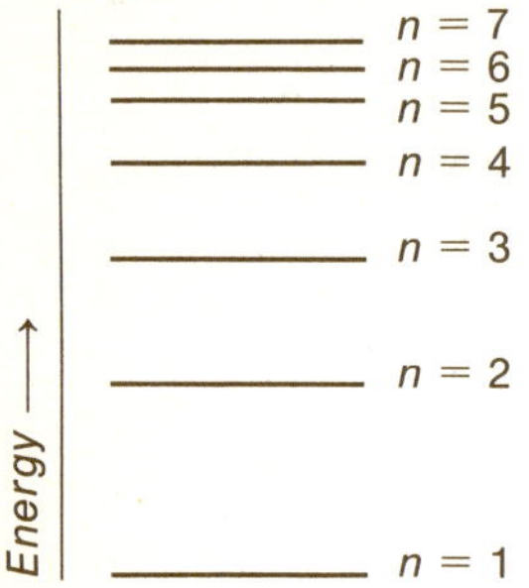

quantum theory) will be delayed until Chapter 13. The discussion here gives an elementary introduction to some of the results of the quantum theory in order that you will have a basis for understanding oxidation-reduction and so that you can predict the nature and reactions of substances encountered in the examples and problems presented in the intervening chapters. In addition, you will find the quantum theory in Chapter 13 much easier to grasp because of this earlier exposure.

An awareness of the electronic configurations of atoms is essential for being able to predict the kinds of reactions that substances will undergo, for the reactivity of an atom is determined in large part by its electrons and their configurations. When a chemical reaction occurs, the atom loses, gains, or shares electrons; therefore, we consider first some elementary aspects of electronic structure.

Energy Levels

An electron in an atom may exist in any of several energy levels (see Figure 2.7). If the atom is not excited (as by some external event such as an electric discharge), it is said to be in the ground state, and the electron will exist in the lowest available energy level. These energy levels are numbered $1, 2, 3, \ldots$, and are identified by the letter n. That is, $n = 1$ represents the lowest energy level, $n = 2$ represents the second, and so forth. The maximum population (number of electrons) that can be accommodated by an energy level is given by $2n^2$. The maximum populations of the first and second energy levels are, therefore, $2 \times (1)^2 = 2$ electrons and $2 \times (2)^2 = 8$ electrons, respectively.

Immediately, we begin to see a correlation between the populations of energy levels and the numbers of elements in the rows of the periodic table. The first period has just two members, hydrogen and helium. The helium atom, having atomic number 2, contains two protons and two electrons. This is just the number of electrons that can inhabit the first energy level. Thus, the element at the end of the first period in the periodic table has just enough electrons to fill the first energy level. The second period has eight members, and its final member, $_{10}Ne$, has ten protons and ten electrons. Two of these electrons are in the first energy level, and the remaining eight fill the second main energy level. Again we see that the period ends just when the main energy level is completely filled. We could feel very satisfied with the correlation if we stopped here. The third period, however, has only eight members even though one predicts from the simple rule that there can be $2 \times (3)^2 = 18$ electrons in the third main energy level. This lack of agreement between the number of elements in the third period and the maximum number of electrons in the third main energy level brings us to a discussion of energy sublevels, for through this discussion we can resolve the apparent discrepancy.

These energy levels should not be thought of as representing fixed positions of the electrons relative to the nucleus. In 1913 Niels Bohr (1885–1962) proposed a model of the atom in which the electrons were imagined to be in fixed circular orbits about the nucleus. This model allowed some properties of the hydrogen atom to be explained, but then it had to be abandoned when it failed to account for other observations. Bohr's model was a great step toward understanding the electronic configuration of atoms.

Spectroscopic experiments give accurate measurements of the different energy levels in an atom. By means of a spectroscope, one measures the energy of light that is emitted when an electron in an atom falls from an excited to a lower state. From measurements of these spectral lines, the energy levels are calculated.

Energy Sublevels

Experiments show, and quantum mechanics leads one to predict, that in an atom that has many electrons, not all electrons having $n = 2$ have the same energy, and, moreover, that this is true for all higher values of n. That is, there are sublevels within an energy level. Some rules based on quantum mechanics and spectroscopy allow us to predict the number of sublevels and the population of each.

(1) For energy level n there are n sublevels. Therefore, there is just one sublevel if $n = 1$, there are two if $n = 2$, and so forth.
(2) The sublevels within an energy level are identified in the order of increasing energy as *s*, *p*, *d*, *f*. The lowest sublevel is always an *s* sublevel, the next a *p* sublevel, and so on.
(3) The maximum number of electrons that can inhabit various sublevels is shown below:

These letters had their origin in spectroscopy; they no longer have their original significance.

SUBLEVEL	MAXIMUM NUMBER OF ELECTRONS
s	2
p	6
d	10
f	14

Note that each sublevel can accommodate just four more electrons than can its predecessor.

Making use of these rules, we can represent the sublevels in each main energy level, as shown in Figure 2.8. The sublevels are identified by a number indicating the main energy level and a letter indicating the sublevel. Thus, the second lowest sublevel in the fourth main energy level is designated 4*p*.

In Figure 2.7, we saw that in addition to the increasing numbers of sublevels with increasing n, there is also a decreasing separation of the main energy levels. The result of these two factors is shown in Figure 2.9, where the first four main energy levels and their sublevels are shown. As n increases, the probability of

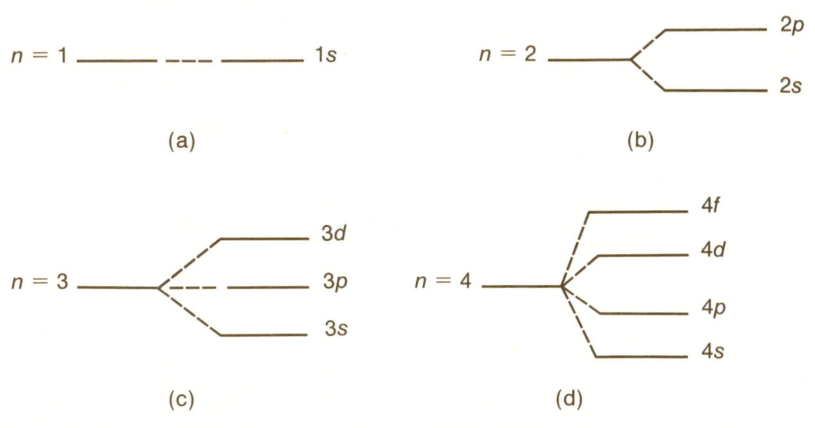

Figure 2.8
Sublevels of the first four main energy levels.

Figure 2.9
Energy sublevels from different levels overlap.
The 3*d* sublevel is higher in energy than the 4*s* sublevel.

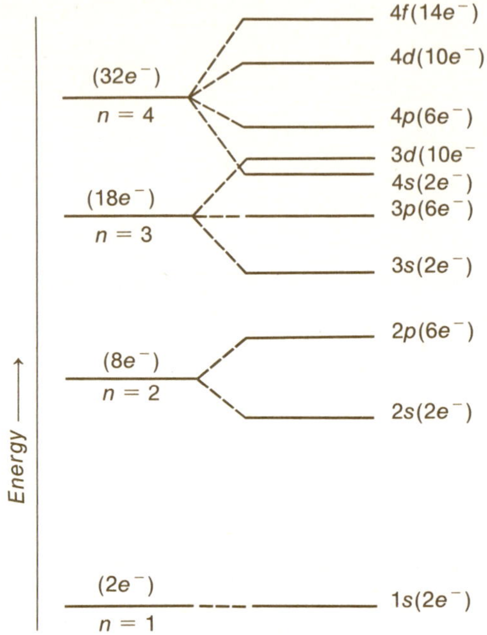

overlap of the sublevels increases, and in the abbreviated energy-level diagram of Figure 2.9, we see the first example of this overlap. Since the 4*s* sublevel lies lower in energy than the 3*d* sublevel, the third energy level is only partially filled before filling of the 4*s* sublevel begins. This lets us account for the unexpected short length of the third period in the periodic table, for a period ends when a *p* sublevel is just filled, and a new period begins when an electron first appears in an *s* sublevel. The overlap of the 4*s* and 3*d* sublevels, therefore, causes the third period to be shortened to eight elements, since the 3*d* sublevel is not filled until the buildup of the fourth period has begun.

Figure 2.10 emphasizes the relationships of periods and energy levels by including marked regions that show the electrons added during the buildup of uncharged atoms of the first three periods. Similar information is given in Figure 2.11, which is a frequently used mnemonic device for showing the order of the filling of sublevels. The best guide of all, however, is the periodic table itself. The division of the table into regions of representative, transition, and inner transition elements is based directly upon the population of the sublevels. The representative elements are those in which *s* and *p* sublevels are partially filled; the transition elements are characterized by partially filled *d* sublevels; and the inner transition elements have electrons added to the *f* sublevels. Figure 2.12 is a periodic table in which these relationships are more specifically shown. By examining the periodic table and making use of your knowledge of the maximum populations of sublevels, you can complete the energy level diagram of Figure 2.9 by indicating the appropriate overlap of sublevels and thus accounting for the lengths of the periods.

The imaginary scheme that we are following of "building up" atoms is known as the *aufbau principle*. An electron and a proton (plus necessary neutrons) are imagined to be added simultaneously to an atom, for example to $_{18}Ar$, to form the next atom, $_{19}K$. The scheme provides a very convenient means for a chemist to retain information concerning the atomic structure of elements. When we discuss this subject in more detail (Chapter 13) we will see that there are several exceptions to the aufbau principle, but also that there are reasonable explanations for most of the important deviations.

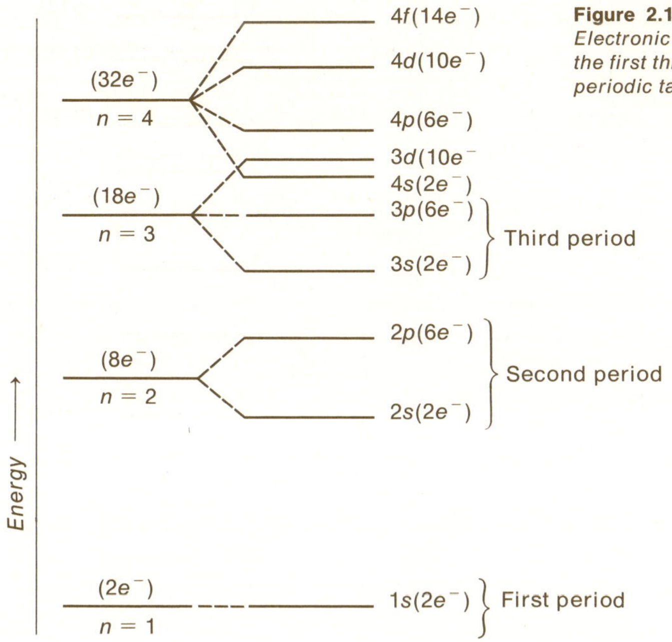

Figure 2.10
Electronic sublevels used in the first three periods of the periodic table.

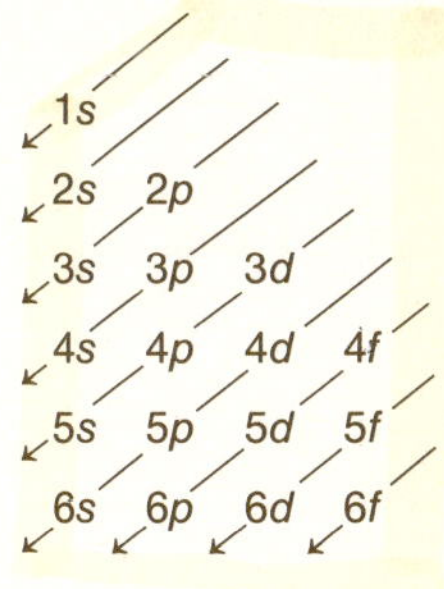

Figure 2.11
Mnemonic device for showing the order in which electronic sublevels are filled. Start at the top and follow each arrow in order. The sublevels are encountered in the same order as they lie in Figures 2.9 and 2.10.

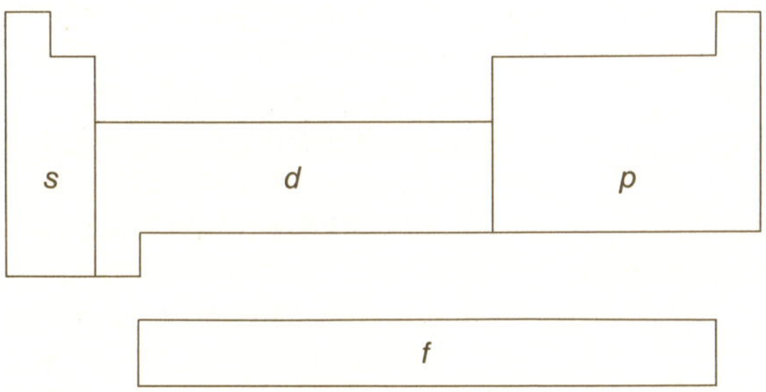

Figure 2.12
Periodic table showing the regions with partially filled s, p, d, *and* f *sublevels.*

EXAMPLE 2.1 In the aufbau concept, which lies lower, 6*s* or 4*f*?

SOLUTION Each period begins after a *p* sublevel is just filled and as *s* electrons are added. The sixth period, therefore, begins with two elements in which *s* electrons are added (Cs and Ba). The 14 4*f* electrons are added in the lanthanum series (cerium to lutetium), and these elements have higher atomic numbers than barium. Therefore, the 6*s* sublevel lies lower than, and is filled before, the 4*f* sublevel.

A simple notation is used for electronic configurations. The element is represented by its symbol (sometimes including the atomic number, which specifies the number of protons, or the number of electrons in the neutral atom) followed by an orderly listing of sublevels with the number of electrons in each sublevel shown by a superscript number:

$$_1\text{H};\ 1s^1$$
$$_6\text{C};\ 1s^2 2s^2 2p^2$$

Thus, the hydrogen atom has a single $1s$ electron while the carbon atom has two $1s$ electrons, two $2s$ electrons, and two $2p$ electrons. The inclusion of the atomic number is often convenient, but it is not necessary. The notation becomes rather unwieldy for elements of high atomic numbers:

$$\text{Sb};\ 1s^2 2s^2 2p^6 3s^2 3p^6 3d^{10} 4s^2 4p^6 4d^{10} 5s^2 5p^3$$

In general, since interest is centered on only the outer electrons, a simpler notation that accounts readily for the lower energy electrons in terms of a noble-gas configuration can be used. The antimony atom, for example, has the krypton configuration plus 15 electrons and can be represented in the following way:

$$\text{Sb};\ [\text{Kr}]4d^{10} 5s^2 5p^3$$

Antimony is spoken of as having the krypton core plus the $4d$, $5s$, and $5p$ electrons that are indicated.

The order in which the sublevels are written in an electronic configuration requires some explanation. In the examples above, we listed the sublevels of the third energy level together even though the aufbau principle shows the $3d$ sublevel to be filled only after the $4s$ electrons are added. It is acceptable to emphasize the aufbau principle by listing the sublevels in the order in which they are filled:

$$\text{Sb};\ 1s^2 2s^2 2p^6 3s^2 3p^6 4s^2 3d^{10} 4p^6 5s^2 4d^{10} 5p^3$$

Notice that the former writing of the electronic configuration emphasizes the energy levels by keeping related sublevels together whereas the latter emphasizes the aufbau principle. Either method is acceptable.

The electronic configurations of ions are written similarly. Additional examples of both atoms and ions are given below.

$$_{11}\text{Na};\ 1s^2 2s^2 2p^6 3s^1$$
$$_{11}\text{Na}^+;\ 1s^2 2s^2 2p^6$$
$$_{10}\text{Ne};\ 1s^2 2s^2 2p^6$$
$$_9\text{F}^-;\ 1s^2 2s^2 2p^6$$
$$_9\text{F};\ 1s^2 2s^2 2p^5$$

A caution: This imaginary system (aufbau) for the building up of atoms must not be confused with what happens when atoms lose electrons to become positively charged ions. In the aufbau concept a proton, an electron, and necessary neutrons are imagined to be added to an atom to make an atom of the next element in the periodic table. In the ionization of an atom electrons are lost, and there is no change in the nucleus. By considering the difference in these two processes, you can see why it is probable that different rules would apply. In the ionization process the electrons lost first are those that are, on the average, farther from the nucleus, and these are not necessarily the electrons imagined to have been added last by the aufbau process. Hence, if Cu loses an electron to become Cu^+, the electron lost is a 4*s* electron and not one of the 3*d* electrons. The electrons in the fourth energy level tend on the average to be farther from the nucleus than those in the 3*d* energy level. Therefore, these outermost electrons are more readily removed from the atom.

Electronic Configuration and Reactivity

The outer electrons, frequently termed the **valence electrons**, profoundly affect the chemical reactivity of an element. These electrons have the highest probability of being at relatively great distances from the nucleus; consequently, the valence electrons are involved in chemical reactions. We will defer until Chapter 13 a discussion of the regions of space occupied by the electrons in various energy levels and sublevels; quantum mechanics allows one to predict shapes and sizes of these regions. For the present, it is sufficient that you realize that a 2*s* electron, on the average, is farther from the nucleus than is a 1*s* electron in the same atom. In the magnesium atom, for example, where the electronic configuration is

$$_{12}\text{Mg};\ 1s^2 2s^2 2p^6 3s^2$$

the two 3*s* electrons have a spatial distribution about the nucleus such that they are more exposed for reaction with other atoms than are the electrons of the first and second energy levels. Moreover, the 3*s* electrons are highest in energy and, therefore, are most easily involved in reaction. Earlier, on page 15, we listed a few properties of certain groups and series of elements. Let us review these properties, now correlating them with the electronic configuration.

The Noble Gases (Group VIIIA)

The noble gases are characterized by having filled *s* and *p* sublevels. (Helium is an exception in that no *p* sublevel is available.) There are no valence electrons; the extraordinary resistance to reaction to form either diatomic molecules (comparable to H_2 or O_2) or compounds with other elements attests to the stability of that configuration. We will find it convenient to refer to the stable electronic configuration of these elements as the **noble-gas configuration.** Table 2.1 shows the number of electrons in each of the various energy levels for the noble gases. Displayed this way the numbers clearly show the similarities in the electronic configurations of the noble gases. Notice that each noble gas after helium has eight electrons in the outermost energy level.

TABLE 2.1

Electrons in Each Main Energy Level in the Noble Gases

	1	2	3	4	5	6	7
He	2						
Ne	2	8					
Ar	2	8	8				
Kr	2	8	18	8			
Xe	2	8	18	18	8		

TABLE 2.2

Electrons in Each Main Energy Level in the Alkali Metals†

	1	2	3	4	5	6	7
Li	2	(1)					
Na	2	8	(1)				
K	2	8	8	(1)			
Rb	2	8	18	8	(1)		
Cs	2	8	18	18	8	(1)	

† The valence electrons have been circled.

When the sodium atom loses an electron, it is said to achieve the neon configuration. Notice that this does not suggest that it has become neon (the number of protons is different) or even that it has the properties of neon. It means simply that it has achieved the electronic distribution that is characteristic of neon. We will see that there are changes in the spatial and energy distributions that result from the higher nuclear charge of the sodium ion.

The Alkali Metals (Group IA)

Each alkali metal has one valence electron, which is an *s* electron (see Table 2.2 and compare it with Table 2.1). The close similarity of behavior observed among the various alkali metals—despite the large differences in atomic number and atomic weight—is the result of the single *s* valence electron in each atom. The high electrical and thermal conductivities are attributed to that loosely held electron, as we shall see in Chapter 13. The ease with which that electron is lost, since the atom achieves the stable noble-gas configuration by its loss, gives rise to much of the observed chemistry of the alkali metals. It is not surprising that +1 ions are formed readily by all Group IA elements. The ions Li^+, Na^+, K^+, Rb^+, and Cs^+ will be encountered constantly in all your study of chemistry.

The Alkaline Earth Metals (Group IIA)

The properties of the alkaline earth metals are largely the result of the two valence electrons that characterize the elements (see Table 2.3). The +2 ion is formed by each element, and the loss of the two valence electrons leaves the noble-gas configuration.

TABLE 2.3

Electrons in Each Main Energy Level in the Alkaline Earth Metals†

	1	2	3	4	5	6	7
Be	2	(2)					
Mg	2	8	(2)				
Ca	2	8	8	(2)			
Sr	2	8	18	8	(2)		
Ba	2	8	18	18	8	(2)	

† The valence electrons have been circled.

TABLE 2.4

Electrons in Each Main Energy Level in the Halogens†

	1	2	3	4	5	6	7
F	2	(7)					
Cl	2	8	(7)				
Br	2	8	18	(7)			
I	2	8	18	18	(7)		

† The valence electrons have been circled.

The Halogens (Group VIIA)

Each halogen atom has two *s* electrons and five *p* electrons beyond a noble-gas configuration. In addition, bromine, iodine, and astatine have filled *d* sublevels, and astatine also has a filled *f* sublevel. The *d* and *f* sublevels have lower n (the principal energy level number), tend to lie deeper in the atom, and make far less contribution to the reactivity of the halogens than do the *s* and *p* valence electrons. Each halogen lacks only a single electron of having the noble-gas configuration and can achieve that configuration by gaining an electron:

As a general rule, the most probable distance of the electrons from the nucleus tends to increase with the value of n, the principal energy level number.

$$Cl + e^- = Cl^-$$

Thus F^-, Cl^-, Br^-, and I^- are common ions observed frequently in compounds and reactions. (Table 2.4 lists the electrons in each main energy level in the halogens.)

Inner Transition Elements

Elements of the lanthanum series are particularly similar to each other chemically despite a consistent increase within that series in the number of electrons in an atom. The basis for the similarity is that the electrons added are f electrons, lying deep in the atoms and not, therefore, contributing appreciably to the chemical reactivity of the element. That is, the increase from a total of 58 to 71 electrons from cerium to lutetium is not considered to be an increase in valence electrons, for the added electrons do not appreciably affect the reactivities of the elements.

We can predict the charge on stable ions for the alkali metals, the alkaline earths, and the halogens from their electronic configurations. However, we must be careful at this point not to conclude that we can now correctly list probable ions for any atom merely by noting its position in the periodic table. We will start, therefore, with a somewhat cautious statement and then work to expand it in the remainder of this and the next chapter: All elements in Groups IA, IIA, VIA, and VIIA form ions in which electrons have been gained or lost in just sufficient number to give the noble-gas configuration. Hence, Na^+, Mg^{2+}, Cl^-, S^{2-}, F^-, Br^-, and O^{2-} are ions that are produced in chemical reactions. Moreover, for Groups IA and IIA, only such ions are formed in ordinary chemical reactions. That is, if barium forms an ion in a chemical reaction, that ion is Ba^{2+}. If cesium forms an ion, it is Cs^+.

It is further observed that in chemical reactions, no element gains or loses more electrons than would be required to produce the noble-gas configuration. Tellurium (Te) will gain two electrons to achieve the xenon configuration, thus forming Te^{2-}, but will not gain more than two in chemical reaction.

2.4 The Electronegativity Concept

In the remainder of this chapter, we will give a very preliminary introduction to chemical bonding, which will be enlarged upon in subsequent chapters. In order to provide a basis for bond character, we discuss first the electronegativity concept.

Ionization Energy

The energy required to remove an electron from a neutral gaseous atom is called **ionization energy.** The magnitude of this energy varies in a periodic manner, as is shown in Figure 2.13. An alkali metal atom readily yields its s electron that lies beyond filled energy levels; consequently, the ionization energy is low. A noble gas, on the other hand, has a very high ionization energy, for the electronic structure of the neutral atom is particularly stable; the removal of an electron requires changing from that very stable to a less stable configuration. In any period, an element between the alkali metal (Group IA) and the noble gas (Group VIIIA) has an intermediate ionization energy. In no case other than the alkali metal does the removal of a single electron from a neutral atom result in the noble-gas configuration. Also, in no case other than the noble gas does the removal of a single electron result in the disruption of that exceptionally stable configuration. It is seen that the general trend from Group IA to Group VIIIA elements is toward higher ionization energies.

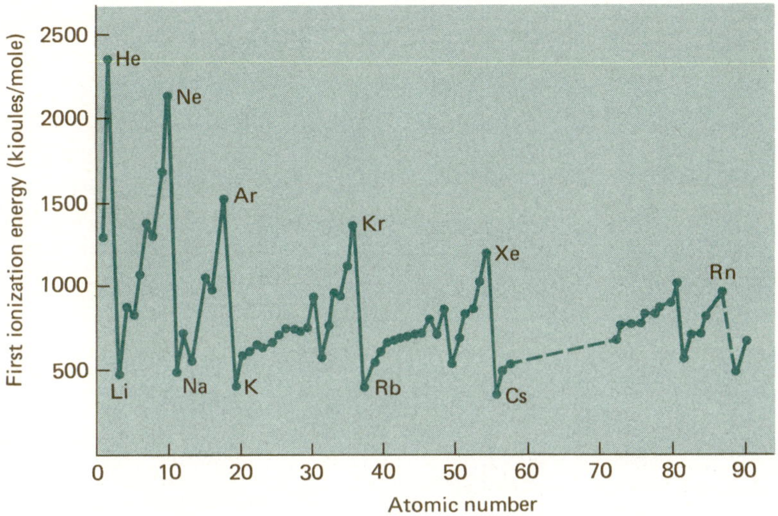

Figure 2.13
First ionization energy of the elements as a function of atomic number.

Electron Affinity

For some elements it is possible to measure the energy released when an electron is added to the neutral gaseous atom; this energy is known as the **electron affinity.** This can be thought of as a measure of the attraction the neutral atom has for an additional electron, in contrast with the ionization energy, which is a measure of the attraction the neutral atom has for its most easily removable electron. Table 2.5 shows the electron affinities for several elements. The trend is generally as expected; the halogens have high values because the addition of a single electron results in the noble-gas configuration, while in no other case does one more electron cause such a stable structure. The table is very incomplete because of the difficulty of measuring this property.

TABLE 2.5

Electron Affinities (kjoules/mole)	
H	72.4
C	121
O	142
F	333
S	197
Cl	349
Br	323
I	295

Electronegativity Values

Ionization potential and electron affinity values can be combined to give a number termed the **electronegativity,** which is a measure of the attraction that an atom has for electrons shared with another atom. As discussed in the next section, a pair of electrons shared by two atoms constitutes a **covalent bond;** the electronegativity values provide chemists with information they need to predict the character of that bond.

In Figure 2.14, electronegativity values of elements are shown in a periodic table arrangement. Fluorine, the most electronegative element, is assigned the value 4.0. On this scale the alkali metals, the elements having the lowest electronegativities, have values approximating unity. While the memorization of many of these values would be a tiresome task carrying small reward, you should notice that several of the numbers are easily kept in mind. The second-row

IA	IIA	IIIB	IVB	VB	VIB	VIIB	VIII	VIII	VIII	IB	IIB	IIIA	IVA	VA	VIA	VIIA	Noble gases
H 2.1																	
Li 1.0	Be 1.5											B 2.0	C 2.5	N 3.0	O 3.5	F 4.0	
Na 0.9	Mg 1.2											Al 1.5	Si 1.8	P 2.1	S 2.5	Cl 3.0	
K 0.8	Ca 1.0	Sc 1.3	Ti 1.6	V 1.6	Cr 1.6	Mn 1.5	Fe 1.8	Co 1.8	Ni 1.8	Cu 1.9	Zn 1.6	Ga 1.6	Ge 1.8	As 2.0	Se 2.4	Br 2.8	
Rb 0.8	Sr 1.0	Y 1.2	Zr 1.4	Nb 1.6	Mo 1.8	Tc 1.9	Ru 2.2	Rh 2.2	Pd 2.2	Ag 1.9	Cd 1.7	In 1.7	Sn 1.8	Sb 1.9	Te 2.1	I 2.5	
Cs 0.7	Ba 0.9	La 1.1	Hf 1.3	Ta 1.5	W 1.7	Re 1.9	Os 2.2	Ir 2.2	Pt 2.2	Au 2.4	Hg 1.9	Tl 1.8	Pb 1.8	Bi 1.9	Po 2.0	At 2.2	

Figure 2.14
Table showing the electronegativities of the elements as a function of position in the periodic table.

elements have a simple trend of electronegativities, each differing by just 0.5 unit from its neighbor. Hence, since the electronegativity of fluorine is 4.0, the values for other second-row elements can be readily remembered. We will have frequent interest in compounds that include hydrogen; therefore, its value (2.1) is worth remembering. Another observation helps one to obtain approximate values when a table is not at hand: With very few exceptions, the electronegativities in a family decrease with increasing atomic number. The decrease from the second to the third period tends to be much greater than subsequent decreases. Each of the elements carbon, nitrogen, oxygen, and fluorine has electronegativity approximately one unit higher than the element immediately below. Silicon, phosphorus, sulfur, and chlorine, on the other hand, differ by only 0.2 electronegativity unit or less from the adjacent elements below.

Here we see first members of groups or families as quite different from subsequent members. This is often observed.

With a good introduction to the concept of electronegativity established, we are now ready to discuss elementary aspects of chemical bonding.

2.5 Chemical Bonds

A simple view of chemical bonding that is sufficient for our immediate purposes is that there are two extreme kinds of bonds between atoms: ionic and covalent. The ionic bond is electrostatic in character and results from the attraction between two ions of opposite charge. The covalent bond is viewed as the result of the sharing of electrons by two atoms.

The Ionic Bond

Two oppositely charged bodies are attracted by a force equal to the product of their charges and divided by the square of the distance separating them. This force is the binding force of the **ionic bond** (known also as the electrovalent bond or the electrostatic bond). The ionic bond is nondirectional; that is, the bond has no angular dependence. It is of equal strength in all directions. A familiar example of an ionically bonded substance is table salt, or sodium chloride. A neutral sodium atom has its first and second energy levels completely filled and has in addition a single 3*s* electron. A neutral chlorine atom also has its first and second energy levels, as well as its 3*s* sublevel, filled, and lacks just one electron of having its 3*p* sublevel filled. The transfer of an electron from sodium to chlorine gives to both atoms the very stable noble-gas configuration. We can represent the reaction in various ways:

It is meaningless to talk of the directional character of a bond if only two atoms are bonded together. But if an atom, A, is bonded to two other atoms, B and C, the question of direction becomes important. For example, are the bonded atoms on opposite sides of A or at right angles or in some other arrangement?

$$Na(1s^22s^22p^63s^1) + Cl(1s^22s^22p^63s^23p^5) = Na^+(1s^22s^22p^6) + Cl^-(1s^22s^22p^63s^23p^6)$$

or

$$\text{sodium atom} + \text{chlorine atom} = \text{sodium ion} + \text{chloride ion}$$

or

$$Na + Cl = Na^+ + Cl^-$$

In the sodium chloride crystal (Figure 2.15) each sodium ion is surrounded by six equidistant chloride ions, and each chloride ion is surrounded by six

Figure 2.15
The sodium chloride structure. The small sphere represents the Na^+ ion and the large sphere the Cl^- ion.

equidistant sodium ions. A particular sodium ion is bonded to each of the surrounding chloride ions in precisely the same way; no single chloride ion is bonded to that sodium ion in particular. When the sodium chloride dissolves in water, it dissolves to give sodium ions and chloride ions and not to give sodium chloride molecules.

Chemists avoid the use of the term *molecule* except where discrete molecular units occur. The sodium chloride crystal described here has the regular pattern of sodium and chloride ions extending throughout the entire crystal, but it has no NaCl molecules. Only at high temperature in the gas phase does sodium chloride exist as discrete NaCl units. In only that case can we speak of sodium chloride molecules. The term **formula unit** is often applied to ionic substances such as NaCl; this is the smallest unit that shows the composition of the compound.

Ionic bonds form between elements of large electronegativity difference. As we saw in the sodium chloride example, an ionic substance is formed between atoms of low ionization energy and atoms of high electron affinity. Thus, binary compounds consisting of elements from the extreme left and right sides of the periodic table have high probability of being ionic.

Formulas of Ionic Compounds

Ions are combined in ionic compounds in such a ratio that the charges neutralize one another and an electrically neutral compound is obtained. Earlier in this chapter we learned how to predict the charge on monatomic ions of the representative elements, and in Chapter 1 we listed some common ions formed from a combination of atoms, such as SO_4^{2-}, PO_4^{3-}, NO_3^-, CO_3^{2-}, ClO_4^-, and NH_4^+. We can now put this information together to predict the formulas of a large number of ionic compounds, as the following example illustrates.

For the transition metals and some of the representative metals in the middle groups of the periodic table, ions are formed that do not have the noble-gas configuration. For example Fe^{2+}, Fe^{3+}, Cu^{2+}, and Sn^{2+}. For compounds containing these elements, we cannot predict the charge on the ion, but we can predict the formula of the compound when the charge on the ion is known. For example,

$$Cu^{2+} + 2\,Cl^- \rightarrow CuCl_2$$
$$2\,Fe^{3+} + 3\,SO_4^{2-} \rightarrow Fe_2(SO_4)_3$$

EXAMPLE 2.2 Predict the formulas of the ionic compounds that result when the positive rubidium and aluminum ions are combined with the negative selenide, fluoride, and phosphate ions.

SOLUTION We note from their positions in the periodic table that rubidium and aluminum achieve the noble-gas configuration by losing one or three electrons, respectively. Hence the formulas of the ions are Rb^+ and Al^{3+}. In like manner, the formulas for selenide and fluoride ions are Se^{2-} and F^-. We also know the formula of the phosphate ion to be PO_4^{3-}. Combining the ions to give electrically neutral compounds gives Rb_2Se, RbF, Rb_3PO_4, Al_2Se_3, AlF_3, and $AlPO_4$.

The Covalent Bond

The ideal covalent (shared electron pair) bond results when two identical atoms are joined into a molecule. For example, in one simple view of bonding two chlorine atoms may be imagined to share one pair of electrons, in effect giving each atom one more electron than it has as a neutral separate atom. Thus, by equal sharing of the electron pair, each of the atoms achieves the noble-gas configuration.

In the Cl_2 molecule the two atoms have identical electronegativities; consequently, there is no inequality in the sharing of the electrons. The centers of the positive charges (the nuclei) are coincident with the centers of negative charges (the electrons). Hence, there is no polar character to the bond (that is, the electrons are not pulled in closer to one atom than to the other).

Bonding theories are discussed in Chapter 14. The oversimplified view of the chlorine molecule does not effectively account for many

The hydrogen chloride (HCl) molecule, on the other hand, consists of two atoms of different electronegativities. Chlorine, the atom with higher electronegativity, has a greater attractive force for the shared electrons than has the hydrogen atom, and while there is not a complete transfer of an electron as in ionic bonding, there is significant polar character to the molecule. Relative to the hydrogen end of the molecule, the chlorine end is distinctly negative. The bond in this molecule is intermediate in character between that in the purely covalent Cl_2 and that in the ionic NaCl; the HCl bond is referred to as a covalent bond with ionic character, or as a **polar covalent bond.** The purely covalent (nonpolar) bond is found only between identical atoms, while the ionic bond requires a very large electronegativity difference.

observations of chemical bonds, but it is still often used by chemists because of its convenience. This is just one of many cases of a model in the sciences that is known to be inadequate, that has been replaced by more involved and more accurate models, but that is still widely used for qualitative purposes because it conveniently covers many observations.

Lewis Structures

A system of representing molecules by means of electron-dot formulas was proposed in 1916 by the American chemist G. N. Lewis. Our understanding of chemical bonding now reaches far beyond the information implicit in these Lewis structures, but the system remains useful as a way of representing many compounds. The outermost *s* and *p* electrons of an atom are shown as dots around the symbol of the element. Thus, the elements of the second period have the following Lewis structures:

Li·   Be: ·B: ·C:·   ·N:·   :O:·   :F:·   :Ne:

The formation of a covalent bond is indicated by the sharing of a pair of electrons:

H· + ·F: → H:F: (or H—F:)

A short line between two atoms is often used to represent a shared pair of electrons.

The Lewis structure for HF shows hydrogen with the helium configuration and fluorine with the neon configuration, both having reached this through the sharing of the electron pair.

The bonding of identical atoms to form diatomic molecules is similarly accounted for by the Lewis structures. For example,

H· + ·H = H:H   or   H—H

and

:I· + ·I: = :I:I: or   :I—I:

The Lewis structure for diatomic nitrogen suggests a triple bond, since this gives each nitrogen atom the neon configuration:

:N· + ·N: = :N:::N: or   :N≡N:

The experimental observation that the bond in N_2 is exceptionally strong gives support to the Lewis representation.

The Lewis structures for such oxygen-containing anions as SO_4^{2-} (sulfate ion) and ClO_4^- (perchlorate ion) are written with the recognition that the oxygen atoms are individually bonded to the other atoms in the anion:

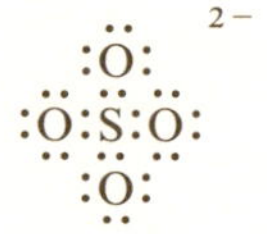

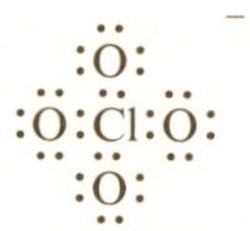

Notice that the total number of outer *s* and *p* electrons in the sulfate ion is 32. There are six from each neutral oxygen and sulfur atom, and in addition there is a net charge on the ion of negative two. This indicates that there is a surplus of two electrons beyond those in the neutral atoms. Similarly, the perchlorate ion has a total of 32 outer *s* and *p* electrons. Twenty-four are from the four neutral oxygen atoms, seven are from the neutral chlorine atom, and there is one more indicated by the single negative charge.

In acids formed from such ions as SO_4^{2-} and ClO_4^-, the hydrogen atoms are bonded to oxygen atoms. To draw the Lewis structure for H_2SO_4, we observe that there is no net charge. Hence, the molecule has 32 electrons, six from each oxygen and sulfur atom and one from each hydrogen atom. A noble-gas configuration can be achieved for each atom by the following arrangement of electrons:

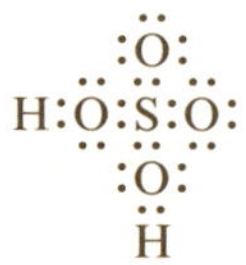

Each element, except the very light elements, achieves an octet of electrons in these simple Lewis structures, and thus the structures are often spoken of as meeting the "octet rule."

As we mentioned above, while there are many molecules for which the Lewis representation is inadequate, this system provides such a simple means for correlating much information about bonding that chemists still make frequent use of it.

Molecular Geometry

Although Lewis electron-dot formulas are useful for describing bonding in covalent molecules, the electron-dot structures by themselves do not describe the geometry of the molecules. However, application of a very simple rule called **electron-pair repulsion** to the information given in the electron-dot formula allows one to make at least qualitative predictions about the shapes of molecules. Simply stated, the rule says that since electrons repel each other, the molecule assumes the shape that will keep the electron pairs (both bonding and nonbonding) as far apart as possible.

We will give several illustrations of the rule. For a first example, consider mercuric chloride, which when vaporized exists as discrete covalent $HgCl_2$

molecules. Since the mercury atom has only two valence electrons ($6s^2$), and the chlorine atom is one electron short of having an octet, the electron-dot structure would show two Hg—Cl bonds and no other electron pairs around the central Hg atom. The electron-pair repulsion rule would predict that the molecule will take the shape that will keep the two sets of electron pairs as widely separated as possible, which is on opposite sides of the mercury atom. Thus the structure for $HgCl_2$ (and all other molecules in which there are only two pairs of electrons around the central atom) is linear:

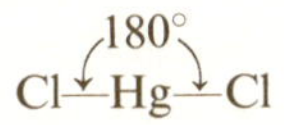

As an example of a molecule in which three electron pairs surround the central atom, let us consider BF_3. Since boron has three valence electrons and fluorine has seven, the Lewis structure would show three B—F bonds with no other electron pairs around the central B atom. The obvious structure that will keep the three bonds as widely separated as possible is the trigonal shape:

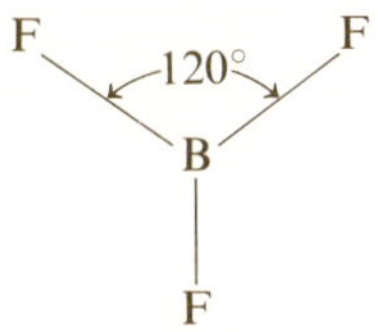

As an example of a compound that has four bonds around the central atom, consider methane, CH_4. Four bonds are separated as widely as possible when the bonds are directed toward the corners of a regular tetrahedron:

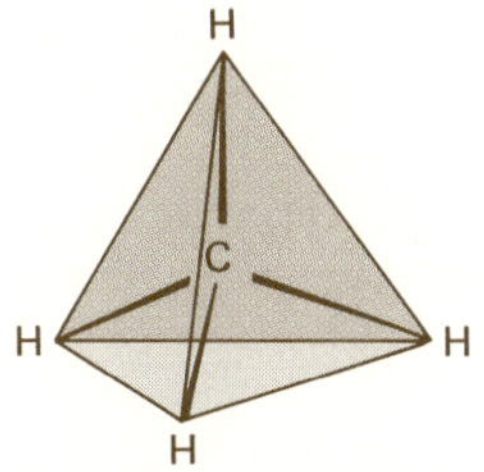

The H—C—H bond angles are all equivalent and equal to 109°28′. This is a very common structure in nature.

Up to this point we have considered examples of molecules in which two, three, and four pairs of electrons (all bonded) surround the central atom, and we have seen that the corresponding molecular shapes are linear, trigonal, and tetrahedral, respectively. In Chapter 14 we will return to the subject of molecular shape and extend our coverage to include five and six electron pairs. At that time we will present the subject in detail from another (somewhat more sophisticated) point of view.

Before we leave this discussion, however, let us consider the effect that nonbonding pairs of electrons have on the geometry of the molecule. Consider the ammonia molecule, NH_3. Its electron-dot formula shows it to have four sets of electron pairs (three pairs in the three N—H bonds and one nonbonded pair) surrounding the central nitrogen atom:

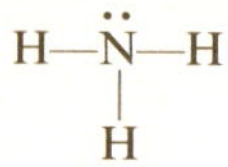

All four electron pairs must be considered in electron-pair repulsion theory. Hence, the **electronic geometry** of the molecule would be predicted to be tetrahedral, with hydrogen atoms occupying three corners of the tetrahedron, and a nonbonded pair of electrons on the fourth corner:

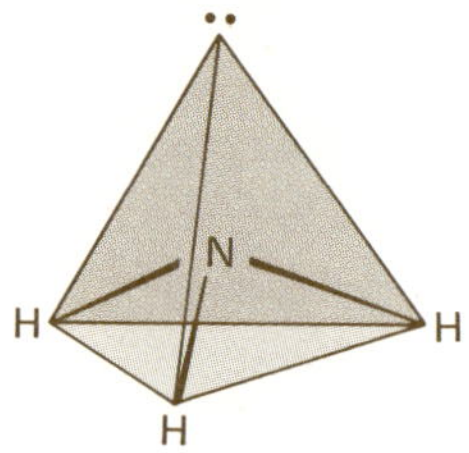

Experimental measurements, however, show that the H—N—H bond angle is approximately 107°, somewhat less than the tetrahedral angle. It appears, therefore, that the nonbonded pair of electrons requires more room than the electrons in a bond, and the effect of the presence of a nonbonded pair is to squeeze the bonds closer together. Hence, the electronic geometry of NH_3 is a somewhat distorted tetrahedron.

Notice that we are making a distinction between the *electronic geometry*, which is the structure obtained by including the nonbonded electrons, and the *shape* of the molecule, which is determined only by the arrangement of the atoms. The electronic geometry of NH_3 is thus described as a distorted tetrahedron; the shape of the molecule could be described as a symmetrical top:

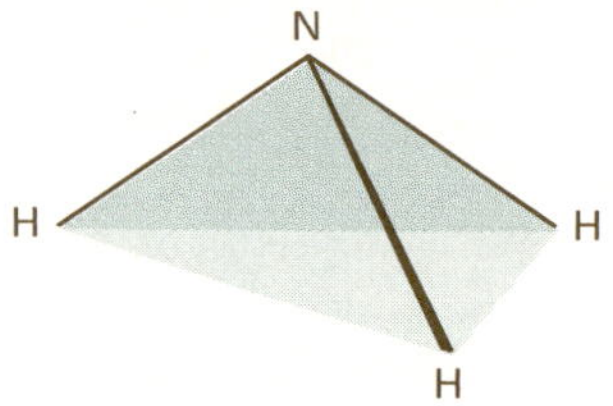

The water molecule is another example of the effect of nonbonded pairs on the geometry of the molecule. The Lewis structure for water shows two nonbonded and two bonded electron pairs surrounding the central atom:

With four electron pairs surrounding the oxygen atom, we would again expect tetrahedral electronic geometry:

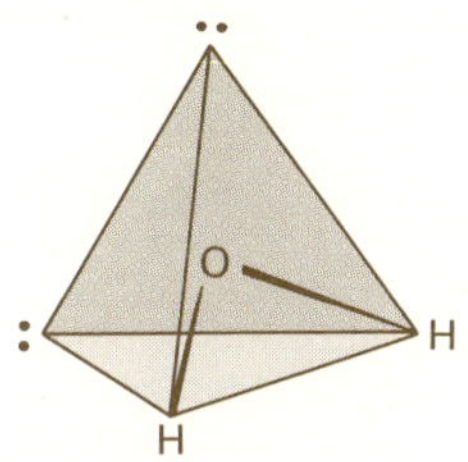

Again, however, the nonbonded pairs crowd the bonds closer together, and a somewhat distorted tetrahedron results. As you would expect, because of the presence of two pairs of nonbonded electrons, the distortion is greater in the water molecule than in ammonia. The bond angle H—O—H is approximately 105°. The electronic geometry of the water molecule is tetrahedral; the shape of the molecule is *bent:*

In Chapter 14 we will be able to predict molecular shapes more accurately by making use of a more adequate theory.

Summary

The periodic table is an arrangement of the elements in order of increasing atomic number. Similarities in properties and trends in properties are clearly seen in the periodic table. Among the representative elements, family similarities are especially pronounced, while among the transition elements there is more similarity along a row or period.

While the origins of the periodic table were based on observed reactions and properties of the elements, the structure of the atoms provides a sound theoretical basis for the table. The experimental and theoretical work of Faraday, Thomson, Millikan, Rutherford, Bohr, and others has led to our present understanding of the nuclear atom and its electronic structure.

The electrons in an atom have quantized energies; that is, only certain energy levels are available to them. Transitions of electrons between energy levels lead to the known spectra of the elements. Since the energy levels are not identical for different elements, the transitions involve different amounts of energy, and the spectra also are different. The chemical reactivity of an atom depends to a large extent upon the electronic configuration of the atom. Noble-gas configurations have low energy, and atoms with one or two electrons deficient or in excess of the noble-gas configuration easily gain or lose electrons to achieve that configuration.

The electronegativity of an atom is a measure of the attraction that the atom has for electrons shared with another atom. Elements at the upper right of the periodic table (noble gases excepted) have the highest electronegativities, while elements at the lower left have the lowest.

The bond between atoms of equal electronegativity is covalent, whereas that between atoms of very large electronegativity difference is ionic.

We will return to this subject and treat it in more detail in Chapters 13 and 14.

Important Terms and Concepts

Periodic properties
Periods
Families
Alkali metals
Alkaline earth metals
Noble gases (inert gases)
Halogens
Lanthanum series (rare earths)
Actinium series
Metals
Nonmetals
Metalloids
Representative elements
Transition elements
Inner transition elements
Monatomic gases
Cathode rays
Mass spectrometer
Valence electrons
Noble-gas configuration
Ionization energy
Electron affinity
Electronegativity
Covalent bond
Ionic bond
Formula unit
Polar covalent bond
Electron-pair repulsion
Electronic geometry

Exercises

2-1.

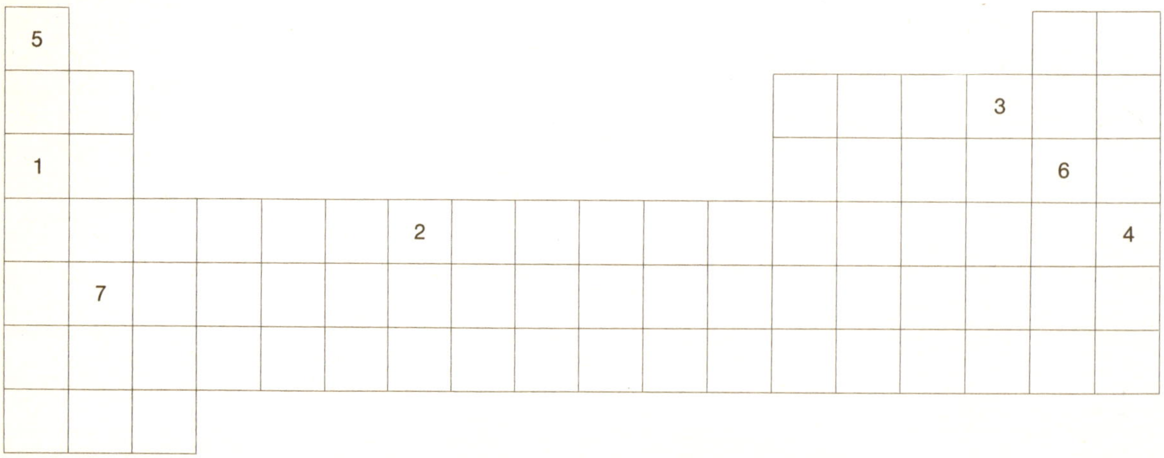

In the periodic table shown on page 40, choose the number that is in the position of the element that will give the correct answer to the question.

(a) Which element is an alkaline earth metal?
(b) Which element is a halogen?
(c) Which element has the smallest ionization energy?
(d) Which element is a noble gas?
(e) Which element is most electronegative?
(f) Which element would be most closely related to bromine in its physical and chemical properties?
(g) Which element is a transition metal?
(h) Which element has the electronic configuration $1s^2 2s^2 2p^4$?
(i) Which elements are nonmetals?
(j) Which element has the highest electron affinity?
(k) Which element easily loses an electron to form a +1 ion?

2-2.

The densities of metallic calcium and barium are 1.55 g/ml and 3.5 g/ml, respectively, at 20° C. Which of the three values 1.4, 2.6, or 3.8 would you choose as the most likely density of strontium?

2-3.

Before element 85 (At) was discovered, many of its physical and chemical properties were predicted correctly. From the known properties of other elements in the periodic table, predict the following properties of At. (See page 567 and Table 19.12 for properties of the halogens.)

(a) Melting point and normal boiling point
(b) Color
(c) Elemental formula (will it be At or At_2 or At_3 or some other formula?)

2-4.

Write the electronic configurations of the following elements:

(a) Na
(b) S
(c) F
(d) Ar
(e) Al

2-5.

Using the aufbau principle, predict the electronic configurations of the following elements:

(a) Ca
(b) Tl
(c) Br
(d) Zr
(e) Xe

2-6.

Using the aufbau principle, predict the electronic configurations of the following elements:

(a) Ba
(b) W
(c) Nd
(d) Pb
(e) At

2-7.

Of the group of compounds listed below, which have bonds that are essentially ionic, and which have bonds that are essentially covalent?

(a) CsCl
(b) CaF_2
(c) HI
(d) NF_3
(e) BaO
(f) CCl_4
(g) KBr
(h) LiH

2-8.

Write the electron-dot formulas (Lewis structures) for the following compounds:

(a) Cl_2
(b) H_2O
(c) N_2
(d) NH_3
(e) CF_4

2-9.

Write the electron-dot formulas (Lewis structures) for the following compounds:

(a) $CH_2{=}CH_2$
(b) CN^-
(c) ClO_3^-
(d) PH_3
(e) NO_3^-

2-10.

Write the electron-dot formulas (Lewis structures), and then describe the geometrical arrangement of atoms in the following molecules:

(a) $SiCl_4$
(b) CO_2
(c) H_2S
(d) NF_3
(e) PO_4^{3-}
(f) NH_4^+

2-11.

Write formulas for the following compounds:

(a) lithium sulfide
(b) aluminum oxide
(c) calcium sulfate
(d) strontium perchlorate
(e) magnesium nitrate
(f) ammonium carbonate

3

Stoichiometry

The term **stoichiometry** (derived from the Greek words meaning *element* and *to measure*) refers to weight relations in chemical formulas and equations. In this chapter we will consider weight relations first in formulas, then in equations, and finally as they apply in the reaction of substances in solution. You should recognize that the fundamental approach is the same in every problem; when the basis for stoichiometry is understood, a single approach should solve any problem involving weight relations.

3.1 Mole Relationships

Atomic Mass, Dimensions, and Mole

A number called the **atomic mass**† is associated with each element. The atomic mass is a relative mass; an atom of a particular isotope of carbon (^{12}C) is assigned the exact value 12, and the masses of all other elements are determined relative to that value of the carbon atom. Thus, ordinary hydrogen atoms have a mass (on the average) of 1.0079, and oxygen atoms have an average mass of 15.9994 on the same scale. Dimensions or units are needed for these numbers, and one set often used for the atomic mass is **atomic mass units per atom.** The atomic mass unit (amu) is thus defined in terms of the relative masses of the elements and is 1/12 the mass of a ^{12}C atom.

Atomic masses of the elements are listed alphabetically in the table of atomic weights and in the periodic table. Both tables are found just inside the front cover of the book.

Isotopes, remember, are atoms of the same element that have different mass numbers.

The dimensions atomic mass units per atom are useful when one's interest is centered upon an atom or a small number of atoms, but they lead to the use of very cumbersome numbers when one works with laboratory-sized quantities. Therefore, we choose a unit that is of such size that only a few of the units need be used for an ordinary reaction in the laboratory. If we take a large number of hydrogen atoms and an identical number of oxygen atoms, the ratio of the total mass of hydrogen to oxygen is the same as for one atom of each, 1.00797/15.9994. Now, if a certain very large number of hydrogen atoms is considered, the total mass is 1.00797 g and the mass of the identical number of oxygen atoms is 15.9994 g. The number of atoms required in each case is 6.0220×10^{23} and is called **Avogadro's number.** Thus, Avogadro's number of platinum atoms weighs 195.09 g while that number of ^{12}C atoms weighs exactly 12 g. Now we can write another set of dimensions for the atomic mass of hydrogen:

When chemists use the dimensions *atomic mass units/atom* with an atomic mass from the periodic table, they do not assume that a particular atom has just that mass. They recognize that the element, as it exists, is a mixture of isotopes and that the value used is the average mass of a very large number of atoms.

$$1.00797 \frac{\text{grams}}{\text{Avogadro's number of hydrogen atoms}}$$

A more convenient and more usual way of writing the same relationship is

$$1.00797 \frac{\text{grams}}{\text{mole of hydrogen atoms}}$$

† The term *atomic weight* is more commonly used than is the precise *atomic mass*. We will use both terms in this text, emphasizing *mass* where apparatus other than a balance is used for determination of the mass. Mass and weight are discussed in more detail in the footnote on page 45. (See the inside front cover for a table of atomic weights.)

Note that the atomic mass of hydrogen is not expressed completely as 1.00797; it is 1.00797 amu/atom, or it is 1.00797 g/mole. Chemists avoid many errors by using dimensions consistently.

for the **mole** is defined as Avogadro's number of particles and, in this context, is free from any connotation of molecules as distinct from atoms or ions. One can speak of a mole of hydrogen atoms (H), a mole of hydrogen molecules (H_2), or a mole of hydrogen ions (H^+), and in each case clearly mean 6.022×10^{23} particles of the kind specified. It is immediately evident that a mole of H_2 molecules weighs 2.01594 g while a mole of H atoms weighs only half as much.

EXAMPLE 3.1 Calculate the mass in grams of 1.00×10^{20} lithium atoms.

SOLUTION To make the conversion from a certain number of atoms to the mass of that number of atoms, we use a set of dimensioned conversion factors. The dimensions guide us in making the conversion.

$$(1.00 \times 10^{20}\ \cancel{\text{atoms}})\left(\frac{1\ \cancel{\text{mole}}}{6.02 \times 10^{23}\ \cancel{\text{atoms}}}\right)\left(6.94\frac{\text{g}}{\cancel{\text{mole}}}\right) = 1.15 \times 10^{-3}\ \text{g}$$

We first wrote the data given in the problem—1.00×10^{20} atoms—and then made use of Avogadro's number to change from number of atoms to number of moles. Multiplying by the atomic mass of lithium changed this to number of grams. Notice that since the number of atoms is known with an accuracy of 1 part in 100, there is no point in using the atomic mass and Avogadro's number to greater accuracy. Hence, they are rounded off in the calculation.

In working this problem, we have considered Avogadro's number and the atomic weight to be conversion factors. Since there are 6.02×10^{23} atoms/mole, two factors can be written that will convert moles to atoms or atoms to moles. The factors are

$$\frac{1\ \text{mole}}{6.02 \times 10^{23}\ \text{atoms}} \quad \text{and} \quad \frac{6.02 \times 10^{23}\ \text{atoms}}{1\ \text{mole}}$$

Similarly, the atomic weight provides two factors

$$\frac{6.94\ \text{g}}{1\ \text{mole}} \quad \text{and} \quad \frac{1\ \text{mole}}{6.94\ \text{g}}$$

that will make conversions between grams and moles. Each of the four factors is equal to unity; the numerator and denominator are equivalent to each other. One mole consists of 6.02×10^{23} atoms, and 6.02×10^{23} atoms make 1 mole. There are 6.94 g in 1 mole of lithium, and 1 mole of lithium weighs 6.94 g. Therefore, the solution to the problem is made by writing the data given in the problem and using a series of factors to convert this to the desired units.

Many other dimensioned numbers can serve similarly as conversion factors. You are never left in doubt whether to multiply or divide by a number if you make careful use of the dimensioned conversion factors, for only if the factor is used correctly will dimensions cancel correctly to give the solution.

EXAMPLE 3.2 Calculate the mass in grams of an atomic mass unit.

SOLUTION We want here a conversion factor from atomic mass units to grams. The answer will have the dimensions *grams/amu.* We can solve the problem by the use of any element (or, indeed, without specific consideration of an element). Let us use beryllium, with atomic mass 9.0122. The number 9.0122 has two sets of dimensions:

$$9.0122 \text{ amu/atom or } 9.0122 \text{ g/mole}$$

If we start with 9.0122 g/mole, we can convert that to g/amu by the use of two factors:

$$\left(9.0122 \frac{\text{g}}{\cancel{\text{mole}}}\right)\left(\frac{1 \cancel{\text{mole}}}{6.0220 \times 10^{23} \cancel{\text{atoms}}}\right)\left(\frac{1 \cancel{\text{atom}}}{9.0122 \text{ amu}}\right) = 1.6606 \times 10^{-24} \frac{\text{g}}{\text{amu}}$$

EXAMPLE 3.3 Which has the greatest weight:† (a) 2.50 moles of aluminum; (b) 68.0 g of aluminum; or (c) 1.51×10^{24} aluminum atoms?

SOLUTION It is not, of course, necessary to calculate the weights in order to answer the question, since the numbers of moles or the numbers of atoms would serve as well. Let us calculate the number of moles in each case.

(a) 2.50 moles

(b) $(68.0 \cancel{\text{g}})\left(\frac{1 \text{ mole}}{26.98 \cancel{\text{g}}}\right) = 2.52 \text{ moles}$

(c) $(1.51 \times 10^{24} \cancel{\text{atoms}})\left(\frac{1 \text{ mole}}{6.02 \times 10^{23} \cancel{\text{atoms}}}\right) = 2.51 \text{ moles}$

It is clear that the quantity of aluminum in (b) is greater than that in either (a) or (c).

EXAMPLE 3.4 What is the weight in grams of an average calcium atom?

SOLUTION A mole of calcium weighs 40.08 g. Hence an atom weighs

$$(1 \cancel{\text{atom}})\left(\frac{1 \cancel{\text{mole}}}{6.0220 \times 10^{23} \cancel{\text{atoms}}}\right)\left(\frac{40.08 \text{ g}}{\cancel{\text{mole}}}\right) = 6.656 \times 10^{-23} \text{ g}$$

Now, let us calculate the atomic weight of an element from the percentage composition of the isotopes.

† *Mass versus weight.* The mass of an object is defined in terms of the object's inertia; it is a measure of the amount of material in the body and not of the force (mass times acceleration) exerted upon that body by the earth. The force just mentioned is the weight and is variable. A piece of iron weighs less on a high mountain than it does at sea level, and it weighs still less when immersed in water because of the buoyant force exerted by the water. For an object "weighed" in a vacuum on an equal-arm balance, not only is the weight of the object equal to that of the comparison standards, but also the masses are equal. There is the same gravitational attraction on each body, and there is no buoyant force on either. Ideally, then, with an equal-arm balance one compares masses as well as weights. Common usage has the operation spoken of as "weighing" and not as "massing." We will bow to tradition and speak of the "weight of aluminum" in the problem, even though we mean mass.

EXAMPLE 3.5 From the percentages of the two isotopes of boron, calculate the atomic weight (the average atomic weight) of the element.

ISOTOPE	ISOTOPIC MASS	% IN NATURAL BORON
^{11}B	11.01	81.17
^{10}B	10.01	18.83

SOLUTION A simple approach to such a problem is to consider 100.00 g of boron, so that 81.17 g of that sample is ^{11}B and 18.83 g is ^{10}B. We can calculate the number of moles of each. Then from the total number of moles and the total weight of the sample, we calculate the atomic weight.

$$(81.17\ \text{g})\left(\frac{1\ \text{mole}}{11.01\ \text{g}}\right) = 7.372\ \text{mole}$$

$$(18.83\ \text{g})\left(\frac{1\ \text{mole}}{10.01\ \text{g}}\right) = 1.881\ \text{mole}$$

There are 9.253 moles in 100.00 g.

$$(100.00\ \text{g})\left(\frac{1}{9.253\ \text{mole}}\right) = 10.81\ \frac{\text{g}}{\text{mole}}$$

Calculations Involving Formulas and Equations

With a clear understanding of the meaning of the term *mole* and with a very slight exposure to the meaning of chemical formulas, one can make stoichiometric calculations in a straightforward manner. Consider, for example, the substance calcium chloride, which is represented by the formula $CaCl_2$. There are two atoms of chlorine for each atom of calcium. Therefore, in a mole of $CaCl_2$ there are two moles of chlorine atoms and one mole of calcium atoms.

The fact that calcium chloride does not exist in discrete molecules need not deter us from speaking of moles of $CaCl_2$; a mole is just Avogadro's number of $CaCl_2$ units.

Because of the simple mole relationship we can readily calculate the weight percentage composition of the substance. In one mole $CaCl_2$, which weighs 110.98 g [40.08 + (2 × 35.45)], the mole of calcium atoms weighs 40.08 g and the two moles of chlorine atoms weigh 70.90 g. The percent calcium is

$$\frac{40.08\ \text{g}}{110.98\ \text{g}} \times 100 = 36.11\%$$

and the percent chlorine is

$$\frac{70.90\ \text{g}}{110.98\ \text{g}} \times 100 = 63.89\%$$

The reverse calculation is also useful. If we know the percentage composition of a chemical compound, we can determine the **simplest** or **empirical formula.** For example, analysis of an oxide of phosphorus gives 56.36 weight percent oxygen and 43.64 weight percent phosphorus. To determine the simplest formula, we take the most convenient amount of the compound, 100.00 g. This sample contains 56.36 g of oxygen and 43.64 g of phosphorus.

We next find the number of moles of each element.

$$(56.36\text{ g O})\left(\frac{1\text{ mole O}}{16.00\text{ g O}}\right) = 3.523\text{ moles O}$$

$$(43.64\text{ g P})\left(\frac{1\text{ mole P}}{30.97\text{ g P}}\right) = 1.409\text{ moles P}$$

The ratio of moles of oxygen to moles of phosphorus is

$$\frac{3.523\text{ moles O}}{1.409\text{ moles P}} = 2.500\text{ moles O per mole P}$$

From this information we could write the formula $PO_{2.5}$, but we choose instead to write it in terms of whole numbers of atoms. Therefore, P_2O_5 is the empirical formula. Our information does not let us know whether P_2O_5, P_4O_{10}, P_6O_{15}, or some other formula in which the ratio is 1:2.5 is the correct **molecular formula,** that is, the formula of the molecules that actually exist. Other information is essential to make that determination. In Chapters 5 and 8 we will see that measurements on gases and measurements of freezing points can be used to determine molecular weights. For the phosphorus-oxygen compound such measurements show that the molecular weight is 284 g/mole. Hence, the correct molecular formula is P_4O_{10}.

Finally, one can calculate the weights of substances that react with each other. When silver and sulfur are heated together, the black compound silver sulfide is produced, and in this compound the silver and sulfur atoms combine in the ratio 2:1 (Ag_2S). Let us calculate how much sulfur is required for complete reaction with 50 g of silver. The equation for the reaction is

$$2Ag + S = Ag_2S$$

so that

$$(50\text{ }\cancel{\text{g Ag}})\left(\frac{1\text{ }\cancel{\text{mole Ag}}}{108\text{ }\cancel{\text{g Ag}}}\right)\left(\frac{1\text{ }\cancel{\text{mole S}}}{2\text{ }\cancel{\text{moles Ag}}}\right)\left(\frac{32\text{ g S}}{\cancel{\text{mole S}}}\right) = 7.4\text{ g S}$$

We could also have calculated the weight of silver sulfide produced. Since two moles of silver give one mole of silver sulfide, the weight of silver sulfide is

$$(50\text{ }\cancel{\text{g Ag}})\left(\frac{1\text{ }\cancel{\text{mole Ag}}}{108\text{ }\cancel{\text{g Ag}}}\right)\left(\frac{1\text{ }\cancel{\text{mole Ag}_2\text{S}}}{2\text{ }\cancel{\text{moles Ag}}}\right)\left(\frac{248\text{ g Ag}_2\text{S}}{\cancel{\text{mole Ag}_2\text{S}}}\right) = 57\text{ g Ag}_2\text{S}$$

Now, since a chemical equation shows the mole relationship of the reactants and products, one can make stoichiometric calculations involving any of the substances that appear in the equation. It should be realized, however, that such calculations indicate only the quantities of substances required or produced if the equation is a true representation of what occurs in the reaction. If the reaction does not proceed to completion—and many do not—this fact would need to be considered in the calculations. Similarly, if the reaction yields products in addition to those that appear in the equation, calculations that are based on the equation are of little value except to show the maximum quantity of a substance that could be produced under ideal conditions.

EXAMPLE 3.6 Methane, CH_4, burns completely in an excess of oxygen to yield carbon dioxide and water. What weight of water is formed in the combustion of 100 g of methane in oxygen?

SOLUTION The equation

$$CH_4 + 2\,O_2 = CO_2 + 2\,H_2O$$

represents the combustion reaction. We are interested in the methane-water relationship, and we note that their mole ratio is 1:2. Therefore, the reaction yields two moles of water for each mole of methane burned.

$$(100\,\cancel{g\,CH_4})\left(\frac{1\,\cancel{mole\,CH_4}}{16.0\,\cancel{g\,CH_4}}\right)\left(\frac{2\,\cancel{moles\,H_2O}}{\cancel{mole\,CH_4}}\right)\left(\frac{18.0\text{ g }H_2O}{\cancel{mole\,H_2O}}\right) = 225\text{ g }H_2O$$

A point that should become apparent here is the simplicity of mole relationships. For instance, if you wanted to know the number of moles of water that would result from the complete combustion of 6.2 moles of methane, a glance at the balanced equation would tell you that 12.4 moles would be produced. The relationship in grams is much less obvious and requires that we first calculate the number of moles in 100 g of methane, relate that to the number of moles of water, and, finally, calculate the weight of that number of moles of water. Mole relationships are much easier to work with than are mass or weight relationships in chemical reactions.

When substances are mixed and allowed to react there will be a residue of one or the other of the reactants unless just stoichiometric amounts were mixed. Often one must first determine which reactant limits the reaction before calculating how much product is formed. The following example illustrates this.

EXAMPLE 3.7 Gaseous ammonia, NH_3, reacts with gaseous hydrogen chloride, HCl, to give solid ammonium chloride, NH_4Cl. What weight of ammonium chloride results from the reaction when 100 g of $NH_3(g)$ and 100 g of $HCl(g)$ are mixed?

SOLUTION In solving this problem, we must determine which of the two reactants limits the reaction. That is, we recognize that in the reaction

$$HCl(g) + NH_3(g) = NH_4Cl(s)$$

one mole of HCl reacts with one mole of NH_3, but that the weight relationship will not be so simple.

First we calculate the number of moles of each reactant:

$$(100 \text{ g } \cancel{HCl})\left(\frac{1 \text{ mole HCl}}{36.5 \text{ g } \cancel{HCl}}\right) = 2.74 \text{ moles HCl}$$

$$(100 \text{ g } \cancel{NH_3})\left(\frac{1 \text{ mole } NH_3}{17.0 \text{ g } \cancel{NH_3}}\right) = 5.88 \text{ moles } NH_3$$

There is an excess of NH_3, and the quantity of NH_4Cl produced is limited by the quantity of HCl. Since one mole of NH_4Cl results from each mole of HCl, the weight of ammonium chloride is

$$(2.74 \cancel{\text{moles HCl}})\left(\frac{1 \cancel{\text{mole } NH_4Cl}}{1 \cancel{\text{mole HCl}}}\right)\left(\frac{53.5 \text{ g } NH_4Cl}{1 \cancel{\text{mole } NH_4Cl}}\right) = 147 \text{ g } NH_4Cl$$

In many chemical reactions, less product results than would be predicted from the equation. This results from an incomplete reaction or from the occurrence of other reactions (side reactions). Example 3.8 illustrates how the percentage yield is taken into account in the calculation.

EXAMPLE 3.8 Naphthalene, $C_{10}H_8$, can be oxidized to phthalic anhydride ($C_8H_4O_3$) by the reaction

$$2\ C_{10}H_8 + 9\ O_2 = 2\ C_8H_4O_3 + 4\ CO_2 + 4\ H_2O$$

In the oxidation of 500 g of naphthalene the yield of phthalic anhydride was just 65% of the theoretical yield. What weight of phthalic anhydride was obtained?

SOLUTION We will calculate the theoretical yield from the equation, and then calculate the actual yield.

$$(500 \text{ g } \cancel{C_{10}H_8})\left(\frac{1 \cancel{\text{mole } C_{10}H_8}}{128 \text{ g } \cancel{C_{10}H_8}}\right)\left(\frac{2 \cancel{\text{moles } C_8H_4O_3}}{2 \cancel{\text{moles } C_{10}H_8}}\right)\left(\frac{148 \text{ g } C_8H_4O_3}{1 \cancel{\text{mole } C_8H_4O_3}}\right) = 578 \text{ g } C_8H_4O_3$$

However, since the yield is 65% of the theoretical yield,

$$(578 \text{ g})(0.65) = 376 \text{ g}$$

are produced.

3.2 Concentrations of Solutions

Most chemical reactions with which we will be concerned are carried out in solutions; that is, the reactants are dissolved in a solvent and react there to give products. We will discuss solutions in some detail in Chapter 8, but for now a few definitions will suffice.

A **solution** is defined as a homogeneous mixture of two or more substances. Later, we will consider solid and gaseous solutions, but our immediate interest is with liquid solutions. One of the components of a liquid solution is referred to—rather arbitrarily—as the **solvent,** and the other components are referred to as **solutes.** In the case of a solution made up of two mutually soluble liquid components, such as alcohol and water, either component could be considered as the solvent, although the common practice is to assign that name to the predominant substance. Thus a liquid solution with water as solvent is called an *aqueous solution.* When the solvent is not specified for a solution, it is to be understood that an aqueous solution is being discussed.

Water and ethyl alcohol are soluble in each other in all ratios. Liquids having this property are described as being miscible.

Weight Percent, Mole Fraction, Molality, and Molarity

Many different systems for expressing the composition of a solution are possible, and several of these are in use. Probably the simplest concentration unit is *weight percent:*

$$\text{Weight percent} = \frac{\text{weight of solute}}{\text{weight of solution}} \times 100 \qquad \text{(dimensionless)}^{\dagger}$$

In its simplicity, however, this concentration unit has the inherent awkwardness of being related to weight rather than to the number of moles. The chemist's work is much easier in moles than in grams. Therefore, the **mole fraction** is more useful:

$$\text{Mole fraction} = \frac{\text{moles of solute}}{\text{moles of solute} + \text{moles of solvent}} \qquad \text{(dimensionless)}^{\dagger}$$

Now we have an expression of concentration that has the great convenience of being based on the number of moles of the components of the solution. For many experiments, mole fraction is the preferred concentration unit.

Closely related to the mole fraction is the *molal concentration* or **molality:**

$$\text{Molal concentration} = \frac{\text{moles of solute}}{\text{kg of solvent}} \qquad \left(\frac{\text{moles}}{\text{kg}}\right)$$

You should notice that while the mole fraction has the total moles of substances in the solution for its denominator, molal concentration has only the mass of *solvent* as the denominator.

† All dimensions (mass, volume, moles, and so on) cancel in the definition of weight percent and mole fraction. These terms are therefore without dimension.

All three of the methods of expressing concentration given so far have the convenience of being invariant with temperature changes. A 1.0-molal solution of sulfuric acid is 1.0 molal at 0° C, at 50° C, and at 150° C, for neither the number of moles of sulfuric acid nor the mass of water changes when the temperature changes, even though the volume does change. On the other hand, all three concentration units suffer from an inconvenience. If a chemist has a solution that is 1.0 molal in sulfuric acid, for example, and he wishes to take just enough of that solution so that he will have 0.50 mole of H_2SO_4, he must first make calculations and then weigh out the desired quantity.† His calculations would go this way: The desired quantity of H_2SO_4 (0.50 mole) weighs

$$(0.50\ \cancel{\text{mole}})\left(\frac{98\ \text{g}}{1\ \cancel{\text{mole}}}\right) = 49\ \text{g}$$

To get this quantity of H_2SO_4 the chemist must also take the water in the solution, the weight of which can be calculated by use of a conversion factor obtained from the molal concentration:

$$(0.50\ \cancel{\text{mole } H_2SO_4})\left(\frac{1\ \cancel{\text{kg } H_2O}}{1.0\ \cancel{\text{mole } H_2SO_4}}\right)\left(\frac{1000\ \text{g } H_2O}{1\ \cancel{\text{kg } H_2O}}\right) = 500\ \text{g } H_2O$$

The weight of solution to be taken is

$$49\ \text{g } H_2SO_4 + 500\ \text{g } H_2O = 549\ \text{g}$$

A similarly time-consuming approach would have been necessary had the concentration of this solution been expressed as mole fraction or as weight percent. Therefore, yet another system of commonly used concentration units is one called *molar concentration* or **molarity:**

$$\text{Molar concentration} = \frac{\text{moles of solute}}{\text{liters of solution}} \qquad \left(\frac{\text{moles}}{\text{liters}}\right)$$

The reference here is to the volume of the *solution* and not to some function of the solvent. Since the solution does undergo volume changes as the result of changes in temperature, and the molarity depends upon the solution volume, the molarity is dependent upon the temperature. Nevertheless, there are two conveniences in having the concentration defined in this way: (1) Volume measurements are easily and quickly made; and (2) numbers of moles are easily translated into volumes of solution. Thus if a chemist wants 0.25 mole of ammonia and the ammonia is available in 1.0-molar solution, she would measure out just

$$(0.25\ \text{mole})\left(\frac{1\ \text{liter}}{1.0\ \text{mole}}\right) = 0.25\ \text{liter of the solution}$$

† Alternatively, he could determine the density of the solution and then measure out the desired volume. Density, or *d*, is the mass per unit volume, for example, g/cm^3.

TABLE 3.1

Concentration Units

NAME	SYMBOL	DEFINITION	DIMENSIONS	NOTES
Weight percent	%	$\frac{\text{g of solute}}{\text{g of solute + g of solvent}} \times 100$	none	Inconvenient if solution is to be used in a reaction, since it is based on weight and not on moles. Not temperature dependent.
Mole fraction	x	$\frac{\text{moles of solute}}{\text{moles of solute+moles of solvent}}$	none	Not temperature dependent.
Molal concentration or molality	m	$\frac{\text{moles of solute}}{\text{kg of solvent}}$	$\frac{\text{moles}}{\text{kg}}$	Not temperature dependent.
Molar concentration or molarity	$M^{\dagger}$	$\frac{\text{moles of solute}}{\text{liter of solution}}$	$\frac{\text{moles}}{\text{liter}}$	Convenient for measurement. Temperature dependent.

† Square brackets around a species refer to the molar concentration of that species. Thus, $[NH_3] = 1.0$ means that the concentration of NH_3 in the solution is 1.0 *M*.

Table 3.1 shows the concentration units we have discussed. You should make the definitions and dimensions of that table part of your working vocabulary.

To understand what various concentration units mean, it is helpful to consider how solutions of known concentration are prepared. The following simple example shows how solutions can be made.

A 10% solution of a simple sugar ($C_6H_{12}O_6$) in water can be prepared by dissolving 10 g $C_6H_{12}O_6$ in 90 g H_2O.

An aqueous solution in which the mole fraction of the sugar is 0.10 can be prepared by dissolving 1.0 mole $C_6H_{12}O_6$ in 9.0 moles H_2O:

$$x_{C_6H_{12}O_6} = \frac{1.0 \text{ mole}}{(1.0 + 9.0) \text{ moles}} = 0.10$$

This requires that one calculate the weight of 1.0 mole of the sugar and the weight of 9.0 moles of water:

$$(1.0 \text{ mole})\left(180 \frac{\text{g}}{\text{mole}}\right) = 180 \text{ g sugar}$$

$$(9.0 \text{ mole})\left(18 \frac{\text{g}}{\text{mole}}\right) = 162 \text{ g water}$$

Therefore, 180 g of sugar are dissolved in 162 g of water to make a solution whose mole fraction of sugar is 0.1.

A 1.0 molal solution of $C_6H_{12}O_6$ in water is prepared by dissolving 1.0 mole (180 g) of $C_6H_{12}O_6$ in 1.0 kg of water.

A 1.0 molar solution is prepared by dissolving 1.0 mole of $C_6H_{12}O_6$ (180 g) in water and then diluting this to a total volume of 1.0 liter. The usual practice is to weigh the sugar into a volumetric flask (a flask calibrated to contain a certain volume of liquid) and add water until the flask is filled to the calibration mark.

Several examples will help illustrate the matter of solution concentrations. In Example 3.9 a solution of known molarity and fixed volume is to be prepared.

EXAMPLE 3.9 How can one prepare just 1.25 liters of 0.200 *M* $Ca(NO_3)_2$ from solid $Ca(NO_3)_2 \cdot 4\,H_2O$?

SOLUTION The $Ca(NO_3)_2$ solution is to contain 0.200 mole of $Ca(NO_3)_2$ per liter and is to have a total volume of 1.25 liters. First, we will calculate the number of moles of $Ca(NO_3)_2$ required:

$$(1.25 \text{ liters})\left(0.200 \frac{\text{mole}}{\text{liter}}\right) = 0.250 \text{ mole } Ca(NO_3)_2$$

$Ca(NO_3)_2 \cdot 4\,H_2O$ is an example of a hydrate salt. Such salts crystallize from aqueous solution with definite quantities of water included in the crystal. The amount of water is usually shown connected by a dot to the formula of the compound; in our case, there are four moles of water per mole of $Ca(NO_3)_2$. When the hydrate salt dissolves, this water of hydration becomes part of the solvent. However, we must include the water of hydration as we calculate the molecular weight of the solute. Moreover, the parentheses in the formula tell us we have two nitrate groups for each calcium in the compound. Hence, the molecular weight of 236.15 is calculated by adding the atomic weight of one calcium (40.08) plus the molecular weights of two nitrates (124.01) and four waters (72.06).

With this molecular weight, we now calculate the weight of 0.250 moles of $Ca(NO_3)_2 \cdot 4\,H_2O$:

$$(0.250 \text{ moles})(236 \text{ g/mole}) = 59.0 \text{ g}$$

Therefore, to prepare the solution one would dissolve 59.0 g $Ca(NO_3)_2 \cdot 4\,H_2O$ in water and dilute the solution to 1.25 liters. Note that the final volume of the solution, and not the volume of the water added, is 1.25 liters.

The next example demonstrates the interrelationship of the various concentration units. Notice that the density of the solution is essential for the calculation of molarity from the weight percent.

EXAMPLE 3.10 A 65.0% (by weight) solution of sulfuric acid has a density of 1.553 g/ml. Calculate the mole fraction, the molality, and the molarity of H_2SO_4.

SOLUTION A simple approach to such a problem is to consider a convenient quantity of the solution and make all calculations on the basis of that quantity. Let us consider 1.000 liter of the 65.0% solution. The weight of that portion is

$$(1.000 \text{ liter})\left(10^3 \frac{\text{ml}}{\text{liter}}\right)\left(1.553 \frac{\text{g}}{\text{ml}}\right) = 1553 \text{ g}$$

Of this, 65.0% is H_2SO_4 while the rest is water. There are $0.350 \times 1553 \text{ g} = 544 \text{ g } H_2O$, and thus there are 1009 g H_2SO_4. The numbers of moles are

$$(0.650)(1553 \text{ g})\left(\frac{1 \text{ mole}}{98.08 \text{ g}}\right) = 10.3 \text{ moles } H_2SO_4$$

and

$$(0.350)(1553 \text{ g})\left(\frac{1 \text{ mole}}{18.01 \text{ g}}\right) = 30.2 \text{ moles } H_2O$$

The mole fraction of H_2SO_4 is

$$x_{H_2SO_4} = \frac{10.3 \text{ moles}}{(10.3 + 30.2) \text{ moles}} = 0.254$$

The molal concentration is

$$m_{H_2SO_4} = \frac{10.3 \text{ moles } H_2SO_4}{0.544 \text{ kg } H_2O} = 18.9$$

and the molar concentration is

$$M_{H_2SO_4} = \frac{10.3 \text{ moles } H_2SO_4}{1.00\text{-liter solution}} = 10.3$$

Example 3.11 illustrates the calculation of the molar concentration of a solution after dilution.

EXAMPLE 3.11 A 50.0 ml portion of 1.00 *M* $AgNO_3$ solution is diluted to 475 ml. What is the molar concentration of $AgNO_3$ in the new solution?

SOLUTION We need only calculate the number of moles of $AgNO_3$ in the new solution and divide this by the new volume, for molar concentration is just the moles per liter.

$$(0.0500 \not{\text{liter}})\left(1.00 \frac{\text{mole}}{\not{\text{liter}}}\right) = 0.0500 \text{ mole } AgNO_3$$

The new volume is 0.475 liter, and the concentration is

$$(0.0500 \text{ mole})\left(\frac{1}{0.475 \cancel{\text{liter}}}\right) = 0.105 \frac{\text{mole}}{\cancel{\text{liter}}} \quad \text{or} \quad 0.105\ M$$

This can all be done in a single expression:

$$(0.0500 \cancel{\text{liter}})\left(1.00 \frac{\text{mole}}{\cancel{\text{liter}}}\right)\left(\frac{1}{0.475 \text{ liter}}\right) = 0.105\ M$$

The following example requires you to consider two sources of the solute: the solid and an existing solution. Since all the latter must be used, according to the statement of the problem, you must be careful to keep an accurate accounting of steps taken in the calculation. The careful use of complete dimensions will prevent many errors.

EXAMPLE 3.12 A student wishes to prepare 0.700 liter of 0.220 *M* KI solution. She has 115 ml of 1.00 *M* KI and only 10.0 g of solid KI. Calculate how she can prepare the desired solution, using all her 1.00 *M* KI solution and some of the solid KI.

SOLUTION She needs

$$(0.700 \cancel{\text{liter}})\left(0.220 \frac{\text{mole}}{\cancel{\text{liter}}}\right) = 0.154 \text{ mole KI}$$

in the final solution. The 1.00 *M* solution contains

$$\left(1.00 \frac{\text{mole}}{\cancel{\text{liter}}}\right)(0.115 \cancel{\text{liter}}) = 0.115 \text{ mole KI}$$

Therefore, she needs to use $0.154 - 0.115 = 0.039$ mole of the solid KI. This weighs

$$(0.039 \cancel{\text{mole}})\left(\frac{166 \text{ g}}{\cancel{\text{mole}}}\right) = 6.5 \text{ g}$$

To prepare the new solution, she should add 6.5 g of solid KI to the 1.00 *M* KI solution and dilute this to 0.700 liter.

Summary

The mole, Avogadro's number of particles or units, is the convenient basis for relating quantities of reactants and products in chemical reactions. The mole concept makes all stoichiometric calculations straightforward.

The dimensions of atomic weight are amu/atom or g/mole.

The dimensions of molecular weight are amu/molecule or g/mole.

The dimensions of Avogadro's number are particles or units of the specified kind per mole.

Several concentration units are in common use among chemists. Each is best defined by its dimensions:

$$\text{mole fraction } (x) = \frac{\text{moles of solute}}{\text{moles of solute + moles of solvent}} \quad \text{(dimensionless and temperature independent)}$$

$$\text{molality } (m) = \frac{\text{moles of solute}}{\text{kg of solvent}} \quad \text{(temperature independent)}$$

$$\text{molarity } (M) = \frac{\text{moles of solute}}{\text{liter of solution}} \quad \text{(temperature dependent)}$$

All stoichiometry problems can be solved by the simple technique of using dimensioned factors accurately.

Important Terms and Concepts

Stoichiometry
Atomic mass
Atomic mass units
Avogadro's number
Mole
Simplest formula or empirical formula
Molecular formula
Solution
Solvent
Solute
Mole fraction
Molality
Molarity

Exercises

3-1.
Units of length on the planet Mars are the garf, ylf, and haw. If 1 garf = 10^3 ylf and 456 ylf = 23 haws, how many garfs are there in 50 haws?

3-2.
If 5.10 g of the compound M_2O_3 contains 2.40 g of oxygen, what is the atomic weight of M?

3-3.
Which amount of matter contains the largest number of atoms: 20.0 g of Na, 72.0 g of Br_2, or 10.0 g of B?

3-4.
Naturally occurring neon has the following isotopic composition:

ISOTOPE	% ABUNDANCE	ISOTOPIC MASS
Ne^{20}	90.51	19.998
Ne^{21}	0.28	20.999
Ne^{22}	9.21	21.997

Calculate the "average" atomic weight of naturally occurring neon.

3-5.
Calculate the number of carbon atoms in each of the following:

(a) 10.0 g of diamond (pure carbon)
(b) 10.0 g of methane (CH_4)
(c) 10.0 g of sucrose ($C_{12}H_{22}O_{11}$)

3-6.
In 200 g of sucrose ($C_{12}H_{22}O_{11}$),

(a) how many moles of sucrose are present?
(b) how many sucrose molecules are present?
(c) how many hydrogen atoms are present?

3-7.
In 1.08 g of quinine ($C_{20}H_{24}N_2O_2$),

(a) how many moles of quinine are there?
(b) how many total atoms are there?

3-8.
What is the mass in grams of 25.0 moles of bromine (Br_2)?

3-9.
Calculate the weight percentage composition of each of the following:

(a) Na_2CO_3
(b) $CaSO_4$
(c) $Na_2S_2O_3 \cdot 5\,H_2O$

3-10.
(a) How many moles of Fe will give the same number of atoms as 45.0 g of carbon?
(b) How many moles of P_4 will give the same number of atoms as 45.0 g of carbon?

3-11.
What is the weight in grams of one molecule of

(a) CH_3OH?
(b) $C_{60}H_{122}$?

3-12.
(a) How many moles of (1) Ba and (2) Cl are contained in 107.0 g of $Ba(ClO_3)_2 \cdot H_2O$?
(b) How many molecules of water of hydration are in this same amount?

3-13.
Hemoglobin, the blood protein responsible for oxygen transport and exchange in the body, contains 0.333% Fe. If each molecule contains four iron atoms, what is the molecular weight of hemoglobin?

3-14.
A given sample of a compound contains 40.1 g of C, 6.67 moles of H atoms, and 2.01×10^{24} atoms of oxygen. Determine the simplest formula of the compound.

3-15.
25.0 g of a hydrocarbon (compound containing only carbon and hydrogen) contains 2.59×10^{24} hydrogen atoms. What is the simplest formula of the compound?

3-16.
A 0.578 g sample of pure tin was treated with gaseous fluorine until the weight of the resulting compound was constant at 0.944 g. What is the simplest formula of the tin fluoride that was formed?

3-17.
Calculate the simplest formula for each of the following compounds:

(a) An oxide of manganese that contains 63.19 weight percent Mn.
(b) An oxide of manganese in which manganese and oxygen are combined in the weight ratio of 1.717 to 1.000.
(c) 50.00 g of a compound of sodium, phosphorus, and oxygen that contains 21.04 g Na and 9.45 g P.
(d) A compound of carbon, hydrogen, and oxygen that is 52.14% C, 13.13% H, and 34.73% O, by weight.

3-18.
A compound of barium and oxygen, when heated, loses part of its oxygen as O_2 gas and is completely converted into BaO. If 1.693 g of the compound produce 1.533 g of BaO, what is the empirical formula of the compound?

3-19.
If 21.3 g (0.0500 mole) of a compound that consists solely of tantalum and oxygen is decomposed completely, 1.09 cm^3 of tantalum (density = 16.6 g/cm^3) is obtained. What is the molecular formula of the compound?

3-20.
14 g of CO gas is mixed with 64 g of oxygen and ignited to form CO_2.

(a) How many moles of CO_2 are formed?
(b) How many grams of carbon are present in the CO_2?
(c) How many oxygen atoms are present in the CO_2?

3-21.
Calculate the weight of H_2O formed when 25.0 g of C_6H_6 are burned. The *unbalanced* equation is $C_6H_6 + O_2 = CO_2 + H_2O$.

3-22.
When sodium hydrogen carbonate is heated, it decomposes according to the equation

$$2\ NaHCO_3(s) = Na_2CO_3(s) + H_2O(g) + CO_2(g)$$

If 0.50 mole $NaHCO_3$ is completely decomposed,

(a) how many grams of Na_2CO_3 are produced?
(b) how many H_2O molecules are formed?

3-23.
0.560 g of ethyl alcohol (C_2H_5OH) is burned completely in O_2 to produce CO_2 and H_2O.

(a) How many grams of CO_2 are produced?
(b) How many O_2 molecules are consumed?

3-24.
How many grams of Ce_2S_3 can be obtained from 10.0 g of Ce and 10.0 g of S?

3-25.
Calculate the weight of H_2O formed when 30.0 g of C_3H_8 are burned. The *unbalanced* equation is

$$C_3H_8 + O_2 = CO_2 + H_2O$$

3-26.
Butane gas (C_4H_{10}) is burned in a container with oxygen gas (O_2). The reaction is

$$2\ C_4H_{10}(g) + 13\ O_2(g) = 8\ CO_2(g) + 10\ H_2O(g)$$

After reaction, the remaining gaseous mixture consists of 24.5 g of H_2O, 9.7 g of O_2, and some CO_2. How many grams each of butane and oxygen were in the flask before combustion?

3-27
A water solution is 38.0% HNO_3 by weight. The density of the solution is 1.23 g/ml.

(a) Calculate the molarity of the solution.
(b) Calculate the mole fraction HNO_3 in the solution.

3-28.
A 10.0% solution of $C_6H_{12}O_6$ in H_2O has a density of 1.25 g/ml.

(a) Calculate the molarity.
(b) Calculate the molality.

3-29.
A 10.00 g portion of glycerine ($C_3H_8O_3$), $d = 1.260$ g/ml, is added to 50.00 g of H_2O, $d = 0.999$ g/ml.

(a) What is the molality of the glycerine in the resultting solution?
(b) What is the mole fraction of water in the resulting solution?

3-30.
A solution of C_2H_5OH (solute) in H_2O is 10.00 *m* and 6.41 *M*. What is the density of the solution?

3-31.
You are given 10.34 g of a solution that contains 0.0543 mole fraction Na in liquid NH_3. How many grams of NH_3 must you evaporate off so that the solution concentrates to 0.1000 mole fraction Na?

3-32.
What volume of dilute nitric acid (19.0% HNO_3 by weight, $d = 1.11$ g/ml) contains 10.0 g HNO_3?

3-33.
What volume of concentrated H_2SO_4 (93.2% H_2SO_4 by weight, $d = 1.835$ g/cm^3) is required to make 500 ml of a 3.00 *M* H_2SO_4 solution?

3-34.
Calculate the volume of water that must be added to 250 ml of a 1.25 *M* solution to make a 0.500 *M* solution. (Assume that the volumes are additive.)

3-35.
How many liters of 0.200 *M* H_3PO_4 solution can be prepared from 33.0 g of H_3PO_4?

3-36.
A solution is made by mixing 57.5 ml of ethyl alcohol (C_2H_5OH) and 300 ml of benzene (C_6H_6).

(a) What is the molality of the solution? (Densities are 0.80 g/ml for C_2H_5OH and 0.90 g/ml for C_6H_6.)
(b) The final volume of the solution is 358.7 ml. What is the molarity of the solution (assuming C_6H_6 is the solvent)?

3-37.
Ammonium sulfate $(NH_4)_2SO_4$, a common fertilizer, is produced by the following series of chemical reactions:

$$3\ H_2 + N_2 = 2\ NH_3$$
$$2\ NH_3 + H_2SO_4 = (NH_4)_2SO_4$$

How many grams of H_2 are required to produce 1.0 kg of $(NH_4)_2SO_4$?

Problems

3-1.
A mixture of KBr and NaBr weighing 0.560 g was treated with aqueous Ag^+, and all the bromide ion was recovered as 0.970 g of pure AgBr. What was the weight of NaBr in the original sample?

3-2.
A 3.50 g sample of a mixture of $NaHCO_3$ and $CaCO_3$ is heated. The compounds decompose as follows:

$$CaCO_3(s) = CaO(s) + CO_2(g)$$
$$2\ NaHCO_3(s) = Na_2CO_3(s) + CO_2(g) + H_2O(g)$$

What percentage of the original mixture was $CaCO_3$ if the solid sample remaining after heating weighs 2.00 g?

3-3.
A commercial preparation of sulfuric acid begins with the burning of sulfur:

$$S(s) + O_2(g) = SO_2(g)$$

The SO_2 is then oxidized by passing SO_2 and O_2 over a catalyst:

$$SO_2(g) + O_2(g) = SO_3(g) \qquad \text{(not balanced)}$$

Dissolving the SO_3 in H_2O produces sulfuric acid:

$$SO_3(g) + H_2O(l) = H_2SO_4(l)$$

In this process:

(a) How many moles of O_2 are required to burn 160 g of S?
(b) How many grams of SO_3 are produced from the complete conversion of 160 g of sulfur into SO_3?
(c) How many grams of sulfur are required to produce 1000 g of H_2SO_4?
(d) How many H_2O molecules are required to produce 1000 g of H_2SO_4?
(e) If the density of SO_2 gas is 2.9 g/liter, how many liters of SO_2 gas are produced from burning 160 g of S?

3-4.
30 g (0.25 mole) of A_3 react completely with B_2 to yield 105 g of A_2B_4 as the only product.

(a) What is the atomic weight of B?
(b) What is the mass of 0.50 moles of A_2B_4?

3-5.
Acetylsalicylic acid (aspirin) can be made by reacting salicylic and acetic acids according to the equation

$$C_7H_6O_3 + C_2H_4O_2 = C_9H_8O_4 + H_2O$$
$$\text{salicylic acid} + \text{acetic acid} = \text{aspirin} + \text{water}$$

(a) Determine the minimum weights of salicylic acid and acetic acid required to make 100.0 g of aspirin.
(b) If this problem has given you a headache, calculate the number of 5-grain (the normal size) aspirin tablets that could be made from this 100.0 g of aspirin (1 grain = 0.0648 g).

3-6.
Suppose the atomic weight scale had been established with the atomic weight of ^{12}C set at 100 instead of 12. A mole of ^{12}C atoms then would weigh 100 g.

(a) What would be the numerical value for Avogadro's number?
(b) What would be the atomic weight of hydrogen?
(c) How many moles of CH_4 would there be in 25 g of CH_4?
(d) Can you think of some good reasons for assigning ^{12}C the atomic weight of 12 rather than 100?

3-7.

It was his first day in the chemistry laboratory. Horatio LeChatelier Smith made an impression. When he pulled his equipment drawer open it fell, breaking everything it contained except the wooden test-tube rack, the nickel spatula, and the box of matches. The blood from his finger, cut while picking up broken glass, appeared ugly on the pan of the analytical balance. He had the largest flame in class from his Bunsen burner, though this was smaller than the flame from the reagent shelf the second time he set it on fire. The heat from the flame did help dry his clothes from the soaking they got in the safety shower after he knocked the nitric acid bottle over, spilling the acid down his pant leg onto his bare foot. He felt tired, so sat on the laboratory bench briefly. Then it was necessary for him to wear his apron backwards because of the sulfuric acid he had spilled on the bench earlier. Still he felt warm inside, partly from his having smelled the concentrated ammonia solution and partly from the obvious consideration and attention he received from the other students, the laboratory instructor, and a latecomer, who proved to be the campus safety officer. It was while the safety officer was talking to him that Horatio, standing with his back to the bench, swung his pony tail over the burner. The fire was quickly extinguished, but the odor of the burned protein and burned greasy kid stuff lingered. The odor gave most students a headache but gave Horatio an idea. Since hair consists of the protein keratin, which contains approximately two sulfur atoms per protein molecule, he thought he should be able to estimate the molecular weight of the protein if he could determine the weight of sulfur in a known weight of hair. He washed his face thoroughly, and with a sense of sacrificing for science he shaved his beard and moustache, carefully saving each whisker. The beard was Horatio's greatest pride. He had begun it 16 months earlier and had watched with joy as it grew. Alas, although the balance had a sensitivity of 0.1 mg, it showed no deflection when the entire beard was added to the pan. Therefore, Horatio washed his hair, even though it was only September, and had it cut. He weighed a quarter of the clippings. He ignited this 500 g sample in an abundance of oxygen and collected the SO_2 produced. This he found to weigh just 4.00 g. Help Horatio by calculating the approximate molecular weight of the molecules in the protein of the hair.

3-8.

(a) A fine young lad, Horatio LeChatelier Smith, filled with an optimism based on innocence, considered the possibility of helping his fellow chemistry students comprehend the magnitude of the mole. His plan was to mark the length of a mole of inches on the wall of each handball court in the physical education building. When the wall proved to be somewhat too short, he decided that it was time that he join the rest of the world and use a unit of length from the metric system. He obtained a piece of glass rod that was 1.0 mm diameter, sharpened his pencil, and using the thickness of the rod for his guide, made careful marks 1.0 mm apart on all four maple walls of the court. You can imagine his disappointment after counting three times to find that he had only 36,576 marks. Therefore, he went to the football field and started on the 0-yard line. Where should Horatio draw the line at the end of a mole of millimeters? The football field is 100 yards long from goal line to goal line. The great circle route from his college town to Tokyo is 8,030 miles. The Milky Way (our galaxy) is estimated to be 1×10^5 light years in diameter. The faintest extragalactic systems photographed with a 100-inch telescope are at a distance of about 5×10^8 light years. The speed of light is 3.0×10^{10} cm/sec.

(b) Horatio felt disappointed that his representation of a mole of millimeters was so outlandish and offearthish. An experience he had in a speech class suggested a better approach. In that class the upper limit of the ear's sensitivity to high frequencies was demonstrated. Most students could detect sounds up to only about 15,000 hertz (cycles/sec). Two students could clearly hear sounds at 35,000 hertz, the limit of the frequency generator. One of these was Horatio and the other was a female springer spaniel named Violet that had sat in the front row for four days without being noticed by the teacher. Eureka! Horatio would determine the time required for either of the two outstanding students, him or Violet, to hear a mole of cycles at the maximum frequency of the generator. What is the length of time required, expressed in millenia? (A millenium is a thousand years.)

4

Chemical Reactions

In Chapter 3 we described how to derive stoichiometric calculations from chemical equations, and, further, we discussed the concentrations of solutions and their use in stoichiometric calculations. With this background we can now consider two major classes of chemical reactions: *acid-base reactions* and *oxidation-reduction reactions*. We will also make a first brief exploration of the field of *chemical equilibrium*.

4.1 Acids and Bases

There are several definitions of acids and bases in common use among chemists, but we will discuss just one in this chapter. Later, in Chapter 10, we will consider other definitions.

Arrhenius or Classical Theory of Acids and Bases

In 1887, the Swedish physical chemist Svante Arrhenius proposed definitions in which an **acid** was said to be a hydrogen-containing substance that dissolves in water to give hydrogen ions (H^+), and a **base** was said to be a substance that contains the OH group and dissociates in water to give hydroxide ions (OH^-).

Note that these definitions are restricted to the behavior of substances in aqueous solutions.

The presence of hydrogen does not qualify a substance as an acid. Neither NH_3 nor CH_4 is an acid under the Arrhenius definition, since neither dissociates to give hydrogen ions in aqueous solution. Similarly, $SO_2(OH)_2$ (which is just H_2SO_4 written in an unexpected way) fails to qualify as a base, since it dissociates to give H^+ and not OH^-.

Do not confuse *strength* with *concentration*. A solution of a strong acid may be dilute or concentrated, as may a solution of a weak acid.

An acid is called a **strong acid** if it dissociates completely in water to give H^+. A **weak acid** is one that does not dissociate completely. Thus, sulfuric acid is a strong acid because the reaction

$$H_2SO_4 = H^+ + HSO_4^-$$

is complete in dilute aqueous solution. Since the second step does not take place completely,

$$HSO_4^- = H^+ + SO_4^{2-}$$

we refer to HSO_4^- (hydrogen sulfate ion) as a weak acid.

An aqueous solution of sulfuric acid contains H^+, HSO_4^-, and SO_4^{2-}, but no H_2SO_4.

Some common strong acids are

H_2SO_4	sulfuric acid
HNO_3	nitric acid
HCl	hydrochloric acid
$HClO_4$	perchloric acid

and some common weak acids are

CH_3COOH	acetic acid
HCOOH	formic acid
H_3PO_4	phosphoric acid
H_3AsO_3	arsenious acid

Similarly, a **strong base** is one that ionizes completely to give hydroxide ions. Examples are

NaOH	sodium hydroxide
KOH	potassium hydroxide
LiOH	lithium hydroxide
$Ba(OH)_2$	barium hydroxide

The most common **weak bases** are ammonia and its derivatives, and in the Arrhenius theory these are imagined to have reacted with water to give a hydroxide:

$$NH_3 + H_2O = NH_4OH$$

$$CH_3NH_2 + H_2O = CH_3NH_3OH$$

However, these reactions are not good representations of what occurs, because "ammonium hydroxide," NH_4OH, has never been isolated or even detected in solution. We mention the name and formula because of the historical interest, and because aqueous ammonia solutions are commonly sold and labeled as ammonium hydroxide. A better representation of the reactions of aqueous ammonia will be given later.

Neutralization

When an acid and a base react, they neutralize each other and produce water and **salt.** Salts are often given names that reflect the acids from which they are formed. The relationship between the name of an acid and the salt that results from its **neutralization** is quite consistent for oxygen-containing acids, as Table 4.1 shows.

For the common binary acids the relationship of names is less simple. When, for example, the hydrogen halides (HF, HCl, HBr, HI) are dissolved in water, they form the hydrohalic acids:

HF(*aq*)	hydrofluoric acid	HBr(*aq*)	hydrobromic acid
HCl(*aq*)	hydrochloric acid	HI(*aq*)	hydroiodic acid

The salts of these acids are the fluorides, chlorides, bromides, and iodides.

TABLE 4.1

Some Oxygen Acids and Their Salts

ACID		SALT	
H_2SO_4	sulfuric acid	Na_2SO_4	sodium sulfate
H_3PO_4	phosphoric acid	K_3PO_4	potassium phosphate
HNO_3	nitric acid	KNO_3	potassium nitrate
H_2SO_3	sulfurous acid	Na_2SO_3	sodium sulfite
H_3PO_3	phosphorous acid	Na_2HPO_3	sodium monohydrogen phosphite
CH_3COOH	acetic acid	CH_3COONa	sodium acetate

The oxygen acid with an *ic* ending gives, upon neutralization, a salt with an *ate* ending. Similarly, the *ous* acid yields an *ite* salt.

Consider now the neutralization reaction between hydrochloric acid and sodium hydroxide:

$$HCl + NaOH = H_2O + NaCl$$

That this is a strong acid-strong base reaction can be emphasized by use of an ionic representation:

$$H^+Cl^- + Na^+OH^- = H_2O + Na^+Cl^-$$

Much more commonly used is the net ionic equation that is obtained if the nonreacting species are omitted:

$$H^+ + OH^- = H_2O \qquad [4.1]$$

This shows that the fundamental strong acid-strong base reaction in the Arrhenius theory is just the formation of water from hydrogen ions and hydroxide ions. Equation [4.1] is the neutralization reaction if both the acid and the base are strong.

If the weak acid acetic acid reacts with the strong base sodium hydroxide, the molecular equation is written

$$CH_3COOH + NaOH = H_2O + CH_3COONa$$

Since acetic acid is weak, it is only slightly dissociated, and CH_3COOH is the predominant species in an aqueous solution.

and the net ionic equation is

$$CH_3COOH + OH^- = H_2O + CH_3COO^- \qquad [4.2]$$

It is evident that a significant change resulted from the substitution of a weak acid for the strong acid of equation [4.1]. Water is again a product, but a proton had to be removed from acetic acid in order for the water to be formed.

The Arrhenius theory serves as a model that fits many observations and thus helps to simplify chemists' understanding of a large group of reactions. Most simple acid-base reactions that occur in water solution can be conveniently discussed in terms of the Arrhenius definitions. After you have had an opportunity to develop confidence in the use of this theory we will discuss other broader acid-base theories. The following examples show the application of stoichiometric calculations to acid-base reactions.

EXAMPLE 4.1 Calculate the volume of 0.20 *M* HCl that just reacts with a solution that contains 0.30 g of $Ba(OH)_2$.

SOLUTION The reaction between the strong acid HCl and the strong base $Ba(OH)_2$ is

$$2\,HCl + Ba(OH)_2 = 2\,H_2O + BaCl_2$$

A dimensioned expression can be written:

$$(0.30 \text{ g Ba(OH)}_2)\left(\frac{1 \text{ mole Ba(OH)}_2}{171 \text{ g Ba(OH)}_2}\right)\left(\frac{2 \text{ mole HCl}}{1 \text{ mole Ba(OH)}_2}\right)\left(\frac{1 \text{ liter}}{0.20 \text{ mole HCl}}\right) = 0.018 \text{ liter}$$

Therefore 0.018 liter (or 18 ml) of 0.20 *M* HCl just reacts with 3.0 g $Ba(OH)_2$.

EXAMPLE 4.2 What weight of H_3PO_4 just reacts with 1.0 mole of NaOH, if PO_4^{3-} is the product?

SOLUTION The reaction is

$$H_3PO_4 + 3\ NaOH = 3\ H_3O + Na_3PO_4$$

or

$$H_3PO_4 + 3\ OH^- = 3\ H_2O + PO_4^{3-}$$

The dimensioned expression by which one calculates the answer is

$$(1.0 \text{ mole NaOH})\left(\frac{1 \text{ mole H}_3\text{PO}_4}{3 \text{ mole NaOH}}\right)\left(\frac{98 \text{ g H}_3\text{PO}_4}{1 \text{ mole H}_3\text{PO}_4}\right) = 33 \text{ g H}_3\text{PO}_4$$

EXAMPLE 4.3 What volume of 0.25 *M* HCl just reacts with 0.050 liter of 0.21 *M* KOH?

SOLUTION The equation for the reaction is

$$HCl + KOH = H_2O + KCl$$

or

$$H^+ + OH^- = H_2O$$

The dimensioned expression for solving the problem is

$$(0.050 \text{ liter KOH})\left(\frac{0.21 \text{ mole KOH}}{1 \text{ liter KOH}}\right)\left(\frac{1 \text{ mole HCl}}{1 \text{ mole KOH}}\right)\left(\frac{1 \text{ liter HCl}}{0.25 \text{ mole HCl}}\right) = 0.042 \text{ liter}$$

Thus we see that the dimensional approach to making stoichiometric calculations fits the acid-base reactions perfectly. A simple dimensioned expression can be written for the problem, and this expression yields the correct result.

4.2 Oxidation-Reduction Reactions

An important class of reactions is called **oxidation-reduction reactions** or **redox reactions.** The name oxidation had its origin with reactions in which oxygen was added to a substance. Reduction then meant the removal of oxygen.

We will see that the meanings of the terms are now expanded to cover a much broader class of reactions.

Oxidation Number

Numbers called **oxidation numbers** are assigned to atoms in a consistent way. In the case of atoms or simple ions the oxidation numbers are related to the charge, as can be seen in the following examples:

SPECIES	OXIDATION NUMBER	SPECIES	OXIDATION NUMBER
Fe	0	Na	0
Fe^{2+}	+2	Na^{+}	+1
Fe^{3+}	+3	Ba	0
Cl_2	0	Ba^{2+}	+2
Cl^{-}	−1	H_2	0
He	0	H^{+}	+1

You will notice that each substance in its elementary state is assigned the oxidation number 0. This is true regardless of whether the substance exists as a single atom (He) or as a polyatomic molecule (H_2). An atom that has lost or gained electrons to become an ion has an oxidation number that is the same magnitude and sign as the charge on the ion.

Two additional rules enable one to assign oxidation numbers to elements in compounds and complex ions.

(1) The sum of the oxidation numbers of all the atoms in a molecule or complex ion is equal in sign and magnitude to the net charge on that molecule or ion.

(2) Many elements *in compounds* can be assigned oxidation numbers by a priority system. In the sequence that follows, rule (a) takes precedence over (b), and so forth.

Increasing Priority of Rule ↑

(a) The *alkali metals* (Group IA of the periodic table; Li, Na, K, Rb, Cs, Fr) all have oxidation number *+1*.

The *alkaline earth metals* (Group IIA; Be, Mg, Ca, Sr, Ba, Ra) all have oxidation number *+2*.

Fluorine has oxidation number *−1*. The remaining *halogens* (Group VIIA; Cl, Br, I) have oxidation number *−1* in binary compounds with metals or hydrogen.

(b) *Hydrogen* has oxidation number *+1*.

(c) *Oxygen* has oxidation number *−2*. Similarly, in binary compounds with metals or hydrogen, *sulfur* has oxidation number *−2*.

These rules cover the majority of cases that we will encounter and will allow us to proceed with a consideration of oxidation-reduction reactions.

In determining oxidation numbers, one does well to notice that metals in compounds have only positive oxidation numbers. Moreover, an element in a given complex ion always has the same oxidation number. In NO_3^- the nitrogen is always +5; in SO_4^{2-} the sulfur is +6; in PO_4^{3-} the phosphorus is +5; and so on. Thus it is not necessary to recalculate the oxidation number of an element in a complex ion each time it is encountered.

You should be aware that all oxidation numbers specified in rule (2) above are the oxidation numbers that would be expected if the atoms gained or lost enough electrons to achieve the noble-gas configuration. For example, an alkali metal atom, M, becomes M^+ (oxidation number +1) when it loses an electron and reaches the configuration of the nearest noble gas. Similarly, the loss of two electrons from an alkaline earth atom gives it oxidation state +2 and the noble-gas configuration. As we apply the rules we will point out other correlations between oxidation numbers and electronic configurations.

The application of the rules is shown below:

$KMnO_4$ Potassium and oxygen have oxidation numbers +1 and −2 [rules (2a) and (2c)]. The oxidation number of manganese is obtained by noting that there is no net charge on the molecule and that, therefore, the sum of oxidation numbers of all the atoms is zero. If we let x be the oxidation number of manganese, then $1(+1) + 1(x) + 4(-2) = 0$, and $x = +7$. This is the maximum oxidation number predicted for manganese from its position in Group VIIB of the periodic table. By losing (through unequal sharing) seven electrons, manganese achieves the argon configuration.

$FeCl_2$ The chlorine is −1 [rule (2a)], and the iron must therefore be +2.

MnO_2 Oxygen is −2 [rule (2c)], and manganese is +4. In both this and the preceding example we observe a metal atom with an oxidation number lower than the maximum allowed value. It is particularly true of transition elements that several such lower oxidation numbers are observed.

Na_2O_2 Sodium is +1 [rule (2a), which takes precedence over (2c)], and oxygen is −1. (Compounds in which oxygen has oxidation state −1 are called **peroxides.**)

ZnS Sulfur is −2 [rule (2c)], and zinc is +2.

KH Potassium is +1 [rule (2a), which takes precedence over (2b)], and hydrogen is −1. (Compounds in which hydrogen has oxidation state −1 are called **hydrides.**)

HPO_4^{2-} Hydrogen is +1 [rule (2b)], oxygen is −2 [rule (2c)], and phosphorus must be +5 in order that the sum of all oxidation numbers be equal to the net charge of 2− on the ion.

Prediction of Oxidation Numbers from the Periodic Table

In a molecule with a polar covalent bond, oxidation numbers are assigned as if the shared electrons were transferred completely to the more electronegative element. Thus, in the CO_2 molecule (O═C═O), the more electronegative oxygen atoms are assigned oxidation number −2 as if they had increased from six to eight valence electrons each, and the carbon atom has oxidation number +4 as if it had lost its four valence electrons.

The rules given above allow one to calculate oxidation numbers for elements in compounds and complex ions. It is often helpful to check after having calculated an oxidation number to see if the value obtained is reasonable. A simple system that permits such a check is based on the electronic configurations of the elements. The noble-gas configuration is particularly stable; we will see that oxidation numbers can often be predicted on the assumption that the elements have attained the noble-gas configuration by loss or gain of electrons. (In ions the electrons have been transferred, while in polar covalent bonds the electrons are shared unequally.) If we keep in mind the restriction that metals do not display negative oxidation numbers, and add the observation that in chemical reactions elements do not lose or gain more electrons than would be required to give the noble-gas configuration, we can readily predict oxidation numbers for the elements of periods 2 and 3:

Li	Be	B†	C	N	O‡	F‡
+1	+2	+3	+4	+5	−2	−1
			−4	−3		
Na	**Mg**	**Al**	**Si†**	**P**	**S**	**Cl**
+1	+2	+3	+4	+5	−2	−1
				−3	+6	+7

We observe that in periods 4 and higher the presence of *d* electrons would require the loss by elements at the right side of the table of very large numbers of electrons to attain the noble-gas configuration. However, if only the *s* and *p* electrons of the highest energy level are lost, a *pseudo-noble-gas configuration* results—that is, the filled electron energy levels typical of the noble gas plus a filled *d* sublevel. (In period 6 there is also the filled 4*f* sublevel.) This pseudo-noble-gas configuration is sufficiently stable that it can be used for predicting positive oxidation numbers of elements of the A groups in periods 4, 5, and 6. Thus As, Sb, and Bi in Group VA all form compounds in which their oxidation number is +5; Ge, Sn, and Pb in Group IVA display the +4 oxidation number; and so on. Moreover, the members of these A groups have a marked tendency to have the oxidation numbers that would be represented by the loss of only their *p* valence electrons. For an element, this oxidation number is two units less positive than that which is characteristic of the pseudo-noble-gas configuration (the two *s* electrons remain) and is referred to as the *pseudo-noble-gas plus 2 configuration*.

EXAMPLE 4.4 Assuming noble-gas, pseudo-noble-gas, and pseudo-noble-gas plus 2 configurations, predict oxidation numbers for the following elements: Tl, N, As, I, Pb.

† We used caution in the case of these metalloids and did not predict the negative oxidation number.
‡ Because fluorine and oxygen are the most electronegative elements, we did not predict positive oxidation numbers.

SOLUTION A tabular arrangement is convenient for presenting the solution:

ELEMENT	NOBLE-GAS OXIDATION STATE	PSEUDO-NOBLE-GAS OXIDATION STATE	PSEUDO-NOBLE-GAS +2 OXIDATION STATE
Tl	None	+3	+1
N	+5, −3	None	None
As	−3	+5	+3
I	−1	+7	+5
Pb	None	+4	+2

Figure 4.1 is a periodic table in which common oxidation numbers are listed along with the symbols of the elements. You should study this figure to see that a large number of the oxidation numbers can be predicted correctly by use of the periodic table.

Oxidation-Reduction

Oxidation is most readily defined as an *increase in oxidation number* or as a change to a more positive oxidation number. In each of the following steps an oxidation is shown.

> There is no requirement that in oxidation the final oxidation number be positive. For example, ammonia (NH_3) is oxidized to become hydroxylamine (NH_2OH) and the oxidation number changes from −3 to −1.

$$\underset{+2}{Mn^{2+}} \text{ to } \underset{+7}{MnO_4^-} \qquad \underset{0}{I_2} \text{ to } \underset{+5}{IO_3^-}$$

$$\underset{-1}{Cl^-} \text{ to } \underset{+5}{ClO_3^-} \qquad \underset{-3}{NH_3} \text{ to } \underset{-1}{NH_2OH}$$

$$\underset{-1}{Br^-} \text{ to } \underset{0}{Br_2}$$

Reduction is defined as a *decrease in oxidation number* or as a change to a less positive oxidation number. The reverse of each step shown in the examples above is a reduction.

The changes called oxidation and reduction can be seen by examination of a series of species involving the same element with different oxidation numbers. For example, the oxidation number of nitrogen in the following series of compounds ranges all the way from +5 to −3:

HNO_3	+5
NO_2	+4
N_2O_3	+3
NO	+2
N_2O	+1
N_2	0
NH_2OH	−1
N_2H_4	−2
NH_3	−3

Figure 4.1
Common oxidation states of the elements.

IA	IIA	IIIB	IVB	VB	VIB	VIIB	VIII			IB	IIB	IIIA	IVA	VA	VIA	VIIA	VIIIA
H −1 1																	**He**
Li 1	**Be** 2											**B** 3	**C** 4	**N** −3 1 2 3 4 5	**O** −1 −2	**F** −1	**Ne**
Na 1	**Mg** 2											**Al** 3	**Si** 4	**P** 3 5	**S** −2 2 4 6	**Cl** −1 1 3 5 7	**Ar**
K 1	**Ca** 2	**Sc** 3	**Ti** 2 3 4	**V** 2 3 4 5	**Cr** 2 3 6	**Mn** 2 3 4 7	**Fe** 2 3	**Co** 2 3	**Ni** 2	**Cu** 1 2	**Zn** 2	**Ga** 3	**Ge** 4	**As** 3 5	**Se** 4 6	**Br** −1 1 3 5 7	**Kr**
Rb 1	**Sr** 2	**Y** 3	**Zr** 3 4	**Nb** 5	**Mo** 3 4 6	**Tc**	**Ru** 3 5	**Rh** 3	**Pd** 2 3	**Ag** 1 2	**Cd** 2	**In** 3	**Sn** 2 4	**Sb** 3 5	**Te** 4 6	**I** −1 1 3 5 7	**Xe**
Cs 1	**Ba** 2	**La** 3	**Hf** 4	**Ta** 5	**W** 6	**Re** 4 6	**Os** 4 6 8	**Ir** 4 6	**Pt** 2 4	**Au** 1 3	**Hg** 1 2	**Tl** 1 3	**Pb** 2 4	**Bi** 3 5	**Po**	**At**	**Rn**

Any change from a substance in the series to a substance above it is an oxidation. Conversely, any change downward in the series is a reduction.

In a complete reaction that involves the oxidation of one substance, there must also be a reduction of a substance. Consider the following reaction, showing the oxidation numbers of substances oxidized and reduced:

$$\underset{+7}{MnO_4^-} + 5\,\underset{+2}{Fe^{2+}} + 8\,H^+ = \underset{+2}{Mn^{2+}} + 5\,\underset{+3}{Fe^{3+}} + 4\,H_2O \qquad [4.3]$$

Manganese in permanganate ion is reduced from oxidation number +7 to +2, while iron is oxidized from +2 to +3. Permanganate ion is called the *oxidizing agent* because it oxidizes the ferrous ion. Ferrous ion, on the other hand, is called the *reducing agent* because it reduces the manganese in permanganate ion to manganous ion (or, more commonly, because it reduces the permanganate ion to manganous ion). The reducing agent is oxidized, and the oxidizing agent is reduced.

Oxidation-Reduction Equivalents

It is useful to think of a unit of reaction as something that always reacts with another unit in a one-to-one ratio. Such a unit of reaction, called an **equivalent,** is defined in terms of a *transfer of one mole of electrons*. Therefore, *an equivalent of an oxidizing agent is that quantity of the substance that will consume 1 mole (Avogadro's number) of electrons. An equivalent of a reducing agent will produce 1 mole of electrons*. In equation [4.3], 1 mole of Fe^{2+} releases 1 mole of electrons in becoming Fe^{3+}; therefore, there is 1 equivalent per mole of Fe^{2+}. One mole of MnO_4^- consumes 5 moles of electrons in being reduced to Mn^{2+}; therefore, there are 5 equivalents per mole of MnO_4^-.

Returning to equation [4.3], we see that we can readily determine the number of equivalents per mole of each reactant directly from that equation. The oxidation number of manganese changes from +7 to +2 for a total change of 5 oxidation-number units. The permanganate ion contains just 1 manganese atom so there are 5 equivalents per mole of permanganate ion. The oxidation number of iron changes from +2 to +3 for a total change of 1 oxidation-number unit. Therefore, there is 1 equivalent per mole of ferrous ion (Fe^{2+}).

We can now make use of the equivalent in balancing oxidation-reduction equations. First, we will consider the reaction of permanganate ion and ferrous ion discussed above. If we know that in acidic solution (hence, an abundance of H^+ and H_2O) the permanganate ion is reduced to manganous ion (Mn^{2+}) and the ferrous ion is oxidized to ferric ion (Fe^{3+}), we can balance the equation. First, we write the known reactants and products in the appropriate places, and then we write the oxidation number of each substance oxidized or reduced below it. Below that, we write the number of equivalents per mole of each of the reactants.

$$\overbrace{MnO_4^- + Fe^{2+} = Mn^{2+}}^{\text{reduction}} + Fe^{3+} \qquad \text{(unbalanced)}$$

$$\begin{array}{cccc} +7 & +2 & +2 & +3 \end{array}$$

(oxidation: Fe^{2+} (+2) → Fe^{3+} (+3))

$$\frac{5\ \text{eq}}{\text{mole}} \qquad \frac{1\ \text{eq}}{\text{mole}}$$

Now, since 1 equivalent of MnO_4^- reacts with 1 equivalent of Fe^{2+} (by the definition of the equivalent), it is evident that 5 moles of Fe^{2+} are required to provide as many equivalents as there are in 1 mole of MnO_4^-:

$$(1\ \text{mole}\ MnO_4^-)\left(\frac{5\ \text{eq}}{\text{mole}}\right) = 5\ \text{eq}\ MnO_4^-$$

$$(5\ \text{moles}\ Fe^{2+})\left(\frac{1\ \text{eq}}{\text{mole}}\right) = 5\ \text{eq}\ Fe^{2+}$$

Therefore, the partially balanced equation becomes

$$MnO_4^- + 5\ Fe^{2+} = Mn^{2+} + 5\ Fe^{3+} \qquad \text{(partially balanced)}$$

We have now adjusted the oxidation-reduction equivalents so that they are equal; we have as many manganese atoms on one side of the equation as on the other, and the same is true of the iron atoms.

Next the charges are balanced. There are 9 positive charges on the left and 17 on the right, indicating a deficit of 8 positive charges on the left. The species that are available in an acidic aqueous solution are H^+ and H_2O, and since H^+ carries a positive charge, we add 8 H^+ to the left. Then we add water as required to balance the hydrogen atoms in the equation:

$$MnO_4^- + 5\ Fe^{2+} + 8\ H^+ = Mn^{2+} + 5\ Fe^{3+} + 4\ H_2O$$

Do not confuse *charge* with *oxidation number*. The oxidation numbers are balanced by having the same number of equivalents of oxidizing agent and reducing agent. The electrical charges on the ions are balanced on the two sides of the equation with H^+ or OH^-, depending on whether the solution is acidic or alkaline.

The charge balance is now complete, as is the balance of hydrogen. A check shows that there are four oxygen atoms on each side, enabling us to conclude that the equation is balanced.

An alternative and equally acceptable approach to balancing equations makes use of **half-reactions.** Half-reactions separate the part of the equation that involves oxidation from the part that involves reduction. Thus, in the reaction of MnO_4^- and Fe^{2+} that we just balanced, the part of the equation in which MnO_4^- is reduced to Mn^{2+} is separated from the part in which Fe^{2+} is oxidized to Fe^{3+}. We will start with the unbalanced equation, separate this into half-reactions, balance the half-reactions, and then combine them to give the balanced equation. The unbalanced equation is

$$MnO_4^- + Fe^{2+} = Mn^{2+} + Fe^{3+}$$

The iron and manganese species are separated into two unbalanced half-reactions, and each is written with oxidation numbers and the number of equivalents per mole:

$$\underset{+7}{MnO_4^-} = \underset{+2}{Mn^{2+}}$$

$$\frac{5 \text{ eq}}{\text{mole}}$$

$$\underset{+2}{Fe^{2+}} = \underset{+3}{Fe^{3+}}$$

$$\frac{1 \text{ eq}}{\text{mole}}$$

Now, since the equivalent is defined in terms of a mole of electrons, we write in the number of electrons required:

$$MnO_4^- + 5e^- = Mn^{2+}$$
$$Fe^{2+} = Fe^{3+} + e^-$$

Notice that the electrons must be written on the side of the half-equation that contains the oxidized species. The charge balance is achieved by adding H^+ (the reaction was specified in acid solution) as required. Then hydrogen and oxygen are balanced by adding water.

$$MnO_4^- + 8\,H^+ + 5e^- = Mn^{2+} + 4\,H_2O$$
$$Fe^{2+} = Fe^{3+} + e^-$$

Now that the half-reactions are balanced, the two are combined to give the balanced equation. The second half-reaction is first multiplied by 5 in order that the electrons cancel upon addition:

$$MnO_4^- + 5\,Fe^{2+} + 8\,H^+ = Mn^{2+} + 5\,Fe^{3+} + 4\,H_2O$$

You should use whichever of the methods of balancing you prefer; if you understand the principles you should be able to balance any equation using either method.

It is clear from the half-reactions that MnO_4^- gains electrons in being reduced and that Fe^{2+} loses electrons in being oxidized. For this reason, oxidation is sometimes defined as a loss of electrons and reduction as a gain of electrons. Since we will usually be working with complete balanced equations in which no electrons appear, and since the word *reduction* ordinarily connotes loss or removal, it will probably be easier for you to use the definitions given earlier: Oxidation is an increase in oxidation number, and reduction is a decrease in oxidation number.

We will make further use of half-reactions in Chapter 11 when working with electrochemical reactions. The following examples give practice in balancing oxidation-reduction equations.

EXAMPLE 4.5 Iodate ion (IO_3^-) is reduced in acidic solution to iodine (I_2) by iodide ion (I^-), which is oxidized to iodine. Write a balanced equation for the reaction.

SOLUTION We can write the reactants and products, their oxidation numbers, and equivalents per mole:

reduction

$$\underset{+5}{IO_3^-} + \underset{-1}{I^-} = \underset{0}{I_2} \qquad \text{(unbalanced)}$$

oxidation

$$\frac{5\text{ eq}}{\text{mole}} \qquad \frac{1\text{ eq}}{\text{mole}}$$

Therefore, 5 moles of I^- react with 1 mole of IO_3^-, and the mass balance of iodine atoms requires that 3 moles of I_2 be formed.

$$IO_3^- + 5\,I^- = 3\,I_2 \qquad \text{(partially balanced)}$$

The charge balance is then performed:

$$IO_3^- + 5\,I^- + 6\,H^+ = 3\,I_2 + 3\,H_2O$$

and a check on the oxygen on each side shows that the equation is balanced.

EXAMPLE 4.6 In alkaline solution, Cl_2 oxidizes thiosulfate ion ($S_2O_3^{2-}$) to sulfate ion. The chlorine is reduced to chloride ion. Write the balanced equation for the reaction.

SOLUTION The reactants, products, oxidation numbers, and equivalents per mole are determined and written in proper order:

reduction

$$\underset{0}{Cl_2} + \underset{+2}{S_2O_3^{2-}} = \underset{-1}{Cl^-} + \underset{+6}{SO_4^{2-}}$$

oxidation

$$\frac{2\text{ eq}}{\text{mole}} \qquad \frac{8\text{ eq}}{\text{mole}}$$

Notice that although chlorine changes oxidation numbers only from 0 to -1, there are 2 equivalents per mole of Cl_2 because of the diatomic molecule. The same argument leads to 8 equivalents per mole of thiosulfate ion. Hence, 4 moles of Cl_2 react with 1 mole of $S_2O_3^{2-}$, and the balance of chlorine and sulfur can be made:

$$4\,Cl_2 + S_2O_3^{2-} = 8\,Cl^- + 2\,SO_4^{2-} \qquad \text{(partially balanced)}$$

We now balance the charges, observing that in the alkaline solution the species that are available in abundance are OH^- and H_2O. The net charge on the left of the partially balanced equation is -2 while that on the right is -12. We therefore add 10 OH^- to the left and 5 H_2O to the right (we chose 5 H_2O to keep the hydrogen atoms balanced):

$$4\,Cl_2 + S_2O_3^{2-} + 10\,OH^- = 8\,Cl^- + 2\,SO_4^{2-} + 5\,H_2O$$

A check of the oxygen shows that it, too, is balanced and that therefore the equation is balanced.

When a reacting species in an oxidation-reduction reaction contains more than one atom of the element that is oxidized or reduced (as Cl_2 and $S_2O_3^{2-}$ in Example 4.6), one must be careful to use the total change in oxidation number units. In Example 4.7, $Cr_2O_7^{2-}$ is reduced and requires the same care.

EXAMPLE 4.7 The dichromate ion, $Cr_2O_7^{2-}$, oxidizes Fe^{2+} to Fe^{3+} in acid solution and is reduced to Cr^{3+}. Write the balanced equation for this reaction.

SOLUTION The partial equation is

$$\underset{+2}{Fe^{2+}} + \underset{+6}{Cr_2O_7^{2-}} = \underset{+3}{Cr^{3+}} + \underset{+3}{Fe^{3+}}$$

$$\frac{1\text{ eq}}{\text{mole}} \qquad \frac{6\text{ eq}}{\text{mole}}$$

The $Cr_2O_7^{2-}$ has 6 equivalents per mole, because chromium is reduced from $+6$ to $+3$ and there are two chromium atoms in the ion. Balancing the iron and chromium we obtain:

$$6\,Fe^{2+} + Cr_2O_7^{2-} = 2\,Cr^{3+} + 6\,Fe^{3+}$$

Now we balance the charge, oxygen, and hydrogen, and have the completed equation:

$$6\,Fe^{2+} + Cr_2O_7^{2-} + 14\,H^+ = 2\,Cr^{3+} + 6\,Fe^{3+} + 7\,H_2O$$

The next two examples introduce a reaction in which there is disproportionation and a reaction in which oxygen is present as the peroxide. In *disproportionation* a single substance is both oxidized and reduced. In peroxides, oxygen has oxidation number -1.

EXAMPLE 4.8 In basic solution, hypochlorite ion (OCl^-) disproportionates to chloride ion and chlorate ion (ClO_3^-). Write a balanced equation for the reaction.

SOLUTION The incomplete equation for the reaction given is

$$\underset{+1}{OCl^-} = \underset{-1}{Cl^-} + \underset{+5}{ClO_3^-}$$

and the oxidation numbers of chlorine are as shown. One sees immediately that it is not possible to assign a discrete number of equivalents per mole to OCl^-, for this substance is reduced to Cl^- and is also oxidized to ClO_3^-. An easy approach is to balance the equation from *right to left*. Assignment of numbers of equivalents per mole is easily made:

$$\underset{+1}{OCl^-} = \underset{-1}{Cl^-} + \underset{+5}{ClO_3^-}$$

$$\frac{2\text{ eq}}{\text{mole}} \quad \frac{4\text{ eq}}{\text{mole}}$$

and the balancing of chlorine gives

$$3\,OCl^- = 2\,Cl^- + ClO_3^-$$

The charges are balanced and oxygen is also balanced. Therefore, the equation is balanced.

EXAMPLE 4.9 In acid solution, hydrogen peroxide oxidizes nitrous acid (HNO_2) to nitrate ion (NO_3^-). Write a balanced equation for the reaction.

SOLUTION We begin by writing the known reactants and products in an incomplete equation, and by noting the oxidation numbers and the numbers of equivalents per mole:

$$\underset{-1}{H_2O_2} + \underset{+3}{HNO_2} = NO_3^- \quad (+5,\ -2)$$

$$\frac{2\text{ eq}}{\text{mole}} \quad \frac{2\text{ eq}}{\text{mole}}$$

The oxygen in hydrogen peroxide is in the -1 oxidation state. Since hydrogen peroxide is the oxidizing agent in this reaction the oxygen must be reduced, and the only lower oxidation state available is -2. Hence, we need not be concerned with the question of where the oxygen from H_2O_2 is found in the product; regardless of the compound or ion, any oxygen must be in the -2 oxidation state after reaction.

There are two equivalents per mole of both H_2O_2 and HNO_2, and these species therefore react in a one-to-one ratio. The equation is already balanced in nitrogen and only needs balancing in charge, hydrogen, and oxygen. Approaching those steps in this order gives

$$H_2O_2 + HNO_2 = NO_3^- + H^+ + H_2O$$

The key to balancing this equation is the recognition that all oxygen in the product must be in the -2 oxidation state. Knowing this, we are able to get the number of equivalents per mole of H_2O_2 and balance the equation. Although we are still unsure of the disposition of particular oxygen atoms, that problem need not be resolved in order to balance the equation.

The following examples require the application of several of the principles discussed so far in this chapter; no new concepts are introduced.

EXAMPLE 4.10 A 0.250 g sample of iron is put into acidic solution as Fe^{2+}. To this solution is added 0.0200 *M* $Cr_2O_7^{2-}$ solution until the iron is all oxidized to Fe^{3+}. In the reaction the chromium is reduced to Cr^{3+}. What volume of the $Cr_2O_7^{2-}$ solution is required?

SOLUTION We will show two convenient approaches to this problem. In the first we will write a balanced equation for the reaction and make the calculation from that equation. In the second we will make the calculation from the number of equivalents without balancing the equation.

(a) Use of the balanced equation.
The reaction is between $Cr_2O_7^{2-}$ and Fe^{2+}:

$$\begin{array}{cccc} Cr_2O_7^{2-} + & Fe^{2+} = & Fe^{3+} + & Cr^{3+} \\ +6 & +2 & +3 & +3 \\ \frac{6\text{ eq}}{\text{mole}} & \frac{1\text{ eq}}{\text{mole}} & & \end{array}$$

When we balance the chromium and iron we obtain

$$Cr_2O_7^{2-} + 6\,Fe^{2+} = 6\,Fe^{3+} + 2\,Cr^{3+}$$

and the balanced equation is

$$Cr_2O_7^{2-} + 6\,Fe^{2+} + 14\,H^+ = 6\,Fe^{3+} + 2\,Cr^{3+} + 7\,H_2O$$

From the equation we see the mole ratio: 6 moles of Fe^{2+} react with 1 mole of $Cr_2O_7^{2-}$. Therefore, we can write an expression to get the volume:

$$(0.250\,\cancel{\text{g Fe}})\left(\frac{1\,\cancel{\text{mole Fe}}}{55.9\,\cancel{\text{g Fe}}}\right)\left(\frac{1\,\cancel{\text{mole Fe}^{2+}}}{1\,\cancel{\text{mole Fe}}}\right)\left(\frac{1\,\cancel{\text{mole Cr}_2\text{O}_7^{2-}}}{6\,\cancel{\text{moles Fe}^{2+}}}\right)$$

$$\times\left(\frac{1\text{ liter}}{0.0200\,\cancel{\text{mole Cr}_2\text{O}_7^{2-}}}\right) = 0.0373\text{ liter}$$

(b) Use of equivalents without balancing the equation.
From the statement of the problem we know that $Cr_2O_7^{2-}$ is reduced to Cr^{3+}, and Fe^{2+} is oxidized to Fe^{3+}. Hence, $Cr_2O_7^{2-}$ has 6 eq/mole and Fe^{2+} has 1 eq/mole. With this information we can write a single expression to calculate the volume of $Cr_2O_7^{2-}$ solution:

$$(0.250\,\cancel{\text{g Fe}})\left(\frac{1\,\cancel{\text{mole Fe}}}{55.9\text{ g Fe}}\right)\left(\frac{1\,\cancel{\text{mole Fe}^{2+}}}{1\text{ mole Fe}}\right)\left(\frac{1\,\cancel{\text{eq Fe}^{2+}}}{1\text{ mole Fe}^{2+}}\right)\left(\frac{1\,\cancel{\text{eq Cr}_2\text{O}_7^{2-}}}{1\text{ eq Fe}^{2+}}\right)$$

$$\times\left(\frac{1\,\cancel{\text{mole Cr}_2\text{O}_7^{2-}}}{6\,\cancel{\text{eq Cr}_2\text{O}_7^{2-}}}\right)\left(\frac{1\text{ liter}}{0.0200\,\cancel{\text{mole Cr}_2\text{O}_7^{2-}}}\right) = 0.0373\text{ liter}$$

Notice the dimensioned factor 1 eq $Cr_2O_7^{2-}$/1 eq Fe^{2+}, which is an expression of the fact that the equivalent is defined in such a way that 1 equivalent of a reactant always requires just 1 equivalent of the other reactant. Both approaches are good ways to solve the problem. In each case the dimensions guide one to the correct operations.

EXAMPLE 4.11 What is the molarity of a thiosulfate ($S_2O_3^{2-}$) solution if 21.20 ml of the solution just react with 25.00 ml of 0.1065 M I_3^- solution? In the reaction, $S_2O_3^{2-}$ is oxidized to $S_4O_6^{2-}$ and I_3^- is reduced to I^-.

SOLUTION Again we will approach the problem two ways:

(a) Use of the balanced equation.
The unbalanced equation is

$$\begin{array}{cccc} S_2O_3^{2-} + & I_3^- = & I^- + & S_4O_6^{2-} \\ +2 & -\frac{1}{3} & -1 & +\frac{5}{2} \\ \frac{1\text{ eq}}{\text{mole}} & \frac{2\text{ eq}}{\text{mole}} & & \end{array}$$

and the balanced equation is

$$2\,S_2O_3^{2-} + I_3^- = 3\,I^- + S_4O_6^{2-}$$

Having the balanced equation we can write an expression for molarity of the $S_2O_3^{2-}$ solution.

$$(0.02500\text{ liter})\left(\frac{0.1065\text{ mole } I_3^-}{\text{liter}}\right)\left(\frac{2\text{ moles } S_2O_3^{2-}}{1\text{ mole } I_3^-}\right)\left(\frac{1}{0.02120\text{ liter}}\right) = 0.2512\,\frac{\text{mole } S_2O_3^{2-}}{\text{liter}}$$

(b) Use of equivalents.
If we choose not to balance the equation but work instead with the equivalents, we first determine the numbers of equivalents per mole of $S_2O_3^{2-}$ and of I_3^-. (This operation was necessary to balance the equation in part (a); but here we do not need to balance the equation in order to get the equivalents per mole.) The iodine goes from oxidation number $-1/3$ to -1. Therefore, I_3^- has 2 equivalents per mole. The sulfur goes from $+2$ to $+5/2$, and $S_2O_3^{2-}$ has 1 equivalent per mole. The expression for the molarity of the $S_2O_3^{2-}$ solution is, therefore

$$(0.2500\text{ liter})\left(\frac{0.1065\text{ mole } I_3^-}{\text{liter}}\right)\left(\frac{2\text{ eq } I_3^-}{1\text{ mole } I_3^-}\right)\left(\frac{1\text{ eq } S_2O_3^{2-}}{1\text{ eq } I_3^-}\right)\left(\frac{1\text{ mole } S_2O_3^{2-}}{1\text{ eq } S_2O_3^{2-}}\right) \times \left(\frac{1}{0.02120\text{ liter}}\right) = 0.2512\,\frac{\text{mole } S_2O_3^{2-}}{\text{liter}}$$

4.3 Chemical Equilibrium: An Introduction

In Chapter 3 and up to this point in this chapter we have considered chemical reactions as going to completion. That is, starting with the equation for a reaction, we have been able to calculate the quantity of product formed from given amounts of reactants, and we have done that with the assumption that the reaction went completely to the right (products) as written. There are many reactions, however, for which that would be a bad assumption. Instead, under a given set of conditions, some reactions are observed to proceed to a certain extent and no farther. Indeed, one can mix the products and find that they react somewhat to give the reactants. Such reactions are spoken of as reaching *equilibrium;* that is, they reach the condition of no longer having any tendency to go forward or backward. The principle of chemical equilibrium was observed long before there was a theoretical base from thermodynamics for understanding the principle. Here we will use that historical approach for convenience, showing the equilibrium relationship presented as an empirical observation. Later, in Chapters 9 and 10, we will derive the relationship so that you can see that it has a solid theoretical base. At that time we will make much more extensive use of the principle.

This short section introducing **chemical equilibrium** is included for use in classes in which this subject is needed early for laboratory work or in which the instructor prefers a first empirical approach to equilibrium. The subject of chemical equilibrium as presented in Chapters 9 and 10 does not depend upon this introduction; hence this section may be skipped if the instructor prefers.

The original observation of chemical equilibrium was that the concentrations of substances in a reaction come to a fixed relationship at a given temperature, and that this fixed relationship is independent of whether the reaction starts with reactants or products, or with some mixture of these. Specifically, in the equilibrium reaction

$$A + 2B = C$$

the equilibrium condition is given by the expression

$$\frac{c_C}{(c_A)(c_B)^2} = K$$

in which K is a constant at a given temperature and is called the **equilibrium constant,** and c_A, c_B, and c_C are the concentrations of the species in the reaction. The form of the equilibrium constant expression is related to the chemical reaction equation. The concentration of each product species is written in the numerator while the concentration of each reactant is written in the denominator, and each concentration is raised to the power given by the coefficient that appears in the balanced equation. In the above equation both A and C have the coefficient 1, and the concentration of each is, therefore, raised to the first power in the equilibrium constant expression. The coefficient of B is 2, and its exponent is, therefore, also 2.

The following are examples of equilibrium expressions:

$$HF(aq) = H^+(aq) + F^-(aq)$$

$$K = \frac{(c_{H^+})(c_{F^-})}{(c_{HF})}$$

$$N_2(g) + 3\,H_2(g) = 2\,NH_3(g)$$

$$K = \frac{(c_{NH_3})^2}{(c_{N_2})(c_{H_2})^3}$$

$$Cu(NH_3)_4^{2+}(aq) = Cu^{2+}(aq) + 4\,NH_3(aq)$$

$$K = \frac{(c_{Cu^{2+}})(c_{NH_3})^4}{(c_{Cu(NH_3)_4^{2+}})}$$

(*aq* indicates that the reaction occurs in water solution.)

Acetic acid, CH_3COOH, dissociates only slightly in water to give H^+ and acetate ion, CH_3COO^-. (Recall that acetic acid is in the list of common weak acids.) The equation for the partial dissociation is

$$CH_3COOH = CH_3COO^- + H^+$$

and the equilibrium constant expression is

$$\frac{[CH_3COO^-][H^+]}{[CH_3COOH]} = K$$

The concentration used for solutes is molarity. The square bracket indicates molar concentration. Thus, $[H^+]$ is the moles of hydrogen ion per liter of solution.

The value of K for this reaction at 25° C is 1.8×10^{-5}. The individual concentration of species will be very different in different solutions, but the value of $[CH_3COO^-][H^+]/[CH_3COOH]$ remains the same.

The equilibrium relationship gives the chemist a powerful means for determining the concentrations of species in a solution. Let us consider some solutions that contain acetic acid, acetate ion, and hydrogen ion.

EXAMPLE 4.12 What is the ratio of $[CH_3COO^-]/[CH_3COOH]$ in a solution containing these species and in which $[H^+]$ is (a) 1.8×10^{-5} mole/liter; (b) 1.0×10^{-10} mole/liter; (c) 1.0×10^{-1} mole/liter?

SOLUTION First we see from the equilibrium constant expression that the ratio of concentrations of acetate ion to acetic acid is controlled by the concentration of H^+. This is apparent if both sides of the expression are divided by $[H^+]$:

$$\frac{[CH_3COO^-]}{[CH_3COOH]} = \frac{K}{[H^+]}$$

With this expression we can calculate the desired ratios.

$$\text{(a)}\ \frac{[CH_3COO^-]}{[CH_3COOH]} = \frac{K}{[H^+]} = \frac{1.8 \times 10^{-5}}{1.8 \times 10^{-5}} = 1$$

$$\text{(b)}\ \frac{[CH_3COO^-]}{[CH_3COOH]} = \frac{K}{[H^+]} = \frac{1.8 \times 10^{-5}}{1.0 \times 10^{-10}} = 1.8 \times 10^{5}$$

$$\text{(c)}\ \frac{[CH_3COO^-]}{[CH_3COOH]} = \frac{K}{[H^+]} = \frac{1.8 \times 10^{-5}}{1.0 \times 10^{-1}} = 1.8 \times 10^{-4}$$

From this example we see that the desired ratio is easily calculated, but we also see that the ratio can be changed greatly by changing the $[H^+]$.

EXAMPLE 4.13 Calculate $[H^+]$ in a solution that contains 0.20 mole of acetic acid per liter.

SOLUTION We know the concentration of acetic acid if none dissociated, and we know that any acetic acid that dissociates results in the formation of equal concentrations of acetate ion and hydrogen ion. This can all be expressed as follows:

$$CH_3COOH = CH_3COO^- + H^+$$

Concentrations: $\quad 0.20 - x \qquad x \qquad x$

The concentration of CH_3COOH would be 0.20, had none dissociated; however, since x mole/liter dissociate, then x mole/liter of H^+ and x mole/liter of

CH_3COO^- are formed. These substitutions are made in the equilibrium constant expression:

$$\frac{(x)(x)}{0.20 - x} = 1.8 \times 10^{-5}$$

Rearranging gives

$$x^2 + (1.8 \times 10^{-5}x) - (3.6 \times 10^{-6}) = 0$$

This is solved by the quadratic equation

$$x = \frac{-b \pm \sqrt{b^2 - 4ac}}{2a} = \frac{-1.8 \times 10^{-5} \pm \sqrt{(1.8 \times 10^{-5})^2 + 4(3.6 \times 10^{-6})}}{2}$$

$$x = 1.9 \times 10^{-3} = [H^+]$$

In Chapter 10 you will see that there is a much more convenient approach to such a problem by a method of approximations.

This is the concentration of H^+ in a 0.20 M CH_3COOH solution.

For the dissociation of acetic acid, $K = 1.8 \times 10^{-5}$. The small value of K tells us that this reaction does not go very far. In Example 4.13 we see that the percent dissociation of the acetic acid is

$$\% = \frac{1.9 \times 10^{-3}}{0.20} \times 100 = 0.95\%$$

We recognize that acetic acid is a weak acid since only about 1% of the acetic acid in this solution is dissociated. Equilibrium constants for chemical reactions range from very large to very small numbers. Examination of the equilibrium constant helps one estimate quickly the extent of a chemical reaction. A large value for K tells that the reaction goes almost to completion, a small K indicates that very little reaction occurs, while a K around unity suggests that significant amounts of both products and reactants are present at equilibrium.

Water and the pH Scale

Water is slightly dissociated to give H^+ and OH^-:

$$H_2O(l) = H^+ + OH^- \qquad [4.4]$$

H^+ in water solution is hydrated (associated with water molecules) and is often written as H_3O^+ to show the hydration. We will see in later chapters that this complicating factor does not affect the equilibrium calculations. For the present we will use the simple form H^+.

The equilibrium constant expression for this reaction is an important relationship that is of great consequence in living systems as well as in many chemical reactions. There is, however, a complication that we must face before writing that expression—namely that different concentration units are used for a species, depending upon whether it is a solute or the solvent. For a solute we use the molar concentration; for the solvent we use the mole fraction. This choice is made for reasons of convenience, since the mole fraction of the solvent in a dilute solution is very nearly unity. Now, let us write the equilibrium constant expression:

$$K = \frac{[H^+][OH^-]}{x_{H_2O}}$$

However, since $x_{H_2O} = 1$, the expression simplifies to

$$K = [H^+][OH^-] \qquad [4.5]$$

Some frequently used constants are given names for easy identification. Later we will encounter more of these named equilibrium constants.

This constant is called the water constant or K_w, and the numerical value is 1.0×10^{-14} at 25° C. This tells us that the product of the molar concentrations of H^+ and OH^- is 1.0×10^{-14}. It also tells us that if $[H^+]$ is high, then $[OH^-]$ is low. We can easily calculate $[H^+]$ and $[OH^-]$ in pure water from equation [4.5], for in pure water the only source of H^+ or OH^- is through the reaction shown in equation [4.4]. Therefore, in this example, $[H^+] = [OH^-]$. That is, a hydrogen ion is formed only when there is dissociation of H_2O, which must also give a hydroxide ion. Since these are equal, we can make the following substitution: Let

$$x = [H^+] = [OH^-]$$
$$[H^+][OH^-] = 1.0 \times 10^{-14} = x^2$$
$$x = 1.0 \times 10^{-7} = [H^+] = [OH^-]$$

In pure water at 25° C the concentrations of H^+ and OH^- are both 1.0×10^{-7}.

The following example illustrates some other calculations that can be made from this relationship.

EXAMPLE 4.14 Calculate $[OH^-]$ in (a) 0.20 *M* H^+, (b) 0.20 *M* CH_3COOH, (c) a solution containing 0.20 mole of HCl per liter, (d) 1.0×10^{-10} *M* H^+.

SOLUTION

(a) Since $[H^+] = 0.20$, $[OH^-]$ is readily calculated from equation [4.5]:

$$[OH^-] = \frac{1.0 \times 10^{-14}}{[H^+]} = \frac{1.0 \times 10^{-14}}{0.20} = 5.0 \times 10^{-14}\ M$$

(b) In Example 4.13 we calculated $[H^+]$ in 0.20 *M* CH_3COOH to be 1.9×10^{-3}. Therefore,

$$[OH^-] = \frac{1.0 \times 10^{-14}}{[H^+]} = \frac{1.0 \times 10^{-14}}{1.9 \times 10^{-3}} = 5.3 \times 10^{-12}\ M$$

(c) HCl is a strong acid. Hence $[H^+] = 0.20\ M$, and the $[OH^-]$ is the same as in part (a):

$$[OH^-] = \frac{1.0 \times 10^{-14}}{0.20} = 5.0 \times 10^{-14}$$

(d) $$[OH^-] = \frac{1.0 \times 10^{-14}}{[H^+]} = \frac{1.0 \times 10^{-14}}{1.0 \times 10^{-10}} = 1.0 \times 10^{-4}$$

Example 4.14 illustrates the large range of values that $[H^+]$ and $[OH^-]$ can have in aqueous solutions. Since this range is so very large, chemists find it convenient to compress the range by means of logarithms (logs) to what is known as the **pH scale.** For our present purposes the following definition is sufficiently accurate:

$$pH = -\log [H^+]$$

If $[H^+] = 1.0 \times 10^{-1}$, then pH = 1.0. If $[H^+] = 1.0 \times 10^{-7}$ (as in pure water), then pH = 7.0. If $[OH^-] = 1.0 \times 10^{-1}$, then $[H^+] = 1.0 \times 10^{-13}$ and pH = 13. Verify your understanding by making your own calculations and checking them against each of these pH values. Examples 4.15 and 4.16 provide further practice in calculating pH values.

EXAMPLE 4.15 Calculate the pH of the following solutions: (a) $[H^+] = 1.0 \times 10^{-3}$, (b) $[H^+] = 4.7 \times 10^{-5}$, (c) $[OH^-] = 2.5 \times 10^{-3}$

SOLUTION

(a) $\text{pH} = -\log 1.0 \times 10^{-3} = 3.0$

(b) $\text{pH} = -\log 4.7 \times 10^{-5} = 4.3$

(c) $$[H^+] = \frac{1.0 \times 10^{-14}}{[OH^-]} = \frac{1.0 \times 10^{-14}}{2.5 \times 10^{-3}} = 4.0 \times 10^{-12}$$

$$\text{pH} = -\log 4.0 \times 10^{-12} = 11.4$$

EXAMPLE 4.16 Calculate $[H^+]$ of solutions having the following pH values: (a) 4.7, (b) 9.5, (c) 7.0

SOLUTION

(a) $[H^+] = 10^{-\text{pH}} = 10^{-4.7} = 2.0 \times 10^{-5}$

(b) $[H^+] = 10^{-9.5} = 3.2 \times 10^{-10}$

(c) $[H^+] = 10^{-7.0} = 1.0 \times 10^{-7}$

Equilibria Involving Solids

In this short introduction we will consider equilibria involving just one other kind of species, slightly soluble salts. An example is the equilibrium between solid silver chloride and an aqueous solution that contains silver and chloride ions:

$$AgCl(s) = Ag^+ + Cl^-$$

The equilibrium constant for this reaction is

$$K = \frac{(c_{Ag^+})(c_{Cl^-})}{(c_{AgCl(s)})}$$

Here we must be careful how we express the concentration of the solid AgCl. It is convenient to use molar concentrations for Ag^+ and Cl^-, since these species are in solution. However, it is unreasonable to attempt to express the concentration of AgCl(*s*) in moles per liter. As in the example of water, it is especially convenient to use the mole fraction, for if the solid silver chloride is pure, the mole fraction is exactly equal to 1. With such choices of concentration units the equilibrium constant expression becomes

$$K = \frac{(c_{Ag^+})(c_{Cl^-})}{(c_{AgCl(s)})} = \frac{[Ag^+][Cl^-]}{x_{AgCl(s)}} = \frac{[Ag^+][Cl^-]}{1}$$

The solution in such an equilibrium mixture is called a **saturated solution.** It contains as much of the dissolved species as can dissolve under the given conditions. In the AgCl example, any Ag^+ or Cl^- that is added to the solution will result in the formation of some solid AgCl that is precipitated from the solution. If by some other reaction either Ag^+ or Cl^- is removed from the solution, more AgCl will dissolve until the solution is again saturated.

The numerical value of this constant is 1.8×10^{-10}:

$$K = [Ag^+][Cl^-] = 1.8 \times 10^{-10}$$

By the use of the equilibrium constant for the dissolving of AgCl(*s*), we can determine the solubility of the substance in water, but of much more interest is the fact that we can determine its solubility in a solution that already contains Ag^+ or Cl^-. The following examples illustrate these calculations.

EXAMPLE 4.17 Calculate the solubility (expressed in moles/liter) of AgCl(*s*).

SOLUTION The reaction by which AgCl(*s*) dissolves is

$$AgCl(s) = Ag^+ + Cl^-$$

Let $y = [Ag^+]$, but since one Cl^- is formed for each Ag^+, $y = [Cl^-]$. The equilibrium constant for the reaction is

$$K = [Ag^+][Cl^-] = 1.8 \times 10^{-10}$$

Therefore,

$$y^2 = 1.8 \times 10^{-10}$$

and

$$y = 1.3 \times 10^{-5} = [Ag^+] = [Cl^-]$$

Since 1 mole of Ag^+ and 1 mole of Cl^- appear in solution for each mole of AgCl(*s*) that dissolves, the concentration of either of these ionic species is also the solubility (in moles/liter) of AgCl(*s*). The solubility, therefore, is 1.3×10^{-5} mole/liter.

EXAMPLE 4.18 Calculate the solubility of AgCl(*s*) in 1.0×10^{-4} *M* Ag^+.

SOLUTION In this problem there is already some Ag^+ in solution. Because of the presence of this Ag^+, the concentration of Ag^+ and Cl^- will not be the same, but will differ by 1.0×10^{-4} mole/liter. However, the equilibrium constant still holds, and

$$[Ag^+][Cl^-] = 1.8 \times 10^{-10}$$

Again we consider the reaction by which AgCl(*s*) dissolves:

$$AgCl(s) = Ag^+ + Cl^-$$

Let z = solubility of AgCl(*s*) in this solution. If z mole/liter of AgCl dissolve, then both $[Ag^+]$ and $[Cl^-]$ increase by z mole/liter. Hence, at equilibrium,

$$[Ag^+] = 1.0 \times 10^{-4} + z$$
$$[Cl^-] = 0 + z$$

and when these values are substituted into the equilibrium constant expression we obtain

$$(1.0 \times 10^{-4} + z)(z) = 1.8 \times 10^{-10}$$

$$z^2 + (1.0 \times 10^{-4}z) - (1.8 \times 10^{-10}) = 0$$

This is solved by the quadratic equation to give

$$z = 1.8 \times 10^{-6}$$

Therefore, the solubility of AgCl(*s*) is only 1.8×10^{-6} mole/liter in 1.0×10^{-4} *M* Ag^+. It was over seven times that soluble in water. Thus, the presence of the Ag^+ decreased the solubility.

With a good understanding of this brief introduction to the subject of chemical equilibrium, you will be able to make calculations of the concentrations of species in many equilibrium mixtures. In Chapters 9 and 10 we will return to the subject with much more rigor and detail.

Summary

In Chapters 3 and 4 you have learned a technique for solving all stoichiometric problems and have, in addition, gained an introduction to both acid-base and oxidation-reduction reactions. You should be able to balance redox equations confidently and solve simple chemical equilibrium problems. It is the nature of the study of chemistry that one constantly makes use of concepts and techniques learned earlier. The topics covered in these two chapters will be needed at every step in the remainder of your study.

In Arrhenius theory an acid is a substance that contains H and ionizes in water to give H^+. A base is a substance that contains OH and ionizes in water to give OH^-. This old and restricted theory is still useful in considering aqueous solutions.

Oxidation-reduction equations can be balanced by (a) balancing the substances oxidized and reduced in terms of the number of equivalents of each, (b) balancing the charges, and (c) balancing the hydrogen and oxygen.

Chemical reactions typically reach a state of equilibrium at which there is no longer a tendency to react in either direction. This condition is reproducible, and for the reaction

$$A + 2B = C$$

as an example, the equilibrium condition can be represented by the expression

$$K = \frac{c_C}{(c_A)(c_B)^2}$$

Knowing the equilibrium constant K, we can calculate concentrations of the species in the equation under any desired conditions. The concentration is expressed as molar concentration for species in solution and as mole fraction for solids, for liquids, and for the solvent in a dilute solution.

Important Terms and Concepts

Acid (Arrhenius definition)
Base (Arrhenius definition)
Strong acids
Weak acids
Strong bases
Weak bases
Neutralization
Salt
Oxidation-reduction reactions or redox reactions
Oxidation number
Peroxides
Hydrides
Oxidation
Reduction
Equivalents (oxidation-reduction)
Half-reactions
Chemical equilibrium
Equilibrium constant
pH scale
Saturated solution

Exercises

4-1.
When Cl_2O_7 is dissolved in water, it reacts according to the equation

$$Cl_2O_7 + H_2O = 2\,H^+ + 2\,ClO_4^-$$

Four (4.00) grams of Cl_2O_7 are dissolved in enough water to make 100.0 ml of solution.

(a) Calculate the concentration of H^+ in the solution in moles per liter.
(b) What volume of 0.10 M KOH is required to just react with all the H^+ of part (a)?

4-2.
How many grams of $H_2C_2O_4$ can be completely neutralized by 30.0 ml of 0.20 M NaOH solution? (Both hydrogens in $H_2C_2O_4$ are acidic, capable of being neutralized.)

4-3.
Calculate the volume in ml of 0.100 M $Ba(OH)_2$ solution that would be needed to neutralize completely 20.00 ml of 0.0500 M H_3PO_4 solution. (All three hydrogens in H_3PO_4 are capable of being neutralized.)

4-4.
How many ml of 0.100 M HCl solution can be neutralized by 1.00 g of $Al(OH)_3$?

4-5.
If 15.30 ml of 0.100 M HCl are required to completely neutralize 20.00 ml of NaOH solution, what is the molarity of the NaOH solution?

4-6.
The molarity of a hydrochloric acid solution was determined by reacting it with sodium carbonate. The net ionic reaction is

$$2\,H^+ + CO_3^{2-} = CO_2(g) + H_2O$$

Calculate the molarity of the hydrochloric acid solution if 16.5 ml of the acid are required to react completely with 0.500 g of sodium carbonate (Na_2CO_3).

4-7.
The molarity of a sodium hydroxide solution was determined by reacting it with the acid-salt $KHC_8H_4O_4$. If 35.70 ml of the sodium hydroxide solution is required to neutralize 0.4500 g of $KHC_8H_4O_4$, calculate the molarity of the NaOH solution. (Only one hydrogen in $KHC_8H_4O_4$ is capable of being neutralized.)

4-8.
Determine the oxidation number of the underlined element:

$\underline{Ba}^{2+}$	$\underline{O}_2$
$\underline{Br}^-$	$\underline{Cu}$
$\underline{Mn}O_2$	$H_2\underline{O}_2$
$\underline{Fe}_2O_3$	$H\underline{O}_2^-$
$\underline{Mn}O_4^-$	$\underline{N}H_3$
$\underline{Cr}_2O_7^{2-}$	$Na_3\underline{N}$
$\underline{Cr}(OH)_4^-$	$\underline{N}_2O_5$
$Mg_2\underline{P}_2O_7$	$H\underline{N}O_3$
$Na_2H\underline{P}O_4$	$\underline{Cr}^{3+}$

4-9.
Complete (add H^+ and H_2O) and balance the following oxidation-reduction equations representing reactions occurring in acid solution.

(a) $MnO_4^- + Fe^{2+} = Fe^{3+} + Mn^{2+}$
(b) $Zn + As_2O_3 = Zn^{2+} + AsH_3$
(c) $H_2O_2 + Cr_2O_7^{2-} = O_2 + Cr^{3+}$
(d) $IO_3^- + I^- = I_3^-$
(e) $Cu + NO_3^- = Cu^{2+} + NO_2 + H_2O$

4-10.
Complete (add OH^- and H_2O) and balance the following oxidation-reduction equations representing reactions occurring in neutral or basic solution.

(a) $MnO_4^- + SO_3^{2-} = SO_4^{2-} + MnO_2$
(b) $Cl_2 = ClO_3^- + Cl^-$
(c) $HO_2^- + Cr(OH)_4^- = CrO_4^{2-}$
(d) $Ag + CN^- + O_2 = Ag(CN)_2^-$
(e) $MnO_4^- + H_2O = MnO_2 + O_2$

4-11.
What volume of 0.1000 M $FeSO_4$ is required to reduce 4.000 g of $KMnO_4$ in a solution acidified with sulfuric acid? (In acid solution, Fe^{2+} is oxidized to Fe^{3+} and MnO_4^- is reduced to Mn^{2+}.)

4-12.
A ferrous sulfate solution was standardized by titration with ceric sulfate. The reaction is

$$Ce^{4+} + Fe^{2+} = Ce^{3+} + Fe^{3+}$$

A 25.00 ml portion of the ferrous sulfate solution required 44.61 ml of 0.0600 M ceric sulfate for complete oxidation. What is the molarity of the ferrous sulfate?

4-13.
How many grams of CO_2 gas could be produced by the reaction of 250.00 ml of 1.00 M MnO_4^- solution with excess oxalic acid ($H_2C_2O_4$)? The reaction takes place in sulfuric acid solution. The products of the reaction are Mn^{2+} and $CO_2(g)$.

4-14.
In the reaction of an HCl solution with excess MnO_2, 13.0 g of Cl_2 were produced. The system was acidified with excess H_2SO_4. The products of the reaction are $Cl_2(g)$ and Mn^{2+}. If the reaction was done with 100.0 ml of HCl solution, what was the molarity of the HCl solution?

4-15.
Write equilibrium constant expressions for the following equilibrium reactions taking place in dilute solutions:

(a) $HF(aq) = H^+(aq) + F^-(aq)$
(b) $H_2O = H^+(aq) + OH^-(aq)$
(c) $Ag_2CrO_4(s) = 2Ag^+(aq) + CrO_4^{2-}(aq)$
(d) $H_2S(aq) = H^+(aq) + HS^-(aq)$
(e) $NH_3(aq) + H_2O = NH_4^+(aq) + OH^-(aq)$

4-16.
Calculate the molar concentration of (a) H^+, (b) CH_3COO^-, and (c) undissociated CH_3COOH in a solution labeled 0.10 M CH_3COOH. The equilibrium constant for ionization of CH_3COOH is $K = 1.8 \times 10^{-5}$.

4-17.
(a) Calculate the concentration of H^+ in a solution that is labeled 0.10 M HF ($K = 7.4 \times 10^{-4}$).
(b) Calculate the $[H^+]$ of a solution that is prepared by adding 0.10 mole of HF and 0.05 mole of NaF to enough water to make 1.00 liter of solution. The salt NaF is completely dissociated into Na^+ and F^- ions.

4-18.
Calculate the molar concentration of OH^- in a 0.20 *M* solution of $NH_3(aq)$. The ionization reaction is

$$NH_3 + H_2O = NH_4^+ + OH^-$$

K for this reaction is 1.8×10^{-5}.

4-19.
Calculate the concentration of acetate ion in a solution that is 0.10 *M* in CH_3COOH (*K* for ionization = 1.8×10^{-5}) and also 0.10 *M* in hydrochloric acid. Hydrochloric acid is a strong acid, fully ionized into H^+ and Cl^-.

4-20.
Calculate $[OH^-]$ in a solution that is 0.10 *M* in $NH_3(aq)$ and also 0.10 *M* in NH_4Cl. The ionization constant for $NH_3(aq)$ is 1.8×10^{-5}. The salt NH_4Cl is fully dissociated into NH_4^+ and Cl^-.

4-21.
Calculate (a) $[H^+]$, (b) $[OH^-]$, and (c) pH of each of the following solutions of strong (fully ionized) acids and bases:

(1) 0.001 *M* HCl
(2) 0.001 *M* NaOH
(3) 0.025 *M* HNO_3
(4) 0.00015 *M* NaOH

4-22.
Calculate (a) $[H^+]$ and (b) $[OH^-]$ concentration in solutions with pH values of (1) 7.00; (2) 3.50; (3) 9.25; (4) 1.00.

4-23.
(a) Calculate the pH of 0.10 *M* CH_3COOH solution (*K* for ionization reaction = 1.8×10^{-5}).
(b) Compare your answer in (a) with the pH of a 0.10 *M* solution of the strong acid hydrochloric acid.

4-24.
(a) Calculate the pH of a 0.10 *M* $NH_3(aq)$ solution. *K* for the reaction

$$NH_3 + H_2O = NH_4^+ + OH^-$$

is 1.8×10^{-5} at 25° C.
(b) Compare your answer in (a) with the pH of a 0.10 *M* NaOH solution. The latter is a fully ionized (strong) base.

4-25.
Calculate the solubility of AgCl in (a) pure water, (b) 0.10 *M* $AgNO_3$ solution ($AgNO_3$ is fully dissociated into Ag^+ and NO_3^-), and (c) 0.10 *M* HCl (fully ionized). The equilibrium constant for the solution process

$$AgCl(s) = Ag^+(aq) + Cl^-(aq)$$

is 1.8×10^{-10}.

4-26.
Calculate the solubility of $Ag_2CrO_4(s)$ in (a) pure water, (b) 0.10 *M* Ag^+ solution, (c) 0.10 *M* CrO_4^{2-} solution. The equilibrium constant for

$$Ag_2CrO_4(s) = 2\,Ag^+ + CrO_4^{2-}$$

is 2.0×10^{-12}.

4-27.
The solubility of Ag_2S in water is 5×10^{-18} moles/liter. Calculate the equilibrium constant for the solution process

$$Ag_2S(s) = 2\,Ag^+(aq) + S^{2-}(aq)$$

Problems

4-1.
Brass is often analyzed for its copper content by the following process: First the brass is dissolved in nitric acid and the copper goes into solution as Cu^{2+}:

$$Cu + H^+ + NO_3^- = Cu^{2+} + NO_2 + H_2O$$

Nitrogen oxides and the excess nitric acid are removed and KI is added. The following reaction occurs:

$$Cu^{2+} + I^- = CuI(s) + I_3^-$$

The I_3^- generated is then titrated with a standard thiosulfate solution.

$$S_2O_3^{2-} + I_3^- = S_4O_6^{2-} + I^-$$

(None of the above reactions is balanced.)

A 0.5432 g sample of brass is dissolved in nitric acid and the various steps accomplished to produce the I_3^-. It requires 24.32 ml of 0.2612 *M* $S_2O_3^{2-}$ to react with all the I_3^-. What is the weight percent copper in the brass sample?

4-2.
Will a precipitate of $PbCl_2(s)$ form if 500 ml of 0.10 *M* Pb^{2+} solution is added to 500 ml of 0.10 *M* Cl^- solution? *K* for the solution process

$$PbCl_2(s) = Pb^{2+}(aq) + 2\ Cl^-(aq)$$

is 2.0×10^{-4}. (*Hint:* The term $[Pb^{2+}][Cl^-]^2$ will not exceed its equilibrium value of 2.0×10^{-4}. Any Pb^{2+} and Cl^- present in excess of the equilibrium values will precipitate as $PbCl_2(s)$. On the other hand, if the concentrations of the ions in the solution are less than the equilibrium values, the solution is undersaturated, and solid will *not* form.)

4-3.
Balance the following nonaqueous reactions. In one respect this is simpler than balancing reactions in aqueous solutions since the step involving the balancing of a charge need not be taken. First balance the oxidation-reduction equivalents, and then balance the number of each type of atom present.

(a) $KClO = KCl + KClO_3$
(b) $Li + N_2 = Li_3N$
(c) $NaNO_3 + Zn = ZnO + NO + Na_2O$
(d) $Fe + N_2O = Fe_2O_3 + N_2$
(e) $Fe_3O_4 + CO = Fe + CO_2$
(f) $Cu_2S + O_2 = CuO + SO_2$

5

Description of Gases

The gaseous state is in many ways simpler than either the solid or liquid state and as such provides a convenient introduction to the study of matter. In this chapter we will give a qualitative description of gases in general, an exact mathematical description of an idealized gas, and some approximate mathematical descriptions of real gases. Then in Chapter 6 we will extend the discussion to a molecular theory that leads to very successful explanations for the macroscopic properties of gases.

5.1 Qualitative Description

A gas has familiar properties that distinguish it from the solid and liquid states. It fills its containing vessel completely, mixes with other gases in all proportions, compresses readily with increased pressure, and expands greatly when its temperature is raised. The pressure and temperature effects are especially important in the development of the description of a gas.

Pressure and Pressure Effects

Pressure is defined as force per unit area, shown by the mathematical equation

$$P = \frac{F}{A} \qquad [5.1]$$

where F is the force exerted on an area A. The units of pressure commonly used in the study of gases are **atmospheres** and **torr.** The relationship of these units can be seen when the techniques of pressure measurement are considered. Gas pressure is measured with a **barometer** or a **manometer.** A barometer is prepared by inverting a tube, sealed at one end and completely filled with mercury, so that its open end is immersed in a vessel of mercury (Figure 5.1a). It is used to measure the pressure of the atmosphere. The height, h, is a measure of the pressure upon the surface of the mercury in an open vessel, for the space above the liquid in the sealed tube contains no gas except a small quantity of mercury vapor.

The manometer is constructed from a "U" tube partially filled with mercury (Figure 5.1b). The manometer is used to measure the gas pressure in a closed container. One end of the U tube is connected to the container of gas. If the other end is open, the height h is a measure of the pressure difference between the gas and the atmosphere. Frequently this end is evacuated and sealed, in which case h is the absolute pressure of the gas. With both the barometer and manometer, h is usually measured in millimeters, and in the past the pressure was expressed in *mm of Hg*. The present practice is to replace that awkward term by *torr*, named for Evangelista Torricelli (1608–1647), who invented the barometer. Hence, if h is 720 mm, the pressure is 720 torr. At sea level the atmospheric pressure is near 760 torr. A pressure of exactly 760 torr has been defined as one atmosphere.[†]

[†] In terms of other units of pressure, 1 atm equals 1.013×10^6 dynes/cm^2, 101.3 kilopascals, or 14.70 lb/in^2. The pascal is the SI unit of pressure. See Appendix I for a description of this system of units.

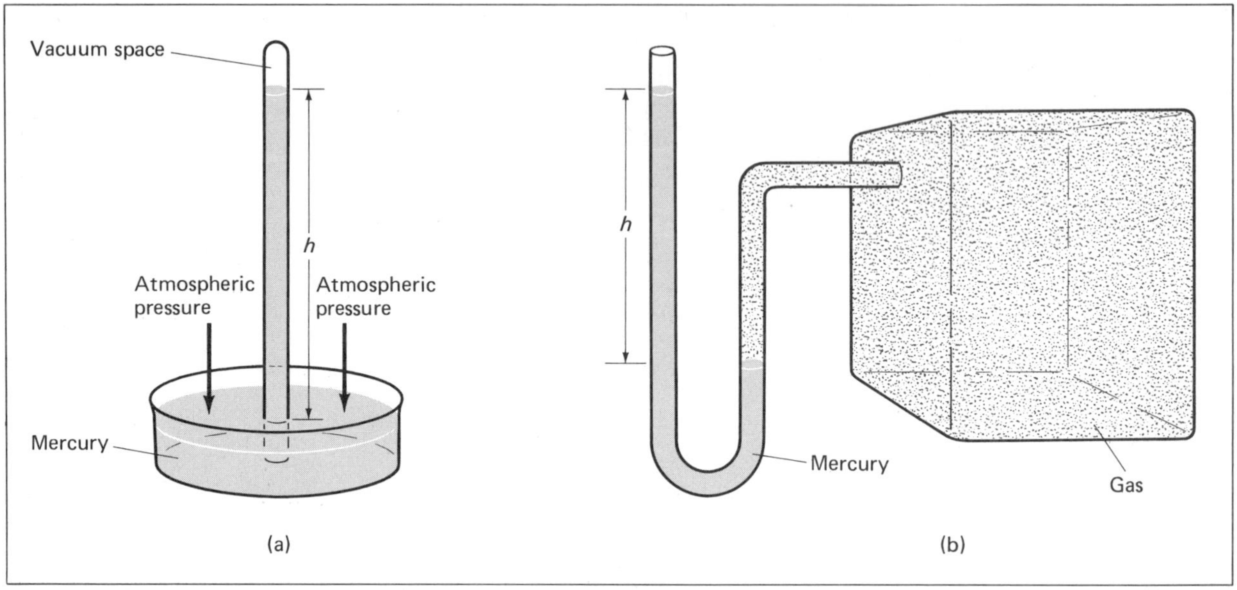

Figure 5.1a
The barometer, used to measure atmospheric pressure. The force from the atmosphere pushes the mercury up in the tube. The height *h* measured in millimeters gives pressure in torr units.

Figure 5.1b
The manometer, used to measure gas pressure. The height *h* gives the pressure difference between the two sides of the manometer. If the end of the manometer is closed and evacuated, *h* is the absolute pressure of the gas. If *h* is measured in millimeters, the pressure is in torr units.

Robert Boyle, an Irishman, first published his experiments on the compressibility of a gas in 1662 in the second edition of his book *New Experiments, Physico-Mechanical, Touching the Spring of Air and Its Effects.*

If Boyle had had apparatus capable of more accurate measurements, he would have recognized that his discovery was only a very good approximation, which became less valid with increased pressure.

Robert Boyle (1627–1692) discovered that at a constant temperature the volume of a sample of gas varies inversely with the pressure on the gas. This observation is known as **Boyle's law** and is expressed mathematically as

$$P = \frac{k'}{V}$$

or

$$PV = k' \qquad [5.2]$$

where k' is a constant. Boyle made his experiments with a tube such as that shown in Figure 5.2. He measured the pressure on the gas in the closed portion of the tube by the difference in mercury levels, and he measured the gas volume by means of the graduations on the tube. To the extent that his measurements were accurate, the PV product was constant for a gas sample. Figure 5.3 depicts two plots of Boyle's law, one showing the constancy of PV versus P and the other showing the hyperbolic shape of the P versus V curve.

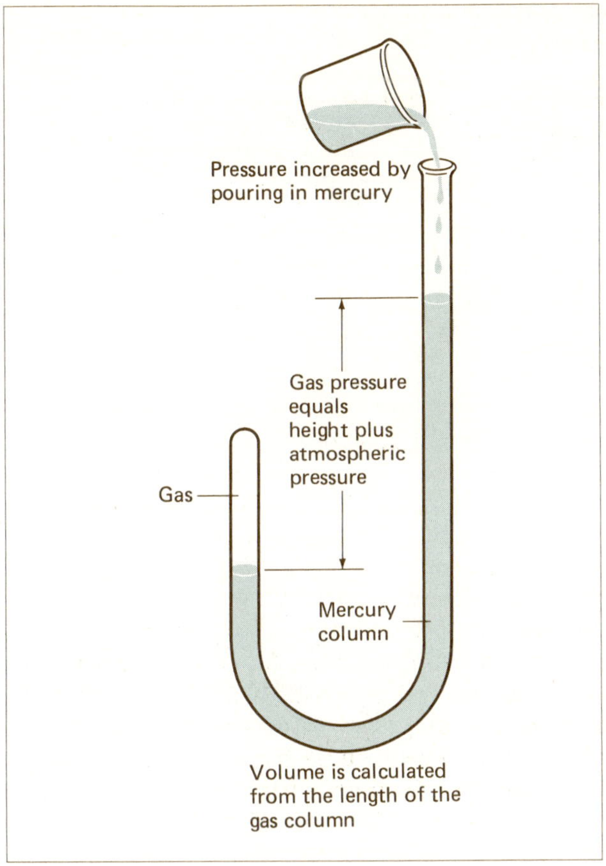

Figure 5.2
Boyle's J-tube, used to determine the effect of pressure on the volume of a gas.

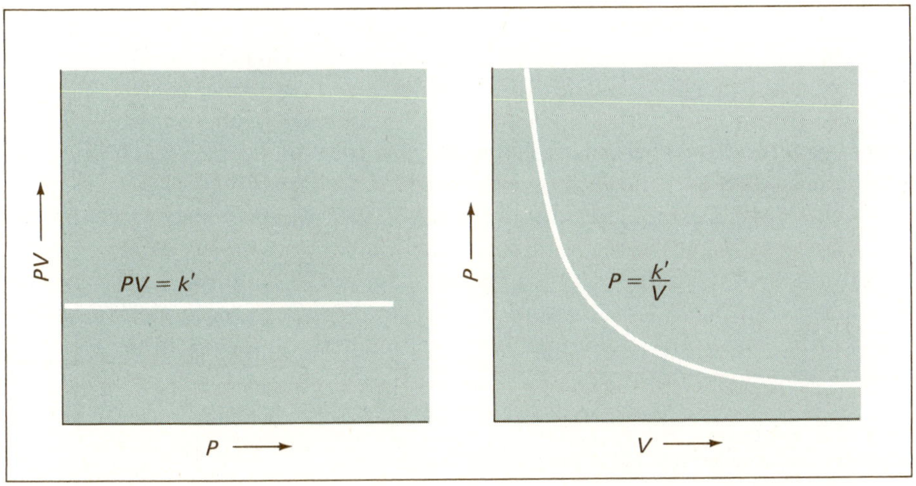

Figure 5.3
Plots of Boyle's law.

Daniel Gabriel Fahrenheit (1686–1736) made the first mercury thermometer. His thermometers marked the beginning of reliable temperature measurements.

Temperature and Temperature Effects

Temperature is measured in terms of some physical property of a substance that is assumed to vary in a simple way as the substance is heated or cooled. Volume is often used for liquid-filled thermometers because volume changes are easily measured. Liquid thermometers (usually containing mercury or alcohol) are constructed with a large bulb filled with the liquid and sealed to a capillary tube.

The thermometer was invented by Galileo Galilei between 1592 and 1597. The following account by an associate of Galileo describes this thermometer:

Galileo took a glass vessel about the size of a hen's egg, fitted to a tube the width of a straw and about two spans long; he heated the glass bulb in his hands and turned the glass upside down so that the tube dipped in water held in another vessel; as soon as the ball cooled down the water rose in the tube to the height of a span above the level in the vessel; this instrument he used to investigate degrees of heat and cold.

Galileo's thermometer suffered from one serious disadvantage. The reading varied with atmospheric pressure.

The liquid level in the capillary tube is a measure of the temperature of the bulb. **Gas thermometers,** on the other hand, usually measure pressure instead of volume. These are constructed with a bulb filled with gas and attached to a manometer. The pressure reading of the manometer is a measure of the temperature of the bulb.

There is no upper limit to temperature, but there is a lower limit. This lowest possible temperature is called **absolute zero.** Later in this chapter we will show that absolute zero is at $-273.15°$ C. It is the temperature at which the volume of an idealized gas becomes zero. In the next chapter the kinetic molecular theory describes absolute zero as the temperature at which all possible thermal energy is removed from the sample.

We learn from thermodynamics that it is impossible to cool a sample to the absolute zero of temperature. Scientists have, however, used some unusual refrigeration techniques to cool a substance to within 10^{-6}° C of this temperature.

The size of the unit of temperature (degree) is established by defining reference temperatures. The **Celsius (or centigrade) scale** sets the normal melting point of ice at zero (0° C) and the normal boiling point of water at one hundred (100° C).† The **absolute temperature scale** has the same size degree as the Celsius scale, but the zero is shifted to the absolute zero of temperature. A mercury thermometer is calibrated on the Celsius scale by marking the capillary tube at the level of the mercury when the bulb is immersed in ice water and again when the bulb is immersed in boiling water (both at one atmosphere pressure). The region between these marks is then divided into 100 equal spaces, each 1 degree, and equal spaces are extended on either side of the liquid water region.

A mercury thermometer and an alcohol thermometer, calibrated as described above, do not agree at intermediate temperatures. This shows that the expansion characteristics of the liquids are different and suggests that the expansion of a *liquid* is not a very fundamental property on which to base a temperature scale. The response of *gas* thermometers, on the other hand, is very nearly independent of the gas used. For this reason, the gas thermometer serves

† Melting and boiling points change with total pressure. The *normal* melting or boiling point is the value at one atmosphere total pressure.

as the basis for our temperature scales. The Celsius temperature scale is defined with the pressure of the gas proportional to the temperature. Mathematically, the relation is

The agreement improves at low pressure and becomes exact in the limit as P becomes zero.

$$P = P_0(1 + \alpha t) \quad \text{(constant volume)} \qquad [5.3]$$

where t is the Celsius temperature, P_0 is the pressure at 0° C, and α is a constant. An equivalent expression can be obtained by letting the volume of the gas change instead of the pressure:

$$V = V_0(1 + \alpha t) \quad \text{(constant pressure)} \qquad [5.4]$$

where V_0 is the volume at 0° C.

Equation [5.4] shows that the volume of a gas is linearly related to the temperature, and thus a plot of volume versus temperature should be a straight line. We can use this relationship to extrapolate to the absolute zero of temperature, that is, the temperature at which the volume of the gas is zero. The method is shown in Example 5.1.

EXAMPLE 5.1 Table 5.1 at right gives the volume of one mole of helium gas at one atmosphere pressure and at various Celsius temperatures. Use the data to determine the number required to convert the centigrade temperature to absolute temperature.

SOLUTION Figure 5.4 is the temperature-volume plot of the data given above. The intercept at $V = 0$ gives $t = -273°$ C.

TABLE 5.1

Volume of He Gas (1 mole)

V (liters)	t (°C)
22.41	0
26.52	50
30.62	100
38.82	200

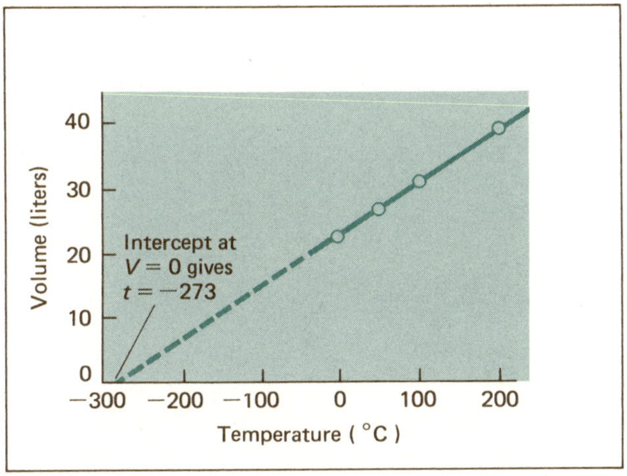

Figure 5.4
Graph of temperature versus volume for helium gas (one mole). The temperature intercept of −273° C at $V = 0$ is the absolute zero of temperature.

More accurate measurements establish the value as $t = -273.150°$ C and the absolute and Celsius scales differ only by this number. Hence

$$T° = t° + 273.150 \qquad [5.5]$$

where T is the symbol for absolute temperature and t for Celsius temperature.

Equations [5.4] and [5.5] can be combined to relate the volume of a gas to the absolute temperature. From [5.4], when $V = 0$, $t = -1/\alpha$. Hence, $1/\alpha = 273.15$, which, when combined with [5.5] and substituted into [5.4], gives

The accurate volume-temperature measurements for a gas that resulted in the law of Charles and Gay-Lussac were made by Joseph Louis Gay-Lussac. In his report, Gay-Lussac gives J.A.C. Charles the credit for obtaining similar results several years earlier. The work of Charles was never published, but on the strength of Gay-Lussac's generous acknowledgment, the temperature-volume relationship is often called the law of Charles.

$$V = V_0 \alpha T = k'' T \quad \text{(constant pressure)} \qquad [5.6]$$

Equation [5.6] states that the volume of a gas is directly proportional to the absolute temperature. This expression is known as the **law of Charles and Gay-Lussac** for Jacques Charles (1746–1832) and Joseph Louis Gay-Lussac (1778–1850), who independently established the temperature-volume relationship. It now serves as the definition of the absolute temperature scale.

In Chapter 7 we will describe another scale of temperature, which is identical with the absolute temperature scale. It is called the **Kelvin (K) scale** and is based on thermodynamics. The relationship existing among the Celsius, absolute, and Kelvin scales is

The definition of absolute temperature includes the stipulation of very low pressures. Equation [5.6] (the law of Charles and Gay-Lussac) is an approximation except at very low pressure.

$$T° A = T° \mathrm{K} = t° \mathrm{C} + 273.150 \qquad [5.7]$$

The two laws that are expressed by equations [5.2] and [5.6] provide simple if approximate statements of the pressure and temperature dependence of the volume of a gas. Later in the chapter we will consider the limitations of these laws. For now, we will simply state that at pressures up to a few atmospheres and at temperatures of the order of room temperature and higher, many gases obey these equations very closely.

5.2 The Ideal Gas

Derivation of the Ideal Gas Equation

The laws of Boyle and of Charles and Gay-Lussac apply only approximately to a real gas. It is convenient to define an idealized gas that obeys these laws under all conditions of pressure and temperature. Calculations made for this **ideal gas** can be applied to real gases once the limitations of the ideal gas model are understood.

We combine the law of Boyle with the law of Charles and Gay-Lussac to obtain a single equation that describes the PVT properties of an ideal gas. The result is

$$\frac{PV}{T} = k \qquad [5.8]$$

The law of Boyle is obtained from equation [5.8] by holding T constant:

$$V = \frac{(kT)}{P} = \frac{k'}{P}$$

and the law of Charles and Gay-Lussac is obtained by holding P constant:

$$V = \left(\frac{k}{P}\right)T = k''T$$

Elaborate mathematical techniques can be used to rigorously combine equations [5.2] and [5.6] to obtain equation [5.8]. The fact that the equation predicts the law of Boyle at constant temperature and the law of Charles and Gay-Lussac at constant pressure is sufficient for our purpose.

In equation [5.8], k is constant for a particular sample of gas. However, P and T do not depend upon the amount of gas (mass or moles), while V varies directly with the amount. Therefore, k must be directly related to mass or moles. This relationship is expressed in terms of the number of moles of gas, n, by the equation

At a given T and P, two moles of gas will occupy twice the volume of one mole. In general, n moles will occupy n times the volume of one mole at a given T and P.

$$k = nR \qquad [5.9]$$

where R is called the **ideal gas constant** and is independent of P, V, T, and n. Substituting equation [5.9] into [5.8] gives

$$\boxed{PV = nRT} \qquad [5.10]$$

Equation [5.10] is known as the **ideal gas equation.** Solved for V, it becomes

$$V = \frac{nRT}{P} \qquad [5.11]$$

which clearly indicates that the volume of a gas is determined by the temperature, pressure, and number of moles (R, of course, is constant). This functional relationship is written mathematically

$$V = f(T, P, n)$$

An expression that relates volume, temperature, pressure, and composition is an example of an **equation of state.** Thus, equation [5.10] is the equation of state for the **ideal gas.** This gas obeys the laws of Boyle and of Charles and Gay-Lussac at all temperatures and pressures.

The Gas Constant

The numerical value for R can be calculated from experimental data for real gases if these data are taken at temperature and pressure conditions at which the gases very nearly obey the laws of Boyle and of Charles and Gay-Lussac, and if the data are then extrapolated to the limit at which the laws are obeyed.

Figure 5.5
Plots of PV/nT versus P for several gases. The value of PV/nT at $P = 0$ is the gas constant R.

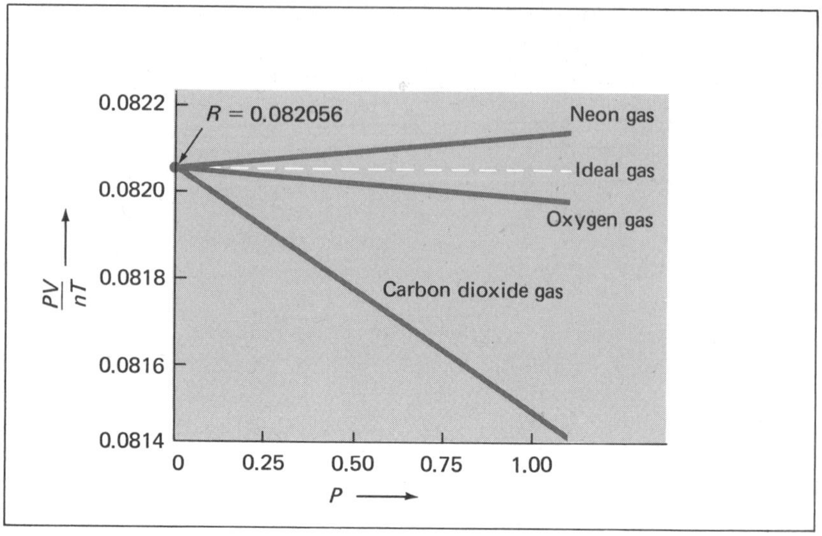

Figure 5.5 is a plot of PV/nT versus P for one mole of several real gases. The deviations from ideality are seen by comparing the curves for the real gases with the horizontal line for the ideal gas. These deviations become smaller at low pressures, and the extrapolated value at $P = 0$ (at which pressure the real gas achieves ideal gas behavior for PV/nT) is 0.082056 liter-atm/°K-mole for each of the real gases. Therefore,

$$R = 0.082056 \frac{\text{liter-atm}}{°\text{K-mole}}$$

Liter-atmospheres is an energy unit (check the dimensions). The gas constant, therefore, has the units of energy/degree-mole.

Table 5.2 gives the value of R expressed in several sets of dimensions. This table will be useful for reference in the study of later chapters where the gas constant often appears in equations in which it is more conveniently expressed in units other than liter-atm/°K-mole.

TABLE 5.2

Values for *R*, the Ideal Gas Constant

UNITS	VALUE
liter-atm/°K-mole	0.082056
cc-atm/°K-mole	82.056
cal/°K-mole	1.9872
joule/°K-mole	8.3144
erg/°K-mole	8.3144×10^7

Application of the Ideal Gas Equation

Some examples will illustrate the use of the ideal gas equation.

EXAMPLE 5.2 Calculate the volume in liters of 1.0000 mole of ideal gas at precisely one atmosphere pressure and 0° C.

SOLUTION Equation [5.11] applies:

$$V = \frac{nRT}{P}$$

Substituting the values of n, R, P, and T (remember that T is the absolute temperature) gives

$$V = \frac{(1.0000 \text{ mole})(0.082056 \text{ liter-atm/°K-mole})(273.15° \text{ K})}{1.0000 \text{ atm}}$$

$$V = 22.414 \text{ liters}$$

The temperature (0° C) and pressure (1 atm) used in Example 5.2 are referred to as standard temperature and pressure (STP). Hence, the volume of one mole of ideal gas at STP is 22.414 liters.

EXAMPLE 5.3 (a) Calculate the volume at STP of ideal gas that occupies a volume of 20.0 liters at 1000 torr pressure and 25° C. (b) Calculate the number of moles of gas present.

SOLUTION

(a) The total effect of a change in temperature and pressure upon the volume of an ideal gas may be seen by writing the ideal gas equation in the form

$$nR = \frac{P_1V_1}{T_1} = \frac{P_2V_2}{T_2} \qquad [5.12]$$

where the subscript 1 refers to the initial condition and the subscript 2 to the final condition. Rearranging this gives

$$V_2 = V_1\left(\frac{P_1}{P_2}\right)\left(\frac{T_2}{T_1}\right)$$

Substituting the following values:

$$V_1 = 20.0 \text{ liters}$$
$$P_1 = (1000 \text{ torr})(1 \text{ atm}/760 \text{ torr}) = 1.32 \text{ atm}$$
$$T_1 = 25 + 273 = 298° \text{ K}$$
$$T_2 = 273° \text{ K}$$
$$P_2 = 1.00 \text{ atm}$$

we find that

$$V_2 = (20.0 \text{ liters})\left(\frac{1.32 \text{ atm}}{1.00 \text{ atm}}\right)\left(\frac{273^\circ \text{ K}}{298^\circ \text{ K}}\right) = 24.2 \text{ liters}$$

(b) Solving $PV = nRT$ for n we obtain

$$n = \frac{PV}{RT}$$

Substituting the original pressure, volume, and temperature into this expression gives

$$n = \frac{(1000 \text{ torr})\left(\dfrac{1 \text{ atm}}{760 \text{ torr}}\right)(20.0 \text{ liter})}{\left(0.0821 \dfrac{\text{liter-atm}}{^\circ\text{K-mole}}\right)(298^\circ \text{ K})} = 1.08 \text{ moles}$$

We could have obtained the same result, of course, by substituting the final pressure, volume, and temperature into the expression for n.

Application of the Ideal Gas Equation to Real Gases

The ideal gas equation is of little value unless it can be applied to real gases. To do this, we must first understand the conditions under which the ideal gas equation approximates real gas behavior. From the ideal gas equation, we can write

$$\frac{PV}{nRT} = 1 \qquad \text{(ideal gas)}$$

Figure 5.6 shows plots of PV/nRT versus P for N_2 gas at different temperatures. It is apparent that temperature and pressure have considerable effect on the ideality of the gas. It is also apparent from Figure 5.7, which is a graph at 0° C of PV/nRT versus P for the three gases, H_2, N_2, and CH_4, that the nature of the gas has an important effect on its ideality. Examination of Figures 5.6 and 5.7 leads to the following conclusions:

(1) *The ideal gas equation approximates well real gas behavior at low pressures.* This approximation improves as the pressure is lowered. At $P = 0$, the gas behaves exactly ideally.
(2) *The higher the temperature, the larger the pressure range over which the ideal gas equation is a good approximation of real gas behavior.*

For temperatures near room temperature and above, and pressures of less than about 10 atmospheres, the ideal gas model can be successfully applied to such commonly encountered gases as He, Ne, Ar, H_2, O_2, N_2, CO_2, and CH_4. The calculations made in Examples 5.2 and 5.3 are applicable to these gases. The following examples illustrate other applications.

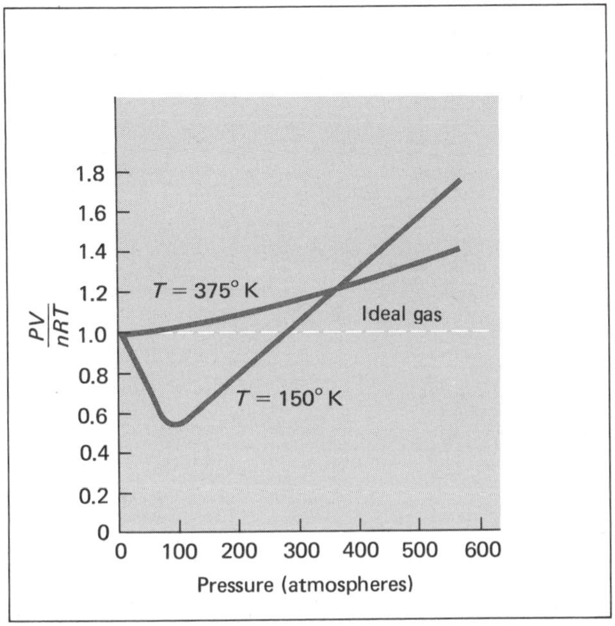

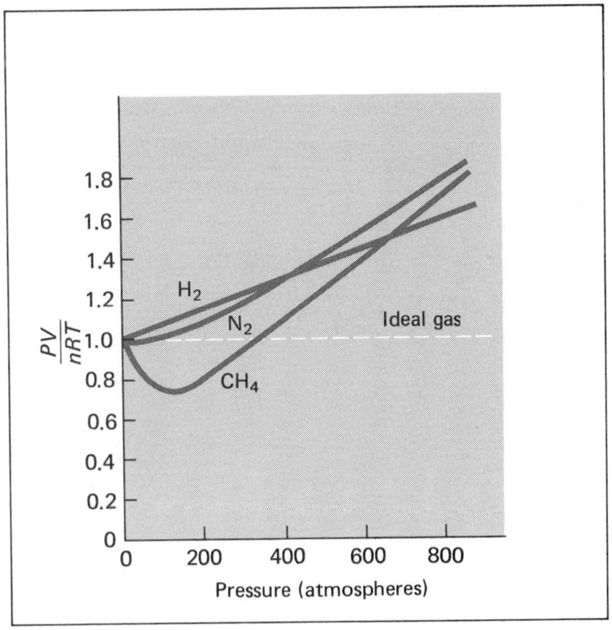

Figure 5.6
Graph showing the effect of pressure and temperature on the ideality of N_2 gas. Serious deviations occur at high pressures and low temperatures.

Figure 5.7
Graph showing deviations from ideal gas behavior for three different gases at 0° C.

EXAMPLE 5.4 Calculate the number of He atoms per cubic centimeter in He gas at 1.50 atm pressure and 300° K.

SOLUTION Under the *TP* conditions of the problem, the *PVT* behavior of helium gas can be approximated by the ideal gas equation. First we will solve for the number of moles of He gas:

$$n = \frac{PV}{RT}$$

$$P = 1.50 \text{ atm}$$
$$V = (1.00 \text{ cm}^3)(10^{-3} \text{ liter/cm}^3) = 1.00 \times 10^{-3} \text{ liter}$$
$$T = 300^\circ \text{ K}$$
$$R = 0.0821 \text{ liter-atm/}^\circ\text{K-mole}$$

$$n = \frac{(1.50 \text{ atm})(1.00 \times 10^{-3} \text{ liter})}{(0.0821 \text{ liter-atm/}^\circ\text{K-mole})(300^\circ \text{ K})} = 6.10 \times 10^{-5} \text{ moles}$$

Therefore, there are

$$6.10 \times 10^{-5} \text{ moles} \times 6.02 \times 10^{23} \text{ atoms/mole} = 3.67 \times 10^{19} \text{ atoms}$$

EXAMPLE 5.5 Calculate the density of O_2 gas at 500 torr and 100° C.

SOLUTION The number of moles, n, can be expressed in terms of the mass, w, and the molecular weight, M_w:

$$n = \frac{w}{M_w}$$

Therefore, the ideal gas equation becomes

$$PV = nRT = \frac{w}{M_w}RT$$

which is rearranged to

$$\frac{w}{V} = \frac{PM_w}{RT}$$

The mass per unit volume (w/V) is the density d; consequently

$$\boxed{d = \frac{PM_w}{RT}} \qquad [5.13]$$

For O_2 gas at 500 torr and 100° C, the ideal gas equation is a good approximation. Hence

$$d = \frac{(500 \text{ torr})(1 \text{ atm}/760 \text{ torr})(32.0 \text{ g } O_2/\text{mole } O_2)}{(0.0821 \text{ liter-atm}/°\text{K-mole})(373° \text{ K})}$$

$$d = 0.687 \text{ g/liter}$$

Equation [5.13] is often used to determine the molecular weight of a gas from an experimental determination of its density at a known temperature and pressure.

EXAMPLE 5.6 Calculate the molecular weight (M_w) of a gas if 2.00 liters of the gas measured at $P = 1.00$ atm and $T = 300°$ K weigh 2.44 g.

SOLUTION

$$PV = nRT$$

$$= \frac{w}{M_w}RT$$

$$\boxed{M_w = \frac{wRT}{PV}}$$

$$= \frac{(2.44 \text{ g})\left(\frac{0.0821 \text{ liter-atm}}{°\text{K-mole}}\right)(300° \text{ K})}{(1.00 \text{ atm})\ (2.00 \text{ liters})}$$

$$= 30.0 \text{ g/mole}$$

Mixture of Gases

The ideal gas equation can be applied to a mixture of gases. With ideal gas behavior, each gas in the mixture behaves independently. **Dalton's law** is the result.

Dalton's Law

This law, expressed by John Dalton (1766–1844), states that in a gas mixture the total pressure is the sum of the partial pressures of the individual gases. It is applicable to a mixture of gases, each showing ideal behavior, in which case the partial pressure of each gas is the pressure it would exert if it were alone in the container in the same quantity present in the mixture. That is,

$$\boxed{P_T = P_1 + P_2 + \cdots + P_N} \qquad \text{(Dalton's law)} \qquad [5.14]$$

where P_T is the total pressure and $P_1, P_2, \ldots, P_N$ are the partial pressures of the individual gases in the mixture.

EXAMPLE 5.7 Very pure N_2 gas can be obtained by heating ammonium nitrite

$$NH_4NO_2(s) \rightarrow N_2(g) + 2H_2O(g)$$

The water is easily removed. A 1.00 g sample of $NH_4NO_2(s)$ is placed in a 1.00-liter evacuated flask and heated to 500° C. Calculate the pressure in the flask (assume complete decomposition of the NH_4NO_2).

SOLUTION First we calculate the moles of N_2:

$$1.00 \text{ g } NH_4NO_2 \left(\frac{1 \text{ mole } NH_4NO_2}{64.0 \text{ g } NH_4NO_2}\right)\left(\frac{1 \text{ mole } N_2}{1 \text{ mole } NH_4NO_2}\right) = 0.0156 \text{ moles } N_2$$

Assuming ideal gas behavior gives

$$P_{N_2} = \frac{n_{N_2}RT}{V} = \frac{(.0156 \text{ moles})\left(.0821 \dfrac{\text{liter-atm}}{°\text{K-mole}}\right)(773° \text{ K})}{1.00 \text{ liter}} = 0.990 \text{ atm}$$

$$P_{H_2O} = 2P_{N_2} = 1.98 \text{ atm}$$

From Dalton's law

$$P = P_{N_2} + P_{H_2O} = 2.97 \text{ atm}$$

Pressure and Concentration

Equation [5.14] and the ideal gas equation can be combined to give a useful relationship between the pressure of a gas in a mixture and its concentration. Assume that a mixture of N gases, each having ideal behavior, is in a container of volume V maintained at temperature T. The pressure of each gas can be expressed by the ideal gas equation:

$$P_1 = n_1 \frac{RT}{V}$$

$$P_2 = n_2 \frac{RT}{V}$$

$$\vdots$$

$$P_N = n_N \frac{RT}{V}$$

or, in general

$$P_i = n_i \frac{RT}{V} \qquad [5.15]$$

where i refers to any of the components of the mixture. Substitution for $P_1 \ldots P_N$ into equation [5.14] gives

$$P_T = n_1 \frac{RT}{V} + n_2 \frac{RT}{V} + n_3 \frac{RT}{V} + \cdots + n_N \frac{RT}{V}$$

or

$$P_T = (n_1 + n_2 + n_3 + \cdots + n_N) \frac{RT}{V}$$

However,

$$n_T = n_1 + n_2 + n_3 + \cdots + n_N$$

when n_T is the total number of moles. Hence

$$P_T = n_T \frac{RT}{V} \quad \text{(ideal gas)} \qquad [5.16]$$

This is as expected, for the mixture was of gases, each behaving as if it were ideal gas. Now, dividing equation [5.15] by [5.16], we obtain

$$\frac{P_i}{P_T} = \frac{n_i}{n_T}$$

The ratio $n_i/n_T = x_i$ is the mole fraction of the ith component, and

$$\boxed{P_i = x_i P_T} \quad \text{(ideal gas)} \qquad [5.17]$$

That is, in a mixture of gases displaying ideal behavior, the pressure of each gas is the product of the mole fraction of that gas and the total pressure. For real gases (except at low temperatures and high pressures), equation [5.17] is a valid approximation.

Notice that equation [5.17] shows a very close relationship between the pressure of a gas and its concentration (mole fraction) in a mixture. A simple *pressure-molar concentration* relationship also exists for the ideal gas. This is seen when the ideal gas equation is written in the form

$$P = \frac{n}{V}RT$$

At constant temperature, P and n/V are related linearly. Since n/V is the concentration expressed in moles/liter, P can be looked upon as a concentration term when the temperature is constant. We will find it convenient to refer to pressure as a concentration term in subsequent chapters. The following example shows the usefulness of equation [5.17].

EXAMPLE 5.8 Calculate the pressure of oxygen gas in a sample of air that is 20% by weight O_2 and 80% by weight N_2. The total pressure is one atmosphere. Assume ideal gas behavior.

SOLUTION The partial pressure will be the same for any size example. Let us take 100 g of air, which will contain 20 g of O_2 and 80 g of N_2.

$$x_{O_2} = \frac{\text{moles } O_2}{\text{moles } O_2 + \text{moles } N_2}$$

$$x_{O_2} = \frac{\dfrac{20\text{ g } O_2}{32\text{ g } O_2/\text{mole } O_2}}{\dfrac{20\text{ g } O_2}{32\text{ g } O_2/\text{mole } O_2} + \dfrac{80\text{ g } N_2}{28\text{ g } N_2/\text{mole } N_2}} = 0.18$$

From equation [5.17]

$$P_{O_2} = x_{O_2}P_T = (0.18)(1.0\text{ atm}) = 0.18\text{ atm}$$

5.3 Real Gases

Equations of State for Real Gases

We saw on pages 100–101 that **real gases** approach ideal behavior at low pressures, but that there are significant deviations at higher pressures. There are two implications of the ideal gas equation that can help us recognize the kinds

We will better understand these implications in the next chapter when we describe the kinetic-molecular theory of gases.

of modifications that must be made on that equation to make it fit real gas behavior more closely.

(1) The volume, V, of the ideal gas is compressible volume; that is, there is no part of its volume that cannot be compressed by increased pressure.
(2) There are no attractive or repulsive intermolecular forces in the ideal gas.

In real gases, however, there are intermolecular forces and the molecules of these gases do have volume. Equations of state must account for these facts. Although many equations of state have been written for real gases, we will consider only the **van der Waals equation,** which is selected because of its simplicity and its theoretical basis.

Van der Waals Equation

Johannes D. van der Waals (1837–1923) modified the ideal gas equation by introducing additional terms in an effort to account for the volume of the gas molecules and the intermolecular forces. He argued that the pressure of a real gas would be less than that of an ideal gas under identical conditions because of attractive forces between the molecules of the real gas. To correct for this pressure effect, he added the term $a(n/V)^2$, where a is a constant specific for each gas, to the measured pressure in an attempt to obtain an *effective pressure*. The term $a(n/V)^2$, sometimes called the *internal pressure*, represents the effect on pressure produced by attraction between molecules. Van der Waals added the correction term $a(n/V)^2$ to the measured pressure to compensate for the decrease in pressure brought about by attraction between molecules; hence, P in the ideal gas equation was replaced by $(P + (n^2a/V^2))$.

The $(n/V)^2$ dependence for this correction term was obtained empirically by van der Waals. The kinetic molecular theory explains from a molecular point of view the validity of this dependence. The term (n/V) is a measure of the number of molecules per unit volume, and the square of the term is a measure of the probability of collision between molecules. Since the attractive forces between molecules are significant only when the molecules are close together, the extent of attraction, and hence the decrease in pressure, should be proportional to the number of collisions, which is given by $(n/V)^2$.

Van der Waals also made a correction on the volume term. His line of reasoning was as follows:

(1) The volume of a real gas would be expected to be larger than that of the ideal gas (under the same conditions) because the finite volume of the real gas molecules contributes to the total volume of the gas.
(2) To arrive at a volume term equivalent to the V in the ideal gas equation, a correction for the "incompressible volume" of the real gas must be made. Accordingly, van der Waals subtracted a term nb from the measured volume and replaced the V of the ideal gas equation with $(V - nb)$. The term b is the volume correction per mole. It is related to the molar volume of the molecules in the gas.

The van der Waals equation is

$$\left(P + \frac{n^2a}{V^2}\right)(V - nb) = nRT \qquad [5.18]$$

For one mole, the equation becomes

$$\left(P + \frac{a}{V^2}\right)(V - b) = RT \qquad [5.19]$$

This is the form in which we will apply the equation. Values for the a and b constants are determined empirically for each gas from measured PVT data. The values thus obtained (See Appendix A) can then be used to make calculations under other PVT conditions.

EXAMPLE 5.9 Calculate the pressure of CO_2 at 300° K if the density of the gas is 44.0 g/liter. Assume van der Waals behavior.

SOLUTION The molecular weight of CO_2 is 44.0 g/mole. For one mole of CO_2 ($n = 1$),

$$V = \frac{44.0 \text{ g/mole}}{44.0 \text{ g/liter}} = 1.00 \text{ liter/mole}$$

Solving the van der Waals equation for P gives

$$P = \frac{RT}{V - b} - \frac{a}{V^2}$$

From Appendix A, $a = 3.59$ liter^2-atm/mole^2 and $b = 0.0427$ liter/mole for CO_2. Substitution into the above equation gives

$$P = \frac{(0.0821 \text{ liter-atm/°K-mole})(300° \text{ K})}{(1.00 \text{ liter/mole} - 0.0427 \text{ liter/mole})} - \frac{(3.59 \text{ liter}^2\text{-atm/mole}^2)}{(1.00 \text{ liter/mole})^2}$$

$$P = 22.1 \text{ atm}$$

In the next chapter we will describe the attractive forces that occur between molecules. The van der Waals constant a is related directly to these forces. The van der Waals constant b is related to the incompressible volume of the molecules in the gas. For the ideal gas, no attractive forces occur and the molecules are point masses. Hence, a and b are both equal to zero, and equation [5.19] reduces to equation [5.10].

The van der Waals equation is easily solved for P and T, but it is a cubic polynomial in V:

$$PV^3 - (RT + bP)V^2 + aV - ab = 0$$

Since an explicit solution for V in this equation cannot be made, numerical methods are used. One approach is to rearrange the van der Waals equation to

$$V = \frac{RT}{P + \dfrac{a}{V^2}} + b \qquad [5.20]$$

and use successive approximations. A first approximate value of V, obtained from the ideal gas equation, is used to calculate a/V^2, which is then substituted into equation [5.20]. The improved value for V so obtained is used to calculate a new value of a/V^2, and the process is repeated until the calculated V agrees, to the required precision, with the inserted value. The following example illustrates this method.

EXAMPLE 5.10 Calculate the volume of one mole of CO_2 at 40° C and 10.00 atm. Assume van der Waals behavior.

SOLUTION From Appendix A, $a = 3.59$ liter²-atm/mole² and $b = 0.0427$ liter/mole for CO_2. Substitution into equation [5.20] gives for the volume per mole

$$V = \frac{(0.08206 \text{ liter-atm/}^\circ\text{ K-mole})(313.0^\circ \text{ K})}{10.00 \text{ atm} + (3.59 \text{ liter}^2\text{-atm/mole}^2)/V^2} + 0.0427 \text{ liter/mole} \qquad [5.21]$$

An approximate value of V (liter/mole) is calculated by use of the ideal gas equation:

$$V = \frac{RT}{P} = \frac{(0.08206 \text{ liter-atm/}^\circ\text{ K-mole})(313.0^\circ \text{ K})}{10.00 \text{ atm}} = 2.568 \text{ liter/mole}$$

This is substituted into equation [5.21]:

$$V = \frac{25.68}{10.00 + 3.59/(2.568)^2} + 0.043 = 2.479 \text{ liter/mole}$$

This new value for V is substituted into equation [5.21] to give

$$V = \frac{25.68}{10.00 + 3.59/(2.479)^2} + 0.043 = 2.469 \text{ liter/mole}$$

With repeated substitutions and calculations the next two values obtained are 2.468 liter/mole and 2.468 liter/mole. The final value for V is the same as the substituted value, and

$$V = 2.468 \text{ liter/mole}$$

Liquefaction of Real Gases: The Critical Point

We will now examine the pressure-temperature conditions under which a gas can be liquefied. Figure 5.8 shows several isotherms, or constant temperature curves, of pressure versus volume for a one-mole sample of carbon dioxide. Consider curve 1, the isotherm for 29.92° C. Initially the gas is at point A_1, a condition of low pressure, large volume, and low density. As the gas is compressed, the pressure increases along line A_1B_1 until point B_1 is reached. With further compression of the sample, however, the pressure remains constant and the gas condenses to form a liquid. The region from B_1 to C_1 is a two-phase

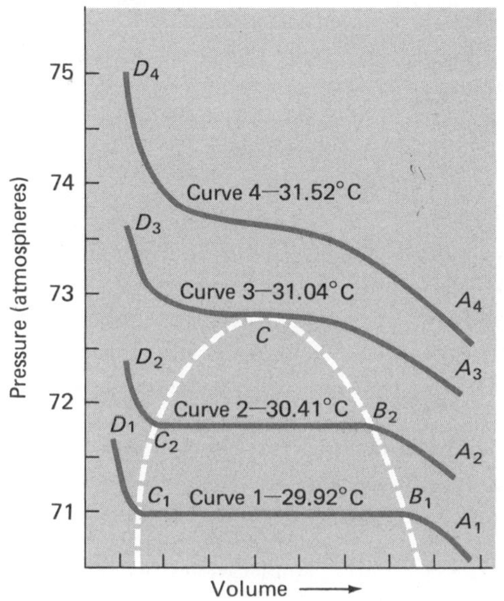

Figure 5.8
Pressure-volume isotherms of carbon dioxide near the critical point.

region; a definite meniscus, or boundary, separates the two phases. At any point between B_1 and C_1 the density of the gas is the same as at point B_1 while that of the liquid is the same as at C_1. The decrease in volume in this region results from the conversion of the gas into the more dense liquid. At point C_1, all the gas has been condensed to liquid. Further compression decreases the volume of the liquid (line C_1D_1). The rapid increase in pressure in this region reflects the very low compressibility of the liquid compared with that of the gas.

Curve 2 is obtained if the compression is carried out at the higher temperature of 30.41° C. The description of the system is much the same as along curve 1. Gas at point A_2 is compressed along A_2B_2 to point B_2. At this point, liquid condenses and the pressure of the system remains constant until all the gas is condensed to liquid at C_2. The line from C_2 to D_2 represents the compression of liquid CO_2 at 30.41° C. In comparing the two isotherms, note that at the higher temperature, the gas represented by point B_2 must be compressed to a higher pressure and a smaller molar volume (or higher density) before liquid forms. On the other hand the liquid represented by point C_1 has a larger molar volume and, hence, is less dense at the higher temperature. That is, the difference in density between gas and liquid is less at this higher temperature.

From consideration of the four isotherms plotted in Figure 5.8, one sees that the dotted line is an envelope that encloses the two-phase region. A horizontal line permits comparison at a particular temperature of the molar volume of the liquid (given by the line's intersection with the left side of the envelope) with the molar volume of the coexisting gas (given by the intersection with the right side of the envelope). It is apparent from the figure that as the temperature is increased, the molar volumes, and hence the densities, of the gas and liquid approach one another. At point C, called the **critical point,** the liquid and gas have the same density. The pressure at this point is the **critical pressure** (P_c), while the volume is the **critical volume** (V_c). The temperature of the isotherm that just touches this point is known as the **critical temperature** (T_c). (For CO_2, $T_c = 31.04°$ C or 304.2° K, $P_c = 72.9$ atm, and $V_c = 94.0$ cm^3/mole.)

The critical point is important, for it represents the maximum temperature at which liquid and gas can coexist. Since the gas and liquid have the same density at the critical point, the two phases are no longer distinguishable, and the phase boundary or separating meniscus disappears. Above the critical temperature on an isotherm, such as curve 4 in Figure 5.8, compression changes the system from a low-density state (point A_4) to the high-density state (point D_4) without a phase transition. Appendix A gives values for the critical pressure and temperature of a number of substances. It is apparent that the so-called permanent gases (N_2, O_2, H_2, He, Ne, . . .) have low critical temperatures. No amount of compression of these gases at room temperature will cause the condensation of liquid. Only if the gas is cooled to a temperature lower than T_c can it be converted to the liquid.

So far in this chapter we have described the macroscopic physical properties of the gaseous state. We now turn to that special and dynamic system of gas surrounding the earth, known as the atmosphere, and describe some of its unique characteristics. In the next chapter, we will return to gases once again and use the kinetic-molecular theory to explain the properties of gas from a molecular point of view.

5.4 The Atmosphere

In terms of mass, volume, and value to humans, the **atmosphere** is a gaseous system of first importance. Since the atmosphere is composed principally of low molecular weight, nonpolar molecules under conditions of relatively low pressure and moderately warm temperatures, its physical state is quite adequately expressed by the equations of state discussed in the previous sections. The example that follows illustrates the kind of calculations that can be used appropriately with gases in the atmosphere.

EXAMPLE 5.11 What volume of air taken at 25° C and 1.00 atm pressure contains just enough oxygen to react with 1.0 g of sulfur according to the equation

$$S + O_2(g) = SO_2(g) \qquad [5.22]$$

Assume that air is 21 mole % O_2.

SOLUTION The gas in the problem is at low pressure and relatively high temperature. Its behavior can be approximated by the ideal gas. Dalton's law is used to calculate the partial pressure of O_2 in the air:

$$P_{O_2} = x_{O_2}P_T = (0.21)(1 \text{ atm}) = 0.21 \text{ atm}$$

The number of moles of O_2 is calculated from the stoichiometry of the reaction:

$$1.0 \text{ g S}\left(\frac{1 \text{ mole S}}{32 \text{ g S}}\right)\left(\frac{1 \text{ mole } O_2}{1 \text{ mole S}}\right) = 0.031 \text{ mole } O_2$$

The ideal gas equation is used to calculate the volume of O_2 (which is also the volume of the air).

$$V = \frac{nRT}{P} = \frac{(0.031 \text{ mole } O_2)(.082 \text{ liter-atm/°K-mole})(298° \text{ K})}{(0.21 \text{ atm})}$$

$$V = 3.6 \text{ liters}$$

Several chemical reactions of great importance to life and its quality take place in the atmosphere. Consider now a few of the chemical properties of this gas mixture.

Chemical Composition

Clean air is a mixture of very simple gaseous molecules as can be seen in Table 5.3, which lists the major components of dry air. Variable amounts of water vapor are also present in the atmosphere. The water vapor concentration at the earth's surface varies from three mole percent in hot wet air to less than one mole percent in cold dry air to even lower values at high altitudes. Additional substances are also present in very small concentrations, usually as a result of chemical reactions occurring in the atmosphere or of the dumping of wastes into the atmosphere.

TABLE 5.3

The Composition of Clean, Dry Air near Sea Level

COMPONENT	CONTENT (mole percent)
Nitrogen	78.084
Oxygen	20.9476
Argon	0.934
Carbon dioxide	0.0318
Neon	0.00182
Helium	0.00052
Krypton	0.00011
Xenon	0.000009

Chemical Reactions with the Atmosphere

Several of the most important reactions of the constituents of the atmosphere are parts of cyclic processes; in other words, a substance consumed in one reaction is produced in another. In this chapter we will limit our discussion of chemical reactions in the atmosphere to a description of several of these important cyclical processes.

The Carbon Dioxide Cycle

Figure 5.9 summarizes important atmospheric reactions that involve carbon dioxide. The combustion, decay, and metabolism reactions generate CO_2, while the photosynthesis reaction consumes CO_2. As a result of these opposing reactions, CO_2 is continuously cycled into and out of the atmosphere. An almost endless variety of carbon-containing compounds participates in the **carbon dioxide cycle.** In the following sections we describe representative examples of the more important reactions.

The cycle is not in balance; at the present time more CO_2 is produced than is consumed. This imbalance is described on pages 116–117.

COMBUSTION REACTIONS Carbon compounds react with O_2 to produce CO_2. These reactions are classified as combustion reactions when they occur at high temperatures (as in a flame). When coal is burned in an excess of oxygen, the principal product is CO_2:

$$C(s) + O_2(g) = CO_2(g) \qquad [5.23]$$

These combustion reactions are highly *exothermic;* that is, they release large amounts of energy. Fossil fuels (coal, natural gas, and petroleum) are important sources of energy. At the present time, approximately three-fourths of the electrical energy produced in the world comes from the burning of fossil fuels.

Hydrocarbons, which are compounds of hydrogen and carbon, burn completely in excess oxygen to give CO_2 and H_2O. These are the products, for example, when methane (CH_4), a principal component of natural gas, is burned:

$$CH_4(g) + 2\,O_2(g) = CO_2(g) + 2\,H_2O(g) \qquad [5.24]$$

The predominant constituents of petroleum are hydrocarbons; these range from methane with one carbon atom through such substances as butane (C_4H_{10}), octane (C_8H_{18}), and octadecane ($C_{18}H_{38}$) to compounds having 40 and more carbon atoms. Crude petroleum is separated by distillation processes into *fractions*, which are mixtures of hydrocarbons of similar molecular weight. Gasoline, for example, is one of these fractions and usually consists mainly of hydrocarbons having from seven to nine carbon atoms. The combustion of

Figure 5.9
The CO_2 cycle.

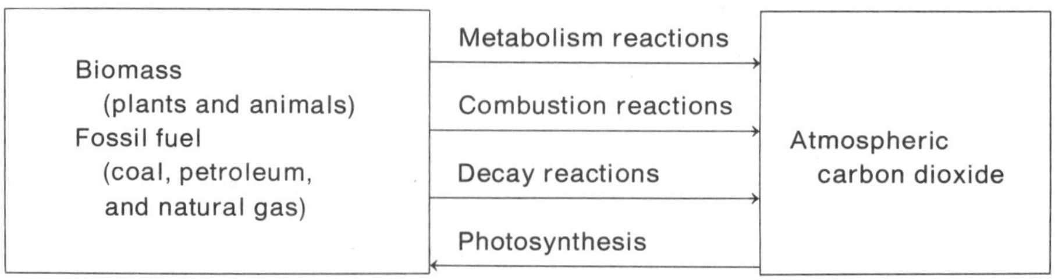

octane is representative of the reactions that occur when gasoline is burned:

$$2\,C_8H_{18}(l) + 25\,O_2(g) = 16\,CO_2(g) + 18\,H_2O(g) \qquad [5.25]$$

As has been indicated, in the complete combustion of hydrocarbons, the carbon is converted to CO_2 and the hydrogen to H_2O. Oxygen and sulfur are frequently present in natural fuels; the oxygen becomes H_2O and the sulfur SO_2 when the fuel is burned.

In a limited supply of oxygen, carbon may be oxidized only to carbon monoxide, CO. For example,

$$2\,C_8H_{18} + 17\,O_2 = 16\,CO + 18\,H_2O \qquad [5.26]$$

EXAMPLE 5.12 Predict the products in the following complete combustion reactions:

(1) $CS_2(l) + O_2(g) \rightarrow$

(2) $C_2H_5OH(l) + O_2(g) \rightarrow$

SOLUTION In reaction (1), the sulfur in carbon disulfide is oxidized to SO_2 and the carbon to CO_2:

$$CS_2(l) + 3\,O_2(g) = CO_2(g) + 2\,SO_2(g)$$

In reaction (2), the carbon in the ethyl alcohol (C_2H_5OH), is oxidized to CO_2 and the hydrogen and oxygen form water:

$$C_2H_5OH(l) + 3\,O_2(g) = 2\,CO_2(g) + 3\,H_2O(g)$$

DECAY REACTIONS In the combustion of carbon compounds, the carbon is rapidly oxidized at high temperatures to CO_2. A similar but very much slower reaction may occur at room temperatures in the decay of plant and animal matter. Wood, for example, is composed principally of cellulose, a high molecular weight compound with a formula that can be represented as $(C_6H_{10}O_5)_n$, where n is a large number. Although the decay of wood involves a number of intermediate products, the final products are CO_2 and H_2O:

$$6n\,O_2 + (C_6H_{10}O_5)_n = 6n\,CO_2 + 5n\,H_2O \qquad [5.27]$$

The same reaction occurs (rapidly) when wood is burned.

CELL METABOLISM REACTION Living organisms obtain the energy necessary for life and growth from metabolic reactions that occur within their cells. A most important example is the oxidation of glucose, $C_6H_{12}O_6$:

$$C_6H_{12}O_6 + 6\,O_2 = 6\,CO_2 + 6\,H_2O \qquad [5.28]$$

Sugar (glucose) and oxygen are carried by the blood to the cells. A reaction occurs, and the CO_2 that is formed is transported away from the cell, again by the blood.

This reaction is highly exothermic; that is, it occurs with the release of energy in the form of heat.

THE PHOTOSYNTHESIS REACTION The photosynthesis reaction is formally the reverse of the cell metabolism reaction; carbon dioxide and water are combined to form oxygen gas and a simple sugar ($C_6H_{12}O_6$). The source of energy for this *endothermic* (energy-absorbing) reaction is the sun; the overall reaction may be written

$$6\,CO_2 + 6\,H_2O + \text{sunlight} = C_6H_{12}O_6 + 6\,O_2 \qquad [5.29]$$

The photosynthesis reaction occurs in plant cells that contain the green pigment chlorophyll. Chlorophyll absorbs radiant energy from the sun and transfers it to the chemical system through a very complex and incompletely understood process. Efforts to understand the details of photosynthesis make up an important area of biochemical research.

The photosynthesis reaction ranks as one of the most important of all chemical reactions, for it is by photosynthesis that radiant energy from the sun is converted to chemical energy. Simple sugar is the starting material for the formation of the more complex carbohydrates that are stored in plants and from which we derive energy for our own metabolic processes.

The Nitrogen Cycle

Nitrogen in the form of the diatomic gas N_2 is the major component of the atmosphere. Small quantities of nitrogen are also present in the atmosphere as nitric oxide, NO, and nitrogen dioxide, NO_2. Nitric oxide results from the direct combination of nitrogen and oxygen:

$$N_2(g) + O_2(g) = 2\,NO(g) \qquad [5.30]$$

The reaction is endothermic and consequently occurs in the atmosphere only at high temperatures, such as in a lightning discharge. When the gases cool, nitric oxide reacts with oxygen to produce nitrogen dioxide:

$$2\,NO(g) + O_2(g) = 2\,NO_2(g) \qquad [5.31]$$

An internal combustion engine causes some formation of the nitrogen oxides in its combustion chamber.

Nitrogen is required for plant and animal growth since it is an essential constituent of protein. Nitrogen is incorporated into living tissue through steps of the **nitrogen cycle** as outlined in Figure 5.10. In step 1 of that cycle elementary nitrogen is converted to usable compounds: nitrates and ammonia. This process is often called *nitrogen fixation.* Nitrates result from the washing of NO_2, which

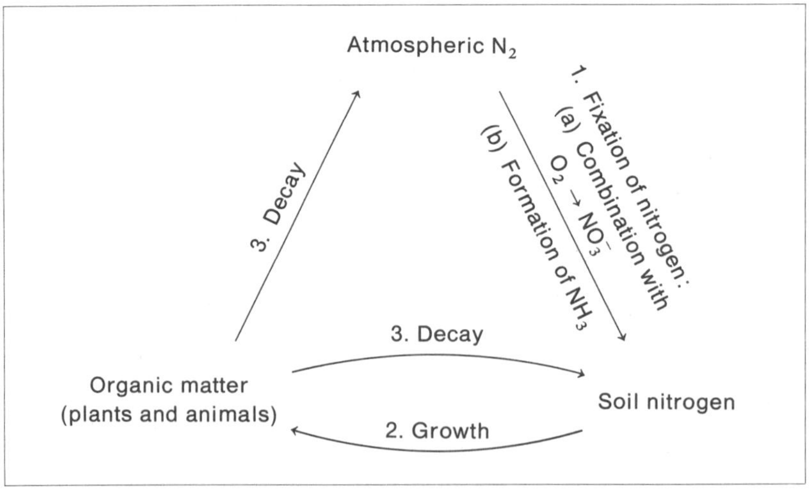

Figure 5.10
The nitrogen cycle.

is formed by equations [5.30] and [5.31], into the soil by rain, with the following reactions occurring:

$$2\ NO_2 + H_2O = H^+ + NO_3^- + HNO_2$$
$$3\ NO_2 + H_2O = 2\ H^+ + 2\ NO_3^- + NO$$

The quantity of NO_2 in the atmosphere is relatively small, and most of the elementary nitrogen that is fixed through natural processes is converted to ammonia by bacterial action. Certain bacteria in the soil convert nitrogen to ammonia in a direct metabolic process. A separate class of bacteria can similarly produce ammonia from elemental nitrogen, but they can do this only through a symbiotic process in the nodules on the roots of leguminous plants, such as alfalfa, beans, peas, clover, and peanuts. The plants provide the nodules in which the bacteria flourish; in return the bacteria provide fixed nitrogen that is essential to the plants' growth. Since the quantity of ammonia produced is greater than the plant requires, these legumes have long been used to enrich soil.

The growth reactions are in step 2 of the cycle. The fixed nitrogen (nitrates and ammonia) in the soil becomes part of the protein of the plant by complicated cell-growth reactions that involve many steps. These processes are a part of the subject of biochemistry.

The third step completes the nitrogen cycle. Decay of organic matter (plant and animal tissue) returns nitrogen in the form of nitrate and ammonia to the soil and N_2 gas to the atmosphere.

The Ozone Cycle

In the upper atmosphere, oxygen undergoes a sequence of reactions of great significance to life on earth. In the **ozone cycle,** molecular oxygen, O_2, is converted to ozone, O_3, which is subsequently changed back to molecular oxygen.

The mechanism (the series of steps that constitute the path of a reaction) is as follows: Ultraviolet radiation from the sun absorbed by an O_2 molecule dissociates the molecule into two oxygen atoms:

$$\text{Ultraviolet radiation} + O_2 = O + O$$

These oxygen atoms are examples of free radicals. They have an unpaired electron, and as a result are very reactive.

Ozone is formed when the reactive oxygen atoms collide with an O_2 molecule:

$$O + O_2 = O_3 + \text{heat}$$

There are two paths by which ozone decomposes:

$$O_3 + O = O_2 + O_2$$
$$\text{light} + O_3 = O_2 + O + \text{heat}$$

Sunburn and, in extreme cases, skin cancer can result from an overexposure to ultraviolet radiation. Without the ozone cycle in the upper atmosphere, exposure to direct sunlight would be dangerous.

Many scientists have been concerned that we would destroy the ozone layer through our irresponsible use of chloroflourocarbons, such as CF_2Cl_2 and $CFCl_3$, which have been used as aerosol propellants (in hair sprays, for example) and as refrigerants. These compounds are very stable as long as they remain in the lower

Figure 5.11
Temperature of the atmosphere as a function of altitude. The increase in temperature above 15 miles is primarily due to the ozone reaction.

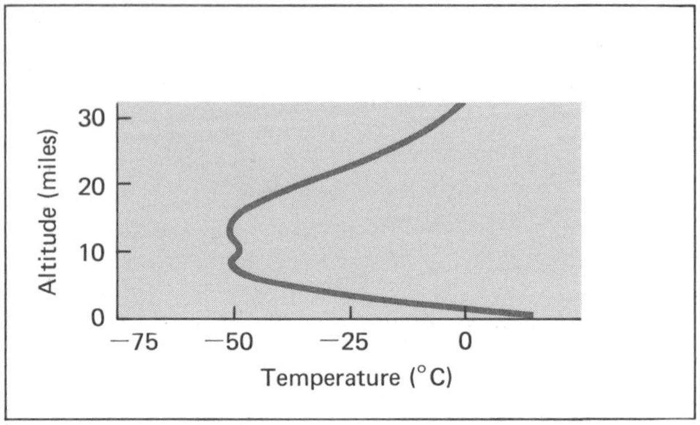

A steady state is established in the upper atmosphere, in which the rate of production of ozone is the same as the rate of its decomposition; consequently, its concentration stays constant. One of the important effects of this reaction is the conversion of ultraviolet radiation into heat. Figure 5.11 shows how the temperature of the atmosphere changes with altitude. The relatively high temperatures 20 to 40 miles above the earth are largely the result of the ozone cycle.

atmosphere. However, as soon as they diffuse into the upper atmosphere, they are decomposed by the intense ultraviolet light to produce chlorine atoms, which appear to be a catalyst in the decomposition of ozone. High-flying supersonic jets may also pose a threat to the ozone layer. The high temperature of the jet engine produces nitric oxide, NO, from the direct combination of N_2 and O_2 gases in the atmosphere. As with the chlorine atom, NO appears to catalyze the decomposition of ozone. Being catalysts, Cl and NO are regenerated in the overall reaction; hence, only small quantities are required to produce a large effect.

Contamination of the Atmosphere

The total mass of the atmosphere is approximately 5×10^{18} kg (6×10^{15} tons). Until the last few years, humans have considered the properties of this massive amount of material to be unalterable. We have indiscriminately dumped large quantities of gases and particulate matter into the atmosphere and expected nature to dilute and remove the contaminants from the environment.

It is now apparent that it is possible to overload the atmosphere with our waste products. The increase in the carbon dioxide content of the atmosphere from the combustion of fossil fuels is an example. The combustion reactions are very important in providing energy, but they add large quantities of carbon dioxide to the atmosphere. It is estimated that within the next 75 years, the total quantity of CO_2 in the atmosphere will double unless the use of fossil fuels is sharply curtailed.

The photosynthesis reaction (which consumes CO_2) cannot work fast enough to consume the CO_2 produced. Measurements show that the quantity of carbon dioxide in the atmosphere is increasing by 5.5×10^9 tons/year. This represents about 37% of the total annual carbon dioxide production.

The temperature and climate of the earth are sensitive to the CO_2 concentration in the atmosphere. The earth is warmed by visible radiation from the sun, which passes through the atmosphere to the earth's surface. Some light is reflected from the earth and passes through the atmosphere back into space. However, much of the incoming light is absorbed by the earth and re-emitted as infrared radiation. The carbon dioxide and water vapor in the atmosphere are opaque to a large fraction of this radiation; consequently, these substances keep some of the radiant energy received from the sun from escaping the earth. We benefit from this energy entrapment because, as a result, the earth's surface is relatively warm and the differences in day and night temperatures are small. However, an increase in the atmospheric CO_2 threatens to upset the delicate temperature balance of the earth. For example, some detailed studies have suggested that the effect of doubling the quantity of atmospheric CO_2 would be an average surface temperature increase of about 2.5° C. This is calculated to be a sufficient increase to cause melting of the polar ice caps. The melting of the southern cap (which is on land rather than floating, as is the northern cap) would raise the level of the oceans about 200 feet, submerging some of the most densely populated areas of the earth. Weather patterns would also be affected. Some dry regions would become wetter, and wet regions would become drier, which would drastically affect food production.

It should be apparent that many uncertainties are involved in estimates such as those in the preceding paragraph. For instance, we have not considered the absorption of CO_2 by the oceans (that tends to decrease, very slowly, the extent of the increases in CO_2 concentration in the atmosphere) or the effect an increased surface temperature has in expelling CO_2 from the oceans. Neither have we considered the effect that increased particulate matter in the atmosphere has on decreasing the amount of radiation received by the earth. This extremely complex and growing problem demands that those who attack it effectively have detailed understanding of chemical and physical equilibria, of biological processes, and of rates and mechanisms of the chemical reactions that occur in the atmosphere. It is important that there be people with concern to awaken others to the problem, and it is essential that there be people with the scientific and technical competence to convert that concern into meaningful action.

The insulating effect of the atmosphere is sometimes referred to as the "greenhouse effect." A greenhouse is made of a substance (glass or plastic) transparent to visible light but relatively opaque to infrared radiation. The visible radiation from the sun passes through the glass walls and roof of the greenhouse and is partially reflected back out as visible light; however, it is also partially absorbed by matter within the enclosure and re-emitted as infrared radiation. Because very little of this invisible radiation is transmitted by the glass, energy is trapped in the greenhouse as heat that keeps the greenhouse warm, thus promoting plant growth.

Local Pollution

In contrast to earthwide pollution problems, such as the increase in atmospheric carbon dioxide, there are serious problems of local air pollution, usually in congested and heavily industrialized areas. The automobile with its internal combustion engine makes a significant contribution to these problems in areas where most commuting is done by private car. While carbon dioxide and water are the major products of the combustion of gasoline, there are other products of much greater consequence to local air pollution. Carbon monoxide is present in the engine exhaust as the result of incomplete combustion; nitrogen oxides are formed in the high temperatures of the engine by the reactions shown in equations [5.30] and [5.31]; and sulfur dioxide results from sulfur compounds in the gasoline.

Photochemical smog is produced by the reaction of sunlight on these latter substances. The generally accepted mechanism for the reactions that occur is as follows:

(1) The NO_2 absorbs radiation from the sun and dissociates to form atomic oxygen:

$$NO_2 + \text{light} = NO + O$$

(2) Atomic oxygen reacts with O_2 in the air to produce ozone. (See the earlier discussion of the ozone reaction on page 115.)

$$O + O_2 = O_3$$

One can no longer ignore the possibility that the atmosphere could become irreversibly poisoned by human carelessness. Even the estimated billion dollars annual damage done by air pollution in one country, the United States, seems inconsequential compared with the destructive effects upon health, environment, and the quality of life.

(3) Atomic oxygen and ozone react with unburned hydrocarbons to produce a series of carbon compounds known as aldehydes and acyl peroxynitrates.

The ozone, aldehydes, and acyl peroxynitrates, along with the NO_2 and SO_2, give smog its objectionable qualities. The aldehydes and acyl peroxynitrates are eye irritants; the SO_2, O_3, and NO_2 attack the mucous membranes and lungs; SO_2 and O_3 in the concentrations present in smog attack and kill plants. In addition, O_3 causes the rapid deterioration of large numbers of organic materials, such as textiles and rubber.

Summary

At low pressures and high temperatures the relationship between pressure, volume, temperature, and number of moles of a gas can be represented to a good approximation by the ideal gas equation of state:

$$PV = nRT$$

where the gas constant R is a universal constant.

The ideal gas equation can be readily solved to obtain any one of the five quantities when the others are known. It can also be used to directly calculate densities of gases at specified temperatures and pressures:

$$d = \frac{PM_w}{RT}$$

or molecular weights from PVT data:

$$M_w = \frac{wRT}{PV}$$

For mixtures of ideal gases, Dalton's law of partial pressures applies:

$$P_T = P_1 + P_2 + \cdots + P_N$$

which can also be stated in the form

$$P_i = x_i P_T$$

The assumptions implicit in the ideal gas equation are (1) that the molecules occupy no volume and (2) that there are no attractive forces between the molecules. At high pressure these assumptions are no longer reasonable for "real" gas. The van der Waals equation is an attempt to describe *PVT* relationships under real gas conditions. Written for one mole it is

$$\left(P + \frac{a}{V^2}\right)(V - b) = RT$$

where *a* and *b* are constants with different values for every gas.

Under the right conditions of temperature and pressure any real gas can be liquefied. However, there is a characteristic temperature for each gas, above which it cannot be liquefied by compression alone. That temperature is known as the critical temperature of the gas.

Taking place in the atmosphere are several cyclic chemical processes that tend to keep that all-important gaseous system in a state of delicate balance. We need to be aware of the possibilities and consequences of upsetting that balance.

Important Terms and Concepts

Pressure
Atmosphere (unit of pressure)
Torr
Barometer
Manometer
Boyle's law
Temperature
Gas thermometer
Celsius (or centigrade) temperature scale
Absolute temperature scale
Absolute zero
Law of Charles and Gay-Lussac
Kelvin temperature scale
Ideal gas
Ideal gas equation
Ideal gas constant
Equation of state
Dalton's law
Real gases
Van der Waals equation
Critical point
Critical pressure
Critical volume
Critical temperature
Atmosphere
Carbon dioxide cycle
Nitrogen cycle
Ozone cycle

Exercises

Assume ideal gas behavior unless instructed otherwise in the exercise.

5-1.
Absolute temperatures can be measured experimentally with a gas-volume thermometer. With a given thermometer, the volume of gas at 0.00° C and a low pressure was 150.0 ml. At the same pressure but at another temperature, the gas volume was found to be 185.6 ml. What is the temperature of the system in °C?

5-2.
A given quantity of gas confined to a constant volume exerts a pressure of 330.0 torr at 0° C and a pressure of 375.9 torr at 38.00° C. From these data, calculate the value of absolute zero on the centigrade scale.

5-3.
A 20.0-liter vessel contains 0.200 g of oxygen at a pressure of 1.00×10^{-2} atm. What is the temperature of the gas?

5-4.
A vessel contains 2.00 moles of ideal gas at a pressure of 5.00 atm and a temperature of 27° C. What is the volume of the vessel in liters?

5-5.
What weight of CO_2 (M.W. = 44.01) is required to fill a 10.0 ml flask to a pressure of 1.00 torr at 30° C?

5-6.
Calculate the density of CO_2 gas at 1.00 atm pressure and 27° C.

5-7.
A 2.00-liter container is filled with N_2 gas at 27° C and 2.00 atm pressure. Calculate the weight of N_2 gas in the container.

5-8.
Which contains more molecules: (a) 500 ml of a gas at 600 torr and 21° C, or (b) 525 ml of a gas at 640 torr and 30° C?

5-9.
At what temperature will the ideal gas at 1.00 atm pressure have a concentration of 1.00 mole/liter?

5-10.
A 10.0-liter vessel contains 5.00 grams of nitrogen (M.W. = 28.01) at a temperature of 20° C. What is the pressure of the gas in atm?

5-11.
The best vacuum pump can generate a vacuum of about 1×10^{-9} torr. At room temperature (25° C), about how many molecules are there per milliliter at this pressure?

5-12.
One of the methods for estimating the temperature of the center of the sun is based on the ideal gas law. If the center is assumed to consist of gases whose average molecular weight is 2.0, and if the density and pressure are 1.4 g/cm^3 and 1.3×10^9 atm, respectively, calculate the temperature in °K.

5-13.
Calculate the number of molecules in 1.00 liter of O_2 gas at 50° C and 900 torr.

5-14.
Ethane gas has the formula C_2H_6. If 10.0 liters of C_2H_6 gas are confined in a vessel at 2.0 atm pressure and 0° C,

(a) how many moles of ethane gas are present in the bulb?
(b) what is the density (in g/liter) of the ethane gas?
(c) how many ethane molecules are in the bulb?
(d) how many grams of ethane are in the bulb?

5-15.
A given quantity of gas at 27° C and 1.00 atm pressure occupies 20.0 liters. What volume will it have if the pressure is raised to 5.00 atm and the temperature is raised to 67° C?

5-16.
A 50.0-liter vessel contains ideal gas at a pressure of 0.500 atm and at an unspecified temperature. When the temperature is increased by 30.0° C, the new pressure is 0.550 atm. Calculate the initial temperature in °C and the number of moles of gas in the vessel.

5-17.
A 3.00-mole portion of ideal gas initially at a pressure of 1 atm and a temperature of 100° C is expanded to twice the initial volume and then heated to 200° C at the new volume. What is the final pressure in atm?

5-18.
One mole of ideal gas initially at 1.00 atm and 100° C is heated to 200° C.

(a) If the process is carried out isobarically (at a constant pressure), what is the final volume?
(b) If the process is carried out isochorically (at a constant volume), what is the final pressure?

5-19.
A container is filled with a gas to a pressure of 10.0 atm at 20° C.

(a) What pressure will develop inside the sealed container when it is warmed to 100° C?
(b) At what temperature would the pressure be 20.0 atm?

5-20.
The density of an unknown gas is 1.00 g/liter at 1.00 atm pressure and 27° C. Calculate the molecular weight of the gas.

5-21.
(a) Analysis of a compound of carbon and hydrogen showed 85.6% C and 14.4% H. What is the empirical formula of the compound?
(b) The density of the gas in part (a) is 3.75 g/liter at STP. What is the *molecular* formula of the compound?

5-22.
A given compound consists of 36.36% C, 6.06% H, and 57.58% F. At 1.00 atmosphere pressure and 0° C, 1.00 liter of the gaseous compound weighs 2.95 g. Calculate the molecular formula of the gas.

5-23.
A mixture of H_2, O_2, and N_2 gas is at a total pressure of 742 torr. If the pressure of the hydrogen is 242 torr and the nitrogen pressure is four times the oxygen pressure, what is the nitrogen pressure?

5-24.
A gas mixture of O_2 and N_2 at STP is 50.0 weight percent O_2. Calculate the pressure of O_2 in the gas.

5-25.
At 27° C, 5.00×10^2 cm^3 of H_2 gas measured under a pressure of 5.00×10^2 torr and 1.000×10^3 cm^3 of N_2 gas at a pressure of 6.00×10^2 torr are introduced into an evacuated 2.00-liter flask. Calculate the final total pressure in the flask.

5-26.
The density of a mixture of CO_2 and SO_2 gases is 2.40 g/liter at 27° C and 1.00 atm pressure. Calculate the percent (by weight) of CO_2 in the mixture.

5-27.
A 5.00-liter bulb containing N_2 gas at 5.00×10^2 torr pressure is connected by a closed stopcock to a 10.0-liter bulb containing O_2 gas at 1.000×10^3 torr pressure. What is the equilibrium pressure in the bulb after the stopcock is opened?

5-28.
A 10.0-liter flask contains 0.500 g each of hydrogen (H_2), nitrogen (N_2), carbon dioxide (CO_2), and methane (CH_4) at 27° C. Calculate the partial pressure of each gas in atmospheres.

5-29.
How many grams of $NaHCO_3$ do you need to produce 10.0 liters of CO_2 gas (measured at 1.00 atm and 27° C) from the reaction

$$NaHCO_3 + H^+ = Na^+ + H_2O + CO_2$$

5-30.
A mixture of $CaCO_3$ and $NaHCO_3$ is heated, and the compounds decompose according to the following equations:

$$CaCO_3(s) \rightarrow CaO(s) + CO_2(g)$$
$$2NaHCO_3(s) \rightarrow Na_2CO_3(s) + CO_2(g) + H_2O(g)$$

(a) If 784 ml of CO_2, measured at STP, and 0.396 g of H_2O were obtained from the decomposition of the sample, what mole percentage of the original sample was $CaCO_3$?
(b) What was the original weight of the sample?

5-31.
In an experiment, 0.202 g of H_2 and 0.800 g of O_2 gas are injected into an evacuated container that has a volume of 2.24 liters. The container and contents are maintained at 546° K.

(a) What is the pressure in the container?
(b) The gaseous mixture is now ignited by a spark. What is the pressure in the container after the temperature is restored to 546° K?

5-32.
What volume of CO_2 at 760 torr pressure and 27° C is produced by the reaction of 200.0 ml of 4.00 molar $Cr_2O_7^{2-}$ solution with excess $C_2O_4^{2-}$ in acidic solution? The reaction is

$$3\,C_2O_4^{2-} + 14\,H^+ + Cr_2O_7^{2-} = 2\,Cr^{3+} + 6\,CO_2 + 7\,H_2O$$

5-33.
Consider the reaction

$$2\,MnO_4^- + 10\,Cl^- + 16\,H^+ \rightarrow 2\,Mn^{2+} + 5\,Cl_2 + 8\,H_2O$$

How many ml of 0.300-molar MnO_4^- solution are required to produce 10.0 liters (measured at 27° C and 380 torr) of Cl_2 gas?

Problems

5-1.
An ideal gas at 700 torr pressure occupies a bulb of unknown volume. A certain amount of the gas is withdrawn and found to occupy 1.50 cc at 1.00 atm pressure. The pressure of the gas remaining in the bulb is 675 torr. Assuming that all measurements were made at the same temperature, calculate the volume of the bulb.

5-2.
Suppose that a hot-air balloon is constructed with a volume of 100,000 liters and that the contained air is maintained at a temperature of 200° C.

(a) If the exterior air has a temperature of 20° C and a density of 1.2 g/liter, what mass could the balloon lift?
(b) What mass could be lifted if the balloon were filled with hydrogen (H_2) at one atm and 20° C

5-3.
PCl_5 gas partially dissociates at high temperatures according to the following reaction:

$$PCl_5(g) = PCl_3(g) + Cl_2(g)$$

In a given experiment, 0.100 mole of PCl_5 was weighed into a 4.00-liter container and heated to 227° C. After equilibrium was established, the total pressure of the system read 1.60 atm. Calculate the number of moles of PCl_5 gas in the system at equilibrium.

5-4.
Given below are *PV* data for one mole of CO_2 and one mole of O_2 gas at 0° C. Plot *PV* versus *P* for each gas, and thereby show that *PV* (which is constant for an ideal gas at constant temperature) is not constant for real gases. From your graph determine (1) the value of *R*, the ideal gas constant, and (2) the *a* and *b* constants for CO_2 gas at 0° C in the equation of state $PV = a + bP$. You will note that this straight line equation adequately describes real gas behavior in this low pressure range.

O_2		CO_2	
P(atm)	*PV* (liter-atm/mole)	*P*(atm)	*PV* (liter-atm/mole)
1.0000	22.3939	1.00000	22.2643
0.7500	22.3978	0.66667	22.3148
0.5000	22.4045	0.50000	22.3397
0.2500	22.4096	0.33333	22.3654
		0.25000	22.3775
		0.16667	22.3897

5-5.
The density of ammonia in grams per liter has been determined at various pressures by weighing the gas in large glass bulbs. The values of d/P (d = density in grams per liter, P = pressure in atmospheres) at 0° C are as follows: 0.77169 g/liter-atm at 1 atm, 0.76773 g/liter-atm at 2/3 atm, 0.76585 g/liter-atm at 1/2 atm, 0.76405 g/liter-atm at 1/3 atm.

(a) Plot d/P versus P and extrapolate to zero pressure, where the gas behaves ideally. From this limiting d/P value, calculate the molecular weight of NH_3.
(b) If the atomic weight of hydrogen is taken as 1.008, what is the atomic weight of nitrogen?

5-6.
Calculate the pressure exerted by 10.0 g of nitrogen in a closed 1.00-liter vessel at 25° C, using (a) the ideal gas equation; (b) the van der Waals equation. See Appendix A for additional data.

5-7.
Calculate the volume of one mole of O_2 gas at 102 atm pressure and 501° K assuming (a) ideal gas behavior; (b) van der Waals gas behavior.

5-8.

Derive a formula for determining the density of a gas that obeys the equation of state

$$PV = nRT + nBP$$

The expression that you obtain should be in terms of T, P, R, B, and M_w (molecular weight).

5-9.

Calculate the percent NO in a mixture of NO plus NO_2 if 0.700 grams of the mixture occupies 5.30 liters at 400° K and a pressure of 0.100 atm.

5-10.

Horatio La Chatelier Smith, a young chemist of sterling character, discovered with dismay that his cat, Fluorine Disprosmia von Boltzmann, had licked the label from a beaker until only the word "alcohol" could be detected, and was now in the process of drinking the contents of the beaker. Remembering that it was either methyl alcohol (CH_3OH) or ethyl alcohol (C_2H_5OH) that he had brought to his room, and remembering further that when imbibed, the former causes blindness and the latter drunkenness, he realized that he must quickly determine whether to take F. Disprosmia v. B. to the Sight-Savers Clinic or to Alcoholics Anonymous. Therefore, he took 10.0 g of the remaining liquid and injected this into a 10.0 liter evacuated flask. (Horatio had been a very sincere Boy Scout. His deep commitment to the principle of being prepared caused him to have an unusual quantity of preparedness items in his room. Since he kept his parachute, iron lung, adrenaline kit, decompression chamber, and elephant gun on his bed in order to have a little floor space, he found it difficult to get prepared to go to bed. But we digress.) He heated the 10.0-liter sealed flask now containing liquid from the beaker to 227° C over the Bunsen burner that had been on top of the snowshoes under the space blanket under the bed. The liquid all evaporated. With his emergency pressure gauge, Horatio found that the pressure within the flask was 0.89 atm. To which agency should he take Fluorine Disprosmia von Boltzmann? Show how you determined this.

6

Kinetic-Molecular Theory

In the last chapter we considered methods for representing the macroscopic *PVT* properties of a gas and very briefly introduced the molecular theory that accounts for these properties. This theory, based on a submicroscopic view of matter, is known as the **kinetic-molecular theory** and is very useful in explaining many of the physical properties of matter. As the name suggests, the principal postulates of the theory are (1) *that matter consists of molecules*[†] and (2) *that the molecules are in motion*. In this chapter we will discuss the application of these postulates to the three states of matter.

6.1 Molecular Arrangement

Figure 6.1 illustrates the general molecular arrangements in the three states of matter. In a crystalline solid, the molecules are arranged in a repeating geometrical array called a *lattice*. The lattice repeats itself in all three directions. As a result, the molecules are lined up in long rows, columns, and planes. The alignment results in a neat, well-ordered arrangement of the molecules. We speak of a solid as having a "long-range order." In a later chapter we will consider in detail a number of the common lattice structures.

In a liquid, the molecules are packed closely together, as they are in a solid. As a result, both phases have relatively high densities and are difficult to compress. A liquid differs from a solid, however, in that the packing is disordered. Clumps of molecules often form into geometric units, like those one would find in a similar amount of solid. However, the structure has no permanence, and it does not repeat itself throughout the liquid. No long rows,

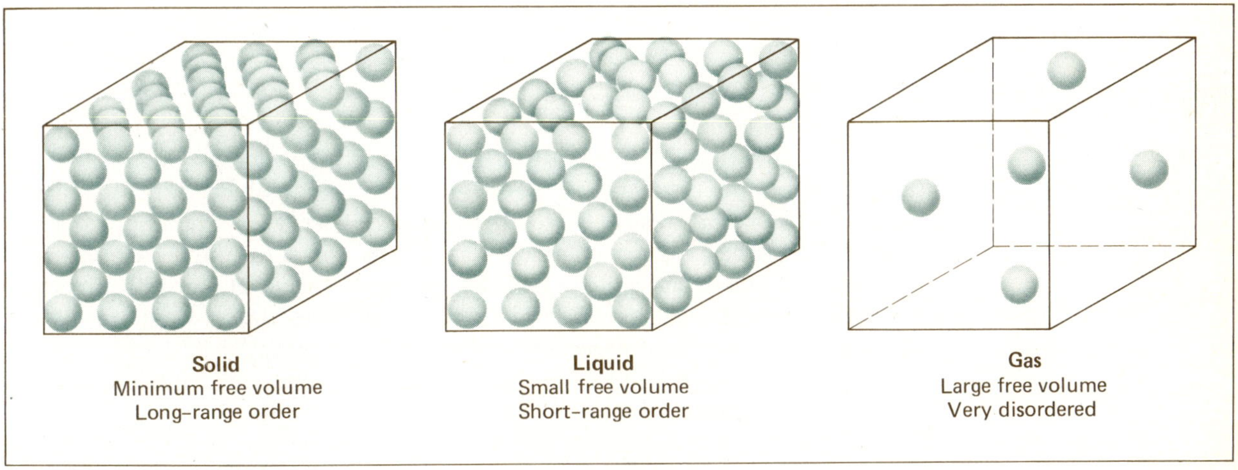

Figure 6.1
Comparison of the atomic structures of solid, liquid, and gas.

† The word *molecule* is used here rather loosely to refer to the unit particle that makes up a substance. In some substances the "molecules" may be atoms or ions.

columns, and planes of molecules are present in a liquid. Thus, a liquid can be described as having only a "short-range order."

The molecular arrangement in a gas, as represented in Figure 6.1, is seen to be quite different from that of a solid or a liquid. In a gas, most of the volume is "empty space," the molecules themselves occupying only a small percentage of the total volume. The fraction of the total space that is empty depends on the temperature and pressure of the gas. Near atmospheric pressures and room temperatures, a gas is usually much less dense than the corresponding liquid or solid; commonly only one part in several thousand of the gaseous volume is occupied by the molecules. The high compressibility of gases becomes understandable when one realizes that the compression merely removes some of the free space. The arrangement of the molecules in a gas is one of general disorder. A real gas may have a small amount of very short-range order; the ideal gas, however, has complete disorder.

6.2 Kinetic Energy

The second postulate of the kinetic theory is that molecules are in motion and, hence, have **kinetic energy.** This "energy of motion" can be present in several forms. **Translational kinetic energy** is the energy associated with motion along a line. For example, a falling rock and a pitched baseball have this type of motion. The translational kinetic energy, E_{Trans}, can be calculated for any moving object from the equation

$$E_{\text{Trans}} = \frac{1}{2}mu^2 \qquad [6.1]$$

where m is the mass of the object and u is the velocity.[†] Thus, the faster the motion and the larger the mass, the greater the kinetic energy.

The motion associated with the rotation of an object around an axis (imaginary or real) results in **rotational kinetic energy.** A spinning propeller, for example, has rotational kinetic energy, as has a turning roulette wheel. **Vibrational kinetic energy** results from the oscillatory motion of an object. A vibrating tuning fork, the vibrating string on a violin, and a weight in motion on the end of a spring, all have vibrational kinetic energy.

In solids, the motion of the molecules, and hence the kinetic energy, is mainly restricted to vibration. The molecules oscillate back and forth around their equilibrium positions in the lattice. Some molecules (spherically shaped molecules such as CH_4 and CCl_4) rotate in a solid as well as vibrate, and occasionally one molecule slips past another in the lattice, resulting in some

† Velocity is a *vector quantity;* that is, it has direction as well as magnitude. However, except where stated otherwise, we will be considering only the magnitude.

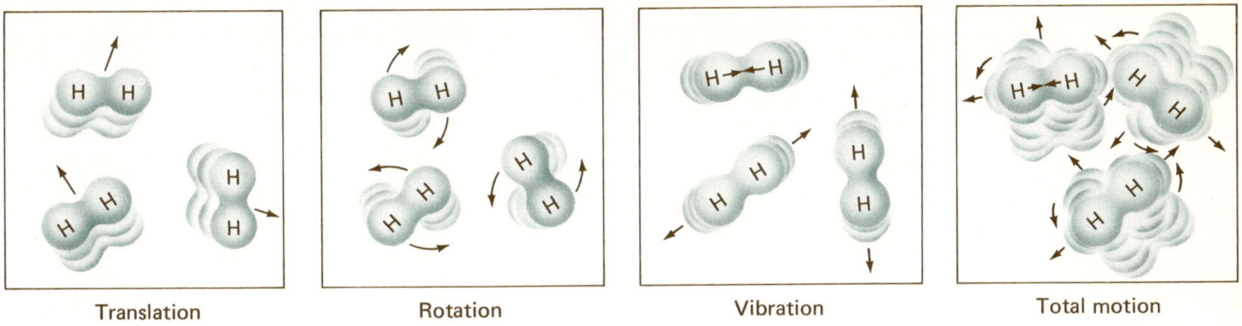

Figure 6.2
The total kinetic energy in hydrogen gas results from a combination of translational, rotational, and vibrational motion.

translational motion. But the principal motion of the molecules in a solid is vibrational. In a liquid, the motion is not so restricted; translational, rotational, and vibrational motion all occur. All gases possess translational kinetic energy, and the molecules are in rapid, random motion. They move in a straight line until they collide with one another or with the walls of the container. In molecular gases, in which the individual molecules can rotate or vibrate, these types of "internal" motions are also present. Figure 6.2 illustrates the motion in hydrogen (H_2) gas.

EXAMPLE 6.1 Calculate the kinetic energy of (a) a 5.00 ounce (142 g) baseball thrown by a pitcher at 100 miles/hr (4470 cm/sec) and (b) a gaseous oxygen molecule moving with a velocity of 1.00×10^5 cm/sec.

SOLUTION

(a) Equation [6.1] applies. For the baseball

$$\begin{aligned} E_{\text{Trans}} &= (1/2)(142\text{ g})(4470\text{ cm/sec})^2 \\ &= 1.42 \times 10^9\text{ ergs} \end{aligned}$$

An *erg* is a g cm^2/sec^2, and is a very small unit of energy. A more convenient unit to use in this calculation is either the joule or calorie. Multiplying by the appropriate conversion factor (see table of conversion factors on the inside back cover) gives

$$E_{\text{Trans}} = (1.42 \times 10^9\text{ ergs})\left(10^{-7}\frac{\text{joules}}{\text{erg}}\right) = 142\text{ joules}$$

or

$$E_{\text{Trans}} = (1.42 \times 10^9\text{ ergs})\left(\frac{1\text{ cal}}{4.18 \times 10^7\text{ ergs}}\right) = 34.0\text{ cal}$$

(b) For the O_2 molecule, we first calculate the molecular mass:

$$m = \frac{32.0 \text{ g/mole}}{6.02 \times 10^{23} \text{ molecules/mole}} = 5.32 \times 10^{-23} \text{ g}$$

Equation [6.1] is then used to calculate the energy:

$$\begin{aligned} E_{\text{Trans}} &= (1/2)(5.32 \times 10^{-23} \text{ g})(1.00 \times 10^5 \text{ cm/sec})^2 \\ &= 2.66 \times 10^{-13} \text{ ergs} \end{aligned}$$

Because this energy is so small, it is appropriate to leave the answer expressed in ergs, although we could convert it to electron volts (eV), which is a unit of energy commonly used when considering atomic-size particles.

$$E_{\text{Trans}} = (2.66 \times 10^{-13} \text{ ergs})\left(\frac{1.00 \text{ eV}}{1.60 \times 10^{-12} \text{ ergs}}\right) = 0.166 \text{ eV}$$

6.3 Potential Energy

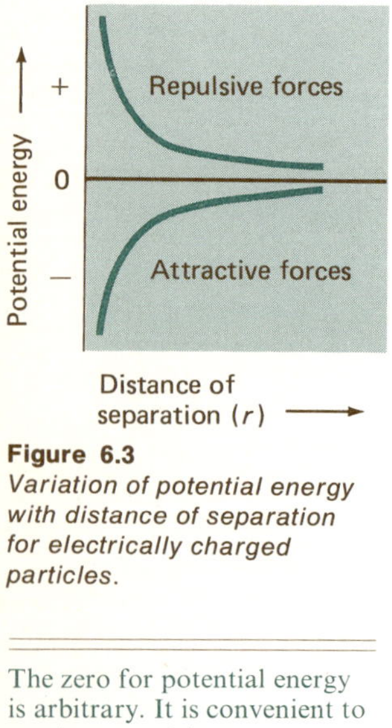

Figure 6.3
Variation of potential energy with distance of separation for electrically charged particles.

The zero for potential energy is arbitrary. It is convenient to set the potential energy at zero when the particles are an infinite distance apart.

Potential energy exists when two objects, either through attraction or repulsion, interact with one another. The potential energy in such a system depends upon the force of attraction or repulsion, which in turn depends upon the distance of separation of the particles. For example, Figure 6.3 shows how the potential energy changes with the distance of separation between two electrically charged particles. The lower curve results when the particles have opposite electrical charges and hence are attracted to one another.† At large separations, where the interaction is minimal, the potential energy is approximately zero. As the particles are brought closer together, the potential energy decreases. *By convention, the potential energy is given a negative sign when attraction occurs.* The upper curve results when the particles are repelled by one another.‡ Again, at large separations, the potential energy is near to zero. As the particles are brought together in this case, however, work must be done to overcome the forces of repulsion between the particles. Adding energy in the form of work increases the potential energy of the system.

When uncharged molecules are brought together, both repulsion and attraction occur. The potential energy curve that results is shown in Figure 6.4. The curve can be explained as follows: At large r values, both the attractive and repulsive forces are close to zero. The potential energy under such a condition would also be near zero. As the particles are brought together, the attractive forces initially increase faster than do the repulsive forces, resulting in a net decrease in poten-

† An electron and a proton in the hydrogen atom would be an example of a system of this type.
‡ Two protons or two electrons would be examples of such a system.

tial energy from A to B in Figure 6.4. Repulsive forces between molecules become important at small separations. The minimum in the curve at B results when the rate of change in the forces of repulsion and attraction is the same. At separations less than that represented by B, the repulsive forces increase faster than the attractive forces. At point C, the two forces are equal, and the potential energy is again zero. At still smaller separations, the repulsive force is larger than the attractive force, and the potential energy is positive.[†]

The potential energy of a system of molecules is highly dependent upon the physical state of the substance. In liquids and solids the molecules are packed closely together. The separation between the particles is small, near to that given by point B in Figure 6.4. A net attractive force exists between the molecules, and the potential energy is close to the minimum value. In other words, *solids and liquids have large attractive forces and consequently large negative potential energies*. In a gas, however, the molecules are, on the average, separated by large distances, and the average potential energy resulting from interaction between molecules is almost zero. The definition of the ideal gas on the molecular scale includes the statement that there are no attractive or repulsive forces between the particles, and hence the potential energy is exactly zero.

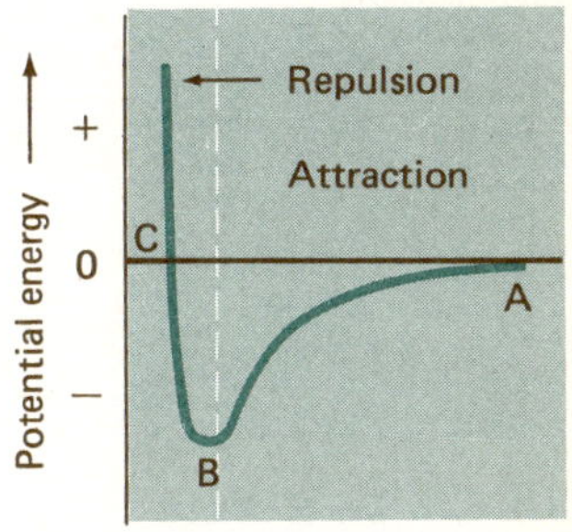

Figure 6.4
Variation of the potential energy with distance of separation for two interacting molecules.

6.4 Distribution of Translational Energy in the Ideal Gas

We have noted that the molecules of a gas are in motion, continually colliding with one another and with the walls of their container. The molecules of any given gas, however, do not all have the same translational velocity, but are moving with a variety of velocities. The range in velocity of the molecules may be from very fast to very slow. Furthermore, because of the transfer of energy from one molecule to another during collisions, every molecule is constantly undergoing changes in velocity.

Because of the large number of molecules and the high frequency of collisions, it would be a monumental task to describe the behavior of each molecule in a mole of gas for even a very short period of time. It *is* possible, however, to make statistical calculations of the most probable distribution of velocities. If we could measure the velocity of all molecules in a sample of gas at any given instant, we could construct a graph similar to Figure 6.5, in which the fraction of the molecules with velocities between u_1 and u_2, u_2 and u_3, and so on, are

† For many molecules, the intermolecular potential energy curve represented in Figure 6.4 is expressed mathematically to a good approximation by the Lennard-Jones equation,

$$E_p = 4\varepsilon\left[\left(\frac{\sigma}{r}\right)^{12} - \left(\frac{\sigma}{r}\right)^{6}\right]$$

where E_p is the potential energy and r is the distance of separation. The quantities ε and σ are constants that depend upon the nature of the molecule. The parameter ε is the value of E_p at the minimum in the curve. The distance parameter σ is the separation between the molecules at point C in Figure 6.4 (roughly the diameter of the molecules). The $(\sigma/r)^6$ term arises from the attractive forces between the molecules, and the $(\sigma/r)^{12}$ term from the repulsive forces. At large separations, $(\sigma/r)^6 > (\sigma/r)^{12}$ and a net attraction (negative potential energy) results. At small separations, the reverse is true.

Figure 6.5
Distribution of molecular speeds in a gas.

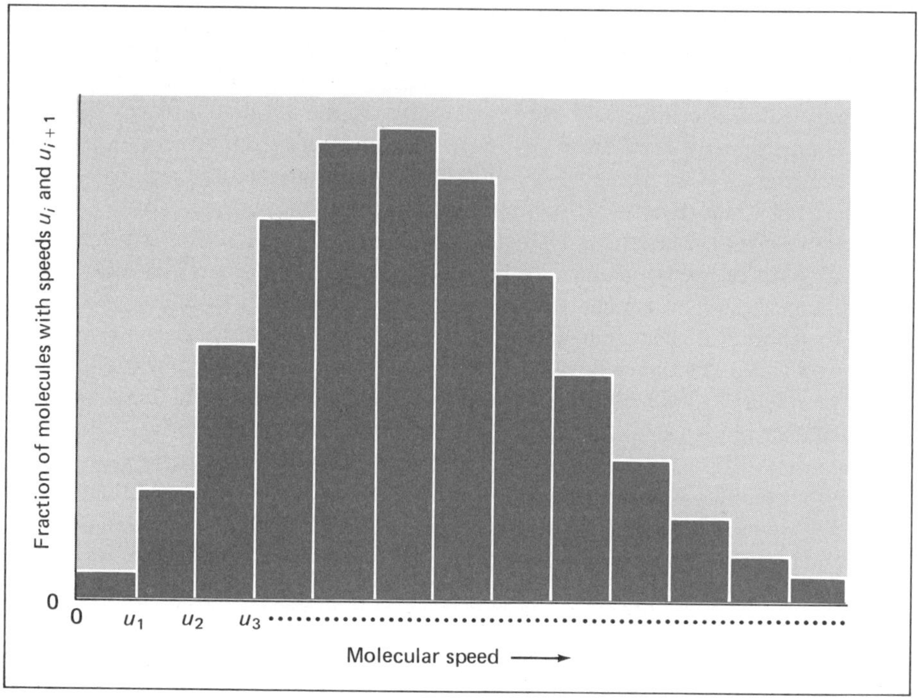

Figure 6.6
The distribution of molecular speeds in a gas spreads out as temperature increases, and the most probable speed shifts to a higher value.

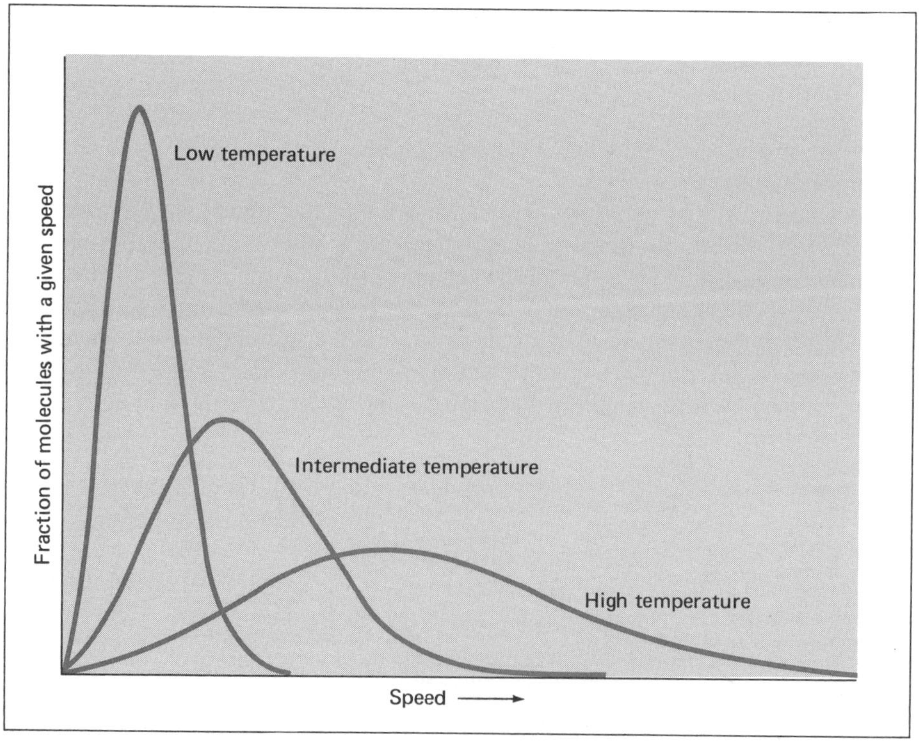

plotted as bar graphs. If the Δu_i increments ($u_2 - u_1$, $u_3 - u_2$, . . .) are very small, the steplike line at the top of the bar graph will approach the line graphs as shown in Figure 6.6. If the Δu_i increments are infinitesimally small, the steplike line will become a smooth curve. Distribution curves at three temperatures are illustrated in Figure 6.6. Note that the most probable velocity (the velocity at which the largest fraction of the molecules will be moving) shifts to the right, and the distribution curve broadens as the temperature is raised.

Root-Mean-Square Velocity

Since the molecules of a gas are not all moving at the same velocity, it becomes useful to define an *average velocity* and an *average translational energy*.

The **average velocity** is obtained by dividing the sum of the magnitudes of the velocities of the individual molecules by the total number of molecules N. Stated mathematically:

$$\bar{u} = \frac{\sum u_i}{N} \qquad [6.2]^{\dagger}$$

A bar over a term, such as $\bar{u}$, denotes the average, or mean, value. The average velocity can be calculated by statistical methods. However, a different average, called the **root-mean-square velocity** and given the symbol u_{rms}, is in general more useful since it is directly related to the average translational kinetic energy (discussed in the next section). It is defined as

$$u_{\text{rms}} = \sqrt{\frac{\sum u_i^2}{N}} \qquad [6.3]$$

It is easy to show from statistical methods that u_{rms} is related to the pressure P and volume V of an ideal gas by the equation

$$PV = \frac{1}{3} n M_w u_{\text{rms}}^2 \qquad [6.4]^{\ddagger}$$

where n is the number of moles of gas of molecular weight M_w.

Equation [6.4] can be combined with the ideal gas equation to derive an equation that enables us to calculate u_{rms} and also $\bar{E}$, the average translational kinetic energy. From the ideal gas equation we write

$$PV = nRT \qquad [5.10]^{\ddagger}$$

Velocity is a vector quantity and hence is positive or negative depending upon the direction of motion. The average velocity, which takes into account the direction of the motion, is zero. The average of the magnitudes of the velocities, often called the *average speed*, is of more interest and is the one we are calculating.

$\bar{u}$ and u_{rms} are not the same, as can be seen from the following calculation. Suppose $u_1 = 1$, $u_2 = 2$, $u_3 = 2$, $u_4 = 3$, and $u_5 = 2$ cm/sec. Then

$$\bar{u} = \frac{(1 + 2 + 2 + 3 + 2)}{5} = 2.00 \text{ cm/sec}$$

$$u_{\text{rms}} = \sqrt{\frac{1^2 + 2^2 + 2^2 + 3^2 + 2^2}{5}} = 2.10 \text{ cm/sec}$$

A third velocity of interest is the most probable velocity α (the velocity at the peak of the distribution curve in Figure 6.6). For a large number of molecules, such as in a gas sample, it can be shown that the velocities are related by

$$u_{\text{rms}} = \sqrt{\frac{3\pi}{8}}\,\bar{u}$$

$$u_{\text{rms}} = \sqrt{\frac{3}{2}}\,\alpha$$

† The Greek letter $\sum$ is used as a symbol to represent the summing operation: $\sum u_i = u_1 + u_2 + u_3 + u_4 + \cdots + u_N$ where $u_1, u_2, u_3, \cdots u_N$ are the velocities of the first, second, third, fourth, ⋯ Nth molecules, respectively In a similar manner

$$\sum u_i^2 = u_1^2 + u_2^2 + \cdots + u_N^2$$
$$\sum E_i = E_1 + E_2 + \cdots + E_N$$

‡ See Appendix J for a derivation of equation [6.4].

‡ See Appendix J for a derivation of equation [5.10] from the kinetic-molecular theory.

Combining equations [5.10] and [6.4] gives

$$nRT = \frac{1}{3} n M_w u_{\text{rms}}^2 \tag{6.5}$$

Solving for u_{rms} gives

$$\boxed{u_{\text{rms}} = \sqrt{\frac{3RT}{M_w}}} \tag{6.6}$$

The Average Translational Energy

The **average translational kinetic energy,** $\bar{E}$, is related to the energies of the individual molecules by the equation

From equation [6.6] one can predict that the average velocity of gas molecules varies inversely with the square root of the molecular weight at a constant temperature. The rate of diffusion of a gas through a hole varies directly with the average velocity, which means that the rate of diffusion of the gas will vary inversely with the square root of the molecular weight. This is a statement of Graham's law of diffusion. (See Problem 6-1 for an application of this law.)

$$\bar{E} = \frac{\sum E_i}{N} \tag{6.7}$$

For any one of the molecules in the gas, the kinetic energy is given by

$$E_i = \frac{1}{2} m u_i^2 \tag{6.8}$$

where again m is the mass of the molecule and u_i is the velocity. Substituting [6.8] into [6.7] gives

$$\bar{E} = \frac{\sum \frac{1}{2} m u_i^2}{N} \tag{6.9}$$

Removing the constant term 1/2 m from the summation gives

$$\bar{E} = \frac{1}{2} m \frac{\sum u_i^2}{N} \tag{6.10}$$

Squaring equation [6.3] and substituting it into equation [6.10] gives equation [6.11], which relates $\bar{E}$ to u_{rms}:

$$\bar{E} = \frac{1}{2} m u_{\text{rms}}^2 \tag{6.11}$$

Equation [6.11] gives the average energy per molecule. The average energy per mole is obtained by multiplying by Avogadro's number N_0 to give

$$\bar{E} \text{ (per mole)} = \frac{1}{2} M_w u_{\text{rms}}^2 \tag{6.12}$$

where $mN_0 = M_w$, the molecular weight. Combining [6.12] with [6.6] gives the equation relating $\bar{E}$ to the temperature T:

$$\bar{E} = \frac{3}{2}RT \quad \text{(per mole)} \qquad [6.13]$$

From equation [6.13] we see that at $T = 0$, $\bar{E} = 0$, and all translational motion ceases at absolute zero temperature.

EXAMPLE 6.2 Calculate the root-mean-square velocity (cm/sec) and the average translational kinetic energy (joules per mole) for O_2 gas at 300° K.

SOLUTION Equations [6.6] and [6.13] apply:

$$u_{rms} = \sqrt{\frac{3RT}{M_w}} = \sqrt{\frac{(3)(8.314 \times 10^7 \text{ ergs/°K-mole})(300.0° \text{ K})}{32.00 \text{ g/mole}}}$$

$$u_{rms} = 4.836 \times 10^4 \text{ cm/sec} \qquad \text{(1080 miles/hr!)}$$

$$\bar{E} = (\tfrac{3}{2})(8.314 \text{ joules/°K-mole})(300° \text{ K})$$

$$= 3741 \text{ joules/mole}$$

Note that R was expressed in erg (centimeter, gram, second) units in calculating u_{rms}, which is necessary if u_{rms} is to be in cm/sec. To calculate $\bar{E}$, the value of R is chosen to give the desired energy units.

6.5 Rotational and Vibrational Energy

Degrees of Freedom

In describing the various forms of energy of a molecule, it is helpful to introduce the concept of the degrees of freedom of a molecule. To do this it is necessary to describe the motion of the molecule in terms of a coordinate system. We begin by analyzing the translational motion of a molecule.

In three dimensional space, it is possible to move "up and down," "back and forth," and "to and fro," that is, along any of the three rectangular coordinate axes. Since motion along any other line in space can be resolved into components along these three axes, we speak of three **degrees of freedom** in translational motion. In general, *the number of coordinates required to specify the position (and hence the motion) of the molecule is the number of degrees of freedom of the molecule.*

Just as translational motion in any direction can be specified in terms of motion along the x, y, and z coordinate axes, the rotational motion of a molecule (regardless of how complex) can be resolved into rotational components around three imaginary axes running through the molecule. Rotation of the nonlinear ethylene molecule around the imaginary axes (dotted lines) of the molecule are shown below. (The axes are at right angles to each other.)

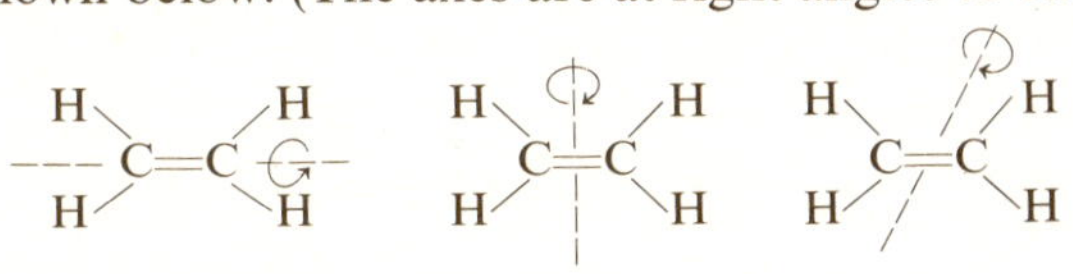

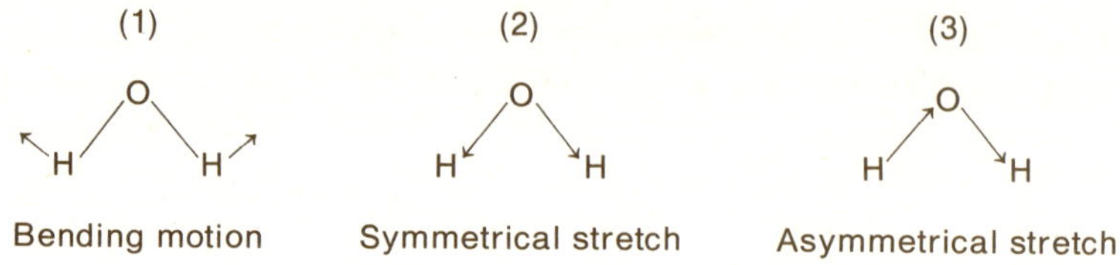

Figure 6.7 *Fundamental vibrational motions in water.* (1) Principally a change in the bond angle is involved. (2) The two bond lengths expand and contract together. (3) One bond expands while the other contracts. All three types of motion are occurring simultaneously, but at different frequencies.

Since three angular coordinates need to be specified to describe the motion, there are three degrees of rotational freedom for ethylene (or any other nonlinear molecule).

For a molecule in which all the atoms are in a straight line (linear molecule) such as H—H, O═C═O, and H—Cl, only two axes of rotation need be taken into account. Using the H_2 molecule as an example, we have

It is not necessary to consider the rotation around the molecular axis:

since this motion does not produce any new configurations; that is, the linear molecule looks just the same at every step in its rotation around the molecular axis. Thus, linear molecules have only two degrees of rotational freedom.

We determine the degrees of freedom for vibrational motion by subtracting those of translation and rotation from the total degrees of freedom of the molecule. To specify completely the position of each atom in a molecule requires a total of three coordinates. A molecule containing η atoms, therefore, requires 3η position coordinates and has 3η degrees of freedom. We have seen above that five coordinates (for linear molecules) or six coordinates (for nonlinear molecules) describe the translation plus rotation. The other $3\eta - 5$ coordinates (for linear molecules) or $3\eta - 6$ coordinates (for nonlinear molecules) describe the vibration in the molecule. These coordinates involve the bond angles and bond lengths in the molecule that change in a vibration. In water, for example, the complex vibrational motion can be resolved into the fundamental vibrations shown in Figure 6.7. Diagram 1 in the figure is principally a change in bond angle, while diagrams (2) and (3) involve changes in bond length.

EXAMPLE 6.3 Describe the rotational and vibrational degrees of freedom for carbon dioxide, a linear molecule.

SOLUTION Carbon dioxide contains three atoms and hence has $3 \times 3 = 9$ total degrees of freedom. Three of these degrees of freedom are translational, and because CO_2 is a linear molecule, two are rotational. The other $9 - 3 - 2 = 4$ degrees of freedom are vibrational. For each vibrational degree of freedom, there is a corresponding vibrational motion.

For CO_2 the motions are as follows:

←O—C—O→

Symmetrical stretch

O→ ←C——O→

Asymmetrical stretch

Energy Content of the Ideal Gas

Earlier we described the ideal gas as one for which there are no attractive or repulsive forces between the molecules and, hence, for which the potential energy is zero. Only translational, rotational, and vibrational energies are present in the ideal gas, and the energy content of the gas depends on the number of different ways the molecules can contain these different types of energy. For example, a monatomic gas, such as He, has no rotation or vibration; thus, translational kinetic energy is the only possibility. The total energy content of such a gas is given by equation [6.13]:

$$\bar{E} = \frac{3}{2} RT \qquad [6.13]$$

A diatomic gas, such as Cl_2 or H_2, has the same three degrees of translational freedom and an amount of translational energy given by equation [6.13]. In addition, however, it has two degrees of rotational freedom and one degree of vibrational freedom. Both rotational and vibrational forms of energy are present, and the total energy content is more than 3/2 RT. A more complex molecule, such as methane, CH_4, has a still larger energy content. Each of the three degrees of translational freedom, the three degrees of rotational freedom, and the nine degrees of vibrational freedom contribute to the total energy content.

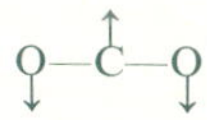

Bending

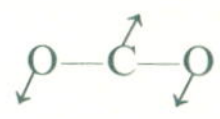

Bending

As in water, the two bond lengths expand and contract together in the symmetrical stretch. In the asymmetric stretch, one bond expands while the other contracts. The two bending motions are perpendicular to each other; that is, one bending motion is "up and down" while the other is "in and out." These two bending vibrations occur at the same frequency; this condition is referred to as *degeneracy*. As with water, all three types of motion (symmetrical stretching, asymmetrical stretching, and bending) occur simultaneously, but at different frequencies. If one could observe a vibrating CO_2 molecule, one would see a complicated pulsating motion. Analysis of this motion would show that it is a combination of the four different fundamental vibrations, each occurring at its own characteristic frequency.

Heat Capacity of the Ideal Gas

When heat is added to a substance, the temperature changes. The **heat capacity** of a substance is defined as the amount of heat required to increase its temperature by one degree. Mathematically, we express this as

$$C = \frac{Q}{\Delta T} \qquad [6.14]$$

where C is the heat capacity and Q is the amount of heat required to raise the temperature by the amount ΔT.

If the heat is added to the ideal monatomic gas in a closed container so that the volume cannot change, we can use the kinetic-molecular theory to calculate C_V, the heat capacity of the gas under this condition of constant volume.

In Chapter 7 we will see that when a substance is heated without allowing its volume to change (constant volume process), the change in the energy of the substance (ΔE) is equal to the quantity of heat (Q) that is added. For a monatomic ideal gas the energy of the gas can be calculated from equation [6.13]. If we let the subscript 1 represent the initial state of a substance (before heating) and subscript 2 represent the final state (after heating), we can make the following argument:

If the gas expands when the heat is absorbed, part of the heat is converted to the work of expansion. If this work does not occur, all heat absorbed ends up as energy content, which we can calculate from the kinetic-molecular theory. In Chapter 7 we will describe more completely the difference between heat capacity at constant volume (C_V) and heat capacity at constant pressure (C_P), where expansion does occur.

$$\Delta E = E_2 - E_1$$

$$E_2 = \frac{3}{2} RT_2$$

$$E_1 = \frac{3}{2} RT_1$$

$$\Delta E = \frac{3}{2} R(T_2 - T_1) = \frac{3}{2} R\Delta T$$

$$Q = \frac{3}{2} R\Delta T$$

Then from equation [6.14], we obtain the result

$$\boxed{C_V = \frac{Q}{\Delta T} = \frac{3}{2} R} \qquad \text{(monatomic, ideal gas, one mole)}$$

Thus we see that for any monatomic ideal gas, C_V equals 3/2 R per mole and

From a rule known as the *principle of equipartition of energy*, one can predict that each translational and rotational degree of freedom in a molecule contributes 1/2 R to the heat capacity while each vibrational degree of freedom contributes R. Thus, He, with three translational degrees of freedom, has $C_V = 3(1/2\ \text{R})$. H_2, with three translational, two rotational, and one vibrational degree of freedom, has $C_V = 3(1/2\ R) + 2(1/2\ R) + R = 7/2\ R$; water vapor, with three translational, three rotational, and three vibrational degrees of freedom, has $C_V = 3(1/2\ R) + 3(1/2\ R) + 3R = 6R$. This rule is an approximation that is obeyed fairly well at high temperatures. At lower temperatures, C_V is less than the value predicted by the rule.

is independent of temperature. For molecular gases with rotational and vibrational energy, C_V is greater than 3/2 R per mole. When heat is absorbed in the latter, it is divided among the different degrees of freedom of the molecule. That is, part of the heat increases the translational motion, part the rotational motion, and part the vibrational motion. The result is that for every quantity of heat absorbed, ΔT is less, and $Q/\Delta T = C_V$ is larger, than for the monatomic gas.

6.6 Intermolecular Attraction

In the ideal gas, no attractive forces exist between the molecules, and the molecules have zero volume. In an actual gas, deviations from ideal gas behavior (especially at high pressure and low temperature) show that attractive forces do exist and can be important.

Liquefaction of gases is possible only because there is **intermolecular attraction.** At high temperatures the molecules have enough kinetic energy

(energy of motion) to overcome the attractive forces and remain gaseous. As the temperature is lowered, however, a point is reached (as was shown in Figure 5.8) at which the kinetic energy is insufficient to overcome the attraction, and liquefaction occurs. Pressure aids the liquefaction process by bringing the molecules closer together so that the intermolecular forces can operate.

Because the ideal gas, by definition, has no attractive forces between molecules, this hypothetical gas could never be liquefied.

To understand the source of these forces of attraction, we will consider three types of interactions: (1) dipole-dipole interactions; (2) van der Waals forces; and (3) hydrogen bonding.

Dipole-Dipole Interactions

One source of attraction between molecules is readily apparent when the molecules are polar, for then there is electrostatic attraction between the positively charged end of one molecule and the negatively charged end of another. This **dipole-dipole interaction** is shown in Figure 6.8. The extent of attraction is a function of the **dipole moment** of the molecules, which is defined by the relationship.

As explained in Chapter 2, the polarity of a diatomic molecule is the result of a difference in electronegativity of the two atoms. The atom with the higher electronegativity draws the shared electrons close to it, making that end of the molecule negative relative to the other end of the molecule.

In Chapter 14 we will return to this subject, and there we will see that in polyatomic molecules the shape of the molecule, as well as the electronegativity of the atoms, is important in determining the extent of polarity of the molecule.

It will be helpful in understanding the attractive forces in gases to know that *symmetrical* molecules, such as CO_2 (linear) and CH_4 (tetrahedral), are nonpolar despite the fact that they are composed of atoms with differing electronegativities. This is because the attractive force of one atom will always be exactly balanced in a symmetrical molecule by the attractive force of another like atom. For example, in carbon dioxide the two oxygen atoms lie on opposite sides of the carbon atom (O=C=O). Since the oxygen atoms attract electrons in exactly opposite directions, no net transfer of charge takes place, and the molecule is nonpolar. Further illustrations of the effect of symmetry on polarity are given in Chapter 14.

$$\mu = \delta l$$

in which μ is the dipole moment, δ is the magnitude of the charge at each end of the dipole, and l is the distance separating the positive and negative charges.

The unit generally used in reporting dipole moments is the **debye,** which is defined as 1.0×10^{-18} esu-cm. The dipole moment obtained when opposite charges, each with the magnitude of an electron (4.8×10^{-10} esu), are separated by one angstrom unit (1×10^{-8} cm) is 4.8×10^{-18} esu-cm. Since these conditions (one angstrom and unit electronic charge) represent the order of magnitude of the values observed in polar bonds, 1×10^{-18} esu-cm (a debye unit) is a convenient way of reporting dipole moments. Table 6.1 lists the dipole moments of a number of commonly encountered molecules of widely varying polarity.

TABLE 6.1

Dipole Moments of a Few Representative Compounds

	DEBYES
HCl	1.03
HBr	0.79
HI	0.38
NO	0.16
CO	0.13
H_2	0.00
H_2O	1.86

Figure 6.8
Polar molecules are attracted to one another by electrostatic forces.

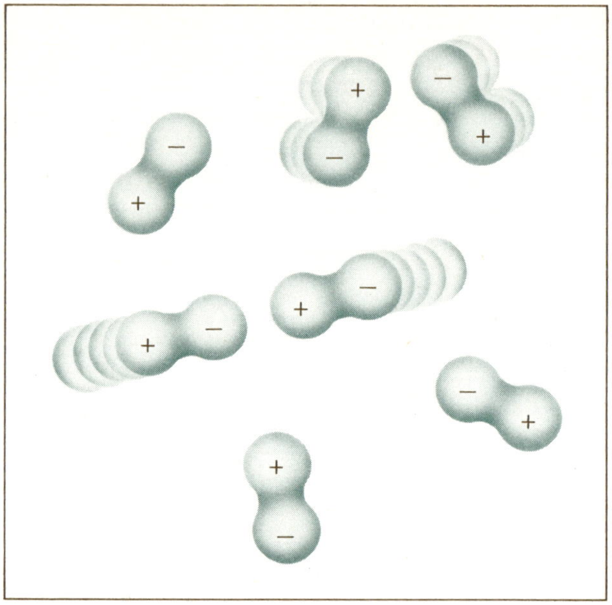

Figure 6.9
An instantaneous, nonsymmetrical distribution of the positive and negative charges in molecule A induces a corresponding effect in molecule B. This instantaneous dipole-induced dipole interaction is the van der Waals force.

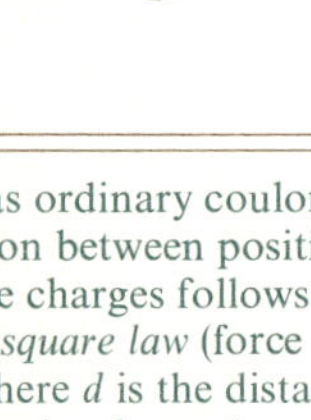

Whereas ordinary coulombic attraction between positive and negative charges follows an *inverse square law* (force $\propto 1/d^2$, where d is the distance between the charges), van der Waals forces follow an *inverse sixth power law* (force $\propto 1/d^6$).

Van der Waals Forces

While dipole-dipole interaction is a significant attractive force between molecules, it obviously is not the only force involved since nonpolar diatomic molecules, such as H_2, N_2, O_2, F_2, He, and Ne, can also be liquefied. This leads us to a second type of attractive force between molecules, the **van der Waals force**. In some respects it is similar to the dipole-dipole effect, but it operates between all molecules regardless of their polarity. Van der Waals forces are best described as momentary, induced dipole-dipole attractions. The key words are *momentary* and *induced*. It is reasonable to expect that the distribution of electrons in a nonpolar molecule is momentarily unsymmetrical; that is, the number of electrons is higher on one side of the molecule than on the other. Thus, for an instant the molecule is a dipole. And during this brief period, the effect of the molecule on a neighboring molecule is to distort its electrical balance (by repelling electrons in the second molecule to its far side). Thus the first molecule *induces* a dipole in the second. This effect is illustrated in Figure 6.9. In the next instant the arrangement of electrons in the first atom will change, and as it does, the effect will be transmitted to the electrons of the second atom. The result of this coupled motion is a fluctuating series of dipole-induced dipole attractions. The larger the number of electrons in a molecule and the greater the mobility of the electrons, the easier it is to distort the electron cloud of that molecule, and the van der Waals forces are more pronounced than in a molecule with few electrons.

Table 6.2 shows the effect that the number of electrons in a molecule has on some properties that one would expect to be related to the degree of their molecular interaction. Since the substances in the table are all nonpolar noble

TABLE 6.2

Correlation Between the Number of Electrons per Molecule in the Noble Gases and Some Physical Properties Related to Intermolecular Attraction

	ELECTRONS PER MOLECULE	VAN DER WAALS a (liter^2 atm/mole^2)	T_b (°K)	T_c (°K)
He	2	0.034	4	5
Ne	10	0.211	27	44
Ar	18	1.35	87	151
Kr	36	2.32	121	209
Xe	54	4.19	164	290

Note: T_b is the normal boiling temperature, and T_c is the critical temperature.

gases, the attractive forces are van der Waals forces only. The *general* trend that is seen in Table 6.2 also exists among other nonpolar molecules. However, one has to use caution. Because of the difference in mobility of electrons in molecules we are not able to make fine distinctions in boiling points, for example, on the basis of the number of electrons alone.

The result is that van der Waals forces are very short-range forces; they are strong while the molecules are close together, but fall off rapidly as the molecules move apart.

Hydrogen Bonding

A third type of intermolecular attraction, **hydrogen bonding,** exists in those substances in which a hydrogen atom is bonded to a very small, very electronegative atom. F, O, and N fit these conditions. Hydrogen compounds of these elements are the principal substances showing appreciable H bonding. In such compounds, a weak bond forms between the electronegative atom of one molecule and the hydrogen bonded to the electronegative atom of another molecule. Water, shown below, is an example.

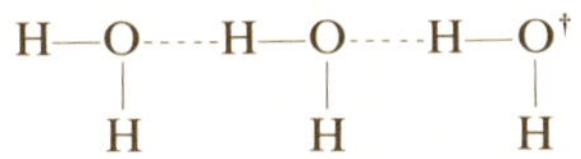

Note that the hydrogen bond in hydrofluoric acid is so stable that when hydrofluoric acid ionizes, the ions formed are not H^+ and F^-, but H^+ and $(FHF)^-$, suggesting the existence of the hydrogen-bonded dimer H_2F_2. When solutions of hydrofluoric acid are half-neutralized with KOH, the salt that forms is KHF_2.

The dotted line, which conveys the implication that it is not a formal chemical bond, represents the hydrogen bond. Compared with the van der Waals force, it is a fairly strong attraction, requiring 20 to 30 kjoule/mole (5 to 7 kcal/mole) of energy to break, but this is still well below the range of ordinary chemical bonding.

The hydrogen bond has a highly significant effect on the properties of the hydrides of the very electronegative elements F, O, and N. Figure 6.10 is a graph of boiling points of the hydrides of Groups IVA, VA, VIA, and VIIA elements. The deviation of NH_3, H_2O, and HF from the behavior of the hydrides of the other members of their groups is attributed to the formation of **intermolecular**

Without hydrogen bonding between water molecules, the properties of water would be quite different. The boiling point, for example, would be very much lower. Thus, extrapolating the data in Figure 6.10 to a molecular weight of 18 would give water a boiling point of −100° C!

† This representation is not meant to portray the geometry, but only the bonding.

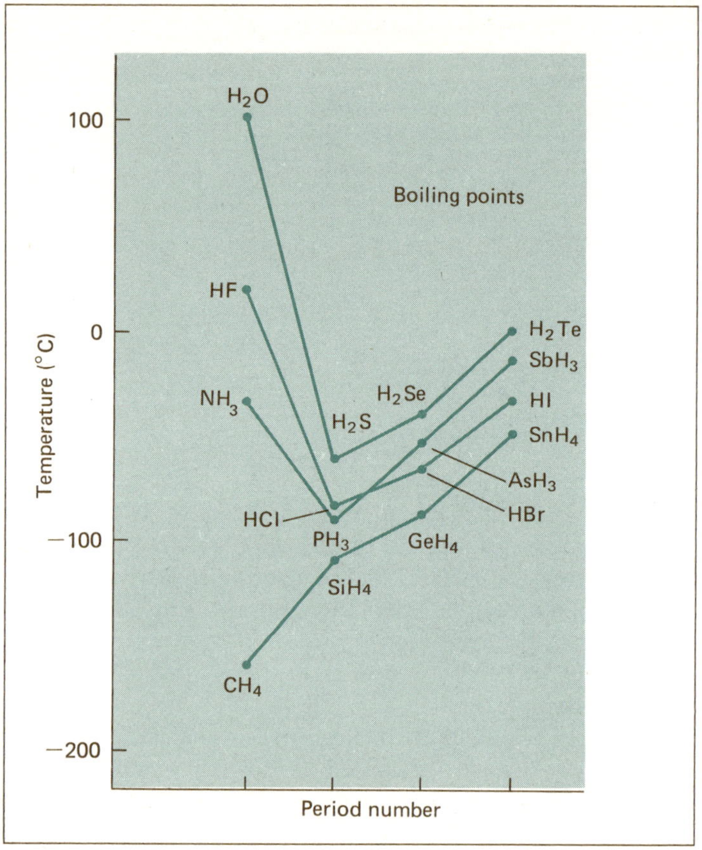

Figure 6.10
Normal boiling points of the Group IVA, VA, VIA, and VIIA hydrides. The abnormally high boiling points of H_2O, HF, and NH_3 are the result of hydrogen bonding.

hydrogen bonds. Note that there is no evidence from Figure 6.10 that hydrogen bonding occurs in CH_4. Carbon is not electronegative enough for its hydrides to form intermolecular hydrogen bonds.

Summary

Kinetic-molecular theory, which is a molecular model of motion, is highly successful in predicting and interpreting macroscopic behavior of matter. The principal postulates are (1) that matter consists of molecules, and (2) that the molecules are in motion. Three types of molecular motion are possible: (1) translational, (2) rotational, and (3) vibrational.

The molecules of a gas have translational motion, are continually colliding with one another and with the walls of the container, and are moving at variable velocities. Several types of "average" velocities can be calculated for ideal gas molecules. The one that especially interests us is the root-mean-square velocity:

$$u_{\text{rms}} = \sqrt{\frac{3RT}{M_w}} \qquad \text{(ideal gas)}$$

The average translational energy of ideal gas molecules is a function solely of the absolute temperature:

$$\bar{E} = \frac{3}{2}RT \qquad \text{(ideal gas, per mole)}$$

If, in addition to being ideal, the gas molecules are monatomic so that translational energy is the only type of energy they possess, we can calculate their heat capacity at constant volume:

$$C_V = \frac{3}{2}R \qquad \text{(per mole, ideal monatomic gas, constant volume)}$$

For polyatomic molecules, additional energy and heat capacity terms have to be considered because of the presence of rotational and vibrational motion.

Attractive forces between molecules contribute potential energy to the total energy content of a system. An ideal gas has zero potential energy; a solid, on the other extreme, has a large negative potential energy. The attractive forces between molecules depend on the nature of the molecule. All molecules attract each other because of van der Waals forces, and for nonpolar molecules these are the only intermolecular forces present. Polar molecules have, in addition, dipole-dipole attractions. A third type of attraction called hydrogen bonding exists between molecules in which a hydrogen atom is bonded to a very electronegative atom (F, O, or N).

Important Terms and Concepts

Kinetic-molecular theory
Kinetic energy
Translational kinetic energy
Rotational kinetic energy
Vibrational kinetic energy
Potential energy
Distribution of translational energy in the ideal gas
Average velocity
Root-mean-square velocity
Average translational kinetic energy
Degrees of freedom
Heat capacity
Intermolecular attraction
Dipole-dipole interactions
Dipole moment
Debye
Van der Waals forces
Hydrogen bonding
Intermolecular hydrogen bonds

Exercises

Assume ideal gas behavior unless instructed otherwise in the exercise.

6-1.
(a) Calculate the root-mean-square velocity of two cars, one traveling at 40 miles an hour and the other at 80 miles an hour.
(b) If the root-mean-square velocity of molecules A, B, and C is 1.00×10^5 cm sec^{-1}, what is the speed of C when the speed of A is 0.40×10^5 cm sec^{-1} and the speed of B is 1.60×10^5 cm sec^{-1}?

6-2.
Calculate the root-mean-square velocity at 25° C of

(a) a free electron ($m = 9.1 \times 10^{-28}$ g).
(b) an H_2 molecule.
(c) a UF_6 molecule.

6-3.
1.00×10^{24} gas particles, each of mass 1.00×10^{-22} g, are contained in a 1.00-liter vessel. The root-mean-square velocity is 1.00×10^5 cm/sec. Calculate

(a) the pressure exerted by the molecules in atmospheres.
(b) the temperature.
(c) the total translational kinetic energy in calories.

6-4.
The following values are given for the speed of sound in air:

T (°C)	SPEED (cm/sec)
20	3.44×10^4
100	3.86×10^4
500	5.53×10^4
1000	7.00×10^4

Compare these values with u_{rms} for N_2 molecules at the above temperatures. Can you suggest a relationship between the velocity of the molecules and the speed of sound in a gas?

6-5.
Calculate the temperature at which u_{rms} for Xe is the same as u_{rms} for He at 300° K.

6-6.
Calculate u_{rms} for N_2, O_2, and H_2 gas at 300° K. The escape velocities from the earth and moon are 1.1×10^4 m/sec and 2.3×10^3 m/sec, respectively. How does this help to explain the lack of gaseous atmosphere on the moon?

6-7.
Consider a system consisting of one mole of helium gas at 1.00 atm pressure and 500° K. Calculate

(a) the *total* kinetic energy in the gas (in joules).
(b) the root-mean-square velocity (cm/sec) of the He atoms.
(c) the density of the gas (in g/liters).

6-8.
Which will have the higher (and how many times higher) root-mean-square velocity, H_2 gas at 27° C or O_2 gas at 927° C?

6-9.
Given one-liter samples of He and O_2 gas at 0° C and 1 atm,

(a) in which gas do the molecules have the larger root-mean-square velocity?
(b) in which gas would the average molecule be exerting the larger force on the walls of the container?
(c) in which gas would the translational energy be the larger?
(d) in which gas would the *total* kinetic energy be the larger?
(e) in which gas would the molecules behave more nearly like an ideal gas?

6-10.
The density of HF at 20° C and 1.00 atm pressure is 0.991 grams liter^{-1}. Compare this density with that calculated from the ideal gas equation, and suggest a reason for the difference.

6-11.
True or False

________ (a) All gases have the same average total kinetic energy when compared at the same temperature. (Ideal gas behavior assumed.)

_______ (b) The molecules of all gases have the same average velocity when compared at the same temperature. (Ideal gas behavior assumed.)

_______ (c) The translational energy of an ideal gas is independent of the molecular weight of the gas.

_______ (d) The root-mean-square velocity of He gas would be 16 times that of a gas whose molecular weight is 196, with both gases at the same temperature and pressure. (Ideal gas behavior assumed.)

_______ (e) The ideal gas equation approximates the behavior of real gases at low temperatures and high pressures.

_______ (f) At the same temperature and pressure, Ne gas would come closer to ideal behavior than Kr gas.

_______ (g) CH_4 gas comes closer to being ideal than does HF gas (same *TP* conditions).

_______ (h) CCl_4 gas is more ideal than CBr_4 (same *T* and *P*). Both molecules are symmetrical tetrahedra.

_______ (i) As temperature is increased, the fraction of molecules in a gas with the "most probable velocity" decreases.

_______ (j) The *b* constant in the van der Waals equation of state has the physical significance of being the potential energy of the gas at 0° K.

_______ (k) The rms velocity of CO_2 at 1 atmosphere and 25° C is the same as the rms velocity of CO_2 at 2 atmospheres and 25° C. (Ideal gas behavior assumed.)

_______ (l) The pressure produced by 10,000 molecules of CH_4 enclosed in a container at constant temperature will be the same as the pressure exerted by 10,000 O_2 molecules in the same volume and at the same temperature. (Assume van der Waal's behavior.)

_______ (m) The dipole moment of NO is greater than that of CO.

_______ (n) The *a* constant in the van der Waals equation is higher for CO gas than for N_2 gas.

_______ (o) The critical temperature of Ar is higher than that of Xe.

_______ (p) Hydrogen bonding is responsible for the difference in the boiling points of CH_4 and CCl_4.

_______ (q) The temperature of the ozone layer is higher than the atmosphere immediately below it.

Problems

6-1.

(a) According to the kinetic-molecular theory, the diffusion of a gas should depend upon the velocity of the molecules. Assuming that the rate of diffusion of a gas is proportional to u_{rms}, show that at a given temperature this rate of diffusion is inversely proportional to the square root of the molecular weight of a gas. Mathematically, we express this relationship as

$$\text{Rate} = \frac{\text{constant}}{\sqrt{M_w}} \quad \text{or} \quad \frac{(\text{Rate})_1}{(\text{Rate})_2} = \sqrt{\frac{M_{w_2}}{M_{w_1}}}$$

where M_{w_1} and M_{w_2} are the molecular weights of two different gases. This relationship is known as Graham's law of diffusion, which was formulated in 1829 by Thomas Graham, a Scottish scientist. The difference in diffusion rate for gases serves as a basis for the separation of gases of different mass. The diffusion of gaseous UF_6 was first used during World War II to separate ^{235}U, the isotope of uranium used in the atomic bomb (and in nuclear reactors), from ^{238}U, the more commonly occurring isotope.

(b) The flow time of a gas through an orifice can be used to estimate its molecular weight. This flow time varies inversely with the rate of diffusion. Hence

$$\frac{t_2}{t_1} = \sqrt{\frac{M_{w_2}}{M_{w_1}}}$$

where t_1 and t_2 are the flow times through an orifice of the same quantity of two gases with molecular weights M_{w_1} and M_{w_2}. At a certain temperature and pressure, it takes 65.3 sec for 500 cm^3 of O_2 gas to diffuse through an orifice. Under the same conditions, it takes 92.3 sec for the same amount of an unknown gas to diffuse through the orifice. Calculate the molecular weight of the unknown gas.

6-2.
In describing the velocity of gas molecules, we usually use the root-mean-square velocity, u_{rms}, defined as

$$u_{rms} = \sqrt{\frac{\sum u_i^2}{N}}$$

It is shown in the text that u_{rms} is calculated from

$$u_{rms} = \sqrt{\frac{3RT}{M_w}}$$

(M_w is the molecular weight of the gas, T is the temperature, and R is the gas constant.) The velocity can also be described in terms of the average velocity $\bar{u}$ defined by

$$\bar{u} = \frac{\sum u_i}{N}$$

It can be shown, again from kinetic theory, that

$$\bar{u} = \sqrt{\frac{8RT}{\pi M_w}}$$

Still another measure of velocity is the most probable velocity α. It is the velocity of the largest fraction of the molecules (velocity at the peak of the distribution curve in Figure 6.6). The most probable velocity can also be expressed in terms of R, T, and M_w. The relation is

$$\alpha = \sqrt{\frac{2RT}{M_w}}$$

(a) Compare u_{rms}, $\bar{u}$, and α for He gas at 1000° K.
(b) To what temperature must He gas be heated so that the most probable velocity is the same as u_{rms} at 1000° K?

6-3.
(a) Acetylene is a linear molecule, H—C≡C—H. Using equipartition principle values for the heat capacity, estimate the amount of energy required to raise the temperature of 5.00 moles of acetylene 200° K. The gas is confined in a constant volume container.
(b) Make use of the equipartition principle to list the following gases in order of increasing heat capacity: O_2, He, CH_4 (nonlinear), CO_2 (linear), BF_3 (nonlinear).

7

Chemical Thermodynamics

In Chapter 1, *chemistry* was defined as the study of matter and the changes that it undergoes. **Thermodynamics** is concerned principally with the energy transformations that accompany those changes. A working knowledge of thermodynamics provides the basis for understanding such important topics as solution behavior, phase equilibrium, chemical reactions, chemical equilibrium, and electrochemistry.

If we define an exact science as one that yields a large body of general information from a very few postulates, then thermodynamics is the preeminent example of an exact science. Thermodynamics is based on two empirical laws (that is, "laws of experience"), which are known as the *first* and *second laws of thermodynamics*. Application of relatively simple mathematics to these two laws results in more than 50 million (a conservative estimate) general mathematical relationships about the behavior of matter and energy. Here then is one of the very powerful tools available to the chemist.

Basically, thermodynamics is concerned with the fundamental quantities of *energy content* (E) and *entropy* (S). These quantities are the subject of the first and second laws, respectively. In this chapter, we will (1) describe the properties of energy content and entropy, (2) state the laws in terms of these quantities, and (3) discuss some of the important deductions from the laws as they apply to chemically reacting systems. We are principally concerned in this course with (1) the amount of heat exchanged in a reaction, and (2) the prediction as to whether any given proposed reaction will take place. In subsequent chapters, we will make extensive application of the equilibrium principles developed from thermodynamics. We will see that it is possible not only to predict *whether* a reaction will take place, but also *to what extent* the reaction will go before equilibrium is reached.

7.1 Energy Content

Energy content is the thermodynamic term used to describe energy that is stored in a substance. In Chapter 6, we found that energy is stored in the form of translational, rotational, and vibrational motions of the atoms and molecules; in the form of potential energy of attraction between molecules; and in the kinetic and potential energy of the electrons.

The word *system* is used to designate the particular portion of the universe that is being subjected to thermodynamic study; for example, 1 mole of N_2 gas, 100 grams of liquid water, or a calorimeter plus its contents. The *surroundings* include all of the rest of the universe.

Two common ways of bringing about a change in the energy content of a system are for the system to exchange heat (Q) with its surroundings and to have work (W) done on (or by) the system. Heat and work may be described as "energy in transit," since neither is stored in the system. Thus, they are the mechanisms for exchanging energy content between *system* and *surroundings*. Once the exchange is made, the energy resides in the motions (kinetic energy) and attractions (potential energy) of the atoms, molecules, and electrons, as previously discussed.

Work

Work can be expressed as the product of an *intensity factor* and a *capacity factor*. The intensity factor in work is *force*, and the capacity factor is *displacement*. Work is thus defined as the product of force times displacement. Some forms of mechanical work are easily visualized. For example, if you stretch a spring, you do work by exerting a force in opposition to the restraining force of the spring. The product of the force exerted and the displacement of the spring is work. Work has the dimensions of energy, and in most applications in chemistry, the unit in which it is recorded is either the calorie or the joule.

Originally, the calorie was defined as the quantity of heat required to raise the temperature of water by one degree Celsius. This definition, in terms of the heat capacity of water, had the built-in defect that the value of the calorie changed each time a more exact measurement of the heat capacity of water became available. To divorce the value of the calorie from an experimental heat capacity measurement, the absolute calorie was defined in 1948 as 4.1840 absolute joules. This was the accepted "best value" of the heat capacity of water at that time, but that is the only connection that now exists between the definition of the calorie and the heat capacity of water.

The joule is the m-k-s (meter-kilogram-second) or SI unit of energy; it is one kg-m^2/sec^2. In terms of electrical units it is one volt-coulomb or one volt-ampere-second.

For chemical systems, more common than stretching a spring is changing the volume of a fluid, such as by compressing or expanding a gas. The intensity factor for this type of work, usually referred to as *PV* work, can be thought of as the external or applied pressure, and the capacity factor, as the change in volume. For a system in which the applied pressure is held constant, work can be calculated from the relationship

$$\boxed{W = -P\Delta V} \quad \text{(constant applied pressure)} \qquad [7.1]$$

The Greek letter Δ (delta) is used to indicate a finite change. Thus, $\Delta V = (V_2 - V_1)$ in a given process is the difference in the volume of the system in the final state and the volume in the initial state.

The negative sign is added because we wish to follow the convention that work done *by* a system has a negative sign, and work done *on* a system has a positive sign. The use of a negative sign in equation [7.1] incorporates this information into the definition of *PV* work. When a system is compressed, work is done on the system, and according to the convention we have adopted, the sign of W should be positive. Since ΔV will be negative for a compression, and the two negative values multiplied together give a positive value, we see that equation [7.1], as written, does give the correct sign as well as the magnitude of the work performed.

EXAMPLE 7.1 Calculate the work done on a sample of gas if it is compressed at a constant pressure of 10.0 atm from a volume of 100.0 liters to 75.0 liters. The change in volume is made possible by cooling the gas.

SOLUTION

$$W = -P\Delta V = -P(V_2 - V_1)$$
$$= -10.0 \text{ atm} (75.0 - 100.0) \text{ liters}$$
$$= 250 \text{ liter-atm}$$

The gas constant R, expressed in both liter-atm/°K-mole and joules/°K-mole, can be used to obtain the conversion factor needed to report the answer in joules:

$$250 \text{ liter-atm} \left(\frac{8.31 \text{ joules/°K-mole}}{0.0821 \text{ liter-atm/°K-mole}}\right) = 2.53 \times 10^4 \text{ joules}$$

To obtain the area graphically, one could "count squares," for example. An interesting graphical method—one that has special appeal to chemists—is to cut out the area from the graph and weigh it on an analytical balance. A comparison of its weight with the weight of a known area (such as a square area cut from the center of the graph) gives the desired area. If the plots are made large, and good quality graph paper is used, fairly satisfactory results can be obtained.

The magnitude of PV work, in general, is the area under a PV graph of the process. If, as in the example shown in Figure 7.1, the pressure remains constant, the area under the PV line will be a rectangle, and W will be simply $-P\Delta V$. If P is not constant, as in the example shown in Figure 7.2, the area under the PV line will not be rectangular and equation [7.1] will not apply. In such cases, the area may be obtained graphically or by the use of an equation of state and calculus.

In the special case of a constant-volume process, there is no area under the PV curve (Figure 7.3) and $W = 0$ (where PV work is the only type of work possible).

One further property of work needs to be described. Since work is related to the area under a PV line, the amount of work involved in a process will depend upon the *path* of the process. Figure 7.4 shows three of the innumerable paths that a system might conceivably follow in going from an initial state a to

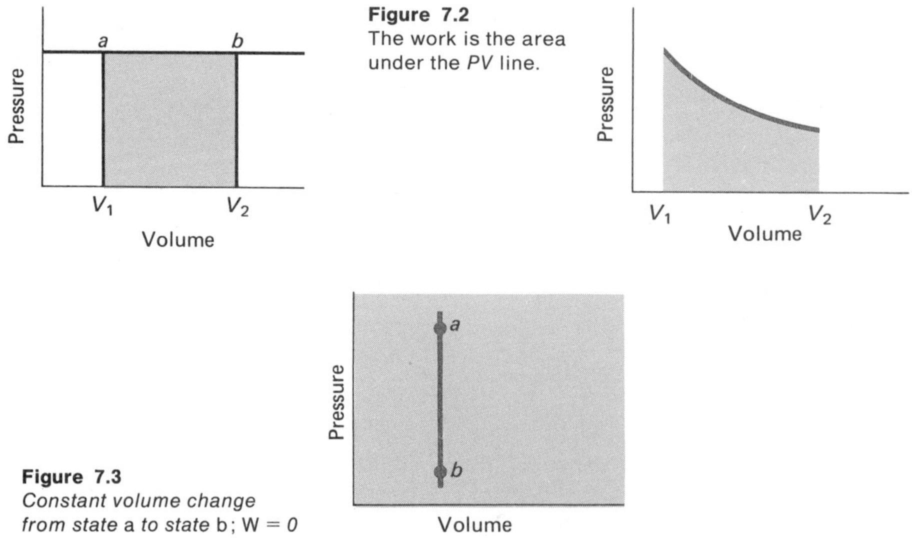

Figure 7.1
Work of expansion for a constant-pressure process.

Figure 7.2
The work is the area under the *PV* line.

Figure 7.3
Constant volume change from state a to state b; W = 0

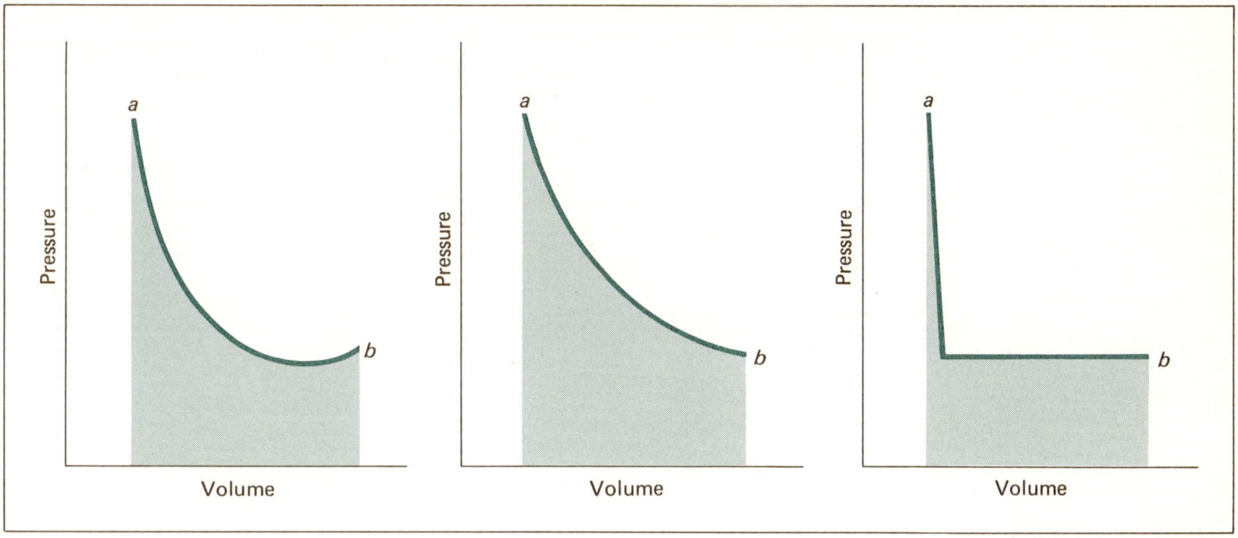

Figure 7.4
Pressure-volume work is equivalent to the area under the *PV* curve. The amount of work done depends upon the path a process follows.

a final state *b*. The shaded area under each curve represents the work done in each of the three expansion processes. Even though the same initial and final states apply in the three examples in Figure 7.4, the amounts of work performed are seen to be very different. We conclude that work is a path-dependent quantity, a fact that can be proved by using more rigorous mathematics. In order to make a calculation for W, therefore, it is not sufficient to know only the initial and final states; one must also have the entire path carefully specified.

Heat and Heat Capacity

Whereas *PV* work is transfer of energy because of a difference in pressure, **heat** is energy that is exchanged between objects because of a difference in temperature. In constant-volume processes, where no work can be done, exchange of heat is the common process by which energy is transferred between a system of constant composition and its surroundings. In an **adiabatic system** (where there is no heat transfer, such as a system enclosed by a perfect heat insulator), work is the mechanism for transfer of energy. In more complex systems, both heat and work come into play.

Heat is not a property of the state of a system but, like work, depends on the path taken by a process. Conventions concerning the sign of Q are the same as those for work: *Heat added to a system has a positive sign; heat given off by a system has a negative sign.*

Heat and work are interconvertible. A steam engine, or even an automobile engine, is an example of a machine designed to convert heat into work. The turning of a paddle wheel in a tank of water to produce heat from friction represents the reverse process, the conversion of work into heat.

In Chapter 6 we defined heat capacity as the quantity of heat required to change the temperature of a system by one degree. Stated mathematically it is

It is necessary that ΔT be kept small, since heat capacities do change with temperature. For example, the heat capacity of liquid water at 25° C is 0.9983 cal/g-°K or 4.178 joules/g-°K, and at 65° C it is 1.0000 cal/g-°K or 4.184 joules/g-°K. The calculations that we consider in this course will generally be made on the assumption that the heat capacity is constant over the temperature range being considered.

$$C = \frac{Q}{\Delta T} \qquad [7.2]$$

Since the amount of heat exchanged depends on the path of the process, heat capacities are measured under specified paths. We will be interested in molar **heat capacities** (heat capacity per mole) **at constant volume** (C_V) and **constant pressure** (C_P). From equation [7.2], we see that the quantities of heat exchanged in (a) a constant-pressure process and (b) a constant-volume process are, respectively,

$$\boxed{Q = nC_P\Delta T} \quad \text{(constant pressure, constant heat capacity)} \qquad [7.3]$$

$$\boxed{Q = nC_V\Delta T} \quad \text{(constant volume, constant heat capacity)} \qquad [7.4]$$

where n is the number of moles.

EXAMPLE 7.2 Calculate the heat required to raise the temperature of 28.0 g of He gas from 300° K to 500° K at constant pressure. The heat capacity of He gas may be considered constant over this temperature range at $C_P = 5/2\ R$ cal/°K-mole.

SOLUTION

$$Q = nC_P\Delta T \qquad \text{(constant pressure)}$$

$$n = (28.0 \text{ g He})\left(\frac{1 \text{ mole He}}{4.00 \text{ g He}}\right) = 7.00 \text{ moles He}$$

$$Q = (7.00 \text{ moles})\left(\frac{5}{2}\right)(8.31 \text{ joules/°K-mole})(500 - 300)° \text{ K}$$

$$= 2.91 \times 10^4 \text{ joules}$$

In general, Q is the area under a heat capacity-temperature curve. If heat capacities are not constant, the areas can be evaluated graphically (refer to Figure 7.5).

Figure 7.5
Heat exchanged at constant pressure is the area under the C_PT curve.

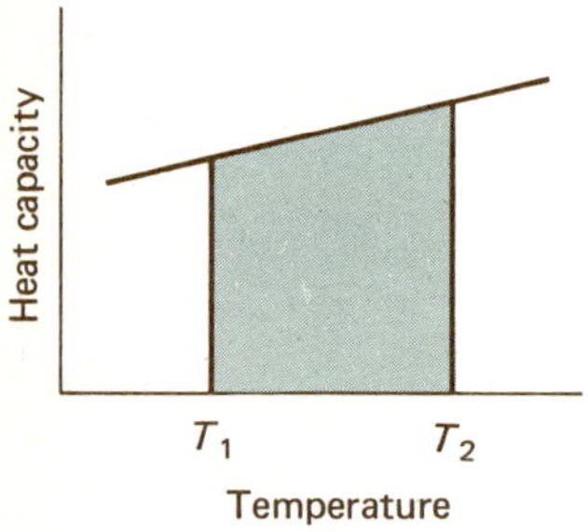

Heat Plus Work: State Functions

The important equation that relates the change in energy content (ΔE) to heat and work is

$$\boxed{\Delta E = Q + W} \qquad [7.5]$$

In essence, equation [7.5] says that there are two ways a system of fixed composition can change its energy content: (1) exchange of heat with the surroundings, and (2) performance of work.

We saw earlier that work and heat taken separately depend on the path of the process. However, it is a matter of experience gained through several hundred years of observations that the sum $(Q + W)$ does not depend on the path, but only on the initial and final states of the system. ΔE, the sum of Q and W, is therefore independent of path. We speak of thermodynamic quantities, such as E, as being *state functions*. We will have more to say on this subject immediately following Example 7.3.

The rationale of the convention that we adopted for the signs of Q and W should now be apparent. Heat and work are given plus signs when they add to the energy content (work done on the system, and heat absorbed by the system) and negative signs when they decrease the energy content of the system.

EXAMPLE 7.3 Calculate the change in energy content when 2.00 moles of He gas is expanded from 10.0 liters to 20.0 liters by raising the temperature of the gas. The pressure is kept constant at 5.00 atmospheres. The molar heat capacity of the He(g) at constant pressure is 5/2 R.

SOLUTION To calculate ΔE, we must first find W and Q.

$$\begin{aligned} W &= -P\Delta V \\ &= (-5.00 \text{ atm})(20.0 - 10.0) \text{ liters} \\ &= (-50.0 \text{ liter-atm})\left(\frac{8.31 \text{ joules/°K-mole}}{0.0821 \text{ liter-atm/°K-mole}}\right) \\ &= -5.06 \times 10^3 \text{ joules} \end{aligned}$$

In order to solve the equation

$$Q = nC_P\Delta T$$

begin by finding the temperature change:

$$T = \frac{PV}{nR}$$

$$T_1 = \frac{(5.00 \text{ atm})(10.0 \text{ liters})}{(2.00 \text{ moles})\left(\frac{0.08206 \text{ liter-atm}}{\text{°K-mole}}\right)} = 305\text{° K}$$

$$T_2 = \frac{(5.00 \text{ atm})(20.0 \text{ liters})}{(2.00 \text{ moles})\left(\frac{0.08206 \text{ liter-atm}}{\text{°K-mole}}\right)} = 609\text{° K}$$

$$\begin{aligned} Q &= (2.00 \text{ moles})\left(\frac{5}{2}\right)(8.31 \text{ joules/°K-mole})(609 - 305)\text{° K} \\ &= 1.26 \times 10^4 \text{ joules} \end{aligned}$$

Now the change in energy content can be determined:

$$\begin{aligned}\Delta E &= Q + W \\ &= 1.26 \times 10^4 \text{ joules} - 5.06 \times 10^3 \text{ joules} \\ &= 7.54 \times 10^3 \text{ joules}\end{aligned}$$

The thermodynamic state of a system is the complete description of its macroscopic properties, such as temperature, pressure, volume, mass, density, composition, and all of the many other observable bulk properties of the system. Fortunately, not all of these properties vary independently, and only a few of the total need to be specified to fix the value of all others. For most systems, the state is adequately and conveniently described by specifying for each phase (a) the composition expressed in terms of the number of moles of each substance and (b) any two of the three properties T, P, and V.

A property is a *state function* if its value depends only on the state the system is in. For example, consider the following process:

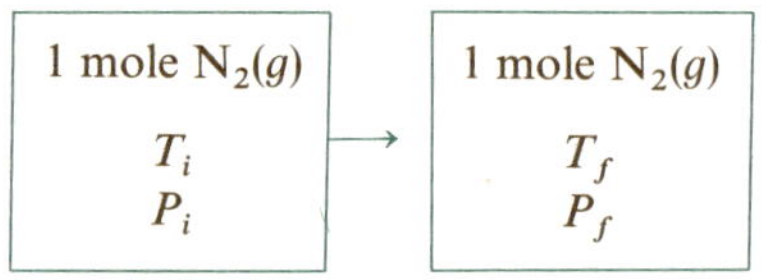

Initial state (i) Final state (f)

The initial and final states are specified by stating the composition, temperature, and pressure of each. Since volume is also a *state function*, there is a single fixed value for the volume of each state. It follows, therefore, that the difference in the volumes of the two states, $\Delta V = V_f - V_i$, is independent of the path by which the system was transferred from state i to state f. In a like manner, E is a *state function*; ΔE (unlike Q and W) depends only on the initial and final states and is not influenced by anything that happens to the system in its trip from the starting point to the final state.

If a system should—by a series of steps—be returned to its initial state as is illustrated below, no net change in any state function is possible.

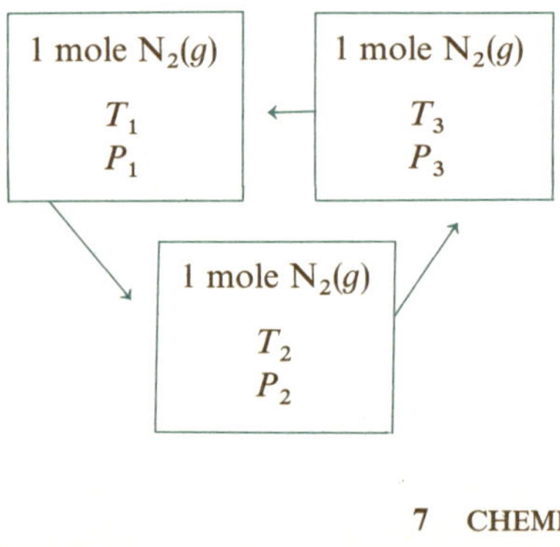

The energy content, for example, would undergo changes from E_1 to E_2 to E_3, and back again to E_1, with the overall change being zero. Thus, the energy content of a system makes no net change in any process that returns the system to its original state. Stated another way, E depends only on the state of a system. Unlike Q and W, ΔE is independent of the path a process follows.

7.2 Entropy

The existence of the highly useful and interesting property of matter called **entropy** (S) was first deduced from a study of a hypothetical heat engine, which had been designed in 1824 by the French engineer Sadi Carnot. Carnot was interested in efficiencies of engines to convert heat into work. His imaginative work, published even before the first law of thermodynamics had been firmly established, proved to be significantly more important than he probably had envisioned. In the years that followed, Kelvin, Planck, and Clausius made use of Carnot's hypothetical engine to (a) define a fundamental temperature scale, (b) formulate the concept of entropy, and (c) arrive at a statement of the second law of thermodynamics. In this brief treatment of thermodynamics, we will not describe the events and arguments that led to the discovery of the quantity called entropy, but will concentrate instead on its physical significance and its applications in chemical processes.

Entropy and Disorder

Once the existence of the property called entropy was established, and, like energy content, it was determined to be a state function, the next step was to discover its physical meaning. It was Ludwig Boltzmann who provided us with a physical picture of entropy by showing it to be a measure of the *disorder* in a system. Boltzmann's contribution was to link entropy, S, to the thermodynamic probability, W, through the relationship

$$S = \text{k} \ln W$$

The definition of the natural logarithm (ln), and a table of four-place common logarithms for solving some later problems, appear in Appendix K.

where k is the Boltzmann constant. Since highly ordered systems have a low probability of existence, they have low entropy. On the other hand, high entropy is associated with high probability, which is the disordered state of random distribution. A coin-tossing example will illustrate these relations among entropy, probability, and disorder.

EXAMPLE 7.4 Calculate the number of ways in which four coins tossed into the air can land in order to obtain (a) the highly ordered state of four heads; (b) the random distribution of two heads and two tails; and (c) an intermediate state of three heads and one tail. Which distribution (a, b, or c) produces the state of highest entropy?

SOLUTION The possible states that can result (using h for heads and t for tails) are as follows:

	COIN 1	COIN 2	COIN 3	COIN 4
State I	h	h	h	h
State II	h	h	h	t
	h	h	t	h
	h	t	h	h
	t	h	h	h
State III	h	h	t	t
	h	t	h	t
	t	h	h	t
	h	t	t	h
	t	h	t	h
	t	t	h	h
State IV	h	t	t	t
	t	h	t	t
	t	t	h	t
	t	t	t	h
State V	t	t	t	t

Since there are six ways to obtain the random distribution of equal number of heads and tails (state III), four ways to obtain three heads and one tail (state II), and only one way to obtain four heads (state I), the random distribution (state III) is the most probable—six times more probable than the highly ordered arrangement of four heads (state I). The entropy of the random distribution (two heads, two tails) is accordingly the largest of any of the possible states.

Another representation of the relationship is shown in Figure 7.6. The rod-shaped objects can be thought of as molecules, match sticks, or any other unit of which the system is composed.

One further example is to consider a deck of cards. An ordered deck (ace, 2, 3, . . . , J, Q, K of hearts; ace, 2, 3, . . . , J, Q, K of diamonds; . . .) is a state of order, low probability, and low entropy. A shuffled deck is a state of disorder, high probability, and high entropy.

Barring sleight-of-hand artistry, shuffling a stacked deck of cards increases the entropy of the deck.

With only this much information, one can make qualitative predictions about the change of entropy in a number of processes. Three examples in which it is very easy to predict the sign of ΔS are (a) phase changes, (b) mixing processes, and (c) temperature changes.

Phase changes. Since a solid represents greater order than a liquid, and a liquid greater order than a gas, phase changes from solid to liquid, solid to gas, and liquid to gas are all accompanied by increases in the entropy of the system.

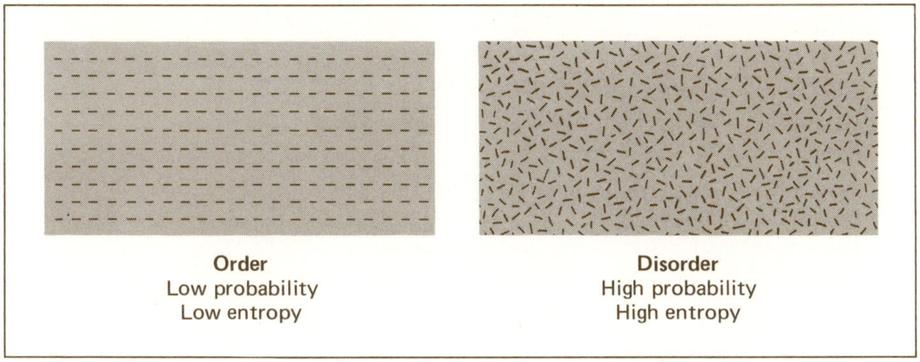

Figure 7.6
Entropy and disorder.

Mixing processes. Other factors remaining constant, any mixing process that results in a random mixing of molecules represents an increase in entropy. The diffusion of gases is an example (see Figure 7.7).

Temperature changes. The effect of temperature on entropy can be understood by examining the extent of order in the distribution of energies of the molecules in a substance. Consider as an example the translational energy of a gas. We saw in Chapter 6, in our discussion of kinetic-molecular theory, that a collection of molecules moves with a wide range of velocities, from very slow to very fast. Figure 7.8 shows this distribution of speeds among the molecules at three different temperatures. At low temperatures, most of the molecules have speeds close to the most probable value (the maximum in the curve). This condition represents a certain degree of order. As the temperature increases, one can see from the graph that the distribution spreads out. This "spreading out" of the energy gives a more disordered (random) distribution of the energy.

Qualitatively, the above discussion also applies to solids and liquids, and to other types of motion (rotation and vibration). In each example, the spreading out of the energy gives a more disordered system. Increasing the temperature then increases the entropy of a substance.

In addition, it is sometimes possible to predict the entropy changes in a chemical reaction, since reactions often change the amount of disorder in a system. In polymerization reactions, many small molecules (monomers) are

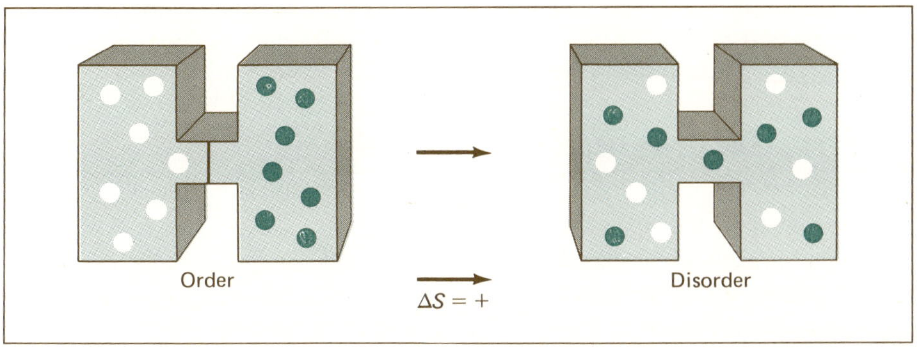

Figure 7.7
Entropy change in a diffusion or mixing process.

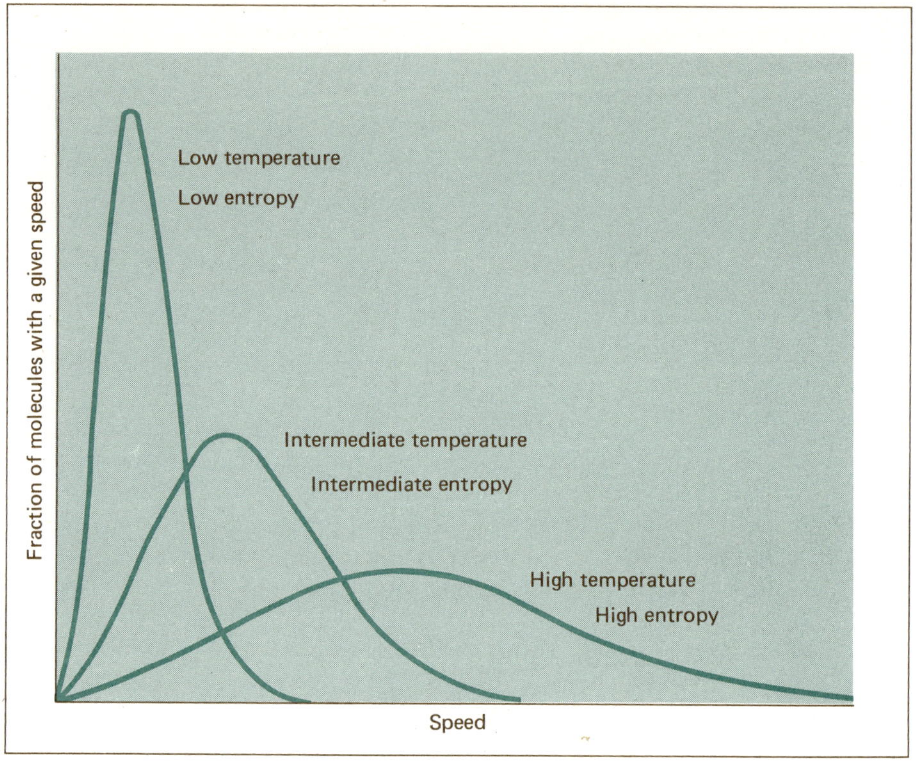

Figure 7.8
The distribution of molecular speeds in a gas spreads out as temperature increases and the gas becomes more disordered.

condensed into one large, well-ordered structure. Such reactions result in a decrease in entropy. Conversion of ethylene into polyethylene is an example. Reactions that change the shape of the molecule also change the order. For example, ΔS is positive for the reaction

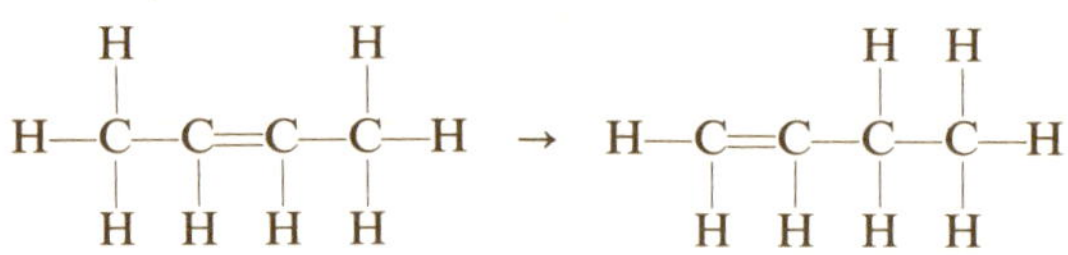

since the symmetry of the structure on the left gives it greater order than the structure on the right.

7.3 Laws of Thermodynamics

Statements of the First and Second Laws

The laws of thermodynamics are brief statements that generalize certain observations about energy and entropy. The laws are not deduced from theoretical principles but are *laws of experience*, the experience going back through all recorded history during which no exceptions have been observed. The laws are based on macroscopic observations that are independent of any theory concerning the microscopic nature of matter and energy. As such, they are entirely general. The laws of thermodynamics have been stated in many different ways, but for sheer elegance it would be difficult to improve on the 1865 version of Clausius:

First law of thermodynamics	Die Energie der Welt ist konstant. *The energy of the universe is constant.*
Second law of thermodynamics	Die Entropie der Welt strebt einem Maximum zu. *The entropy of the universe is increasing.*

The corresponding mathematical statements are:

$$\textbf{First law:} \quad \boxed{\Delta E = 0} \quad \text{(universe)} \qquad [7.6]$$

$$\textbf{Second law:} \quad \boxed{\Delta S > 0} \quad \text{(universe)} \qquad [7.7]$$

Equation [7.6] is seen to be the **law of conservation of energy:** *the energy content of the universe is constant.* Transfer of energy within the universe is, of course, not ruled out. Equation [7.6] requires only that the algebraic sum of all such changes in energy content be zero; that is,

$$\Delta E = \Delta E_1 + \Delta E_2 + \cdots + \Delta E_n = 0$$

Equation [7.7] states that any process that takes place causes an increase in the entropy of the universe. A given system (not isolated) may either increase or decrease in entropy, but the *net* change for the universe (*system plus surroundings*) must, according to equation [7.7], be an increase. The second law thus provides us with the criterion by which we can judge whether a proposed process is thermodynamically possible. If the process would result in an increase in entropy for the universe, it can occur. If the proposed process would cause a net decrease in the entropy of the universe, that process is unnatural and will not occur; in fact, the reverse of the proposed process will take place. Thus, the second law tells us the *direction* of change of a process. A positive value for ΔS (universe) means that a reaction will take place as written, and a negative value means that the reaction will go in the opposite direction.

Thermodynamics, however, tells us nothing about how fast a process will occur. Thermodynamically possible processes can occur at immeasurably slow rates. For example, at room temperature diamond is thermodynamically unstable with reference to graphite. However, the rate of this conversion is slow enough that few would worry that the crown jewels will change to pencil lead.

Intermediate to the natural process and its reverse reaction is the state of equilibrium, in which

$$\Delta S = 0 \quad \text{(universe, equilibrium)} \qquad [7.8]$$

Implications of the Laws

The first law of thermodynamics implies the existence of a finite quantity of energy in the universe. The energy can be changed from one form into another and transferred from one part of the universe to another, but it can be neither created nor destroyed. The table below summarizes the types of energy changes that are possible.

	ΔE				
	System	+	Surroundings	=	Universe
(a)	+		−		0
(b)	−		+		0
(c)	0		0		0

The vaporization of a liquid is an example of (a). In this process, heat flows from the surroundings into the liquid to give the molecules the increased energy content they need to exist as a gas. As the liquid vaporizes, energy is added back into the surroundings through the performance of pressure-volume work on the surroundings. The net quantity of energy removed from the surroundings must equal the increase in energy content of the system.

The adiabatic expansion of a gas is an example of (b). (In an adiabatic process, no heat enters or leaves the system.) In the expansion of gas, work is done on the surroundings, increasing its energy content. Since no heat can be exchanged, the energy content of the gas must decrease by the amount that the energy content of the surroundings has increased. Accordingly, the temperature of the system falls.

Finally, (c) is represented by a cyclic process occurring within an isolated system. With the system returning to its initial state, and with no opportunity for transfer of energy between system and surroundings, all energy changes in the process must add up to zero.

The second law says that when all the entropy changes in the universe that result from a natural process are added together, the sum is greater than zero. In other words, a net amount of entropy is produced in the universe in any naturally occurring process. The four possible ways in which this can be accomplished are listed below:

	ΔS				
	System	+	Surroundings	=	Universe
(a)	+		+		+
(b)	−		+		+
(c)	+		−		+
(d)	+		0		+

Processes represented by (a) are large entropy producers; burning and decay reactions are examples. During such processes the *system* not only becomes more disorganized (an increase in entropy), but the heat produced enters the *surroundings* to add still more entropy to the *universe*.

The freezing of supercooled water is an example of (b). This spontaneous ordering of the water molecules may at first seem to be a violation of the second law. However, when one takes into account the accompanying entropy change in the surroundings, the second law is still found to be valid. While the change of liquid water to ice decreases the entropy of the *system*, the heat of crystallization released in the process causes an increase in the entropy of the *surroundings* that is larger than the decrease in the entropy of the system. The net effect is, as always in a natural process, an increase in the entropy of the *universe*.

Vaporization of a liquid is an example of (c). The *system* becomes more disordered (ΔS is positive), but the *surroundings* decrease in entropy as a result of supplying the heat of vaporization. Once again, the increase in entropy exceeds the decrease, and there is a net gain in the entropy of the *universe*.

Any change occurring in an isolated system is an example of (d). Under this condition, no entropy change can take place in the surroundings. Only changes that will cause an increase in the entropy of the system occur, for it is the experience of all of recorded time that the entropy of the universe increases whenever a process occurs.

To summarize, in all kinds of naturally occurring processes for which the total energy and entropy changes in the universe can be calculated, energy is conserved and entropy is produced. The laws were formulated on the basis of experience, and historically they were proposed at about the same time. The first law gained almost immediate acceptance, probably because it appeared reasonable and in agreement with the comfortable concept that nature is permanent. Not so with the second law, however. Because of its far-reaching implication—that every natural process permanently alters the universe in such a way that no amount of effort can return it to its original state—the second law was slower in receiving general acceptance.

The production of entropy as predicted by the second law might be looked upon as an effect working against the best interests of life. We have seen previously that entropy is related directly to disorder in the universe. Thus, the second law predicts an inevitable state of complete disorder in the future. It is an interesting paradox that as we struggle to improve our surroundings by increasing the complexity of our environment, we speed up the decay of the universe.

A popular student version of the second law: "If you think things are mixed up now, just wait awhile."

Clausius envisioned the "death" of the universe as the state when the universe would reach maximum entropy. In this state all systems would be in equilibrium, and no exchange of energy would be possible. This is referred to as the "heat death" (*Wärme Tod*). In this state the same quantity of energy as ever would be present in the universe, but none of it would be available for work. In other words, the universe simply "runs down" and eventually everything, animate and inanimate alike, comes to a stop.

This is the way the world ends
This is the way the world ends
This is the way the world ends
Not with a bang, but a whimper.†

† From "The Hollow Men" in COLLECTED POEMS 1909–1962 by T.S. Eliot, copyright, 1936, by Harcourt Brace Jovanovich, Inc.; copyright © 1963, 1964 by T.S. Eliot. Reprinted by permission of the publishers, Harcourt Brace Jovanovich, Inc., and Faber and Faber Ltd.

The earliest statements of the laws of thermodynamics were not in terms of E and S. The formulation of the laws was to some extent a by-product of scientists' early interest in perpetual motion. These statements generally described the impossibility of accomplishing certain cyclic processes. The first law denied the possibility of creating energy in a cyclic process that returned both system and surroundings to their initial state (*perpetual motion of the first kind*). The second law denied the possibility of converting heat completely into work in a cyclic process (*perpetual motion of the second kind*). Perpetual-motion machines that would allow a ship to extract heat from the ocean, convert the heat into the work of running a propeller, and finally return the energy to the ocean via heat from the friction of the propeller (none of which violates the first law) have long intrigued some people (those who are unaware of the second law). All schemes to produce perpetual motion of either variety—(a) cyclic processes that create their own energy or (b) cyclic processes that attempt to exchange heat with only one reservoir—have failed without exception.

Popular versions of these statements are "You can't get something for nothing" (First law), and "You not only can't get something for nothing, you can't even break even" (Second Law).

The Third Law, Absolute Entropy, and Applications to Chemical Reactions

We have seen that entropy is the subject of the second law. Another important generalization concerning entropy is known as the **third law of thermodynamics** which states that *the entropy of any element or compound present in the form of a perfectly ordered crystal is zero at the absolute zero of temperature.*

The phrase "perfectly ordered" has to be included because some compounds are capable of crystallizing with the molecules in more than one geometrical orientation; that is, there are different orientations with nearly the same energy that are available to the molecule. The result is some disorder, which, as we have already seen, results in positive entropy. Examples of compounds exhibiting disorder at 0° K are CO, N_2O, and H_2O. Before this disorder was fully understood, discrepancies between entropy changes determined on the assumption that $S^0 = 0$ and non-third-law methods led some to question the validity of the third law. All such apparent discrepancies have now been resolved.

As with the first and second laws, the third law is based on experimental measurements, not deduction. It is easy, however, to rationalize such a law. In a perfectly ordered crystal, every atom is in its proper place in the crystal lattice. At 0° K, no disorder due to molecular motion exists in the crystal. Therefore, such a state would have perfect order, and since entropy is a measure of the disorder in a system, perfect order would mean zero entropy. The third law gives us an absolute reference point from which we can make determinations of the absolute values of S.

The numerical value of the entropy of a system depends on the pressure as well as on the temperature of the system. In order to have a common state in which to report thermodynamic data, scientists set up a *standard state*, defined as the pure element or compound at one atmosphere pressure. A superscript zero is used to designate the standard state, for example, $S^0_{298.15}$. Note that the temperature is not a part of the definition of the standard state; it must be specified separately. Some examples of substances in their standard states are C(graphite, 1 atm), Hg(liquid, 1 atm), CH_4(gas, 1 atm). $S^0_{298.15}$ values have been measured for many substances and collected in tables. Once a set of S^0 data is available, it may be used to calculate ΔS^0 for a chemical change by subtracting the sum of the entropies of the reactants from the sum of the entropies of the products. An example will help clarify this procedure.

Since solids can sometimes exist in more than one crystalline form (such as graphite and diamond [see

Chapter 16]), it is necessary to specify the crystalline form of the solid. To be completely rigorous, we would need to specify that the standard state for a gas is the hypothetical state in which the gas has one atmosphere pressure and ideal behavior. The correction to ideal behavior is usually small and will be omitted in this study.

EXAMPLE 7.5 Given the following entropy data at 25° C, calculate the standard entropy change at 25° C accompanying the reaction

$$CH_4(g) + 2\ O_2(g) = CO_2(g) + 2\ H_2O(l)$$

	$S^0_{298.15°K}$ (joules/°K-mole)
$CH_4(g)$	186.2
$O_2(g)$	205.0
$CO_2(g)$	213.6
$H_2O(l)$	69.9

SOLUTION

$$\begin{aligned}\Delta S^0 &= S^0_{CO_2} + 2\ S^0_{H_2O} - S^0_{CH_4} - 2\ S^0_{O_2}\\ &= 213.6 + 2(69.9) - 186.2 - 2(205.0)\\ &= -242.8 \text{ joules/°K}\end{aligned}$$

Note that S^0 is the standard entropy *per mole* and must be multiplied by the number of moles in the balanced equation.

From the point of view of thermodynamics, the fundamental properties of matter are T, P, V, E, and S. It is convenient to define several other thermodynamic properties in terms of these five. This we will do in the following section.

7.4 Enthalpy

Definition of Some Useful Relationships of Enthalpy

One very useful combination of the fundamental properties is the quantity $E + PV$, to which we assign the single symbol H. Hence, we write by definition

$$\boxed{H = E + PV} \qquad [7.9]$$

It is important to keep in mind that H, *having been defined in terms of three other state functions, must also be a state function.*

The function H is called the **enthalpy**; it was introduced into thermodynamics by the American physicist J. Willard Gibbs. The function is also called the *heat content*, a well-established but rather unfortunate name in that it implies that matter contains heat. This name comes from the important relationship that equates ΔH to the heat exchanged in a process carried out at *constant pressure*, as can be seen in the following argument:

$$\boxed{\Delta E = Q + W} \qquad [7.5]$$

If the work is restricted to a constant-pressure process in which only PV work is taking place, we can substitute

$$\boxed{W = -P\Delta V} \quad \text{(constant pressure)} \qquad [7.1]$$

into equation [7.5] to obtain

$$\Delta E = Q - P\Delta V \quad \text{(constant pressure)} \qquad [7.10]$$

If we write equation [7.9] for two states, an initial state (1) and a final state (2), and then subtract the initial from the final to obtain a difference function, we have

$$\begin{aligned} H_2 &= E_2 + P_2V_2 \\ H_1 &= E_1 + P_1V_1 \\ \hline H_2 - H_1 &= E_2 - E_1 + P_2V_2 - P_1V_1 \end{aligned} \qquad [7.11]$$

or

$$\Delta H = \Delta E + \Delta(PV) \qquad [7.12]$$

For a constant-pressure process, $P_1 = P_2$, and we can factor a P out of the last two terms of equation [7.11] to obtain

$$H_2 - H_1 = E_2 - E_1 + P(V_2 - V_1)$$

or

$$\boxed{\Delta H = \Delta E + P\Delta V} \quad \text{(constant pressure)} \qquad [7.13]$$

Combining equations [7.10] and [7.13], we find

$$\boxed{\Delta H = Q} \quad \text{(constant pressure)} \qquad [7.14]$$

Q, in general, is dependent upon the actual path of a process, but for the specific case of constant pressure, Q is equal to the difference in the values of a state function.

This conclusion is significant because so many of the processes in which we are interested take place at the constant pressure of the atmosphere. For example, the heat produced or absorbed in a chemical reaction carried out in an open vessel occurs at constant pressure. Equation [7.14] applies, and the heat is seen to be a definite quantity for a given system, *dependent only on the initial and final states of the system.*

A parallel relation between Q and ΔE exists. By placing the constraint of constant volume on equation [7.5], we have

$$\boxed{\Delta E = Q} \quad \text{(constant volume)} \qquad [7.15]$$

Since

$$Q = nC_V\Delta T \quad \text{(constant volume)} \qquad [7.4]$$

and

$$Q = nC_P\Delta T \quad \text{(constant pressure)} \qquad [7.3]$$

we can write

$$\boxed{\Delta H = nC_P\Delta T} \quad \text{(constant pressure, constant heat capacity)} \qquad [7.16]$$

and

$$\boxed{\Delta E = nC_V\Delta T} \quad \text{(constant volume, constant heat capacity)} \qquad [7.17]$$

Equations [7.16] and [7.17] make possible the calculation of energy content and enthalpy changes when a system undergoes a temperature change.

Enthalpy of Reaction: Additivity of ΔH Data

We can make use of the relationship between ΔH and Q at constant pressure (equation [7.14]) to obtain values for the **enthalpy of reaction.** The heat involved in the reaction is experimentally measured, corrected to constant pressure, and equated to the ΔH of the reaction. Conversely, if we know the ΔH for a reaction, we know how much heat will be involved in the constant-pressure process. Since chemical reactions taking place in containers open to atmospheric pressure are common, the condition of constant pressure required by equation [7.14] is frequently encountered in practice.

Since H is a state function, the change in H depends only on the initial and final states of the system and is therefore independent of any intermediate states that the system might pass through. Hence, one can add several equations and their corresponding ΔH values together to obtain the ΔH of some other desired reaction.

EXAMPLE 7.6 Calculate ΔH^0 at 25° C from the data given below for the reaction

$$\text{C(graphite)} + CO_2(g) = 2\,CO(g)$$

$$FeO(s) + H_2(g) = Fe(s) + H_2O(g), \qquad \Delta H^0_{25°C} = 24.7 \text{ kjoules}$$

$$FeO(s) + CO(g) = Fe(s) + CO_2(g), \qquad \Delta H^0_{25°C} = -16.3 \text{ kjoules}$$

$$H_2(g) + CO(g) = \text{C(graphite)} + H_2O(g), \qquad \Delta H^0_{25°C} = -131.4 \text{ kjoules}$$

As with entropy, the superscript zero signifies that each reactant and product is in its standard state condition of 1 atm pressure.

SOLUTION Reverse the second and third equations (change the sign of ΔH^0) and add them to the first as written. All quantities cancel except the desired reaction.

C(graphite) + $H_2O(g)$	= $H_2(g)$ + CO(g),	$\Delta H^0 = 131.4$ kjoules
FeO(s) + $H_2(g)$	= Fe(s) + $H_2O(g)$,	$\Delta H^0 = 24.7$ kjoules
Fe(s) + $CO_2(g)$	= FeO(s) + CO(g),	$\Delta H^0 = 16.3$ kjoules
C(graphite) + $CO_2(g)$	= 2 CO(g)	$\Delta H^0 = 172.4$ kjoules

Enthalpy of Formation

It would be desirable to have absolute values of H^0 of the elements and compounds as we do for S^0. If H^0 values could be obtained, ΔH^0 could be calculated from them in the same way that we calculate ΔS^0 from S^0 data. (See Example 7.5.) However, since we have no equivalent of the third law to apply to E (and therefore H), we are unable to obtain absolute values of E and H. For enthalpy data, therefore, we must content ourselves with measurements of *differences* only.

The reference state for the elements is the state stable at 25° C and 1 atm pressure; for example $O_2(g)$, Hg(*l*), Na(*s*), C(*s*, graphite), and S(*s*, rhombic).

We will see that this is not a serious limitation. A convenient system for recording the enthalpy of a compound is in terms of the difference between its enthalpy and that of the elements of which it is composed. With the elements and compounds in their standard states, this quantity is called the standard **enthalpy of formation** of the compound, ΔH_f^0. For example, ΔH_f^0 of CH_4 gas at 25° C is the ΔH^0 of the reaction

$$\text{C(graphite, 1 atm, 25° C)} + 2\,H_2(g, \text{1 atm, 25° C}) = CH_4(g, \text{1 atm, 25° C})$$

There are relatively few reactions in which the enthalpy of formation can be measured directly. Most ΔH_f^0 data are obtained indirectly by addition of two or more reactions. The reaction just written for the formation of CH_4 from its elements is an example of a process in which ΔH is measured by an indirect means. The ΔH_f^0 data for CH_4, as for most carbon compounds, come from the directly measured enthalpies of combustion of these compounds, coupled with the directly measured enthalpies of combustion of carbon and hydrogen.

The enthalpy of combustion for hydrocarbons is the heat of the reaction in which the reactant is completely burned in oxygen to form CO_2 gas and liquid water. This measurement is made with a heat of combustion calorimeter, a schematic diagram of which is shown below:

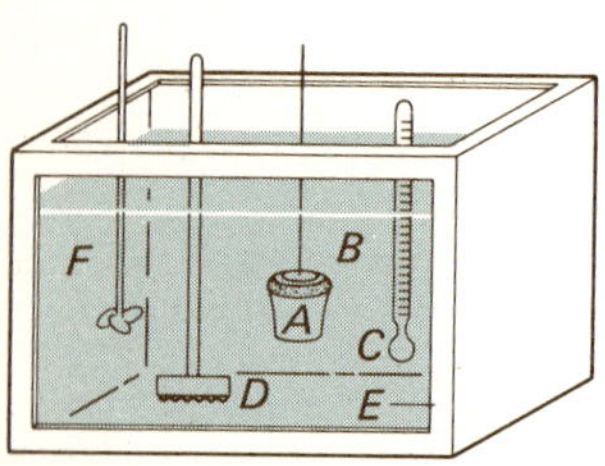

EXAMPLE 7.7 Given the following ΔH^0 data obtained from the combustion reaction, calculate ΔH_f^0 for $CH_4(g)$.

	ΔH^0 (combustion with O_2; kjoules/mole)
$CH_4(g)$	−890.36
C(graphite)	−393.55
$H_2(g)$	−285.85

SOLUTION Addition of the combustion data gives the ΔH_f^0 of CH_4. The first equation is written in the reverse direction of the combustion reaction. Note that the sign of ΔH^0 for this reaction has also been reversed.

$$
\begin{array}{llll}
CO_2(g) + 2\,H_2O(l) & = CH_4(g) + 2\,O_2(g), & \Delta H^0 = 890.36 \text{ kjoules} \\
2\,H_2(g) + O_2(g) & = 2\,H_2O(l), & \Delta H^0 = -571.70 \text{ kjoules} \\
C(\text{graphite}) + O_2(g) & = CO_2(g). & \Delta H^0 = -393.55 \text{ kjoules} \\
\hline
C(\text{graphite}) + 2\,H_2(g) & = CH_4(g), & \Delta H^0 = -74.89 \text{ kjoules} = \Delta H^0_{f(CH_4)}
\end{array}
$$

Standard enthalpies of formation have been experimentally determined for many substances. Values for a representative list of compounds are given in Appendix B. Since ΔH data are additive, we can use the enthalpy of formation data to calculate ΔH for numerous reactions. For example, consider the reaction given below for the dimerization of butadiene:

$$
2\,C_4H_6 \xrightarrow{-2\Delta H^0_{f(C_4H_6)}} 8\,C + 6\,H_2 \xrightarrow{\Delta H^0_{f(C_8H_{12})}} C_8H_{12}
$$

$$
\Delta H = \Delta H^0_{f(C_8H_{12})} - 2\Delta H^0_{f(C_4H_6)}
$$

The upper path is a two-step process in which (1) the reactants are decomposed to the elements and (2) the elements are then recombined to form the products. The lower path represents the direct reaction. Since ΔH is the same by either path, we see that the enthalpy of reaction for the overall process is the molar enthalpy of formation of the product (C_8H_{12}) minus twice the molar enthalpy of formation of the reactant (C_4H_6). In general, we obtain ΔH^0 of a reaction from ΔH_f^0 data in the same way that we obtained ΔS^0 from S^0 data (Example 7.5). ΔH^0 is the sum of the ΔH_f^0 of the products minus ΔH_f^0 of the reactants.

EXAMPLE 7.8 Calculate ΔH^0 at 25° C for the gaseous reaction

$$2\,C_2H_6 + 7\,O_2 = 4\,CO_2 + 6\,H_2O$$

from the following ΔH_f^0 data:

	ΔH_f^0 (25° C; kjoules/mole)
$H_2O(g)$	−241.8
CO_2	−393.5
C_2H_6	−84.7

Is the reaction (a) exothermic, or (b) endothermic?

SOLUTION

$$\Delta H^0 = 4\,\Delta H^0_{f(CO_2)} + 6\,\Delta H^0_{f(H_2O)} - 2\,\Delta H^0_{f(C_2H_6)} - 7\,\Delta H^0_{f(O_2)}$$

The basic features are *A*, the constant volume "bomb" calorimeter that contains both oxygen gas and the sample to undergo combustion; *B*, water to absorb the heat released in the combustion reaction; *C*, a thermometer to record the temperature rise when the explosively fast reaction takes place inside the calorimeter; *D*, an electrical heater of known resistance that can be used to determine the heat capacity of the system; *E*, an insulated jacket to contain the entire apparatus; and *F*, a stirrer to keep the water bath at uniform temperature. Not shown is an electrical lead into the calorimeter that is used to ignite the reaction.

The procedure consists of measuring the rise in temperature produced by combustion of the sample and comparing that rise in temperature with the change in temperature brought about in a calibration "run," where a known quantity of energy is added electrically by means of the heater.

Corrections for a change to the standard state pressure and a temperature of 25° C are required, and since the heat is measured at constant volume, a conversion from $Q_V(\Delta E)$ to $Q_P(\Delta H)$ must be made.

By definition, $\Delta H^0_{f(O_2)}$ is zero; that is, the enthalpy of formation of an element from itself is zero. Hence,

$$\Delta H^0 = 4(-393.5) + 6(-241.8) - 2(-84.7) - 0$$
$$= -2855.4 \text{ kjoules}$$

Since ΔH is negative, the reaction evolves heat (is exothermic).

Effect of Temperature on ΔH

Equation [7.16] can be used to calculate the ΔH of a reaction (or any other process) at one temperature from the known ΔH value at another. Generally it is ΔH^0 at 25° C that we know because it can be calculated from ΔH^0_f tables that are commonly recorded at 25° C. The procedure is to go through a cycle that includes the reaction at both the temperature at which ΔH is known and at the new temperature. Consider the cycle shown below, in which ΔH for step I is the value known at T_1 and ΔH for step III is the value we wish to calculate at the new temperature, T_2. A represents reactants, and B, products.

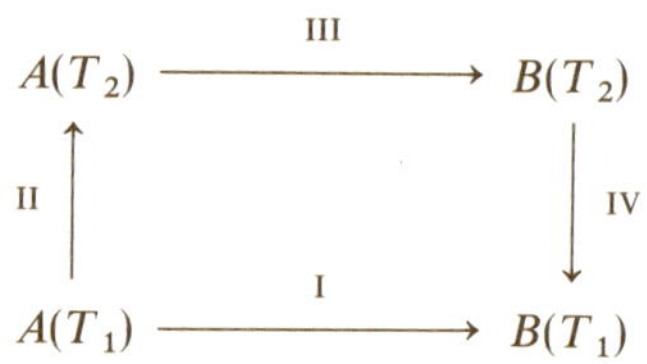

Since enthalpy is a state function, ΔH will depend only on the initial and final states of the system. Step I and the sum of the three steps II, III, and IV start at the same point and end at the same point; hence,

$$\Delta H(\text{I}) = \Delta H(\text{II}) + \Delta H(\text{III}) + \Delta H(\text{IV})$$

or rearranging,

$$\Delta H(\text{III}) = \Delta H(\text{I}) - \Delta H(\text{II}) - \Delta H(\text{IV})$$

ΔH(III) is the quantity that we want to calculate; ΔH(I) is the known value; and ΔH(II) and ΔH(IV) can be calculated from equation [7.16.] An example will help clarify the procedure.

EXAMPLE 7.9 Calculate ΔH^0 at 100° C for the reaction in Example 7.8. Assume that the molar heat capacities will remain constant over the temperature range 25° C to 100° C at the following values:

	C_P (joules/°K-mole)
$CO_2(g)$	37.13
$H_2O(g)$	33.58
$C_2H_6(g)$	52.66
$O_2(g)$	29.36

SOLUTION

$$\begin{array}{ccccccc}
 & & & \text{III} & & & \\
2\,C_2H_6(T_2) & + & 7\,O_2(T_2) & \longrightarrow & 4\,CO_2(T_2) & + & 6\,H_2O(T_2) \\
\text{II}\uparrow & & \text{II}'\uparrow & & \text{IV}\downarrow & & \text{IV}'\downarrow \\
2\,C_2H_6(T_1) & + & 7\,O_2(T_1) & \longrightarrow & 4\,CO_2(T_1) & + & 6\,H_2O(T_1) \\
 & & & \text{I} & & &
\end{array}$$

$$\begin{aligned}
\Delta H(\text{I}) &= \Delta H(\text{II}) + \Delta H(\text{II}') + \Delta H(\text{III}) + \Delta H(\text{IV}) + \Delta H(\text{IV}') \\
\Delta H(\text{I}) &= -2{,}855{,}400 \text{ joules (from Example 7.8)} \\
\Delta H(\text{II}) &= nC_P\Delta T = (2 \text{ moles})(52.66 \text{ joules/°K-mole})(373.15 - 298.15)\text{° K} \\
\Delta H(\text{II}') &= nC_P\Delta T = (7 \text{ moles})(29.36 \text{ joules/°K-mole})(373.15 - 298.15)\text{° K} \\
\Delta H(\text{IV}) &= nC_P\Delta T = (4 \text{ moles})(37.13 \text{ joules/°K-mole})(298.15 - 373.15)\text{° K} \\
\Delta H(\text{IV}') &= nC_P\Delta T = (6 \text{ moles})(33.58 \text{ joules/°K-mole})(298.15 - 373.15)\text{° K}
\end{aligned}$$

$$\begin{aligned}
\Delta H(\text{III}) &= \Delta H(\text{I}) - \Delta H(\text{II}) - \Delta H(\text{II}') - \Delta H(\text{IV}) - \Delta H(\text{IV}') \\
&= -2{,}855{,}400 - 7899 - 15{,}410 + 11{,}140 + 15{,}110 \\
&= -2{,}852{,}500 \text{ joules} = -2{,}852.5 \text{ kjoules}
\end{aligned}$$

7.5 Gibbs Free Energy

Free Energy and Spontaneity

We have seen that the second law gives us the criterion for telling whether a proposed process can take place. In the previous section on enthalpy, we stated that if the entropy of the universe would be increased through the occurrence of a process, that process is *natural* and possible; if the process would result in a decrease in entropy of the universe, then it is unnatural and impossible. Unfortunately, in this form the criterion of natural versus unnatural change is often difficult to apply. In this section, we see that a more convenient approach is to define a new state function and then apply the second-law criteria to it. The function derived for this purpose is called the **Gibbs free energy** in honor of Willard Gibbs, who introduced the function. Its definition in terms of the fundamental quantities is

$$\boxed{G = E + PV - TS = H - TS} \qquad [7.18]$$

Since G is defined in terms of other state functions, it is also a state function. Our purpose in defining it as we have is to give ourselves a criterion for direction of change in terms of the system at constant temperature and pressure. We can see how ΔG fits this requirement from the following arguments.

We start with

The equality sign applies to equilibrium, the inequality sign to a natural process.

$$\Delta S(\text{universe}) \geqq 0$$

which is a combination of equations [7.7] and [7.8]. Since the universe is composed of a system and its surroundings, we can write

$$\Delta S(\text{system}) + \Delta S(\text{surroundings}) \gtreqless 0 \qquad [7.19]$$

Now let us limit the discussion to processes taking place at constant temperature and pressure. At constant temperature, the entropy change produced in the surroundings by exchange of heat with the system can be calculated through the relationship

Equation [7.20] applies only to a special hypothetical path called the *reversible process*. It is the only path by which we can make the calculations for an entropy change. However, since entropy is a state function, independent of path, we are at liberty to use any pathway between specified initial and final states, including the hypothetical reversible one.

$$\Delta S(\text{surroundings}) = \frac{Q(\text{surroundings})}{T} \qquad [7.20]$$

Since

$$Q(\text{surroundings}) = -Q(\text{system})$$

and

$$Q(\text{system}) = \Delta H(\text{system}) \quad (\text{constant pressure}) \qquad [7.14]$$

we can express ΔS(surroundings) in terms of properties of the system. At constant temperature we can write

$$\Delta S(\text{surroundings}) = -\frac{Q(\text{system})}{T} \quad (\text{constant temperature}) \qquad [7.21]$$

and at both constant T and constant P we have

$$\Delta S(\text{surroundings}) = -\frac{\Delta H(\text{system})}{T} \quad (\text{constant temperature, constant pressure}) \qquad [7.22]$$

Substitution of equation [7.22] into [7.19] gives

$$\Delta S(\text{system}) - \frac{\Delta H(\text{system})}{T} \gtreqless 0 \qquad [7.23]$$

Rearranging gives

$$-\Delta H(\text{system}) + T\Delta S(\text{system}) \gtreqless 0 \qquad [7.24]$$

Multiplying both sides of equation [7.24] by -1 gives

Note the change in the inequality sign. Multiplying by a negative number reverses the inequality sign.

$$\Delta H(\text{system}) - T\Delta S(\text{system}) \lesseqgtr 0 \qquad [7.25]$$

The left-hand side of equation [7.25] is equal to ΔG as shown below:

Since

$$G = H - TS \qquad [7.18]$$

we can write

$$\Delta G = \Delta H - \Delta(TS) \qquad [7.26]$$

for a change from an initial to a final state. If the process is carried out at a constant temperature

$$\Delta(TS) = TS_2 - TS_1 = T\Delta S \qquad [7.27]$$

and we obtain

$$\boxed{\Delta G = \Delta H - T\Delta S} \quad \text{(constant temperature)} \qquad [7.28]$$

Combining equation [7.28] with [7.25] gives the desired relation

$$\boxed{\Delta G \lesseqgtr 0} \quad \text{(system; constant temperature, constant pressure)} \qquad [7.29]$$

where the equality refers to equilibrium and the inequality to a natural process. Thus, to predict whether any given process will occur, it is necessary to calculate ΔG (at constant T and P) for the *system alone.* If $\Delta G = 0$, the system is already in the equilibrium state. If the calculation shows that the process would decrease the free energy of the system ($\Delta G < 0$), then the reaction will occur. A calculation that results in $\Delta G > 0$ shows that the process is unnatural and would not take place. In fact, the reverse process would take place. The conditions for equilibrium and change are summarized in Table 7.1.

TABLE 7.1

Conditions for Equilibrium and Change

	EQUILIBRIUM	NATURAL PROCESS	UNNATURAL PROCESS
ΔS(universe)	0	+	−
ΔG(system, constant T, P)	0	−	+

The summary in Table 7.1 is the basis for deriving an important body of information about phase and chemical equilibria that is discussed in later chapters.

Free Energy of Formation

Before proceeding to our study of equilibria, it is necessary to know how free energy data are obtained and in what form these data are recorded. Since the definition of G involves E, we have the same problem with G as we had with H; namely, we are forced to use relative rather than absolute values. We handle this problem in the same way we did with enthalpy, by recording the free energy data in terms of standard **free energies of formation.** Thus, we report differences in the free energy of a compound and the elements of which it is composed. Once we have a table of standard free energies of formation, we can calculate ΔG^0 for a reaction in the same manner as we did ΔH^0. That is, the free energy of the reaction is the sum of the free energies of formation of the products minus the sums of the free energies of formation of the reactants. (See Example 7.8 for the calculations of ΔH^0 from ΔH^0_f.)

One important means of obtaining free energy values is by combining enthalpy and entropy data. Equation [7.28] gives the relationship among these quantities at constant temperature. If we are referring to standard state conditions, equation [7.28] becomes

$$\Delta G^0 = \Delta H^0 - T\Delta S^0 \quad \text{(constant temperature)} \qquad [7.30]$$

EXAMPLE 7.10 Given the following enthalpy and entropy data, calculate ΔG_f^0 of CH_4 gas at 25° C:

	ΔH_f^0 (25° C; kjoules/mole)	S^0 (25° C; joules/°K-mole)
C(graphite)	0	5.69
$H_2(g)$	0	130.59
$CH_4(g)$	−74.85	186.20

SOLUTION Consider the reaction:

$$\text{C(graphite)} + 2\ H_2(g) = CH_4(g)$$

$$\begin{aligned}\Delta H^0 &= \Delta H^0_{f(CH_4)} - \Delta H^0_{f(C)} - 2\ \Delta H^0_{f(H_2)}\\ &= -74.85 - 0 - 0 = -74.85 \text{ kjoules}\end{aligned}$$

$$\begin{aligned}\Delta S^0 &= S^0_{CH_4} - S^0_{C} - 2\ S^0_{H_2}\\ &= 186.20 - 5.69 - 2(130.59) = -80.67 \text{ joules/°K}\end{aligned}$$

$$\begin{aligned}\Delta G^0 &= \Delta H^0 - T\Delta S^0 = -74{,}850 \text{ joules} - (298.15° \text{ K})(-80.67 \text{ joules/°K})\\ &= -50{,}800 \text{ joules}\end{aligned}$$

Since the reaction represents the formation of CH_4 from the elements, ΔG^0 of the reaction is ΔG_f^0 for CH_4. Hence,

$$\Delta G^0_{f(CH_4)} = -50{,}800 \text{ joules}$$

We will make extensive use of ΔG_f^0 data in later chapters. The following example illustrates one application.

EXAMPLE 7.11 Will the reaction

$$N_2O_4(g) = 2\ NO_2(g)$$

occur in a mixture of the two substances, each at a partial pressure of one atmosphere (standard state) and the system at 25° C?

SOLUTION The reaction can occur only if by doing so it lowers the free energy of the system ($\Delta G < 0$). From Appendix B we find

	ΔG_f^0 (25° C)
$NO_2(g)$	51,840 joules/mole
$N_2O_4(g)$	98,286 joules/mole

$$\Delta G^0 = 2\Delta G^0_{f(NO_2)} - \Delta G^0_{f(N_2O_4)}$$
$$= 2(51{,}840) - 98{,}286 = 5394 \text{ joules}$$

The answer is *no*; ΔG^0 is positive. The reverse reaction (dimerization of NO_2 to N_2O_4) is the natural change under these conditions.

Effect of Pressure on Free Energy

We will have need on several occasions later in the course to consider the effect of pressure on free energy. We will summarize in this section the equations that apply, but leave their applications until later.

The effect of pressure on the free energy of a system that is not very compressible (a solid, for example) can be calculated from the relation

$$\Delta G = V\Delta P \quad \text{(constant } T) \qquad [7.31]$$

The effect of pressure on the free energy of an ideal gas is given by

$$\Delta G = RT \ln \frac{P_2}{P_1} \quad \text{(constant } T) \qquad [7.32]$$

This equation, in the form

$$G = G^0 + RT \ln \frac{P}{P^0} \qquad [7.33]$$

will be used as a starting point in the development of the equations that apply to chemical equilibrium.

Derivation of equations [7.31]–[7.33] is reserved for a more advanced course. The superscript used in equation [7.33] refers to the standard state; ln is the symbol for natural logarithm, logarithm to the base *e*.

7.6 Thermodynamics and the Environment

Efficiency of Thermodynamic Engines

In order to sustain life and be productive, we must continually convert other forms of energy into work. It is work energy that runs the machines, pumps the water, and provides the transportation, and it is work energy that will be required to clean the environment so that we may improve the quality of life on our planet. The general procedure used to convert other forms of energy into

Figure 7.9
Procedure for converting other forms of energy into work.

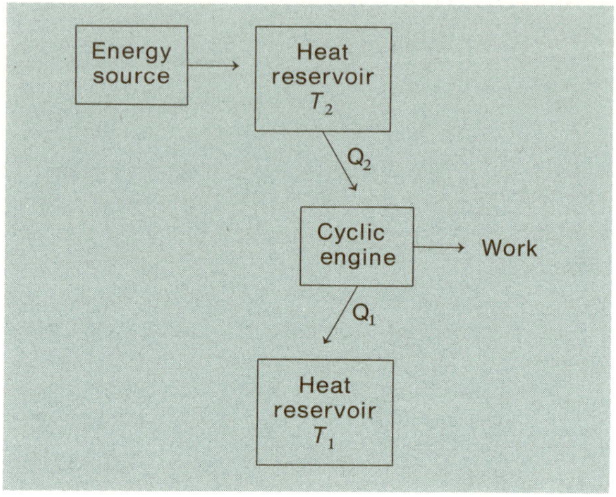

work is illustrated in Figure 7.9. A hot reservoir is maintained at temperature T_2 by an energy source (for example, coal that is burned to release heat). A cyclic engine withdraws Q_2 units of heat from the hot reservoir, converts part of this heat energy into work, and discharges the remainder (Q_1) into a lower-temperature (T_1) reservoir.

The laws of thermodynamics restrict the amount of work that can be obtained from this process. The obvious restriction of the first law is that we cannot get more work out of the process than the energy we put in. The second law restricts the process further in that only part of the heat that flows into the engine can be converted into work. The remainder must be removed in order for the engine to operate (the reservoir at T_1 is essential to the operation of the engine).

The equation, first derived by Carnot in 1824, to calculate the maximum efficiency of conversion of heat into work is

$$\boxed{\eta = \frac{T_2 - T_1}{T_2}} \qquad [7.34]$$

where η is the efficiency of the engine for converting heat into work, and T_2 and T_1 are the temperatures on the Kelvin scale for the hot and cold reservoirs, respectively. From [7.34] one can see that an efficiency of 1 (100% conversion) would be possible only if the temperature of the cold reservoir were at absolute zero ($T_1 = 0$). It is instructive to see what the limit of efficiency is under reasonable operating conditions. For example, if the hot reservoir is operated at the boiling point of water and the cold reservoir at the freezing point of water, we would have

$$\eta = \frac{T_2 - T_1}{T_2} = \frac{(373 - 273)^\circ \mathrm{K}}{373^\circ \mathrm{K}} = 0.27$$

That is, a maximum of only 27% conversion would be possible, and 73% of Q_2 would be added to the lower reservoir.

In discussing equation [7.34], no reference was made to the nature of the work machine. It is easy to visualize the engine as a steam engine, with the hot reservoir as the boiler, the cold reservoir as the condenser, and the machinery in between as the device for performing the work. The same principle applies to any heat engine, from an internal combustion engine performing mechanical work to an atomic reactor producing electrical energy. It is the latter that we want to consider in this section. Before we do that, however, let us digress briefly to discuss present energy requirements, our future needs for energy, and its possible sources. Then we will return to the subject of heat engines (particularly nuclear reactors) and their relationship to thermal pollution.

Energy Requirements and Sources

Highly industrialized countries use large amounts of energy. An interesting way of expressing this rate of energy conversion is to make comparisons in terms of the rate at which a 100-watt light globe uses energy. Living on a 2500 kcal/day diet provides energy to the body at about the same rate that energy is used in a 100-watt light globe. In primitive societies, the power consumption is roughly 100 watts per capita. In more advanced (but not industrialized) societies, where fire and some water power are used, the per-capita power consumption might be 200 watts. For a striking comparison, the present rate of energy conversion in the United States is 10,000 watts per person!

Power is energy per unit time, and the unit of power is the *watt*. One *watt* is one joule of energy per second. Hence, 100 watts of power represents the consumption of 100 joules of energy every second.

Requirements for energy can be expected to increase greatly in the future. The present rate of increase is 3.3% per year, and forecasts for the future give skyrocketing curves. Even if people were to curtail their demands for a higher standard of living (which requires large expenditures of energy), the expanding population and the energy requirements to control and clean up pollution are enormous. It is becoming abundantly clear that preserving the life of the planet will require very large expenditures of energy. The questions we now face are (1) where do we get the energy? and (2) what are the consequences of its use?

The current world average rate of energy use is estimated to be 1500 watts per capita.

The principal sources of energy at the present time are fossil fuels and water power. Neither of these is likely to be the important source in the future. We are fast running out of fossil fuels, and while some further expansion of hydroelectric facilities is possible, the potential is still small compared with projected needs.

Other major sources we can look at for future needs are solar radiation and controlled nuclear reactions. Intercepting and collecting solar radiation has some very appealing aspects because of the large quantity of energy available and because of its pollution-free quality. There are drawbacks, however. Since the sun does not shine all the time, elaborate provisions for storage and transmission of energy would be required. In addition, fairly large areas of the earth's surface would be needed to collect the energy. Hence, we come to nuclear reactions as probably the most likely future source of energy.

One can calculate that 15 to 20 square miles of collecting devices would be required to produce the same amount of energy as a modern steam power plant.

The nuclear power plants in operation now derive their energy from fission of ^{235}U, which is the isotope that makes up only 0.7% of natural uranium. Breeder reactors that are capable of converting ^{238}U, the most abundant isotope of naturally occurring uranium, into fissionable material are being seriously considered. The controlled fusion of hydrogen nuclei (the reaction of the H-bomb) has not yet been achieved, but it is a very promising source of energy for the future. If the nuclear fusion reaction can be controlled, enough energy could be made available from the deuterium (^2_1H) in the ocean to last for billions of years.

The one out of three units of energy that becomes electrical energy will also eventually end up as heat. Since the electrical energy can be transmitted long distances from the generator and dissipated gradually in varied ways (lighting, heating, and operating mechanical devices), it is not as much of a pollution problem as the waste heat discharged at the site of the generator.

We find ourselves with an interesting dilemma: The energy that is needed to restore the environment may have to be derived by power plants that are themselves polluters.

The energy from a nuclear reactor is in the form of heat, which is used to form superheated steam. The steam drives a turbine in the same way that a coal-burning power plant operates. The force that pushes the turbine comes from the difference in pressure of the steam in the hot and cold reservoirs, which are on opposite sides of the turbine. At the cold reservoir, the steam is condensed into liquid water, making pressure drops of the order of 2000 pounds per square inch.

Thermal Pollution

If our conclusions are correct that future needs for energy will be large and that nuclear reactions will be the likely source of supply, then we need to examine the consequences of large energy production by this means. The performance of the nuclear-powered electric generating plant, like all heat engines, is bound by the restrictions embodied in equation [7.31]. Only a fraction of the heat produced in the reactor can be converted to electrical energy. In practice, the efficiency obtained is about 33%, which means that for each unit of electrical energy produced, two units of heat are discharged into the environment. If this is to be done on a large scale, we have reasons to be concerned with indiscriminately adding waste heat to the environment. This special type of pollution is appropriately called **thermal pollution.**

One way of expressing energy consumption is to compare it with the net amount of energy that reaches the surface of the earth from the sun. This energy is referred to as the **insolation.** Expressed as a percentage of the insolation, the total energy consumption from fossil fuels, water power, and fissionable materials is presently less than .005%. Working on the basis of a population twice the number living on the earth today, and with each person enjoying the 10,000-watt power consumption that now exists in the United States, one calculates that global consumption would approach 0.05% of the insolation. Many people who have studied the problem believe that the planet could withstand a heating effect amounting to 0.5% of the insolation without serious climatic and ecological changes. Hence, we are currently two orders of magnitude away from a global problem of thermal pollution. Considering the projected expansion in energy consumption in the future, we can estimate that we are as much as 100 years (not a comfortably long period of time) away from worldwide thermal pollution problems.

The problem of *local thermal pollution* caused by the heat from electrical power plants being discharged into rivers, lakes, and coastal waters is already with us. Near the vicinity of a modern 1000-megawatt power plant, the discharge of waste heat into the environment can be a sizable fraction of the insolation in the region. The most convenient way of dissipating the heat is to add it to a fast-flowing stream, but unless the quantity of water is large, the temperature of the stream rises enough to cause adverse ecological changes in the plant and fish life.

Other ways of dissipating the heat are cooling by (1) evaporating water either in wet cooling towers or in ponds, and (2) discharging the heat directly into the atmosphere by means of a dry cooling tower. As one might guess, the quantity of water used in evaporative cooling for a 1000-megawatt plant is very large. It seems likely that many such plants concentrated in a small area (such as the eastern seacoast of the United States, where energy consumption is high) could put enough moisture in the air to affect the local climate. The dry tower has the advantage of discharging the heat directly into the atmosphere, where it is effectively dispersed by natural convection. Dry cooling has the disadvantage, however, of being very expensive.

To summarize, our energy needs for the future will be necessarily large and will likely have to be met from nuclear power. Because of the limitations placed on conversion of heat into work by the second law of thermodynamics, electrical generating plants do *now* produce local thermal pollution and could conceivably in the not too distant future cause serious problems on a global scale.

If the temperature of the cooling water is not allowed to rise more than 20° F, a flow rate of about one-half million gallons of water per minute are required in a 1000-megawatt plant. This warm water must then be mixed with enough cold water to prevent the water downstream from heating more than 3 to 5° F, or else pronounced ecological changes will result.

Summary

Thermodynamics is an exact science that is based on two well-tested laws of experience, the first and second laws of thermodynamics.

The subject of the first law is energy content, E. A change in energy content is related to the heat and work involved in a process:

$$\boxed{\Delta E = Q + W}$$

The subject of the second law is entropy. This interesting quantity has the physical significance of being a measure of disorder. The higher the disorder in a system, the greater is the entropy.

The first law of thermodynamics is a statement of the conservation of energy in the universe:

First law: $\boxed{\Delta E = 0}$ (universe)

The second law tells us whether or not a process is possible. Only those processes that increase the entropy of the universe can take place. These are natural (spontaneous) processes.

Second law: $\boxed{\Delta S > 0}$ (universe)

A "third law," sometimes considered as a corollary of the second law, states that the entropy of a perfectly ordered crystal at 0° K is zero. This sets a reference point from which absolute values of entropies can be measured.

Two very important derived quantities are defined in terms of E and S. One of these is called enthalpy, H, the definition of which is

$$\boxed{H = E + PV}$$

Under the condition of constant pressure

$$\boxed{\Delta H = Q} \qquad \text{(constant pressure)}$$

Thus ΔH is a measure of the heat exchanged in a chemical reaction carried out in a container open to the atmosphere, a common practice. If the size of ΔH is negative, the reaction is exothermic; if positive, endothermic.

A second derived quantity is the Gibbs free energy, defined as

$$\boxed{G = E + PV - TS}$$

G is defined to give a quantity that will include the predictive property of the second law and thus give a criterion for direction of change. Under the condition of constant T and P, a negative value for ΔG(system) is the condition for a natural change. If ΔG is positive, the reverse reaction will take place. The condition for equilibrium is

$$\boxed{\Delta G = 0} \qquad \text{(constant temperature, constant pressure)}$$

Values of ΔS^0, ΔH^0, and ΔG^0 for chemical reactions can be obtained from tabulated values of S^0, ΔH_f^0, and ΔG_f^0 by subtracting the sum of the thermodynamic quantity for the reactants from that of the products.

At constant temperature, ΔG^0, ΔH^0, and ΔS^0 are related by

$$\boxed{\Delta G^0 = \Delta H^0 - T\Delta S^0} \qquad \text{(constant temperature)}$$

Thermodynamics limits the extent to which heat can be converted into work in a thermodynamic engine. The maximum fraction allowable is given by Carnot's relation

$$\boxed{\eta = \frac{T_2 - T_1}{T_2}}$$

This limitation has important implications in the solution of our present and future energy problems.

Important Terms and Concepts

Thermodynamics
Energy content
Work
Heat
Adiabatic system
Heat capacity at constant volume
Heat capacity at constant pressure
State function

Entropy	Enthalpy of reaction
First law of thermodynamics	Enthalpy of formation
Second law of thermodynamics	Gibbs free energy
Law of conservation of energy	Free energy of formation
Third law of thermodynamics	Thermal pollution
Enthalpy	Insolation

Exercises

7-1.
One thousand grams of liquid water at 0.00° C are heated to 100.00° C. In this process, the water expands and does work as it pushes back the atmosphere. Assuming that atmospheric pressure is 1.0000 atm, calculate W (in calories) for the process. The density of liquid water is 0.99973 g/cm^3 at 0.00° C and 0.95838 g/cm^3 at 100.00° C.

7-2.
Two moles of hydrogen gas are produced at 25° C and a pressure of 1.00 atmosphere by the action of acid on a metal. Calculate the work (in joules) done by the gas in pushing back the atmosphere. Assume ideal gas behavior.

7-3.
How many joules of heat are required to raise the temperature of 10.0 g of neon (a monatomic gas, $C_V = 3/2\ R$, $C_P = 5/2\ R$) from 0° C to 100° C at (a) constant volume? (b) constant pressure?

7-4.
One hundred grams of Ar gas with $C_V = 3/2\ R$ is heated in a sealed container from 200° K to 400° K. Assume that the container does not expand, and calculate Q and W for the process.

7-5.
The freezing point of benzene (C_6H_6) at 1 atm pressure is 5.45° C. The density of solid benzene is 0.978 g/cm^3, and the density of the liquid is 0.874 g/cm^3 at the melting point. $\Delta H_{fus} = 9832$ joules/mole for benzene. For the process

$$C_6H_6(l,\ 1\ \text{atm},\ 5.45°\ \text{C}) \rightarrow C_6H_6(s,\ 1\ \text{atm},\ 5.45°\ \text{C})$$

calculate in joules (a) W and (b) Q.

7-6.
Calculate in joules (a) W, (b) Q, and (c) ΔE for the isobaric (constant pressure) expansion at 1.00 atm pressure of 1.00 mole of an ideal gas ($C_V = 3/2\ R$, $C_P = 5/2\ R$) from a volume of 20.0 to 30.0 liters.

7-7.
One mole of an ideal gas is expanded isobarically (at constant pressure) at 2.00 atm pressure from a volume of 11.2 liters to a volume of 22.4 liters. $C_P = 5/2\ R$. Calculate (a) W; (b) Q; (c) ΔE; and (d) ΔH.

7-8.
For Hg gas, $C_P = 5/2\ R$. Calculate ΔH for the process

$$\text{Hg}(g,\ 1\ \text{atm},\ 800°\ \text{K}) \rightarrow \text{Hg}(g,\ 1\ \text{atm},\ 1000°\ \text{K})$$

7-9.
The ΔH^0 of combustion of normal butane and isobutane have been measured at 25° C.

$$n\text{-}C_4H_{10}(g) + 6\tfrac{1}{2}\ O_2(g) \rightarrow 4\ CO_2(g) + 5\ H_2O(l), \quad \Delta H^0 = -2878\ \text{kjoules}$$

$$i\text{-}C_4H_{10}(g) + 6\tfrac{1}{2}\ O_2(g) \rightarrow 4\ CO_2(g) + 5\ H_2O(l), \quad \Delta H^0 = -2870\ \text{kjoules}$$

Calculate the heat of isomerization of n-C_4H_{10} at 25° C and 1 atm pressure.

7-10.
Use the following data to determine the standard enthalpy of formation of tungsten carbide, WC(s).

$$\text{C(graphite)} + O_2(g) = CO_2(g),\ \Delta H^0_{298} = -393.51\ \text{kjoules}$$

$$\text{WC}(s) + \frac{5}{2}\ O_2(g) = \text{WO}_3(s) + CO_2(g), \quad \Delta H^0_{298} = -1195.79\ \text{kjoules}$$

$$\text{W}(s) + \frac{3}{2}\ O_2(g) = \text{WO}_3(s),\ \Delta H^0_{298} = -837.47\ \text{kjoules}$$

7-11.
Given the following information at 25° C.

$$2\ NH_3(g) + \frac{7}{2}\ O_2(g) = 3\ H_2O(g) + 2\ NO_2(g), \quad \Delta H^0 = -565.4 \text{ kjoules}$$

$$N_2(g) + 2\ O_2(g) = 2\ NO_2(g),\ \Delta H^0 = 67.70 \text{ kjoules}$$

$$H_2(g) + \tfrac{1}{2}O_2(g) = H_2O(g),\ \Delta H^0 = -241.8 \text{ kjoules}$$

Calculate ΔH^0 at 25° C for the reaction

$$3\ H_2(g) + N_2(g) = 2\ NH_3(g)$$

7-12.
The combustion of diamond and of graphite are given by the equations

$$\text{C(graphite)} + O_2(g) = CO_2(g),\ \Delta H^0 = -393.5 \text{ kjoules}$$

$$\text{C(diamond)} + O_2(g) = CO_2(g),\ \Delta H^0 = -395.4 \text{ kjoules}$$

Calculate the heat that must be added to 10.0 grams of graphite at 25° C to convert it into diamond (constant pressure of 1 atm).

7-13.
The ΔH^0 of combustion of cyclopropane is −2091.2 kjoules/mole at 298.15° K. Use the ΔH^0 of combustion of carbon and hydrogen to calculate the heat of formation of cyclopropane (C_3H_6) at 298.15° K.

7-14.
The ΔH^0 of combustion of benzene, C_6H_6, at 25° C is −3267 kjoules/mole. Calculate the enthalpy of formation of benzene from the heats of formation of CO_2 and H_2O given in Appendix B.

7-15.
Using data on enthalpies of formation (Appendix B), calculate the heats of combustion (ΔH) at 25° C of the following substances to $H_2O(l)$ and $CO_2(g)$:

(a) propane (C_3H_8)
(b) methanol (CH_3OH)
(c) acetic acid (CH_3COOH)

7-16.
ΔH for the reaction

$$A = 2B$$

is −42,050 joules at 300° K. Calculate ΔH for this reaction at 400° K. $C_p = 41.8$ joules/°K-mole for A, and 62.8 joules/°K-mole for B.

7-17.
(a) Use the enthalpy of formation data in Appendix B to calculate ΔH^0 at 298.15° K for the decomposition reaction

$$2\ O_3 = 3\ O_2$$

(b) Now use the C_p data in Appendix B together with your answer in (a) to calculate ΔH^0 for the reaction in (a) at 400° K.

7-18.
Calculate ΔH^0 for the reaction

$$10\ HNO_3(l) + C_2H_2(g) = 2\ CO_2(g) + 6\ H_2O(l) + 10\ NO_2(g)$$

See Appendix B for necessary data. Is the reaction exothermic or endothermic when carried out at standard conditions and 25° C?

7-19.
Using Appendix B, calculate ΔS^0 at 298° K for each of the following reactions. Explain the sign of each of the entropy changes by qualitatively comparing the molecular disorder in the reactants and products.

(a) $Cl_2(g) = 2\ Cl(g)$
(b) $H_2O(l) = H_2O(g)$
(c) $C(s) + O_2(g) = CO_2(g)$

7-20.
Using atomic and molecular entropies, calculate ΔS for the following processes at 25° C:

(a) $H_2(g) + \frac{1}{2}O_2(g) = H_2O(l)$

(b) $H_2(g) + Cl_2(g) = 2\ HCl(g)$

(c) $H_2O(g) = H_2O(l)$

7-21.
Calculate the free energy of formation of urea, $CO(NH_2)_2(s)$, at 25° C from the following data:

$$CO_2(g) + 2\ NH_3(g) = H_2O(g) + CO(NH_2)_2(s), \quad \Delta G^0_{298} = 1908 \text{ joules}$$

$$H_2O(g) = H_2(g) + \frac{1}{2}O_2(g),\ \Delta G^0_{298} = 228{,}590 \text{ joules}$$

$$\text{C(graphite)} + O_2(g) = CO_2(g),\ \Delta G^0_{298} = -394{,}380 \text{ joules}$$

$$N_2(g) + 3\ H_2(g) = 2\ NH_3(g),\ \Delta G^0_{298} = -33{,}280 \text{ joules}$$

7-22.
Using Appendix B calculate ΔH^0 and ΔG^0 at 25° C for the following processes:

(a) $H_2O(l) = H_2O(g)$

(b) $2\ SO_2(g) + O_2(g) = 2\ SO_3(g)$

(c) $C_2H_4(g) + H_2(g) = C_2H_6(g)$

(d) $C_2H_4(g) + H_2O(g) = CH_3CH_2OH(l)$

(e) $F_2O(g) + H_2O(l) = 2\ HF(g) + O_2(g)$

(f) $C_2H_2(g) + \frac{5}{2} O_2(g) = 2\ CO_2(g) + H_2O(l)$

(g) $Mg(s) + CaO(s) \quad MgO(s) + Ca(s)$

7-23.
Consider the chemical reaction

$$C_2H_6(g) + \frac{3}{2} O_2(g) = CH_3COOH(l) + H_2O(l)$$

Use the table in Appendix B to calculate (a) $\Delta H^0_{298.15}$; (b) $\Delta G^0_{298.15}$; and (c) $\Delta S^0_{298.15}$. (d) The negative ΔG^0 value as calculated in (b) indicates that the reaction is spontaneous. How do you reconcile this with the fact that $\Delta S^0_{298.15}$ as calculated in (c) is also negative, but the second law states $\Delta S > 0$ (natural process)?

7-24.
The following information at 500° K is given.

	ΔH^0_f (kjoules/mole)	ΔG^0_f (kjoules/mole)	S^0 (joules/°K-mole)
A	−20.0	—	40.0
B	−10.0	−30.0	20.0

(a) For the reaction 2A → 3B, calculate ΔH^0 at 500° K. Is the reaction (as written) exothermic or endothermic?

(b) Calculate ΔG^0_f for A. Will the reaction take place under the conditions above?

7-25.
The following information at 300° K is given.

	ΔH^0_f (kjoules/mole)	ΔG^0_f (kjoules/mole)	S^0 (joules/deg-mole)
A	0	0	30.0
B	−20.0	−30.0	50.0
C	−50.0	−40.0	—

For the reaction 2A + B → 2C, calculate (a) ΔH^0; (b) ΔG^0; (c) ΔS^0; and, (d) the standard entropy of C. (e) Will the reaction take place (as written) under standard state conditions at 300° K? (f) Is the reaction exothermic or endothermic under standard conditions at 300° K?

7-26.
The following information at 25° C is given.

	ΔH^0_f (joules/mole)	ΔG^0_f (joules/mole)	S^0 (joules/°K-mole)
$H_2(g)$	0	0	130.58
$CO(g)$	−110,525	−137,269	197.90
$CH_3OH(l)$	−238,638	−166,314	—

(a) Calculate ΔS^0 at 25° C for the reaction

$$CO(g) + 2\ H_2(g) = CH_3OH(l)$$

(b) Calculate S^0 for $CH_3OH(l)$ at 25° C.

(c) Will the reaction take place under the conditions above?

(d) Is the reaction exothermic or endothermic?

(e) Is there more or less order in the product than in the reactants?

(f) Would the formation of CH_3OH under the conditions above increase the entropy of the universe?

7-27.
The following information at 25° C is given.

	ΔH^0_f (joules/mole)	ΔG^0_f (joules/mole)	S^0 (joules/°K-mole)
$O_2(g)$	0	0	205.0
$O_3(g)$	142,300	—	237.7

(a) Calculate ΔG^0 for the reaction $3\ O_2(g) = 2\ O_3(g)$ at 25° C.

(b) Calculate ΔG^0_f for O_3 at 25° C.

(c) Is the reaction exothermic or endothermic?

(d) Will the reaction go as written under standard-state conditions and 25° C?

7-28.
The following data are given.

	$NH_4NO_3(s)$	$H_2O(g)$
ΔH^0_f (kjoules/mole)	−365.3	−241.8
ΔG^0_f (kjoules/mole)	−188.7	−228.4

Calculate ΔH^0 and ΔG^0 for the reaction

$$NH_4NO_3(s) = N_2(g) + \frac{1}{2} O_2(g) + 2\ H_2O(g)$$

and discuss the danger of storing large quantities of $NH_4NO_3(s)$.

7-29.
Consider the process

$$COCl_2(g) = CO(g) + Cl_2(g)$$

(a) Will the reaction take place as written when each chemical species is present in its *standard state* at 298° K? Show calculations and give rationale for your answer.
(b) Will increasing the temperature of the system favor the formation of the products or of the reactants? For endothermic reactions, increasing T favors the formation of products. For exothermic reactions the reverse is true.
(c) Do the products represent more or less order than the reactant? Show calculations and give reasoning.

Problems

7-1.
One mole of an ideal gas with $C_V = 3/2$ R and $C_P = 5/2$ R undergoes the following change of state:

$$P_1 = 1.00 \text{ atm} \qquad P_2 = 0.250 \text{ atm}$$

$$V_1 = 25.0 \text{ liters} \qquad V_2 = 80.0 \text{ liters}$$

The change may be brought about in an infinite number of ways. Consider for this problem two paths, each consisting of two parts:

I. (1) A constant pressure expansion to 80.0 liters followed by (2) a constant volume drop in pressure to 0.250 atm.
II. (1) A constant volume drop in pressure to 0.250 atm followed by (2) a constant pressure expansion to 80.0 liters.

(a) Determine P, V, and T for 1.00 mole of the gas after the initial step of each of the two paths. Plot the paths on a PV diagram.
(b) Calculate W, Q, and ΔE for each part [(1) and (2)] of each path (I and II) and for the total path. Tabulate your results and compare values for the two paths.

7-2.
Calculate the entropy change of the universe when 2.0 moles of water at 90° C are mixed at constant atmospheric pressure with 3.0 moles of water at 40° C. The heat capacity of water may be considered to be constant at 75.3 joules/°K-mole over the temperature range of the process.

[*Hints:* Consider the 2.0 moles of hot water to be the system and the 3.0 moles of cooler water to be the surroundings. The sum of the two entropy changes will be the change for the universe.

Calculate the final temperature of the water after mixing by assuming that the heat lost by the hot water is equal to the heat gained by the cold water.

The change in entropy resulting from a change in temperature (pressure held constant) can be calculated from the relation

$$\Delta S = nC_P \ln \frac{T_2}{T_1} \qquad \text{(constant pressure)}$$

7-3.
The spontaneous process (1 mole of water)

$$H_2O(l, -10°\text{ C}) \rightarrow H_2O(s, -10°\text{ C})$$

$$C_P = 75.3 \text{ joules/°K-mole} \qquad C_P = 37.7 \text{ joules/°K-mole}$$

is given.

(a) Assuming that the heat capacities of liquid and solid are constant, calculate ΔS for the process. [*Hint:* ΔS must be calculated by a reversible path. A convenient path for this problem is to heat the liquid to 0° C, freeze the liquid reversibly at 0° C, then cool the solid to −10° C.]
(b) Calculate the amount of heat absorbed into the surroundings by calculating ΔH for the process. Using this value for Q, calculate ΔS(surroundings) and show that $\Delta S > 0$ (universe) for this process even though $\Delta S < 0$ for the system. ΔS as a function of temperature (constant pressure) is given by the relation

$$\Delta S = nC_P \ln \frac{T_2}{T_1} \qquad \text{(constant pressure)}$$

8

Phase Equilibria and Solutions

In Chapter 6 we discussed briefly the three physical states—solid, liquid, and gas—and in Chapter 7 we derived expressions that define the equilibrium condition. In this chapter these subjects are brought together as we apply the equilibrium principle to changes of a substance from one physical state to another. Such transitions are known as **phase changes.**

8.1 Phase Change and Kinetic-Molecular Theory

Figure 8.1 is obtained when the substance is heated at a constant atmospheric pressure. The liquid vaporizes by boiling when the vapor pressure is the same as this atmospheric pressure.

Heating and Cooling Curves

When heat is added at a low and constant rate to a solid, the temperature changes in the manner shown by the **heating curve** in Figure 8.1. The added heat initially causes increased vibration of the particles of the solid about their lattice points, and this increased vibrational kinetic energy is evidenced by a higher temperature. The temperature-time line (x_1 to x_2) approaches linearity over short regions, but in general it has curvature because the heat capacity of a substance is a function of temperature. When the melting temperature of the solid is reached, a break in the heating curve is observed, and the temperature remains constant until all the solid has melted (x_3). Along the portion of the curve from x_2 to x_3 the kinetic energy of the substance does not increase (hence there is no temperature increase), but all heat added increases the potential energy. When the last of the solid melts (x_3), the added heat again increases the kinetic energy of the substance, now liquid, and the temperature increases at an approximately constant rate. At x_4, vaporization begins, and the added heat increases the potential energy of the substance. The temperature remains

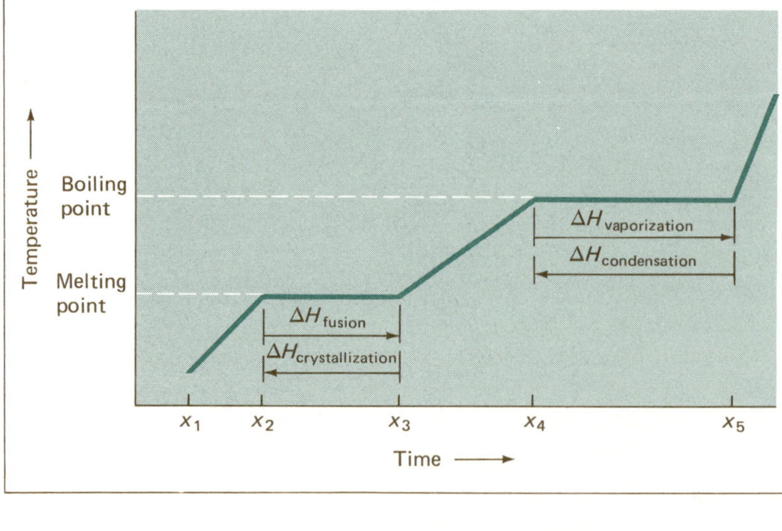

Figure 8.1
Time-temperature warming curve for a pure substance. The temperature halts occur at a phase change. The vaporization halt is usually longer than the fusion because the length of the halt depends upon the magnitude of ΔH, and ΔH_{vap} is usually larger than ΔH_{fus}. The slopes of the heating curves depend upon the heat capacity of the phase that is being warmed.

constant until the vaporization is complete (x_5), whereupon the kinetic energy again begins to increase.

On a heating curve the abrupt changes in slope that occur at x_2, x_3, x_4, and x_5 are known as *breaks*, and the horizontal segments x_2 to x_3 and x_4 to x_5 are *temperature halts*. As indicated in Figure 8.1, the heat added in the region x_2 to x_3 is the heat of fusion **(enthalpy of fusion),** and that added from x_4 to x_5 is the heat of vaporization **(enthalpy of vaporization).**

Similar but reversed plots, known as **cooling curves,** are obtained when heat is removed at a constant rate from a substance initially in the gaseous state. Heating and cooling curves are valuable in the study of phase changes because they yield information concerning the temperature and ΔH values of such changes. We will refer to data from heating and cooling curves occasionally throughout this chapter.

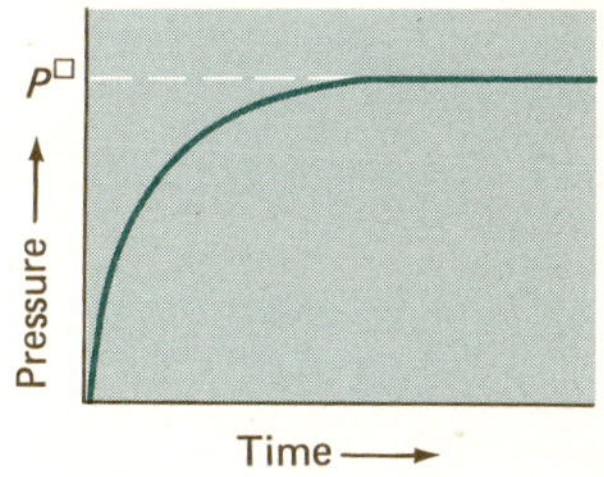

Figure 8.2
Change of pressure with time in the vapor-pressure experiment. $P^{\square}$ is the equilibrium vapor pressure.

Vapor Pressure and Dynamic Equilibrium

Consider an experiment in which a sample of liquid is injected into an evacuated flask, and the change in gas pressure is observed. The sample of liquid must be large enough that it does not all vaporize. As the liquid evaporates, the pressure will at first increase in the bulb in a manner similar to that shown in Figure 8.2. After a short time, however, the pressure levels off at a value represented in the graph as $P^{\square}$. The leveling-off occurs when the rate of vaporization of the liquid is exactly balanced by the rate of condensation of the gas. The pressure $P^{\square}$ is known as the **vapor pressure** of the liquid. Further experimentation will show that $P^{\square}$ is unaffected by the size and surface area of the liquid sample. It is affected rather markedly, however, by the temperature, composition, and nature of the substance.

This equilibrium gas pressure is easily explained from kinetic-molecular theory. First, we recall that the molecules in a liquid are in motion. Usually, as a molecule approaches the surface of the liquid, it changes direction and is drawn back into the liquid by the attractive forces of the other molecules. Occasionally, however, a fast-moving molecule reaches the surface with enough kinetic energy to overcome the attractive forces and escape into the gas phase.

Removing the fast-moving molecules from the liquid or solid removes energy from the system. The temperature will be lowered unless heat energy is added.

As long as the temperature remains constant, molecules evaporate from a liquid at a constant rate. However, the number of molecules in the gas phase, and hence the gas pressure above the liquid, does not increase indefinitely. As molecules in the gas phase strike the surface of the liquid, the attractive forces of the liquid cause some of the gas molecules to rejoin the liquid phase. As a substance evaporates, the gas pressure increases. This increase in pressure, in turn, increases the rate of return of the gas molecules to the liquid, since at the higher pressure, more molecular collisions occur per unit time. Eventually, a pressure is reached at which the rate of condensation just balances the rate of evaporation. This condition of two opposing processes occurring at the same rate is an example of **dynamic equilibrium.** Thus, gas and liquid are in a state of dynamic **phase equilibrium** at a pressure equal to the vapor pressure.

A similar equilibrium can be established between solids and gases, since solids also exhibit vapor pressures. The explanation is similar to that for a liquid. An occasional molecule on the surface vibrates with sufficiently high frequency to overcome the attractive forces in the solid and escape into the gas phase. Equilibrium is established when the rate of condensation from gas to solid equals the rate of sublimation.

Two or more condensed phases can also exist in equilibrium. Solid and liquid, for example, are in equilibrium at the melting point of a substance. If a mixture of solid and liquid at the melting point of a substance is surrounded by a bath maintained at this same temperature, no change will occur in the relative amounts of the two phases. In the mixture, solid is continuously changing to liquid, and liquid is changing to solid. Because the two processes are occurring at the same rate, no net change in the distribution of the substance between the phases will occur.

Two solid phases are in equilibrium at the transition temperature. The situation is analogous to the melting process. As an example, rhombic and monoclinic sulfur (two different crystalline forms of sulfur) are in equilibrium at 95.5° C and one atmosphere pressure. Lowering the temperature converts monoclinic sulfur to rhombic, while raising the temperature converts rhombic to monoclinic.

Pressure-Temperature Relationships in Phase Equilibria

Temperature has a large effect on the vapor pressure of a solid or liquid because it affects the rate of evaporation of the condensed phase. The distribution of kinetic energy in a liquid or solid is much like that for a gas. The graph shown in Figure 8.3 is representative of this distribution at two temperatures. At the higher temperature, more of the molecules have energies in excess of the escape energy; hence, a larger fraction of the molecules can exist in the gas phase. Similar arguments apply to solid-vapor equilibria.

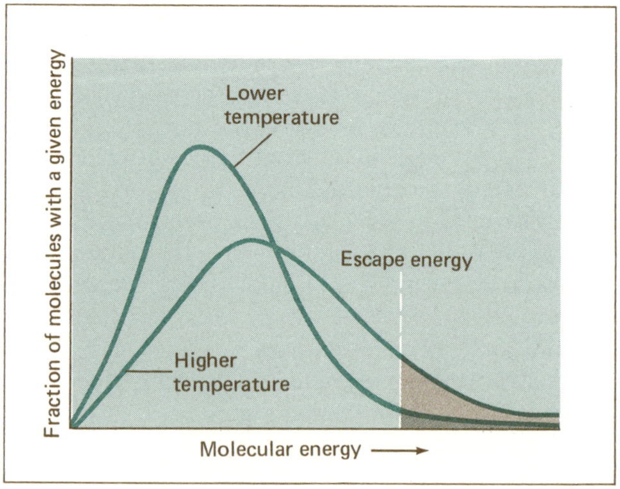

Figure 8.3
Distribution of molecular energy in a liquid. At the higher temperature, a larger fraction of the molecules has an energy greater than the escape energy.

While these kinetic-molecular explanations are helpful, thermodynamics is our best tool for examining systems of phases in equilibrium. Application of thermodynamics to phase equilibria produces important quantitative temperature-pressure relationships.

8.2 Phase Equilibria, Single Component

Consider an equilibrium phase change such as the evaporation of a liquid or solid at a pressure equal to its vapor pressure, the melting of a solid at its melting point, or a solid-state transition at the equilibrium transition temperature. Each is an example of an equilibrium process occurring under the condition of constant temperature and pressure. Let us represent this process by the equation

$$A = B$$

where A and B are any two phases of a substance in equilibrium. In Chapter 7 we saw that the change in the Gibbs free energy for an equilibrium process at constant temperature and pressure is zero. Hence, for the above process

$$\Delta G = 0 = G_B - G_A \quad \text{(constant temperature, constant pressure)} \qquad [8.1]$$

or

$$G_B = G_A \qquad [8.2]$$

In other words, phases in equilibrium have the same free energy per mole.

Solid-Liquid Equilibria

As an application of equation [8.2] consider the process

$$A(\text{solid}) = B(\text{liquid})$$

Suppose we change the pressure on this equilibrium system by an amount ΔP. To maintain equilibrium, the temperature must change by an amount ΔT. Starting with equation [8.2], it can be shown that ΔP and ΔT are related by the equation

$$\boxed{\frac{\Delta P}{\Delta T} = \frac{\Delta S}{\Delta V}} \quad \text{(solid-liquid equilibrium)} \qquad [8.3]$$

where ΔS and ΔV are the change in entropy and volume, respectively, for the melting of solid to liquid.

Equation [8.3] is an approximate version of a more rigorous equation known as the Clapeyron equation. It is most useful since it will quantitatively as well as qualitatively predict the effect of pressure on the melting point of a substance. Since ΔS (melting) will always be positive (for melting leads to greater disorder), the sign of $\Delta P/\Delta T$ will depend on the sign of ΔV (melting). If, as commonly occurs, the volume increases on melting, ΔV will be positive. The result is a positive sign for $\Delta P/\Delta T$; that is, T and P will change in the same direction. Thus, an *increase* in pressure on a solid-liquid system will cause an *increase* in the equilibrium temperature (melting point), and a *decrease* in pressure will cause a *decrease* in the equilibrium temperature.

If, on the other hand, melting is accompanied by a decrease in volume, as it is in the somewhat unusual case of ice, ΔV is negative, $\Delta P/\Delta T$ is negative, and an *increase* in pressure *lowers* the melting point.

EXAMPLE 8.1 Calculate the melting point of ice at 10.00 atm pressure. The specific volume (reciprocal of density) of liquid water is 1.0002 cm^3/g at 0° C and that of ice at the same temperature is 1.0908 cm^3/g.

SOLUTION Equation [8.3] is applicable. Solving for ΔT gives

$$\Delta T = \frac{\Delta V}{\Delta S}\Delta P$$

ΔV can quickly be determined:

$$\begin{aligned}\Delta V &= [(1.0002 - 1.0908)\ \text{cm}^3/\text{g}][(10^{-3}\ \text{liters/cm}^3)(18.0\ \text{g/mole})]\\ &= -1.63 \times 10^{-3}\ \text{liters/mole}\end{aligned}$$

ΔS is obtained from the enthalpy of fusion data in Appendix C. From equation [7.28]

$$\Delta G = \Delta H - T\Delta S$$

At equilibrium, $\Delta G = 0$ and

$$\Delta S = \frac{\Delta H}{T} \qquad [8.4]$$

From Appendix C, $\Delta H_{\text{fus}} = 6008$ joules/mole. Hence,

$$\Delta S = \left(\frac{6008\ \text{joules/mole}}{273.1°\ \text{K}}\right)\left(\frac{0.08206\ \text{liter-atm/°K-mole}}{8.314\ \text{joules/°K-mole}}\right) = \frac{0.2171\ \text{liter-atm}}{\text{°K-mole}}$$

The gas constant expressed in two appropriate sets of units gives a convenient means of converting between joules, calories, and liter-atmospheres. Thus, in Example 8.1, ΔS in joules/°K-mole was converted to liter-atm/°K-mole by multiplying by the ratio of the gas constant.

When $P = 1.00$ atm, $T = 273.150°$ K. When $P = 10.00$ atm,

$$\Delta P = (10.00 - 1.00)\ \text{atm} = 9.00\ \text{atm}$$

Substitution into equation [8.3] gives

$$\Delta T = \frac{(-1.63 \times 10^{-3}\ \text{liter/mole})}{(0.2171\ \text{liter-atm/°K-mole})}(9.00\ \text{atm}) = -0.0675°\ \text{K}$$

The new melting point is

$$T = (273.150 - .068)°\ \text{K} = 273.082°\ \text{K}$$

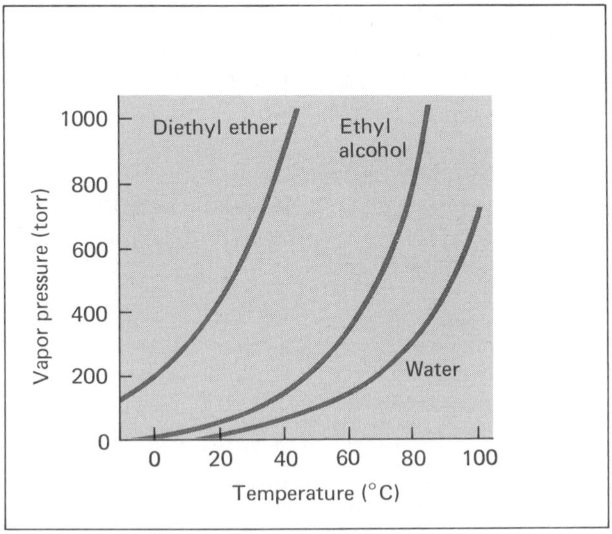

Figure 8.4
Vapor pressure as a function of temperature.

Liquid-Vapor Equilibria

Plots of vapor pressure versus temperature are curved lines similar to those shown in Figure 8.4. Experience shows that a plot of the logarithm of the vapor pressure against the reciprocal of the absolute temperature, however, produces a *straight* line, such as the examples illustrated in Figure 8.5. Thermodynamics not only confirms that the ln P versus $1/T$ plot should be a straight line, but it also identifies the slope and intercept in terms of some of the physical properties of the liquid. The equation derived from thermodynamics and usually referred to as the Clausius-Clapeyron equation is

$$\ln P^{\square} = -\frac{\Delta H}{R}\left(\frac{1}{T}\right) + \frac{\Delta H}{RT_b} \qquad [8.5]$$

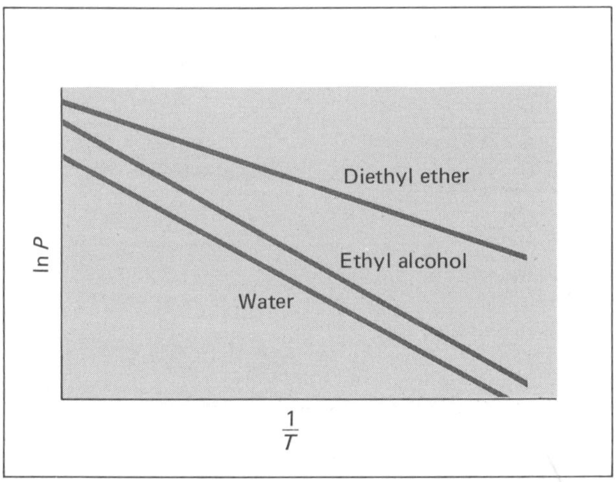

Figure 8.5
The logarithm of the vapor pressure plotted against $1/T$ is a straight line.

Boiling occurs when a substance is heated in an open container until the vapor pressure equals atmospheric pressure. At this point, the pressure of the vapor is large enough to push back the atmosphere and escape from the liquid. The normal boiling point is the temperature at which the vapor pressure is one atmosphere.

where $P^{\square}$ is the vapor pressure of the pure liquid in atmospheres at temperature T, ΔH is the enthalpy of vaporization, and T_b is the *normal* (one atmosphere) boiling point. The equation is seen to be in the straight-line form

$$y = mx + b$$

with $y = \ln P$, $x = 1/T$, and m (the slope) $= -\Delta H/R$.

Having information on the normal boiling point and the enthalpy of vaporization allows one to (1) calculate the vapor pressure of a liquid at any given temperature, or (2) calculate the boiling point at any specified pressure.

EXAMPLE 8.2 Calculate the boiling point of water on top of Mt. Everest on a clear day (barometric pressure = 0.312 atm). From Appendix C, we know that the enthalpy of vaporization of water is 40,660 joules/mole at its boiling point of 373.1° K.

SOLUTION Water will boil when its vapor pressure equals the atmospheric pressure. Our problem thus reduces to a calculation of T at $P = 0.312$ atmospheres. The solution is as follows:

$$\ln P^{\square} = -\frac{\Delta H_{\text{vap}}}{R}\left(\frac{1}{T}\right) + \frac{\Delta H_{\text{vap}}}{RT_b}$$

$$T = \frac{(\Delta H_{\text{vap}})(T_b)}{\Delta H_{\text{vap}} - RT_b \ln P}$$

$$T = \frac{(40{,}660 \text{ joules/mole})(373.1^\circ \text{ K})}{40{,}660 \text{ joules/mole} - (8.314 \text{ joules/}^\circ\text{ K-mole})(373.1^\circ \text{ K})(\ln 0.312)}$$

$$T = 342.7^\circ \text{ K} = 69.6^\circ \text{ C}$$

There are still other applications of the Clausius-Clapeyron equation. For example, one method for determining enthalpies of vaporization is to measure vapor pressures as a function of temperature, plot the data in the form of $\ln P$ versus $1/T$, and calculate ΔH from the slope of the resulting line.

EXAMPLE 8.3 Given the following vapor-pressure data for liquid N_2, calculate (a) the normal boiling point, and (b) the enthalpy of vaporization of liquid N_2.

Vapor Pressure of N_2

TEMPERATURE (°K)	VAPOR PRESSURE (torr)
63.14	94.01
67.37	193.9
70.36	305.7
73.42	466.7
76.53	691.3
78.01	823.2

SOLUTION Figure 8.6 is a graph of ln $P^{\square}$ versus $1/T$ for N_2 using the above data. The normal boiling point is obtained when $\ln P^{\square} = \ln 760 = 6.633$. This occurs when $1/T = 0.01294$. T, therefore, is 77.3° K.

From equation [8.5] we know that a graph of ln P versus $1/T$ should have a slope of $-\Delta H/R$. The slope from the graph is $-718.5°$ K; hence,

$$\Delta H = -(-718.5° \text{ K})(8.314 \text{ joules/°K-mole}) = 5974 \text{ joules/mole}$$

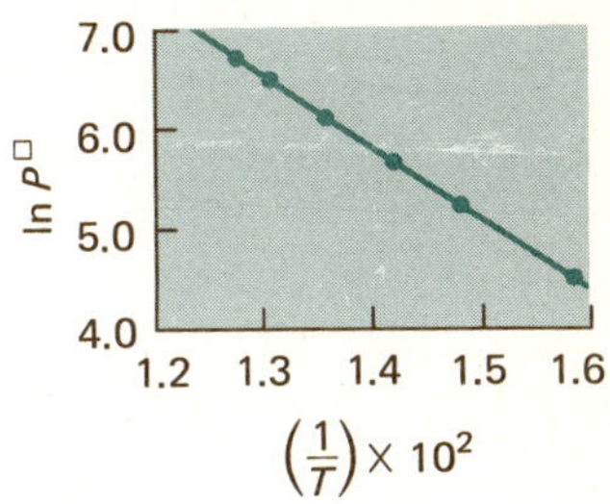

Figure 8.6
Graph of ln $P^{\square}$ versus $1/T$ for N_2. $P^{\square}$ is the vapor pressure. The slope of the straight line is $-\Delta H_{vap}/R$.

Equation [8.5] can also be applied to equilibrium between solid and gas phases. In such systems, ΔH is the molar enthalpy for sublimation, and T_b is the normal sublimation temperature (the temperature when the vapor pressure of the solid reaches one atmosphere).

Phase Diagrams, Single Component

Temperature-Pressure Diagrams

The pressure-temperature relationships between the different phases of a substance are summarized in a **phase diagram.** Figure 8.7, for example, is the

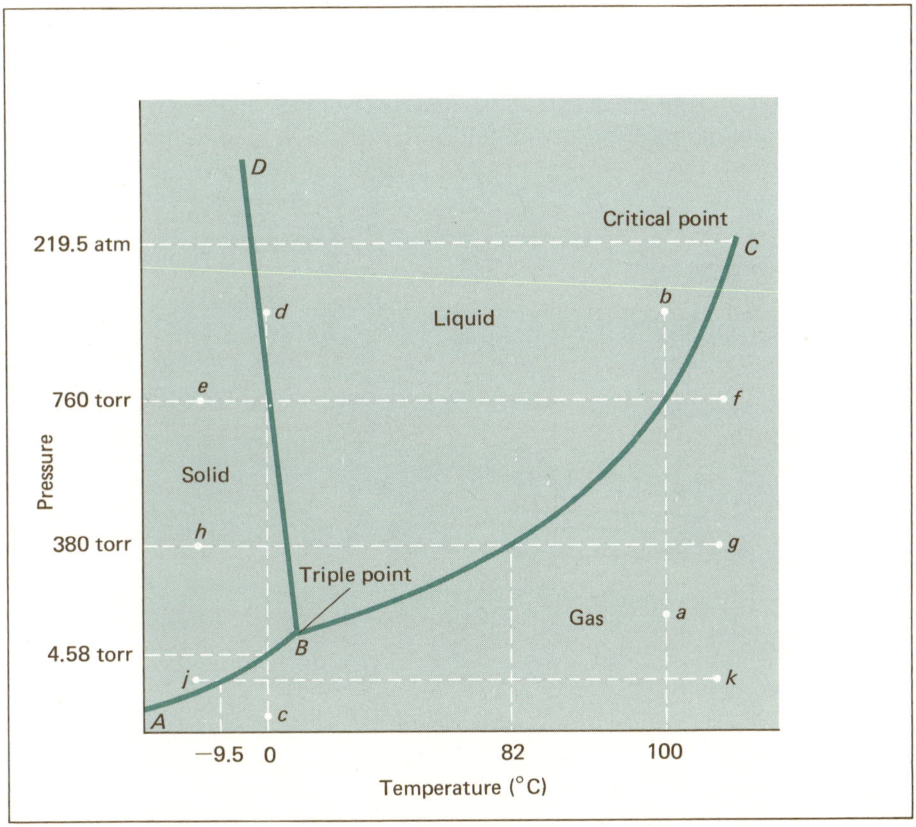

Figure 8.7
Phase diagram for water. Line *BD* is the melting-point line while *AB* and *BC* are solid and liquid vapor-pressure lines, respectively. The position of *B* and the slope of *BD* have been exaggerated in order to show the important features of the diagram.

phase diagram for water, in which *BC* is the vapor-pressure line for liquid water. Only along this line can liquid and gaseous water exist together in equilibrium. Point *C* represents the critical point for water, meaning that liquid water cannot exist above this temperature. Similarly, *AB* is the vapor-pressure line for ice. Only along this line can solid and vapor exist together in equilibrium. The temperature at point *B*, the intersection of the solid and liquid vapor-pressure lines, is the melting temperature of ice at the pressure equal to its vapor pressure. This intersection, called the **triple point** of water, gives the temperature and pressure at which pure solid, liquid, and gas can be in equilibrium.

Line *BD* shows how the melting point of ice changes with pressure. Only along this line can solid and liquid exist together in equilibrium. From the phase diagram it is evident that the melting point of ice decreases as the pressure is increased, which is in agreement with equation [8.3]. Thus, the normal melting point (0° C at 1 atm) is lower than the triple-point temperature (0.010° C at 4.579 torr).

Notice that at the triple point, solid and liquid water have the same vapor pressure. Figure 8.8 shows a representative vapor-pressure curve for a substance in the region of the triple point. The dotted lines represent the vapor pressure of a superheated solid (line *BE*) above the melting point or a supercooled liquid (line *DB*) below the melting point. Above the melting point, the liquid is stable. Below the melting point, the reverse is true.

The different regions of the phase diagram (refer again to Figure 8.7) are labeled to designate the pressure-temperature conditions under which a given phase is stable. To the left of and above line *ABD*, only solid ice can exist; above and to the right of line *DBC*, only liquid water can exist; and to the right of and below line *ABC*, only water vapor can be present.

Figure 8.8
Vapor pressures of solid and liquid at temperatures near the melting point. T_0 is the triple-point temperature. Below T_0, the solid has the lower vapor pressure and is the stable form. Above T_0, the reverse is true.

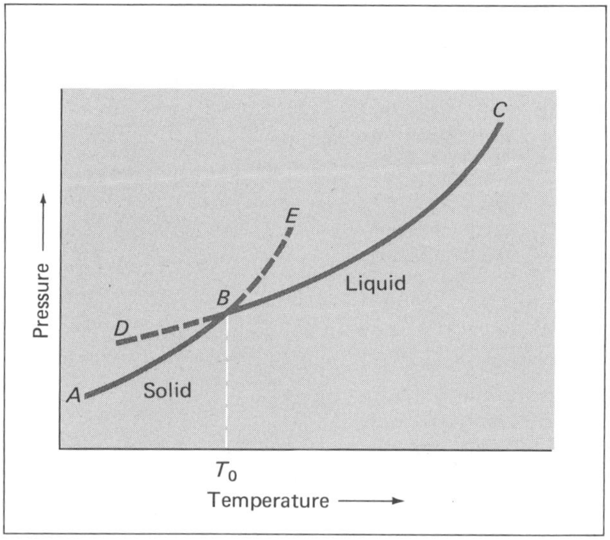

EXAMPLE 8.4 Describe the phase changes that will occur in water during the following processes:

(a) Isothermal (100° C) change of pressure from 10 torr to 1000 torr.
(b) Isothermal (0° C) change in pressure from 1 torr to 1000 torr.
(c) Isobaric (380 torr) temperature change from 110° C to −10° C.
(d) Isobaric (2 torr) temperature change from −10° C to 110° C.

SOLUTION

(a) We are changing the system along the line *ab* in Figure 8.7. Under the initial conditions (point *a*), we see that water is a gas. An isothermal compression of the gas decreases its volume until a pressure of 760 torr is reached, at which pressure the gas liquefies. A further increase in pressure to 1000 torr compresses the liquid.
(b) Once again, the water is initially present as a gas. We are changing the conditions along the dotted line marked *cd*. When the pressure is increased to 4.57 torr, the gas condenses directly to the solid. The pressure is increased until one atmosphere is reached, at which pressure the solid melts to form liquid. A further increase in pressure compresses the liquid.
(c) The water, initially present as a gas, follows line *gh*. Liquid water forms at 82° C; further cooling produces ice at 0.005° C.
(d) The system follows the dotted line *jk*. Initially the water is present as ice. At −9.5° C, the ice sublimes. Further heating increases the temperature of the gas.

LeChatelier's Principle

It is convenient to be able to make a qualitative prediction of the effect of a change in the conditions of a system at equilibrium. Such qualitative predictions can be made from **LeChatelier's principle:**

When a stress is applied to a system at equilibrium, the system shifts, if possible, to reduce the stress.

Consider an isothermal experiment in which ice at 0° C is in equilibrium with water. As long as the pressure on the system is 1 atm, no net change occurs. If, however, the pressure is increased, the ice melts. In terms of LeChatelier's principle, this is the predicted effect. An increase in pressure was the stress. This stress can be partially relieved or reduced by a decrease in volume, and such a decrease occurs when ice melts to form the more dense liquid water.

Ice is rather unusual in that the solid has a larger molar volume than the liquid. For most substances, the reverse is true, and we would predict, from LeChatelier's principle, that an increase in pressure would cause liquid to crystallize.

In an isobaric experiment with a water-ice mixture, the qualitative prediction is also easily made. Consider the addition of heat to such a mixture initially

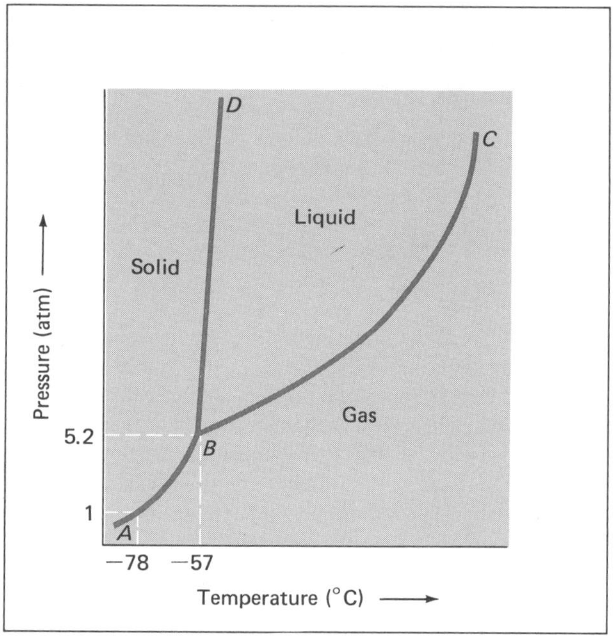

Figure 8.9
The phase diagram for carbon dioxide. It has been necessary to exaggerate the slope of line *BD* in order to show the important features of the diagram.

Many phase diagrams are not as simple as those discussed above. Even the water diagram, when extended to high pressures, has a number of different solid phase transitions. Seven different solid crystalline phases have been identified for water. A more complete description of solid-liquid phase equilibria for water is left for a more advanced course.

at equilibrium. The stress, heat, is relieved by the melting of ice. LeChatelier's principle can also be applied to other phase-equilibrium systems. In Chapter 10 we will use it to make qualitative predictions concerning systems in chemical equilibrium.

Phase diagrams have been determined for a large number of substances. Figure 8.9 is the phase diagram for CO_2. In CO_2, the solid is more dense than the liquid. The positive slope of the solid-liquid equilibrium line follows from the Clapeyron equation (or LeChatelier's principle). The triple point of CO_2 occurs at 216.7° K and 5.11 atm. If solid CO_2 is heated at a constant pressure above 5.11 atm, it will melt at a temperature given by line *BD*. If, however, the pressure is less than 5.11 atm, solid CO_2 sublimes rather than melts. At 1 atm pressure, the temperature at which sublimation occurs is −78° C.

8.3 Phase Equilibria Involving Solutions

To this point we have limited the discussion to phase equilibria in a single substance. But now we are ready to expand the discussion to include equilibria between a two-component liquid solution and its vapor and between a liquid solution and a solid component. In this section we will be interested in composition as well as temperature and pressure effects.

Vapor Pressure of Solutions

The vapor pressures of solutions vary with composition as well as with temperature. Under ideal conditions the relation of the vapor pressures of solute and solvent to the concentration (at constant temperature) can be expressed mathematically by two simple equations, Raoult's law and Henry's law, which are discussed below.

Raoult's Law

Figure 8.10a shows how the vapor pressure changes with concentration (mole fraction) in solutions of CCl_4 and $SiCl_4$. P_1 is the vapor-pressure curve for CCl_4; it goes from zero, where the mole fraction of CCl_4 is zero, to $P_1^{\square}$, the vapor pressure of pure CCl_4. P_2 is a similar curve for $SiCl_4$. Curve $P_1 + P_2$ represents the total vapor pressure above the solution; it is the sum of the vapor pressures of the two components in the solution.

Good approximations for the lines in Figure 8.10a are the linear equations

$$\boxed{P_1 = x_1 P_1^{\square}} \qquad [8.6]$$

$$\boxed{P_2 = x_2 P_2^{\square}} \qquad [8.7]$$

where $P_1^{\square}$ and $P_2^{\square}$ are the vapor pressures of pure components, and the x_1 and x_2 are mole fractions.

Equations [8.6] and [8.7] are statements of **Raoult's law.** Solutions that obey these equations over the entire composition range are called **ideal solutions.** The CCl_4-$SiCl_4$ system is nearly ideal.

Real solutions may have vapor pressures either higher (Figure 8.10b), which is very common, or lower (Figure 8.10c) than predicted by Raoult's law. In the region where a substance is almost pure, however, its vapor pressure approaches closely the behavior predicted by Raoult's law, and under these conditions (mole fraction approaching unity), Raoult's law can be used as a satisfactory approximation. Stated in another way, the vapor pressure of the solvent in a *dilute* solution can be approximated by Raoult's law, and the more dilute the solution, the better the approximation.

Henry's Law

Figure 8.11 is an enlargement of the vapor-pressure curves of acetone in chloroform (curve A) and acetone in carbon disulfide (curve B) in the region of low acetone concentration. (Refer back to Figures 8.10b and 8.10c for the entire curve.) The straight line C is drawn from Raoult's law. It is evident that even in very dilute solutions, the vapor pressure of the solute does not approach that predicted by Raoult's law. However, the vapor-pressure curves, at very low concentrations, do approach straight lines. The dotted lines D and E are straight lines that have the same slopes as the vapor-pressure curves in the limiting case of infinite dilution. In the very dilute region, the vapor pressure of the solute can be approximated well by the equation of the dotted line, which is

$$\boxed{P_2 = k_x x_2} \qquad [8.8]$$

Figure 8.10a
Vapor pressure in CCl_4-$SiCl_4$ solutions, an example of a nearly ideal system.

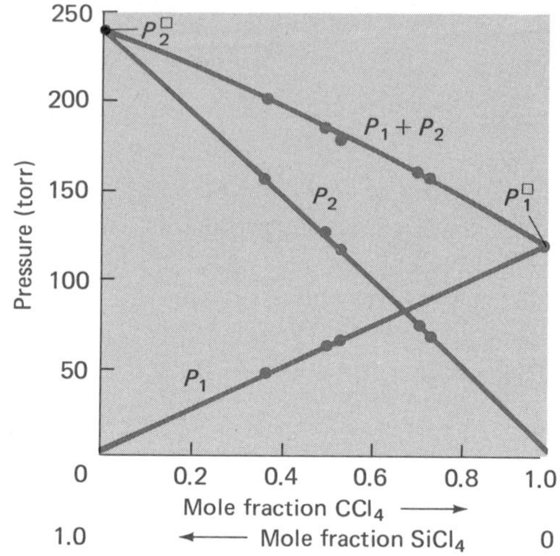

Figure 8.10b
Vapor pressure in acetone-carbon disulfide solutions, an example of positive deviations from Raoult's law. Although the system is far from being ideal, the vapor pressure of the solvent approaches the Raoult's law value in dilute solutions.

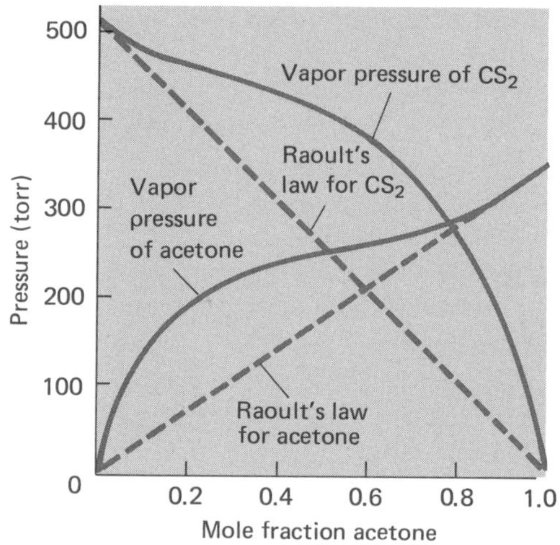

Figure 8.10c
Vapor pressure in acetone-chloroform solutions, an example of negative deviations from Raoult's law. In dilute solutions, the vapor pressure of the solvent approaches the Raoult's law value.

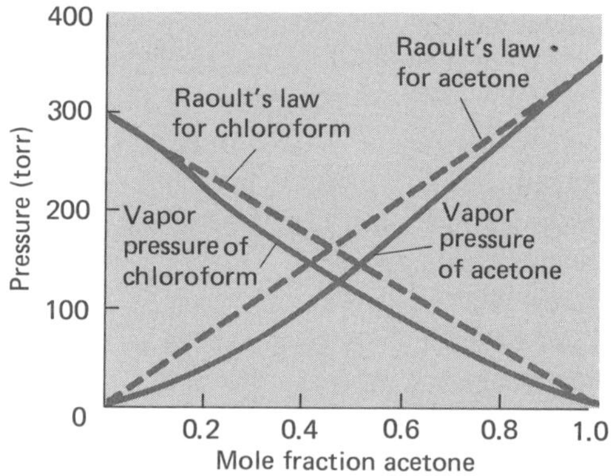

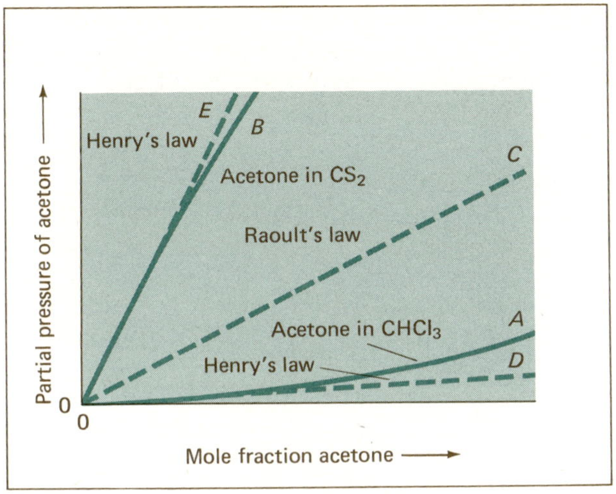

Figure 8.11
Representation of Henry's law and Raoult's law in dilute solutions of acetone in carbon disulfide and in chloroform. The vapor pressure of the solutes deviates widely from Raoult's law, but approximates closely Henry's law in dilute solutions.

where P_2 is the vapor pressure of the solute above the dilute solution, x_2 is its mole fraction, and k_x is the slope of the straight line. This equation can also be expressed with the concentration in molality m or molarity $[s]$ instead of mole fraction. Hence we can write

In discussing solutions, we use the subscript 1 to denote the solvent and the subscripts 2, 3, . . . to denote the solutes.

$$P_2 = k_m m \qquad [8.9]$$

and

$$P_2 = k_c[s] \qquad [8.10]$$

The relationship among the Henry's law constants in dilute solutions is

$$k_c = \frac{k_m}{d_0} = \frac{k_x M_1}{1000\, d_0} \qquad [8.11]$$

where M_1 is the molecular weight of the solvent, and d_0 is the density of the pure solvent.

Henry's law is stated in these words: *In a dilute solution, the vapor pressure of the solute is proportional to its concentration (mole fraction, molality, or molarity).* Table 8.1 gives the values of the Henry's law constants for a number of substances in water solution.

TABLE 8.1

Henry's Law Constants for Solutes in Water at 25° C

	k_x(torr/mole fraction)	k_m[torr/(moles/kg)] or k_c[torr/(moles/liter)]†
Nitrogen	6.3×10^7	1.1×10^6
Oxygen	3.4×10^7	6.1×10^5
Hydrogen	5.4×10^7	9.7×10^5
Carbon dioxide	1.2×10^6	2.2×10^4
Hydrogen sulfide	4.1×10^6	7.4×10^4
Bromine	5.2×10^4	9.4×10^2

† For water solution, d_0 in equation [8.11] is very nearly unity. Hence, k_c and k_m have the same value within the accuracy given in the table.

Henry's law, like Raoult's law, is a limiting law; that is, it is exact only in the limit of infinite dilution. It is, however, a good approximation in dilute solutions.

In summary, Raoult's law is not obeyed by most solutions over the entire range of concentration. However, in dilute solutions, the behavior of the solvent approaches Raoult's law, and the behavior of the solute approaches Henry's law. The more dilute the solution, the better are these approximations.

EXAMPLE 8.5 Estimate the total vapor pressure at 30° C above a solution of ethyl alcohol (C_2H_5OH) and *n*-propyl alcohol (C_3H_7OH) in which the mole fraction of ethyl alcohol is 0.750. The vapor pressures at 30° C of pure ethyl alcohol (1) and *n*-propyl alcohol (2) are 79.1 and 27.6 torr, respectively.

SOLUTION Assuming that the alcohols form an ideal solution, we have

$$P_1 = x_1 P_1^{\square} = (0.750)(79.1) = 59.3 \text{ torr}$$
$$P_2 = x_2 P_2^{\square} = (0.250)(27.6) = 6.90 \text{ torr}$$
$$P_T = P_1 + P_2 = 59.3 + 6.9 = 66.2 \text{ torr}$$

EXAMPLE 8.6 Calculate the solubility (g solute/kg solvent) of carbon dioxide in water that is in contact with CO_2 gas at 1 atm pressure.

SOLUTION The solubility of gases in liquid solvents is usually small unless chemical reaction occurs. It is a good approximation to assume that CO_2 obeys Henry's law in the solution:

Henry's law governs the "fizz" in soda pop. In the sealed bottle, the pressure of CO_2 is greater than the partial pressure of CO_2 in the atmosphere; hence, the solubility of CO_2 in the sealed bottle is greater than it is when open to the atmospheric pressure. When the bottle is opened, the pressure of CO_2 above the liquid decreases, and the liquid is supersaturated. The excess dissolved CO_2 then escapes as bubbles.

$$P_2 = k_m m$$

$P_2 = 1$ atm $= 760$ torr, and, from Table 8.1, $k_m = 2.2 \times 10^4$ torr/(mole/kg). Hence,

$$m = \frac{P_2}{k_m} = \frac{760 \text{ torr}}{2.2 \times 10^4 \text{ torr/(mole/kg)}} = 0.035 \text{ molal}$$

$$\text{Solubility} = \left(\frac{0.035 \text{ moles } CO_2}{\text{kg } H_2O}\right)\left(\frac{44 \text{ g } CO_2}{\text{mole } CO_2}\right) = \frac{1.5 \text{ g } CO_2}{\text{kg } H_2O}$$

This calculated value is in good agreement with the experimentally determined value.

EXAMPLE 8.7 Calculate the total vapor pressure at 100° C above a solution made by mixing 0.1000 mole of sucrose with 800.0 g of water. Assume that the vapor pressure of sugar is negligible.

SOLUTION The solution is sufficiently dilute that the vapor pressure of the water can be approximated by Raoult's law. It is immaterial for this calculation whether the sucrose obeys Henry's law, Raoult's law, or neither, since sucrose has such a small vapor pressure above the solution that its contribution to the total pressure is negligible. Only the vapor pressure of the water needs to be considered.

For H_2O at 100° C ($P_1^{\square} = 760.0$ torr), we have

$$x_1 = \frac{(800.0 \text{ g } H_2O)/(18.01 \text{ g } H_2O/\text{mole } H_2O)}{(800.0 \text{ g } H_2O)/(18.01 \text{ g } H_2O/\text{mole } H_2O) + 0.100 \text{ mole sucrose}}$$

$$x_1 = 0.9978$$

$$P_1 = x_1 P_1^{\square} = 0.9978(760.0) = 758.3 \text{ torr}$$

The vapor pressure of the solution in the above example is less than that of pure water. The vapor pressure of a solution will always be lower than the vapor pressure of the pure solvent if the solute is nonvolatile and the solvent follows Raoult's law.

Boiling-Point Elevation of Solutions Containing a Nonvolatile Solute

In Example 8.7, it was shown that the addition of a nonvolatile solute to a solvent lowers the vapor pressure, which has the result of raising the boiling point of the solution. This consequence is diagramed in Figure 8.12, which shows the solid-liquid-gas phase diagram for a typical solvent. The solid curve BC represents the vapor pressure as a function of temperature for a pure solvent. Adding a nonvolatile solute lowers the vapor pressure of the solvent (which is the total vapor pressure above the solution). The vapor pressure of the solution is represented by the broken line, curve EF. It is evident from the graph that the solution must be heated to a higher temperature (T_s) than the pure solvent (T_b) before the vapor pressure reaches atmospheric pressure, where it will boil.

You should be aware that adding a *volatile* solute does not necessarily lower the vapor pressure of a solution and raise its boiling point. For example, since ethyl alcohol is more volatile than water, adding ethyl alcohol to water raises the vapor pressure of the solution and lowers its boiling point.

For a dilute solution of a nonvolatile solute in which the solvent obeys Raoult's law, thermodynamics can be used to derive an equation relating the

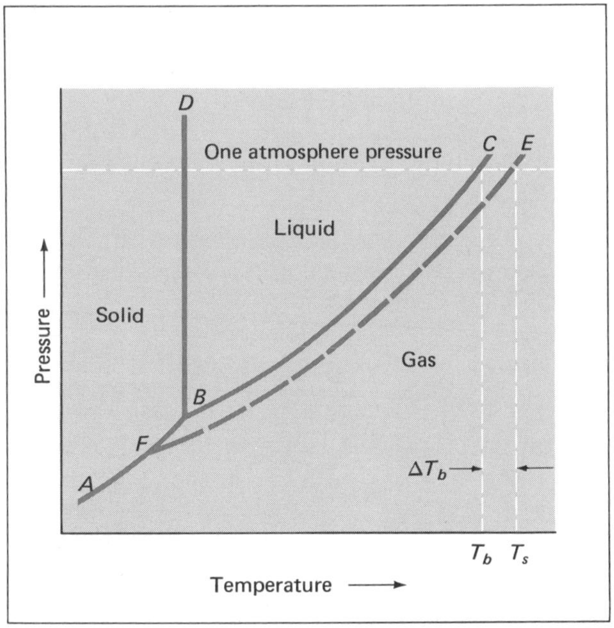

Figure 8.12
Vapor pressure of solvent and solution. Adding a solute decreases the vapor pressure and increases the boiling point.

TABLE 8.2

Boiling-Point Elevation Constants

SOLVENT	K_b[°C/(moles/kg)]
Acetone	1.71
Benzene	2.53
Carbon tetrachloride	5.03
Chloroform	3.63
Diethyl ether	2.06
Ethyl alcohol	1.20
Water	0.512

boiling-point rise (ΔT_b in Figure 8.12) to the molal concentration of solute *m*. The resulting equation is

$$\Delta T_b = K_b m \tag{8.12}$$

The constant K_b, known as the molal boiling-point elevation constant, is related to properties of the solvent through equation [8.13]:

$$K_b = \frac{RT_b^2 M_1}{1000\,\Delta H_{vap}} \tag{8.13}$$

where R is the gas constant, T_b is the normal boiling point[†] for the solvent, M_1 is the molecular weight of the solvent, and ΔH_{vap} is the enthalpy (or heat) of vaporization, again for the solvent. The constant K_b can be either calculated from equation [8.13] or measured experimentally. Table 8.2 gives values for K_b for a number of solvents, and Appendix C gives values for ΔH_{vap} and T_b, from which K_b values can be calculated for additional substances.

Equation [8.12] has several useful applications. The assumption of dilute solution and nonvolatile solute made in the derivation of this equation, however, should be kept in mind.

EXAMPLE 8.8 Calculate the normal boiling point of a 0.25 *m* pyrene solution in benzene. Assume that pyrene is a nonvolatile solute.

SOLUTION The normal boiling point of benzene is 80.10° C, and from Table 8.2 we find that $K_b = 2.53$ for benzene. Therefore, using equation [8.12] we have

$$\Delta T_b = K_b m = [2.53^\circ\text{ C/(moles/kg)}](0.25\text{ moles/kg}) = 0.63^\circ\text{ C}$$
$$T_s = 80.10^\circ\text{ C} + 0.63^\circ\text{ C} = 80.73^\circ\text{ C}$$

† The normal boiling point is the boiling temperature at one atmosphere total pressure.

Notice that a 0.25 m solution of any nonvolatile solute in benzene would boil at 80.73° C. In other words, the boiling point of a *dilute* solution of a nonvolatile solute is independent of the nature of the solute.

EXAMPLE 8.9 The normal boiling point of a solution of carbon tetraiodide in carbon tetrachloride is 77.95° C. Calculate the mole fraction CI_4 in the solution. The normal boiling point of pure CCl_4 is 76.54° C. The vapor pressure of CI_4 is negligible.

SOLUTION From Table 8.2 we find that K_b for CCl_4 is 5.03° C/(moles/kg). Since K_b is expressed in terms of molality, it is easiest to calculate the molality of the solution first, then convert the concentration to a mole fraction:

$$m = \Delta T_b/K_b$$
$$\Delta T_b = T_s - T_0 = 77.95^\circ\text{ C} - 76.54^\circ\text{ C} = 1.41^\circ\text{ C}$$
$$m = 1.41^\circ\text{ C}/[5.03^\circ\text{ C}/(\text{mole/kg})] = 0.280\text{ molal}$$

The mole fraction is related to molality through the equation

$$x_2 = \frac{m}{m + 1000/M_1}$$

which in dilute solution becomes (since m is negligible compared with $1000/M_1$)

$$x_2 = \frac{M_1 m}{1000} = \frac{(153.84\text{ g/mole})(0.280\text{ moles/kg})}{1000\text{ g/kg}} = 0.0431$$

Freezing-Point Depression of Solutions

We have seen in the preceding section that increasing the concentration of a solute in a solution lowers the vapor pressure of the solvent, which in turn raises the boiling point of the solution. We now show that the lowering of the vapor pressure of a solvent also affects the *freezing point* of a solution. This may be seen by examining Figure 8.13, which shows a graph of the vapor pressure of a solvent near the freezing point. The lines CE and ABC give the vapor pressure of a pure liquid and a pure solid solvent, respectively. The intersection of these two lines at point C is the freezing point of the solvent, since at the freezing point the solid and liquid must have the same vapor pressure. Line BH represents the vapor pressure of the liquid solvent after a solute has been added. If a solid crystallizes from a solution as a pure substance, the vapor pressure of this solid is given by line ABC. The intersection of the vapor-pressure curve for the solution (line BH) with line ABC at point B gives the freezing point of the solution, T_m. At this point, the vapor pressures of the solvent in both the solid and liquid are the same. As can be seen from the graph, T_m is less than T_0, and the addition of a solute lowers the freezing point of the solvent by an amount $\Delta T_f = T_m - T_0$.

Since the solid and liquid are in equilibrium with each other, they must be in equilibrium with the same vapor phase; that is, their vapor pressures must be equal.

Figure 8.13
Vapor pressures near the freezing point. Adding a solute lowers the freezing point of the solvent.

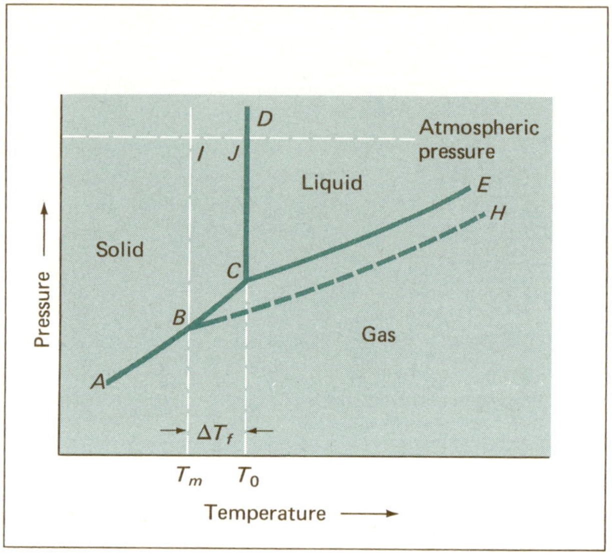

The argument we have just used gives the freezing-point lowering at an external pressure equal to the vapor pressure. The melting points of pure solvent and of the solution at atmospheric pressure are points J and I, respectively. To a very good approximation, lines IB and JC are parallel, and the freezing-point *lowering* is the same at atmospheric pressure as at the vapor pressure.

The change in the freezing point of a solvent depends upon the concentration of the solute. With the aid of Raoult's law, the dependence of the freezing point upon concentration can be calculated. We assume that (1) the solid that freezes out is pure solvent, and (2) the concentration of solute is sufficiently low that the vapor pressure of the solvent can be approximated by Raoult's law. The result is

$$\Delta T_f = K_f m \qquad [8.14]$$

in which

$$K_f = -\frac{RT_0^2 M_1}{1000\,\Delta H_{\text{fus}}} \qquad \text{(dilute solutions; the solid in equilibrium with the solution is pure solvent)} \qquad [8.15]$$

In equation [8.15], T_0 is the freezing point of pure solvent; M_1 and ΔH_{fus} are the molecular weight and enthalpy of fusion of the solvent. K_f in equation [8.14] is called the molal freezing-point lowering constant. Equation [8.14] states that in a dilute solution, in which the approximations used in its derivation are valid and where no solid solutions form, the freezing-point lowering of the solution is proportional to its molal concentration.

Table 8.3 gives values of K_f for a number of substances. Some applications of these data are illustrated in the examples that follow.

TABLE 8.3

Freezing-Point Lowering Constants and Freezing Points of Solvents

SOLVENT	K_f[°C/(moles/kg)]	T_0(°C)
Acetic acid	−3.73	16.63
Benzene	−5.13	5.533
Camphor	−37.7	178.4
Cyclohexane	−20.5	6.554
Naphthalene	−6.9	80.21
Nitrobenzene	−7.00	5.7
Water	−1.86	0.00

EXAMPLE 8.10 The freezing-point lowering of a dilute solution of a nonvolatile solute in benzene is −1.00° C. Calculate the boiling point of this solution.

SOLUTION From equations [8.12] and [8.14],

$$\Delta T_b = K_b m$$

$$m = \frac{\Delta T_f}{K_f}$$

Combining these equations gives

$$\Delta T_b = \frac{K_b}{K_f}\Delta T_f$$

For benzene, we see from Tables 8.2 and 8.3 that $K_b = 2.53°$ C/(moles/kg), $K_f = -5.13°$ C/(moles/kg). Since $\Delta T_f = -1.00°$ C,

$$\Delta T_b = \frac{2.53}{-5.13}(-1.00° \text{ C}) = 0.493° \text{ C}$$

The normal boiling point of pure benzene is 80.100° C. The boiling point of the solution, then, is 80.100 + 0.493 = 80.593° C.

EXAMPLE 8.11 0.5160 g of a solute is dissolved in 50.0 g of benzene. The freezing point of the resulting solution is 5.139° C. Calculate the molecular weight of the solute.

SOLUTION From Table 8.3, we know that the freezing point of pure benzene is 5.533° C; therefore

$$\Delta T_f = 5.139 - 5.533 = -0.394° \text{ C}$$

For benzene, $K_f = -5.13°$ C/(moles/kg). Hence,

$$m = -0.394° \text{ C}/[-5.13° \text{ C/(moles/kg)}]$$

But, by definition,

$$m = \left(\frac{0.5160 \text{ g solute}/(M_2 \text{ g solute/mole solute})}{50.0 \text{ g solvent}}\right)\left(\frac{10^3 \text{ g solvent}}{\text{kg solvent}}\right)$$

where M_2 is the molecular weight of the solute. Equating the two expressions for m and solving for M_2 gives

$$M_2 = \frac{0.5160(10^3)}{(50.0)(-0.394)/(-5.13)} = 134 \text{ g/mole}$$

This is an example of a very common method for estimating the molecular weight of a solute. Camphor is often used as the solvent because it has a very large K_f value [$-37.7°$ C/(moles/kg)]. Even a very small amount of solute dissolved in camphor (the very dilute solution should closely approximate Raoult's law) gives a freezing-point lowering large enough to be easily measured.

Vapor-pressure lowering, freezing-point depression, and boiling-point elevation of dilute solutions depend only on the concentration and not on the nature of the solute. The term *colligative* (depending on the number of particles, not their nature) is therefore applied to these properties. Equations [8.12] and [8.14] express this dependence on concentration. Note that no term involving the nature of the solute appears in the equations. Keep in mind, however, that these equations were derived for *dilute* solutions. At high concentrations, the nature of the solute does affect the vapor pressure and the freezing and boiling points of solutions.

EXAMPLE 8.12 Calculate the freezing point of a water solution that is 0.10 m in ethyl alcohol (C_2H_5OH) and 0.10 m in methyl alcohol (CH_3OH).

SOLUTION The freezing-point lowering depends upon the total molality:

$$m = m_{CH_3OH} + m_{C_2H_5OH} = 0.20 \frac{\text{mole}}{\text{kg}}$$

$$\Delta T_f = K_f m = \left(\frac{-1.86° \text{ K}}{\text{mole/kg}}\right)\left(0.20 \frac{\text{mole}}{\text{kg}}\right)$$

$$\Delta T_f = -0.37° \text{ C}$$

Accounting for the total concentration of dissolved species in colligative property calculations is important in electrolyte solutions, which we now consider.

Freezing-Point Lowering and Boiling-Point Elevation Calculations in Electrolyte Solutions

Electrolytes are substances that dissociate into ions in solution. The name electrolyte comes from the electrical conductivity of such solutions. When the colligative properties of these solutions are calculated, the total concentration of dissolved species must be used in the equations. For example, 0.1 m Na_2SO_4 is 0.2 m in Na^+ (two Na^+ in a formula unit of Na_2SO_4) and 0.1 m in SO_4^{2-}, or 0.3 m in total solute particles.

EXAMPLE 8.13 Calculate the freezing-point lowering of a 0.10 molal solution of NaCl in water. Assume that the solvent obeys Raoult's law.

SOLUTION NaCl is dissociated completely into Na^+ and Cl^- ions. Each ion lowers the freezing point. Since the solvent obeys Raoult's law and the solution is dilute,

$$\Delta T_f = -1.86(m_{Na^+} + m_{Cl^-})$$

where m_{Na^+} and m_{Cl^-} are the molality of the Na^+ and Cl^- ions, respectively. Since the electrolyte consists only of Na^+ and Cl^- ions,

$$m_{Na^+} = m_{Cl^-} = 0.10$$
$$\Delta T_f = -1.86(0.10 + 0.10)$$
$$\Delta T_f = -0.37^\circ\,C$$

In general, ionic solutions must be more dilute than nonionic solutions before a reasonable approximation of Raoult's law behavior is obtained.

A nonelectrolyte would have lowered the freezing point by only 0.19° C. An electrolyte, composed of two ions (NaCl, for example), lowers the freezing point by twice that expected for a nonelectrolyte. As another example, $CaCl_2$ exists as three ions in solution, and the boiling-point elevation is three times as large as would be predicted for a nonelectrolyte. For dilute solutions in general, an electrolyte that exists as n particles in solution increases by a factor of n the freezing-point lowering or boiling-point elevation predicted for a nonelectrolyte.

Some electrolytes do not completely dissociate in solution. The freezing point and boiling point can then be used to estimate the percentage dissociation in solutions of these substances.

EXAMPLE 8.14 A 0.100 molal solution of $NaHSO_4$ lowers the freezing point of water by 0.426° C. Assume that $NaHSO_4$ dissociates completely into Na^+ and HSO_4^-:

$$NaHSO_4(aq) = Na^+ + HSO_4^- \qquad \text{(completely)}$$

and that the weak acid HSO_4^- dissociates partially to H^+ and SO_4^{2-}:

$$HSO_4^- = H^+ + SO_4^{2-} \qquad \text{(partially)}$$

Calculate from the observed freezing-point lowering, the molality of H^+ in the solution.

SOLUTION The total molality of the solution is calculated from equation [8.14]:

$$m = \frac{\Delta T_f}{K_f} = \frac{-0.426^\circ\,C}{-1.86^\circ\,C/(\text{moles/kg})} = 0.229\text{ molal}$$

But

$$m = m_{Na^+} + m_{HSO_4^-} + m_{H^+} + m_{SO_4^{2-}}$$

$$m_{Na^+} = 0.100$$

$$m_{H^+} = m_{SO_4^{2-}}$$

$$m_{HSO_4^-} = 0.100 - m_{H^+}$$

Combining gives

$$0.229 = 0.100 + (0.100 - m_{H^+}) + m_{H^+} + m_{H^+}$$

Solving for m_{H^+} gives

$$m_{H^+} = 0.229 - 0.200 = 0.029 \text{ molal}$$

Summary

Vapor pressure is the equilibrium pressure of a gas above a liquid or a solid. In addition to liquid-vapor and solid-vapor equilibria, liquid-solid and solid-solid phase equilibria also can occur. The phase diagram is a graph describing the equilibrium conditions under which two or more phases can coexist. For a pure substance, the diagram is usually a graph of temperature versus pressure. Solid-vapor and liquid-vapor equilibria are represented as lines that give the vapor pressure as it changes with temperature. Solid-liquid and solid-solid equilibria are represented by the change of melting point or transition temperature, respectively, with pressure.

Approximate equations, which apply when pressure and temperature changes are not extreme, can be used to calculate the effect of a change in pressure or temperature on the equilibrium condition. The Clapeyron equation applies to solid-liquid and solid-solid equilibria:

$$\boxed{\frac{\Delta P}{\Delta T} = \frac{\Delta S}{\Delta V}}$$

where ΔT is the change in temperature that must be made to maintain equilibrium when the pressure is changed by ΔP. ΔS and ΔV are the entropy and volume changes that occur during the phase change. The Clausius-Clapeyron equation gives the effect of temperature on vapor pressure:

$$\boxed{\ln P^{\square} = -\frac{\Delta H}{R}\left(\frac{1}{T}\right) + \frac{\Delta H}{RT_b}}$$

ΔH is the enthalpy of vaporization for liquid-vapor phase equilibrium and the enthalpy of sublimation for solid-vapor phase equilibrium.

The vapor pressure of the components in a solution vary with composition. In an ideal solution, both components obey Raoult's law over the entire range of composition:

$$P_1 = x_1 P_1^{\square}$$

$$P_2 = x_2 P_2^{\square}$$

Most real solutions do not obey Raoult's law over the entire range of composition. However, the behavior of the solvent is approximated by Raoult's law in dilute solutions, and in this same concentration range, the behavior of the solute is approximated by Henry's law:

$$P_2 = k_x x_2$$

or

$$P_2 = k_m m$$

or

$$P_2 = k_c[s]$$

A solution prepared from a nonvolatile solute has a higher boiling point than the pure solvent. In a dilute solution, the increase in boiling temperature is given by

$$\Delta T_b = K_b m$$

where

$$K_b = \frac{RT_b^2 M_1}{1000\ \Delta H_{\text{vap}}}$$

When a pure solid freezes from solution, the freezing-point lowering is

$$\Delta T_f = K_f m$$

where

$$K_f = \frac{-RT_0^2 M_1}{1000\ \Delta H_{\text{fus}}}$$

Important Terms and Concepts

Phase changes	Phase equilibrium
Heating curve	Phase diagram
Enthalpy of fusion	Triple point
Enthalpy of vaporization	LaChatelier's principle
Cooling curves	Raoult's law
Vapor pressure	Ideal solutions
Dynamic equilibrium	Henry's law

Exercises

8-1.
Predict whether ΔS and ΔG are positive, negative, or zero for the following processes:

(a) Vaporization of a liquid at its boiling point.
(b) Solidification of a supercooled liquid.
(c) Melting of a solid at its triple point.

8-2.
For the transition

$$\text{HgS(red)} = \text{HgS(black)}$$
$$\Delta G^0 = (17{,}200 - 25.5T) \text{ joules/mole}$$

(a) What is the stable form of HgS at room temperature (25° C) and atmospheric pressure?
(b) What is the transition temperature at exactly one atm pressure?

8-3.
The normal (one atmosphere) boiling point of *n*-butane, *n*-C_4H_{10}, is −0.50° C. Its enthalpy of vaporization at its boiling point is 22,400 joules/mole. Assuming that ΔH remains constant, calculate the vapor pressure of *n*-C_4H_{10} at 20.0° C.

8-4.
The normal (one atmosphere) boiling point of mercury (Hg) is 357° C. Calculate to what value one would need to reduce the external pressure on liquid Hg to cause it to boil at 100° C. The enthalpy of vaporization of Hg may be considered to be constant at 59,160 joules/mole. (*Hint:* A liquid boils when the vapor pressure equals or exceeds the external pressure.)

8-5.
Benzene and toluene form solutions that are nearly ideal. At 80° C, the vapor pressures of pure benzene and pure toluene are 753 torr and 290 torr, respectively. Calculate the vapor pressure of a solution made by mixing 100 g benzene (C_6H_6) with 100 g toluene ($C_6H_5CH_3$).

8-6.
At 25° C, the vapor pressure of pure CCl_4 is 0.151 atm. If you dissolve 2.97 g I_2 in 29.7 g CCl_4, what will the vapor pressure of CCl_4 become? Assume ideal solution behavior.

8-7.
At 60° C, the vapor pressure of water and glycerine ($C_3H_8O_3$) are 149.4 torr and 0.0071 torr, respectively. Calculate the total vapor pressure above an aqueous solution, 10% by weight glycerine. Assume ideal solution behavior.

8-8.
At 140° C the vapor pressure of pure chlorobenzene (C_6H_5Cl) is 939.4 torr and that of pure bromobenzene (C_6H_5Br) is 495.8 torr. What would be the mole fraction of C_6H_5Cl in a mixture of the two that boils at 140° C under 1 atm pressure?

8-9.
At 24° C, 0.00178 g N_2 gas will dissolve in 1.00×10^2 g H_2O if the nitrogen pressure is 737 torr. Calculate the Henry's law constant: (a) k_m, (b) k_x.

8-10.
At 20° C, the Henry's law constant for neon in water is 9.14×10^7 torr/mole fraction. How many grams of neon can you dissolve in 1000 g H_2O if the pressure of the neon gas above the water is 724 torr?

8-11.
How many grams of O_2 are dissolved in a round lake that is 1 mile in diameter and an average of 20 feet deep? Assume that O_2 obeys Henry's law when dissolved in water, that atmospheric pressure is 1 atm, and that air is 20 mole % O_2.

8-12.
A solution is made by dissolving 1.00 g of Br_2 in 100 g of water. Assuming that the water obeys Raoult's law and that the Br_2 obeys Henry's law, calculate the total vapor pressure above the solution at 25° C. The vapor pressure of pure water at 25° C is 23.76 torr.

8-13.
A solution of benzoic acid (C_6H_5COOH), a nonvolatile solute, in acetic acid (CH_3COOH) boils at 120.00° C. The boiling point of pure acetic acid is 118.11° C, and K_b (molal boiling-point elevation constant) is 3.07 for acetic acid. Assume ideal solution behavior and calculate the molality of the solution.

8-14.
What is the freezing point of a solution made from 1.00 g methyl alcohol (CH_3OH) and 10.00 g H_2O.

8-15.
The boiling point of pure carbon disulfide (CS_2) is 46.26° C. When 3.795 g of sulfur are dissolved in 100 g of CS_2, the boiling point increases to 46.64° C. The enthalpy of vaporization of CS_2 is 331.4 joules/gram. What are the molecular weight and molecular formula for sulfur?

8-16.
A compound with the empirical formula C_8H_5 is dissolved in benzene at 25° C. When 10.0 g of the compound are dissolved in 100 g of benzene, the vapor pressure above the resulting solution is 91.64 torr. If the vapor pressure of pure benzene is 95.18 torr at 25° C, calculate the molecular weight and determine the molecular formula of the compound. You may assume the compound is nonvolatile and that benzene obeys Raoult's law in the solution.

8-17.
A new compound between selenium and sulfur has just been discovered. Its formula is Se_yS_z. From the following data, calculate y and z: The compound consists of 45% selenium by weight. It dissolves in benzene at 20° C to the extent of 12 g Se_yS_z per 880.0 g C_6H_6, and it lowers the vapor pressure of C_6H_6 from 75.50 to 75.27 torr.

8-18.
4.68 g of a solute are dissolved in 287 g H_2O. If the measured freezing point of the solution is -0.153° C, what is the apparent molecular weight of the solute?

8-19.
It is suspected that selenium is a polymer of the type Se_y. Some solid selenium is dissolved in benzene and the freezing-point depression of the benzene is measured. With 3.25 g selenium dissolved in 225 g benzene, the observed freezing point is 0.117° C lower than for pure benzene. What is the molecular formula of selenium?

8-20.
When 0.555 g of a solute (Mol. Wt. = 110.0 g/mole) is dissolved in 100 g of a solvent (Mol. Wt. = 94.10 g/mole, $T_0 = 318.1$° K), a freezing-point lowering for the solvent of 0.382° C occurs. Calculate (a) K_f, the molal freezing-point lowering constant, and (b) the enthalpy of fusion for the solvent.

8-21.
Sodium sulfate (Na_2SO_4) dissociates completely into Na^+ and SO_4^{2-} ions when dissolved in water. 3.00 g of solid $Na_2SO_4 \cdot 10\,H_2O$ are dissolved in 100.0 g of H_2O at 25° C. At this temperature, the vapor pressure of pure water is 23.756 torr. Assume ideal solution behavior and calculate

(a) the vapor pressure above the solution at
(b) the normal boiling point of the solution.
(c) the normal freezing point of the solution.

8-22.
Zinc sulfate, $ZnSO_4$, is a weak electrolyte. That is, the reaction

$$ZnSO_4 = Zn^{2+} + SO_4^{2-}$$

does not go to completion. The freezing point of a solution that is made by dissolving 0.5811 g $ZnSO_4$ in 180.0 g of water, is -0.0530° C. Calculate the apparent percent dissociation of $ZnSO_4$ into Zn^{2+} and SO_4^{2-}.

8-23.
Acetic acid, $HC_2H_3O_2$, dissociates in water to give H^+ and $C_2H_3O_2^-$. A solution labeled 0.100 *m* $HC_2H_3O_2$ shows an observed freezing point that is 0.190° C below that of pure water. Calculate the apparent percent dissociation.

8-24.
$CaCl_2$ dissociates completely in water to form Ca^{2+} and Cl^- ions. Assuming that Raoult's law holds, calculate the vapor pressure of water at 25° C over an aqueous solution that is 0.55 *m* in $CaCl_2$. The vapor pressure of pure water at 25° C is 23.756 torr.

Problems

8-1.
Our friend, Horatio LeChatelier Smith was involved in a traffic accident in which his automobile, traveling at 50 mph, became intimately associated with a pole that supported a mercury-vapor street lamp. The wavelength of the brightest spectral line from the lamp was 4358.35 Å. (Horatio's 3000 lb car was red in color and was indeed a pitiful sight bent around the pole in the blue-violet light of the lamp.) A blood analysis indicated that there was more ethanol in Horatio's blood than would result from consumption of even very large quantities of bread made from stoneground flour. The total quantity of alcohol was estimated at 49 g. (Horatio was not of voting age but was able to name 800% more members of the Senate from the minority party than from the other party, which led some people to give special credence to the story he told.) This was Horatio's story: "I was asked—since I am a teetotaler—to act as bartender at a party being held by some friends. I spent three hours breathing the fumes of the liquor and apparently inhaled enough alcohol to become intoxicated." (Many comments regarding this story were made by the police officers who were present.) If ethanol (C_2H_5OH) and water formed an ideal solution (which they do not); if all the solutions opened during the three-hour period had been 100-proof alcohol (50% water and 50% ethanol by volume, or 42.5% ethanol by weight); and if Horatio breathed 250 ml of air every four sec, could he have breathed in enough ethanol in three hours to account for the quantity estimated to be in his blood at the time of his accident? (The vapor pressure and density of pure ethanol are 60 torr and 0.789 g/ml, respectively, at 20° C. The molecular weight of C_2H_5OH is 46 g/mole.)

8-2.
Figure 8.14 shows the phase diagram for CCl_4 that exists in three different solid crystalline phases.

(a) Draw the time-temperature warming curves that would result if CCl_4 were heated uniformly from point *a* to *b* and from point *c* to *d*. (Remember that a temperature halt occurs whenever two phases are in equilibrium.)

(b) Which phase has the higher density, liquid or solid I? Use LeChatelier's principle or the Clapeyron equation to justify your answer.

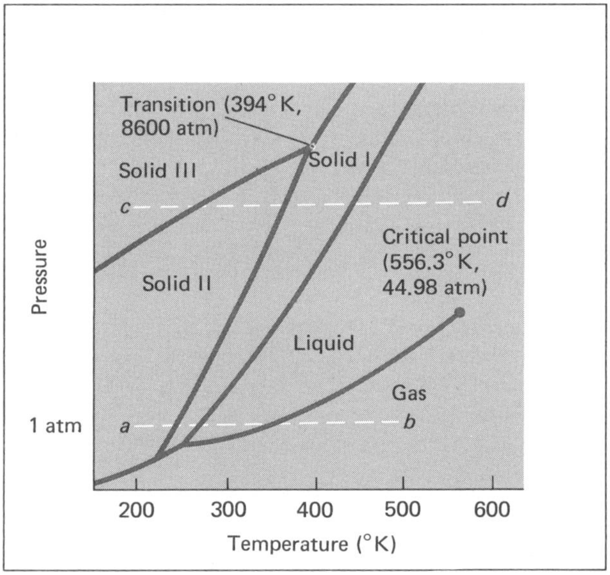

Figure 8.14
Phase diagram for carbon tetrachloride. (Note that the *y* axis is distorted.)

8-3.
The isopiestic (equal pressure) method for determining molecular weights uses an apparatus similar to that shown in the drawing on page 209. In a given experiment, 0.365 g of biphenyl ($C_{12}H_{10}$) was placed in arm A and 0.273 g of an unknown substance was placed in arm B. Some benzene was then poured into each arm, and the apparatus was sealed off. After placing the apparatus in a constant temperature

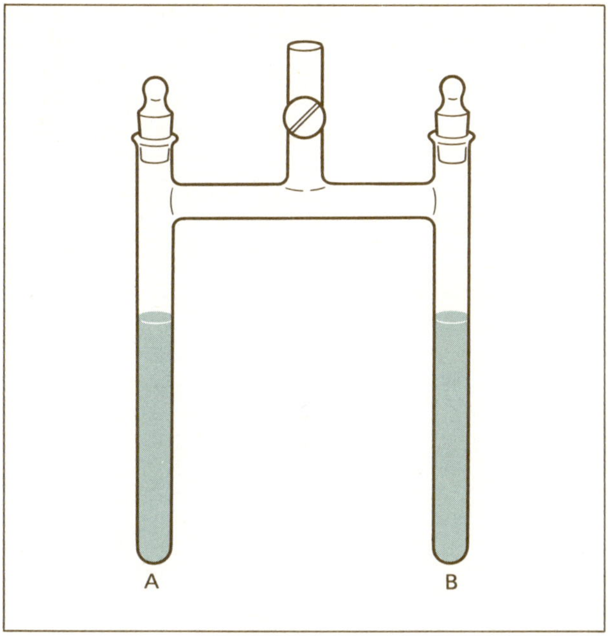

bath for a few days to establish equilibrium, the quantity of benzene in each arm was measured and found to be 36.20 g in arm A and 42.65 g in arm B. From these data, calculate the molecular weight of the unknown substance. (*Hint:* When equilibrium is established, the vapor pressure of benzene must be the same in each arm.)

8-4.

The partial pressures at 25° C for solutions of *n*-propanol with water are as shown below:

$x_{n\text{-PROPANOL}}$	P_{H_2O}(torr)	$P_{n\text{-PROPANOL}}$(torr)
0	23.76	0.00
0.020	23.5	5.05
0.050	23.2	10.8
0.100	22.7	13.2
0.200	21.8	13.6
0.400	21.7	14.2
0.600	19.9	15.5
0.800	13.4	17.8
0.900	8.13	19.4
0.950	4.20	20.8
1.000	0.00	21.76

Draw the complete pressure-composition diagram. Show the deviations from ideal solution behavior by drawing dotted lines representing the ideal solution.

8-5.

(a) Prove that for an ideal liquid solution of two volatile liquids A and B,

$$P = P_A + P_B = P_A^{\square} + x_B(P_B^{\square} - P_A^{\square})$$

where x_B is the mole fraction of component B, $P_A^{\square}$ and $P_B^{\square}$ are the vapor pressures of pure A and B, respectively, and P is the total vapor pressure above the solution.

(b) With $P_A^{\square} = 200$ torr and $P_B^{\square} = 400$ torr, make a graph of P against x_B. The graph should illustrate that in an ideal solution, the total vapor pressure is a linear (straight-line) function of mole fraction.

8-6.

After paying his debt to society, Horatio returned home. "Ah, yes; Thanksgiving! Over the river and through the woods!" It was a romantic thought. But his grandmother lived in Sun City along with 52,027 other retirees, abandoned to the riverless and woodless ghetto, and Horatio was too young to obtain visiting privileges. (Horatio wished that her home were in Havasu City, for then at least he could cross over the transplanted London Bridge.) Thinking of roasted turkey, the fuel shortage, pumpkin pie, smoked ham, and colligative properties in general, he decided that even if he could not enjoy Thanksgiving dinner with his grandmother, he could think about cooking. He decided to measure the difference in boiling temperature of the water in two open pots, one of which contains 500 g cranberries and the other of which contains 500 g rice. (The berries and rice have just been added; each berry and each grain of rice exists as a discrete unit.) Each pot contains 1000 g H_2O. Help the poor lad by determining the temperature difference that his thermometer must be able to detect if he is to measure the difference in the two boiling points.

The cranberries weigh 5.0 g each. The average weight of the rice grains is 0.10 g. The standard enthalpy of vaporization of water is 9717 cal/mole at the normal boiling point of 373° K. The molecular weight of water is 18 g/mole.

9

Chemical Equilibrium

In Chapter 7 we saw that equilibrium is attained in a system when $\Delta G = 0$ (constant temperature and pressure). With this criterion as a starting point, we went on, in Chapter 8, to discuss some useful relationships that apply to phases in equilibrium with one another. In this chapter we will be considering equilibrium in *chemically reacting* systems.

During the course of a chemical reaction, the products that are formed can react with one another in a reverse reaction. Since rates of reaction are proportional to concentrations of the reacting substances, the rate of the forward reaction slows down as the reactants are consumed, and the rate of the reverse reaction speeds up as the products are formed. Eventually, a point is reached at which the forward and reverse reactions proceed at the same rate and no net change occurs in the system. At this point, **chemical equilibrium** is established. Thermodynamics can be used to predict this point of equilibrium in a chemical reaction. We saw in Chapter 7 that the sign of ΔG for a proposed reaction tells us whether or not the reaction can take place. When you master the present chapter you will have all the equations necessary to tell not only *whether* a reaction will take place, but to what *extent*. That is, you will be able to calculate the concentration of all chemical species that will be present in the system when it reaches the equilibrium state.

Before a reaction can take place, the reacting substances must first encounter one another. The frequency of these encounters is proportional to the number of molecules of each reacting species per unit volume. Hence, the rate of reaction is proportional to the concentration of the reactants.

9.1 Free Energy Relations and the Equilibrium Constant

To develop the equations that apply to chemical equilibrium, we begin with an equation given in Chapter 7 that relates pressure and free energy change:

$$G - G^0 = nRT \ln \frac{P}{P^0} \qquad [9.1]$$

To apply the equation to a gas, the gas must be ideal; G is the free energy of the gas at pressure P, and G^0 is the free energy in a standard state where the pressure is P^0. Equation [9.1] also applies to liquids and solids, and for these, the pressure is the *vapor* pressure of the condensed phase. Thus, equation [9.1] can be applied to changes in the state of solids and liquids simply by letting P be the vapor pressure of the solid or liquid in the final state, and P^0 the vapor pressure in the standard state.

The standard state of a gas is chosen as the gas at one atmosphere pressure so that $P^0 = 1$ atm. To keep the dimensions consistent, the pressure P must also be expressed in atmospheres.

We are now ready to define a new term called **activity** (a) as

$$a = \frac{P}{P^0} \qquad [9.2]$$

where P and P^0 are the same quantities as in [9.1] above. We will describe more of the characteristics of activity shortly, and show how it can be approximated by a concentration term. For now, it is enough to keep in mind that when we speak of the *activity of a gas*, we mean that the pressure terms are the *partial pressures of the gas*, and when we refer to the *activity of a solid or liquid*, we are speaking of the *vapor pressures*. So, substituting our definition of activity into equation [9.1] gives

$$G = G^0 + nRT \ln a \qquad [9.3]$$

Equation [9.3] can now be used to express the free energy change in a chemical reaction in terms of activities. Below is a chemical reaction in the general form

$$a\text{A} + b\text{B} + \cdots = l\text{L} + m\text{M} + \cdots$$

where *a* moles of reactant A combine with *b* moles of reactant B and so forth to form *l* moles of product L, *m* moles of product M, ... The free energy change for this reaction would be

$$\Delta G = (G_\text{L} + G_\text{M} + \cdots) - (G_\text{A} + G_\text{B} + \cdots) \qquad [9.4]$$

where G_L is the *total* free energy of component L, G_M is the *total* free energy of component M, and similarly for all other products and reactants. But, from equation [9.3] we can write

$$G_\text{L} = G_\text{L}^0 + lRT \ln a_\text{L}$$
$$G_\text{M} = G_\text{M}^0 + mRT \ln a_\text{M}$$
$$G_\text{A} = G_\text{A}^0 + aRT \ln a_\text{A}$$
$$G_\text{B} = G_\text{B}^0 + bRT \ln a_\text{B}$$

Substituting these relationships into [9.4] and rearranging gives

$$\begin{aligned}\Delta G = {} & (G_\text{L}^0 + G_\text{M}^0 + \cdots) - (G_\text{A}^0 + G_\text{B}^0 + \cdots) \\ & + (lRT \ln a_\text{L} + mRT \ln a_\text{M} + \cdots) \\ & - (aRT \ln a_\text{A} + bRT \ln a_\text{B} + \cdots)\end{aligned}$$

Recognizing that

$$(G_\text{L}^0 + G_\text{M}^0 + \cdots) - (G_\text{A}^0 + G_\text{B}^0 + \cdots) = \Delta G^0$$

where ΔG^0 is the free energy change for the reaction with the reactants and products in the standard state, and combining the logarithm terms, we obtain

Equation [9.5] is valuable because it allows us to use free energy data for standard state concentrations to calculate ΔG for a reaction under any set of concentration conditions. We will return to its application after we cover the next section on activities.

$$\boxed{\Delta G = \Delta G^0 + RT \ln \frac{(a_\text{L}^l)(a_\text{M}^m)\cdots}{(a_\text{A}^a)(a_\text{B}^b)\cdots}} \qquad [9.5]$$

Under the conditions of equilibrium ($\Delta G = 0$), equation [9.5] becomes

$$\Delta G = \Delta G^0 + RT \ln \frac{(a_L^l)(a_M^m)\cdots}{(a_A^a)(a_B^b)\cdots} = 0 \quad \text{(equilibrium)} \qquad [9.6]$$

Rearranging, we can write

$$-\frac{\Delta G^0}{RT} = \ln \frac{(a_L^l)(a_M^m)\cdots}{(a_A^a)(a_B^b)\cdots} \qquad [9.7]$$

For a given reaction at a specified temperature, all quantities on the left-hand side of the equation are constants; hence, the right-hand side is also constant. For convenience, we designate the constant as ln K; hence,

$$\ln \frac{(a_L^l)(a_M^m)\cdots}{(a_A^a)(a_B^b)\cdots} = \ln K \qquad [9.8]$$

Taking the antilog of both sides of equation [9.8], we have

$$\boxed{K = \frac{(a_L^l)(a_M^m)\cdots}{(a_A^a)(a_B^b)\cdots}} \qquad [9.9]$$

The term K in equations [9.8] and [9.9] is called the **equilibrium constant** for the reaction. We can calculate K from the standard free energy change of the reaction by combining equations [9.9] and [9.7] to obtain

$$\boxed{\Delta G^0 = -RT \ln K} \qquad [9.10]$$

This is a very important relation because it allows one to obtain numerical values of K from standard free energy data.

It is important to note that the activity quotient term

$$\frac{(a_L^l)(a_M^m)\cdots}{(a_A^a)(a_B^b)\cdots}$$

in equation [9.9] is a constant (the equilibrium constant) only when the system has reached equilibrium ($\Delta G = 0$). The activity quotient term in equation [9.5] is *not* the equilibrium constant, since the value of the activities will not, in general, be the equilibrium values. In order to emphasize this distinction, some chemists use the symbol J for the activity quotient term in equation [9.5]. Using such notations, we have

$$\Delta G = \Delta G^0 + RT \ln J \qquad [9.5]$$

where

$$J = \frac{(a_L^l)(a_M^m)\cdots}{(a_A^a)(a_B^b)\cdots}$$

but

$$\Delta G^0 = -RT \ln K \qquad [9.10]$$

where

$$K = \frac{(a_L^l)(a_M^m)\cdots}{(a_A^a)(a_B^b)\cdots} \quad \text{(equilibrium)}$$

EXAMPLE 9.1 (a) Write the equilibrium-constant expression, and (b) calculate K at 298.15° K for the reaction

$$F_2(g) = 2\ F(g)$$

SOLUTION

(a) Using equation [9.9]

$$K = \frac{a_F^2}{a_{F_2}}$$

(b) The value of K is obtained from equation [9.10]:

$$\Delta G^0 = 2\ \Delta G^0_{f(F)} - \Delta G^0_{f(F_2)}$$

From Appendix B

$$\Delta G^0_{f(\mathrm{F})} = 118.0 \text{ kjoules/mole}$$

$$\Delta G^0_{f(\mathrm{F_2})} = 0$$

$$\Delta G^0 = (2 \text{ moles})(118.0 \text{ kjoules/mole})(10^3 \text{ joules/kjoule}) - 0$$

$$= 235{,}800 \text{ joules}$$

$$\ln K = -\frac{\Delta G^0}{RT}$$

$$= \frac{-235{,}800 \text{ joules}}{(8.314 \text{ joules/}^\circ\text{K})(298.2^\circ \text{ K})}$$

$$= -95.13$$

$$K = 4.85 \times 10^{-42}$$

The dimensions of R, the gas constant, appear here as joules/°K instead of joules/mole-°K. The reason for this is that in the derivation of equation [9.10] the *mole* in the denominator cancels with the *mole* of n, the number of moles. (See equation [9.3].)

Equation [9.9] is the expression that will allow us to calculate the concentration of each product and each reactant at equilibrium. In order to make use of the equation, however, we need to know how activities are related to concentrations. In the next section, we show that by proper selection of the standard state we can obtain very simple relationships between activities and concentrations.

9.2 Activity and Standard States

Standard State and Activity of Gases

The **standard state for a gas** is ideal gas at one atmosphere pressure. This hypothetical, but convenient, selection for the reference state allows us to obtain very readily a relationship between activity and concentration (pressure, in the case of a gas). Thus we have

$$a = \frac{P}{P^0} = \frac{P}{1} = P \quad \text{(gases)} \qquad [9.11]$$

We saw in Chapter 5 that pressure can be related to concentration from the ideal gas equation written in the form

$$P = \frac{n}{V} RT$$

At constant temperature, pressure is seen to be directly proportional to n/V (moles/liter).

Hence, where gases are involved, we can substitute their partial pressures for their activities in equations such as [9.5] and [9.9].

Shortly we will see what approximations can be made for the activities of solids, liquids, and solutions. First, however, let us restrict ourselves to reactions in which only gases appear, and illustrate a few applications of equations [9.5] and [9.9]. In Chapter 7 we demonstrated how to calculate ΔG^0 for a chemical reaction by subtracting the standard free energies of formation of the reactants from those of the products. Since thermodynamic data are tabulated for the standard state, ΔG^0 for the reaction is easily calculated. Equation [9.5] is important because it allows one to calculate ΔG for a reaction in any state from a knowledge of the free energy change in the standard state.

Ideal gas behavior is assumed in equation [9.11]. If the pressures are so high that the ideal gas assumption is not reasonable, the pressure term must be multiplied by a correction factor. In all of our applications, however, we will be working at pressures low enough that equation [9.11] can be used without modification.

EXAMPLE 9.2 Calculate ΔG for the following reaction at 25° C:

$$N_2(P = 0.10 \text{ atm}) + 3\,H_2(P = 0.10 \text{ atm}) = 2\,NH_3(P = 10 \text{ atm})$$

SOLUTION

$$\Delta G^0 = 2\,\Delta G^0_{f(NH_3)} - \Delta G^0_{f(N_2)} - 3\,\Delta G^0_{f(H_2)}$$

From Appendix B, ΔG^0_f for $NH_3 = -16{,}640$ joules/mole. Since $\Delta G^0_{f(H_2)} = \Delta G^0_{f(N_2)} = 0$,

$$\Delta G^0 = (-16{,}640 \text{ joules/mole})(2 \text{ moles}) - 0 - 0 = -33{,}280 \text{ joules}$$

Combining equations [9.5] and [9.11], we can write

$$\Delta G = \Delta G^0 + RT \ln \frac{P^2_{NH_3}}{(P_{N_2})(P^3_{H_2})}$$

$$\Delta G = -33{,}280 \text{ joules} + (8.314 \text{ joules/°K})(298.15° \text{ K}) \ln \frac{(10)^2}{(0.10)(0.10)^3}$$

$$\Delta G = -33{,}280 + 1286 \ln (1.0 \times 10^6) = 970 \text{ joules}$$

The reaction under the standard state conditions

$$N_2(P = 1 \text{ atm}) + 3\,H_2(P = 1 \text{ atm}) = 2\,NH_3(P = 1 \text{ atm})$$

is spontaneous, since ΔG^0 has a negative value (−33,280 joules). Under the conditions used in Example 9.2, however, the reaction is not spontaneous, since the value of ΔG is positive. It is evident that the direction or spontaneity of a reaction depends upon the concentrations of the reactants and products. In a chemical reaction it is the concentrations at equilibrium that interest us. We shall now see how thermodynamics permits one to predict these equilibrium conditions.

EXAMPLE 9.3 For the reaction

$$2\,CO(g) + O_2(g) = 2\,CO_2(g)$$

write the equilibrium expression, calculate the value for K at 25° C, and predict the extent of reaction.

SOLUTION The equilibrium-constant expression for this reaction is

$$K = \frac{a^2_{CO_2}}{(a^2_{CO})(a_{O_2})} = \frac{P^2_{CO_2}}{(P^2_{CO})(P_{O_2})}$$

The relation between K and ΔG^0 is

$$\ln K = -\frac{\Delta G^0}{RT}$$

$$\Delta G^0 = 2\,\Delta G^0_{f(CO_2)} - 2\,\Delta G^0_{f(CO)} - \Delta G^0_{f(O_2)}.$$

$$\Delta G^0 = 2\text{ moles}(-394.4\text{ kjoules/mole}) - 2\text{ moles}(-137.3\text{ kjoules/mole}) - 0$$ (data from Appendix B)

$$\Delta G^0 = -514.2\text{ kjoules}$$

$$\ln K = -\frac{(-514.2\text{ kjoules})(10^3\text{ joules/kjoule})}{(8.314\text{ joules/}^\circ\text{K})(298.15^\circ\text{ K})} = 207.4$$

$$K = 1.2 \times 10^{90}$$

Even without making quantitative calculations we can use the equilibrium constant to predict whether or not a reaction will occur to an appreciable extent. A large constant like that shown in this example means that at equilibrium, the pressures of the products (since they appear in the numerator of the equilibrium-constant expression) will be much larger than the pressures of the reactants. In such a case, the reaction will go essentially to completion. Conversely, a small equilibrium constant (such as the one for the reaction in Example 9.1) means that the pressures of the reaction products will be much smaller than the pressures of the reactants. Such a reaction would produce very little product at equilibrium.

Quantitative calculations are easily made from the equilibrium expression when we know the equilibrium constant.

EXAMPLE 9.4 Nitrogen dioxide gas partially dimerizes according to the equation

$$2\,NO_2(g) = N_2O_4(g)$$

Calculate the equilibrium partial pressures of NO_2 and N_2O_4 gases in a mixture of the two at a temperature of 25° C and a total pressure of 1 atm.

SOLUTION Again, using free energy of formation data from Appendix B, we have

$$\Delta G^0 = 98{,}286\text{ joules} - 2(51{,}840\text{ joules}) = -5394\text{ joules}$$

K can then be calculated from ΔG^0:

$$\ln K = -\frac{\Delta G^0}{RT} = -\frac{-5394\text{ joules}}{(8.314\text{ joules/}^\circ\text{K})(298.15^\circ\text{ K})} = 2.176$$

$$K = \frac{a_{N_2O_4}}{a^2_{NO_2}} = \frac{P_{N_2O_4}}{P^2_{NO_2}} = 8.81$$

Since the total pressure is 1.00 atm,

$$P_{N_2O_4} = 1.00 - P_{NO_2}$$

Substituting this relationship in the above gives

$$\frac{1.00 - P_{NO_2}}{P^2_{NO_2}} = 8.81$$

Rearranging gives the equation

$$8.81\, P^2_{NO_2} + P_{NO_2} - 1.00 = 0$$

which can then be solved for P_{NO_2} by using the quadratic formula:

$$P_{NO_2} = \frac{-1 + \sqrt{1 - 4(8.81)(-1)}}{2(8.81)}$$

$$= 0.285 \text{ atm}$$

Therefore,

$$P_{N_2O_4} = 1.000 - 0.285 = 0.715 \text{ atm}$$

Activities and Standard States for Pure Solids, Pure Liquids, and Solutions

The definition of activity is

$$a_i = \frac{P_i}{P_i^0} \qquad [9.2]$$

The subscript i is added to take care of solutions in which there is more than one component to identify.

Previously we set the standard state for a gas and found that pressure was a substitution for the activity of a gas. Now we define the standard state for substances in other phases and find approximations that we can use for their activities.

(a) *Pure solids and liquids.* The **standard state for a pure liquid or solid** is chosen as the pure substance under 1 atm pressure. This makes P_i^0 identical with $P_i^{\square}$, the vapor pressure of the pure substance at 1 atm total pressure. Therefore, for the activity of a pure solid or liquid, we have

$$a_i = \frac{P_i}{P_i^0} = \frac{P_i^{\square}}{P_i^{\square}} = 1 \qquad [9.12]$$

which is an even simpler relationship than we had for gases.

(b) *Solutions.* In dilute solutions, the vapor pressure of a solvent can be approximated by Raoult's law and the vapor pressure of the solute by Henry's law (Chapter 8, pages 193–196). These laws are the bases for setting standard states in solutions.

Raoult's law, as was shown in equation [8.6], is $P_1 = x_1 P_1^{\square}$.

(1) *Solvent.* The usual choice of standard state for a solvent is a solution in which the mole fraction of the solvent is unity. This is referred to

as **Raoult's law standard state.** It is identical to the standard state chosen for a pure liquid for which $P_1^0 = P_1^\square$. The activity of the solvent is given by

$$a_1 = \frac{P_1}{P_1^0} \qquad [9.13]$$

which by our choice of standard state becomes

$$a_1 = \frac{P_1}{P_1^\square} \quad \text{(solvent, Raoult's law standard state)} \qquad [9.14]$$

If we write Raoult's law in the form

$$x_1 = \frac{P_1}{P_1^\square} \qquad [8.6]$$

and compare this expression with equation [9.14], it becomes apparent that a_1 and x_1 are related. Where Raoult's law applies, $a_1 = x_1$. Hence, in dilute solutions we have

$$a_1 \cong x_1 \quad \text{(solvent, dilute solution, Raoult's law standard state)}^\dagger \qquad [9.15]$$

In the limit, when the concentration of solute becomes zero, the activity of the solvent becomes exactly equal to its mole fraction, which is 1 in the pure solvent. The result we obtained in the preceding section for the activity of pure solid and pure liquid is thus seen to be just a special case of the activity of the solvent in a solution. In summary, the activity of solvent is given approximately by its mole fraction. When the mole fraction of solvent becomes 1, the activity becomes exactly 1.

EXAMPLE 9.5 Figure 9.1 depicts the vapor pressure of water versus mole fraction in dilute aqueous solutions of sucrose. Indicate the concentration region for which the activity of the water can be well approximated by the mole fraction, hence showing that equation [9.15] is valid.

SOLUTION The experimental vapor pressure line follows the Raoult's law line for $x_{H_2O} = 1$ to $x_{H_2O} = 0.985$. In that concentration region, equation [9.15] is valid. For $x_{H_2O} = 0.985$, m, the molality of sucrose, is 0.85. We can say that Raoult's law holds for water in aqueous solutions of sucrose less than 0.85 m.

† In solutions too concentrated for Raoult's law to apply, activities can be related to concentrations through the relationship $a_1 = \gamma_1 x_1$, where γ_1 is the **activity coefficient** of the solvent. There are various experimental ways for determining γ_1. Its numerical value approaches unity as the mole fraction of solvent approaches unity.

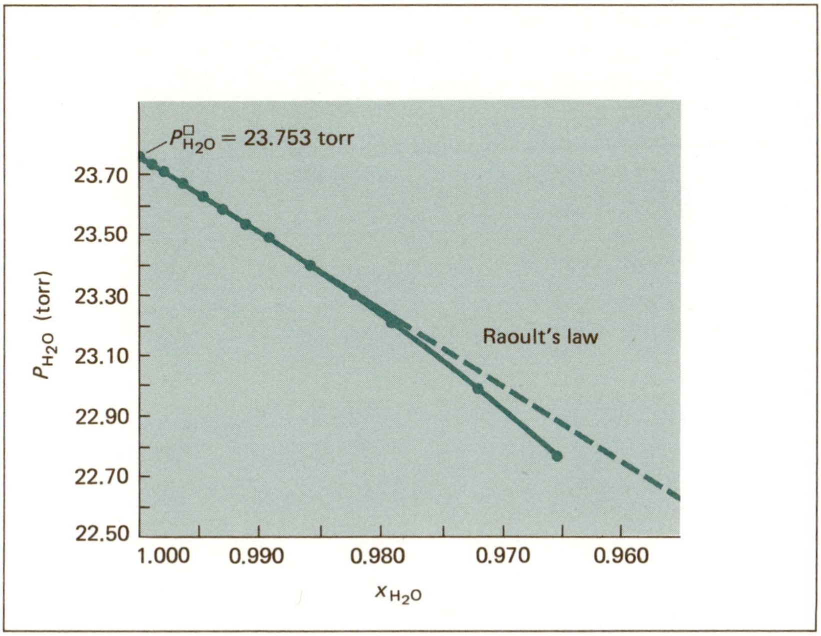

Figure 9.1
Vapor pressure of water above aqueous solutions of sucrose at 25° C.

(2) *Solute.* It is convenient to choose the standard state of a solute as a 1-molar solution in which Henry's law is obeyed. However, since Henry's law generally does not apply at a concentration that high, this state needs some explaining. Figure 9.2 shows the vapor pressure of solute plotted against the molar concentration for a

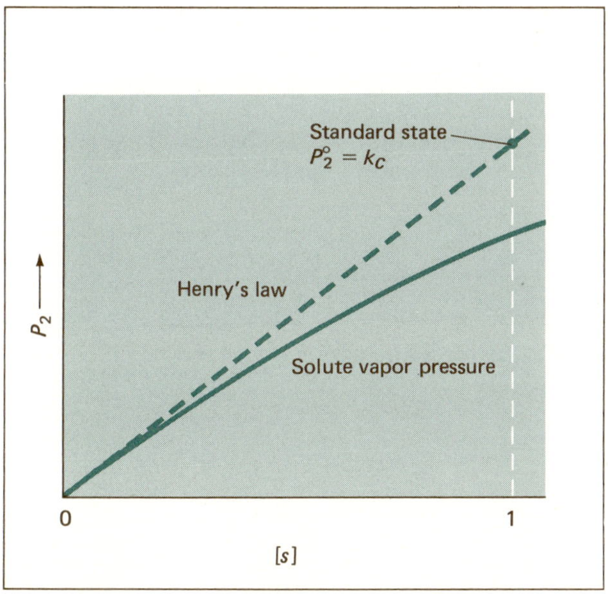

Figure 9.2
The standard state for the solute is a hypothetical 1-molar solution that obeys Henry's law.

typical solution. The solid line represents the measured vapor pressures; the dotted line is a plot of Henry's law, which is seen to be a valid description of the vapor pressure of the solute in the dilute range. By extrapolating the dilute range behavior to $[s] = 1$, we arrive at the Henry's law standard state, denoted by a bullet (•) in Figure 9.2. At this point, the vapor pressure of the solute is P_2^0 by definition, and since Henry's law is

$$P_2 = k_c[s] \qquad [8.10]$$

we have at $[s] = 1$ (standard state)

$$P_2^0 = k_c \qquad [9.16]$$

The activity of a solute is given by

$$a_2 = \frac{P_2}{P_2^0} \qquad [9.17]$$

which by our choice of standard state becomes

$$a_2 = \frac{P_2}{k_c} \quad \text{(solute, Henry's law standard state)} \qquad [9.18]$$

Writing Henry's law in the form

$$[s] = \frac{P_2}{k_c} \qquad [8.10]$$

and comparing this expression with equation [9.18] shows that molarity and activity are related. In solutions dilute enough for Henry's law to be a valid approximation, we see that the activity of a solute can be approximated by molarity, and we have

As an example, experimental data verify that Henry's law is valid for solutions of O_2 in water when the concentration of O_2 is less than 5×10^{-3} molal. The equilibrium pressure of O_2 above a 5×10^{-3} molal solution is approximately 4 atm. [9.19] is valid, therefore, for O_2 solutions when P_{O_2} is less than 4 atm.

$$a_2 \cong [s] \quad \text{(solute, dilute solution, Henry's law standard state)}^\dagger \qquad [9.19]$$

A summary of standard state pressures and activity-concentration relations appears in Table 9.1.

† In solutions too concentrated for Henry's law to apply, concentration and activity can be related through an activity coefficient, γ_2, by the expression

$$a_2 = \gamma_2[s] \qquad [9.20]$$

The numerical value of γ_2, which can be experimentally determined, is a measure of the extent to which the solute obeys Henry's law. In dilute solutions, the numerical value of γ_2 will be near unity.

TABLE 9.1

Relations between Activities and Concentrations	
STATE	ACTIVITY–CONCENTRATION RELATION
Gas	$a = P$ (ideal gas)
Pure liquid or solid	$a = 1$ (atmospheric pressure)
Solvent in solution	$a_1 = x_1$ (dilute solution)
Solute in solution	$a_2 = [s]$ (dilute solution)

Equilibrium Calculations

We are now able to extend equilibrium calculations to substances in any phase. Several representative calculations follow. These examples will help you become familiar with the equilibrium-constant expression and its simplification by the substitutions for activities discussed above.

The entire next chapter is devoted to detailed applications of the equilibrium-constant expression to a variety of reactions involving gases, solids, liquids, and solutions.

EXAMPLE 9.6 Barium chloride monohydrate dissociates according to the reaction

$$BaCl_2 \cdot H_2O(s) = BaCl_2(s) + H_2O(g)$$

Calculate the equilibrium vapor pressure of water above $BaCl_2 \cdot H_2O(s)$ at 25° C.

SOLUTION First we calculate ΔG^0 for the reaction. From Appendix B,

$$\Delta G^0 = (-228{,}600 \text{ joules} - 810{,}900 \text{ joules}) - (-1{,}059{,}000 \text{ joules}) = 19{,}500 \text{ joules}$$

From ΔG^0 we can obtain the equilibrium constant:

$$\ln K = -\frac{\Delta G^0}{RT} = -\frac{19{,}500 \text{ joules}}{(8.314 \text{ joules/°K})(298.15° \text{ K})} = -7.7$$

$$K = 4 \times 10^{-4}$$

Finally, using K, we can calculate the equilibrium pressure of the water vapor:

$$K = \frac{(a_{H_2O})(a_{BaCl_2})}{a_{BaCl_2 \cdot H_2O}}$$

Since $BaCl_2$ and $BaCl_2 \cdot H_2O$ are pure solids, their activities are unity, and since the H_2O is a gas, its activity is the same as the pressure of the water vapor. Making these substitutions, we find

Refer to Table 9.1.

$$K = P_{H_2O} = 4 \times 10^{-4} \text{ atm}$$

The units of pressure must be in atmospheres since the standard state was defined in atmospheres.

EXAMPLE 9.7 Water reacts with itself and dissociates slightly according to the reaction

$$2\,H_2O(l) = H_3O^+(aq) + OH^-(aq)$$

The equilibrium constant for this reaction, commonly designated as K_w, has the numerical value of 1.00×10^{-14} at 25° C. Calculate $[H_3O^+]$ and $[OH^-]$ in pure water at 25° C.

SOLUTION

$$K_w = \frac{(a_{H_3O^+})(a_{OH^-})}{a_{H_2O}^2}$$

Since the water is pure, we can write

$$a_{H_2O} = 1$$

and since the solution is dilute (as evidenced by the small value of the equilibrium constant), we can write for the activities of the solutes

$$a_{H_3O^+} = [H_3O^+]$$
$$a_{OH^-} = [OH^-]$$

Substitution into the expression above for K_w gives

$$K_w = [H_3O^+][OH^-] = 1.00 \times 10^{-14}$$

From the stoichiometry of the reaction, we see that

$$[H_3O^+] = [OH^-]$$

Substituting this into the previous expression gives

$$[H_3O^+]^2 = 1.00 \times 10^{-14}$$

and taking the square root, we have

$$[H_3O^+] = 1.00 \times 10^{-7} = [OH^-]$$

EXAMPLE 9.8 Silver chloride dissolves slightly in water to form a saturatea solution according to the reaction

$$AgCl(s) = Ag^+ + Cl^-$$

In the saturated solution, equilibrium is established between the solid AgCl and Ag^+ and Cl^- in solution.

The equilibrium constant for this type of reaction is commonly referred to as a **solubility product constant** and is given the notation K_{sp}. Calculate the concentration of Ag^+ in a saturated solution of AgCl.

$$K_{sp(AgCl)} = 1.8 \times 10^{-10} \text{ at } 25° \text{ C}$$

SOLUTION For the reaction we have

$$K_{sp} = \frac{(a_{Ag^+})(a_{Cl^-})}{a_{AgCl}}$$

Since AgCl is a pure solid, it follows that

$$a_{AgCl} = 1$$

and since the solutes will be in low concentrations ($K = 1.8 \times 10^{-10}$), we can write

$$a_{Ag^+} = [Ag^+]$$

and

$$a_{Cl^-} = [Cl^-]$$

Substitution into the equilibrium expression gives

$$K_{sp} = \frac{[Ag^+][Cl^-]}{1} = 1.8 \times 10^{-10}$$

From the stoichiometry of the reaction, we see that

$$[Ag^+] = [Cl^-]$$

Substituting this into the previous expression gives

$$[Ag^+]^2 = 1.8 \times 10^{-10}$$
$$[Ag^+] = 1.3 \times 10^{-5}$$

9.3 Effect of Temperature on Chemical Equilibrium

The **temperature effect on K** (the equilibrium constant) is related to the ΔH of the reaction. The equation is

$$\ln \frac{K_2}{K_1} = \frac{\Delta H^\circ}{R}\left(\frac{1}{T_1} - \frac{1}{T_2}\right) \qquad [9.21]$$

where K_1 and K_2 are the equilibrium constants at T_1 and T_2, respectively. Note that if ΔH^0 is positive (endothermic), $\ln (K_2/K_1) = \ln K_2 - \ln K_1$ will be positive when $T_2 > T_1$. Thus, raising the temperature of an exothermic reaction pushes the point of equilibrium farther toward the products. Conversely, raising the temperature of an exothermic reaction (ΔH^0 is negative) decreases the yield of the products at equilibrium.

If the equilibrium constant for a reaction is known at one temperature, equation [9.21] can be used to calculate K at a second temperature.

EXAMPLE 9.9 Calculate K at 100° C for the reaction

$$2\ NO_2(g) = N_2O_4(g)$$

SOLUTION From Example 9.4, $K = 8.81$ at 298.15° K. Enthalpies of formation can be used to calculate ΔH^0 for the reaction. Using Appendix B we have

$$\Delta H^0 = 9661 \text{ joules} - 2(33{,}850 \text{ joules}) = -58{,}040 \text{ joules}$$

Equation [9.21] can be written in the form

$$\ln K_2 = \ln K_1 + \frac{\Delta H^0}{R}\left(\frac{1}{T_1} - \frac{1}{T_2}\right)$$

Substituting in the values $\Delta H^0 = -58{,}040$ joules, $K_1 = 8.81$, $T_1 = 298.15°$ K, and $T_2 = 373.15°$ K gives

$$\ln K_2 = \ln 8.81 + \frac{-58{,}040 \text{ joules}}{8.314 \text{ joules/°K}}\left(\frac{1}{298.15° \text{ K}} - \frac{1}{373.15° \text{ K}}\right)$$

Therefore,

$$K_2 = 0.0796$$

9.4 Chemical Equilibrium and the Principle of LeChatelier

The equilibrium constant provides a solid base from which one can predict quantitatively the results of changes made on an equilibrium system. In addition, **LeChatelier's principle,** introduced in Chapter 8, can often be used to predict qualitatively the effect of changes made on a system at equilibrium. This principle, first published in 1884 by the French chemist Henri LeChatelier, states that *when a stress is applied to a system at equilibrium, the system shifts, if possible, to reduce the stress.*

A few examples will demonstrate the application of this principle. As we saw in Example 9.4, NO_2 and N_2O_4 exist in equilibrium in the gas phase:

$$2\ NO_2(g) = N_2O_4(g)$$

Let us consider separately the results of two stresses applied to this system:

(a) NO_2 is added to an equilibrium mixture of NO_2 and N_2O_4. This stress (added NO_2) can be reduced by a shift to produce more N_2O_4, thus decreasing the added NO_2. On the other hand, adding more N_2O_4 to an equilibrium mixture of the two gases will cause more NO_2 to form. The phrase often used is "the equilibrium shifts." This must not be supposed to imply that the equilibrium constant has changed. More NO_2 has combined to form N_2O_4; the constant has not changed.

(b) The total pressure on the equilibrium system is increased by compressing the system. The stress here is increased pressure. The principle of LeChatelier states that the system will shift in such a way that the stress is reduced. Since the conversion of some NO_2 to N_2O_4 decreases the total number of moles and, hence, the total pressure, such a change is predicted on the basis of LeChatelier's principle and, indeed, does occur. (Note that if the total pressure had been increased by the addition of an inert gas, no shift in the equilibrium would have resulted nor would a shift be predicted by LeChatelier's principle. Such an increase in pressure is not a stress on the *system*—which is $NO_2(g) + N_2O_4(g)$; rather, it is extraneous to the system.)

The formation of gaseous HI from H_2 and I_2

$$H_2(g) + I_2(g) = 2\ HI(g)$$

is an exothermic reaction:

$$\Delta H^0_{298} = -10{,}400 \text{ joules}$$

If a mixture of H_2, I_2, and HI is at equilibrium, what is the effect of an increase in the temperature (all other factors, including pressure, held constant)? A simple method of applying LeChatelier's principle follows: Since the reaction is exothermic, heat may be considered to be a product of the reaction. The stress applied was an increase in the temperature, and this stress can be decreased by the shifting of the reaction to the left, since this shift consumes heat.

Hence, the effect of an increase in temperature upon the equilibrium system will be a decreased pressure of HI and an increased pressure of H_2 and I_2. It should be noted that the equilibrium constant is changed by the increase in temperature. Equation [9.21] permits a quantitative calculation of the new value of the equilibrium constant.

Recall that LeChatelier's principle includes the phrase *if possible*. Thus, if a stress in the form of increased total pressure is applied to an equilibrium gaseous mixture of H_2, I_2, and HI, then no change in the equilibrium will occur. The equation

$$H_2(g) + I_2(g) = 2\ HI(g)$$

represents the reaction. Since the total number of molecules is not changed by the reaction, the increased total pressure of the system cannot be relieved by the formation or the decomposition of HI. Again, LeChatelier's principle leads us to a correct prediction. Each of the above effects can be predicted from the equilibrium equations derived earlier in this chapter. You can see the usefulness of LeChatelier's principle in allowing one to make qualitative predictions very easily.

EXAMPLE 9.10 Consider the exothermic reaction

$$N_2(g) + 3\ H_2(g) = 2\ NH_3(g);\ \Delta H \text{ is negative}$$

What will be the effect on the point of equilibrium of the reaction (shift right, shift left, no effect) of making the following changes on the system when it is initially at equilibrium?

(a) Adding additional H_2 to the system, all other factors held constant.
(b) Adding additional NH_3 to the system, all other factors held constant.
(c) Removing NH_3 from the system, all other factors held constant.
(d) Compressing the system at constant temperature.
(e) Adding an inert gas to the system.

SOLUTION The stress caused by adding additional reactants or removing products is relieved by shifting the composition further in the direction of the products. The stress caused by adding more products (right-hand side of the equation) is relieved by some of those products changing to reactants (left side). Hence, the answers deduced by LeChatelier's principle are

(a) shift right (toward products)
(b) shift left (toward reactants)
(c) shift right (toward products)
(d) shift right (toward products)
(e) no effect

An example of an equilibrium reaction that can be made to go to completion by removing a product is

$CaCO_3(s) = CaO(s) + CO_2(g)$

If the reaction is carried out in an open container, the $CO_2(g)$ escapes into the atmosphere, and the $CaCO_3(s)$ decomposes completely. If, however, the $CO_2(g)$ is contained in the reaction vessel, an equilibrium state will be reached, and there will be only partial decomposition of the $CaCO_3(s)$.

One can reach the same conclusions by examining the equilibrium constant expression for the reaction:

$$K = \frac{(P_{NH_3})^2}{(P_{N_2})(P_{H_2})^3}$$

K cannot change as long as the temperature is constant; hence, any increase in the reactants (an increase in the size of the denominator) will be compensated by a shift toward the products (to increase the size of the numerator). Removing products, as in part (c), as fast as they are formed is an effective way of driving the reaction to completion.

Summary

Starting with the thermodynamic principles developed in Chapter 7, we obtain the very important relations for chemical reactions

$$\boxed{\Delta G = \Delta G^0 + RT \ln \frac{(a_L^l)(a_M^m)\cdots}{(a_A^a)(a_B^b)\cdots}}$$

$$\boxed{K = \frac{(a_L^l)(a_M^m)\cdots}{(a_A^a)(a_B^b)\cdots}} \qquad \text{(equilibrium)}$$

and

$$\boxed{\Delta G^0 = -RT \ln K}$$

The first of these lets us know (from the sign of ΔG) whether a reaction will take place under a given set of concentration conditions from information about the free energy change in the standard state conditions. The second equation is the highly important equilibrium-constant expression, through which we can calculate the concentration of reactants and products in a reaction mixture that has attained equilibrium. The principle of equilibrium expressed by this equation is the most important concept of the chapter. The entire next chapter is devoted to its applications. The third equation provides an important means by which equilibrium-constant data are obtained.

The effect that temperature has on the equilibrium constant is a function of the enthalpy change of the reaction, as shown below:

$$\ln \frac{K_2}{K_1} = \frac{\Delta H^0}{R}\left(\frac{1}{T_1} - \frac{1}{T_2}\right)$$

It is convenient when studying equilibrium in chemical systems to define the term *activity*. By proper choice of the reference conditions that we refer to as the *standard state*, we can see that various concentration terms can be used as approximations of activities, thus making it possible to calculate concentrations at equilibrium.

Important Terms and Concepts

Chemical equilibrium
Activity
Equilibrium constant
Standard state for gas
Standard state for pure liquid or solid
Raoult's law standard state
Henry's law standard state
Activity coefficient
Solubility product constant
Temperature effect on K
LeChatelier's principle

Exercises

Refer to the Appendix for additional data.

9-1.
Gaseous hydrogen peroxide decomposes according to the equation

$$H_2O_2(g) = H_2O(g) + \frac{1}{2} O_2(g)$$

for which $\Delta G^0 = -125$ kjoules at 25° C. Calculate K at 25° C.

9-2.
Calculate the equilibrium constant at 25° C of the following reactions:

(a) $H_2(g) + Cl_2(g) = 2HCl(g)$
(b) $CH_4(g) + 2\ O_2(g) = CO_2(g) + 2\ H_2O(g)$
(c) $2\ Ag(s) + Cl_2(g) = 2\ AgCl(s)$
(ΔG_f^0 for AgCl = −109.7 kjoules/mole.)

9-3.
What are the possibilities of producing (1) methanol and (2) ethanol by the following reactions at 25° C:

(a) $CO(g) + 2\ H_2(g) \rightarrow CH_3OH(l)$
(b) $C_2H_4(g) + H_2O(g) \rightarrow C_2H_5OH(l)$

9-4.
For the reaction at 25° C

$$CH_3OH(l) = CO(g) + 2\ H_2(g)$$

(a) calculate ΔH^0, ΔS^0, and ΔG^0.
(b) calculate K and then the equilibrium pressure of CO above $CH_3OH(l)$ at 25° C.

(*Hint:* Note the relationship between the number of moles of CO and H_2 produced, and then express the pressure of hydrogen in terms of the pressure of CO.)

9-5.
Calculate the equilibrium pressure of oxygen at 25° C for the reaction

$$KClO_3(s) = KCl(s) + \frac{3}{2} O_2(g)$$

ΔG_f^0 for $KClO_3$ = −289.9 kjoules/mole.

9-6.
When platinum is heated in the presence of chlorine gas, the following reaction takes place:

$$Pt(s) + Cl_2(g) = PtCl_2(g)$$

At 1000° K, $\Delta G^0 = 59$ kjoules. If the pressure of Cl_2 is 1 atm, what will be the partial pressure of $PtCl_2$?

9-7.
For the reaction

$$2\ CaSO_4(s) = 2\ CaO(s) + 2\ SO_2(g) + O_2(g)$$

$K = 1.45 \times 10^{-5}$ at 1625° K. What is the dissociation pressure of $CaSO_4$ in torr at 1625° K? (The dissociation pressure is the total gas pressure above the solid. *Hint:* Note from the balanced equation the relationship that exists between P_{SO_2} and P_{O_2}.)

9-8.
For the reaction

$$Br_2(g) = 2\ Br(g)$$

$\Delta H^0 = 230$ kjoules and $K = 0.0071$ at 1000° C.

(a) calculate ΔG^0 and ΔS^0 at 1000° C.
(b) calculate P_{Br} when $P_{Br_2} = 1.000$ atm at 1000° C.

9-9.
At 3000° K, the equilibrium pressures of CO_2, CO, and O_2 are 0.6, 0.4, and 0.2 atm, respectively. Calculate (a) K and (b) ΔG^0 at 3000° K for the reaction

$$2\ CO_2(g) = 2\ CO(g) + O_2(g)$$

(c) Calculate ΔG for the reaction under the conditions that $P_{CO_2} = 0.10$ atm, $P_{CO} = 2.0$ atm, $P_{O_2} = 1.0$ atm, and $T = 3000°$ K. Will the reaction occur under the latter conditions?

9-10.
(a) Using the data in Appendix B, calculate K at 25° C and at 100° C for the reaction

$$2\ NO(g) + O_2(g) = 2\ NO_2(g)$$

(b) Calculate ΔG for the reaction when $P_{NO} = 10.0$ atm, $P_{O_2} = 1.00$ atm, and $P_{NO_2} = 0.010$ atm at 25° C. Will the reaction occur under these conditions?

9-11.
For the reaction

$$SbCl_5(g) = SbCl_3(g) + Cl_2(g)$$

$K = 1.064$ at 248° C and 0.271 at 206° C. Calculate (a) ΔH^0, (b) ΔG^0, and (c) ΔS^0 for the reaction at 227° C.

9-12.
For the reaction (at 298.2° K)

$$2\ NaHSO_4(s) = Na_2S_2O_7(s) + H_2O(g)$$
$$\Delta H^0 = 82.8 \text{ kjoules}$$
$$\Delta G^0 = 37.7 \text{ kjoules}$$

Assuming that ΔH^0 is constant with temperature, calculate the dissociation pressure (total gaseous pressure) of $NaHSO_4$ at 700° K.

9-13.
Consider the endothermic reaction

$$COCl_2(g) = CO(g) + Cl_2(g);\ \Delta H \text{ is positive}$$

What will be the effect on the point of equilibrium (shift left, shift right, have no effect) of the following changes on a system initially at equilibrium, assuming that all other factors are held constant in each case?

(a) Adding additional $Cl_2(g)$ to the system.
(b) Removing $CO(g)$.
(c) Removing $COCl_2(g)$.
(d) Compressing the system.
(e) Adding 10.0 atm of $Ne(g)$.
(f) Cooling the system.

9-14.
Consider the exothermic reaction

$$2\ CO(g) + O_2(g) = 2\ CO_2(g);\ \Delta H \text{ is negative}$$

What effect will the following changes have on the point of equilibrium (shift left, shift right, have no effect) of the system that is initially at equilibrium? Consider each factor separately, assuming that all other factors are constant.

(a) Adding O_2.
(b) Removing CO.
(c) Removing CO_2.
(d) Heating the system.
(e) Expanding the system.
(f) Compressing the system.
(g) Adding an inert gas.

9-1.
NH_4HS dissociates as follows:

$$NH_4HS(s) = NH_3(g) + H_2S(g)$$

At 25° C the dissociation pressure of the pure solid is 500 torr. Calculate (a) K and (b) the total pressure at equilibrium when 300 torr of NH_3 are introduced into a flask containing solid NH_4HS.

9-2.
Calcium carbonate decomposes according to the reaction

$$CaCO_3(s) = CaO(s) + CO_2(g)$$

(a) For $CaCO_3$, $\Delta G^0_{formation} = -1128.8$ kjoules/mole at 25° C. Calculate the pressure of CO_2 in equilibrium with $CaCO_3$ and CaO at 25° C.
(b) Measurement of the pressure of CO_2 in equilibrium with a CaO-$CaCO_3$ mixture at various temperatures gives the following results.

P_{CO_2}(atm)	0.029	1.05	3.9	11.5	28.6
T(°K)	973	1173	1273	1373	1473

Calculate ΔH^0 for the reaction by a graphical method.

9-3.
Consider the reaction

$$CO_2(g) + C(s) = 2\ CO(g)$$

(a) What will be the effect on the point of equilibrium of the reaction (shift right, shift left, no effect) of making the following changes on a system that is initially at equilibrium? Assume that all other factors are held constant.
   (1) Adding additional CO(g) to the system.
   (2) Adding additional C(s) to the system.
   (3) Removing $CO_2(g)$ from the system.
   (4) Compressing the system.
(b) Will the reaction take place to the right at 25° C with all products and reactants in their standard states?
(c) Will an increase in temperature, all other factors held constant, cause the point of equilibrium to shift right, shift left, or be unaffected?
(d) Calculate P_{CO} at equilibrium (25° C) if the equilibrium pressure of the CO_2 is 1.00 atm, and the only source of CO is from the decomposition of CO_2.
(e) Which of the following will happen to a mixture (at 25° C) of CO_2(g, 0.899 atm), CO(g, 0.00114 atm), and C(s)?
   (1) The partial pressure of the CO(g) will decrease.
   (2) The partial pressure of the $CO_2(g)$ will decrease.
   (3) No change will take place, since these are the equilibrium conditions for the reaction.
   (Show calculations, and give rationale for answer.)
(f) Calculate ΔG^0 at 600° K. Assume that ΔH^0 is constant over the temperature range 298–600° K.

9-4.
The equations below are from important industrial processes. What will be the effect on the yield (extent of conversion to products) of these reactions of (a) increasing the pressure on the system at constant temperature, and (b) increasing the temperature at constant pressure? Report answers as (1) increase, (2) decrease, or (3) have no effect on the yield. Necessary data are given in the Appendix. Assume that all reactions are carried out in closed containers; that is, nothing escapes from the reaction vessel.

$$S(s) + O_2(g) = SO_2(g)$$
$$2\ SO_2(g) + O_2(g) = 2\ SO_3(g)$$
$$N_2(g) + 3\ H_2(g) = 2\ NH_3(g)$$
$$4\ NH_3(g) + 5\ O_2 = 4\ NO(g) + 6\ H_2O(g)$$
$$C(s) + H_2O(g) = CO(g) + H_2(g)$$
$$2\ NaHCO_3(s) = Na_2CO_3(s) + H_2O(g) + CO_2(g)$$

10

Applications of Chemical Equilibrium

In Chapter 9 we saw that by the selection of appropriate standard states we could conveniently apply the equilibrium principle to systems consisting of gases, liquids, solids, undissociated species, ions, and any combination of these. The equilibrium principle is completely general and does not require modification or adjustment to fit a particular system; in this chapter we will apply it to many systems. As you proceed through the chapter, you will notice that some equilibrium constants are given identifying names and that a preponderance of the examples involve ions in aqueous solutions. The names are for convenience in discussing chemical equilibrium, and the many examples concerning aqueous solutions are to emphasize that a great deal of interesting chemistry occurs in such solutions.

10.1 Chemical Equilibria in Gas Phases

In the section titled "Activity and Standard States" in Chapter 9, it was shown that the equilibrium constant for a reaction can be expressed in terms of the activities of the species in the reaction. It was shown also that for gas-phase reactions the partial pressure closely approximates the activity. Therefore, for the general gas-phase reaction

$$aA + bB + \ldots = lL + mM + \ldots$$

the equilibrium constant can be written

$$K = \frac{(P^l_L)(P^m_M)\ldots}{(P^a_A)(P^b_B)\ldots}$$

and its numerical value is calculated from free energy data by use of the expression

$$\Delta G^0 = -RT \ln K$$

Since the standard free energy data are based on pressures in atmospheres, the equilibrium constants calculated directly from these must also be based on pressures in atmospheres. Example 9.3 of the preceding chapter and Examples 10.1, 10.2, and 10.3 that follow all illustrate the calculation of equilibrium conditions in gas mixtures. Notice that although the data are given in different forms, the equilibrium principle is applicable to each example.

EXAMPLE 10.1 The equilibrium constant for

$$2\,NO_2(g) = N_2O_4(g)$$

is 8.81 at 25° C. If 3.0 moles of N_2O_4 are placed in a 4.0-liter container, and the system is maintained at 25° C until equilibrium is attained, how many moles of each gas are finally present?

SOLUTION The equilibrium constant for the reaction is

$$\frac{P_{N_2O_4}}{P^2_{NO_2}} = 8.81$$

If we assume that the behavior of the gases is ideal, the initial and equilibrium conditions can be expressed as follows (x is the number of moles of N_2O_4 that decompose; the pressures are calculated from the number of moles by use of the ideal gas equation):

	MOLES	PRESSURES
Initially:	3.0 moles N_2O_4	$\frac{3.0RT}{V}$ atm N_2O_4
	0 moles NO_2	0 atm NO_2
Finally:	$(3.0 - x)$moles N_2O_4	$\frac{(3.0 - x)RT}{V}$ atm N_2O_4
	$2x$ moles NO_2	$\frac{2xRT}{V}$ atm NO_2

These values substituted into the expression for the equilibrium constant give

$$\frac{(3.0 - x)RT/V}{(2xRT/V)^2} = 8.81$$

From this,

$$x = 0.12 \text{ mole}$$

Therefore,

$$2x = 0.24 \text{ mole } NO_2$$

and

$$3.0 - x = 2.9 \text{ moles } N_2O_4$$

are present at equilibrium.

EXAMPLE 10.2 At 395° C, $K = 0.0444$ for the dissociation of carbonyl chloride, $COCl_2$, to CO and Cl_2. A sample of $COCl_2$ is placed in a container and the system is heated to 395° C. The sample size and volume are such that the pressure before decomposition begins is 0.100 atm. Calculate the equilibrium partial pressures of the three components of the gas mixture.

SOLUTION The dissociation reaction is

$$COCl_2 = CO + Cl_2$$

and, for this,

$$K = \frac{(P_{CO})(P_{Cl_2})}{P_{COCl_2}} = 0.0444$$

The pressure and the number of moles are proportional if volume and temperature are constant. This is seen if the ideal gas equation is rearranged to

$$P = n\frac{RT}{V} = nk$$

Therefore, we can express the initial and equilibrium conditions as well in terms of pressures or moles.

	MOLES	PRESSURES
Initially:	n_i moles $COCl_2$	0.100 atm $COCl_2$
	0 moles CO	0 atm CO
	0 moles Cl_2	0 atm Cl_2

At equilibrium, x moles of $COCl_2$ dissociated, or, stated differently, the pressure of $COCl_2$ has decreased by y atmospheres.

	MOLES	PRESSURES
At equilibrium:	$n_i - x$ moles $COCl_2$	$0.100 - y$ atm $COCl_2$
	x moles CO	y atm CO
	x moles Cl_2	y atm Cl_2

Hence,

$$\frac{(P_{CO})(P_{Cl_2})}{P_{COCl_2}} = \frac{y^2}{0.100 - y} = 0.0444$$

or

$$y^2 + 0.0444y - 0.00444 = 0$$

This is solved by the quadratic formula to give $y = 0.0481$. Therefore,

$$P_{CO} = P_{Cl_2} = 0.0481 \text{ atm}$$
$$P_{COCl_2} = 0.0519 \text{ atm}$$

An alternative solution makes use of the concept of the *degree of dissociation*. We can write a parallel set of terms for number of moles and pressures based on the degree of dissociation defined as

$$\alpha = \frac{n_d}{n_i}$$

where n_i is the initial number of moles of a substance and n_d is the number of moles dissociated. Hence,

$$\alpha n_i = n_d$$

Similarly, αP_i is the decrease in pressure of a substance as the result of its partial dissociation. Therefore, for the $COCl_2$ dissociation,

	MOLES	PRESSURES
Initially:	n_i moles $COCl_2$	0.100 atm $COCl_2$
	0 moles CO	0 atm CO
	0 moles Cl_2	0 atm Cl_2
At equilibrium:	$n_i - \alpha n_i$ moles $COCl_2$	$(0.100 - 0.100\alpha)$ atm $COCl_2$
	αn_i moles CO	0.100α atm CO
	αn_i moles Cl_2	0.100α atm Cl_2

If we substitute the pressure terms into the equilibrium-constant expression, we obtain

$$\frac{(0.100\alpha)^2}{(0.100 - 0.100\alpha)} = 0.0444$$

This is rearranged to

$$\alpha^2 + 0.444\alpha - 0.444 = 0$$

which, solved, gives

$$\alpha = 0.481$$

Hence

$$P_{Cl_2} = P_{CO} = 0.100 \times 0.481 = 0.0481 \text{ atm}$$

and

$$P_{COCl_2} = 0.0519 \text{ atm}$$

In the preceding two examples, initial conditions were already known. In the laboratory, however, a type of problem that is more commonly encountered is one in which the *final* (equilibrium) pressure of the system has been measured, and we need to calculate the partial pressures of each species present. An example follows.

EXAMPLE 10.3 Elemental diatomic gases (H_2, N_2, O_2, F_2, Cl_2, Br_2, and I_2) dissociate somewhat to give the atoms. The chemical equation for the dissociation of chlorine is

$$Cl_2(g) = 2\,Cl(g)$$

Using the free energy data in Appendix B, calculate the partial pressure of atomic chlorine in chlorine gas at 25° C and 2.00 atmospheres total final pressure.

SOLUTION K is obtained from free energy data:

$$\Delta G^0 = 2\Delta G^0_{f(\mathrm{Cl})} - \Delta G^0_{f(\mathrm{Cl_2})}$$

$$\Delta G^0 = (2)(105{,}400) - 0 = 210{,}800 \text{ joules}$$

$$\ln K = \frac{-\Delta G^0}{RT} = \frac{-210{,}800 \text{ joules}}{(8.314 \text{ joules/}^\circ\text{K})(298.15^\circ \text{ K})}$$

$$\ln K = -85.04$$

$$K = 1.2 \times 10^{-37}$$

From the statement of the problem,

$$P_{\mathrm{Cl}} + P_{\mathrm{Cl_2}} = 2.00 \text{ atm}$$

or

$$P_{\mathrm{Cl_2}} = 2.00 - P_{\mathrm{Cl}}$$

Therefore

$$K = \frac{P^2_{\mathrm{Cl}}}{P_{\mathrm{Cl_2}}} = \frac{P^2_{\mathrm{Cl}}}{2.00 - P_{\mathrm{Cl}}} = 1.2 \times 10^{-37}$$

which in quadratic form is

$$P^2_{\mathrm{Cl}} + 1.2 \times 10^{-37} P_{\mathrm{Cl}} - 2.4 \times 10^{-37} = 0$$

This expression can be solved for P_{Cl} by use of the quadratic formula. The computations are relatively simple in this case, since—as can be seen below—several terms are too small to add:

$$P_{\mathrm{Cl}} = \frac{-1.2 \times 10^{-37} + \sqrt{1.4 \times 10^{-74} + 9.6 \times 10^{-37}}}{2}$$

$$P_{\mathrm{Cl}} = 4.9 \times 10^{-19} \text{ atm}$$

This extremely low pressure of atomic chlorine shows that the dissociation of Cl_2 at 25° C is very slight.

10.2 Chemical Equilibria in Mixed Phases

In Chapter 8 we discussed equilibria between different phases of a single substance. In this section we consider chemical equilibria that reach across phase boundaries.

EXAMPLE 10.4 The dissociation of mercuric oxide is represented by the equation

$$2\ \mathrm{HgO}(s) = 2\mathrm{Hg}(g) + \mathrm{O_2}(g)$$

When HgO is maintained at 450° C, the equilibrium pressure is found to be 1.07 atm. Calculate the equilibrium constant for the reaction at 450° C.

SOLUTION The equilibrium constant is

$$K = (P_{Hg}^2)(P_{O_2})$$

since the activity of HgO(*s*) is 1. The total pressure is

$$P_{Hg} + P_{O_2} = 1.07 \text{ atm}$$

and the stoichiometry of the reaction shows that

$$P_{Hg} = 2\,P_{O_2}$$

These two equations are solved simultaneously to give $P_{O_2} = 0.357$ atm and $P_{Hg} = 0.714$ atm. Therefore,

$$K = (0.714)^2(0.357) = 0.182$$

EXAMPLE 10.5 At 100° C, $K = 0.23$ for the reaction

$$2\ NaHCO_3(s) = Na_2CO_3(s) + CO_2(g) + H_2O(g)$$

Calculate the weight of Na_2CO_3 formed if a 5.0 g sample of $NaHCO_3$ is placed in a 1.0-liter container and the system is maintained at 100° C until equilibrium is reached.

SOLUTION Let us first plan the method of attacking the problem:

(a) From the stoichiometry of the reaction and K we can calculate the pressures of CO_2 and of H_2O.
(b) The number of moles of each of these can be calculated from the ideal gas equation.
(c) One mole of Na_2CO_3 is produced for each mole of CO_2 (or for each mole of H_2O).
(d) The weight of Na_2CO_3 is calculated from the number of moles.

The equilibrium constant is

$$K = (P_{CO_2})(P_{H_2O}) = 0.23$$

Since the pressures of CO_2 and H_2O are equal,

$$P_{CO_2} = \sqrt{0.23} = 0.48 \text{ atm}$$

The number of moles of CO_2 is given by the ideal gas equation:

$$n_{CO_2} = \frac{P_{CO_2}V}{RT} = \frac{(0.48 \text{ atm})(1.0 \text{ liter})}{(0.082 \text{ liter-atm/}^\circ\text{K-mole})(373^\circ \text{ K})}$$

$$n_{CO_2} = 0.016 \text{ mole}$$

Therefore, 0.016 mole of Na_2CO_3 is produced, and this weighs

$$(0.016 \text{ mole})\left(\frac{106 \text{ g}}{\text{mole}}\right) = 1.7 \text{ g}$$

The initial weight of $NaHCO_3$ does not enter into the calculation, for the equilibrium depends upon that substance being present and not upon the quantity. If, however, less than 2 × 0.016 mole had been present, all would have decomposed and the system would not have been at equilibrium. The minimum initial weight of $NaHCO_3$ that would come to equilibrium under the conditions of this problem is

$$(0.032 \text{ mole})\left(\frac{84 \text{ g}}{\text{mole}}\right) = 2.7 \text{ g}$$

It is often a help in solving problems to map a method of attack as was done in this example. This procedure keeps one from becoming lost in unproductive calculations.

10.3 Acid-Base Equilibria

In Chapter 4, while considering the stoichiometry of reactions, we discussed the Arrhenius theory of acid-base reactions. In the Arrhenius theory an acid is a hydrogen-containing substance that dissolves in water to give hydrogen ions, and a base is a substance that contains the OH group and dissociates in water to give hydroxide ions.

We are now ready to introduce a theory of acids and bases that is more general in that it is not restricted to water solutions. Such a theory was proposed independently in 1923 by both T. M. Lowry, an English chemist, and J. N. Brønsted, a Danish chemist. This theory, sometimes called the proton theory, defines an acid as a proton (H^+) donor and a base as a proton acceptor. Notice that no restriction is made concerning the solvent and, further, that no requirement is made of the base except that it will take up a proton. Brønsted-Lowry acids and bases may be molecules or ions, and the reactions between them can occur in various solvents or independent of any solvent.

The **Brønsted-Lowry acid-base reaction** is the transfer of a proton from the acid (proton donor) to the base (proton acceptor). Many reactions that qualified as Arrhenius acid-base reactions also fit as Brønsted-Lowry reactions. For example, in equation [10.1] acetic acid and hydroxide ion react by the transfer of a proton from acetic acid to the hydroxide ion:

$$CH_3COOH + OH^- = H_2O + CH_3COO^- \qquad [10.1]$$

The result of this reaction is the formation of the **conjugate acid,** water, and the **conjugate base,** acetate ion. Let us rewrite equation [10.1], this time labeling

There are several striking historical examples of two or more scientists presenting the same proposal or theory independently and almost simultaneously. This is not as surprising as it may first seem, for there is much interchange of information among workers in a field of study. At any particular time, many investigators are studying the same problem, and each is aware of past work in the field through the scientific literature, conferences, and meetings. Thus, by 1923, problems with the classical theory caused many people to consider alternative models. Within the space of a few months, Brønsted and Lowry proposed virtually identical solutions to the problems.

the reactants and products:

$$\underset{\text{acid}}{CH_3COOH} + \underset{\text{base}}{OH^-} = \underset{\substack{\text{conjugate}\\ \text{acid}}}{H_2O} + \underset{\substack{\text{conjugate}\\ \text{base}}}{CH_3COO^-}$$

In generalized form, the Brønsted-Lowry acid-base reaction is

$$\text{acid} + \text{base} = \text{conjugate acid} + \text{conjugate base}$$

or

$$\text{acid}_1 + \text{base}_2 = \text{acid}_2 + \text{base}_1 \qquad [10.2]$$

As we will see, each species may be uncharged, positively charged, or negatively charged.

If in equation [10.2] acid_1 is strong and base_2 is strong, it follows that their conjugates (base_1 and acid_2) are weak, for the strong Brønsted-Lowry acid donates its proton readily, and the strong base accepts a proton readily. If the acid HA has a strong tendency to donate its proton, it follows that its conjugate base A^- will not be a strong base. Since HNO_3 is a strong acid, NO_3^- is a weak base.

The gas-phase reaction of hydrogen chloride and ammonia illustrates the additional breadth of the Brønsted-Lowry definition:

$$HCl(g) + NH_3(g) = NH_4^+ + Cl^-$$

A proton from HCl (the acid) goes to NH_3 (the base) to form NH_4^+ (the conjugate acid) and Cl^- (the conjugate base). This reaction does not qualify as an Arrhenius acid-base reaction since neither water nor a base containing the OH group is present, but it does meet all the Brønsted-Lowry requirements. Ammonia is a base without any pretense of NH_4OH formation, and the ammonium ion appears as an acid. In the Arrhenius (classical) theory, NH_4^+ would be considered only as a positive ion (cation).

Positive ions are called *cations* and negative ions are called *anions*. These names had their origin in electrochemistry, in which reactions occur at a pair of electrodes, one called the *cathode* and the other the *anode*.

An interesting aspect of the Brønsted-Lowry theory is that a substance may be both an acid and a base. For example, water acts as a proton donor in its reaction with ammonia:

$$H_2O + NH_3 = NH_4^+ + OH^- \qquad [10.3]$$

Water also reacts as a proton acceptor or base: The addition of perchloric acid results in the formation of protonated water, the **hydronium ion:**

$$HClO_4 + H_2O = H_3O^+ + ClO_4^- \qquad [10.4]$$

Because the reactions in equations [10.3] and [10.4] are both clearly acid-base reactions in the Brønsted-Lowry system, water qualifies in that system as either an acid or a base.

The interesting conclusion reached by chemists after considering reactions such as those in equations [10.3] and [10.4] was that the apparent strength of an acid (HA) or of a base depends not only upon the nature of the acid or base but also upon the solvent in which it is dissolved. Thus, if the solvent (Sol) is a relatively strong base (proton acceptor), the extent of the reaction

$$HA + Sol = HSol^{+} + A^{-}$$

will be greater than if the solvent has little tendency to accept a proton. Similarly, the strength of a base depends in part upon the readiness with which the solvent gives up a proton.

We will use the Brønsted-Lowry system in most of the acid-base equilibria in this chapter because of the clarity it gives to the effect of the solvent. Frequently, however, we will show how the reaction appears in the Arrhenius theory, so that you will be acquainted with both systems. Subsequently in the text, and especially where the emphasis is not upon the solvent interaction as an acid or a base, we will represent the proton in aqueous solution as H^{+}, meaning $H^{+}(aq)$ or H_3O^{+}, in order to avoid having large numbers of water molecules appearing in the equations.

Strengths of Acids and Bases

A convenient and useful table of Brønsted-Lowry acids and bases can be made in which these are listed in conjugate pairs and in order of strength. By the definitions of acids and bases it is seen that a strong acid must have a conjugate base that is weak. If the base were strong, it would be a good proton acceptor, which would be incompatible with the fact that the acid is strong and hence is an effective donor of its proton. Therefore, if a list is made in which acids are shown in order of *decreasing* strength, the conjugate bases will fall into the order of *increasing* strength. Table 10.1 is such a listing. The common strong acids are included to help you become familiar with them. Next appears the hydronium ion followed by a selection of increasingly weak acids.

An acid tends to transfer a proton to any base below it in the table; if more than one base is present, that which is lowest is the strongest proton acceptor and will react most with the added acid. Thus, HSO_4^{-} tends to transfer its proton to CO_3^{2-}, and the overall reaction is

$$HSO_4^{-} + CO_3^{2-} = HCO_3^{-} + SO_4^{2-}$$

or, to OH^{-},

$$HSO_4^{-} + OH^{-} = H_2O + SO_4^{2-}$$

If a limited quantity of HSO_4^{-} is added to a solution that contains larger quantities of both OH^{-} and CO_3^{2-}, the reaction will occur with OH^{-} to the virtual exclusion of the reaction with CO_3^{2-}, since OH^{-} is the stronger base. This fact leads to an interesting conclusion. Table 10.1 is based on water as the common solvent; consequently, the species H_2O is always present in abundance.

Lewis acids and bases. In 1923 the American chemist G. N. Lewis proposed a theory of acids and bases that is even more general than the Brønsted-Lowry theory. He defined a base as a substance that can donate a pair of electrons and an acid as a substance that can accept a pair of electrons. In the reaction of a proton with ammonia,

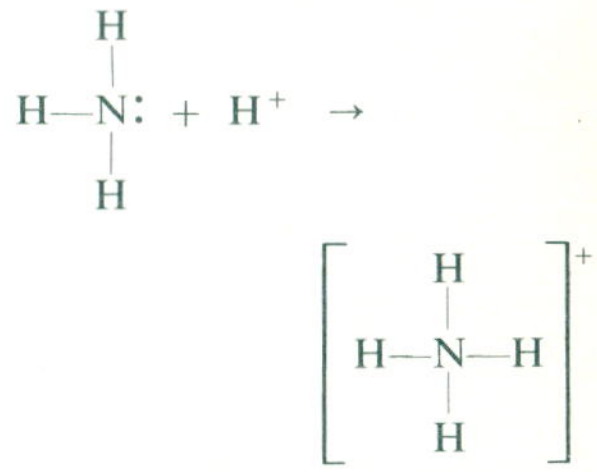

the proton accepts a pair of electrons from the ammonia molecule and is, therefore, an acid. The ammonia is a base, for it donates a pair of electrons. Similarly, complex ion formations such as

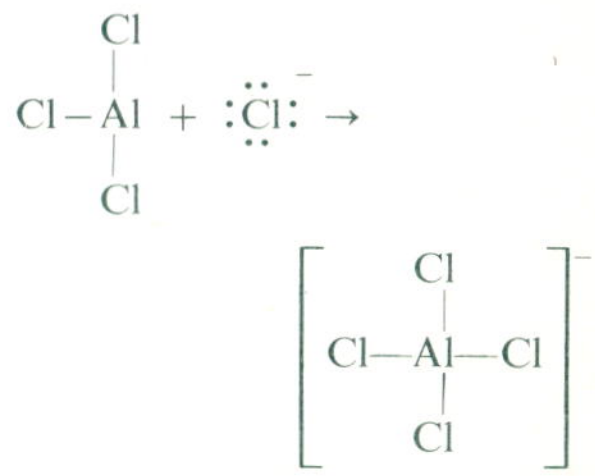

and

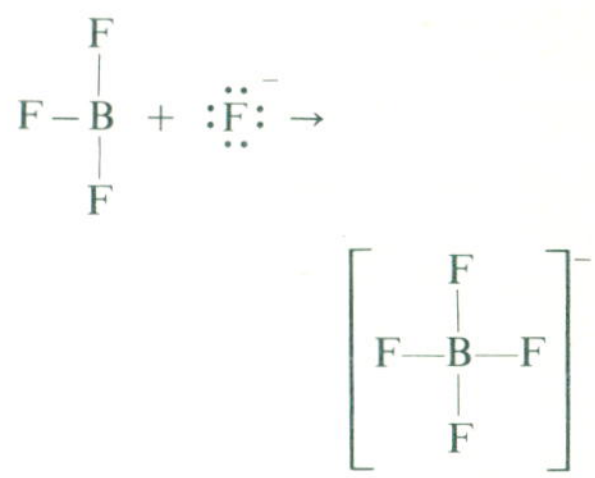

can be thought of as reactions of Lewis acids and Lewis bases.

TABLE 10.1

Brønsted-Lowry Conjugate Pairs

ACID		CONJUGATE BASE	
perchloric acid	$HClO_4$	ClO_4^-	perchlorate ion
hydrochloric acid	HCl	Cl^-	chloride ion
nitric acid	HNO_3	NO_3^-	nitrate ion
sulfuric acid	H_2SO_4	HSO_4^-	hydrogen sulfate ion
hydronium ion	H_3O^+	H_2O	water
hydrogen sulfate ion	HSO_4^-	SO_4^{2-}	sulfate ion
phosphoric acid	H_3PO_4	$H_2PO_4^-$	dihydrogen phosphate ion
formic acid	$HCOOH$	$HCOO^-$	formate ion
acetic acid	CH_3COOH	CH_3COO^-	acetate ion
dihydrogen phosphate ion	$H_2PO_4^-$	HPO_4^{2-}	monohydrogen phosphate ion
hydrogen sulfide (hydrosulfuric acid)	H_2S	HS^-	hydrogen sulfide ion
ammonium ion	NH_4^+	NH_3	ammonia
hydrogen carbonate ion	HCO_3^-	CO_3^{2-}	carbonate ion
water	H_2O	OH^-	hydroxide ion
hydroxide ion	OH^-	O^{2-}	oxide ion
ammonia	NH_3	NH_2^-	amide ion

Any acid that is stronger than the hydronium ion, H_3O^+, will have a pronounced tendency to react with the plentiful H_2O to form H_3O^+. If $HClO_4$ is added to an aqueous solution of $NaHSO_4$, the proton-transfer reaction is

$$HClO_4 + H_2O = H_3O^+ + ClO_4^-$$

rather than

$$HClO_4 + HSO_4^- = H_2SO_4 + ClO_4^-$$

because water is a much stronger base than is the hydrogen sulfate ion. No matter what strong acid is added to water, the result is that a proton is transferred to water to form H_3O^+, and one concludes that *the hydronium ion is the strongest acid that can exist in high concentration in aqueous solution.* Similarly, the addition to water of a base stronger than OH^- invariably results in a proton transfer from the abundant acid water to form OH^-, and it is seen that *OH^- is the strongest base that can exist in high concentration in aqueous solution.* The two dotted lines in Table 10.1, therefore, enclose the acids and bases that are found in water systems. Because of the apparent identical strengths of very strong acids and very strong bases in aqueous solution, water is said to have a *leveling effect.* Other solvents have leveling effects also, in each case dependent upon the proton-donating and proton-accepting ability of the solvent. For

These proton transfers are equilibrium reactions. The strong acids so completely transfer their protons to water, however, that the equilibrium concentration of such species as $HClO_4$ or HCl in a dilute aqueous solution is negligible.

example, when liquid ammonia is used as a solvent, the addition of a strong acid produces the ammonium ion,

$$HA + NH_3 = NH_4^+ + A^-$$

and a strong base produces the amide ion,

$$B + NH_3 = BH^+ + NH_2^-$$

The strongest acid in liquid ammonia is NH_4^+, and the strongest base is NH_2^-.

One should gain familiarity with Table 10.1 in order to avoid the pitfalls of writing species such as O^{2-} or $HClO_4$ in equations that represent equilibria in aqueous solution. Example 10.6 shows the possibility of predicting the species in aqueous solutions.

EXAMPLE 10.6 Predict, with the aid of Table 10.1, the predominant species in each of the following solutions:

(a) HCl in water.
(b) Na_2O in water. (Na_2O is soluble and ionizes completely.)
(c) CH_3COOK in water. (Potassium acetate is soluble and ionic, $K^+ + CH_3COO^-$.)

SOLUTION

(a) From its position in Table 10.1 we see that HCl is so strong that H_3O^+ and Cl^- are the predominant species.
(b) The oxide ion is a stronger base than OH^-; the result is

$$O^{2-} + H_2O = 2\,OH^-$$

and the predominant species are Na^+ and OH^-.
(c) Water as an acid is below acetate ion; hence there is only slight reaction by

$$CH_3COO^- + H_2O = CH_3COOH + OH^-$$

to give hydroxide ion and acetic acid. The predominant species are, therefore, acetate ion and potassium ion.

We have spoken of the strong tendency for reaction to occur when an acid is added to a base that is lower in Table 10.1. But what of the case where an acid is added to a base that is higher in the table? As long as we stay within the bounds of the acids and the bases that exist in water solution, we can make the following statement: Any acid will react with any base, but the tendency for reaction is very small when the base is higher than the acid in the table; the greater that separation, the smaller the reaction. For example, consider the reactions of HSO_4^- and CH_3COOH with the base water:

(a) $HSO_4^- + H_2O = H_3O^+ + SO_4^{2-}$

(b) $CH_3COOH + H_2O = H_3O^+ + CH_3COO^-$

The separation (given in Table 10.1) is greater for acetic acid and water than for hydrogen sulfate ion and water. Therefore, we predict that reaction (b) will not proceed as far as will reaction (a). We observe that in either case the extent of the reaction with water is indicated by the concentration of H_3O^+. The prediction is borne out by the fact that 0.20 *M* HSO_4^- has $[H_3O^+] = 4.3 \times 10^{-2}$, while 0.20 *M* CH_3COOH has $[H_3O^+] = 1.9 \times 10^{-3}$. (We will see in Example 10.9 how to calculate $[H_3O^+]$ in such solutions.)

Self-Ionization of Water

Since we have observed that water can act either as an acid or a base, it is not surprising that to a small but measurable extent it reacts with itself:

$$H_2O + H_2O = H_3O^+ + OH^- \qquad [10.5]$$

Here the acid water reacts with the base water to give the conjugate acid and base. Such reactions are called *autoprotolysis* or *self-ionization* and occur to a limited extent in solvents capable of acting both as acids and bases. The equilibrium condition for equation [10.5] is

$$K = \frac{(a_{H_3O^+})(a_{OH^-})}{a_{H_2O}^2}$$

Referring to the discussion of standard states in Chapter 9, we conclude that if we consider the application of equation [10.5] to pure water, the liquid water is in its standard state and that the term a_{H_2O} has precisely the value of 1. The denominator, $a_{H_2O}^2$, therefore, also has the value unity, and we can write, with no approximation whatever,

$$K = \frac{(a_{H_3O^+})(a_{OH^-})}{a_{H_2O}^2} = (a_{H_3O^+})(a_{OH^-}) \qquad [10.6]$$

This equilibrium constant is of such great importance and frequent use in work involving aqueous solutions that it is given an identifying name, the **water constant,** K_w. The term water constant is applied always to the expression of equation [10.6], and the term carries the full implication of the more detailed name for that expression, "the equilibrium constant for the self-ionization of water."

In a following section it is shown that the activity of water is approximately 1 in dilute aqueous solutions. Therefore, in such solutions it is a good approximation that $K = (a_{H_3O^+})(a_{OH^-})$.

As was indicated in Chapter 9, on page 220, the activity of a solute can be approximated by its molarity. The lower the concentration, the better the quality of this approximation. The extent of the self-ionization of water is very small. As we saw in Example 9.7, less than one molecule of water in 100 million has undergone such reaction at 25° C; consequently, it is a good approximation to use molarity in place of activity.

Therefore, equation [10.6] becomes

$$K_w = (a_{H_3O^+})(a_{OH^-}) \cong [H_3O^+][OH^-] \qquad [10.7]$$

Before making numerical calculations regarding the self-ionization of water, we should make one other observation. Chemists are often inclined to write the expression of the preceding paragraphs in a short-hand version, remembering that the proton is hydrated in aqueous solution but writing the self-ionization reaction of equation [10.5] in the simpler Arrhenius form

$$H_2O = H^+ + OH^-$$

The equilibrium-constant expression for this is

$$K_w = \frac{(a_{H^+})(a_{OH^-})}{a_{H_2O}} = (a_{H^+})(a_{OH^-})$$

The same argument as was used above is made for the activity of water having the value unity. Continuing, then, we obtain

$$K_w \cong [H^+][OH^-] \qquad [10.8]$$

which is identical with equation [10.7] except that in the former equation the hydration of the proton is explicit whereas here it is implicit.

Careful measurements of the water constant, K_w, show that it has the numerical value 1.00×10^{-14} at 25° C:

$$K_w = [H_3O^+][OH^-] = 1.00 \times 10^{-14} \qquad [10.9]$$

Notice that this time the approximation sign, $\cong$, was replaced by an equal sign; for calculations that appear in the remainder of this text, it will be assumed that molarities can be used for activities of solutes. We will ordinarily work with low concentrations, for which this approximation is reasonably valid. However, we will mention the approximation from time to time to remind you to be aware of it each time you make it.

We saw in Chapter 9 that equilibrium constants are temperature dependent. The water constant is 1.00×10^{-14} at 25° C, but decreases to 1.0×10^{-15} at 0° C and increases to 1.0×10^{-13} at 60° C. One must be careful to use the correct value of the constant for the temperature of the system.

Strong Acids and Strong Bases, and the Concept of pH

In pure water the concentrations of H_3O^+ and OH^- are equal, and each has the value 1.00×10^{-7} *M* (Example 9.7). If a strong acid is added, $[H_3O^+]$ increases and $[OH^-]$, as seen in equation [10.9], decreases.

EXAMPLE 10.7 A 2.0 ml portion of a solution that contains 0.10 mole of HCl per liter is diluted to 100 ml. Calculate $[H_3O^+]$ and $[OH^-]$.

The symbol mmole is used for millimole. Notice that a solution that is 0.10 mole/liter is also 0.10 mmole/ml. Calculations involving volumes of solutions of the order of milliliters are often most conveniently made if the concentrations are expressed in mmole/ml.

SOLUTION There are two acids in this solution, HCl and H_2O. Since one, HCl, is strong and the other, H_2O, is weak, we are justified in using the approximation that the concentration of HCl determines the $[H_3O^+]$. The concentration of hydronium ion in the diluted solution is given by

$$[H_3O^+] = (2.0 \text{ ml})\left(\frac{0.10 \text{ mmole}}{\text{ml}}\right)\left(\frac{1}{100 \text{ ml}}\right) = 2.0 \times 10^{-3}\ M$$

The hydroxide ion concentration is calculated from the water constant, equation [10.9]:

$$[OH^-] = \frac{K_w}{[H_3O^+]} = \frac{1.0 \times 10^{-14}}{2.0 \times 10^{-3}} = 5.0 \times 10^{-12}\ M$$

pH is defined properly in terms of the activity of H^+. In Chapter 4 we made use of the approximate definition.

It is apparent from Example 10.7 that the concentrations of hydronium ion and hydroxide ion can vary over many orders of magnitude. Therefore, a common practice is to use a compact logarithmic scale, called the *pH scale*, for expressing these concentrations. The term **pH** is defined as

$$\text{pH} = -\log a_{H_3O^+} \cong -\log [H_3O^+]$$

A neutral solution (pure water) has a pH of 7.00 whereas the solution in Example 10.7 has a pH of 2.70. Similar negative logarithmic terms are used for other small values in chemistry, such as $[OH^-]$ and K. Thus, **pOH** is $-\log [OH^-]$ and pK is $-\log$ K. The pOH of pure water is 7.00; that of the HCl solution in Example 10.7 is 11.30. This latter value can be obtained either by taking the negative logarithm of 5.0×10^{-12} or by noting that pK_w (the negative logarithm of the water constant) is 14.00 and that pH + pOH = pK_w. Thus, pOH = 14.00 − 2.70 = 11.30.

EXAMPLE 10.8 Calculate the pH of a solution obtained by mixing 1.0 ml of a solution containing 0.10 mole of HCl per liter, 1.0 ml of a solution containing 0.20 mole of NaOH per liter, and enough water to make the volume 200 ml.

SOLUTION A strong acid and a strong base are mixed. The quantity of hydronium ion added is 1.0 ml × 0.10 mmole/ml = 0.10 mmole, and the quantity of hydroxide ion is 1.0 ml × 0.20 mmole/ml = 0.20 mmole. The reaction of H_3O^+ and OH^- to produce water consumes 0.10 mmole of each and leaves 0.10 mmole of OH^-. The concentration is, therefore,

$$[OH^-] = (0.10 \text{ mmole})\left(\frac{1}{200 \text{ ml}}\right) = 5 \times 10^{-4}\ M$$

Hence,

$$[H_3O^+] = \frac{K_w}{[OH^-]} = \frac{1.0 \times 10^{-14}}{5 \times 10^{-4}} = 2.0 \times 10^{-11}\ M$$

and

$$\text{pH} = 10.70$$

We need to consider one problem that was introduced in the two examples above. The activity of water is precisely *one* in pure water, but the question is whether chemists are justified in continuing to use *one* for the activity of water when they are working with a solution and not with pure water. A simple method of answering this question is to consider actual activity values for water in solutions of various concentrations. Table 10.2 shows some such data using HCl as the solute. Notice that even in 1 M HCl, the error caused by using $a_{H_2O} = 1$ is less than 4%. Most equilibrium constants are not known so accurately as ±4%; therefore, one concludes that in dilute aqueous solutions, it is a good approximation to use unity for the activity of water.

TABLE 10.2

Activity of Water in Hydrochloric Acid Solutions of Various Concentrations

$\frac{\text{Moles HCl}}{\text{Liter}}$	a_{H_2O}
0.0010	0.9999+
0.010	0.999+
0.10	0.997
1.0	0.963

Weak Acids and Weak Bases

When a weak acid is added to water, the reaction to form hydronium ion is not complete, and the calculation of the pH of the resulting solution is somewhat more complex than it was in the case of a strong acid. Consider a weak acid, HA. The equilibrium constant for its reaction with water

$$HA + H_2O = H_3O^+ + A^- \qquad [10.10]$$

is

$$K_a = \frac{(a_{H_3O^+})(a_{A^-})}{(a_{HA})(a_{H_2O})} \cong \frac{[H_3O^+][A^-]}{[HA]x_{H_2O}} \cong \frac{[H_3O^+][A^-]}{[HA]} \qquad [10.11]$$

This constant is called the **acid constant** or the acid ionization constant or the acid dissociation constant. These latter names have their origin in Arrhenius theory, in which the acid, HA, is said to ionize or dissociate,

$$HA = H^+ + A^-$$

and the interaction with water is not apparent. The names persist, and the proton-transfer reaction shown in equation [10.10] is commonly called the dissociation or ionization reaction of HA.

In 0.10 M HA solution, the concentration of H_3O^+ is greater than it is in pure water, and equation [10.10] indicates that $[H_3O^+]$ and $[A^-]$ are produced in equal quantities. The concentrations of these species would appear, therefore, to be identical. There is, however, a source of H_3O^+ other than the dissociation of HA; the water undergoes self-ionization (equation [10.5]) and makes a very small contribution to the concentration of H_3O^+ in the solution. In most of the examples that we will consider, this contribution of H_3O^+ by the water is so small that it is negligible when compared with that from the weak acid; however, one needs to be aware always of the ionization of the water. If we let $x = [A^-]$, then $[H_3O^+]$ is essentially equal to x, and $[HA] = 0.10 - x$. These values substituted into equation [10.11] give

$$\frac{x^2}{0.10 - x} = K_a$$

If K_a is known, the expression can be solved and the concentrations of all species calculated. Some numerical examples will demonstrate this. Numerical values of acid and base constants are given in Appendix E.

EXAMPLE 10.9 Calculate $[H_3O^+]$, $[CH_3COO^-]$, $[CH_3COOH]$, and $[OH^-]$ in 0.20 *M* CH_3COOH. K_a of acetic acid is 1.8×10^{-5}.

SOLUTION The dissociation of the acetic acid is

$$CH_3COOH + H_2O = H_3O^+ + CH_3COO^-$$

and the acid constant is

$$\frac{[CH_3COO^-][H_3O^+]}{[CH_3COOH]} = K_a = 1.8 \times 10^{-5}$$

Let $y = [CH_3COO^-] = [H_3O^+]$, and $0.20 - y = [CH_3COOH]$. Therefore,

$$\frac{y^2}{0.20 - y} = 1.8 \times 10^{-5}$$

This equation can be solved by the quadratic formula, but an approximation method is more commonly applied to such expressions. Consider the denominator of the left side of the equation. If y is less than 0.005, it is insignificant relative to 0.20. Therefore, let us assume that $y < 0.005$ (or, in other words, that $y \ll 0.20$, where $\ll$ means *much less than*) and complete the calculation on that assumption. We can then check to see if the assumption was valid. Therefore,

$$\frac{y^2}{0.20} \cong 1.8 \times 10^{-5}$$

and

$$y \cong \sqrt{3.6 \times 10^{-6}} = 1.9 \times 10^{-3}$$

We see, upon checking this number against the approximation, that y is much less than 0.20; thus, the assumption was valid. We will make frequent use of this method of solving problems. The concentrations asked for are

$$[CH_3COO^-] = [H_3O^+] = 1.9 \times 10^{-3}\ M$$

$$[CH_3COOH] = 0.20 - 1.9 \times 10^{-3} = 0.20\ M$$

and

$$[OH^-] = \frac{K_w}{[H_3O^+]} = \frac{1.0 \times 10^{-14}}{1.9 \times 10^{-3}} = 5.3 \times 10^{-12}$$

We should now check the approximation, made early in the problem that $[H_3O^+] = [CH_3COO^-]$. Since the $[OH^-]$ calculated above came from the ionization of water, the same $[H_3O^+]$ also would come from that source. This very low concentration, 5.3×10^{-12} *M*, indicates how much the ionization of water was repressed by the H_3O^+ from the acetic acid. The concentration of

H_3O^+ from this latter source is seen to be almost 10^9 times as great as that from the water. Therefore, the statement that $[H_3O^+] = [CH_3COO^-]$ is an excellent approximation.

The equilibrium constant for the reaction of a base with water is called the **base constant,** K_b:

$$B + H_2O = BH^+ + OH^-$$

$$K_b = \frac{[BH^+][OH^-]}{[B]}$$

Calculations of $[H^+]$ and $[OH^-]$ can be made from a base constant in the same way as from an acid constant, as the following example shows.

EXAMPLE 10.10 Calculate the pH of 0.40 *M* NH_3. The base constant of NH_3 is 1.8×10^{-5}.

SOLUTION The reaction of ammonia with water is shown by the equation

$$NH_3 + H_2O = NH_4^+ + OH^-$$

and the equilibrium constant for this is

$$K_b = 1.8 \times 10^{-5} = \frac{[NH_4^+][OH^-]}{[NH_3]}$$

Let $z = [NH_4^+] = [OH^-]$. (Notice that this is parallel to the equality of $[H_3O^+]$ and $[A^-]$ in the dissociation of the acid HA. The equality in both cases is an approximation, but a very good one.) Then $[NH_3] = 0.40 - z$, and

$$\frac{z^2}{0.40 - z} = 1.8 \times 10^{-5}$$

If we assume that $z \ll 0.40$, then

$$z = \sqrt{7.2 \times 10^{-6}} = 2.7 \times 10^{-3}$$

The assumption is seen to be good; therefore,

$$[OH^-] = 2.7 \times 10^{-3}$$

$$[H_3O^+] = \frac{1.0 \times 10^{-14}}{2.7 \times 10^{-3}} = 3.7 \times 10^{-12}$$

and

$$pH = 11.43$$

Notice that the ionization of water was repressed so that only 3.7×10^{-12} mole/liter H_3O^+ was formed. This same concentration of OH^- came from the water's ionization, and the approximation that $[NH_4^+] = [OH^-]$ is valid.

Hydrolysis

When NaCl is added to water, the pH remains at 7.00, but when NH_4Cl is added, the resulting solution is found to be acidic, and when CH_3COONa is added to water, the solution is alkaline. The explanation for these observations lies in the nature of the ions and other species formed when the various salts are dissolved. The NaCl gives Na^+ and Cl^-. Na^+ has no tendency to form NaOH since this is a strong base, and Cl^- has no tendency to form HCl since this is a strong acid. On the other hand, the NH_4^+ in solution that results from dissolving NH_4Cl does have the tendency to form NH_3 since NH_3 is a weak base (if it were strong, it would hold firmly to its additional proton and remain as NH_4^+). The reaction

As was discussed in Chapter 2, page 33, such compounds as NaCl, NH_4Cl, and CH_3COONa are ionic in the solid state as well as in aqueous solution.

$$NH_4^+ + H_2O = NH_3 + H_3O^+$$

occurs to a small extent, and therefore the solution is more acidic than is pure water.

Similarly, the CH_3COO^- in solution that results from dissolving sodium acetate forms some of the weak acid, CH_3COOH, by the reaction

$$CH_3COO^- + H_2O = CH_3COOH + OH^- \qquad [10.12]$$

and the solution is more alkaline than is pure water. These reactions are commonly spoken of as **hydrolysis** reactions of salts because of Arrhenius theory, although NH_4^+ is an acid and CH_3COO^- a base in the Brønsted-Lowry system.

The equilibrium constant for a hydrolysis reaction is readily obtained from known constants; for this reason hydrolysis constants are seldom tabulated. For example, consider the hydrolysis reaction of NH_4^+ shown above. The equilibrium constant for this reaction is called the **hydrolysis constant:**

$$\frac{[NH_3][H_3O^+]}{[NH_4^+]} = K_h$$

If numerator and denominator are multiplied by $[OH^-]$ the result is seen to be a product of known constants:

$$\frac{[NH_3][H_3O^+][OH^-]}{[NH_4^+][OH^-]} = \frac{[NH_3]}{[NH_4^+][OH^-]}[H_3O^+][OH^-] = \frac{1}{K_b}K_w = K_h$$

An example will show how calculations of the concentrations of species are made for solutions of salts that hydrolyze.

EXAMPLE 10.11 Calculate the pH of 0.10 *M* sodium acetate. For acetic acid, $K_a = 1.8 \times 10^{-5}$.

SOLUTION The equilibrium constant for equation [10.12] is called the hydrolysis constant,

$$K_h = \frac{[CH_3COOH][OH^-]}{[CH_3COO^-]}$$

and can be expressed in terms of known constants. This is seen readily if the numerator and denominator are multiplied by $[H_3O^+]$:

$$K_h = \frac{[CH_3COOH][OH^-][H_3O^+]}{[CH_3COO^-][H_3O^+]} = \frac{K_w}{K_a} = \frac{1.0 \times 10^{-14}}{1.8 \times 10^{-5}}$$

Therefore,

$$K_h = 5.6 \times 10^{-10}$$

Let $x = [CH_3COOH] = [OH^-]$; consequently, $0.10 - x = [CH_3COO^-]$. Substitution into the expression for K_h gives

$$\frac{x^2}{0.10 - x} = 5.6 \times 10^{-10}$$

from which $x = 7.5 \times 10^{-6}$. Therefore $[H_3O^+] = 1.3 \times 10^{-9}$ and pH = 8.87.

A similar calculation would apply to the determination of the pH of an ammonium chloride solution, or, indeed, a solution of any simple salt of a weak acid or a weak base.

Buffers

A solution that contains a weak acid and its salt or a weak base and its salt will show much smaller changes in pH when a strong acid or base is added than will water. Such solutions are said to be **buffered.** The reason for the resistance to pH change is made clear when one considers the species in the buffered solution. If a buffer is prepared from a weak acid, HA, and its salt, NaA, the species in solution are HA, A^-, and Na^+. If a strong base is added, it reacts with HA, which is the strongest acid available:

$$OH^- + HA = A^- + H_2O$$

If a strong acid is added, it reacts with A^-, which is the strongest base in the solution:

$$H_3O^+ + A^- = HA + H_2O$$

The significant point is that the addition of a strong acid or base causes a change in the ratio of [HA] to $[A^-]$ but does not cause a large increase in $[H_3O^+]$ or $[OH^-]$ as would such an addition to water. The $[H_3O^+]$ of the solution is determined by the ratio of the acid and its salt, as can be seen clearly when the expression for the acid constant is rearranged:

$$[H_3O^+] = K_a \frac{[HA]}{[A^-]} \qquad [10.13]$$

Therefore, the addition of a strong acid or base changes the $[H_3O^+]$ of the solution by changing the ratio $[HA]/[A^-]$, but this change in $[H_3O^+]$ is very small compared with the change in an unbuffered solution. The following three examples illustrate the effectiveness of a buffer.

EXAMPLE 10.12 Calculate the pH of a solution 0.50 *M* in CH_3COOH and 0.50 *M* in CH_3COONa. K_a for CH_3COOH is 1.8×10^{-5}.

SOLUTION The equilibrium equation is

$$CH_3COOH + H_2O = CH_3COO^- + H_3O^+$$

If we let $x = [H_3O^+]$, then $[CH_3COOH] = 0.50 - x$, and $[CH_3COO^-] = 0.50 + x$. The calculation is easily made if one assumes that $x \ll 0.50$:

> Note that $[CH_3COO^-] = 0.50 + x$. The total concentration of each species from all sources must be included in the equilibrium-constant expression.

$$K_a = \frac{[H_3O^+][CH_3COO^-]}{[CH_3COOH]} = \frac{x(0.50 + x)}{(0.50 - x)} \approx \frac{x(0.50)}{0.50}$$

Therefore $x \approx K_a = 1.8 \times 10^{-5}$; the assumption is seen to be good. The pH is 4.74.

It can be seen that the assumption should be very good in such a case, for the presence of A^- represses the dissociation of HA, whereas the presence of HA represses the hydrolysis of A^-. As long as K_a is small (about 10^{-4} or less) and the concentrations of HA and A^- are of the order of 0.1 *M* or greater, one can be confident that the assumption made above will be valid.

EXAMPLE 10.13 Calculate the pH of 100 ml of the solution in Example 10.12 after 1 ml of 12 *M* HCl is added.

SOLUTION 1 ml × 12 mmoles/ml = 12 mmoles H_3O^+ added. The effect of this is to convert 12 mmoles of CH_3COO^- to CH_3COOH. Initially there were 50 mmoles of CH_3COO^- and 50 mmoles of CH_3COOH; therefore, there are now 38 mmoles of CH_3COO^- and 62 mmoles of CH_3COOH. From equation [10.13]

$$[H_3O^+] = K_a \frac{62 \text{ mmoles}/101 \text{ ml}}{38 \text{ mmoles}/101 \text{ ml}}$$

The volumes cancel,

$$[H_3O^+] = (1.8 \times 10^{-5})\frac{62}{38} = 2.9 \times 10^{-5}$$

and pH = 4.54. The change was 0.20 pH units, and there was a 60% increase in the $[H^+]$. Compare this finding with Example 10.14, which follows.

EXAMPLE 10.14 Calculate the pH change that results from the addition of 1 ml of 12 *M* HCl to 100 ml H_2O.

SOLUTION Initially the pH is 7.00 and $[H_3O^+] = 1.0 \times 10^{-7}$. After the HCl is added, $[H_3O^+] = 12$ mmoles/101 ml = 0.12 *M* and pH = 0.92. The pH changed by 6.8 units, and the final $[H_3O^+]$ is over a million times as high as the original $[H_3O^+]$.

Buffers are of extraordinary importance to living systems. The delicate control of the pH of body fluids, so essential to good health, is accomplished through buffer systems. Industrial processes, from the leaching of metals from waste dumps to the control of conditions in the production of an antibiotic, depend upon buffers. Through the remaining chapters of this book we will present practical and essential applications of buffers.

You should familiarize yourself with the results of these three examples to make certain you clearly understand the effect of a buffer on a solution.

A few general statements can be made concerning buffered solutions:

(a) At a given total concentration of HA + A^-, the most effective buffer is that in which $[HA] = [A^-]$. The reason is that when a strong acid or base is added, the ratio $[HA]/[A^-]$ changes least when it is initially equal to 1. A convenient rule is that if either [HA] or $[A^-]$ is more than about ten times the other, the buffering action is poor. Thus, for effective buffering,

$$\frac{10}{1} > \frac{[HA]}{[A^-]} > \frac{1}{10}$$

For the best buffering, choose a system such that $pK_a \approx pH$.

(b) Buffering effectiveness increases with the concentration of the buffering species. Thus, if $[HA] = [A^-] = 1$ *M*, the buffer will change pH much less with a strong acid added than would the same volume of a buffer in which $[HA] = [A^-] = 0.001$ *M* with the same quantity of a strong acid added.

(c) A solution of a strong acid and its anion does not make a buffered solution because the dissociation of the acid is complete, or essentially so.

Analogous statements are applicable to buffers formed from bases and their salts.

Acid-Base Reactions

We have seen that when a strong acid reacts with a strong base, the reaction is the formation of water:

$$H_3O^+ + OH^- = 2\,H_2O$$

The equilibrium constant for this neutralization[†] reaction is

$$K_n = \frac{1}{[H_3O^+][OH^-]} = \frac{1}{K_w} = 1.0 \times 10^{14}$$

If acetic acid ($K_a = 1.8 \times 10^{-5}$) reacts with NaOH, the reaction is

$$CH_3COOH + OH^- = CH_3COO^- + H_2O$$

and

$$K_n = \frac{[CH_3COO^-]}{[CH_3COOH][OH^-]}$$

The numerical value for such a neutralization constant will seldom be found in tabulations of equilibrium constants, for it can be easily calculated from known constants. Notice that K_a and K_n bear a simple relationship to each other.

$$K_a = \frac{[CH_3COO^-][H_3O^+]}{[CH_3COOH]}$$

If K_n had $[H_3O^+]$ in its numerator, it would contain all the factors of K_a. To obtain this $[H_3O^+]$ we need only multiply the numerator and denominator of K_n by $[H_3O^+]$:

$$K_n = \frac{[CH_3COO^-]}{[CH_3COOH][OH^-]}\frac{[H_3O^+]}{[H_3O^+]}$$

and we see that $K_n = K_a/K_w$. Therefore

$$K_n = \frac{[CH_3COO^-]}{[CH_3COOH][OH^-]} = \frac{K_a}{K_w} = \frac{1.8 \times 10^{-5}}{1.0 \times 10^{-14}} = 1.8 \times 10^9$$

This neutralization constant is not so large as was that for a strong acid and a strong base (10^{14}), and examination of the form of K_n shows that the weaker the acid, the smaller the neutralization constant.

If a weak acid and a weak base react, the neutralization constant is found to be affected by both the acid and the base constants. Consider, for example,

---

† The term *neutralization* has long been used for the reaction of an acid and a base. The term is appropriate for cases in which both acid and base are strong because the solution is neutral (neither acidic nor basic) when precisely equivalent quantities of the two are mixed. However, as we shall see, the solution is not neutral if equivalent quantities of a strong acid and a weak base are mixed. Moreover, when $HClO_4$ is added to water, the reaction that occurs is a Brønsted-Lowry acid-base reaction, but it hardly seems to qualify as a neutralization reaction:

$$HClO_4 + H_2O = H_3O^+ + ClO_4^-$$

In common usage the term neutralization is applied to the reaction of acids and bases of widely differing strengths, but not to such extreme cases as the $HClO_4$-H_2O reaction. Thus the reaction of HCl and NH_3 is often called a neutralization reaction even though NH_3 is a weak base.

the reaction of formic acid, HCOOH ($K_a = 2.0 \times 10^{-4}$), with ammonia ($K_b = 1.8 \times 10^{-5}$):

$$NH_3 + HCOOH = NH_4^+ + HCOO^-$$

for which the neutralization constant is

$$K_n = \frac{[NH_4^+][HCOO^-]}{[NH_3][HCOOH]}$$

Multiplication of the numerator and denominator by $[H_3O^+][OH^-]$ gives

$$K_n = \frac{[NH_4^+][OH^-]}{[NH_3]} \times \frac{[HCOO^-][H_3O^+]}{[HCOOH]} \times \frac{1}{[H_3O^+][OH^-]} = \frac{K_b K_a}{K_w}$$

Therefore,

$$K_n = (1.8 \times 10^{-5})(2 \times 10^{-4})\left(\frac{1}{10^{-14}}\right) = 3.6 \times 10^5$$

Titrations: Strong Acid-Strong Base

Titration is the process by which the quantity of a substance can be determined by measuring the volume of a solution of known concentration that just reacts with it. For example, the quantity of base in an unknown sample can be determined by a titration in which a standard (known concentration) hydrochloric acid solution is added until a just-equivalent quantity has been added. From the quantity of HCl used, the quantity of base originally present can be readily calculated. Of course, in order for such a procedure to be effective, there must be some means of determining when the equivalant quantity of titrant (HCl, in this case) has been added.

It is interesting to follow the changes that occur in the concentrations of species as the acid is added. Of particular interest is the **equivalence point**—the point at which just-equivalent quantities of the unknown substance and the titrant have been added. Just below, we calculate a few points on a titration curve for sodium hydroxide with hydrochloric acid; then, Problem 10-9 at the end of the chapter gives practice in calculating additional points.

If 50.0 ml of 0.10 *M* NaOH are titrated with 0.10 *M* HCl, we can calculate the pH at various points in the titration and make a plot of the pH versus the fraction titrated.

(a) 25 ml HCl added (the fraction titrated is 0.5). There were 50.0 ml $\times$ 0.10 mmole/ml = 5.0 mmoles OH^- initially present. Since the reaction goes essentially to completion ($K_n = 10^{14}$), 2.5 mmoles have reacted with H^+ and 2.5 mmoles of OH^- remain. Therefore, $[OH^-] = 2.5$ mmoles/75 ml = 0.033 *M*, and $[H_3O^+] = 1.0 \times 10^{-14}/0.033 = 3.0 \times 10^{-13}$. pH = 12.52.

(b) 49.5 ml HCl added (the fraction titrated is 0.99). Only 0.05 mmole of the OH^- remains unreacted, and the volume is essentially 100 ml; consequently $[OH^-] = 0.05$ mmole/100 ml $= 5 \times 10^{-4}$. $[H_3O^+] = 1.0 \times 10^{-14}/5 \times 10^{-4} = 2 \times 10^{-11}$. pH = 10.7.

(c) 50.0 ml HCl added (the fraction titrated is 1.0). Here at the equivalence point there is no excess of either OH^- or H_3O^+; thus, the concentrations are equal to each other and to the square root of the water constant:

$$K_w = [H_3O^+][OH^-] = 1.0 \times 10^{-14}$$
$$[H_3O^+] = [OH^-] = 1.0 \times 10^{-7}$$
$$pH = 7.00$$

(d) 50.5 ml HCl added (the fraction titrated is 1.01). $[H_3O^+] = 0.05$ mmole/100 ml $= 5 \times 10^{-4}$, and $[OH^-] = 2 \times 10^{-11}$. pH = 3.3.

Figure 10.1 is a plot of pH versus the fraction titrated for this neutralization reaction and has features that are characteristic of all titration curves. The pH

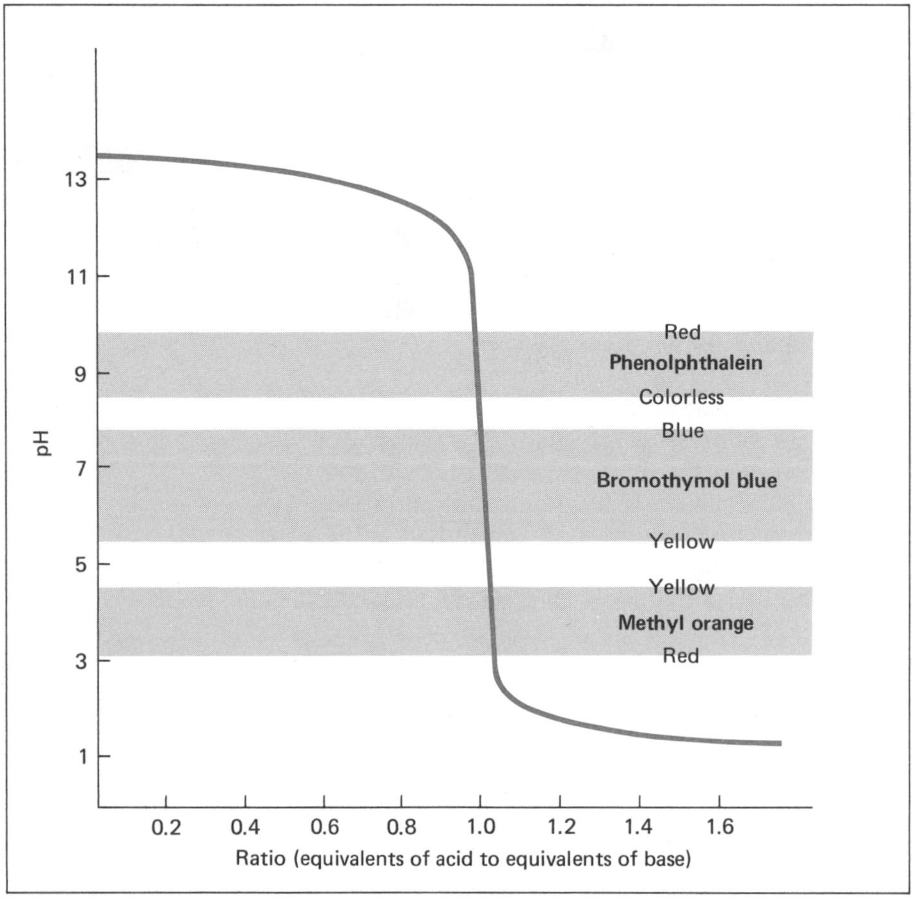

Figure 10.1
Change in pH during the titration of a strong base with a strong acid.

changes only very slowly through most of the titration, then changes rapidly in the region near the equivalence point, and again changes slowly after this region. If there were no volume change during the course of the titration, the curve of Figure 10.1 would be symmetrical. It is the rapid change of concentration near the equivalence point that makes possible the determinations of the equivalence point; even on the plot of pH versus fraction titrated the curve appears to be essentially vertical in that region.

Acid-Base Indicators

An **acid-base indicator** is used to determine the end point in a titration. It is a substance that is itself a weak acid and has different colors in the acidic and basic forms. The very large change in $[H_3O^+]$ near the equivalence point of the titration of NaOH with HCl makes it easy to find such an indicator that will change color very close to the equivalence point. The titration is stopped when the color of the indicator changes. This point is called the **end point** of the titration and should be clearly distinguished from the equivalence point. At the equivalence point an equivalent quantity of titrant has been added. The end point is the stopping point of the titration. Any difference between the two is the error in that titration.

Several common indicators and their transition ranges are shown on the plot in Figure 10.1. Each indicator shown in the figure can be used successfully for the titration of the strong base with a strong acid. This is true even though the color changes occur at widely different pH values. Because of the long, almost-vertical portion of the titration curve, the titration can be stopped anywhere in the range of about pH 9 to pH 4 with satisfactory results. We will see that this range becomes smaller if a weak acid or a weak base is used.

Titrations: Weak Acid or Weak Base

When a weak base such as NH_3 is titrated with HCl, the titration curve has a shorter inflection region near the equivalence point than has the curve of Figure 10.1. We will now make just a few calculations for the titration curve when 50 ml of 0.10 *M* NH_3 are titrated with 0.10 *M* HCl. These will show the general shape of the curve and indicate the differences from the strong acid-strong base titration curve in Figure 10.1. The base constant for NH_3 is 1.8×10^{-5}. The titration reaction is

$$NH_3 + H_3O^+ = NH_4^+ + H_2O$$

(a) 25 ml HCl added (the fraction titrated is 0.5). Now 2.5 mmoles HCl have been added to 5.0 mmoles NH_3. The result is that 2.5 mmoles NH_4^+ are formed and 2.5 mmoles NH_3 remain. This is a buffered solution with equal concentrations of the base NH_3 and its salt NH_4^+.

Any equilibrium expression that involves both NH_4^+ and NH_3 and either H_3O^+ or OH^- could be used for the calculation of the concentrations of species in the solution. Let us use the base constant,

$[NH_4^+][OH^-]/[NH_3] = 1.8 \times 10^{-5}$. Let $x = [OH^-]$. Then $[NH_3] =$ (2.5 mmoles/75 ml) $- x$, and $[NH_4^+] =$ (2.5 mmoles/75 ml) $+ x$:

$$\frac{(0.033 + x)x}{(0.033 - x)} = 1.8 \times 10^{-5}$$

If we assume that $x \ll 2.5/75$, then $x = 1.8 \times 10^{-5}$, and the assumption is seen to be very good. $[H_3O^+] = 5.6 \times 10^{-10}$, and pH = 9.25.

(b) 50.0 ml HCl added. This is the equivalence point, and here the solution is the same as 0.05 *M* NH_4Cl. Therefore, we use the hydrolysis constant for NH_4^+. (The hydrolysis and acid constants are equally true in representing the relationship of the concentrations of species.)

$$K_h = \frac{[NH_3][H_3O^+]}{[NH_4^+]}\frac{[OH^-]}{[OH^-]} = \frac{K_w}{K_b} = \frac{1.0 \times 10^{-14}}{1.8 \times 10^{-5}} = 5.6 \times 10^{-10}$$

Let $z = [NH_3] = [H_3O^+]$, and $0.05 - z = [NH_4^+]$. Therefore, $z^2/(0.05 - z) = 5.6 \times 10^{-10}$, and $z = 5.3 \times 10^{-6}$. pH = 5.28. Thus, the equivalence-point solution in this titration is distinctly acidic.

(c) 55 ml HCl added. There is now an excess of H_3O^+, and this excess strong acid alone controls the pH of the solution. The much weaker acid NH_4^+ is so undissociated in the presence of H_3O^+ that it need not be considered in the calculation. Hence, after the equivalence point the titration curve is identical to that of Figure 10.1. With 55 ml HCl, there is 0.5 mmole excess H_3O^+, and

$$[H_3O^+] = \frac{0.5 \text{ mmole}}{105 \text{ ml}} = 4.8 \times 10^{-3}$$

$$\text{pH} = 2.32$$

Figure 10.2 is a plot of the titration curve for this titration of 0.10 *M* NH_3 with 0.10 *M* HCl. Observe how the nearly vertical inflection region has been shortened relative to that in Figure 10.1, thus decreasing the range of indicators that could be used for this titration.

One sees clearly that the use of phenolphthalein would lead to a very serious error in this titration even though it was a good indicator in the strong acid-strong base titration. The two titration curves (Figure 10.1 and Figure 10.2) should be compared carefully until the differences are recognized and understood.

EXAMPLE 10.15 If a titration of formic acid, HCOOH, with NaOH requires 25.0 ml to reach the equivalence point, (a) what is the pH after 20.0 ml of NaOH have been added? (b) what is the pH after 24.0 ml of NaOH have been added? K_a of formic acid is 2.0×10^{-4}.

SOLUTION

(a) After 20.0 ml have been added, 20.0/25.0 of the formic acid has been converted to the formate ion. Therefore the ratio of formic acid to formate ion is 5.0/20.0, and from K_a,

$$[H^+] = \frac{[HCOOH]}{[HCOO^-]} 2.0 \times 10^{-4} = \left(\frac{5.0}{20}\right) 2.0 \times 10^{-4}$$

$$[H^+] = 5.0 \times 10^{-5}$$

$$pH = 4.30$$

(b) Now the ratio of formic acid to formate ion is 1.0/24.0, and

$$[H^+] = \left(\frac{1.0}{24}\right) 2.0 \times 10^{-4} = 8.3 \times 10^{-6}$$

$$pH = 5.08$$

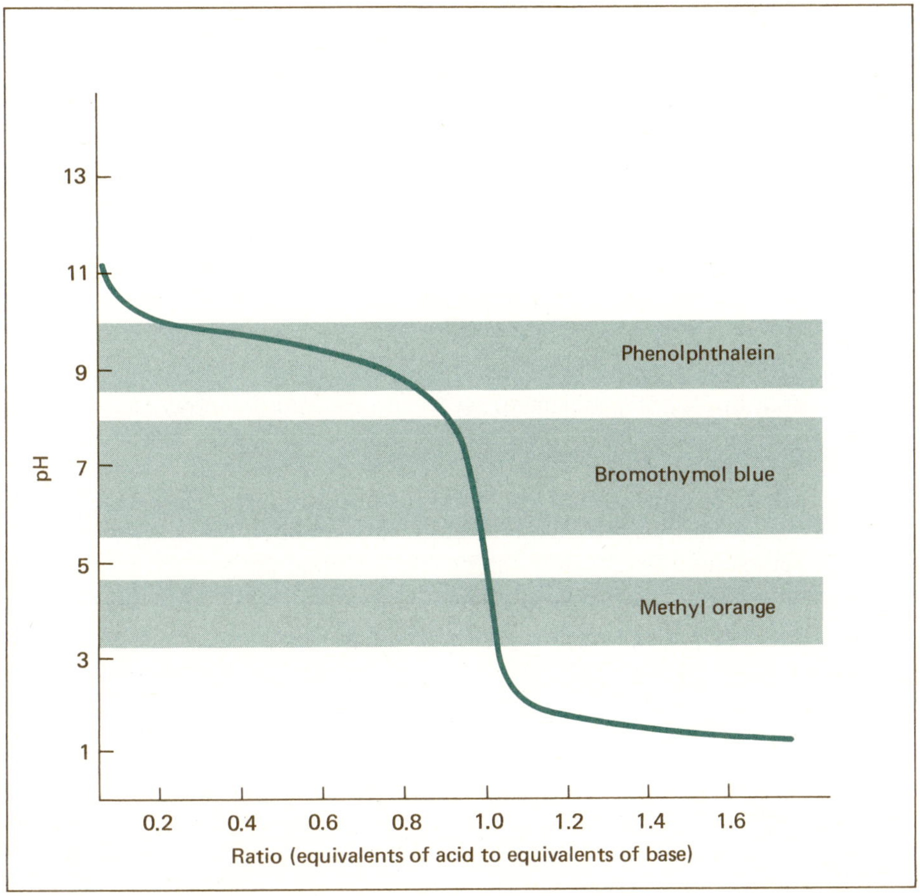

Figure 10.2
Change in pH during the titration of a weak base (NH_3) with a strong acid.

10.4 Ionization Equilibria: Complex Systems

In many ionic solutions, two or more reactions produce the same species. When that happens the equilibria are not independent. For example, the H_3O^+ produced from the dissociation of 0.20 *M* CH_3COOH suppresses the dissociation of water, and $[OH^-]$ is only 5.3×10^{-12} instead of 1.0×10^{-7} as it would be in pure water (see Example 10.9). The presence of CH_3COOH profoundly affects the dissociation of water. The following section illustrates methods used in solving simultaneous equilibria problems.

Multistage Ionizations

In the equilibrium problems considered so far, the acids have had a single ionizable hydrogen (monoprotic acids, such as CH_3COOH and HCOOH).

In this section we will treat the ionization of weak polyprotic acids. These are representative of systems that dissociate in several steps or stages; hence, the name *multistage equilibria.* Examples of polyprotic acids are phosphoric acid, H_3PO_4, arsenic acid, H_3AsO_4, carbonic acid, H_2CO_3, hydrogen sulfide, H_2S, and oxalic acid, $H_2C_2O_4$.

Let us consider a 0.10 *M* solution of oxalic acid. The dissociation reactions and constants are

$$H_2C_2O_4 = HC_2O_4^- + H^+$$
$$HC_2O_4^- = C_2O_4^{2-} + H^+$$
$$K_1 = \frac{[HC_2O_4^-][H^+]}{[H_2C_2O_4]} = 6 \times 10^{-2}$$
$$K_2 = \frac{[C_2O_4^{2-}][H^+]}{[HC_2O_4^-]} = 6 \times 10^{-5}$$

It is possible in such a problem to write enough simultaneous equations to permit rigorous solution, but two factors make that approach seem unwise here. First, while the equations are easily written, their rigorous solution is tedious and their approximate solution offers no apparent advantage in this case. Second, when the two acid constants are separated by a factor as large as 10^3, a very simple approach to the problem is possible.

If x mole/liter of $H_2C_2O_4$ dissociate to give $HC_2O_4^-$ and H^+, and y moles/liter of $HC_2O_4^-$ dissociate to $C_2O_4^{2-}$ and H^+, the concentrations of all species can be expressed in terms of x and y and the molar concentration of oxalic acid.

$$[H_2C_2O_4] = 0.10 - x$$
$$[HC_2O_4^-] = x - y$$
$$[H^+] = x + y$$
$$[C_2O_4^{2-}] = y$$

Let us assume that since K_2 is just a thousandth of K_1, the concentrations of H^+ and $HC_2O_4^-$ are equal; that is, $y \ll x$. Therefore,

$$\frac{x^2}{0.10 - x} = 6 \times 10^{-2}$$

From this,

$$x = 5.3 \times 10^{-2} = [H^+] = [HC_2O_4^-]$$

and

$$[H_2C_2O_4] = 0.10 - 0.053 = 0.047$$

The expression for K_2 shows that if $[H^+] = [HC_2O_4^-]$, then $y = [C_2O_4^{2-}] = 6 \times 10^{-5}$. That is, because of the second step in the dissociation, $[H^+]$ has been increased by 6×10^{-5} *M* and $[HC_2O_4^-]$ has been decreased by the same amount. These concentration changes are much smaller than the 0.05 *M* concentrations of the species calculated from the first dissociation. Hence, the assumption made in the calculation is valid, and the concentrations are as shown.

A slightly different way of looking at the oxalic acid example may help you see the basis for the assumptions. Both the first and the second dissociations yield H^+. Therefore, the H^+ from the first dissociation tends to repress the second step. Consequently, if the magnitudes of K_1 and K_2 differ by a factor of 100 or more, it is probable that the second dissociation occurs only slightly relative to the first.

One might well ask if the oxalic acid problem just considered could be made easier by the use of a constant for the complete dissociation of the acid. This dissociation is

$$H_2C_2O_4 = 2\,H^+ + C_2O_4^{2-} \qquad [10.14]$$

and the constant is

$$K = \frac{[H^+]^2[C_2O_4^{2-}]}{[H_2C_2O_4]} = K_1K_2$$

If we substitute the calculated values of the concentrations into the expression for the equilibrium constant, the equation is satisfied. However, equation [10.14] does not represent the stoichiometry of the reaction. A molecule of oxalic acid does not dissociate in a single step to give two hydrogen ions and an oxalate ion. Consequently there is not a simple relationship between the concentrations of the species such as we found above when we considered the step-by-step dissociation. Remember, the equilibrium constants obtained by combining valid equilibrium relationships are valid also. However, the overall equation obtained by combining the step-by-step reactions often does not represent what happens and may, therefore, be quite misleading as one attempts to set up the relationships between quantities or concentrations.

10.5 Solubility Equilibria

The Solubility Product

Barium sulfate, $BaSO_4$, is an example of a slightly soluble salt. When $BaSO_4(s)$ is added to water, a small amount dissolves to form the species Ba^{2+} and SO_4^{2-} in solution. A **saturated solution** results when as much solid $BaSO_4$ is added to the water as will dissolve. If excess $BaSO_4$ is added, this remains as a separate phase in contact with the solution. At a fixed temperature the concentrations of Ba^{2+} and SO_4^{2-} in a saturated solution are found to be independent of the quantity of solid $BaSO_4$ added and of the volume of water. That is, when the solution is saturated, equilibrium is established between the solid $BaSO_4$ and the ions in solution. The chemical equation representing this equilibrium is written

This is analogous to the phase equilibria described in Chapter 8. A saturated solution is one in which the $BaSO_4$ dissolves to form Ba^{2+} and SO_4^{2-} at the same rate as these ions recombine to form solid $BaSO_4$.

$$BaSO_4(s) = Ba^{2+} + SO_4^{2-}$$

and its equilibrium constant is

$$K = \frac{(a_{Ba^{2+}})(a_{SO_4^{2-}})}{a_{BaSO_4}}$$

However, $a_{BaSO_4} = 1$(pure solid), and

$$K = (a_{Ba^{2+}})(a_{SO_4^{2-}})$$

In general, a solubility reaction can be represented by the equation

$$A_nB_m(s) = nA^{m+} + mB^{n-}$$

for which the equilibrium constant is

$$K = \frac{(a_{A^{m+}})^n(a_{B^{n-}})^m}{a_{A_nB_m(s)}} \qquad [10.15]$$

The activity of the pure solid A_nB_m is 1. Therefore, this general equilibrium constant becomes

$$K = (a_{A^{m+}})^n(a_{B^{n-}})^m \qquad [10.16]$$

No approximation was involved in the writing of this latter expression; equations [10.15] and [10.16] are identical statements. A word statement for equation [10.16] might be that it is "the equilibrium constant for the dissolving reaction of a slightly soluble salt." This rather long statement is too awkward for convenient use and is commonly replaced by the short phrase **solubility product** (or, sometimes, solubility-product constant; we will use the shorter phrase throughout the text). That is, *the solubility product is the equilibrium constant for the dissolving reaction of a slightly soluble salt.*

The term *solubility product* is applied only to dissolving reactions in which ionic species are formed in solution. When

The symbol used for the solubility product is K_{sp} (just as K_a identifies the acid constant and K_b the base constant). Thus, equation [10.16] is rewritten

$$K_{sp} = (a_{A^{m+}})^n(a_{B^{n-}})^m \qquad [10.17]$$

iodine dissolves slightly in water by the reaction

$$I_2(s) = I_2(aq)$$

the equilibrium constant is

$$K = \frac{a_{I_2(aq)}}{a_{I_2(s)}} = a_{I_2(aq)} \approx [I_2(aq)]$$

This is an acceptable and functional equilibrium constant for a dissolving reaction, but it is not labeled a solubility product.

The solubility is small for the slightly soluble salts for which we will calculate the equilibrium conditions. As in the earlier equilibrium considerations, we will work in low concentrations where the assumption is good that the activity coefficients are unity and, therefore, the approximate form of equation [10.17] can be used. We will write

$$K_{sp} = [A^{m+}]^n[B^{n-}]^m \qquad [10.18]$$

with the recognition that in any equilibrium expression the substitution of concentrations for activities involves an approximation.

If solid $BaSO_4$ is equilibrated with pure water at 25° C, the concentrations of Ba^{2+} and SO_4^{2-} in the aqueous solution are equal and are found experimentally to have the value 1.0×10^{-5} *M*. As seen earlier, the dissolving reaction is

$$BaSO_4(s) = Ba^{2+} + SO_4^{2-}$$

and the solubility product is

$$K_{sp} = [Ba^{2+}][SO_4^{2-}] \qquad [10.19]$$

The numerical value of K_{sp} is readily calculated from the solubility to be

$$K_{sp} = (1.0 \times 10^{-5})(1.0 \times 10^{-5}) = 1.0 \times 10^{-10}$$

Thus, in Appendix F, which lists the common slightly soluble salts and their solubility products, the K_{sp} for $BaSO_4$ is given as 1.0×10^{-10}. Some of the values reported in Appendix F were obtained by direct measurement of the concentrations of species in equilibrium with the solid, but a more usual method is an electrochemical approach that will be described in Chapter 11.

It is apparent from equation [10.18] that the solubility product will not always be the simple product of two concentrations, and that, indeed, direct comparison of the numerical values of two solubility products may not give a valid indication of the relative solubilities of the two substances. An example will illustrate this.

EXAMPLE 10.16 The solubility of silver chromate, Ag_2CrO_4, in water at 25° C is found to be 8.0×10^{-5} mole/liter. Calculate the solubility product of Ag_2CrO_4. Compare the solubilities and the solubility products of Ag_2CrO_4 and of $BaSO_4$.

SOLUTION The dissolving reaction of silver chromate is

$$Ag_2CrO_4(s) = 2\ Ag^+ + CrO_4^{2-}$$

and

$$K_{sp} = [Ag^+]^2[CrO_4^{2-}]$$

Since 8.0×10^{-5} mole/liter of the Ag_2CrO_4 dissolves and two Ag^+ and one CrO_4^{2-} result from each Ag_2CrO_4 that dissolves, then $[Ag^+] = 1.6 \times 10^{-4}$ and $[CrO_4^{2-}] = 8.0 \times 10^{-5}$. (Notice that $[Ag^+]$ under these circumstances is double the $[CrO_4^{2-}]$ because of the stoichiometry of the dissolving reaction; $[Ag^+]$ itself, and not some multiple of it, is substituted into the equation for the equilibrium constant.) Therefore

$$K_{sp} = (1.6 \times 10^{-4})^2(8.0 \times 10^{-5}) = 2.0 \times 10^{-12}$$

The solubility of Ag_2CrO_4 is seen to be eight times as great as that of $BaSO_4$ (8×10^{-5} mole/liter versus 1×10^{-5} mole/liter), yet the K_{sp} of Ag_2CrO_4 is the much smaller of the two. This apparent reversal results from the difference in form of the constants, one involving the concentrations of just two ions and the other involving three (one chromate and two silver ions).

Having calculated the solubility product from the solubility, let us reverse the process and calculate the solubility from the K_{sp}.

EXAMPLE 10.17 Calculate the solubility of $PbCl_2$ in grams/liter.

SOLUTION The K_{sp} of $PbCl_2$ is seen from Appendix F to be 2.0×10^{-4}. The dissolving is represented by

$$PbCl_2 = Pb^{2+} + 2Cl^-$$

and

$$K_{sp} = [Pb^{2+}][Cl^-]^2 = 2.0 \times 10^{-4}$$

If we let x be the solubility of $PbCl_2$ in moles/liter, then $x = [Pb^{2+}]$ and $2x = [Cl^-]$. Therefore,

$$x(2x)^2 = 2.0 \times 10^{-4}$$

and

$$x = 3.7 \times 10^{-2} \text{ mole/liter}$$

The solubility in grams/liter is

$$\left(3.7 \times 10^{-2} \frac{\text{mole}}{\text{liter}}\right)\left(278 \frac{\text{g}}{\text{mole}}\right) = 10 \frac{\text{g}}{\text{liter}}$$

The Effect of Common Ions upon Solubility

Thus far we have considered only the dissolving of slightly soluble salts in pure water. The equilibrium relationship indicates that less of the salt will dissolve in a solution that already contains a low concentration of one of the ions of the salt. Hence, if solid $BaSO_4$ is equilibrated with a solution of Na_2SO_4 in water, less of the solid must dissolve in reaching the equilibrium condition than dissolved in pure water. The solubility product is the same as before (see equation [10.19]),

> Care must be used in extrapolating this concept to a solution that contains a high concentration of one of the ions, since species that are more complex may form and *increase* the solubility of the salt. We will meet this issue later, but for now we will work only with low concentrations in which complex-species formation does not make a large contribution to the solubility.

$$[Ba^{2+}][SO_4^{2-}] = K_{sp}$$

but the concentration of sulfate is the sum of that from the dissolving of $BaSO_4$ plus the original concentration from Na_2SO_4. If the original solution was 0.0010 M Na_2SO_4, and if x mole/liter of $BaSO_4$ dissolve, then

$$[Ba^{2+}] = x$$

and

$$[SO_4^{2-}] = 0.0010 + x$$

The numerical value of K_{sp} is 1.0×10^{-10} and

$$x(0.0010 + x) = 1.0 \times 10^{-10}$$

This is solved to give $x = 1.0 \times 10^{-7}$ M. That is, the solubility of $BaSO_4$ in 0.0010 M Na_2SO_4 is only 1.0×10^{-7} mole/liter, while in pure water the solubility is 1.0×10^{-5} M.

The precise effect of the presence of a common ion on the solubility of a salt of the form A_nB_m is less obvious than for such a substance as $BaSO_4$. However, the calculation is straightforward, as the next two examples show.

EXAMPLE 10.18 What weight of Ag_2CrO_4 will dissolve in a liter of 2.0×10^{-3} M $AgNO_3$?

SOLUTION The solubility product of Ag_2CrO_4 is 2.0×10^{-12} (Appendix F); therefore,

$$[Ag^+]^2[CrO_4^{2-}] = 2.0 \times 10^{-12}$$

If we let y represent the quantity of Ag_2CrO_4 that dissolves, expressed in moles/liter, then $[CrO_4^{2-}] = y$ and $[Ag^+] = 2y + 2.0 \times 10^{-3}$. Therefore,

$$(2y + 2.0 \times 10^{-3})^2(y) = 2.0 \times 10^{-12}$$

and

$$y = 5.0 \times 10^{-7} \frac{\text{mole}}{\text{liter}}$$

The solution of this cubic equation would be laborious if it were not possible to use approximation methods. The assumption that $2y \ll 2.0 \times 10^{-3}$ proves to be valid, and this permits the equation to be solved as a linear equation. The weight of Ag_2CrO_4 that dissolves in 1 liter is

$$\left(5.0 \times 10^{-7} \frac{\text{mole}}{\text{liter}}\right)\left(332 \frac{\text{g}}{\text{mole}}\right) = 1.7 \times 10^{-4} \text{ g}$$

The solubility of Ag_2CrO_4 is 8×10^{-4} mole/liter in water. This is over 10^3 times as great as the solubility calculated in this example for Ag_2CrO_4 in 2.0×10^{-3} M $AgNO_3$.

EXAMPLE 10.19 Calculate the solubility in moles/liter of Ag_2CrO_4 in 2.0×10^{-3} M K_2CrO_4, and compare this with the result obtained in Example 10.18.

SOLUTION Let z represent the solubility of Ag_2CrO_4 in this solution expressed in moles/liter. Then

$$[CrO_4^{2-}] = 2.0 \times 10^{-3} + z$$

and

$$[Ag^+] = 2z$$

Therefore,

$$(2z)^2(2.0 \times 10^{-3} + z) = 2.0 \times 10^{-12}$$

and

$$z = 1.6 \times 10^{-5} \frac{\text{mole}}{\text{liter}}$$

This is about 30 times the solubility calculated in Example 10.18 for a silver nitrate solution of the same concentration. These two examples show the relative effects of the presence of the two ions upon the solubility of a salt of the form A_2B.

The solubility product for a substance is a statement of the equilibrium condition in a saturated solution. Thus, in an equilibrium mixture that consists of solid AgCl and an aqueous solution of Ag^+ and Cl^-, the activity of the AgCl is fixed, as is the product

$$[Ag^+][Cl^-] = K_{sp}$$

When equilibrium obtains, the product of $[Ag^+]$ and $[Cl^-]$ cannot exceed K_{sp}. If no solid AgCl is present, the product of $[Ag^+]$ and $[Cl^-]$ can have any value equal to or less than K_{sp}, but if the value is less than K_{sp}, the solution is unsaturated. That is, the product $[Ag^+][Cl^-]$ is the solubility product only in an equilibrium-saturated solution.

Consider a very dilute solution of NaCl to which $AgNO_3$ solution is added in small increments. If concentrations and increments are small enough, the equilibrium condition is not met when the first portions are added, and the concentration of Ag^+ increases with each addition. Eventually, with more additions, $[Ag^+]$ becomes high enough that $[Ag^+][Cl^-]$ is equal to the solubility product. The solution is saturated, even though no AgCl(*s*) is present. Solid AgCl could be added with no change in the ionic concentrations. If more $AgNO_3$ is added, there will be precipitation of AgCl(*s*) until again the product $[Ag^+][Cl^-]$ is equal to the solubility product. The following examples illustrate these matters.

EXAMPLE 10.20 75 ml of 0.67 *M* NaCl are added to 25 ml of 0.20 *M* $Pb(NO_3)_2$. Will a precipitate of $PbCl_2$ form?

SOLUTION The solubility product of $PbCl_2$ is

$$[Pb^{2+}][Cl^-]^2 = 2.0 \times 10^{-4}$$

Let us calculate the concentrations of Pb^{2+} and Cl^- on the assumption that no precipitate forms. We can then examine the product $[Pb^{2+}][Cl^-]^2$ and compare this with the solubility product.

$$25 \text{ ml}\left(\frac{0.20 \text{ mmole Pb}^{2+}}{\text{ml}}\right)\left(\frac{1}{100 \text{ ml}}\right) = 0.050\ M\ \text{Pb}^{2+}$$

$$75 \text{ ml}\left(\frac{0.067 \text{ mmole Cl}^{-}}{\text{ml}}\right)\left(\frac{1}{100 \text{ ml}}\right) = 0.050\ M\ \text{Cl}^{-}$$

$$[Pb^{2+}][Cl^-]^2 = (0.050)(0.050)^2 = 1.3 \times 10^{-4}$$

This is less than K_{sp}, and the solution is unsaturated.

EXAMPLE 10.21 Calculate the volume of 0.010 *M* $AgNO_3$ solution that must be added to 100 ml of 0.010 *M* NaCl in order to saturate the solution with AgCl.

SOLUTION Since no AgCl is to be precipitated and only a very small volume of $AgNO_3$ solution needs to be added, the concentration of Cl^- remains essentially 0.010 *M*. Hence the $[Ag^+]$ can be calculated from $[Cl^-]$ and K_{sp}.

$$[Ag^+] = \frac{K_{sp}}{[Cl^-]} = \frac{1.8 \times 10^{-10}}{0.010} = 1.8 \times 10^{-8}\ M$$

In 100 ml there is

$$\left(1.8 \times 10^{-8}\ \frac{\text{mmole}}{\text{ml}}\right)(100 \text{ ml}) = 1.8 \times 10^{-6} \text{ mmole Ag}^+$$

This requires

$$(1.8 \times 10^{-6} \text{ mmole})\left(\frac{1 \text{ ml}}{0.010 \text{ mmole}}\right) = 1.8 \times 10^{-4} \text{ ml}$$

of the 0.010 *M* $AgNO_3$ solution. (An eye dropper with a small tip emits drops that are about 0.05 ml. The volume of the $AgNO_3$ solution required is about 1/300 the volume of such a drop.)

EXAMPLE 10.22 Given 100 ml of 0.010 *M* KSCN, calculate the volume of 1.0 *M* $AgNO_3$ that would be required to lower the concentration of the thiocyanate ion to 1.0×10^{-3} *M*. Disregard the change in volume.

SOLUTION The solubility product of AgSCN is

$$[Ag^+][SCN^-] = 1.0 \times 10^{-12}$$

Therefore, if $[SCN^-]$ is to be 1.0×10^{-3}, then

$$[Ag^+] = \frac{1.0 \times 10^{-12}}{1.0 \times 10^{-3}} = 1.0 \times 10^{-9}$$

It is evident from this result that an inconsequential volume of 1.0 *M* $AgNO_3$ will be required to provide the Ag^+ in solution. On the other hand, the $[SCN^-]$ went from 0.010 *M* to 0.0010 *M* because of the precipitation of AgSCN. The quantity precipitated is

$$(100\ \text{ml})\left(\frac{0.010\ \text{mmole}}{\text{ml}}\right) - (100\ \text{ml})\left(\frac{0.0010\ \text{mmole}}{\text{ml}}\right) = 0.9\ \text{mmole}$$

Therefore,

$$(0.9\ \text{mmole})\left(\frac{1\ \text{ml}}{1.0\ \text{mmole}}\right) = 0.9\ \text{ml of } 1.0\ M\ AgNO_3$$

is required. Since the volume change from 100 to about 101 ml is within the accuracy limits of the data, it was proper to disregard this change.

Multiple Solubility Equilibria

If a solution contains several species, all of which form slightly soluble salts with a given ion, the various concentrations can often be calculated through use of the solubility products. It is important to recognize that a species can have only one concentration in a solution and that this single concentration must be in equilibrium with all appropriate species in the solution. Thus, if an aqueous solution is in equilibrium with AgI(*s*), AgBr(*s*), and AgCl(*s*), the Ag^+ must simultaneously be in equilibrium with the I^-, Br^-, and Cl^-.

EXAMPLE 10.23 The following solutions are mixed: 25 ml 0.020 *M* NaI, 25 ml 0.010 *M* NaBr, 25 ml 0.030 *M* NaCl, and 25 ml 0.050 *M* $AgNO_3$. Calculate the concentrations of I^-, Br^-, Cl^-, and Ag^+ in the equilibrium solution. Assume that the volumes are additive.

SOLUTION The solubility products are

$$[Ag^+][Cl^-] = 1.8 \times 10^{-10}$$
$$[Ag^+][Br^-] = 5.0 \times 10^{-13}$$

and

$$[Ag^+][I^-] = 1.0 \times 10^{-16}$$

The quantities of substances mixed are

$$25 \text{ ml} \times 0.020 \frac{\text{mmole}}{\text{ml}} = 0.50 \text{ mmole } I^-$$

$$25 \text{ ml} \times 0.010 \frac{\text{mmole}}{\text{ml}} = 0.25 \text{ mmole } Br^-$$

$$25 \text{ ml} \times 0.030 \frac{\text{mmole}}{\text{ml}} = 0.75 \text{ mmole } Cl^-$$

and

$$25 \text{ ml} \times 0.050 \frac{\text{mmole}}{\text{ml}} = 1.25 \text{ mmole } Ag^+$$

Since AgI is the least soluble of the silver halides, it is precipitated, leaving 0.75 mmole Ag^+. AgBr is next least soluble, and after it is precipitated, 0.50 mmole of Ag^+ remains. Therefore, 0.50 mmole of AgCl is precipitated, and the $[Cl^-]$ is

$$\frac{0.25 \text{ mmole}}{100 \text{ ml}} = 2.5 \times 10^{-3}\ M$$

This very simple calculation was made with the assumption that the excess chloride ion is large compared to the quantity of AgCl that dissolves. That this is a valid assumption can be seen when $[Ag^+]$ is calculated from the chloride ion concentration:

$$[Ag^+] = \frac{1.8 \times 10^{-10}}{2.5 \times 10^{-3}} = 7.2 \times 10^{-8}$$

The other two halide concentrations are

$$[Br^-] = \frac{5.0 \times 10^{-13}}{7.2 \times 10^{-8}} = 6.9 \times 10^{-6}$$

and

$$[I^-] = \frac{1.0 \times 10^{-16}}{7.2 \times 10^{-8}} = 1.4 \times 10^{-9}$$

EXAMPLE 10.24 A 50 ml portion of 0.020 *M* NaBr is added to 1.0 g of AgCl, and the mixture is stirred until equilibrium is reached. Calculate the weight of AgCl that is converted to AgBr.

SOLUTION The reaction that occurs is

$$AgCl(s) + Br^- = AgBr(s) + Cl^-$$

and the equilibrium constant for this is

$$K = \frac{[Cl^-]}{[Br^-]}$$

This can be expressed in terms of known constants if both numerator and denominator are multiplied by $[Ag^+]$:

$$K = \frac{[Cl^-][Ag^+]}{[Br^-][Ag^+]} = \frac{K_{sp(AgCl)}}{K_{sp(AgBr)}}$$

The numerical value of the constant is

$$\frac{1.8 \times 10^{-10}}{5.0 \times 10^{-13}} = 3.6 \times 10^2$$

That is,

$$\frac{[Cl^-]}{[Br^-]} = 3.6 \times 10^2$$

If we let x be the equilibrium $[Cl^-]$, then $[Br^-] = 0.020 - x$, and

$$\frac{x}{0.020 - x} = 3.6 \times 10^2$$

From this, $x \approx 0.020$ and $[Br^-]$ is calculated from K to be $0.020/360 = 5.6 \times 10^{-5}$. Notice that $[Br^-]$ cannot be calculated from the equation $[Br^-] = 0.020 - x$, because 0.020 and x are not known with sufficient accuracy to permit the subtraction.

We have now a 50 ml solution that is 0.020 *M* in Cl^-, all of which came from the original solid AgCl. Therefore, the weight of AgCl that was converted to AgBr is

$$(50\text{ ml})\left(\frac{0.020\text{ mmole}}{\text{ml}}\right)\left(\frac{143\text{ mg}}{\text{mmole}}\right) = 143\text{ mg}$$

The units of molecular weight can be expressed as mg/mmole as well as g/mole.

Hence, in the case of a solution that contains a mixture of halides, as in Example 10.23, the addition of Ag^+ may initially cause local precipitation of AgCl; however, if there is still a significant concentration of Br^- or I^- present, the AgCl will be converted to AgBr or AgI.

Solubility Increase Through Reaction

Salt of a Weak Acid

In Example 10.24 we saw that a slightly soluble salt can be converted to one that is less soluble by adding an appropriate substance. In a related reaction one of the ions of the slightly soluble salt reacts with an added species to form a soluble substance. The result is an increase in solubility. $BaSO_4$ is more soluble in an acidic solution than in a neutral solution because of the formation of HSO_4^- in the former:

$$BaSO_4(s) = Ba^{2+} + SO_4^{2-}$$
$$SO_4^{2-} + H^+ = HSO_4^-$$

The conversion of some of the SO_4^{2-} to HSO_4^- requires that additional $BaSO_4$ dissolve in order to satisfy the solubility product. Such an effect requires that the anion of a *weak* acid be involved in the solubility product. For example, the solubility of AgCl is not increased as the solution's acidity is increased, for HCl is a strong acid.

EXAMPLE 10.25 Calculate the solubility of $BaSO_4$ in 0.40 *M* HCl. Compare this with the solubility in water.

SOLUTION K_a for HSO_4^- is

$$\frac{[H^+][SO_4^{2-}]}{[HSO_4^-]} = 1.2 \times 10^{-2}$$

Remember that H^+ is often used by chemists as shorthand for H_3O^+ even though such use tends to obscure the significance of the water in acid-base reactions.

and K_{sp} for $BaSO_4$ is

$$[Ba^{2+}][SO_4^{2-}] = 1.0 \times 10^{-10}$$

In 0.40 *M* HCl, $[H^+] = 0.40$ *M* and

$$\frac{[SO_4^{2-}]}{[HSO_4^-]} = \frac{1.2 \times 10^{-2}}{0.40} = 0.030$$

That is, the concentration of HSO_4^- is about 33 times the concentration of SO_4^{2-}. Observing this, we conclude that the $BaSO_4$ dissolves much more through the formation of HSO_4^-,

$$BaSO_4(s) + H^+ = Ba^{2} + HSO_4^- \qquad [10.20]$$

than through the more familiar reaction

$$BaSO_4(s) = Ba^{2+} + SO_4^{2-}$$

We could set up and solve simultaneous equations for these two dissolving reactions. However, the approximation method used here is not only much simpler but gives a result that is as accurate as the data justify.

If we neglect the latter reaction, we will introduce an error of only 3% (there is 33 times as much HSO_4^- as SO_4^{2-}), which is less than the uncertainty in the equilibrium constants. The equilibrium constant for equation [10.20] is

$$K = \frac{[Ba^{2+}][HSO_4^-]}{[H^+]}$$

and if numerator and denominator are multiplied by $[SO_4^{2-}]$, this becomes

$$K = \frac{[Ba^{2+}][SO_4^{2-}][HSO_4^-]}{[H^+][SO_4^{2-}]} = \frac{K_{sp}}{K_a} = \frac{1.0 \times 10^{-10}}{1.2 \times 10^{-2}}$$

Therefore,

$$\frac{[Ba^{2+}][HSO_4^-]}{[H^+]} = 8.3 \times 10^{-9}$$

Either $[Ba^{2+}]$ or $[HSO_4^-]$ represents the solubility. Let $x = [Ba^{2+}] = [HSO_4^-]$. Therefore $0.40 - x = [H^+]$ and

$$\frac{x^2}{0.40 - x} = 8.3 \times 10^{-9}$$

$$x = 5.7 \times 10^{-5}\ M = [Ba^{2+}] = [HSO_4^-]$$

The $[H^+]$ is virtually unchanged from its initial value of 0.40 *M*, and $[SO_4^{2-}]$ can be calculated from either K_{sp} or K_a to be 1.7×10^{-6} *M*. The solubility of $BaSO_4$ has been increased by a factor of 5.7 as a result of the change from water to 0.40 *M* HCl as solvent.

The approach made in this example is satisfactory if one of the reactions is predominant, and if the approximations are checked carefully.

Complex Ion Formation

The preceding example demonstrated the increase in solubility of $BaSO_4$ in an acid solution as the result of the formation of HSO_4^-. A similar increase in solubility is observed when one of the ions of a slightly soluble compound forms complexes with a substance in the solution. For example, AgCl is more soluble in a solution of ammonia than in water, because NH_3 forms a very stable complex ion with Ag^+:

$$AgCl(s) + 2\ NH_3 = Ag(NH_3)_2^+ + Cl^-$$

We introduce a few complex ions in this chapter as examples of solubility effects, and then discuss them in much more detail in Chapter 20.

EXAMPLE 10.26 What weight of AgCl can be dissolved by 25 ml of 0.10 *M* NH_3? The dissociation constant, K_d, for $Ag(NH_3)_2^+ = Ag^+ + 2\ NH_3$ is 6×10^{-8}.

SOLUTION Again, as in Example 10.25, we will assume that there is a predominant dissolving reaction. In this case the two reactions are

$$AgCl(s) = Ag^+ + Cl^-$$

and

$$AgCl(s) + 2\ NH_3 = Ag(NH_3)_2^+ + Cl^- \qquad [10.21]$$

The low dissociation constant for $Ag(NH_3)_2^+$,

$$Ag(NH_3)_2^+ = Ag^+ + 2\ NH_3$$

$$K_d = \frac{[Ag^+][NH_3]^2}{[Ag(NH_3)_2^+]} = 6 \times 10^{-8}$$

indicates that the complex ion is only slightly dissociated. Let us assume that this dissociation is so small that equation [10.21] is the predominant dissolving reaction. The effect of a small value for K_d is seen when we calculate the equilibrium constant for reaction [10.21]

$$K = \frac{[Ag(NH_3)_2^+][Cl^-]}{[NH_3]^2} = \frac{K_{sp}}{K_d} = \frac{1.8 \times 10^{-10}}{6 \times 10^{-8}} = 3 \times 10^{-3}$$

This K is much larger than K_{sp} for AgCl, which again suggests that [10.21] is the important reaction. If we let $y = [Ag(NH_3)_2^+] = [Cl^-]$, then $[NH_3] = 0.10 - 2y$, for two molecules of NH_3 are lost for each $Ag(NH_3)_2^+$ formed. Hence,

$$\frac{y^2}{(0.10 - 2y)^2} = 3 \times 10^{-3}$$

and

$$y = 4.9 \times 10^{-3}\ M$$

This shows that the solubility of AgCl in this NH_3 solution is about 400 times that in water ($4.9 \times 10^{-3}\ M$ versus $1.3 \times 10^{-5}\ M$). Therefore, the assumption that the reaction of equation [10.21] is predominant is valid. The weight dissolved by 25 ml of the NH_3 solution is given by

$$\left(4.9 \times 10^{-3}\ \frac{\text{mmole}}{\text{ml}}\right)(25\ \text{ml})\left(143\ \frac{\text{mg}}{\text{mmole}}\right) = 18\ \text{mg}$$

Summary

The principle of chemical equilibrium was derived in Chapter 9 from thermodynamics. We saw that for the general reaction

$$aA + bB + \cdots = lL + mM + \cdots$$

the equilibrium constant is

$$K = \frac{(a_L^l)(a_M^m) \cdots}{(a_A^a)(a_B^b) \cdots} \qquad [9.9]$$

and that this general expression fits all systems of chemical equilibrium. Standard states had to be chosen with care and understanding; beyond that, the application of equation [9.9] has been the same in all problems.

In all cases the activity is given by the expression

$$a = \frac{P}{P^0}$$

Now by the appropriate selection of standard states, the expression for the activity simplifies as follows:

For gases	$a = P$	(ideal gas behavior assumed)
For pure solids and pure liquids	$a = 1$	(at atmospheric pressure)
For the solvent in a dilute solution	$a_1 = x_1$	(Raoult's law assumed)
For the solute in a dilute solution	$a = [s]$	(Henry's law assumed)

For the student who desires to understand equilibrium principles and to have competence in solving equilibrium problems, we know of no substitute for experience. Therefore, this chapter has included a large number of examples that are solved in detail. In addition, there is a large set of exercises and problems. We urge you to master the examples and problems, for the subject of equilibrium is of great importance in chemistry and all chemistry-related disciplines. We will make frequent use of equilibrium calculations in the remainder of the book, especially as we discuss the descriptive chemistry of the elements.

Important Terms and Concepts

- Brønsted-Lowry acid-base reaction
- Conjugate acid
- Conjugate base
- Hydronium ion
- Water constant
- pH scale
- pOH
- Acid constant
- Base constant
- Hydrolysis
- Hydrolysis constant
- Buffered solution
- Titration
- Equivalence point
- Acid-base indicator
- End point
- Saturated solution
- Solubility product

Exercises

All reactions are assumed to occur at 25° C unless the temperature is specifically stated otherwise. Data from the tables of equilibrium constants and the tables of thermodynamic data in the Appendix are required for many of the problems.

10-1.
For the reaction

$$Br_2(g) = 2\ Br(g)$$

$\Delta H^0 = 55$ kcal and $K = 0.0071$ at 1000° C,

(a) calculate ΔG^0 and ΔS^0 at 1000° C.
(b) calculate P_{Br} at 1000° C when $P_{Br_2} = 1.00$ atm.

10-2.
For the reaction $2\ O_3(g) = 3\ O_2(g)$,

(a) calculate K at 25° C from thermodynamic data.
(b) calculate P_{O_3} in equilibrium with O_2 at a total pressure of 1.00 atm.
(c) what change would you make in the temperature and pressure to make O_3 more stable?

10-3.
Given the chemical reaction at 25° C

$$2\ HCl(g) + Br_2(g) = 2\ HBr(g) + Cl_2(g)$$

(a) calculate ΔG^0 for the reaction from thermodynamic data.
(b) calculate K for the reaction.
(c) calculate the concentration (moles/liter) of HBr(g) if 1.00 mole of $Br_2(g)$ and 2.00 moles of HCl(g) are placed in a 100-liter flask and allowed to react to equilibrium.

10-4.
Following are free energy of formation data at 25° C:

COMPOUND	$\Delta G^0_{\text{FORMATION}}$ (kcal/mole)
$CdCl_2 \cdot H_2O(s)$	−140.13
$CdCl_2(s)$	−81.88
$H_2O(g)$	−54.64

Calculate P_{H_2O}, the equilibrium vapor pressure of water above $CdCl_2 \cdot H_2O(s)$, at 25° C.

10-5.
At 30° C, $K = 2.9 \times 10^{-2}$ for the reaction

$$SO_2Cl_2(g) = SO_2(g) + Cl_2(g)$$

(a) calculate the equilibrium pressure of chlorine in a sample of SO_2Cl_2 if the total pressure is one atmosphere.
(b) calculate the total pressure at which the equilibrium pressures of Cl_2, SO_2, and SO_2Cl_2 are all equal in a sample of SO_2Cl_2.

10-6.
For the reaction

$$2\ HBr(g) = H_2(g) + Br_2(g)$$

$K = 6.4 \times 10^{-20}$ at 25°C. If 0.010 mole of HBr and 0.010 mole of H_2 are placed in a 1.00-liter flask and allowed to react at 25° C, calculate the moles of Br_2 present at equilibrium.

10-7.
At 986° C, $K = 1.60$ for the reaction

$$H_2(g) + CO_2(g) = H_2O(g) + CO(g)$$

In several experiments, the following numbers of moles of the different gases were mixed in a 10.0-liter container at 986° C. Calculate, in each case, the equilibrium number of moles of each kind of gas.

(a) 1.00 mole of H_2 and 1.00 mole of CO_2.
(b) 1.00 mole of H_2O and 1.00 mole of CO.
(c) 1.00 mole each of H_2, CO_2, CO, and H_2O.
(d) 1.00 mole of H_2, 2.00 moles of CO_2, 3.00 moles of H_2O, and 4.00 moles of CO.

10-8.
A 3.0-liter flask containing 10.0 moles of NH_3 gas was heated to a certain temperature. When equilibrium was attained, 4.0 moles of NH_3 remained. Calculate K for the reaction

$$2\ NH_3 = N_2 + 3\ H_2$$

at the temperature of the experiment.

10-9.
In a given experiment, 1.20 moles of $CO(g)$ were mixed with 2.12 moles of $H_2(g)$ in a 2.00-liter container. The reaction that takes place is

$$CO(g) + 2\ H_2(g) = CH_3OH(g)$$

After equilibrium was established, 0.60 mole of $CH_3OH(g)$ was found in the container. Calculate K for the reaction at the temperature of the experiment.

10-10.
Consider the gaseous reaction

$$CO + Cl_2 = COCl_2$$

To a 7.5-liter container were added 2.00 moles of CO and 1.00 mole of Cl_2, and these were allowed to equilibrate at a given constant temperature. At equilibrium under the above conditions, 0.50 mole of Cl_2 was present. The contents of the container were then compressed to 1.00 liter (same temperature as before), and the system was allowed to reach equilibrium. What is the concentration of Cl_2 in the 1.00-liter container after the new equilibrium is established?

10-11.
For the equilibrium

$$C(s) + CO_2(g) = 2\ CO(g)$$

$K = 98.4$ at 1250° K. Calculate P_{CO} at 1250° K when $P_{CO_2} = 20.0$ torr.

10-12.
For the reaction

$$2\ CaSO_4(s) = 2\ CaO(s) + 2\ SO_2(g) + O_2(g)$$

$K = 1.45 \times 10^{-5}$ at 1625° K. What is the total dissociation pressure of $CaSO_4$ at 1625° K? (The dissociation pressure is the sum of the partial pressures of the gaseous products.)

10-13.
A 1.00-liter flask that contains 10.00 grams of $NaHCO_3$ is heated to 100° C. At this temperature, $K = 0.150$ for the reaction

$$2\ NaHCO_3(s) = Na_2CO_3(s) + CO_2(g) + H_2O(g)$$

(a) What is the equilibrium pressure of CO_2 gas in the container?
(b) How many grams of $NaHCO_3$ remain undissociated?

10-14.
For the reaction

$$2\ Au_2O_3(s) = 4\ Au(s) + 3\ O_2(g)$$

$K = 2.5 \times 10^{57}$ at 25° C.

(a) Calculate the equilibrium oxygen pressure above Au_2O_3 at 25° C. (Make the ridiculous assumption that O_2 is an ideal gas at the equilibrium pressure.)
(b) From the above calculations explain why gold does not oxidize (tarnish) at 25° C in air.

10-15.
For the reaction at 25° C

$$H_2(g) = 2\ H(g)$$

calculate (a) K, and, (b) the partial pressure of H in hydrogen gas at a total pressure of 1.00 atm.

10-16.
Calculate $[CH_3COO^-]$ in 0.25 M acetic acid.

10-17.
Calculate $[H^+]$, $[F^-]$, and $[HF]$ in a solution labeled 0.0010 M HF.

10-18.
Glycine, $HC_2H_4O_2N$ (HG), is one of the amino acids that make up body proteins. The ionization constant for

$$HG = H^+ + G^-$$

is 1.7×10^{-10}. What are $[H^+]$ and $[G^-]$ in a solution 1.00×10^{-4} molar in HG?

10-19.
Calculate the pH of a 0.40 M HCN solution.

10-20.
What concentration of fluoroacetic acid is needed to give a solution with $[H^+] = 2 \times 10^{-3}$ moles/liter? $K_a = 2.6 \times 10^{-3}$ for fluoroacetic acid.

10-21.
The pH of a 0.37 M solution of a weak acid, HA, is 4.20. What is K_a for the acid?

10-22.
The $[H^+]$ concentration in a 0.30 M solution of hyponitrous acid ($H_2N_2O_2$) is 1.7×10^{-4} mole/liter. Calculate K_a for hyponitrous acid. Assume that only one hydrogen ionizes.

10-23.
A 0.25 M solution of propanoic acid, $HC_3H_5O_2$, is 0.72% ionized. What is K_a for propanoic acid?

10-24.
For a 0.50 M NH_3 solution, calculate (a) $[OH^-]$; (b) $[H^+]$; and (c) pH and pOH.

10-25.
Calculate the pH of a solution 0.10 M in HCN and 0.10 M in NaCN.

10-26.
What is the molar concentration of formate ion ($HCOO^-$) in a solution 0.020 M in formic acid (HCOOH) and 0.010 M in HCl?

10-27.
A 0.20 M solution of HNO_2 containing some $NaNO_2$ has a pH of 3.90. What is the molar concentration of $NaNO_2$ in the solution?

10-28.
Calculate $[H^+]$ in a solution 0.20 M in NH_3 and 0.40 M in NH_4Cl.

10-29.
Calculate the pH of a 0.20 M KNO_2 solution.

10-30.
A 0.25 M solution of pyridinium chloride ($C_5H_6N^+Cl^-$), the salt of the weak base pyridine, has a pH of 2.89. What is K_b for pyridine, C_5H_5N?

10-31.
What is $[CO_3^{2-}]$ in a 0.0010 M Na_2CO_3 solution (after hydrolysis)?

10-32.
Calculate the pH of the resulting solution when 20.0 ml of 0.10 M NaOH is added to 20.0 ml of 0.10 M HCN. Assume that the volumes are additive.

10-33.
To 750 ml of 0.010 M HF are added 250 ml of 0.010 M NaOH. Calculate the pH of the resulting solution.

10-34.
Calculate $[H^+]$, $[F^-]$, and $[HF]$ in a solution prepared by adding 10 ml of 0.50 M NaF to 10 ml of 0.30 M HCl. Assume that the volumes are additive.

10-35.
Calculate the pH of the following solutions:

(a) A 0.010 M CH_3COOH solution.
(b) A solution 0.010 M in CH_3COOH and 0.010 M in CH_3COONa.
(c) A solution 0.010 M in CH_3COONa.

10-36.
Calculate the Cu^+ concentration in a saturated solution of CuCl.

10-37.
Calculate the solubility of PbF_2 in water at 25° C.

10-38.
In a saturated solution of a slightly soluble salt, MX, the $[M^+]$ concentration is 1.50×10^{-8} molar. Calculate K_{sp} for MX.

10-39.
The solubility of calcium fluoride (CaF_2) is 2.0×10^{-4} moles/liter. Calculate the K_{sp} for CaF_2.

10-40.
Calculate the solubility of $BaSO_4$ in a 1.00 M solution of $BaCl_2$.

10-41.
Calculate the number of grams of solid MgF_2 that can be dissolved in one liter of (a) pure water; (b) 0.020 M NaF solution.

10-42.
When solid AgI is added to a 0.050 M NaI solution, the Ag^+ concentration after equilibrium is established is 2.0×10^{-15} M. Calculate K_{sp} for AgI.

10-43.
Solid $NaIO_3$ is added to 100 ml of a 0.010 M solution of $Pb(NO_3)_2$. What weight of $NaIO_3$ must be added before $Pb(IO_3)_2$ will start to precipitate?

10-44.
Calculate the Ag^+ concentration in each of the following:

(a) A saturated solution of Ag_2SO_4.
(b) A solution saturated with AgCl and 0.10 *M* in KCl.

10-45.
Calculate the Mn^{2+} concentration in a water solution buffered at pH = 10.00 and in contact with solid $Mn(OH)_2$.

Problems

All reactions are assumed to occur at 25° C unless the temperature is specifically stated otherwise. Data from the tables of equilibrium constants and the tables of thermodynamic data in the Appendix are required for many of the problems.

10-1.
For the reaction

$$N_2O_3(g) = NO_2(g) + NO(g)$$

$K = 10$ at a certain temperature. Some of the NO_2 associates according to the reaction

$$2\ NO_2(g) = N_2O_4(g)$$

$K = 1$ at the same temperature. Calculate P_{NO_2} in the equilibrium gas mixture resulting from the dissociation of N_2O_3 at a total pressure of one atmosphere.

10-2.
Gaseous PCl_5 decomposes into PCl_3 and Cl_2:

$$PCl_5(g) = PCl_3(g) + Cl_2(g)$$

At 546° K, the equilibrium mixture of PCl_5, PCl_3, and Cl_2 has a density of 2.48 g/liter when the total pressure is 1.00 atm. Calculate *K* for the decomposition.

10-3.
For the reaction

$$2\ SO_2(g) + O_2(g) = 2\ SO_3(g)$$

$K = 42.8$ at 900° K and 3.44 at 1000° K.

(a) What is the average value of ΔH^0 in this temperature range?
(b) Calculate *K* at 950° K.

10-4.
A 3.00-molar water solution of sucrose is allowed to react at 25° C according to the equation

$$C_{12}H_{22}O_{11} + H_2O = 2\ C_6H_{12}O_6$$

(a) Write the equilibrium constant expression in terms of activities.
(b) This solution is too concentrated to assume that the activity of water is unity. If the vapor pressure of water above the solution is 22.226 torr, what is the activity of water? $P^{\square}_{H_2O} = 23.756$ torr.
(c) Assume that $C_{12}H_{22}O_{11}$ and $C_6H_{12}O_6$ obey Henry's law in the solution, and calculate the equilibrium concentrations of each. Assume $K = 10.0$.

10-5.
Liquid ammonia ionizes to a slight extent according to the reaction

$$2\ NH_3 = NH_4^+ + NH_2^-$$

At $-50°$ C, $K_{NH_3} = 10^{-30}$ for this self-ionization. How many NH_4^+ ions are present per ml of pure liquid ammonia at $-50°$ C?

10-6.
(a) 50.00 ml of a 0.0500 *M* NaOH solution is added to 50.00 ml of a 0.1000 *M* HCN solution. Calculate the pH of the resulting solution.
(b) 50.00 ml of a 0.1000 *M* NaOH solution is added to 50.00 ml of a 0.1000 *M* HCN solution. Calculate the pH of the resulting solution.

10-7.
Calculate the pH of a solution in which 50.00 ml of 0.0500 *M* HCl is added to 50.00 ml of a 0.1000 *M* KCN solution.

10-8.
The base imidazole has a K_b of 8.1×10^{-8}. Calculate $[OH^-]$ in a solution made by mixing 50.0 ml of 0.050 *M* HCl with 50.0 ml of 0.100 *M* imidazole. Assume that the volumes are additive.

10-9.
Tetramethylammonium hydroxide, $(CH_3)_4NOH$, is a strong organic base.

(a) Exactly 100 ml of a 0.10 *M* solution of $(CH_3)_4NOH$ is titrated with 0.10 *M* HNO_3. Calculate the pH of the $(CH_3)_4NOH$ solution before any HNO_3 is added and after just 10, 50, 90, 99, 100, 101, 110, and 150 ml of HNO_3 are added. Display your results by making a graph of pH versus volume of HNO_3 added. You may assume the volumes of acid and base are additive.
(b) Repeat the above calculation for novocain (Nvc), a weak organic base. For the reaction,

$$Nvc + H_2O = NvcH^+ + OH^-; K_b = 9.0 \times 10^{-6}$$

10-10.
What concentrations of NH_3 and NH_4^+ would you use to prepare a buffer solution at a pH of 9.50? Can more than one set of concentrations give this pH?

10-11.
What volume of 0.60 *M* NaOH solution should be added to 100 ml of a 0.60 *M* benzoic acid (C_6H_5COOH) solution to prepare a buffer with a pH of 5.00? Assume that the volumes are additive.

10-12.
How many grams of solid NaOH must be added to a liter of 0.010 *M* H_3BO_3 to make a buffered solution of pH = 10.00? Assume that the volume of the NaOH added is negligible. H_3BO_3 is a monoprotic acid with $K_a = 5.8 \times 10^{-10}$.

10-13.
A buffered solution is prepared by adding 0.10 mole of formic acid and 0.10 mole of sodium formate to enough water to make one liter of solution.

(a) What is the pH of the solution?
(b) What is the pH after 0.02 mole of HCl is added?
(c) What is the pH after 0.02 mole of NaOH is added?
(d) What is the pH after 0.20 mole of HCl is added?

10-14.
You want to prepare a buffered solution at pH = 7.00. Choose an acid-conjugate base pair from the table of dissociation constants that would work well in a buffered solution at this pH, and calculate the ratio of the substances you would use to prepare the buffer.

10-15.
Calculate $[H^+]$, $[HSO_4^-]$, and $[SO_4^{2-}]$ in a 0.0050 *M* H_2SO_4 solution.

10-16.
Fumaric acid is a diprotic acid with the formula $H_2C_4H_2O_4$. The ionization constants are $K_I = 9.0 \times 10^{-5}$ and $K_{II} = 4.0 \times 10^{-5}$. Calculate the pH of a 1.00-molar solution of fumaric acid.

10-17.
Calculate the $[S^{2-}]$ concentration in a 0.100 *M* H_2S solution.

10-18.
Calculate $[CN^-]$ in a solution 0.10 *M* in HF and 0.10 *M* in HCN.

10-19.
A solution buffered at pH = 4.50 is saturated with H_2S (equilibrium concentration of H_2S = 0.10 *M*). Calculate $[S^{2-}]$ and $[HS^-]$ in the solution.

10-20.
Calculate the concentrations of all the species present in a 0.10 *M* H_3AsO_4 solution.

10-21.
Calculate the pH of a solution 0.20 *M* in H_3PO_4 and 0.10 *M* in NaH_2PO_4.

10-22.
Calculate the concentrations of Na^+, S^{2-}, HS^-, and OH^- in a solution labeled 0.020 *M* Na_2S.

10-23.
HPO_4^{2-} ion can either dissociate as a weak acid or hydrolyze as the salt of a weak acid. The two reactions are

$$HPO_4^{2-} + H_2O = H_3O^+ + PO_4^{3-}$$
$$HPO_4^{2-} + H_2O = H_2PO_4^- + OH^-$$

(a) Which of the above reactions is the principal one that would occur if 0.100 mole of Na_2HPO_4 is added to a liter of water? Justify your choice.
(b) Calculate the pH of the solution of part (a) above.

10-24.
(a) Calculate the pH of a 0.10 M Na_2CO_3 solution.
(b) Calculate the pH of a 0.10 M $NaHCO_3$ solution.

10-25.
Calculate the pH of a (a) 0.010 M NaH_2PO_4 solution; (b) 0.0010 M Na_3PO_4 solution.

10-26.
Calculate the pH of 1.00 liter of a 0.200 M $H_2C_2O_4$ solution to which 0.400 mole of solid NaOH has been added. Assume that the volume of the added NaOH is negligible.

10-27.
Enough of a mixture of solid AgCN and AgBr is put in water to form a saturated solution of each. Calculate $[Ag^+]$, $[CN^-]$, and $[Br^-]$ in the solution. K_{sp} for AgBr $= 5 \times 10^{-13}$; K_{sp} for AgCN $= 1.6 \times 10^{-14}$.

10-28.
One mole of solid $PbCl_2$ and one mole of solid AgCl are added to a liter of water. What will be the weights of dissolved $PbCl_2$ and AgCl in the solution?

10-29.
One mole of $AgNO_3$ is added to a liter of solution containing one mole of sodium acetate (CH_3COONa) and one mole of sodium nitrite ($NaNO_2$). Given K_{sp} for $AgNO_2 = 6 \times 10^{-4}$ and K_{sp} for $CH_3COOAg = 4 \times 10^{-3}$, calculate the composition (wt % $AgNO_2$ and wt % CH_3COOAg) of the precipitate.

10-30.
Potassium hydrogen tartrate ($KHC_4H_4O_6$), a slightly soluble salt, dissolves in water according to the reaction

$$KHC_4H_4O_6(s) = K^+ + HC_4H_4O_6^-$$

for which $K_{sp} = 3.8 \times 10^{-4}$. The $HC_4H_4O_6^-$ ion then dissociates slightly according to the reaction

$$HC_4H_4O_6^- = H^+ + C_4H_4O_6^{2-}$$

for which $K_{diss} = 4.5 \times 10^{-5}$. Calculate the hydrogen ion concentration $[H^+]$ of a saturated solution of $KHC_4H_4O_6$.

10-31.
A solution 0.050 M in Cu^{2+} and 0.050 M in Fe^{2+} is buffered at a pH of 1.00, then saturated with H_2S gas. (The resulting concentration of H_2S in the solution is 0.10 M.)

(a) Will CuS precipitate?
(b) Will FeS precipitate?

10-32.
Calculate the solubility of $CaCO_3$ in a solution buffered at a pH of (a) 11; (b) 6.

10-33.
Calculate the solubility of AgI in a solution 0.10 M in NH_3.

10-34.
$HgCl_2$ is a weak electrolyte that dissociates in the two steps

$$HgCl_2 = HgCl^+ + Cl^- \qquad K_I = 3.3 \times 10^{-7}$$
$$HgCl^+ = Hg^{2+} + Cl^- \qquad K_{II} = 1.8 \times 10^{-7}$$

while $Hg(NO_3)_2$ is a strong electrolyte (completely dissociated). Calculate $[Cl^-]$ in (a) a 0.10 M $HgCl_2$ solution; (b) a solution 0.10 M in $HgCl_2$ and 0.15 M in $Hg(NO_3)_2$.

10-35.
The complex ion $CuCl_4^{2-}$ dissociates by the following steps:

$$CuCl_4^{2-} = CuCl_3^- + Cl^- \qquad K_I = 1.9 \times 10^{-1}$$
$$CuCl_3^- = CuCl_2 + Cl^- \qquad K_{II} = 3.3 \times 10^{-1}$$
$$CuCl_2 = CuCl^+ + Cl^- \qquad K_{III} = 2.5 \times 10^{-2}$$
$$CuCl^+ = Cu^{2+} + Cl^- \qquad K_{IV} = 1.6 \times 10^{-3}$$

Calculate the concentrations of all the species present in a solution made by adding 0.10 mole of $CuSO_4$ and 1.40 moles of NaCl to enough water to make 1.00 liter of solution.

10-36.
At 900° K, $\Delta H^0 = 102.2$ kjoules/mole and $\Delta G^0 = 22.4$ kjoules/mole for the reaction

$$C_2H_6(g) = C_2H_4(g) + H_2(g)$$

(a) Calculate the mole % H_2 present at equilibrium if pure C_2H_6 is passed over a dehydrogenation catalyst at 900° K and one atmosphere pressure.
(b) Estimate the mole % present at equilibrium at 1000° K. Again, assume that the pressure is one atmosphere.

11

Electrochemistry

Electrochemistry divides logically into two subjects: (1) the production of electrical energy from chemical change, and (2) the converse of (1), the production of chemical change by means of electrical energy.

The first subject is concerned with the **voltaic cell,** a simple, ingenious device that physically separates the two half-reactions of an oxidation-reduction reaction. This division makes it possible for the electrons that are produced in the oxidation half-reaction to be intercepted as they flow—through an external circuit—to the reduction half of the cell. The flow of electrons can be used to run a motor, heat a filament, open a solenoid valve, or perform any other such electrical operation before the electrons reach their destination at the reduction electrode. Thus, the energy of the reaction, which otherwise would appear as heat, can often be extracted in the form of electrical energy.

The second subject of electrochemistry is called **electrolysis.** Here we supply electrical energy to a chemical system, and thereby make chemical reactions occur that require a net absorption of energy. Many important industrial operations make use of this procedure.

11.1 Electrical Energy from Chemical Change

The Voltaic Cell

An electrochemical cell designed to produce electrical energy from a chemical reaction is called a *voltaic cell.*

To illustrate the principle of the voltaic cell, let us consider the cell illustrated in Figure 11.1, which from the point of view of theory of operation is a very simple one. Basically, a voltaic cell consists of two chemically different

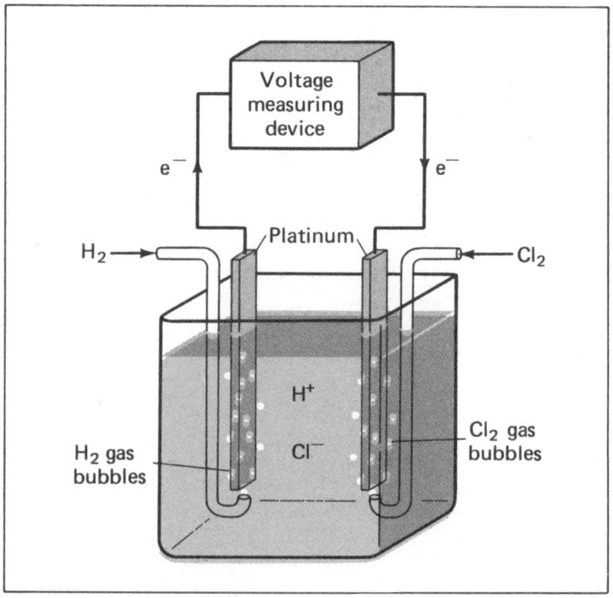

Figure 11.1
An example of an electrochemical cell. Electrical work is obtained from the chemical reaction $H_2 + Cl_2 = 2\,H^+ + 2\,Cl^-$

electrodes in a solution of electrolyte. In our example, the electrodes are H_2 and Cl_2 gases bubbling over platinum strips. The inert platinum serves simply as an electrically conducting support for the chemically active H_2 and Cl_2. The electrolyte of the cell is hydrochloric acid. When the electrodes are connected through an external electrical conductor, chemical reactions occur at the electrodes. By definition, the **anode** is the electrode at which oxidation takes place, and the **cathode** is the one at which the reduction half of the reaction occurs. For our example, the anode reaction is

$$H_2(g) = 2\,H^+ + 2\,e^- \qquad \text{(anode)}$$

The electrons generated at the anode then travel through the external circuit to the cathode, where they are taken up in the reduction half of the reaction,

$$Cl_2(g) + 2\,e^- = 2\,Cl^- \qquad \text{(cathode)}$$

The overall reaction (the sum of the two half-reactions) is thus

$$H_2(g) + Cl_2(g) = 2\,H^+ + 2\,Cl^-$$

During the anode reaction, a positive ion (H^+) appears each time an electron is released, giving, momentarily, an excess of positively charged ions in the region of the anode. During the cathode reaction, a negatively charged ion (Cl^-) is produced for each electron consumed, giving, momentarily, an excess of negative ions in the region of the cathode. These electrical imbalances are immediately corrected by migration of H^+ away from the anode toward the cathode, and by migration of Cl^- away from the cathode and toward the anode. This migration of ions is the mechanism by which an electrical charge is transported through the solution. Thus, while the flow of current through the external circuit is by transfer of electrons along a metallic conductor, the flow of current through the internal circuit is by migration of ions.

The flow of electrons in the external circuit represents electrical energy capable of doing work (running a motor, ringing a bell, or forcing electrons through a resistance such as a light-globe filament to produce heat and light). In theory, at least, any oxidation-reduction reaction can be set up in a cell to produce electrical energy. The maximum amount of electrical work available from a voltaic cell is easily calculated. If, during the discharge of a cell, a quantity of electricity $\mathcal{Q}$ flows through the circuit, the amount of electrical work W_e produced is given by the relation

$$W_e = -\mathcal{E}\mathcal{Q} \qquad [11.1]$$

where $\mathcal{E}$ is the **electromotive force** or voltage of the cell. This is the force that drives the electrical charge (electrons in the external circuit and ions in the internal circuit) through the cell. The electromotive force is measured in volts (V), and its value depends on the nature of the chemical reaction taking place in the cell, and on the conditions (temperature, pressure, and concentration) of operation of the cell.

Work, in general, is given by the product of a potential term and a capacity term. In this example, the potential term is $\mathcal{E}$ and the capacity term is $\mathcal{Q}$. The negative sign is included to make the equation agree with our previous definition of positive and negative work.

The quantity of electrical charge, $\mathscr{Q}$, can be calculated from the stoichiometry of the reaction. If n moles of electrons are transferred from cathode to anode during the chemical reaction, and $\mathscr{F}$ **(faraday)** is the quantity of electricity contained in a mole of electrons, then

$$\mathscr{Q} = n\mathscr{F}$$

and

$$W_e = -n\mathscr{F}\mathscr{E} \qquad [11.2]$$

When one mole of H_2 reacts with one mole of Cl_2 as shown in the cell in Figure 11.1, the number of moles of electrons transferred is two ($n = 2$). This is most easily seen by examining the two half-reactions

$$H_2 = 2\,H^+ + 2\,e^- \qquad \text{(oxidation)}$$
$$Cl_2 + 2\,e^- = 2\,Cl^- \qquad \text{(reduction)}$$

In the oxidation half reaction, one mole of H_2 is seen to produce two moles of electrons, which are transferred through the external circuit to the cathode, where they cause the reduction of one mole of Cl_2.

The numerical value of $\mathscr{F}$, the faraday, can easily be calculated. Since the electronic charge is 1.6022×10^{-19} coulombs/electron, we have

$$\mathscr{F} = \left(6.0220 \times 10^{23}\,\frac{\text{electrons}}{\text{mole}}\right)\left(1.6022 \times 10^{-19}\,\frac{\text{coulombs}}{\text{electron}}\right)$$

$$\mathscr{F} = 96{,}485\,\frac{\text{coulombs}}{\text{mole}}$$

Substituting this value $\mathscr{F}$ into equation [11.2] would give the work in joules. To calculate work in calories, the value of $\mathscr{F}$ in coulombs (joules/volt) is divided by the factor that converts joules to calories (4.1840 joules/cal) to give

$$\mathscr{F} = 23{,}060\,\frac{\text{cal}}{\text{V-mole}}$$

Relationship Between Electromotive Force and Free Energy

The electromotive force of the cell can be related to the free energy change in the chemical reaction through equation [11.2]. Free energy was defined in Chapter 7 as

$$G = E + PV - TS \qquad [7.18]$$

In a constant-pressure and constant-temperature process, a change in free energy is given by

$$\Delta G = \Delta E + P\,\Delta V - T\,\Delta S \quad (\text{constant } T \text{ and } P) \qquad [11.3]$$

We have seen earlier that

$$\Delta E = Q + W \qquad [7.5]$$

When we discussed *work* in Chapter 7, we restricted the discussion to PV work only. Now that we are considering the subject of electrochemistry, we must broaden the discussion of work to include electrical as well as PV work. Equation [11.4] is written on the assumption that the only kinds of work involved in the process are pressure-volume and electrical. If other kinds of work (such as work against a magnetic field or work of forming a surface) are involved, more terms must be included in equation [11.4].

and W, the total work done on the surroundings, is given by W_e (electrical work) plus W_{PV} (pressure-volume work at constant pressure, $-P\Delta V$):

$$W = W_e + W_{PV} = W_e - P\Delta V \qquad [11.4]$$

If the process is reversible and isothermal, the heat Q added to the surroundings is given by

$$Q = T\Delta S \qquad [7.20]$$

Substituting [7.5], [11.4], and [7.20] into [11.3] gives

$$\Delta G = Q + W_e - P\Delta V + P\Delta V - Q = W_e \qquad [11.5]$$

Canceling terms and substituting for W_e, from equation [11.2], yields

$$\Delta G = W_e = -n\mathcal{F}\mathcal{E} \qquad [11.6]$$

For a change under standard state conditions, equation [11.6] can be written

$$\boxed{\Delta G^0 = -n\mathcal{F}\mathcal{E}^0} \qquad [11.7]$$

where $\mathcal{E}^0$ is the electromotive force of the cell with all the reactants and products in their standard states.

In reporting electromotive force data it is also necessary to report the temperature, which is commonly 25° C. It is interesting to note how reference temperatures have changed over the years. It is convenient to choose the reference temperature near the laboratory temperature. In the very early tables, the reference was 15° C. This temperature from the days of "wooden ships and iron men" and unheated laboratories gave way at successively later periods to 17° C and then 20° C. Now in our decadent age of central heating and air-conditioning, we find 25° C to be a comfortable laboratory, and therefore reference, temperature.

Standard Cell Potentials

We have just seen that the electromotive force of a voltaic cell depends on the ΔG of the cell reaction, that is, on the driving force of the reaction. This, in turn, depends not only on the nature of the reaction, but also—as we will see in more detail later—on the *concentration* of the reactants and products of the reaction. Because of this concentration dependence, electromotive force values are recorded at a common reference concentration. This reference state is the thermodynamic standard state discussed in Chapter 9 (1 M concentration for solutes, and 1 atm pressure for gases).

Values of the **standard potential** (standard electromotive force, $\mathcal{E}^0$) have been measured for the combination of a number of half-cells *in which one of the electrodes is a standard* H_2 *electrode*. The H_2-Cl_2 cell previously described is one example. The calomel-H_2 cell shown in Figure 11.2 is another example. A list of $\mathcal{E}^0$ values for such cells is given in Table 11.1, Standard Reduction Potentials. In each case, the $\mathcal{E}^0$ value is that obtained by coupling the half-cell given in the table with the standard H_2 half-cell.

Figure 11.2
A schematic drawing of the calomel (Hg_2Cl_2)-hydrogen voltaic cell.

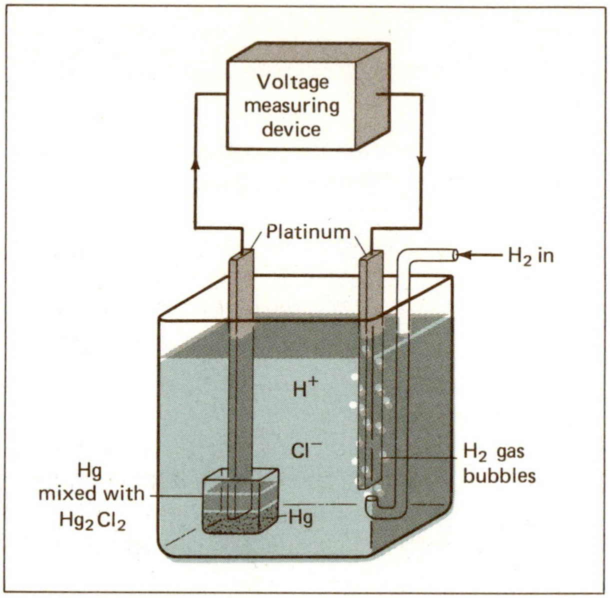

The data in Table 11.1 can be used to calculate $\mathscr{E}^0$ for a large number of chemical reactions by use of equation [11.7] in the form

$$\mathscr{E}^0 = -\frac{\Delta G^0}{n\mathscr{F}} \qquad [11.8]$$

Since ΔG's are additive, the net ΔG^0 values for an overall reaction can be obtained by summing the ΔG^0 values of a series of reactions that add up to the overall reaction.

EXAMPLE 11.1 Calculate $\mathscr{E}^0$ for the reaction

$$Cu^{2+} + Zn = Zn^{2+} + Cu$$

SOLUTION From Table 11.1:

	Reaction	$\mathscr{E}^0$	ΔG^0
	$Cu^{2+} + H_2 = Cu + 2\,H^+$,	$\mathscr{E}_1^0 = 0.337$ V,	$\Delta G_1^0 = -2\mathscr{F}(0.337\text{ V})$
	$Zn + 2\,H^+ = Zn^{2+} + H_2$,	$\mathscr{E}_2^0 = 0.763$ V,	$\Delta G_2^0 = -2\mathscr{F}(0.763\text{ V})$
Adding:	$Zn + Cu^{2+} = Zn^{2+} + Cu$		$\Delta G^0 = -2\mathscr{F}(1.100\text{ V})$

$$\mathscr{E}^0 = -\frac{\Delta G^0}{2\mathscr{F}} = \frac{(1.100\text{ V})2\mathscr{F}}{2\mathscr{F}} = 1.100\text{ V}$$

Observe that $\mathscr{E}_2^0$ is opposite in sign from the value given in the table, since the value in the table is for the reverse reaction.

TABLE 11.1

Standard Reduction Potentials at 298.15° K

HALF–REACTION	$\mathcal{E}^0$ (V)	HALF–REACTION	$\mathcal{E}^0$ (V)
$Li^+ + e^- = Li$	−3.045	$Cu^+ + e^- = Cu$	0.521
$K^+ + e^- = K$	−2.925	$I_2 + 2\,e^- = 2\,I^-$	0.535
$Ba^{2+} + 2\,e^- = Ba$	−2.90	$I_3^- + 2\,e^- = 3\,I^-$	0.536
$Sr^{2+} + 2\,e^- = Sr$	−2.89	$Cu^{2+} + Cl^- + e^- = CuCl$	0.538
$Ca^{2+} + 2\,e^- = Ca$	−2.87	$MnO_4^- + e^- = MnO_4^{2-}$	0.564
$Na^+ + e^- = Na$	−2.714	$Cu^{2+} + Br^- + e^- = CuBr$	0.640
$Mg^{2+} + 2\,e^- = Mg$	−2.37	$O_2 + 2\,H^+ + 2\,e^- = H_2O_2$	0.682
$\frac{1}{2}\,H_2 + e^- = H^-$	−2.25	$Fe^{3+} + e^- = Fe^{2+}$	0.771
$Be^{2+} + 2\,e^- = Be$	−1.85	$Hg_2^{2+} + 2\,e^- = 2\,Hg$	0.789
$Al^{3+} + 3\,e^- = Al$	−1.66	$Ag^+ + e^- = Ag$	0.799
$Mn^{2+} + 2\,e^- = Mn$	−1.18	$Cu^{2+} + I^- + e^- = CuI$	0.86
$V^{2+} + 2\,e^- = V$	−1.18	$2\,Hg^{2+} + 2\,e^- = Hg_2^{2+}$	0.920
$TiO^{2+} + 2\,H^+ + 4\,e^- = Ti + H_2O$	−0.89	$NO_3^- + 4\,H^+ + 3\,e^- = NO + 2\,H_2O$	0.96
$H_3BO_3 + 3\,H^+ + 3\,e^- = B + 3\,H_2O$	−0.87	$HNO_2 + H^+ + e^- = NO + H_2O$	1.00
$SiO_2 + 4\,H^+ + 4e^- = Si + 2\,H_2O$	−0.86	$Br_2\,(l) + 2\,e^- = 2\,Br^-$	1.065
$Zn^{2+} + 2\,e^- = Zn$	−0.763	$N_2O_4 + 2\,H^+ + 2\,e^- = 2\,HNO_2$	1.07
$Cr^{3+} + 3\,e^- = Cr$	−0.74	$ClO_4^- + 2\,H^+ + 2\,e^- = ClO_3^- + H_2O$	1.19
$H_3PO_2 + H^+ + e^- = P + 2\,H_2O$	−0.51	$IO_3^- + 6\,H^+ + 5\,e^- = \frac{1}{2}\,I_2 + 3\,H_2O$	1.195
$H_3PO_3 + 2\,H^+ + 2\,e^- = H_3PO_2 + H_2O$	−0.50	$ClO_3^- + 3\,H^+ + 2\,e^- = HClO_2 + H_2O$	1.21
$Fe^{2+} + 2\,e^- = Fe$	−0.440	$O_2 + 4\,H^+ + 4\,e^- = 2\,H_2O$	1.229
$Cr^{3+} + e^- = Cr^{2+}$	−0.41	$MnO_2 + 4\,H^+ + 2\,e^- = Mn^{2+} + 2\,H_2O$	1.23
$Cd^{2+} + 2\,e^- = Cd$	−0.403	$Cr_2O_7^{2-} + 14\,H^+ + 6\,e^- = 2\,Cr^{3+} + 7\,H_2O$	1.33
$PbI_2 + 2\,e^- = Pb + 2\,I^-$	−0.365	$Cl_2 + 2\,e^- = 2\,Cl^-$	1.3595
$PbSO_4 + 2\,e^- = Pb + SO_4^{2-}$	−0.356	$HIO + H^+ + e^- = \frac{1}{2}\,I_2 + H_2O$	1.45
$Co^{2+} + 2\,e^- = Co$	−0.277	$PbO_2 + 4\,H^+ + 2\,e^- = Pb^{2+} + 2\,H_2O$	1.455
$H_3PO_4 + 2\,H^+ + 2\,e^- = H_3PO_3 + H_2O$	−0.276	$Au^{3+} + 3\,e^- = Au$	1.50
$Ni^{2+} + 2\,e^- = Ni$	−0.250	$MnO_4^- + 8\,H^+ + 5\,e^- = Mn^{2+} + 4\,H_2O$	1.51
$CuI + e^- = Cu + I^-$	−0.185	$BrO_3^- + 6\,H^+ + 5\,e^- = \frac{1}{2}\,Br_2 + 3\,H_2O$	1.52
$AgI + e^- = Ag + I^-$	−0.151	$HBrO + H^+ + e^- = \frac{1}{2}\,Br_2 + H_2O$	1.59
$Sn^{2+} + 2\,e^- = Sn$	−0.136	$H_5IO_6 + H^+ + 2\,e^- = IO_3^- + 3\,H_2O$	1.6
$Pb^{2+} + 2\,e^- = Pb$	−0.126	$Ce^{4+} + e^- = Ce^{3+}$	1.61
$2\,H^+ + 2\,e^- = H_2$	0.000	$HClO + H^+ + e^- = \frac{1}{2}\,Cl_2 + H_2O$	1.63
$P + 3\,H^+ + 3\,e^- = PH_3(g)$	0.06	$HClO_2 + 2\,H^+ + 2\,e^- = HClO + H_2O$	1.64
$AgBr + e^- = Ag + Br^-$	0.095	$Au^+ + e^- = Au$	1.68
$CuCl + e^- = Cu + Cl^-$	0.137	$NiO_2 + 4\,H^+ + 2\,e^- = Ni^{2+} + 2\,H_2O$	1.68
$S + 2\,H^+ + 2\,e^- = H_2S$	0.141	$PbO_2 + HSO_4^- + 3\,H^+ + 2\,e^- = PbSO_4 + 2\,H_2O$	1.685
$Sn^{4+} + 2\,e^- = Sn^{2+}$	0.15	$MnO_4^- + 4\,H^+ + 3\,e^- = MnO_2 + 2\,H_2O$	1.695
$Cu^{2+} + e^- = Cu^+$	0.153	$H_2O_2 + 2\,H^+ + 2\,e^- = 2\,H_2O$	1.77
$HSO_4^- + 3\,H^+ + 2\,e^- = H_2SO_3 + H_2O$	0.17	$Co^{3+} + e^- = Co^{2+}$	1.82
$AgCl + e^- = Ag + Cl^-$	0.222	$O_3 + 2\,H^+ + 2\,e^- = O_2 + H_2O$	2.07
$Cu^{2+} + 2\,e^- = Cu$	0.337	$F_2 + 2\,e^- = 2\,F^-$	2.87
$Fe(CN)_6^{3-} + e^- = Fe(CN)_6^{4-}$	0.36	$F_2 + 2\,H^+ + 2\,e^- = 2\,HF(aq)$	3.06
$H_2SO_3 + 4\,H^+ + 4\,e^- = S + 3\,H_2O$	0.45		

Source: W. M. Latimer, THE OXIDATION STATES OF THE ELEMENTS AND THEIR POTENTIALS IN AQUEOUS SOLUTIONS, 2nd Ed.

The numerical value of an *intensive* factor does not depend upon the quantity of substance. Examples of intensive factors are pressure, temperature, and density. However, the numerical values of *extensive* factors such as volume, energy, and entropy do depend upon the quantity of substance.

In Example 11.1 the $\mathscr{E}^0$'s are seen to be additive; that is $\mathscr{E}^0 = \mathscr{E}_1^0 + \mathscr{E}_2^0$. In general, half-cell potentials, being intensive factors, are not additive; however, when two electrode potentials are combined in such a way that the hydrogen half-cells cancel, the resulting $\mathscr{E}^0$ value does turn out to be the sum $(\mathscr{E}_1^0 + \mathscr{E}_2^0)$. This is so because, under these conditions, both the n's and the $\mathscr{F}$'s cancel; for example,

$$\begin{aligned} \Delta G^0 &= \Delta G_1^0 + \Delta G_2^0 \\ -n\mathscr{F}\mathscr{E}^0 &= -n_1\mathscr{F}\mathscr{E}_1^0 - n_2\mathscr{F}\mathscr{E}_2^0 \end{aligned} \quad [11.9]$$

To cancel the hydrogen half-cell, the condition $n_1 = n_2 = n$ must be met. Then

$$\mathscr{E}^0 = \mathscr{E}_1^0 + \mathscr{E}_2^0 \quad [11.10]$$

Equation [11.9] is always valid. Equation [11.10] can be used if the net reaction is a *complete* reaction, one in which the number of electrons cancel. An example in which [11.10] does not apply is discussed later.

Since the reactions in Table 11.1 are written as reduction half-reactions, the $\mathscr{E}^0$'s listed are **standard reduction potentials.** When it is more convenient to speak of the potential of an oxidation half-reaction, we simply need to reverse the reaction and the sign of $\mathscr{E}^0$ to obtain a **standard oxidation potential.**

Up to this point we have interpreted the $\mathscr{E}^0$ values in Table 11.1 as potentials for whole cells consisting of the half-cell given in the table combined with the standard hydrogen half-cell. But there is another way we can interpret those tabulated values. In theory, each electrode in a cell has a standard electromotive force or potential. If two half-cells were put together to form a cell, adding up the **half-cell potentials** would give the voltage of the cell. Since, however, it is not possible to measure the potential of only one-half of a chemical reaction, *absolute* half-cell potentials have never been obtained. The values given in Table 11.1 may be thought of as *relative* half-cell potentials, where the $\mathscr{E}^0$ value for the hydrogen half-cell has been arbitrarily set at zero, and the values of the other half-cells have been measured relative to that value. Fortunately, addition of these relative $\mathscr{E}^0$ values works as well in determining the $\mathscr{E}^0$ for a complete reaction as would the absolute values, since as we have seen earlier, in the addition, the hydrogen half-cells cancel.

EXAMPLE 11.2 Calculate $\mathscr{E}^0$ for the reaction

$$2\,Br^- + Cl_2(g) = 2\,Cl^- + Br_2(l)$$

SOLUTION The two half-cell reactions with corresponding $\mathscr{E}^0$ values from Table 11.1 are given below. (The Br_2-Br^- half-cell is the reverse of that given in the table, hence the change in arithmetic sign.)

$$\begin{array}{rll} & 2\,Br^- = Br_2(l) + 2\,e^-, & \mathscr{E}^0 = -1.065\ V \\ & Cl_2(g) + 2\,e^- = 2\,Cl^-, & \mathscr{E}^0 = 1.360\ V \\ \hline \text{Adding:} & 2\,Br^- + Cl_2(g) = 2\,Cl^- + Br_2(l), & \mathscr{E}^0 = 0.295\ V \end{array}$$

Relationship Between Electromotive Force and Concentration—The Nernst Equation

In Chapter 9, we derived equation [9.5] for calculating the free energy change in a general chemical reaction

$$aA + bB + \cdots = lL + mM + \cdots$$

which is

$$\Delta G = \Delta G^0 + RT \ln \frac{(a_L^l)(a_M^m)\cdots}{(a_A^a)(a_B^b)\cdots} \qquad [9.5]$$

Substituting equations [11.6] and [11.7] into [9.5] gives

$$\mathscr{E} = \mathscr{E}^0 - \frac{RT}{n\mathscr{F}} \ln \frac{(a_L^l)(a_M^m)\cdots}{(a_A^a)(a_B^b)\cdots} \qquad [11.11]$$

Be reminded that the term

$$\frac{(a_L^l)(a_M^m)\cdots}{(a_A^a)(a_B^b)\cdots}$$

in equation [11.11] is not, in the general case, the equilibrium constant (K) since the activities will *not*, in general, be the equilibrium values. However, in *the special case of equilibrium*, $\mathscr{E}$ will be equal to zero, and the activity quotient term *will* be K.

The equation is most often used at 25° C (298.15° K). At this temperature

$$\frac{RT}{\mathscr{F}} = \frac{(8.3143 \text{ joules/°K})(298.15° \text{ K})}{(96{,}485 \text{ joules/V-eq})} = 0.025692 \text{ V-eq} \qquad [11.12]$$

Substitution into [11.11] gives

$$\mathscr{E} = \mathscr{E}^0 - \frac{0.025692}{n} \ln \frac{(a_L^l)(a_M^m)\cdots}{(a_A^a)(a_B^b)\cdots} \quad (25° \text{ C}) \qquad [11.13]$$

Equation [11.11] is commonly referred to as the **Nernst equation.** It allows one to calculate the electromotive force of a cell when the activities (and therefore concentrations) are different from those of the standard state.

EXAMPLE 11.3 The lead storage battery, which performs so well in the electrical system of automobiles, has an anode of lead and a cathode of lead dioxide (Figure 11.3). (The powdered PbO_2 is supported by being tightly packed into small grid compartments in a lead electrode.) The electrolyte is H_2SO_4. As the cell discharges, the Pb(*s*) of the anode and the $PbO_2(s)$ of the cathode are both converted into $PbSO_4(s)$; water is produced, and the concentration of H_2SO_4 decreases. The half-reactions during discharge are

Anode:	$Pb(s) + HSO_4^- = PbSO_4(s) + H^+ + 2\,e^-$,	$\mathscr{E}^0 = 0.356$ V
Cathode:	$PbO_2(s) + HSO_4^- + 3\,H^+ + 2\,e^- = PbSO_4(s) + 2\,H_2O$,	$\mathscr{E}^0 = 1.685$ V

Calculate the voltage at 25° C of a lead storage battery in which the H_2SO_4 concentration is 0.100 *M*.

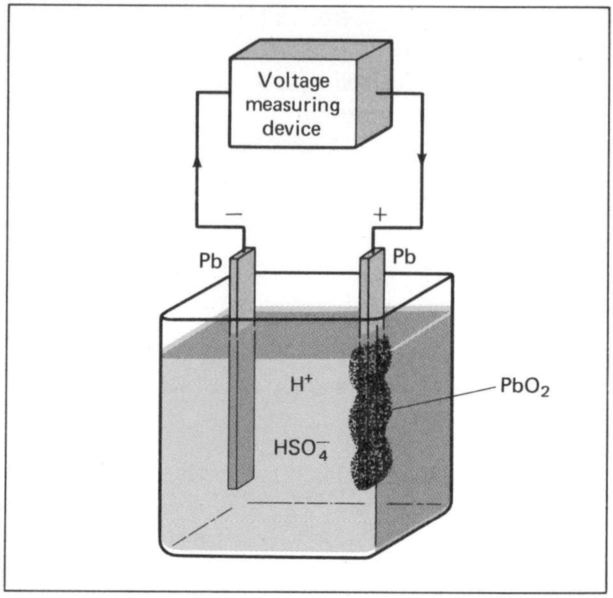

Figure 11.3
The lead storage cell (schematic diagram).

SOLUTION We first calculate $\mathscr{E}^0$ for the cell reaction. Adding the half-reactions gives

$$Pb(s) + PbO_2(s) + 2\,HSO_4^- + 2\,H^+ = 2\,PbSO_4(s) + 2\,H_2O, \qquad \mathscr{E}^0 = 2.041\ \text{V}$$

In this reaction, $n = 2$. Substituting these values into equation [11.11] gives

$$\mathscr{E} = 2.041 - \frac{0.0257}{2} \ln \frac{(a_{H_2O}^2)(a_{PbSO_4}^2)}{(a_{Pb})(a_{PbO_2})(a_{H^+}^2)(a_{HSO_4^-}^2)}$$

But $a_{PbSO_4} = a_{Pb} = a_{PbO_2} = 1$, since $PbSO_4$, Pb, and PbO_2 are pure solid substances. Water is the solvent in the cell; hence, its activity is defined in terms of Raoult's law standard state. The solutes are H^+ and HSO_4^-, whose activities are defined in terms of Henry's law standard state. This solution is dilute enough to assume that Raoult's law and Henry's law are good approximations in the solution. Hence,

$$a_{H_2O} \approx X_{H_2O} \approx 1$$
$$a_{H^+} \approx [H^+]$$
$$a_{HSO_4^-} \approx [HSO_4^-]$$

Substituting into the above expression gives

$$\mathscr{E} = 2.041 - \frac{0.0257}{2} \ln \frac{(1)^2(1)^2}{(1)(1)[H^+]^2[HSO_4^-]^2}$$

or

$$\mathscr{E} = 2.041 + \frac{0.0257}{2} \ln [H^+]^2[HSO_4^-]^2$$

Since sulfuric acid is a strong acid, it is completely dissociated into H^+ and HSO_4^- ions. Hence, $[H^+] = 0.100 = [HSO_4^-]$:

$$\mathscr{E} = 2.041 + \frac{0.0257}{2} \ln [0.10]^2[0.10]^2$$

$$\mathscr{E} = 2.041 + \frac{0.0257}{2} \ln (1.00 \times 10^{-4}) = 1.92 \text{ V}$$

The relatively small dissociation of the HSO_4^- ion to give H^+ and SO_4^{2-} in the acidic solution is neglected in this calculation.

This voltage is less than the 2 V ordinarily found in a lead battery since the sulfuric acid concentration is smaller than is usually encountered. Such a cell needs recharging (which reverses the cell reaction). The 12-volt car battery uses six such cells connected in series.

Calculation of Standard Half-Cell Potentials

The $\mathscr{E}^0$ data in Table 11.1 can be combined in various ways to produce the $\mathscr{E}^0$ values for still other half-cell reactions.

EXAMPLE 11.4 Given

$$Fe^{2+} + 2\,e^- = Fe, \qquad \mathscr{E}_1^0 = -0.440 \text{ V}$$
$$Fe^{3+} + e^- = Fe^{2+}, \qquad \mathscr{E}_2^0 = 0.771 \text{ V}$$

Calculate $\mathscr{E}^0$ for the half-cell

$$Fe^{3+} + 3\,e^- = Fe$$

SOLUTION The correct approach to this problem is to relate the $\mathscr{E}^0$ values to ΔG^0 values, which can then be added:

$$\begin{array}{lll} & Fe^{2+} + 2\,e^- = Fe, & \Delta G_1^0 = -2\mathscr{F}\mathscr{E}_1^0 = -2\mathscr{F}(-0.440 \text{ V}) \\ & Fe^{3+} + e^- = Fe^{2+}, & \Delta G_2^0 = -1\mathscr{F}\mathscr{E}_2^0 = -1\mathscr{F}(0.771 \text{ V}) \\ \hline \text{Adding:} & Fe^{3+} + 3\,e^- = Fe, & \Delta G^0 = -2\mathscr{F}(-0.440 \text{ V}) - 1\mathscr{F}(0.771 \text{ V}) \end{array}$$

Solving for $\mathscr{E}^0$:

$$\Delta G^0 = -n\mathscr{F}\mathscr{E}^0 = -3\mathscr{F}\mathscr{E}^0$$

$$\mathscr{E}^0 = \frac{\Delta G^0}{-3\mathscr{F}} = \frac{\Delta G_1^0 + \Delta G_2^0}{-3\mathscr{F}}$$

$$\mathscr{E}^0 = \frac{-2\mathscr{F}(-0.440 \text{ V}) - 1\mathscr{F}(0.771 \text{ V})}{-3\mathscr{F}}$$

$$\mathscr{E}^0 = \frac{2(-0.440 \text{ V}) + 1(0.771 \text{ V})}{3} = -0.036 \text{ V}$$

Notice that in Example 11.4 the mere adding of $\mathscr{E}_1^0$ and $\mathscr{E}_2^0$ (as one might be tempted to do, since the equations are added) does not give the correct value for $\mathscr{E}^0$. Notice, also, that since the $\mathscr{F}$'s cancel, adding $n\mathscr{E}^0$'s is all that is required in such calculations.

Calculation of Equilibrium Constants

One of the most useful applications of $\mathscr{E}^0$ values is in the calculation of equilibrium constants in chemical reactions. Equation [9.10] gives the relationship between ΔG^0 and K:

$$\Delta G^0 = -RT \ln K \qquad [9.10]$$

Equation [11.7] gives the relationship between ΔG^0 and $\mathscr{E}^0$:

$$\Delta G^0 = -n\mathscr{F}\mathscr{E}^0 \qquad [11.7]$$

The result of combining the equations is

$$\boxed{\ln K = \frac{n\mathscr{F}\mathscr{E}^0}{RT}} \qquad [11.14]$$

or

$$\boxed{K = e^{n\mathscr{F}\mathscr{E}^0/RT}} \qquad [11.15]$$

At 25° C, $\mathscr{F}/RT = 1/0.0257$, so that

$$\ln K = \frac{n\mathscr{E}^0}{0.0257} \quad (25° \text{ C}) \qquad [11.16]$$

EXAMPLE 11.5 Calculate the equilibrium constant at 25° C for the reaction

$$Cu^{2+} + Zn = Zn^{2+} + Cu$$

SOLUTION In Example 11.1 we found that $\mathscr{E}^0 = 1.100$ V and $n = 2$ for this reaction. Hence,

$$\ln K = \frac{(2 \text{ eq})(1.100 \text{ V})}{0.0257 \text{ V-eq}} = 85.6$$

$$K = 1.5 \times 10^{37}$$

The values of $\mathscr{E}^0$ can be used to calculate equilibrium constants even in reactions that do not involve oxidation and reduction.

EXAMPLE 11.6 Calculate K_{sp} for AgCl at 25° C from $\mathscr{E}^0$ values.

SOLUTION We are interested in finding K for the reaction

$$AgCl(s) = Ag^+ + Cl^-$$

The key to solving this problem is finding the half-cell reactions that will add together to give this reaction. From Table 11.1 we have

$$\begin{array}{lrl} & Ag(s) = Ag^+ + e^-, & \mathscr{E}^0 = -0.799\ V \\ & e^- + AgCl(s) = Ag(s) + Cl^-, & \mathscr{E}^0 = 0.222\ V \\ \hline \text{Adding:} & AgCl(s) = Ag^+ + Cl^-, & \mathscr{E}^0 = -0.577\ V \end{array}$$

and since $n = 1$,

$$\ln K_{sp} = \frac{(1)(-0.577)}{.0257} = -22.5$$

$$K_{sp} = 1.8 \times 10^{-10}$$

The use of electrochemical measurements to obtain K_{sp} values is a frequently used, highly satisfactory method.

$\mathscr{E}$ Values and Natural Change

We can use $\mathscr{E}$ values, as well as ΔG values, to predict whether a proposed reaction can occur. In a natural reaction, $\Delta G < 0$ (constant temperature and pressure), and since $\Delta G = -n\mathscr{F}\mathscr{E}$, it follows that $\mathscr{E} > 0$ for the same process. Stated in another way, reactions with positive $\mathscr{E}$ values are natural, and the larger the value of $\mathscr{E}$, the more completely the reaction will proceed to form the products.

From now on as we discuss oxidation-reduction reactions, we will continually refer to the electromotive force data in Table 11.1 to (1) ascertain the strength of a given oxidant or reductant, and (2) determine whether a proposed reaction is thermodynamically possible. Comparing all substances in their standard states, one sees that the best reductant in Table 11.1 is Li, since it has the greatest tendency to give up electrons ($Li \rightarrow Li^+ + e^-$, $\mathscr{E}^0 = 3.045$ V). Aqueous HF is the poorest reducing agent in the table ($2\,HF(aq) = F_2 + 2\,H^+ + 2\,e^-$, $\mathscr{E}^0 = -3.06$ V). For oxidizing ability, we look at the reaction in which electrons are gained, and see that the highest positive voltage is that of F_2 in acid solution ($F_2 + 2\,H^+ + 2\,e^- = 2\,HF(aq)$, $\mathscr{E}^0 = 3.06$ V). The poorest oxidant in the table is Li^+.

EXAMPLE 11.7 If it is assumed that all chemical species are present in their standard states, will Ce^{4+} and/or $Cr_2O_7^{2-}$ be capable of oxidizing Mn^{2+} in acid solution to MnO_4^-?

SOLUTION Ce^{4+} and $Cr_2O_7^{2-}$ are commonly used oxidants in analytical chemistry. If either of them is added to a reasonably strong reducing agent, one would have no hesitation in predicting that a reaction would take place. However, Mn^{2+} is a very poor reducing agent; therefore, we will need to examine the half-cell potentials to answer our question. From Table 11.1 we have

$$\begin{array}{rl} Cr_2O_7^{2-} + 14\,H^+ + 6\,e^- = 2\,Cr^{3+} + 7\,H_2O, & \mathscr{E}^0 = 1.33\ V \\ MnO_4^- + 8\,H^+ + 5\,e^- = Mn^{2+} + 4\,H_2O, & \mathscr{E}^0 = 1.51\ V \\ Ce^{4+} + e^- = Ce^{3+}, & \mathscr{E}^0 = 1.61\ V \end{array}$$

As we have mentioned before (Chapters 7 and 9), thermodynamic arguments tell us only about the state of the system after it reaches equilibrium. Like ΔG^0, $\mathscr{E}^0$ data give us no information about how fast the reaction will occur.

For Ce^{4+} we have

$$\begin{array}{lr} 5\,Ce^{4+} + 5\,e^- = 5\,Ce^{3+}, & \mathscr{E}^0 = 1.61\text{ V} \\ Mn^{2+} + 4\,H_2O = MnO_4^- + 8\,H^+ + 5\,e^-, & \mathscr{E}^0 = -1.51\text{ V} \\ \hline 5\,Ce^{4+} + Mn^{2+} + 4\,H_2O = 5\,Ce^{3+} + MnO_4^- + 8\,H^+, & \mathscr{E}^0 = +0.10\text{ V} \end{array}$$

The positive voltage (+0.10) tells us that the reaction has not yet reached equilibrium and will proceed as written. Thus, under the conditions specified in the problem, Ce^{4+} will tend to oxidize Mn^{2+}. For $Cr_2O_7^{2-}$ we have

$$\begin{array}{lr} 5\,Cr_2O_7^{2-} + 70\,H^+ + 30\,e^- = 10\,Cr^{3+} + 35\,H_2O, & \mathscr{E}^0 = 1.33\text{ V} \\ 6\,Mn^{2+} + 24\,H_2O = 6\,MnO_4^- + 48\,H^+ + 30\,e^-, & \mathscr{E}^0 = -1.51\text{ V} \\ \hline 5\,Cr_2O_7^{2-} + 6\,Mn^{2+} + 22\,H^+ = 10\,Cr^{3+} + 6\,MnO_4^- + 11\,H_2O, & \mathscr{E}^0 = -0.18\text{ V} \end{array}$$

The negative voltage (−0.18) tells us that the reaction will not take place. In fact, the reverse reaction (oxidation of Cr^{3+} by MnO_4^-) will occur under the standard state conditions.

We might have approached the problem a little less formally by simply noting from the reduction potentials that Ce^{4+} is a stronger oxidant than MnO_4^-, and MnO_4^- is a stronger oxidant than $Cr_2O_7^{2-}$; hence, Ce^{4+} will oxidize Mn^{2+} to MnO_4^-, whereas $Cr_2O_7^{2-}$ will not. The first procedure has the advantage of supplying quantitative data. Knowing $\mathscr{E}^0$ for the cell reaction, we could extend the problem to the calculation of K and the equilibrium concentrations. Hence, we would know how far the reaction between Ce^{4+} and Mn^{2+} will go before the equilibrium state is reached.

$\mathscr{E}^0$ Data and Descriptive Oxidation-Reduction Chemistry

From the data in Table 11.1, one can deduce a large body of information about oxidation-reduction reactions. In this section we will select a few important reactions that will serve as an introduction into that large body of information about properties and reactions that is called the descriptive chemistry of oxidation-reduction processes. The data of Table 11.1 form the basis for predicting and correlating this information. We will first consider the relative reactivity of metals acting as reducing agents. This is important factual material that is readily obtainable from half-cell potentials.

EXAMPLE 11.8 Which of the following metals will react with water and/or dilute hydrochloric acid to produce H_2 gas: K, Ca, Na, Zn, Fe, Ni, Sn, Pb, Cu, Ag, Hg, and Au? Consider all reactants and products to be in their standard states.

SOLUTION The reduction of water,

$$2\,H_2O + 2\,e^- = H_2 + 2\,OH^-$$

has a standard potential of −0.829 V. The oxidation $\mathscr{E}^0$'s of K (2.93 V), Na (2.71 V), and Ca (2.87 V) are sufficiently high to give an overall positive voltage when added to −0.829; hence, these metals reduce water to produce H_2 gas and OH^-. All the other metals in the list have oxidation potentials too low to react with water, but the following have oxidation potentials that are high enough to oxidize 1 M H^+ to $H_2(g$, 1 atm): Zn (0.76 V), Fe (0.44 V), Ni (0.25 V), Sn (0.14 V), Pb (0.13 V). The remaining metals have standard oxidation potentials below H_2, and hence do not react with H^+: Cu (−0.34 V), Ag (−0.80 V), Hg (−0.85 V), and Au (−1.50 V).

In summary, we have

Half-reaction	
$K = K^+ + e^-$	Will reduce H_2O to form H_2 gas.
$Ca = Ca^{2+} + 2\,e^-$	
$Na = Na^+ + e^-$	
$Zn = Zn^{2+} + 2\,e^-$	Will not react appreciably with H_2O, but will reduce H^+ to form H_2 gas.
$Fe = Fe^{2+} + 2\,e^-$	
$Ni = Ni^{2+} + 2\,e^-$	
$Sn = Sn^{2+} + 2\,e^-$	
$Pb = Pb^{2+} + 2\,e^-$	
$Cu = Cu^{2+} + 2\,e^-$	Will not react appreciably with either H_2O or H^+.
$Ag = Ag^+ + e^-$	
$Hg = Hg^{2+} + 2\,e^-$	
$Au = Au^{3+} + 3\,e^-$	

It will be to your advantage to remember the order (at least roughly) of the reducing ability of the twelve common metals listed in the example above. A chemist needs to keep a few facts in mind, and the information in this list is in that category. Knowing the order of the reducing power of these common metals allows one to predict (without reference to an $\mathscr{E}^0$ table) that, for example, Zn will reduce Cu^{2+}, or Cu will reduce Ag^+. The reactions are

$$Zn + Cu^{2+} = Zn^{2+} + Cu$$
$$Cu + 2\,Ag^+ = 2\,Ag + Cu^{2+}$$

In general, any metal will reduce the ion of any metal below it in the list.

Table 11.1 allows one to readily identify the strong oxidizing and reducing agents. Those agents with large positive reduction potentials have a strong tendency to become reduced, which means that they are strong oxidizing agents. Conversely, those substances with large positive oxidation potentials are the strong reducing agents. The oxidation potential is found by reversing the half-reaction and changing the sign of the $\mathscr{E}^0$ (reduction) given in Table 11.1.

With thirteen half-reactions (including $H_2 = 2\,H^+ + 2\,e^-$) we can form $13!/2!\,11! = 78$ complete reactions. Contained in a standard potential table of 200 half-reactions is $\mathscr{E}^0$ information on 19,900 reactions!

EXAMPLE 11.9 List four strong oxidants and four strong reductants.

SOLUTION The strongest oxidants are found at the end of Table 11.1, where the reduction potentials are high. F_2, O_3 (ozone), Co^{3+}, and H_2O_2 (hydrogen peroxide) are among the strongest of the oxidants.

The strongest reductants will be those having the highest oxidation potentials. Li, K, Ba, Sr are, thus, among the best reductants.

Some of the common and fairly strong oxidants that you will encounter in laboratory work (especially in analytical chemistry) are H_2O_2 (hydrogen peroxide), HOCl (hypochlorous acid), MnO_4^- (permanganate ion), $Cr_2O_7^{2-}$ (dichromate ion), and NO_3^- (nitrate ion), all in acid solution.

Some of the common reductants (in addition to the active metals) are $H_2S(aq)$, H_2SO_3, I^-, and Fe^{2+}. The half-reactions are

The products shown in these reactions are those that are commonly encountered. Others are possible. The reduction product of HNO_3, for example, is shown only as NO. In Chapter 19 we will see that the reduction product of HNO_3 depends on the strength of the reducing agent and on the conditions (principally the concentration of the acid and the temperature) of the reaction.

OXIDANTS

$$H_2O_2 + 2\,H^+ + 2\,e^- = 2\,H_2O$$
$$HOCl + H^+ + 2\,e^- = Cl^- + H_2O$$
$$MnO_4^- + 8\,H^+ + 5\,e^- = Mn^{2+} + 4\,H_2O$$
$$Cr_2O_7^{2-} + 14\,H^+ + 6\,e^- = 2\,Cr^{3+} + 7\,H_2O$$
$$NO_3^- + 4\,H^+ + 3\,e^- = NO(g) + 2\,H_2O$$

REDUCTANTS

$$H_2S(aq) = S(s) + 2\,H^+ + 2\,e^-$$
$$H_2SO_3(aq) + H_2O = HSO_4^- + 3\,H^+ + 2\,e^-$$
$$2\,I^- = I_2(s) + 2\,e^-$$
$$Fe^{2+} = Fe^{3+} + e^-$$

Once you are familiar with these nine half-reactions you will be on your way to mastering an important section of inorganic chemistry. With this much information you can answer such questions as those asked below.

EXAMPLE 11.10 What are the principal oxidation-reduction products when

(a) MnO_4^- in acid solution is added to Fe^{2+}?
(b) H_2S gas is bubbled through dilute nitric acid solution?
(c) solutions of HOCl and H_2SO_3 are mixed?

SOLUTION

(a) $Mn^{2+} + Fe^{3+}$
(b) $S(s) + NO(g)$
(c) $Cl^- + HSO_4^-$

11.2 Electrolysis

Up to this point in the discussion of electrochemistry we have been concerned with extracting electrical energy from a chemical reaction. We now consider the reverse process, the use of electrical energy to produce chemical change. This latter process is the subject of *electrolysis*. The subject in all its detail can become quite complex, and in the practice of electroplating, it requires almost as much art as science. Our discussion, however, will be limited to a few important examples that will illustrate the general principles.

General Principles of Electrolysis

Molten Salts with Inert Electrodes

Let us begin with a simple system, the electrolysis with inert electrodes of a molten binary salt. Consider, as an example, molten NaCl(l) with carbon electrodes in an arrangement such as that shown schematically in Figure 11.4. When the circuit is closed, the battery transfers electrons from the anode to the cathode. The cathode, now negatively charged, attracts the cations to it, where the reduction half-reaction takes place. Hence, at the cathode we have

$$Na^+ + e^- = Na(l) \qquad \text{(cathode, reduction)}$$

At the same time, anions are being attracted to the anode, where the oxidation half-reaction

$$Cl^- = \tfrac{1}{2}\,Cl_2(g) + e^- \qquad \text{(anode, oxidation)}$$

takes place. The overall reaction is

$$Na^+ + Cl^- = Na(l) + \tfrac{1}{2}\,Cl_2(g) \qquad \text{(molten NaCl)}$$

This is the commercial process for preparing Na and other reactive metals (K, Ca, Mg, Al, for example). The electrolysis cells make provision for keeping the very active metals out of contact with oxygen or moisture. Na, K, and Ca,

These elements have such high oxidation potentials that chemical reduction of their salts is not practicable.

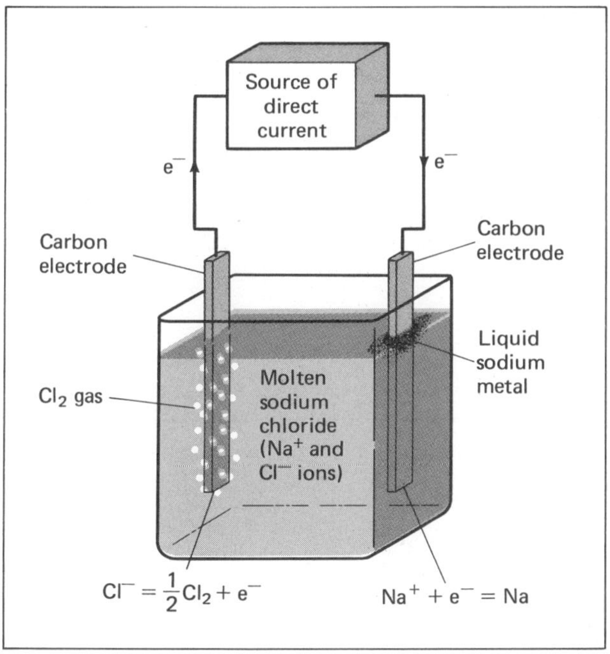

Figure 11.4
Electrolysis of molten sodium chloride (schematic diagram).

for example, are oxidized rapidly when exposed to air, and react vigorously with water.

Aqueous NaCl Solutions

Next, let us consider the electrolysis of an aqueous solution of NaCl, again with inert electrodes. We can write two possible cathodic reactions involving the major constituents of the solution: (a) reduction of Na^+,

$$Na^+ + e^- = Na$$

and (b) reduction of H_2O,

$$2\,H_2O + 2\,e^- = H_2 + 2\,OH^-$$

We can predict the extent to which each contributes to the reaction at the cathode by comparing the free energies of the two reactions, or since ΔG is related to $\mathscr{E}$, by comparing their half-cell potentials. From Table 11.1 we find that the standard potentials for reduction are -2.714 for Na^+ and -0.829 for H_2O. Therefore, in the standard state, the reduction of water is highly favored. In NaCl solutions more dilute than the standard state concentration, the reduction of Na^+ would be even less significant. Hence, we conclude that reaction (b) is by far the predominant cathode reaction.

At the anode, we also have two possibilities to consider: (a) oxidation of Cl^-,

$$2\,Cl^- = Cl_2 + 2\,e^-, \qquad \mathscr{E}^0 = -1.36\text{ V}$$

and (b) oxidation of water,

$$2\,H_2O = O_2 + 4\,H^+ + 4\,e^-, \qquad \mathscr{E}^0 = -1.23\text{ V}$$

Even though the ions of a salt may not enter directly into the electrode reactions, their presence is necessary for the electrolysis of water, since by their migration to the oppositely charged electrode, they are responsible for conducting the electric current through the solution. For example, the cathode reaction in the electrolysis of water produces negative ions. Without the migration of the positive ions of the salt into the cathode region, electrical balance could not be maintained, and the reaction would cease. A similar migration of anions to the cathode region also takes place.

Here we see that the oxidation of water is favored, but only slightly so in the standard state. If the concentration of the salt solution is high, the first reaction becomes important and $Cl_2(g)$ is the principal oxidation product; however, in very dilute solutions reaction (b) predominates. In the electrolysis of dilute solutions of NaCl, therefore, water is both oxidized and reduced. The reactions are

$$\begin{array}{lrr} & 4\,H_2O + 4\,e^- = 2\,H_2 + 4\,OH^- & \text{(cathode)} \\ & 2\,H_2O = O_2 + 4\,H^+ + 4\,e^- & \text{(anode)} \\ \hline \text{Adding:} & 6\,H_2O = 2\,H_2 + O_2 + 4\,H^+ + 4\,OH^- & \end{array}$$

If the solution is stirred so that H^+ and OH^- come into contact, the four sets of these ions on the right of the equation produce four water molecules, leaving as the net overall equation

$$2\,H_2O = 2\,H_2 + O_2$$

The net effect for the electrolysis of most dilute solutions of active metal salts—Na_2SO_4, KNO_3, $Ca(ClO_4)_2$, for example—is simply the electrolysis of water.

Aqueous Solutions of Metallic Ions with High Reduction Potentials

The reduction potentials for Cu^{2+}, Ag^{+}, and Hg^{2+} are

$$Cu^{2+} + 2\,e^- = Cu(s), \qquad \mathscr{E}^0 = 0.337\ V$$
$$Ag^{+} + e^- = Ag(s), \qquad \mathscr{E}^0 = 0.799\ V$$
$$Hg^{2+} + 2\,e^- = Hg(l), \qquad \mathscr{E}^0 = 0.854\ V$$

These are highly favorable reactions relative to the reduction of water ($\mathscr{E}^0 = -.829$). Hence, the cathodic reaction for the electrolysis of aqueous solutions of the salts of these metals is the plating-out of the metal.

If the salt solution is dilute, the anode reaction will almost always be the oxidation of water. The products of electrolysis of $Cu(NO_3)_2$ solution are thus Cu at the cathode, O_2 at the anode, and H^+ in the solution.

Active Electrodes

Consider next the cell pictured in Figure 11.5, the copper anode of which is capable of entering into the reaction. The possibilities for the anode reaction are (a) oxidation of water, $\mathscr{E}^0 = -1.23$ V, and (b) oxidation of the Cu electrode

$$Cu = Cu^{2+} + 2\,e^-, \qquad \mathscr{E}^0 = -0.337\ V$$

It is apparent that alternative (b) is the thermodynamically favorable reaction. The cathode reaction, as we saw in the preceding section, is the electroplating

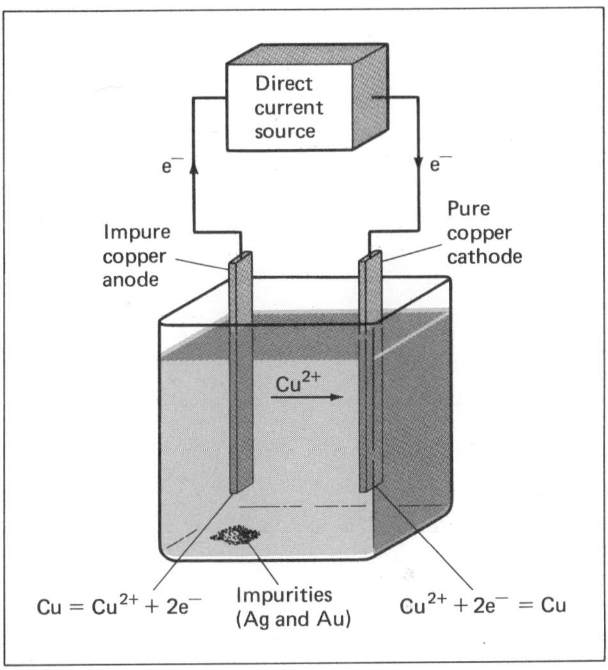

Figure 11.5
Electrolysis cell used in the purification of copper metal (schematic diagram).

of copper metal onto the cathode. The reactions are

$$\begin{array}{lr} \mathrm{Cu} = \mathrm{Cu}^{2+} + 2\,\mathrm{e}^- & \text{(anode)} \\ \mathrm{Cu}^{2+} + 2\,\mathrm{e}^- = \mathrm{Cu} & \text{(cathode)} \\ \hline \mathrm{Cu(anode)} = \mathrm{Cu(cathode)} & \text{(net reaction)} \end{array}$$

The net reaction is simply the transfer of Cu from one electrode to the other. This is the process used in all electroplating and electrorefining of metals. Copper, for example, is refined by using the impure copper as the anode. By controlling the voltage in the electrolysis cell, one can prevent such impurities as Ag and Au, which show less tendency to be oxidized than Cu, from going into solution at the anode. We can see how this works by comparing the standard potentials of Cu, Ag, and Au:

$$\begin{array}{ll} \mathrm{Cu} = \mathrm{Cu}^{2+} + 2\,\mathrm{e}^-, & \mathcal{E}^0 = -0.337\ \mathrm{V} \\ \mathrm{Ag} = \mathrm{Ag}^{+} + \mathrm{e}^-, & \mathcal{E}^0 = -0.799\ \mathrm{V} \\ \mathrm{Au} = \mathrm{Au}^{3+} + 3\,\mathrm{e}^-, & \mathcal{E}^0 = -1.500\ \mathrm{V} \end{array}$$

The oxidation of Cu is seen to take place at a lower voltage than that required for oxidation of either Ag or Au.

On the other hand, such impurities as Fe and Zn are oxidized more readily than Cu; they will enter the solution as ions along with Cu^{2+}:

$$\begin{array}{ll} \mathrm{Zn} \rightarrow \mathrm{Zn}^{2+} + 2\,\mathrm{e}^-, & \mathcal{E}^0 = 0.763\ \mathrm{V} \\ \mathrm{Fe} \rightarrow \mathrm{Fe}^{2+} + 2\,\mathrm{e}^-, & \mathcal{E}^0 = 0.440\ \mathrm{V} \\ \mathrm{Cu} \rightarrow \mathrm{Cu}^{2+} + 2\,\mathrm{e}^-, & \mathcal{E}^0 = -0.337\ \mathrm{V} \end{array}$$

At the cathode, however, Cu^{2+} is reduced more readily than Fe^{2+} or Zn^{2+}:

$$\begin{array}{ll} \mathrm{Cu}^{2+} + 2\,\mathrm{e}^- = \mathrm{Cu}, & \mathcal{E}^0 = 0.337\ \mathrm{V} \\ \mathrm{Fe}^{2+} + 2\,\mathrm{e}^- = \mathrm{Fe}, & \mathcal{E}^0 = -0.440\ \mathrm{V} \\ \mathrm{Zn}^{2+} + 2\,\mathrm{e}^- = \mathrm{Zn}, & \mathcal{E}^0 = -0.763\ \mathrm{V} \end{array}$$

Hence, proper control of the voltage will prevent Ag and Au in the impure anode from going into solution, and will prevent Fe^{2+} and Zn^{2+} from plating-out at the cathode. The net reaction is

$$\mathrm{Cu(impure)} = \mathrm{Cu(pure)}$$

As the Cu of the anode goes into solution in the electrorefining process, the Ag and Au impurities fall to the bottom of the cell, where they are periodically removed. Since the price of refined copper is near to that of crude copper, it appears that the value of the Au and Ag recovered in the refining process approximately offsets the cost of refining.

Stoichiometry in Electrochemistry

Stoichiometry in electrochemical systems is a straightforward application of the principles discussed in Chapter 3. The only additional information that one needs to know is that a mole of electrons has a charge of 96,485 coulombs (1 faraday), as discussed in Section 11.1. Sample calculations are given below.

EXAMPLE 11.11 A solution of $Cu(NO_3)_2$ is electrolyzed by a current of 0.100 amperes (0.100 coulombs/sec) flowing for 10.0 hours. The electrodes are platinum. Calculate the mass of Cu deposited on the cathode.

SOLUTION The reaction at the cathode is $Cu^{2+} + 2\,e^- = Cu$.

$$10.0\text{ hrs}\left(\frac{3600\text{ sec}}{\text{hr}}\right)\left(0.100\,\frac{\text{coulombs}}{\text{sec}}\right)\left(\frac{1\,\mathscr{F}}{96{,}485\text{ coulombs}}\right) = .0373\,\mathscr{F}$$

$$.0373\,\mathscr{F}\left(\frac{1\text{ mole Cu}}{2\,\mathscr{F}}\right)\left(\frac{63.54\text{ g Cu}}{1\text{ mole Cu}}\right) = 1.19\text{ g Cu}$$

EXAMPLE 11.12 Calculate the amperage required to produce 0.020 moles of H_2 gas in 1.00 hour by the electrolysis of a dilute solution of Na_2SO_4. (Note, amps = coulombs/sec.)

SOLUTION The hydrogen is produced by the reaction at the cathode:

$$2\,H_2O + 2\,e^- = H_2 + 2\,OH^-$$

$$(0.020\text{ moles }H_2)\left(\frac{2\,\mathscr{F}}{1\text{ mole }H_2}\right)\left(\frac{96{,}485\text{ coulombs}}{\text{faraday}}\right) = 3860\text{ coulombs}$$

$$\frac{3860\text{ coulombs}}{(1\text{ hr})\left(\frac{3600\text{ sec}}{\text{hr}}\right)} = 1.07\text{ amps}$$

EXAMPLE 11.13 A molten indium salt is electrolyzed with inert electrodes. Indium metal plates-out at the cathode. A second electrolysis cell containing $AgNO_3$ solution and inert electrodes is connected in series with the indium salt cell so that the same current flows through both cells. During the time of the electrolysis, 9.15 g of silver and 3.24 g of indium metal plate-out. The reaction at the cathode of the $AgNO_3$ cell is

$$Ag^+ + e^- = Ag$$

The molecular weight of Ag is 107.9, and that of In is 114.8. Calculate the charge on the indium ion.

SOLUTION The quantity of electrical charge that passed through the $AgNO_3$ cell can be calculated from the mass of the Ag metal that plated-out, the molecular weight of silver, and the charge of the silver ion:

$$(9.15\text{ g of Ag})\left(\frac{1\text{ mole Ag}}{107.9\text{ g}}\right)\left(\frac{1\,\mathscr{F}}{1\text{ mole Ag}}\right) = 0.0848\,\mathscr{F}$$

Knowing that the same quantity of electrical charge had to pass through the indium salt cell (since the cells are connected in series), we can calculate the number of faradays that were required to electroplate 1 mole of In. This is the same as the charge on the indium ion.

$$(3.24 \text{ g In})\left(\frac{1 \text{ mole In}}{114.8 \text{ g In}}\right)\left(\frac{x\,\mathscr{F}}{1 \text{ mole In}}\right) = 0.0848\,\mathscr{F}$$

$$x = 3.00$$

Summary

We have seen that electrochemistry is a two-way road: (1) Through the use of the voltaic cell, we can extract the energy of an oxidation-reduction reaction in the form of electrical energy; and (2) through the electrolysis cell, we can perform the reverse process of using electrical energy to produce chemical change.

The concept of oxidation-reduction potentials is of prime importance. Further and more extensive use of such potentials as a measure of the relative oxidizing or reducing strength of a substance and as a criterion for natural change will be made when we discuss the detailed chemistry of the elements and their compounds.

Application of thermodynamics to oxidation-reduction potentials leads to important quantitative relationships, three of the most useful being

(1) the relationship between free energy and cell potential:

$$\Delta G^0 = -n\mathscr{F}\mathscr{E}^0$$

(2) the Nernst equation, which describes the effect of concentration on the electromotive force of a voltaic cell:

$$\mathscr{E} = \mathscr{E}^0 - \frac{RT}{n\mathscr{F}} \ln \frac{a_L^l \cdot a_M^m \cdots}{a_A^a \cdot a_B^b \cdots}$$

(3) the relationship between the equilibrium constant and the standard cell potential:

$$\mathscr{E}^0 = \frac{RT}{n\mathscr{F}} \ln K$$

Since Avogadro's number of electrons represents 96,485 coulombs of electrical charge and an ampere is a coulomb/sec, the stoichiometry of electrochemical systems is seen to be nothing more than a straightforward application of the principles of stoichiometry that were discussed in Chapter 3.

Important Terms and Concepts

Voltaic cell
Electrolysis
Anode
Cathode
Electromotive force
Faraday
Standard potential
Half-cell potential
Standard reduction potential
Standard oxidation potential
Nernst equation

Exercises

$\mathscr{E}^0$ values from the table of standard reduction potentials (Table 11.1) will be required to work many of the problems.

Assume that the temperature is 25° C unless it is stated otherwise.

11-1.
Calculate $\mathscr{E}^0$ for the following reactions:

(a) $HClO_2 + ClO_4^- = 2\ ClO_3^- + H^+$
(b) $MnO_4^- + 5\ Fe^{2+} + 8\ H^+ = Mn^{2+} + 5\ Fe^{3+} + 4H_2O$
(c) $2\ H_2O_2 = 2\ H_2O + O_2(g)$

11-2.
Using $\mathscr{E}^0$ values in Table 11.1, calculate $\mathscr{E}^0$ for the following half-cell reactions:

(a) $Cr = Cr^{2+} + 2\ e^-$
(b) $Sn = Sn^{4+} + 4\ e^-$
(c) $2\ NO + 2\ H_2O = N_2O_4 + 4\ H^+ + 4\ e^-$

11-3.
Given a cell for which the overall (unbalanced) cell reaction is

$$Cu^{2+} + Cu(s) + Cl^- = CuCl(s)$$

(a) write the two half-cell reactions for this cell and then the balanced overall reaction.
(b) calculate $\mathscr{E}^0$ and $\mathscr{E}$ for the cell if $[Cl^-] = 0.10$ and $[Cu^{2+}] = 0.010$.

11-4.
What is $\mathscr{E}$ for the reaction

$$Pb(s) + 2\ H^+ = Pb^{2+} + H_2(g)$$

under conditions where $[H^+] = 0.010\ M$, $[Pb^{2+}] = 0.10\ M$, and $P_{H_2} = 10^{-6}$ atm?

11-5.
The electromotive force of the cell shown below is 0.695 V.

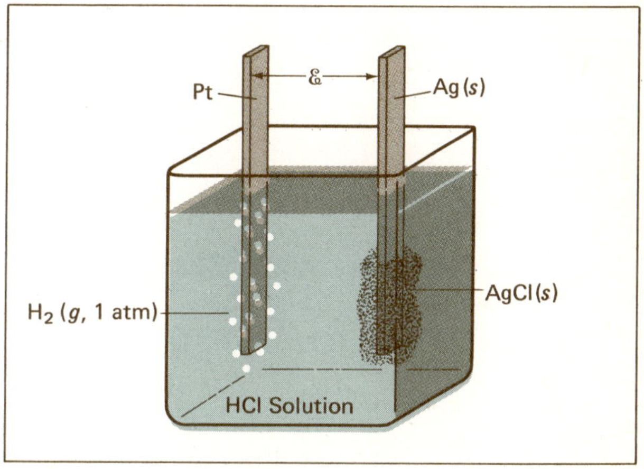

The reaction at the cathode is

$$AgCl(s) + e^- = Ag + Cl^-$$

What is the pH of the HCl solution used as the electrolyte in the cell?

11-6.
A hydrogen electrode and a calomel half-cell

$$Hg_2Cl_2(s) + 2\ e^- = 2\ Hg + 2\ Cl^-, \qquad \mathscr{E}^0 = 0.2676\ V$$

are used to determine the pH of a hydrochloric acid solution on top of a mountain where the hydrogen bubbles out of the solution at 520 torr. If the pH is calculated to be 5.00 when one wrongly assumes that the pressure is one atmosphere, what is the correct pH?

11-7.
Using standard electrode potentials, calculate K for the reaction

$$Pb + Sn^{2+} = Pb^{2+} + Sn$$

11-8.
For the reaction

$$2\,Cu^{+} = Cu^{2+} + Cu(s)$$

(a) write the two half-cell reactions.
(b) calculate the equilibrium constant for the above reaction.
(c) A 0.030 M $Cu(NO_3)_2$ solution is stored in a copper vessel. Calculate the equilibrium Cu^{+} concentration in this solution.

11-9.
(a) Using the standard reduction potential data, calculate ΔG^0 and the equilibrium constant for the reaction

$$Fe(s) + 2\,Fe^{3+} = 3\,Fe^{2+}$$

(b) If 100 g of iron nails are added to 1.00 liter of 1.00 M ferric nitrate solution, and the solution is stirred until equilibrium is attained, what will the concentration of Fe^{3+} be?

11-10.
Use $\mathscr{E}^0$ values to calculate K_{sp} for $PbSO_4$.

11-11.
Use $\mathscr{E}^0$ values to calculate the Cu^{+} concentration in a saturated solution of CuI.

11-12.
Given the following standard oxidation potentials,

$$Cu(s) = Cu^{2+} + 2\,e^{-}, \qquad \mathscr{E}^0 = -0.337\ V$$
$$Cu(s) + 4\,NH_3 = Cu(NH_3)_4^{2+} + 2\,e^{-}, \qquad \mathscr{E}^0 = +0.053\ V$$

calculate the dissociation constant for the $Cu(NH_3)_4^{2+}$ ion:

$$Cu(NH_3)_4^{2+} = Cu^{2+} + 4\,NH_3$$

11-13.
Which of the following could reduce H^{+}(1.0 molar) to H_2(g, 1 atm): Pb, Cu, Cr, Zn?

11-14.
Which of the following could oxidize Cr^{3+} to $Cr_2O_7^{2-}$ in acid solution: Pb^{2+}, Cu^{2+}, Br_2, O_2, Cl_2, MnO_4^{-}? Consider all species to be present in their standard state concentrations.

11-15.
Use the table of standard reduction potentials to specify an oxidizing agent capable of converting

(a) NaI to I_2
(b) Fe to Fe^{2+} (but not to Fe^{3+})
(c) Cr^{3+} to $Cr_2O_7^{2-}$
(d) Mn^{2+} to MnO_4^{-}

and a reducing agent capable of converting

(e) Sn^{4+} to Sn^{2+}
(f) $Cr_2O_7^{2-}$ to Cr^{3+} (but not to Cr^{2+})
(g) Br_2 to Br^{-}
(h) Cu^{2+} to CuI(s) (but not to Cu)

11-16.
Predict the products of the following reactions. If the reaction will not go to at least 50% products under standard conditions, write "no reaction."

(a) $Ce^{4+} + Cr_2O_7^{2-} =$ (acid solution)
(b) $Ca(s) + H_2O =$
(c) $Ag(s) + H^{+} =$
(d) $Hg^{2+} + Cu(s) =$
(e) $HOCl + Fe^{2+} =$ (acid solution)
(f) $H_2O_2 + H_2SO_3 =$

11-17.
Decide what reaction (if any) will occur when the following are mixed at the standard state concentration ($P = 1$ atm for gases and $M = 1$ molar for solutions).

(a) $Cl_2(g)$, Br^{-}, and F^{-}
(b) I^{-}, H^{+}, and Fe^{3+}
(c) $Cr_2O_7^{2-}$, H^{+}, and Fe^{2+}
(d) $H_2(g)$ and Ag^{+}

11-18.
Predict the products when the following solutions are electrolyzed:

(a) 0.10 M $Ca(NO_3)_2$
(b) 0.10 M $CuBr_2$
(c) 0.10 M $ZnSO_4$
(d) 0.10 M KCl

11-19.
Two electrolytic cells containing $CuSO_4$ and $AgNO_3$ were connected in series. A steady current of 1.00 amp was passed through the cells until 1.25 g of Ag was deposited on the cathode of the second cell.

(a) How long did the current flow?
(b) What weight of copper was deposited?

11-20.
A lead storage battery is rated at 100 ampere-hours. This means that the battery is large enough to deliver a current of 100 amperes for one hour. Calculate the minimum mass in grams of the lead anode in the battery. The half-cell reactions are

$$\text{Anode:} \quad Pb(s) + HSO_4^- = PbSO_4(s) + H^+ + 2e^-$$
$$\text{Cathode:} \quad PbO_2(s) + HSO_4^- + 3H^+ + 2e^- = PbSO_4(s) + 2H_2O$$

11-21.
6.30 g of platinum metal are deposited on the cathode of an electrolysis cell containing a solution of a platinum compound when a current of 3.46 amperes is passed through the cell for one hour. What is the oxidation state of platinum in the solution?

Problems

11-1.
Following is a series of oxidation potentials:

$$U \xrightarrow{1.80} U^{3+} \xrightarrow{0.61} U^{4+} \xrightarrow{-0.62} UO_2^+ \xrightarrow{-0.05} UO_2^{2+}$$

(This form of notation is called a potential diagram and is used as a shorthand representation of several half-reactions. That is,

$$U = U^{3+} + 3\,e^-, \qquad \mathscr{E}^0 = 1.80\text{ V}$$
$$U^{3+} = U^{4+} + e^-, \qquad \mathscr{E}^0 = 0.61\text{ V}$$

and so on.)

(a) Which oxidation state of uranium is the best oxidizing agent?
(b) Calculate $\mathscr{E}^0$ for the half-cell $U = U^{4+} + 4\,e^-$.
(c) Calculate the equilibrium UO_2^+ concentration in a solution 0.20 M in U^{4+} and 0.10 M in UO_2^{2+}, and buffered at pH = 1.00.

11-2.
A cell is made by putting a Ag-AgCl and a $H_2(P = 1\text{ atm})$ electrode in 100.0 ml of 0.010 M HCl. (See Exercise 11-5.)

(a) Calculate the electromotive force of the cell.
(b) Calculate the potential of the cell after the addition of 0.80, 0.90, 0.95, 1.00, 1.10, and 1.50 ml of 1.00 M $AgNO_3$. You may neglect the effect of the volume change produced by addition of the $AgNO_3$ solution. Make a plot of $\mathscr{E}$ versus volume added.

11-3.
Given the following single electrode potentials at 25° C, calculate K_{sp} of CuI at 25° C:

$$Cu^{2+} + I^- + e^- = CuI(s), \qquad \mathscr{E}^0 = 0.86\text{ V}$$
$$Cu^+ + e^- = Cu(s), \qquad \mathscr{E}^0 = 0.521\text{ V}$$
$$Cu^{2+} + 2\,e^- = Cu(s), \qquad \mathscr{E}^0 = 0.337\text{ V}$$

11-4.
A terrorist, Alphonzo Gluck, hijacked a jetliner out of Panama City with an old but fiendish scheme. He would force the plane to land at Mexico City, would obtain six million pesos (10^{-17} mole) by threats of violence, and would use a parachute to leave the plane over the Sierra Madre range where his fellow conspirators waited. That way he could avoid the argument in some corner of an airport with a troop of special forces. Alas, the terrorist was not aware that a passenger on the plane was named Horatio LeChatelier Smith. Horatio had considered going into public service, and in preparation he had spent two hot months studying the large watergates in the Panama Canal.

As the plane came within 10,000 meters of Mexico City, Alphonzo made his spine-chilling announcement. Crew and passengers froze with fear, with one exception. Horatio had watched Alphonzo and now made several suggestions regarding the appropriate way to begin a hijacking operation. "It would be much better to wait until lunch is served. It seems discourteous to interrupt the movie. Japanese yen would be much better than Mexican pesos. Your friends will never find you in that mountain range. August is a bad month for landing in that dry country. That really looks silly for you to hold that pistol in both hands and have it wobbling and shaking like that. If a man were planning a first-class hijacking, at least he'd wear a clean shirt. Is that pizza on your tie?" And so on and so on. Alphonzo became so rattled that he forgot to force the plane down at Mexico City. Finally he volunteered to turn over his weapon to the pilot, submit to arrest, and lead the police to his friends in the mountains if Horatio were thrown off the plane. As a token of appreciation for the effects of his curiosity, Horatio was allowed by the pilot to take Alphonzo's parachute (with protest from its owner and several passengers), a package of soda crackers, and a jar of peanut butter. Horatio landed in the desert east of Los Angeles.

In the two days until he reached the Colorado aqueduct and got a drink, he often considered water. Even when he walked under the power lines from Hoover Dam, he thought of water. But as he enjoyed again the feeling of water in his throat, he shifted to humanitarian thoughts. Why not get the water *and* the power to Los Angeles in a single conduit? Horatio's excitement was almost beyond his capacity to contain. He would transport hydrogen in a pipeline from the Colorado River to L.A. and would burn the hydrogen in electric-generating plants, thus obtaining both power and water. This scheme demonstrates that too much sun can harm even a demented mind. Do help Horatio by considering the following question: Is the ratio of power produced to water produced by the combustion of hydrogen appropriate for an average family? Assume that the average water use is about 400 liters per capita per day, that the average electric power use is about 5 kilowatt-hours per capita per day, and that the efficiency of conversion of the heat produced from burning the H_2 into electrical work is 0.25. The heat of combustion of H_2 is -68 kcal/mole. One calorie is 1.16×10^{-6} kwh.

Horatio worried whether the burning of all that hydrogen might seriously deplete the oxygen supply in an area already noted for unusual air quality. Suggest some other matters about the practicality of his scheme that should have caused him to worry.

11-5.

Horatio L. Smith had an idea. He was in a straitjacket in the back seat of the car, being taken to volunteer for Peace Corps service in Patagonia. During the drive he counted the large towers of a power transmission line that ran parallel to the road. This was a challenge to the lad for he was being held in the back seat by his father and his chemistry professor, who were sharing the expense of getting the Peace Corps to accept Horatio, both eager to enjoy the benefits of his being away. As she drove, Horatio's mother sang a happy thinking-of-the-next-twelve-months song. Horatio could not even see his fingers, much less count on them, so he had no way to test his idea. However, it seemed clear to him that the tall, sturdy, ugly, and expensive towers were not made strong merely to support the small mass of the electrons they carried. It was the copper cables that were heavy; the sensible approach would be to carry the electrons separately.

This discovery, which must stand alongside Columbus's discovery of India and Napoleon's discovery of Moscow, lifted Horatio to ecstasy. He came close to having a repulsive thought, but he knew too little of the effects of electrons upon each other to develop the thought. He tried to calculate how many amperes (coulombs/second) he could deliver from a power plant to a city 27.5 miles away using a 0.50-ton pickup truck traveling at the speed limit of 55 miles an hour. He had no idea what the speed limit is in Patagonia, but he could imagine a four-wheel-drive pickup getting over rough country to deliver its load safely. Thinking of the drivers making those regular trips regardless of the Patagonian weather, Horatio tried briefly to find a suitable tune for a brave thought he had memorized from the post office. "Neither snow, nor rain, nor heat, nor gloom of night stays these couriers from the swift completion of their appointed rounds."

His tunes would not fit, so he went back to calculations. Even without available fingers he determined that one pickup load could be delivered each hour, but then his calculations became a shambles. (H.L. had visited York and had biochemical thoughts when considering the butchering that once took place along a street named The Shambles. His mother always thought of Bedlam when she couldn't forget Horatio.)

Help the poor lad by calculating the current in amperes delivered to the city. It will be useful for you to remember that the electron mass is approximately 1/1875 that of a proton or of a hydrogen atom, and that there are 2.2 lbs to a kilogram, and 2000 lbs to a ton.

12

Chemical Kinetics

Chemical kinetics has to do with the rate at which reactions occur (the change in concentration of chemical reactants over time) and with the mechanisms by which they occur. Unlike thermodynamics, kinetics is concerned with the actual *path* of a reaction and with *time*. These variables add complexity and interest to the study of chemical reactions. We will see that knowing how fast a reaction takes place and understanding the mechanism by which it happens affords not only practical information but is intellectually satisfying as well.

12.1 Effects of Concentration on Rates of Reaction

The date of the first publication of *Zeitschrift für Physikalische Chemie* (1887) is generally considered to be the beginning of classical physical chemistry.

The history of chemical kinetics goes back to the very beginning of physical chemistry (1887), with rates being measured and the results being *expressed mathematically* by such intellectual giants as Ostwald, Nernst, and Arrhenius.

Probably the first fundamental contribution to the subject of chemical kinetics was the law of mass action (1863) proposed by Guldberg and Waage, which recognized the significant effect of concentration on the rate of a reaction. Since the very minimum requirement for chemical reaction is that reacting species must encounter one another, and since the rate of these encounters is proportional to the concentrations, it is not surprising that concentration of reactants is at least one of the important factors that affect the rate of a reaction. Temperature, catalysts, and the chemical properties of the reacting molecules are the other important factors. These we will discuss later. In this section we will be concerned with the concentration effect. It will be our objective to formulate a mathematical relationship between the *rate* of a reaction and the *concentration* of the reactants. Such an expression is known as the **rate law.**

Rate Law and Order of Reaction

Consider a chemical reaction of the type

$$A = B + C$$

We can follow the course of the reaction by watching either the disappearance of the reactant or the formation of the products. The rate of reaction can, therefore, be expressed in terms of either the *decrease* in concentration of A per unit time, or the *increase* in concentration of B or C per unit time. Figure 12.1 is a plot of concentration versus time for a reactant in a typical reaction. The rate of reaction (change in concentration with time) is given by the slope of the concentration-time curve. Initially, the rate of reaction is seen to be high (steep slope), but then it falls off as the concentrations of the reactants and products approach their equilibrium values in the reaction. Figure 12.2 is a similar plot for the change in concentration of a *product* with time.

The rate law states that the rate of reaction is proportional to the concentration, raised to an appropriate power, of each reacting species. Expressed mathematically, the rate law for a reaction in which A and B are reactants will

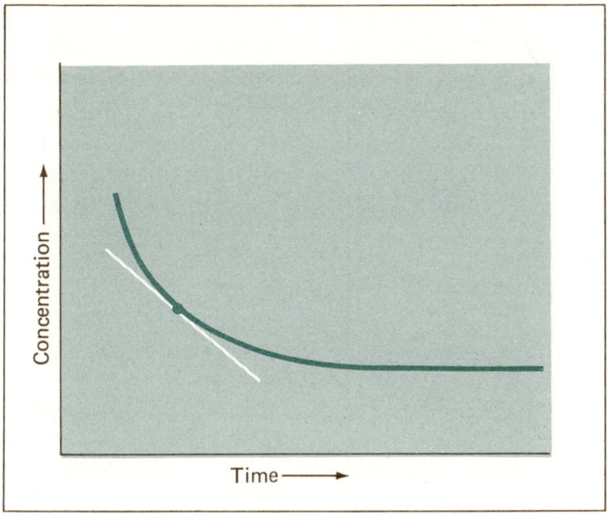

Figure 12.1
Change in concentration with time for a typical reactant in a chemical reaction. The rate of disappearance of the reactant at a specified point on the curve is given by the slope of the tangent to the curve at that point. The rate of disappearance of the reactant is seen to be high (steep slope) initially and then to fall off gradually until the equilibrium value of zero slope is reached. The slopes are negative because the concentration of a reactant decreases with time.

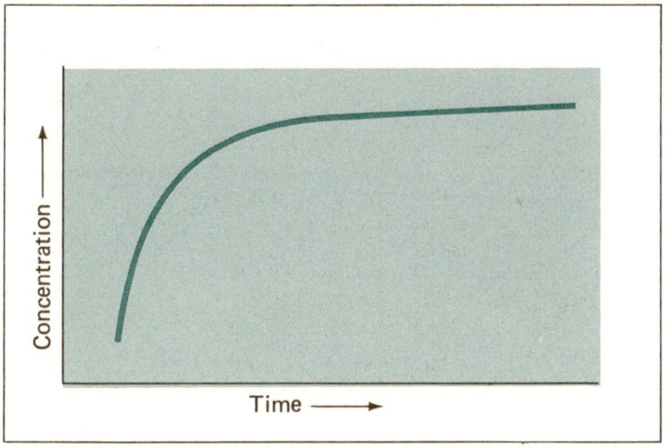

Figure 12.2
Change in concentration with time for a typical product. Initial increase in concentration of the product is rapid. The rate falls off as the equilibrium concentration of the product is approached.

usually be of the form

$$\text{Rate} = k[\text{A}]^m[\text{B}]^n$$

The exponents m and n are referred to as the **order of the reaction** for A and B, respectively. The overall order of the reaction is given by the sum $(m + n)$. The values of m and n are generally integers or half-integers. Their values are determined experimentally, and except in the special case of an elementary process (to be discussed later), they *cannot* be predicted from the stoichiometry of the reaction. To clarify the nomenclature, consider as a specific example the reaction

$$2\,NO + 2\,H_2 = 2\,H_2O + N_2$$

The rate law for this reaction is known from experimental studies to be

$$\text{Rate} = k[H_2][NO]^2$$

Rate of reaction is expressed as a change in concentration per unit of time. The most commonly used concentration unit is moles/liter, and the unit of time is chosen as the second (or as the minute or hour for slower reactions). The dimensions of the rate constant depend not only on the units of concentration and time chosen to express the rate, but also on the order of the reaction. For example, if the rate is expressed in moles/liter-sec for a first-order reaction, k has the dimensions of sec^{-1}:

$$\text{Rate} = k[\text{A}];\ k = \frac{\text{Rate}}{[\text{A}]}$$

Putting in the dimensions only, gives dimensions of

$$k = \frac{\text{moles/liter-sec}}{\text{moles/liter}} = \text{sec}^{-1}$$

For a second-order reaction, k has the dimensions of liter/mole-sec.

Since the concentrations of substances in a system change as the reaction proceeds, the rate also changes. Hence, we speak of the instantaneous rate at a given concentration—the rate at the very moment the concentration is at a specified value.

The reaction is thus first order in H_2, second order in NO, and third order overall.

The proportionality constant k is called the **specific rate constant** (usually shortened to *rate constant*). Its numerical value is characteristic of the properties of the reacting molecules and the temperature.

Knowing the values of $k, m, n, \ldots$ for a given reaction allows one to calculate the instantaneous rate of reaction at any given concentration of reactants.

EXAMPLE 12.1 The reaction

$$\text{A} + \text{B} = \text{products}$$

is known to be first order with respect to A and three-halves order with respect to B. At a given temperature the rate of the reaction is 0.18 moles/liter-sec when [A] = [B] = 0.040. Calculate the rate of reaction when [A] = 0.080 and [B] = 0.010 with the temperature the same as before.

SOLUTION

$$\text{Rate} = k[\text{A}][\text{B}]^{\frac{3}{2}}$$

Substituting the numerical values for the rate and the concentrations from the first set of data we can find the value of k:

$$\frac{0.18 \text{ moles}}{\text{liter-sec}} = k\left(0.040 \frac{\text{moles}}{\text{liter}}\right)\left(.040 \frac{\text{moles}}{\text{liter}}\right)^{\frac{3}{2}}$$

$$k = \frac{0.18 \text{ moles/liter-sec}}{(0.040 \text{ moles/liter})^{\frac{5}{2}}} = 5.6 \times 10^2 \frac{\text{liter}^{\frac{3}{2}}}{\text{mole}^{\frac{3}{2}}\text{-sec}}$$

Knowing k, we simply substitute the new concentration data to find the new rate:

$$\text{Rate} = \left(5.6 \times 10^2 \frac{\text{liter}^{\frac{3}{2}}}{\text{mole}^{\frac{3}{2}}\text{-sec}}\right)\left(\frac{0.080 \text{ moles}}{\text{liter}}\right)\left(\frac{0.010 \text{ moles}}{\text{liter}}\right)^{\frac{3}{2}}$$

$$= 0.045 \frac{\text{mole}}{\text{liter-sec}}$$

Experimental Determination of Rate Laws

There are several experimental procedures for determining the order of a reaction. One of the simplest is the **method of initial rates.**

Method of Initial Rates

Consider a general reaction of the type

$$a\text{A} + b\text{B} + \cdots = \text{products}$$

The rate law is

$$\text{Rate} = k[\text{A}]^m[\text{B}]^n \cdots$$

In order to evaluate $m, n, \ldots$, we make a series of mixtures in which [B] and the concentrations of all other components except A are held constant while [A] is varied, and then measure the rate of the reaction under the known initial conditions. Since the only variable is the concentration of A, the initial rates will be proportional to $[\text{A}_0]^m$, where $[\text{A}_0]$ is the initial concentration of A. If, for example, the initial rate doubles when $[\text{A}_0]$ is doubled, m must be unity, and the reaction is first order with respect to A. On the other hand, if the initial rate quadruples when $[\text{A}_0]$ is doubled, m must be *two*, and the reaction is second order with respect to A. Other orders would be determined similarly.

After the order of A is established, the order of B is deduced from a second series of mixtures in which [A] and the concentration of all other components except B are held constant while [B] is varied. Once the order for each reactant has been determined, k can be evaluated from the measured rate and from the $m, n, \ldots$ values.

EXAMPLE 12.2 The following initial rates of reaction were measured for the reaction

$$\text{A} + \text{B} = \text{products}$$

Calculate the order of the reaction in A and in B.

EXPERIMENT	INITIAL CONCENTRATION (moles/liter) $[\text{A}_0]$	$[\text{B}_0]$	INITIAL RATE (moles/liter-sec)
I	0.10	0.10	2.0×10^{-3}
II	0.20	0.10	4.0×10^{-3}
III	0.10	0.20	8.0×10^{-3}

SOLUTION Note that in experiments I and II, the initial concentration of B is the same value (0.10 moles/liter). Hence, differences in the intial rate of reaction in I and II are due solely to the effect that the concentration of A has on the rate. Under these special conditions we have

$$\text{Rate} = k[\text{A}]^m$$

Substituting the data from I and II into the above expression gives the following equations:

$$2.0 \times 10^{-3} \frac{\text{moles}}{\text{liter-sec}} = k\left(0.10 \frac{\text{moles}}{\text{liter}}\right)^m$$

$$4.0 \times 10^{-3} \frac{\text{moles}}{\text{liter-sec}} = k\left(0.20 \frac{\text{moles}}{\text{liter}}\right)^m$$

which can be solved simultaneously for m. Divide the second equation by the first to get

$$2.0 = (2.0)^m$$
$$m = 1.0$$

Thus, the reaction is first order in A. This result may have been immediately obvious to you without going through the detailed mathematics that we just did. *Doubling* the concentration of A (keeping [B] the same) *doubled* the rate. Hence, in this reaction the rate changes linearly with concentration—a first-order reaction. However, in the case of more complex systems, such as those with fractional orders, you will need the procedure that we outlined in the problem.

In order to establish the effect of [B] on the rate, we compare the rates in experiments I and III. Here the concentration of A is held constant and the concentration of B is varied. The rate law under these conditions is

$$\text{Rate} = k[\text{B}]^n$$

Substituting data from I and III gives

$$2.0 \times 10^{-3} \frac{\text{moles}}{\text{liter-sec}} = k(0.10)^n$$

$$8.0 \times 10^{-3} \frac{\text{moles}}{\text{liter-sec}} = k(0.20)^n$$

If the solution of the equation

$$4.0 = 2.0^n$$

had not been readily apparent, we could have solved it by taking logarithms of both sides of the equation:

$$\ln 4.0 = n \ln 2.0$$

$$n = \frac{\ln 4.0}{\ln 2.0} = 2.0$$

Dividing the second equation by the first gives

$$4.0 = 2.0^n$$
$$n = 2.0$$

which is second order in B. The complete rate law is, therefore,

$$\text{Rate} = k[\text{A}][\text{B}]^2$$

EXAMPLE 12.3 The following initial rates of reaction were measured for the reaction

$$\text{A} + \text{B} = \text{C}$$

at 25° C.

EXPERIMENT	INITIAL CONCENTRATION (moles/liter) $[A_0]$	$[B_0]$	INITIAL RATE (moles/liter-sec)
I	0.10	0.10	2.0×10^{-3}
II	0.20	0.20	8.0×10^{-3}
III	0.10	0.20	8.0×10^{-3}

(a) Write the rate law.
(b) Calculate the specific rate constant k at 25° C.
(c) Calculate the initial rate of formation of C at 25° C when the initial concentrations of A and B are each 0.50 M.

SOLUTION

(a) If we compare the results of experiments II and III, we see that the rate is independent of the concentration of A. (Doubling $[A_0]$ with $[B_0]$ held constant produced no difference in the rate of reaction.)

Examination of the results of experiments I and III shows the reaction to be second order in B. (Doubling $[B_0]$ with $[A_0]$ held constant quadrupled the rate.) Hence, the rate law that applies is

$$\text{Rate} = k[B]^2$$

(b) Substituting a set of data from any of the three experiments into the rate law found in part (a) will give a value for k. From experiment I we have

$$2.0 \times 10^{-3} \frac{\text{moles}}{\text{liter-sec}} = k\left(0.10 \frac{\text{moles}}{\text{liter}}\right)^2$$

$$k = \frac{2.0 \times 10^{-3}\ \text{moles/liter-sec}}{1.0 \times 10^{-2}\ \text{moles}^2/\text{liter}^2} = 0.20 \frac{\text{liters}}{\text{mole-sec}}$$

(c)

$$\text{Rate} = 0.20 \frac{\text{liters}}{\text{mole-sec}} \left(0.50 \frac{\text{moles}}{\text{liter}}\right)^2$$

$$= 0.050 \frac{\text{moles}}{\text{liter-sec}}$$

Integrated Rate Law Equations

The rate law, as we have expressed it up to this point, gives the relationship between *rate* and *concentration*. This is the *differential* form of the rate law. It is possible by means of the calculus to derive from the differential form an *integrated* rate law that expresses the relationship between *concentration* of reactants and *time*.

Let us consider any first-order reaction, and let c stand for the concentration of the reactant. The differential form of the rate law for this first-order process is

$$\text{Rate} = kc$$

The corresponding **integrated rate law equation** is

$$\boxed{\ln \frac{c}{c_0} = -kt} \quad \text{(first order)} \qquad [12.1]$$

where c is the concentration at any given time t, and c_0 is the concentration at the start of the measurement ($t = 0$).

For a second-order reaction involving a single reactant we have

$$\text{Rate} = kc^2$$

for the differential form and

$$\boxed{\frac{1}{c} = \frac{1}{c_0} + kt} \quad \text{(second order, single reactant)} \qquad [12.2]$$

for the corresponding integrated equation.

Through the integrated form of the rate laws, we can calculate the concentration of a reactant (or product) at any given time after the start of a reaction. This is useful information from a practical point of view. An example of the application of equation [12.1] is given below.

EXAMPLE 12.4 The overall reaction for the thermal decomposition of N_2O_5 is

$$N_2O_5(g) = 2\ NO_2(g) + \tfrac{1}{2}O_2(g)$$

The reaction is first order with a rate constant of $4.80 \times 10^{-4}\ \text{sec}^{-1}$. If the initial concentration of N_2O_5 is 0.100 moles/liter, what will be its concentration 100 seconds later? The temperature remains constant.

SOLUTION We use equation [12.1]:

$$\ln \frac{c}{0.100} = (-4.8 \times 10^{-4}\ \text{sec}^{-1})(100\ \text{sec})$$

$$\ln \frac{c}{0.100} = -0.0480$$

Now take the antilog of both sides to get

$$\frac{c}{0.100} = 0.953$$

$$c = 0.0953 \frac{\text{moles}}{\text{liter}}$$

EXAMPLE 12.5 Calculate the concentration of A just 40 minutes after the start of a reaction that is second order in A. The rate constant at the temperature of the reaction is 0.10 liter/mole-minute. The initial concentration of A was 0.20 moles/liter.

SOLUTION

$$\frac{1}{c} = \frac{1}{c_0} + kt$$

$$= \frac{1}{0.20 \text{ moles/liter}} + \frac{0.10 \text{ liter}}{\text{mole-minute}}(40 \text{ minutes})$$

$$= 9.0 \frac{\text{liters}}{\text{mole}}$$

$$c = 0.11 \frac{\text{mole}}{\text{liter}}$$

A useful piece of information to know, especially about decomposition reactions, is the fraction of the sample that will react in a given period of time. The next example illustrates how this calculation is readily made from the integrated equation.

EXAMPLE 12.6 Calculate the fraction of a sample of N_2O_5 that will decompose in 500 seconds. The reaction is first order. Its rate constant is 4.80×10^{-4} sec^{-1}.

SOLUTION The integrated rate equation is

$$\ln \frac{c}{c_0} = -kt$$

The term c/c_0 represents the fraction of the sample remaining at time, t, as shown below:

$$\frac{c}{c_0} = \frac{\text{moles of sample present at time } t\text{/volume of container}}{\text{moles of sample initially present } (t = 0)\text{/volume of container}}$$

$$= \text{fraction of initial quantity still present}$$

$$\ln \frac{c}{c_0} = -(4.80 \times 10^{-4} \text{ sec}^{-1})(500 \text{ sec})$$

$$\frac{c}{c_0} = 0.787$$

Since c/c_0 is the fraction present, $1 - (c/c_0)$ will be the fraction decomposed:

$$1 - \frac{c}{c_0} = 1 - 0.787 = 0.213$$

or

$$\left(1 - \frac{c}{c_0}\right) 100 = (.213)(100) = 21.3\% \text{ decomposed}$$

A descriptive way of expressing the rate of a reaction is to give the time that it takes for one-half of the sample to react. This time is called the **half-life** of a reaction, $t_{1/2}$. It is very commonly used when referring to first-order decomposition reactions. For a first-order reaction, the integrated equation is

Radioactive decay follows a first-order rate law. We will make use of the half-life concept again in Chapter 23, where we discuss nuclear decompositions.

$$\ln\frac{c}{c_0} = -kt$$

When half of the sample is reacted, $c/c_0 = 1/2$, and the elapsed time is the half-life, $t_{1/2}$. Substituting these values into the integrated first-order equation, we get

$$\ln\frac{1}{2} = -kt_{\frac{1}{2}}$$

or, rearranging,

$$t_{\frac{1}{2}} = \frac{-\ln(\frac{1}{2})}{k} = \frac{\ln 2}{k}$$

$$\boxed{t_{\frac{1}{2}} = \frac{0.693}{k}} \qquad \text{(first-order reaction)}$$

EXAMPLE 12.7 Calculate the fraction of the initial sample still remaining after 16 hours for a first-order reaction with a half-life of 46 hours.

SOLUTION

$$t_{\frac{1}{2}} = \frac{0.693}{k}$$

$$k = \frac{0.693}{t_{\frac{1}{2}}} = \frac{0.693}{46 \text{ hours}} = 0.015 \text{ hours}^{-1}$$

$$\ln\frac{c}{c_0} = -kt$$

$$\ln\frac{c}{c_0} = -(0.015 \text{ hours}^{-1})(16 \text{ hours})$$

$$\frac{c}{c_0} = 0.79$$

Hence, 21% has reacted,

Graphing Integrated Rate Equations

The integrated rate law equations are in a convenient form for plotting. The first-order equation [12.1] can be rewritten in the form

$$\ln c = \ln c_0 - kt$$

If you compare this with the equation of a straight line

$$y = mx + b$$

you see that ln c plotted on the ordinate versus t on the abscissa is a straight line for a first-order reaction. The slope of the line is $-k$. Figure 12.3 is an example of such a plot.

The integrated equation for a second-order reaction

$$\frac{1}{c} = \frac{1}{c_0} + kt \qquad [12.2]$$

is also an equation of a straight line, but this time the variables are $1/c$ for the ordinate versus t for the abscissa, and the slope is equal to k. (See Figure 12.4.) Plots similar to those in Figures 12.3 and 12.4 are used to simultaneously determine the correct order of a reaction and the value of the rate constant. Experimental data in kinetics usually consist of measured concentrations of a reactant at various times during the course of the reaction. These data are then plotted in the form dictated by the integrated rate equations. For example, if a plot of ln c versus t produces a straight line, the reaction is first order. Measuring the slope of the line gives the information needed to calculate the rate constant (slope $= -k$).

If the experimental data do not fall on a straight line when ln c is plotted versus t, the reaction is not first order. In that case one would want to examine a plot of $1/c$ versus t, which will be a straight line if the reaction is second order.

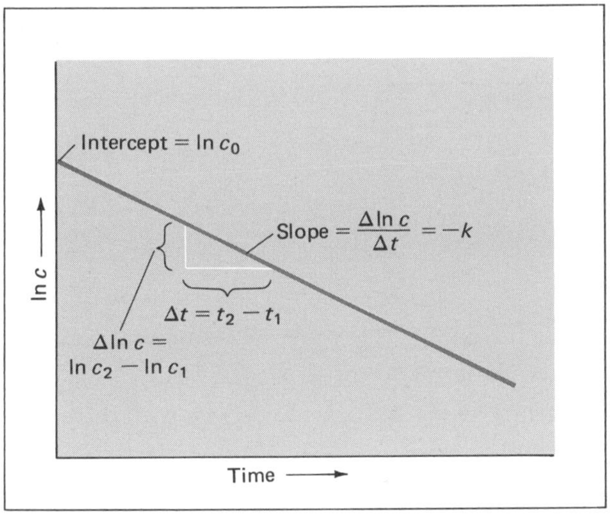

Figure 12.3
A plot of ln c versus time. This is a straight line for a first-order reaction. The slope of the line gives the value of the rate constant.

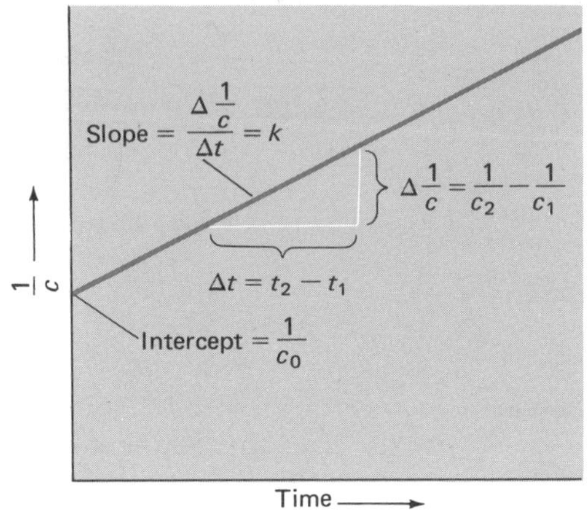

Figure 12.4
A plot of the reciprocal of the concentration of the reactant versus time. This is a straight line for a second-order reaction.

If the data do produce a straight line under these conditions, the slope would be measured to determine the rate constant (slope = k). If neither plot yielded a straight line, then integrated equations for other orders, including fractional orders, would need to be examined.

EXAMPLE 12.8 The reaction

$$A \rightarrow \text{products}$$

is known from a study of initial rates to be first order. Use a graphical method to calculate the rate constant from the following experimental data:

t (seconds)	[A]
10	0.20
20	0.14
30	0.099
40	0.070
50	0.049
60	0.035

SOLUTION Since the reaction is known to be first order, we will use the integrated equation for a first-order reaction:

$$\ln c = \ln c_0 - kt$$

and plot ln [A] versus t.

t	ln [A]
10	−1.61
20	−1.97
30	−2.31
40	−2.66
50	−3.02
60	−3.35

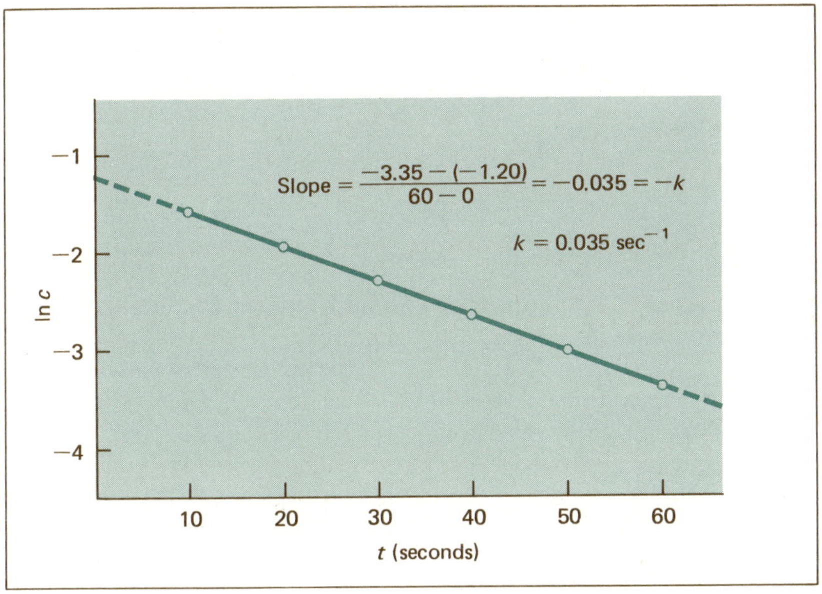

Measurements of Rates of Reaction

It is not within the scope of this course to discuss the experimental details involved in obtaining rate data. It should be mentioned, however, that kinetic studies often require considerable ingenuity on the part of the experimenter. Finding a procedure for following the course of a reaction is generally a separate experimental problem for each system studied. The most convenient methods follow the concentration of a component of the system through the measurement of a physical property. In this way the measurement can be made without disturbing the reaction, and the concentration can be monitored throughout the course of the reaction. For example, gaseous reactions whose stoichiometry shows a different number of product molecules than reactant molecules can be followed by measuring pressure changes. Optical properties, such as the absorption of light, can often be related to concentration, and they can thus be used to follow the course of a reaction.

Sometimes one has to resort to the less elegant procedure of sampling and chemical analysis. Devising a suitable method for following the concentration changes during the reaction is only part of the challenge. Precautions must be taken to guard against unwanted catalytic or inhibiting effects that may be caused by the reaction chamber and measuring system. Many of the standard laboratory procedures involving the use of mercury seals, metallic gauges, stopcock greases, and sometimes even glass surfaces may have to be avoided in the design of the experimental apparatus.

12.2 Transition State Theory

Activated Complexes

Let us consider the detailed molecular picture when the following reaction takes place:

$$H_2 + I_2 \rightarrow 2\,HI \qquad [12.3]$$

One way the reaction could take place would be for the H_2 and I_2 molecules to completely dissociate into atoms, and then for the H and I atoms to combine to form the product. This path seems reasonable until one calculates the energies involved. The dissociation energies of H_2 and I_2 are 435 and 151 kjoules/mole, respectively. Hence, to dissociate a mole of each of H_2 and I_2 completely would require an energy expenditure of 586 kjoules. That is, before the reaction could take place, an energy barrier of 586 kjoules would have to be surmounted, as schematically illustrated in the upper curve of Figure 12.5.

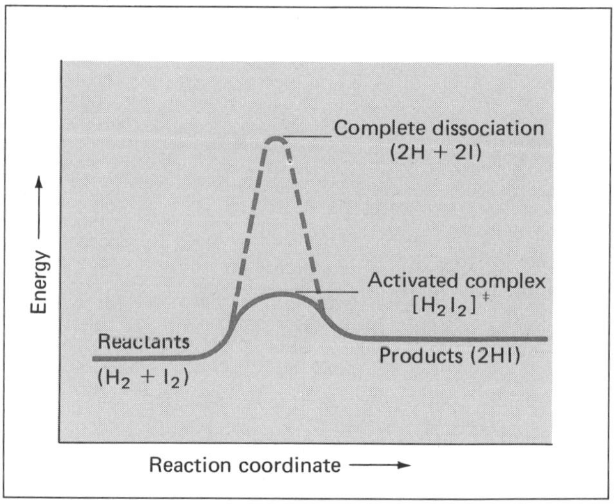

Figure 12.5
Potential energy diagram for the $H_2(g) + I_2(g)$ reaction
The dashed line would be the path of the reaction if complete dissociation of H_2 and I_2 were required for reaction. The actual path—via the activated complex—presents an energy barrier only about one-fourth that of the upper path.

Calculations show that a more favorable path (nature usually takes the low-energy route) is the formation of a complex between H_2 and I_2 molecules.[†] This process is illustrated below:

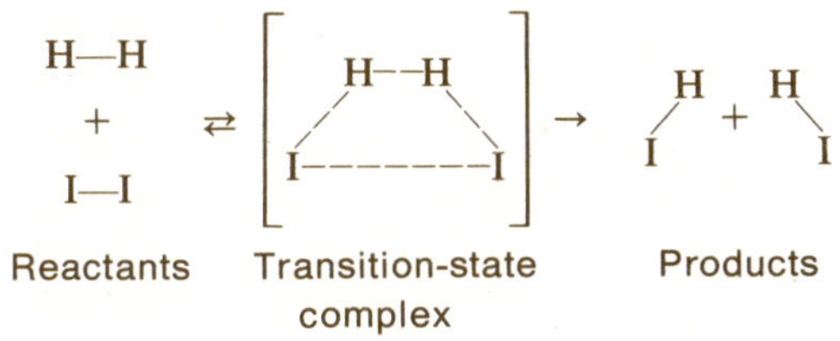

Allowing H—I interactions to take place at the same time that H—H and I—I bonds are being weakened lowers the energy barrier to 167 kjoules, as shown in the lower curve of Figure 12.5. The structure at the top of the barrier is the **activated (or transition-state) complex,** which is identified by the symbol‡. It is a highly unstable (check the bonding), purely hypothetical structure. In general, the structure of the complex formed will be that which will give the most gradual transition from reactants to products that is possible. This minimum change in structure is generally consistent with the lowest energy path.

The activated complex forms when the total kinetic energy of the colliding molecules is as large as the energy required to form the complex.

The energy difference between the *reactants* and the *activated complex* is the **energy of activation.** This is the height of the energy barrier that needs to be surmounted before reaction can occur. The energy of activation is highly important in determining the rate of the reaction. If the energy required to form the activated complex is large, not many collisions will be energetic enough to produce the complex, and the reaction will be slow. Conversely, a low energy of activation means that many collisions will be energetic enough to surmount the barrier. The more frequently this happens, the faster the reaction.

The energy of activation is a function of the chemical nature of the reacting molecules. The experimental measurement of energy of activation is described in the following section.

12.3 Effect of Temperature on Rates of Reaction

The influence of temperature on the rate of reaction is usually quite pronounced. The very crude generalization that the rate of a reaction doubles for each 10° rise in temperature applies (roughly) to a rather large number of reactions.

Early in the development of kinetics, Arrhenius deduced from analogy with the temperature effect on an equilibrium constant that the relationship between k and T should be of the form

$$k = Ae^{-\Delta E^{\ddagger}/RT} \qquad [12.4]$$

† In this discussion we are assuming that the $H_2 + I_2$ reaction is an elementary one, that is, one that takes place in only one step. It is quite possible that the mechanism for this reaction is not a one-step process. A reasonable multi-step mechanism has also been proposed.

where $\Delta E^{\ddagger}$ is the energy of activation for the reaction, and A is a constant referred to as the **frequency factor.**

While this equation, known as the **Arrhenius equation,** has its shortcomings,† it still finds use as a means of estimating the overall energy of activation for a reaction. Example 12.9 illustrates the procedure used.

EXAMPLE 12.9 The rate constant for the vapor-phase decomposition of acetaldehyde (CH_3CHO) has been measured at several temperatures:

T (°K)	RATE CONSTANT (liters/mole-sec)
700	0.00977
725	0.0302
750	0.0851
775	0.229
800	0.575
825	1.38
850	3.09

Calculate from the Arrhenius equation the energy of activation for the decomposition reaction.

SOLUTION

$$k = Ae^{-\Delta E^{\ddagger}/RT}$$

Taking the natural logarithm of both sides gives

$$\ln k = \ln A - \frac{\Delta E^{\ddagger}}{RT}$$

which is the equation of a straight line with $\ln k$ plotted as the ordinate and $1/T$ as the abscissa. (See Figure 12.6.) The intercept is $\ln A$, and the slope equals $-\Delta E^{\ddagger}/R$. $\Delta E^{\ddagger} = -R(\text{slope}) = 190$ kjoules.

Conversely, having an estimate of $\Delta E^{\ddagger}$ and knowing the rate constant at one temperature, one can estimate k at a second temperature.

EXAMPLE 12.10 The rate law for the reaction

$$\text{B} + \text{C} = \text{products}$$

is first order in B and second order in C. Given that the energy of activation is 42,000 joules for the reaction and that the rate constant is 1.5 liters²/moles²-min

† For example, the "constant" A is found in practice to be slightly temperature dependent.

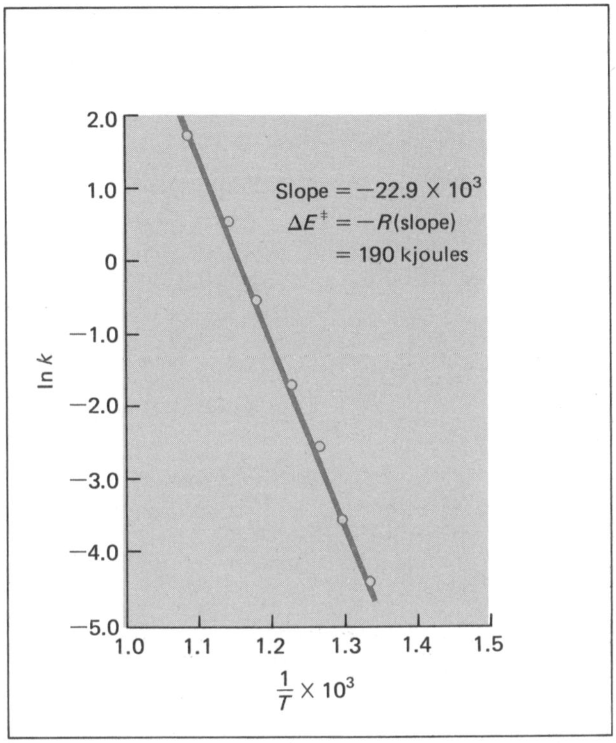

Figure 12.6
Rate data for vapor-phase decomposition of acetaldehyde as a function of temperature.

at 300° K, calculate the rate constant and the rate of the reaction at 400° K at the instant when [B] = 1.0 mole/liter and [C] = 2.0 moles/liter.

SOLUTION

$$k = Ae^{-\Delta E^\ddagger/RT}$$

$$\ln k = \ln A - \frac{\Delta E^\ddagger}{RT}$$

$$\ln k_2 = \ln A - \frac{\Delta E^\ddagger}{RT_2} \qquad \text{(at temperature } T_2\text{)}$$

$$\ln k_1 = \ln A - \frac{\Delta E^\ddagger}{RT_1} \qquad \text{(at temperature } T_1\text{)}$$

Subtracting the expression for $\ln k_1$ at T_1 from that of $\ln k_2$ at T_2 eliminates A to give

$$\ln \frac{k_2}{k_1} = \frac{-\Delta E^\ddagger}{R}\left(\frac{1}{T_2} - \frac{1}{T_1}\right)$$

We then solve for k_2 with $k_1 = 1.5$ liters^2/mole^2-min, $T_1 = 300°$ K, $T_2 = 400°$ K, and $\Delta E^\ddagger = 42{,}000$ joules.

$$\ln\frac{k_2}{1.5} = -\frac{42{,}000 \text{ joules}}{8.31 \text{ joules/°K}}\left(\frac{1}{400° \text{ K}} - \frac{1}{300° \text{ K}}\right)$$

$$\ln\frac{k_2}{1.5} = 4.21$$

$$k_2 = \frac{100 \text{ liter}^2}{\text{moles}^2\text{-min}}$$

Since the rate is first order in B and second order in C, we can write the rate law at 400° K as

$$\text{Rate} = k_2[\text{B}][\text{C}]^2$$

Knowing the value of k_2, we can now find the instantaneous rate at $[\text{B}] = 1$ mole/liter and $[\text{C}] = 2$ moles/liter that was asked for in the problem:

$$\text{Rate} = \left(100\,\frac{\text{liter}^2}{\text{mole}^2\text{-min}}\right)\left(1.0\,\frac{\text{mole}}{\text{liter}}\right)\left(2.0\,\frac{\text{moles}}{\text{liter}}\right)^2 = 400\,\frac{\text{moles}}{\text{liter-min}}$$

EXAMPLE 12.11 Assuming Arrhenius equation behavior, calculate the energy of activation for a reaction that doubles its rate when the temperature is raised from 300° K to 310° K.

SOLUTION Let k represent the rate constant at 300° K. Then the rate constant at 310° K will be $2k$ since the rate at 310° K (and therefore the rate constant) is twice that at the lower temperature. The Arrhenius equations for the two temperatures thus are

$$2k = Ae^{-\Delta E^\ddagger/310R} \qquad (310° \text{ K})$$

and

$$k = Ae^{-\Delta E^\ddagger/300R} \qquad (300° \text{ K})$$

Taking logarithms of both expressions gives

$$\ln 2 + \ln k = \ln A - \frac{\Delta E^\ddagger}{R}\left(\frac{1}{310}\right)$$

$$\ln k = \ln A - \frac{\Delta E^\ddagger}{R}\left(\frac{1}{300}\right)$$

Subtracting gives

$$\ln 2 = \frac{\Delta E^\ddagger}{R}\left(\frac{1}{300} - \frac{1}{310}\right)$$

Finally, solving for $\Delta E^\ddagger$ we obtain

$$\Delta E^\ddagger = 53{,}600 \text{ joules}$$

12.4 Reaction Mechanisms

We are now in a position to investigate the detailed paths of reactions. Of the several hundred common chemical reactions whose kinetics have been studied, only a few are found to be simple one-step processes. Even though their overall stoichiometry may appear simple, most reactions go from reactants to products by more than one step. For example, the conversion of ozone to oxygen:

$$2\,O_3 \rightarrow 3\,O_2$$

is thought to proceed by the following pair of simpler reactions:

$$\begin{array}{ll} (1) & O_3 \rightleftarrows O_2 + O \\ (2) & O_3 + O \rightarrow 2\,O_2 \\ \hline & 2\,O_3 = 3\,O_2 \quad \text{(overall)} \end{array}$$

Together, (1) and (2) constitute the *mechanism* for the overall reaction.

As a further example, consider the thermal decomposition of N_2O_5 into NO_2 and O_2. A mechanism that is in agreement with a large body of experimental facts concerning this reaction is given below:

$$\begin{array}{ll} (1) & 2\,N_2O_5 + 2\,M \rightleftarrows 2\,N_2O_5^* + 2\,M \\ (2) & 2\,N_2O_5^* \rightarrow 2\,NO_2 + 2\,NO_3 \\ (3) & NO_2 + NO_3 \rightarrow NO + NO_2 + O_2 \\ (4) & NO + NO_3 \rightarrow 2\,NO_2 \\ \hline & 2\,N_2O_5 = 4\,NO_2 + O_2 \quad \text{(overall)} \end{array}$$

The asterisk next to N_2O_5 in (1) and (2) indicates an *activated* molecule, one that by collision with another molecule, M, has gained energy equal to, or in excess of, that required to cause dissociation of the molecule. M in reaction (1) stands for any other molecule in the system, including another N_2O_5 molecule. The NO_3 molecule in reactions (2) and (3) is an example of the exotic molecules or fragments that often play an important role in reaction mechanisms. Some of these short-lived intermediates are actually observed spectroscopically, but the existence of others is inferred from less direct evidence. Each of the one-step reactions listed in (1) through (4) is referred to as an **elementary reaction.** The collection of elementary reactions that contributes to the overall reaction is the **reaction mechanism.**

Elementary Reactions and the Rate Law

The elementary reaction is the one process for which the order of the reaction can be predicted from its stoichiometry. Consider the elementary reaction

$$A + B \rightarrow \text{products}$$

Since a necessary condition for reaction is that the A and B molecules must collide, the rate of reaction will be dependent on the frequency of A-B collisions. Let us consider one A molecule in a container with several B molecules. The rate at which the A molecule encounters a B molecule will be directly proportional to the number of B molecules in the container. Stated another way, the frequency of collision of a given A molecule with any of the B molecules is directly proportional to the concentration of B molecules. If we now add a second A molecule to the container, there will be twice the opportunity for A-B collisions, and hence we conclude that the frequency of A-B collisions is proportional to the concentration of A (as well as B) molecules. The rate of reaction is, therefore, proportional to the product of [A] and [B], and the reaction is first order in A, first order in B, and second order overall.

For an elementary reaction of the form

$$A + A + B \rightarrow \text{products}$$

the rate of collision is seen by similar reasoning to be proportional to [A][A][B]. Hence the reaction is second order in A and first order in B. Thus, we see that for *elementary reactions*, the coefficients of the balanced equation become the exponents in the rate law.

EXAMPLE 12.12 One of the steps in the mechanism for the conversion of ozone to oxygen is

$$\text{(a)} \quad O_3 \rightarrow O_2 + O$$

Write the rate law for this one step in the overall mechanism.

SOLUTION This is an elementary reaction since it is a step in the mechanism; hence, its stoichiometry will determine its order. Since the coefficient of the reactant O_3 is one, the order is also one.

$$\text{Rate} = k[O_3]$$

Had the question asked for the order of the overall reaction

$$2\,O_3 = 3\,O_2$$

we could not have given an answer based on its stoichiometry, for the overall reaction is not elementary. As we saw earlier, it is a two-step process.

The Relationship Between Rate Constants and Equilibrium Constants

From Figure 12.7 one readily sees that a relationship exists between the energy of a reaction (a purely thermodynamic quantity) and the difference in the energies of activation of the reverse and forward directions of a reaction. This is not coincidence. Kinetics and thermodynamics do tie together. The connection is best expressed by the relationship that exists between the thermodynamic

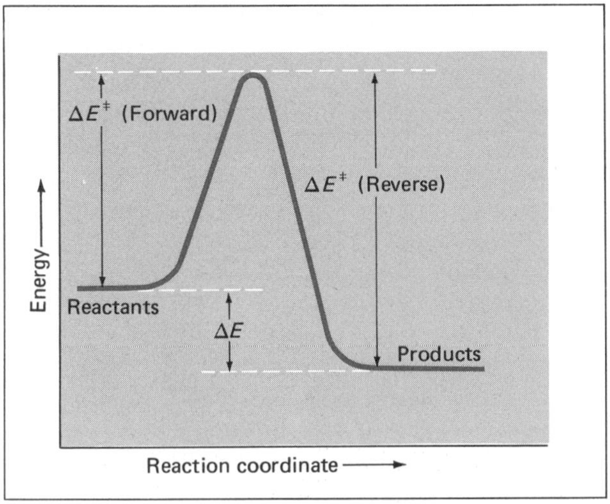

Figure 12.7
Potential energy diagram showing the relationship between the energy of a reaction (exothermic in this example) and the difference in the energies of activation of the forward and reverse reactions.

equilibrium constant and the rate constants for an equilibrium reaction. To obtain this relation, consider the following elementary equilibrium reaction.

$$A + B \underset{k_{-1}}{\overset{k_1}{\rightleftarrows}} C + D$$

where k_1 is the rate constant for the forward reaction (A + B →) and k_{-1} is the rate constant for the reverse direction (C + D →).

Since the reaction is an elementary process, the stoichiometry determines the form of the rate law, and we can write

$$\text{Rate of forward reaction} = k_1[\mathrm{A}][\mathrm{B}]$$
$$\text{Rate of reverse reaction} = k_{-1}[\mathrm{C}][\mathrm{D}]$$

Thermodynamic equilibrium is established when the two opposing rates become equal. Thus, we have

$$\text{Rate (forward)} = \text{Rate (reverse)}$$
$$k_1[\mathrm{A}][\mathrm{B}] = k_{-1}[\mathrm{C}][\mathrm{D}]$$

Rearranging gives

$$\frac{k_1}{k_{-1}} = \frac{[\mathrm{C}][\mathrm{D}]}{[\mathrm{A}][\mathrm{B}]}$$

Since the reaction is at equilibrium, the right-hand side of this equation is the expression for the equilibrium constant. Thus, we have

$$\boxed{K = \frac{k_1\ \text{(forward)}}{k_{-1}\ \text{(reverse)}}} \quad \text{(elementary process)}$$

where K is the equilibrium constant for the system.

While the equation

$$K = \frac{k_1\ \text{(forward)}}{k_{-1}\ \text{(reverse)}}$$

was derived here only for the case of an elementary reaction, it is true in general. The derivation in the general case is more involved and is left for a more advanced course.

Agreement Between Overall Experimental Rate Law and Proposed Mechanism

The most useful single piece of information in deducing a mechanism is the experimentally measured rate law for the reaction. For any proposed mechanism to have validity, it must, as a minimum requirement, predict the same form of the rate law as is found experimentally. As an illustration of finding the rate law predicted for a specified mechanism, let us consider a reaction of the type

$$A + B = C$$

and propose three separate mechanisms for it. We will then obtain the rate law that is consistent with each of our proposed mechanisms.

Let the first mechanism of our example consist of a "slow" (that is, slow relative to the other reactions in the sequence) first step to produce an intermediate (I) followed by any number of "fast" reactions involving additional intermediates (I′) or stable products (C). Thus, our proposed mechanism is

The rate constants, k_1, k_2, and k_3, are written over the arrows. The first rate constant, k_1, is much smaller than k_2 or k_3 in this mechanism.

$$(1) \quad A + B \xrightarrow{k_1} I \qquad \text{(slow)}$$
$$(2) \quad I + A \xrightarrow{k_2} I' \qquad \text{(fast)}$$
$$(3) \quad I' + B \xrightarrow{k_3} 2C \qquad \text{(fast)}$$

The "slow" step is the "bottleneck" that limits the overall rate. Since the overall reaction can go no faster than the slowest step in a sequence, the *relatively* slow step is referred to as the **rate-limiting reaction.** The rate law for the *elementary* reaction (1) can be deduced from its stoichiometry. We have

$$\text{Rate} = k_1[A][B]$$

Since its rate is already expressed in terms of the reactants, and it is considered to be very much slower than any of the succeeding steps, its rate law is the rate law for the overall reaction. Thus, this mechanism leads to a second-order rate law for the overall reaction.

Consider next the same reaction with a mechanism in which a fast equilibrium is followed by one "slow" and any number of "fast" steps:

The symbol k_{-1} is the rate constant for the reverse of reaction (1).

$$(1) \quad A + B \underset{k_{-1}}{\overset{k_1}{\rightleftarrows}} I \qquad \text{(fast)}$$
$$(2) \quad I + A \xrightarrow{k_2} I' \qquad \text{(slow)}$$
$$(3) \quad I' + B \xrightarrow{k_3} 2C \qquad \text{(fast)}$$

The intermediates I and I′ are not necessarily the same as those in the first mechanism. The rate of the overall reaction is again limited by the rate of the "slow" step, whose rate law is

$$\text{Rate} = k_2[I][A]$$

For the overall reaction, however, we want the rate law *not* in terms of an intermediate, but in terms of the original reactants A and B. We can do this by using the equilibrium-constant expression for step (1) to relate [I] to [A] and [B]. Thus,

$$\frac{[\mathrm{I}]}{[\mathrm{A}][\mathrm{B}]} = \frac{k_1}{k_{-1}} = K$$

where K is the equilibrium constant. Solving for [I] gives

$$[\mathrm{I}] = K[\mathrm{A}][\mathrm{B}]$$

Substituting this result into the rate law for the limiting reaction above gives

$$\begin{aligned}\text{Rate} &= k_2K[\mathrm{A}][\mathrm{B}][\mathrm{A}] \\ &= k[\mathrm{A}]^2[\mathrm{B}]\end{aligned}$$

where $k = k_2K$. Hence our second mechanism leads to a third-order rate law.

For a third possibility, let us assume the following reaction mechanism:

$$\left.\begin{array}{ll}(1) & \mathrm{A} + \mathrm{B} \underset{k_{-1}}{\overset{k_1}{\rightleftarrows}} \mathrm{I} \\ (2) & \qquad \mathrm{I} \overset{k_2}{\rightarrow} \mathrm{C}\end{array}\right] \text{(comparable rates)}$$

This mechanism differs from the one discussed just above only in that it contains no single rate-determining step; that is, no reaction in the mechanism can be singled out as being the "slow" reaction. This case can be treated by employing the **steady state approximation** method, which is based on the assumption that an equilibrium (steady state) concentration of the intermediate I is built up soon after the reaction begins and is maintained throughout the course of the reaction.

The steady state treatment applied to the mechanism above is as follows: Since the product C is formed directly from the intermediate I, the rate of the overall reaction depends on the concentration of I; that is,

$$\text{Rate} = k_2[\mathrm{I}] \qquad [12.5]$$

The concentration of I depends upon the balance between the rates of formation and disappearance of I. The intermediate is formed by the forward reaction of step (1), and it disappears in two ways: (a) decomposition into the reactants A and B [reverse reaction of equation (1)], and transformation into the product [reaction (2)]. The rate laws that apply are:

$$\begin{aligned}\text{Rate of formation of I} &= k_1[\mathrm{A}][\mathrm{B}] \\ \text{Rate of disappearance of I} &= k_{-1}[\mathrm{I}] + k_2[\mathrm{I}]\end{aligned}$$

In the steady state, the rate of formation of I will exactly equal its rate of disappearance. Hence, by assuming that we have the steady state, we can write:

$$\begin{aligned}\text{Rate of formation of I} &= \text{Rate of disappearance of I} \\ k_1[\mathrm{A}][\mathrm{B}] &= k_{-1}[\mathrm{I}] + k_2[\mathrm{I}]\end{aligned}$$

from which we obtain the concentration of the intermediate in terms of A and B:

$$[\mathrm{I}] = \frac{k_1[\mathrm{A}][\mathrm{B}]}{k_{-1} + k_2}$$

Substituting this result into equation [12.5] gives us the second-order rate equation

$$\text{Rate} = \frac{k_1 k_2[\mathrm{A}][\mathrm{B}]}{k_{-1} + k_2} = k[\mathrm{A}][\mathrm{B}]$$

where the rate constant for the overall reaction is given by

$$k = \frac{k_1 k_2}{k_{-1} + k_2}$$

Now let us consider actual examples of the three types of mechanisms just discussed. An example of the first mechanism (slow first step followed by any number of rapid steps) is the reaction

$$F_2 + 2\ NO_2 = 2\ NO_2F$$

The experimentally determined rate law is

$$\text{Rate} = k[F_2][NO_2]$$

The mechanism that has been proposed for the reaction is

$$(1)\quad NO_2 + F_2 \xrightarrow{k_1} NO_2F + F \qquad \text{(slow)}$$
$$(2)\quad F + NO_2 \xrightarrow{k_2} NO_2F \qquad \text{(fast)}$$

The slow step limits the rate of the overall reaction; hence, the mechanism leads to the prediction of second-order kinetics, first order in NO_2 and first order in F_2.

Since the F atom is a highly reactive particle, it seems reasonable to expect that the *atom-molecule* encounter of (2) would form an activated complex with less expenditure of energy than the *molecule-molecule* encounter of (1). Hence, (2) would be fast relative to (1). Thus, the mechanism appears to be reasonable in terms of energetics, and it leads to the correct rate law.

An example of the second type of mechanism (one or more fast equilibrium reactions followed by a slow step) is the reaction

$$CO + Cl_2 = COCl_2$$

whose rate law is known from experiment to be

$$\text{Rate} = k[CO][Cl_2]^{3/2}$$

The generally accepted mechanism is

$$\text{(1)} \qquad M + Cl_2 \underset{k_{-1}}{\overset{k_1}{\rightleftharpoons}} 2Cl + M \qquad \text{(fast, equilibrium)}$$

$$\text{(2)} \qquad M + Cl + CO \underset{k_{-2}}{\overset{k_2}{\rightleftharpoons}} COCl + M \qquad \text{(fast, equilibrium)}$$

$$\text{(3)} \qquad COCl + Cl_2 \overset{k_3}{\rightarrow} COCl_2 + Cl \qquad \text{(slow)}$$

The first step of the mechanism is the decomposition of a Cl_2 molecule to form Cl atoms. This occurs when a Cl_2 molecule and any other molecule M (which could be another Cl_2 molecule) collide with sufficient energy to overide the energy barrier for dissociation of Cl_2. The purpose of the "third-body" molecule M in the reverse reaction of (1) is to help absorb the large amount of energy that is released when the two Cl atoms combine. Without a third body to remove the excess energy, the Cl_2 molecule would not exist as a normal, stable molecule, and we could not have the equilibrium represented.† A similar condition occurs in step (2): COCl would not be a stable species without a third body to carry away the excess energy of the reaction. From the rate-limiting step (3) we can write

$$\text{Rate} = k_3[Cl_2][COCl]$$

From steps (1) and (2) we can obtain [COCl] in terms of the original reactants:

$$[COCl] = K_2[CO][Cl] = K_2K_1^{\frac{1}{2}}[CO][Cl_2]^{\frac{1}{2}}$$

and hence obtain for the rate law of the overall reaction

$$\begin{aligned}\text{Rate} &= k_3K_2K_1^{\frac{1}{2}}[CO][Cl_2]^{\frac{3}{2}} \\ &= k[CO][Cl_2]^{\frac{3}{2}}\end{aligned}$$

which agrees with the experimental rate law.

The third type of mechanism (formation of intermediates, comparable rates in all steps, steady state assumption) is common. We use the thermal decomposition of gaseous N_2O_5 (nitrogen pentoxide) as a specific example. The mechanism was described on page 323. The rate of the reaction is dependent on the concentration of the intermediate $N_2O_5^*$, which is formed through the forward reaction of equation (1) and decomposed by two routes—reaction (2) and the reverse reaction of (1). Thus, in the steady state we have

$$\begin{aligned}\text{Rate of formation of } N_2O_5^* &= \text{Rate of deactivation of } N_2O_5^* \\ &\quad + \text{Rate of decomposition of } N_2O_5^* \\ k_1[N_2O_5][M] &= k_{-1}[N_2O_5^*][M] + k_2[N_2O_5^*]\end{aligned}$$

† With no third body, the reverse of reaction (1) would be $Cl_2^* = 2\,Cl$, where Cl_2^* is an activated species.

From this result, we see that the steady state concentration of the intermediate is

$$[N_2O_5^*] = \frac{k_1[N_2O_5][M]}{k_{-1}[M] + k_2}$$

and the rate of the overall reaction is

$$\text{Rate} = k_2[N_2O_5^*] = \frac{k_1k_2[N_2O_5][M]}{k_{-1}[M] + k_2} \quad [12.6]$$

At very low pressures, $[M]$, and therefore $k_{-1}[M]$, is small relative to k_2, and equation [12.6] reduces to

$$\text{Rate} = k_1[N_2O_5][M]$$

giving second-order kinetics. At higher pressures, $k_{-1}[M]$ is large compared with k_2, and equation [12.3] becomes

$$\text{Rate} = \frac{k_1k_2}{k_{-1}}[N_2O_5]$$
$$= k[N_2O_5]$$

which is a first-order rate law. The predicted switch from first to second order as the pressure is lowered is experimentally observed.

The physical interpretation of this phenomenon appears to be clear. At very low pressures where collision frequency is low, the bimolecular activation reaction (1) becomes rate-controlling. At higher pressures, where the frequency of collisions is high, the rate is independent of the activation process and is limited by the unimolecular decomposition reaction (2).

Chain Reactions

A special type of reaction is one that continuously regenerates an intermediate during the course of a reaction to produce a self-sustaining cyclical reaction. This is a **chain reaction,** a highly important mechanism in polymerizations, flames, and explosions.

We illustrate the chain mechanism with the H_2-Br_2 reaction:

$$H_2(g) + Br_2(g) = 2\,HBr(g)$$

The first step in the mechanism of this reaction is thought to be the dissociation of Br_2 molecules into Br atoms:

This is an example of a photochemical reaction. The energy required to dissociate the Br_2 molecule comes from the energy of the light absorbed.

$$Br_2 + \text{light} \rightarrow 2\,Br$$

This is the **chain-initiation reaction.** In the second step, a Br atom produced in the reaction above reacts with an H_2 molecule to form the final product molecule

HBr and an H atom. The atomic H, in turn, reacts with a Br_2 molecule to complete one cycle:

$$Br + H_2 \rightarrow HBr + H$$
$$H + Br_2 \rightarrow HBr + Br$$

These two steps are the **chain-propagation reactions.** The cycle repeats again and again until the chain is broken, which happens whenever two atoms encounter each other and react; for example, the reaction

$$Br + Br \rightarrow Br_2$$

is a significant **chain-termination reaction** in this mechanism.

It is assumed that a "third-body" molecule is present to absorb the excess energy released in the combination of the two Br atoms. Without the third body, Br_2 would be unstable.

The H_2-Br_2 reaction has complex and very interesting kinetics. The experimental rate equation has the unusual form

$$\frac{d[HBr]}{dt} = \frac{k[H_2][Br_2]^{\frac{1}{2}}}{1 + \dfrac{[HBr]}{k'[Br_2]}}$$

It was years after this rate law was obtained experimentally before a mechanism was postulated that predicts this complex rate law. A summary of the mechanism is given below. The step that we have not previously discussed appears in (4), which is included in the mechanism to account for the experimental observation that HBr inhibits the overall reaction. The mechanism requires five steps:

(1)	$Br_2 \xrightarrow{k_1} 2\,Br$	(initiation)
(2)	$Br + H_2 \xrightarrow{k_2} HBr + H$	(propagation)
(3)	$H + Br_2 \xrightarrow{k_3} HBr + Br$	(propagation)
(4)	$H + HBr \xrightarrow{k_4} H_2 + Br$	(inhibition)
(5)	$2\,Br \xrightarrow{k_5} Br_2$	(termination)

The steady state treatment applied to this mechanism leads to a rate law of exactly the same form as the experimental rate law and, hence, lends credibility to the proposed mechanism.

Free-Radical Mechanisms

When a Cl_2 molecule is heated to a high temperature, dissociation of the molecules into atoms occurs as shown below:

$$:\ddot{\underset{..}{Cl}}:\ddot{\underset{..}{Cl}}: \rightarrow :\ddot{\underset{..}{Cl}}\cdot + \cdot\ddot{\underset{..}{Cl}}:$$

Each dissociation fragment retains one of the electrons of the "broken" bond and, therefore, is a neutral atom. Larger molecules are known to dissociate in a similar way. The dissociation products in this case, however, may not be atoms, but molecular fragments. For example, under the right conditions ethane can be dissociated into two particles:

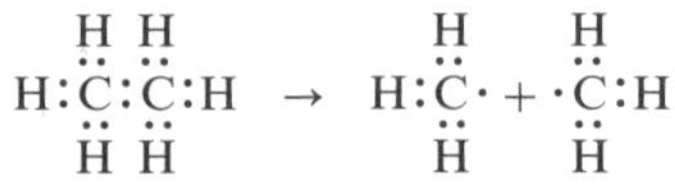

The neutral molecular fragments, each with an unpaired electron, are called **free radicals.** As might be expected, they are generally highly reactive, short-lived species. Their presence, however, is frequently detected in organic reactions.

A very simple procedure that has been used in the past for detecting free radicals is made possible by the ready reaction of free radicals with metals. Disappearance of a metallic mirror deposited on the inside wall of the reaction vessel is indirect evidence of the presence of free radicals. Modern instrumental techniques are now available for not only detecting, but also identifying the composition of, these short-lived fragments.

As an example of a **free-radical mechanism,** consider the decomposition of C_2H_6 into C_2H_4 and H_2. The mechanism that is successful in describing this reaction is given below (the single dot in the formulas denotes an unpaired electron):

(1) $C_2H_6 \rightarrow 2\,CH_3\cdot$

(2) $CH_3\cdot + C_2H_6 \rightarrow CH_4 + C_2H_5\cdot$

(3) $C_2H_5\cdot \rightarrow C_2H_4 + H\cdot$

(4) $H\cdot + C_2H_6 \rightarrow C_2H_5\cdot + H_2$

(5) $H\cdot + C_2H_5\cdot \rightarrow C_2H_6$

Trace amounts of CH_4 and C_4H_{10} are observed in the products of the C_2H_6 decomposition reaction. The presence of CH_4 can be accounted for by reaction (2), which is part of the initiation reaction of the chain mechanism. The presence of small amounts of C_4H_{10} can be accounted for by combination of two $C_2H_5\cdot$ radicals in a chain-termination step, similar to reaction (5). The presence of small amounts of CH_4 and C_4H_{10} thus provide further evidence of the proposed free-radical mechanism.

Generally, one can tell rather easily whether free radicals are playing a part in a mechanism through the addition of *scavengers*, which are chemicals that are exceptionally reactive with free radicals. Nitric oxide, a molecule with an unpaired electron, is very effective in this way. If the addition of a scavenger appreciably alters the rate of a reaction, free radicals in the mechanism are indicated.

The mechanism above leads to a first-order rate law, which is in agreement with experimental observations. However, since the decomposition via the one-step molecular mechanism

(1) $C_2H_6 \rightarrow C_2H_4 + H_2$

would also predict first-order kinetics, the rate law does not distinguish between the free-radical and the molecular mechanism. Abundant evidence for the presence of free radicals during the course of the reaction, however, strongly suggests a free-radical mechanism.

Free-radical chain mechanisms are important in polymerization reactions. The reaction for the formation of polyethylene is an example:

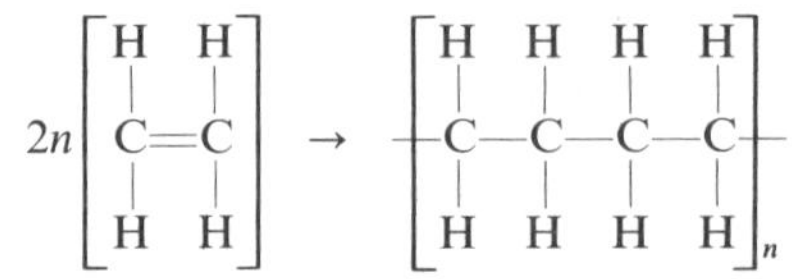

The reaction is initiated by adding benzoyl peroxide, which readily decomposes into free radicals. Letting R stand for any of these free radicals, we can write for the first step

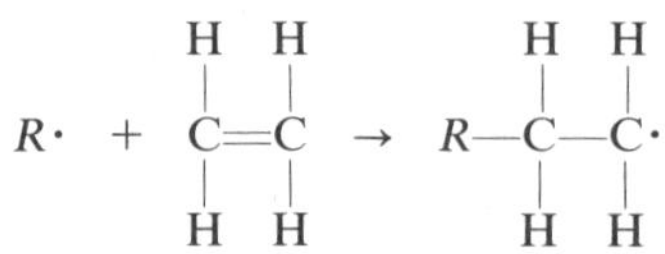

This sets up a chain:

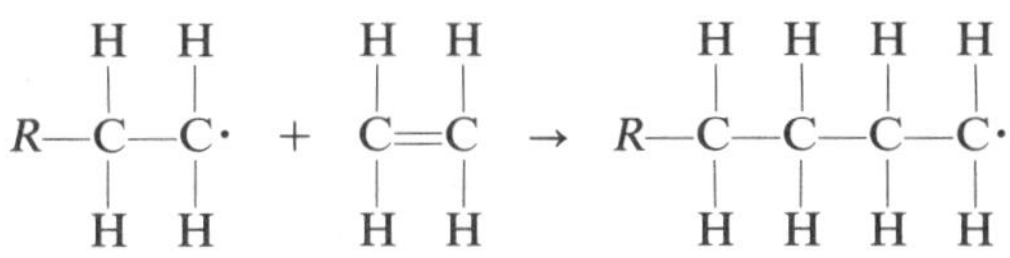

giving first a dimer (two units joined), then a trimer (three units joined), and so on up to macromolecular size.

Catalysis

It is known from experimentation that the rate of many reactions can be dramatically altered by the presence of small quantities of substances that themselves are left unchanged at the end of the reaction. Such substances are called **catalysts.** An example of catalysis that you are probably familiar with is the MnO_2-catalyzed thermal decomposition of $KClO_3$:

$$2\,KClO_3 \xrightarrow[\text{heat}]{MnO_2} 2\,KCl + 3\,O_2$$

Addition of only a small quantity of MnO_2 speeds up considerably the production of O_2 gas, and at the end of the experiment, the original quantity of MnO_2 remains. We will describe several such examples in this section and suggest mechanisms by which the catalyst operates.

Apparently, catalysts increase the rates of reactions by opening up new, faster reaction paths, that is, reaction paths with lower energies of activation (Figure 12.8). How this is accomplished is a separate story for each reaction, and the detailed mechanisms are not always known. An example of a mechanism that is well understood is the catalysis by Mn^{2+} of the oxidation of Tl^{+} by Ce^{4+}.

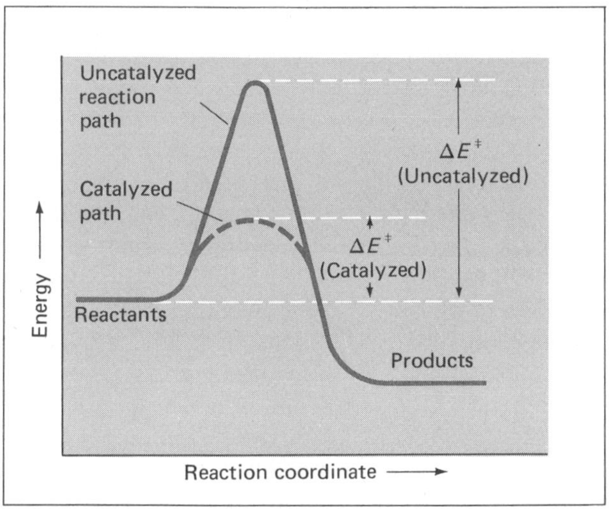

Figure 12.8
A catalyst increases the speed of a reaction by opening up a reaction path with a low energy barrier.

The overall reaction is

$$2\,Ce^{4+} + Tl^{+} \rightarrow 2\,Ce^{3+} + Tl^{3+}$$

The reaction is thermodynamically favorable, but very slow in the absence of a catalyst. A cursory examination of the reaction shows why it is so slow. If the mechanism were the one-step process just written, we would have a termolecular reaction between ions of the same charge, and, hence, we could expect slow kinetics on two counts: (1) low probability of three-body collisions, and (2) repulsion of like charges.

What at first would seem more likely than a single elementary process, however, is a step-by-step oxidation-reduction via bimolecular reactions. As possibilities for a first step, one could write

$$Ce^{4+} + Tl^{+} \rightarrow Ce^{3+} + Tl^{2+}$$

or

$$Ce^{4+} + Tl^{+} \rightarrow Ce^{2+} + Tl^{3+}$$

However, neither of these possibilities is energetically feasible since the high (endothermic) heats of formation of Ce^{2+} and Tl^{2+} would require prohibitively high activation energies. Hence, in both theory and fact, the uncatalyzed oxidation of Tl^{+} to Tl^{3+} by the reduction of Ce^{4+} to Ce^{3+} is very slow.

Add Mn^{2+} to the system, however, and new (and faster) paths for the reaction become available. A reasonable mechanism in the presence of Mn^{2+} is

$$\begin{array}{rl} & Ce^{4+} + Mn^{2+} \rightarrow Ce^{3+} + Mn^{3+} \\ & Ce^{4+} + Mn^{3+} \rightarrow Ce^{3+} + Mn^{4+} \\ & Tl^{+} + Mn^{4+} \rightarrow Tl^{3+} + Mn^{2+} \\ \hline \text{Adding:} & 2\,Ce^{4+} + Tl^{+} \rightarrow 2\,Ce^{3+} + Tl^{3+} \end{array}$$

This mechanism still requires encounters of ions of like charge, but we no longer have a termolecular reaction, and the bimolecular encounters have reasonable energy requirements. Note that the Mn^{2+} is regenerated, as is always true of a catalyst, and, hence, only a small quantity of it is necessary to have a large effect on the rate of reaction.

The example of catalysis just described can be classified as **homogeneous catalysis,** since the catalyst was present in the same phase as the reactants. Another important class of catalytic reactions makes use of the surface properties of solids to catalyze reactions in either gaseous or liquid solutions. With the catalyst and reactants in different phases we have **heterogeneous catalysis.** Many important industrial processes—involving billions of dollars of products annually—are dependent on the use of finely divided solid catalysts. Some of these processes are the catalytic cracking of petroleum, in which large hydrocarbon molecules are "cracked" to gasoline size; catalytic reforming, in which small hydrocarbon molecules are built up to larger sizes; hydrogenation of double bonds, as in the addition of H_2 to unsaturated vegetable oils to form saturated fats; and syntheses of such industrially important products as NH_3, HNO_3, and H_2SO_4.

There is no one general mechanism that describes the action of surface catalysts. The solid catalyst in heterogeneous catalysis appears to function by adsorbing at least one of the reactants on its surface. The adsorption has two effects: (1) A bond in the reactant can be stretched and weakened (in order that the atoms of the reactant can coincide with the adsorption sites of the catalyst), thus making it easier to form an activated complex; and (2) the adsorption of the catalyst may orient a complex molecule in a way that would conveniently expose active groups in the molecule to another reactant. Enzymes are thought to function as catalysts in this way. The field of heterogeneous catalysis is one of the many intriguing fields in chemistry that has need for continued fundamental research.

One further comment about catalysis needs to be made. Since thermodynamic functions of state are independent of reaction path, ΔG^0 (and hence K) for any given reaction is independent of a catalyst being present. In other words, the point of equilibrium in a reaction is *not* affected by a catalyst. Only the length of time required to reach the equilibrium state is shortened if an effective catalyst is present. An alternative way of viewing the picture is to think of the catalyst as something that lowers the energy of activation barrier in a reaction; since the same activated state appears in both forward and reverse reactions, any increase in the rate constant of a forward reaction by use of a catalyst is necessarily accompanied by a compensating increase in the rate constant of the reverse reaction. Thus, $k_1/k_{-1} = K$ remains constant.

Formulation of Mechanisms

If there is any general procedure at all for determining the mechanism for a reaction, it consists of listing all the reasonable paths one can conceive of, and then systematically eliminating (because the proposal is not consistent with known experimental facts) all but one of these. Thus, it is true that mechanisms are only *disproved*, never *proved*. Nevertheless, a proposal that can be shown to be in harmony with a large body of experimental information about the reaction rightfully takes on the stature of a "satisfactory mechanism," and serves as a useful model for making further predictions about the reaction.

We have referred to the type of tests that a proposed mechanism must pass as we discussed individual reactions. Now, we will simply recall these discussions by a summary listing:

(a) *There must be agreement between the experimental rate law and the rate law that is deduced from the mechanism.* This is the basic requirement. (Refer to the discussion of the N_2O_5 and the $H_2 + Br_2$ systems.)

(b) *The mechanism must provide for reasonable collision frequencies.* Four- and five-body collisions are so improbable that they need not be considered at all in a reaction that proceeds at a measurable rate. Termolecular collisions are only 10^{-3} as likely to occur as bimolecular collisions. (Refer to the discussion of the $Ce^{4+} + Tl^{+}$ reaction.)

(c) *The energetics of each elementary process must be reasonable for the reaction.* Reactions between free radicals, for example, have practically zero activation energies, and must be considered relatively fast steps in a process. The same is true for reactions between ions of the opposite sign. On the other hand, reactions between ions of like sign and between molecules often have large energies of activation. (Refer to the Ce^{4+} + Tl^{+} and $F_2 + NO_2$ discussions.)

(d) *The mechanism must be consistent with the experimentally detected intermediates and products, and with already well-established principles of chemical behavior.* (Refer to the thermal decomposition of C_2H_6.)

(e) *The mechanism must be capable of explaining the observed effects of scavengers, surfaces, and inhibitors.* (Refer to the decomposition of C_2H_6 and the polymerization of C_2H_4.)

Summary

Chemical kinetics deals with (1) rates of reaction and (2) mechanisms.

The rate of a reaction is affected by the concentration of reactants, the temperature, and the catalysts. The effect of concentration is summarized in the rate law. For the general reaction

$$aA + bB + \cdots = \text{products}$$

the differential rate law is

$$\boxed{\text{Rate} = k[A]^m [B]^n \cdots}$$

In general, the constants k, and $m, n, \ldots$ must be experimentally determined from concentration versus time measurements. Knowing the value of the proportionality constant k, called the rate constant, and the values of the exponents, referred to as the order of the reaction, one can calculate the instantaneous rate of reaction at any concentration of the reactants [A], [B], ...

The integrated forms of the rate law for first- and second-order reactions are

$$\boxed{\ln c = \ln c_0 - kt} \qquad \text{(first order)}$$

and

$$\boxed{\frac{1}{c} = \frac{1}{c_0} + kt} \qquad \text{(second order)}$$

Through these equations the concentration c of the reactant can be calculated at any given time t after the start of the reaction. The term c_0 is the initial concentration of the reactant, the concentration at $t = 0$. Alternatively, one can calculate c/c_0, the fraction unreacted, or $(1 - c/c_0)$, the fraction reacted after a given time has elapsed.

The integrated rate equations, written in the form that they are above, are equations of a straight line. Plotting $\ln c$ versus t for a first-order reaction gives a straight line of slope $= -k$. Plotting $1/c$ versus t for a second-order reaction gives a straight line of slope $= k$. Such plots are useful both to check on the order of the reaction and to obtain the value of k.

The half-life of a reaction is the time that it takes for half of the original quantity of reactant to disappear. It is a useful way to visualize the speed of a reaction and is very commonly used in discussing rates of first-order decompositions. The relationship between half-life ($t_{1/2}$) and the rate constant for a first-order reaction is

$$t_{\frac{1}{2}} = \frac{\ln 2}{k} \qquad \text{(first order)}$$

Transition-state theory postulates that the reactants first form an intermediate complex, which then decomposes into the products. The activated complex is the highest point on the lowest energy path that connects reactants and products.

The energy barrier that the reactants must surmount before they can change into products is called the energy of activation. The higher the energy of activation, the fewer the reactant molecules that are able to surmount the barrier, and the slower the reaction.

Catalysts increase the rate of a reaction by opening up a new reaction path of lower activation energy. Catalysts are regenerated in a true catalytic reaction and, therefore, need be present in only very small amounts to be effective. Catalysts change the rate of a reaction but have no effect on the point of equilibrium of a reaction.

In logarithmic form the Arrhenius equation is

$$\ln k = \ln A - \frac{\Delta E^{\ddagger}}{RT}$$

which is the equation of a straight line when $\ln k$ is plotted as the ordinate and $1/T$ as the abscissa. The slope is $-\Delta E^{\ddagger}/R$. Energies of activation are obtained from such plots. Knowing the value of the rate constant at one temperature, and knowing $\Delta E^{\ddagger}$, one can calculate from the Arrhenius equation a value of the rate constant at another temperature.

The second main topic of chemical kinetics has to do with mechanisms of reactions. Hardly any reaction goes by a simple one-step mechanism. Even the very simple-appearing reactions usually have several-step mechanisms that often include chain reactions and free-radical mechanisms. The mechanisms are deduced from an analysis principally of data from kinetics studies. The most important test that a proposed mechanism must pass is that it must lead to the same rate law that is found experimentally. Other tests are that the mechanism must provide reasonable collision frequencies, have reasonable energy requirements, be consistent with experimentally detected intermediates and products, and be able to explain observed effects of scavengers, surfaces, and inhibitors.

Important Terms and Concepts

Chemical kinetics
Rate law
Order of the reaction
Specific rate constant
Method of initial rates
Integrated rate law equations
Half-life
Transition-state complex (Activated complex)
Energy of activation
Frequency factor
Arrhenius equation
Reaction mechanism
Elementary reaction
Rate-limiting reaction
Steady state approximation
Chain reaction
Chain-initiation reaction
Chain-propagation reaction
Chain-termination reaction
Free radical mechanisms
Catalysts
Homogeneous catalysis
Heterogeneous catalysis

Exercises

12-1.
For the reaction

$$2\ NO(g) + O_2(g) = 2\ NO_2(g)$$

doubling the O_2 pressure doubles the reaction rate, but doubling the NO pressure increases the reaction rate fourfold. Write the rate law expression.

12-2.
For the reaction

$$H_2(g) + C_2H_4(g) \rightarrow C_2H_6(g)$$
$$\text{Rate} = k[C_2H_4][H_2]$$

Suppose the reacting gas is compressed so that the pressure increases by a factor of four. If the temperature stays constant, how much does the rate increase?

12-3.
For the reaction

$$A + B \rightarrow \text{products}$$

the following data about the rate of the reaction at different concentrations were obtained.

INITIAL MOLAR CONCENTRATION $[A_0]$	INITIAL MOLAR CONCENTRATION $[B_0]$	INITIAL RATE (moles/liter-sec)
.10	.20	5
.10	.40	20
.10	.60	45
.20	.10	1.25
.40	.10	1.25
.60	.10	1.25

(a) What is the rate law expression for the reaction?
(b) Calculate a numerical value for the rate constant.
(c) What would the initial rate of the reaction be if the initial concentration of A is 1.0 and the initial concentration of B is 0.80 molar?

12-4.
For the reaction

$$A + B = C$$

the following data were obtained:

EXPERIMENT	$[A_0]$	$[B_0]$	INITIAL RATE (moles/liter-sec)
1	0.60 M	0.15 M	6.3×10^{-3}
2	0.20 M	0.60 M	2.8×10^{-3}
3	0.20 M	0.15 M	7.0×10^{-4}

(a) What is the rate expression for the reaction?
(b) What is the numerical value for the rate constant k?
(c) What is the rate of the reaction when $[A] = [B] = 0.50\ M$?

12-5.
The following experimental data were obtained for a reaction involving A, B, and C as reactants.

$[A_0]$	$[B_0]$	$[C_0]$	INITIAL RATE (moles/liter-min)
2	4	3	48
2	4	6	96
2	2	3	48
2	4	3	48
1	2	3	12
2	2	3	48

(a) Write the rate law that applies.
(b) Calculate a numerical value for the specific rate constant, k.
(c) Calculate the rate when the concentration is 2 M in A, 3 M in B, and 4 M in C.

12-6.
The rate constant is 0.0170 min^{-1} for the first-order decomposition of CH_3NO_2 at 200° C. Starting with a concentration of 0.100 mole/liter, calculate

(a) the time for the concentration to drop to 0.050 moles/liter.
(b) the concentration of CH_3NO_2 after 30 minutes and after 24 hours.

12-7.
A first-order reaction has a specific rate constant of $4.23 \times 10^{-5}\ \text{sec}^{-1}$. Calculate $t_{\frac{1}{2}}$ for the reaction.

12-8.
Radioactive decay is an example of a first-order reaction. $^{32}_{14}Si$ decays according to the reaction

$$^{32}_{14}Si \rightarrow {}^{32}_{15}P + {}_{-1}^{0}e$$

with a half-life of 700 years.

(a) Calculate the rate constant for the radioactive decay reaction.
(b) How many years would it take for 0.90 g of a 1.00 g sample of $^{32}_{14}Si$ to decompose?

12-9.
The half-life for the first-order decomposition of N_2O_5 is 5.7 hours. Calculate

(a) the specific rate constant.
(b) the time required for 90% of the N_2O_5 to decompose.

12-10.
The mechanism for the reaction

$$2A + B = C$$

has the following steps

$$2A \xrightarrow{k_1} D \qquad \text{(slow)}$$
$$D + B \xrightarrow{k_2} C \qquad \text{(fast)}$$

Write the rate law that this mechanism would predict.

12-11.
The kinetic data listed in the accompanying table were obtained at 25° C for the reaction

$$A(g) + B(g) = AB(g)$$

EXPERIMENT	$[A_0]$ (moles/liter)	$[B_0]$ (moles/liter)	INITIAL RATE (moles/liter-sec)
1	0.10	0.10	0.25
2	0.20	0.20	2.0
3	0.10	0.20	1.0
4	0.20	0.40	8.0

(a) Write the rate law expression for the reaction.
(b) Show that the rate law is compatible with the mechanism

$$2B \underset{k_{-1}}{\overset{k_1}{\rightleftarrows}} B_2 \qquad \text{(fast, equilibrium)}$$
$$B_2 + A \rightarrow AB + B \qquad \text{(slow)}$$

and express k, the experimental constant, in terms of k_1, k_{-1}, and k_2.
(c) Calculate a value from the table for k, and then calculate the initial rate at which A is being used up when $[A] = 0.15\ M$ and $[B] = 0.25\ M$.

12-12.
The experimentally determined rate expressions for several reactions are given below. Assume that these rate expressions hold at equilibrium, and determine the rate expression for the reverse reactions at equilibrium.

(a) $C_2H_2 + H_2 \rightarrow C_2H_4$

$$\text{Rate} = k\frac{[H_2]}{[C_2H_2]}$$

(b) $2\ H_2 + O_2 \rightarrow 2\ H_2O$
$\text{Rate} = k[H_2][O_2]^{4/3}$

(c) $N_2O_5 \rightarrow 2\ NO_2 + \frac{1}{2}O_2$

$\text{Rate} = k[N_2O_5]$

12-13.
At 378.5° C, $t_{1/2} = 363$ min and $\Delta E^{\ddagger} = 52{,}000$ cal/mole for the first-order thermal decomposition of ethylene oxide. From these data, estimate the time required to 75% decompose ethylene oxide at 450° C.

12-14.
The steps in a particular reaction are as follows:

$$A + B \underset{k_{-1}}{\rightleftarrows} C \quad \text{(fast, equilibrium)}$$
$$C + A \xrightarrow{k_2} D \quad \text{(slow)}$$
$$D \xrightarrow{k_3} P \quad \text{(fast)}$$

Therefore, the overall reaction is

$$B + 2A \rightarrow P$$

The graph below shows the energy changes in this chemical reaction.

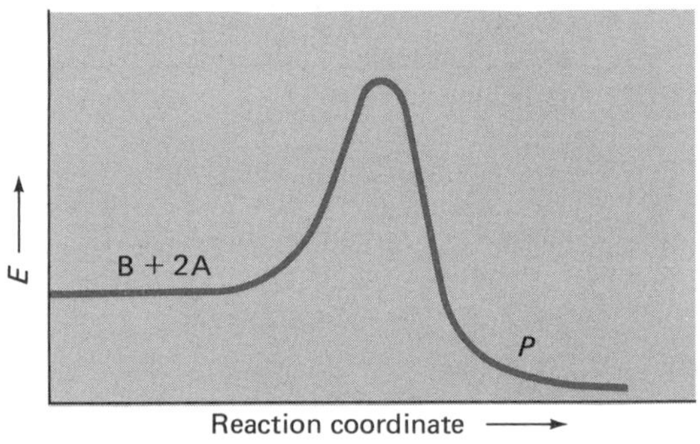

(a) Deduce the rate law that corresponds to the proposed mechanism.
(b) On the above graph, label clearly (1) the energy of activation for the forward reaction, (2) the energy of activation for the reverse reaction, and (3) the energy change corresponding to ΔE for the reaction.
(c) Is the forward reaction exothermic or endothermic?

Problems

12-1.
The following data were obtained for the decomposition of diazomethane, CH_2N_2, at 600° C.

t (minutes)	$[CH_2N_2]$ (moles/liter)	t (minutes)	$[CH_2N_2]$ (moles/liter)
0	0.100	15	0.044
5	0.076	20	0.033
10	0.058	25	0.025

(a) Show graphically that the reaction is first order and determine the rate constant k.
(b) Determine the rate of the reaction at $t = 20$ minutes.

12-2.
Butadiene, C_4H_6, dimerizes according to the reaction

$$C_4H_6(g) \rightarrow \tfrac{1}{2}C_8H_{12}(g)$$

The rate law expression is

$$\text{Rate} = k[C_4H_6]^2$$

At 500° K, $k = 1.42 \times 10^{-2}$ liter/mole-sec for this reaction. Starting with butadiene gas at a concentration of 1.60×10^{-2} moles/liter, how long (in sec) would it take for the concentration to fall to 1.00×10^{-2} moles/liter?

12-3.
For the isomerization reaction

(1) (2)

the following data have been determined

t (minutes)	Moles of (1)
0	1.00
10	0.90
20	0.81
30	0.73

(a) What is the order of the reaction?
(b) What is the half-life for the reaction?

12-4.
The rate expression for the reaction of hydrogen and iodine to form hydrogen iodide is given by

$$\text{Rate} = k[H_2][I_2]$$

For many years, the accepted mechanism for this reaction (as described in the chapter) was a simple binary collision:

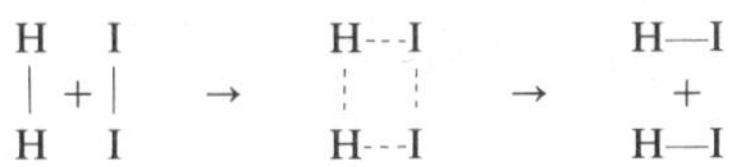

In recent years, new evidence suggests that instead the mechanism may be

$$I_2(g) + M \underset{k_{-1}}{\overset{k_1}{\rightleftarrows}} 2\,I(g) + M \quad \text{(fast, equilibrium)}$$

$$2\,I(g) + H_2(g) \xrightarrow{k_2} 2\,HI(g) \quad \text{(slow)}$$

where M is a third body providing the energy to dissociate the I_2. Show that this mechanism is also consistent with the experimental rate expression, and relate k_1, k_{-1}, and k_2 to k, the experimental rate constant.

12-5.
In acid solution the rate of the reaction

$$NH_4^+ + HNO_2 = N_2 + 2\,H_2O + H^+$$

is consistent with the mechanism

$$HNO_2 + H^+ \underset{k_{-1}}{\overset{k_1}{\rightleftarrows}} H_2O + NO^+ \text{ (rapid equilibrium)}$$

$$NH_4^+ \underset{k_{-2}}{\overset{k_2}{\rightleftarrows}} NH_3 + H^+ \quad \text{(rapid equilibrium)}$$

$$NO^+ + NH_3 \xrightarrow{k_3} NH_3NO^+ \quad \text{(slow)}$$

$$NH_3NO^+ \xrightarrow{k_4} H_2O + H^+ + N_2 \quad \text{(fast)}$$

Write the rate law that is consistent with this mechanism.

12-6.
The following table gives the specific rate constants for the decomposition of N_2O_5 at various temperatures.

t (°C)	k (sec^{-1})
0	7.87×10^{-7}
25	3.46×10^{-5}
35	1.35×10^{-4}
45	4.98×10^{-4}
55	1.50×10^{-3}
65	4.87×10^{-3}

Use a graphical method to

(a) determine the energy of activation.
(b) find k at 50° C.

12-7.
For the reaction

$$2\,I(g) + H_2(g) \rightarrow 2\,HI(g)$$

the Arrhenius factor $A = 6.6 \times 10^7$ liter^2/mole^2-sec, and $\Delta E^{\ddagger} = 5310$ cal/mole. Calculate the specific rate constant for the reaction at 600° K.

13

Atomic Structure

Most of the chemistry that we have described thus far has been concerned with observations of matter on a *macroscopic scale.* For example, we have talked about the effect of pressure on the volume of a gas, the heat capacity of a solid, the freezing point of an aqueous solution, and the vapor pressure of a liquid. Thermodynamics has provided many of the mathematical relationships for correlating and extending our understanding of the observable world.

Generalization from macroscopic observations represents only one approach to science, however. We now consider a radically different way of gaining information, a study of the *submicroscopic* structure of matter. Since these structures cannot be observed directly, any study of them must depend on models. The model in use is the atom, with its nucleus of protons and neutrons and its extranuclear arrangements of electrons. No one has ever directly observed the atom, much less an electron, proton, or neutron. Here, our physical picture of these units may be quite naïve. This is not important in the formation of a model. The important point is that correct predictions about matter *can* be made from the model. The submicroscopic model is "correct" if it leads to correct predictions about the macroscopic, measurable properties of the structure it represents.

The **quantum theory** provides the mathematical relationships we use in describing the atom. The quantum theory is based on a series of postulates, but unlike the laws of thermodynamics, these postulates lack direct experimental verification. They do, however, lead to the correct description and prediction of observable properties, and this is the justification we need for their use.

Our goal in this chapter is to present the postulates of the quantum theory and to use them to describe atomic structure. In the process, we will discover the basis for the electronic structures and chemical periodicity discussed in Chapter 2.

13.1 Quantum Theory

Wave Nature of Electromagnetic Radiation

Many of the properties of light and other forms of **electromagnetic radiation** (for example, γ-rays, x-rays, microwaves, and radio waves) are accurately described by a **wave model.** The wave form of an electromagnetic wave is shown in Figure 13.1a, where the x-axis is the direction of wave propagation, and the y-axis is the displacement of the wave from its mean height. The maximum displacement, represented by A_0, is the *amplitude.*

The distance between successive corresponding points in a wave (for example, the distance between successive maximums—one complete cycle) is the **wave length,** λ. An electromagnetic wave moves through space with a velocity c. In a vacuum, c is independent of λ and has a value of 3.00×10^{10} cm/sec. For any fixed point on the x-axis, the y displacement changes with time as shown in Figure 13.1b. As seen in the figure, displacement at a given location varies periodically with time. The number of complete cycles of the wave passing a

Figure 13.1a
Electromagnetic wave as a function of position. The wavelength, λ, is the distance from crest to crest. A_0 is the maximum amplitude of the wave.

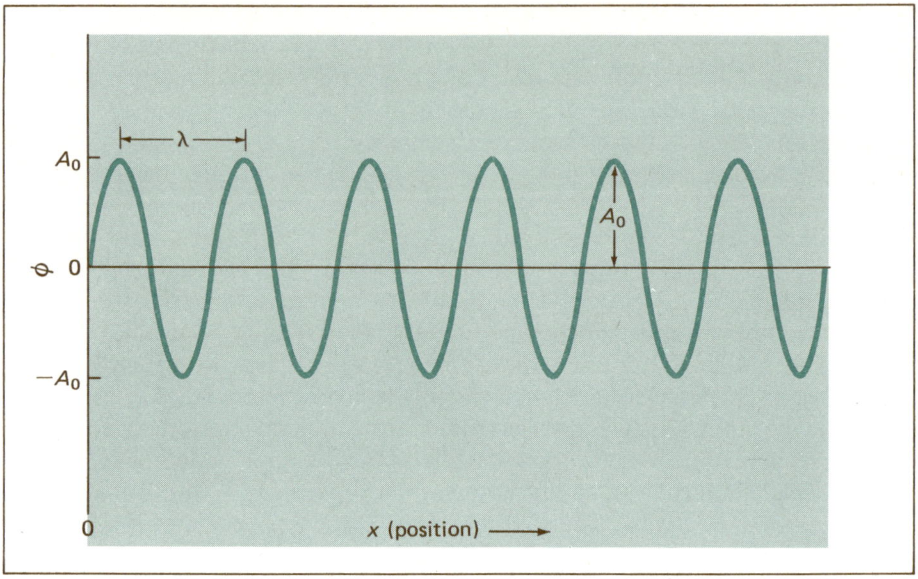

Figure 13.1b
Electromagnetic wave as a function of time. The Greek letter τ is the period of the wave, the length of time of one complete cycle of the wave. The frequency (ν) is the number of cycles that pass a fixed position in unit time. The frequency is the reciprocal of the period.

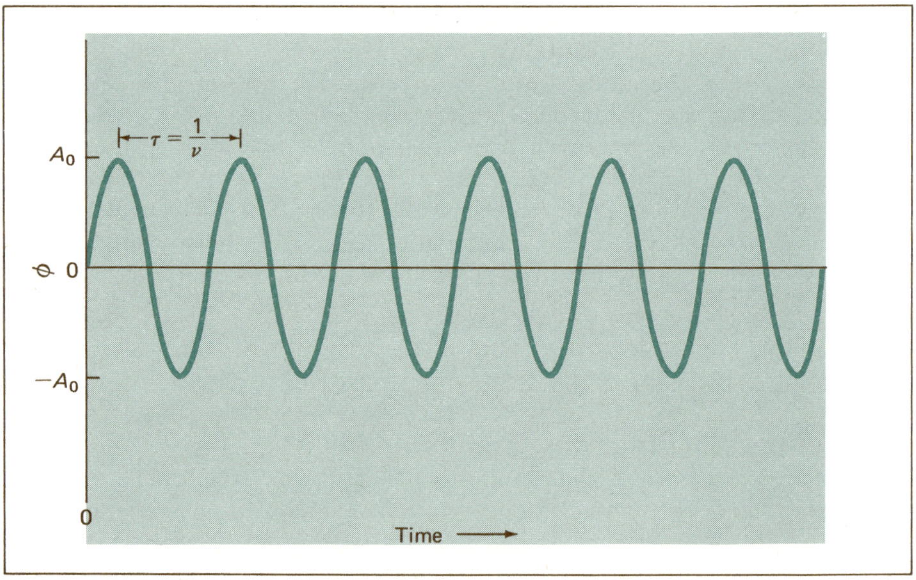

fixed location per unit of time is the **frequency,** ν. It is apparent from the dimensions that ν is inversely proportional to λ, and that the proportionality constant is c:

$$\nu\,(\text{cycles/time}) = \frac{c\,(\text{length/time})}{\lambda\,(\text{length/cycle})} \qquad [13.1]$$

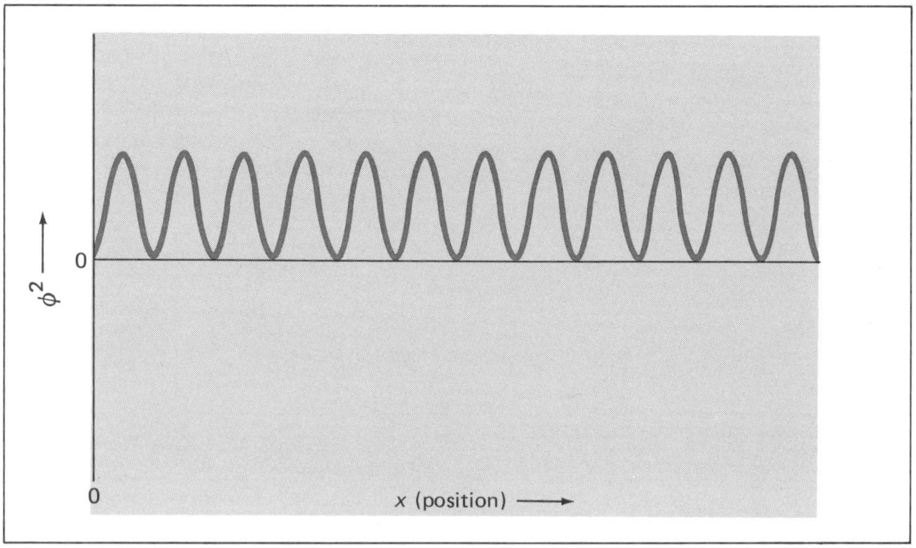

Figure 13.2
Graph of energy density ϕ^2 for an electromagnetic wave as a function of position.

The form of a wave can be represented by a mathematical function; for example, the waves of Figure 13.1a and 13.1b are described by the function

$$\phi = A_0 \sin\left[2\pi\left(\frac{x}{\lambda} - vt\right)\right] \qquad [13.2]$$

Figure 13.1a is obtained with t held constant, and Figure 13.1b is obtained with a constant value for x. The mathematical function ϕ, which denotes the form of this one-dimensional wave, is called the **wave function.** Various properties of the wave can be obtained from the wave function; for example, energy density of the wave is proportional to ϕ^2. Figure 13.2 is a plot of ϕ^2 against x to show graphically how the energy density of an electromagnetic wave varies with position.

Electromagnetic Spectrum

Electromagnetic waves can be generated with wavelengths that vary all the way from fractions of angstroms to kilometers. Radiation of the entire range of wavelengths is known as the **electromagnetic spectrum.** It is convenient to divide the electromagnetic spectrum into wavelength ranges (see Figure 13.3). For example, that range of the spectrum that the human eye responds to is referred to as the visible light range. It extends from $\lambda = 400$ nanometers (violet light) to $\lambda = 800$ nanometers (red light). At wavelengths just shorter than violet light is a range called ultraviolet. In the ultraviolet range the waves are too short to be seen with the eye; however, they can be detected photographically and by various electronic methods. At increasingly shorter wavelengths (higher frequencies) are x-rays and γ-rays. On the other side of visible light are—in

Figure 13.3
The electromagnetic spectrum.

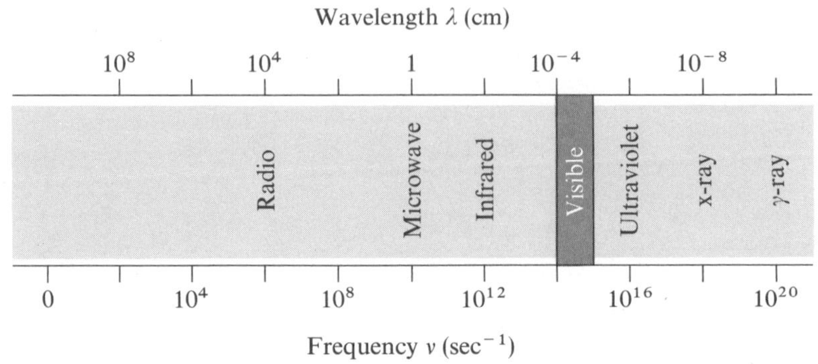

Frequency Ranges

	λ (cm)	λ (usual units)
x-rays	1×10^{-10}–1×10^{-7}	10^{-3}–1 nm*
ultraviolet	1×10^{-7}–4×10^{-5}	1–400 nm
visible	4×10^{-5}–8×10^{-5}	400–800 nm
infrared	8×10^{-5}–3×10^{-2}	0.8–300 μ*
microwave	3×10^{-2}–100	0.03–100 cm
radio waves	10^2–10^5	1–10^3 meters

* nm = nanometer = 10^{-9} meter
μ = microns = 10^{-6} meter

increasingly longer wavelengths—the infrared, microwave (radar), and radio wave regions.

White light is a combination of all the various colors in the visible light range. It can be separated into its constituent colors by being passed through a prism. This process, called dispersion of light, is shown schematically in Figure 13.4. On passing through the prism, the light of the shortest wavelength is bent the most, and that of the longest wavelength is bent the least. The result of white-light dispersion is a rainbow of colors, a continuous spectrum of waves ranging from violet to red.

Figure 13.4
Dispersion of visible electromagnetic radiation by a prism.

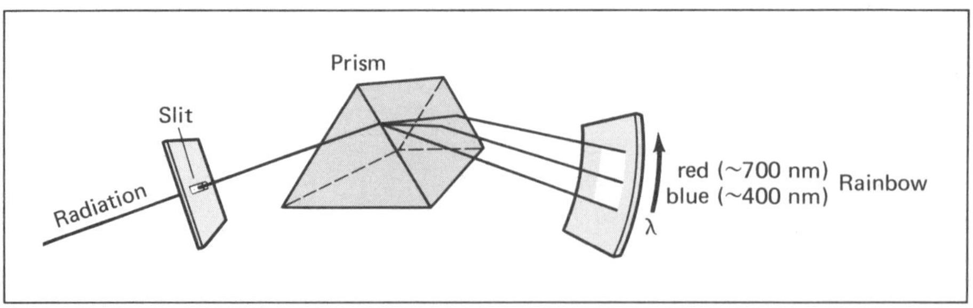

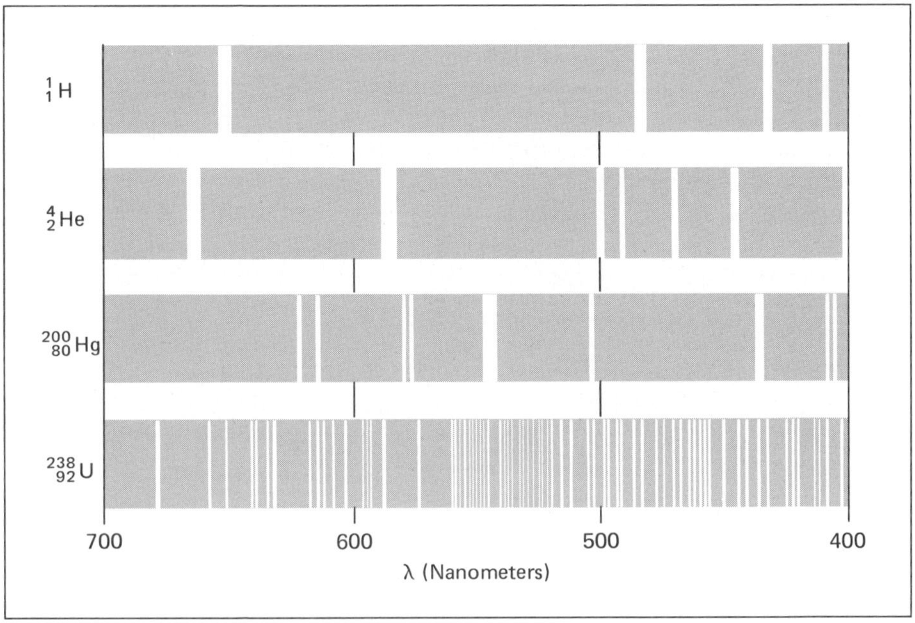

Figure 13.5
Atomic spectra of some of the elements in the visible region.

On the other hand, if an element is heated to incandescence in a flame, arc, or spark, and the emitted light is dispersed, one obtains a **line spectrum** as illustrated in Figure 13.5. A schematic diagram of a spectrograph used to obtain line spectra of gaseous elements is illustrated in Figure 13.6. The electromagnetic radiation is produced by an electrical discharge through the substance being investigated. The radiation is passed through a slit and then into a prism, where it is dispersed into its various wavelengths. Because of the dispersion, each wavelength produces an image of the slit at a different position on the photographic plate. Later we will see that there is an energy value associated with each wavelength; therefore, a line spectrum implies that the energy changes in the atom that generate the light are *restricted to discrete values.* Discovery of the line spectra of elements contributed greatly to the present understanding of atomic structure, and we will refer to this subject again later.

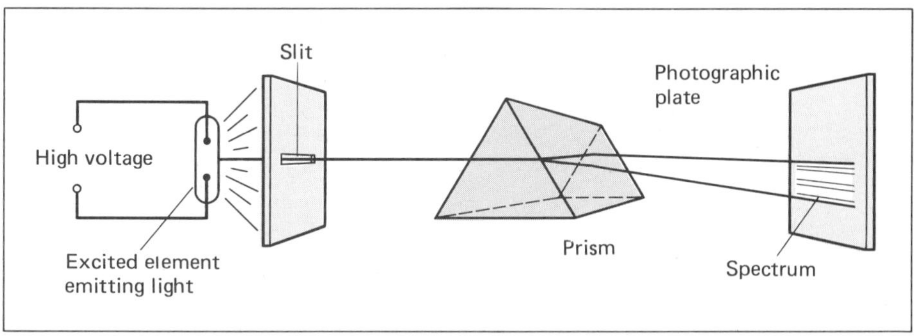

Figure 13.6
Schematic diagram of the essential elements of a spectrograph.

Each element has a different line spectrum. This has great theoretical significance as we will see later, but in addition it forms the basis for an important method for chemical analysis. The position of the spectral lines identifies the elements present, and the intensity of the lines is a measure of the concentrations of the elements.

Thus far the discussion on dispersion has been restricted to visible light and glass prisms for the sake of simplicity. The same general principles apply, however, to other ranges of the electromagnetic spectrum when some modifications are made of the methods of dispersion and of detecting the radiation. Another device (besides the prism) for producing dispersion is a *diffraction grating*, which can be any array of parallel, equidistant lines or slits. Etched lines on a glass or quartz plate, for example, can be used in the ultraviolet, visible, and infrared ranges. Microwaves are dispersed by a wire screen, and x-rays can be dispersed by the uniformly spaced rows and columns of atoms in a crystal.

Particulate Nature of Electromagnetic Radiation

Since Einstein's proposal of the theory of relativity early in this century, scientists have known that matter and energy are closely related. The well-known Einstein equation

$$\boxed{E = mc^2} \qquad [13.3]$$

gives the quantitative relationship, where E is the energy equivalent of mass m, and c is the speed of light (3×10^{10} cm/sec). Since the proportionality constant c^2 is such a large number (9×10^{20} cm^2/sec^2), a small quantity of mass is the equivalent of a very large quantity of energy.

There are interesting analogies that can be drawn between matter and radiant energy. Matter appears continuous on a macroscopic scale, and yet scientists have had great success in explaining the macroscopic properties of matter by means of submicroscopic models in which matter is considered to be discontinuous (the atomic theory, for example). In a like manner, radiant energy also appears continuous on a macroscopic scale. For example, the intensity of an electric light can be made to vary *smoothly* as the quantity of electricity supplying the lamp is varied. On the other hand, many of the properties of radiation can be explained only if we assume that it consists of small indivisible energy packets called **photons,** with the energy of the photon being proportional to the frequency of the electromagnetic wave. Mathematically, this is expressed as

$$\boxed{\epsilon = h\nu} \qquad [13.4]$$

where ϵ is the energy/photon, ν is the frequency, and h is the proportionality constant known as *Planck's constant*. Since Planck's constant is a small number (6.625×10^{-27} erg-sec), photons are very small packets of energy. It is the small size of the particles of energy, just as it is the small size of atoms, that make both matter and energy appear continuous on a macroscopic scale.

EXAMPLE 13.1 Calculate the energy of an x-ray photon with $\lambda = 1.00$ Å and of a photon of green visible light with $\lambda = 500$ nm.

SOLUTION Equations [13.1] and [13.4] can be combined to give

$$\epsilon = h\nu = \frac{hc}{\lambda} \qquad [13.5]$$

For the x-ray,

$$\epsilon = \frac{(6.63 \times 10^{-27}\ \text{erg-sec})(3.00 \times 10^{10}\ \text{cm/sec})}{(1.00\ \text{Å})(10^{-8}\ \text{cm/Å})}$$

$$\epsilon = 1.99 \times 10^{-8}\ \text{ergs}$$

For the green light,

$$\epsilon = \frac{(6.63 \times 10^{-27}\ \text{erg-sec})(3.00 \times 10^{10}\ \text{cm/sec})}{(5.00 \times 10^{2}\ \text{nm})(10^{-7}\ \text{cm/nm})}$$

$$\epsilon = 3.98 \times 10^{-12}\ \text{ergs}$$

Equations [13.3] and [13.5], each written for one photon of energy (ϵ), can be equated to give the relationship

When particles having mass are converted into photons of electromagnetic radiation, equations [13.3] and [13.4] can be used to calculate the energy and also the frequency or wavelength of the photon.

$$\epsilon = mc^2 = \frac{hc}{\lambda}$$

Solving for λ gives

$$\lambda = \frac{h}{mc} \qquad [13.6]$$

which is an interesting relationship that allows one to calculate the momentum for a photon. For a particle that has mass, momentum (p) is the product of mass (m) and velocity (u):

$$p = mu$$

Since a photon has no mass, there is an obvious difficulty in ascribing a momentum in the usual sense to the photon. Equation [13.6] allows one to calculate the product of the velocity (c) and the mass that has been converted to energy, which is the momentum.

$$p = mc$$

This result, substituted into equation [13.6] gives

$$\boxed{\lambda = \frac{h}{p}} \qquad [13.7]$$

Equation [13.7] is known as the **de Broglie equation.** It played a highly important role in the development of the present wave model of the atom by suggesting that since waves have the properties of particles, particles can be described by the equations of waves.

The momentum of a photon is very small and, as a result, is not easily observed; however, there is experimental verification of equation [13.7]. In suitable apparatus, high-energy γ-ray photons can be caused to collide with electrons. The electrons exchange momentum with the photons, and the wavelength of the photon is observed to change. This is known as the *Compton effect*. The new wavelength, experimentally observed, agrees with the value calculated from equation [13.7].

EXAMPLE 13.2 Compare the momentum of a 600 nm photon of light with that of a gaseous O_2 molecule at 300° K that is moving with the root-mean-square velocity.

SOLUTION The momentum of the photon is obtained from equation [13.7]:

$$p = \frac{h}{\lambda} = \frac{6.63 \times 10^{-27} \text{ erg-sec}}{(6.00 \times 10^2 \text{ nm})(10^{-7} \text{ cm/nm})}$$

$$p = 1.11 \times 10^{-22} \text{ g-cm/sec}$$

For the O_2 molecule, we found in Example 6.2 that $u_{rms} = 4.84 \times 10^4$ cm/sec at 300° K. The molecular weight of O_2 is 32.0 g/mole. Hence,

$$p = (m)(u_{rms}) = \left(\frac{32.0 \text{ g/mole}}{6.02 \times 10^{23} \text{ molecules/mole}}\right)(4.84 \times 10^4 \text{ cm/sec})$$

$$p = 2.57 \times 10^{-18} \text{ g-cm/sec-molecule}$$

The momentum of the O_2 molecule is approximately 25,000 times greater than the momentum of the photon.

The Wave Nature of Matter and the First Postulate of the Quantum Theory

We can summarize our comparison of mass and radiant energy as follows:

(1) Mass and energy can be equated:

$$\boxed{E = mc^2}$$

(2) Mass has momentum:

$$\boxed{p = mu}$$

Radiant energy has momentum:

$$\boxed{p = \frac{h}{\lambda}}$$

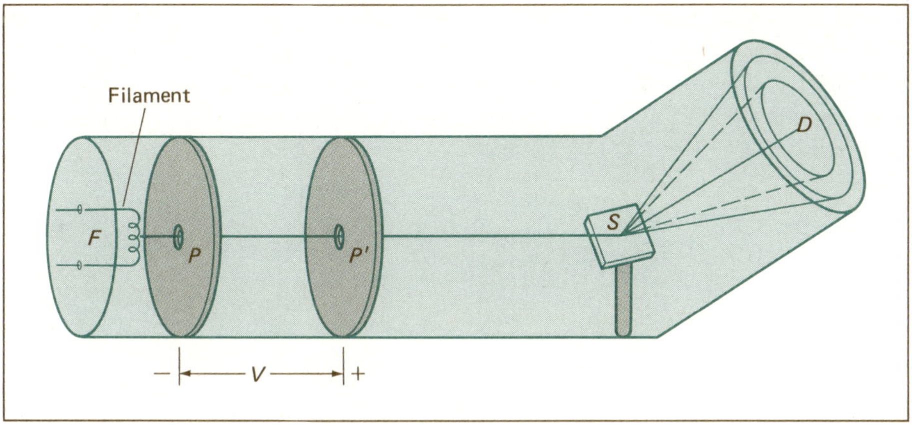

Figure 13.7
Schematic diagram of an electron diffraction device. Electrons from the hot filament *F* are accelerated through a potential drop *V*. The two pinholes *P* and *P′* direct a beam of electrons onto the sample *S*. The intensity of the scattered beam is measured by a detector *D*.

(3) Mass is particulate; it consists of atoms. Radiant energy is particulate; it consists of photons.
(4) Radiant energy has wave properties:

$$\lambda = \frac{c}{\nu}$$

We can complete the analogy between mass and energy by showing that matter has wave properties. Equation [13.7] suggests the possibility. In 1927 (only three years after de Broglie proposed his equation), experimental verification of the wave character of particles was obtained when a beam of electrons was diffracted by the use of the atomic planes of a crystal for a diffraction grating. In such experiments, a beam of electrons is accelerated by passing it through an electrical field (See Figure 13.7). The beam is then directed onto a crystal that serves as a diffraction grating, which enables patterns characteristic of a diffracted wave to be obtained. The velocity of the beam of electrons can be calculated from a knowledge of the voltage difference between the accelerating plates, as shown in Figure 13.7. The wavelength of the electrons, as determined from the diffraction pattern, agrees with the value calculated from equation [13.7].

EXAMPLE 13.3 An electron is accelerated through a voltage difference of 150 volts. Calculate the wavelength of the electron. (The increase in energy of the electron is the product of the charge on the electron e and the voltage difference V through which it is accelerated.)

SOLUTION The energy increase must equal the kinetic energy ($1/2\ mu^2$) of the electron

$$eV = \frac{1}{2}mu^2 \qquad [13.8]$$

(This equation neglects the relatively small velocity the electron may have had before the acceleration.) Combining the relations $p = mu$, $p = h/\lambda$ from equation [13.7], and equation [13.8] gives

$$\lambda = \frac{h}{\sqrt{2meV}}$$

For the electron

$$m = 9.11 \times 10^{-28}\ \text{g}$$

$$e = 1.60 \times 10^{-19}\ \text{coulomb}$$

$$\lambda = \frac{6.63 \times 10^{-27}\ \text{erg-sec}}{\sqrt{(2)(9.11 \times 10^{-28}\ \text{g})(1.60 \times 10^{-19}\ \text{coulombs})(150\ \text{volts})(10^{7}\ \text{ergs/volt-coulomb})^{\dagger}}}$$

$$\lambda = (1.00 \times 10^{-8}\ \text{cm})\left(10^{7}\ \frac{\text{nm}}{\text{cm}}\right) = 0.100\ \text{nm}$$

To produce satisfactory diffraction patterns, the spacing of the lines of the diffraction grating must be the same order of magnitude as the wavelength of the particle being diffracted. The 0.1 nm (1 Å) wavelength of the electrons in Example 13.3 thus requires a grating in which the spacing is as small as the interatomic spacings in crystals.

Diffraction of a neutron beam can also be accomplished with crystals. Both electron and neutron diffraction are now commonly employed in the study of crystal structure. The experimental evidence for the wave nature of small particles is thus well established. It is reasonable to assume, therefore, that the properties of matter can be obtained from a wave function. This is the first of the postulates of the quantum theory.

First postulate of the quantum theory: *Each system has a wave function. The state of that system is fully described by its wave function.*

As discussed earlier, the wave function is a mathematical statement that describes the form of the wave. Earlier in the chapter we wrote the wave function ϕ that describes a simple one-dimensional wave. The wave functions that describe the more complex systems, such as the atoms of the elements, are understandably more complex. The principle, however, is the same. Each system has a wave function, ψ, that describes that system.

The properties of ψ are determined by the other postulates of the quantum theory, which we will discuss later. Before discussing these postulates, however, we need to examine one of the implications of equation [13.7]. According to this equation, the wavelength of a particle is inversely related to its momentum. It is only for atomic-size particles (with very small masses), such as the electron, proton, and neutron, and to a lesser extent atoms and molecules, that the wave

† A volt-coulomb is a joule. One joule = 10^{7} ergs.

properties become important. For larger particles, the wavelength is so small that the wave description is not as useful as the particle description. In other words, *the wave description is useful in describing atomic- and subatomic-size particles.* These are the particles we will be describing in our discussion of atomic and molecular structure.

The problems encountered in applying the wave description to larger particles can be seen by considering the following examples.

EXAMPLE 13.4 Calculate the wavelength of a baseball thrown by a pitcher at 100 miles/hour (4470 cm/sec). The mass of a baseball is 5.00 ounces (142 g).

SOLUTION From equation [13.7]

$$\lambda = \frac{h}{p} = \frac{h}{mu}$$

$$\lambda = \frac{6.63 \times 10^{-27} \text{ erg-sec}}{(142 \text{ g})(4.47 \times 10^{3} \text{ cm/sec})}$$

$$\lambda = 1.04 \times 10^{-32} \text{ cm} = 1.04 \times 10^{-25} \text{ nm}$$

We see from the results of the calculation in Example 13.4 that a grating with an impossibly small spacing (very much smaller than atomic dimensions) would be required to diffract baseball waves.

EXAMPLE 13.5 How slow must a 10.0 g snail be moving to have a wavelength of 700 nm (equivalent to the wavelength of red light)?

SOLUTION From equation [13.7]

$$u = \frac{h}{\lambda m}$$

$$u = \frac{6.63 \times 10^{-27} \text{ erg-sec}}{(7.00 \times 10^{2} \text{ nm})(10^{-7} \text{ cm/nm})(10.0 \text{ g})}$$

$$u = 9.47 \times 10^{-24} \text{ cm/sec} = 3.00 \times 10^{-5} \text{ Å/millenium}$$

The calculation in Example 13.5 makes it clear that wave properties of 10-gram objects are of no practical consequence.

The Heisenberg Uncertainty Principle and the Second Postulate of the Quantum Theory

The **Heisenberg uncertainty principle** is one of the consequences of wave-particle duality. This principle is stated as follows:

> *There is a limit to the accuracy with which one can know both the momentum and position of a particle simultaneously.*

The quantitative statement of this principle is given by the mathematical equation

$$\boxed{(\Delta p)(\Delta x) \geqq \frac{h}{4\pi}} \qquad [13.9]$$

where Δp is the uncertainty in the momentum of the particle and Δx is the uncertainty in the position.

The uncertainty principle does not restrict the accuracy with which we can measure either the position or momentum of a particle. From an examination of equation [13.9], however, we see that the more accurately we measure one of these quantities, the less accurately we know the other.

The uncertainty principle is not important when we deal with macroscopic-size particles. Both the momentum and position of a pendulum or a falling rock can be determined with high accuracy. The uncertainty principle becomes important only when we consider atomic-size particles, as illustrated in the example below:

EXAMPLE 13.6 Compare the uncertainty in the velocity of a one-gram pea and of an electron if the uncertainty in the position in each case is 1.00 Å (1.00×10^{-8} cm).

SOLUTION The minimum uncertainty is

$$(\Delta p)(\Delta x) = \frac{h}{4\pi} \qquad [13.9]$$

$$\Delta p = p_2 - p_1 = mu_2 - mu_1 = m\,\Delta u$$

Combining gives

$$\Delta u = \frac{h}{4\pi m\,\Delta x}$$

For the pea

$$\Delta u = \frac{6.63 \times 10^{-27}\ \text{erg-sec}}{4(3.14)(1.00\ \text{g})(1.00 \times 10^{-8}\ \text{cm})}$$

$$\Delta u = 5.28 \times 10^{-20}\ \text{cm/sec}$$

For the electron

$$\Delta u = \frac{6.63 \times 10^{-27}\ \text{erg-sec}}{4(3.14)(9.11 \times 10^{-28}\ \text{g})(1.00 \times 10^{-8}\ \text{cm})}$$

$$\Delta u = 5.79 \times 10^{7}\ \text{cm/sec}$$

In the example, the uncertainty in the position of the pea is very small in relation to its size. We are usually not interested in knowing the position of a macroscopic-size particle this accurately. Under these conditions, the pea may be moving with a velocity of 5×10^{-20} cm/sec without our knowing it. For the electron, the situation is not so ridiculous. Atomic dimensions are of the order of an angstrom. Specifying the position of an electron to an Å merely restricts it to somewhere within an atom. Even under these conditions, the uncertainty in the velocity of the electron is still very great, about 0.2% of the speed of light.

The proper derivation of equation [13.9] is beyond the scope of this course. We can, however, qualitatively rationalize the uncertainty principle. To determine the position or momentum of an object requires that we interact with it in some manner. This interaction disturbs the object and results in the uncertainty in position or momentum. We can touch the object, throw rocks at it, or bounce sound waves off it. But probably the best way to detect the particle is to examine how the object scatters electromagnetic radiation. Since the momentum of a photon is very small, the electromagnetic radiation would disturb the object less than would the other methods of detection. One limitation in the use of electromagnetic radiation is that it will resolve only those objects that are larger in size than its wavelength. Thus radar waves ($\lambda = 3$ to 20 cm) are used to detect objects such as buildings, airplanes, and mountains; and visible radiation ($\lambda = 4 \times 10^{-4}$ to 8×10^{-4} cm) can be used to detect particles as small as bacteria (10^{-4} cm).

The interaction of microwave or visible radiation with mountains, airplanes, and even microbes is so small that the disturbing effect is of no consequence. This is not the case, however, when a photon is used to determine the position of an atomic-size particle. Here, a very short wavelength photon must be used to give the desired resolution. However, its momentum is large enough to seriously disturb the small particle.

EXAMPLE 13.7 Suppose we try to locate an electron whose velocity is 1.0×10^8 cm/sec to an accuracy of 0.10 Å. To resolve something this small, we use a γ-ray photon with a wavelength of 0.010 Å (0.0010 nm). Compare the momentum of the electron and the γ-ray.

SOLUTION For the photon

$$p = \frac{h}{\lambda} \qquad [13.7]$$

$$p = \frac{6.63 \times 10^{-27}\ \text{erg-sec}}{(0.010\ \text{Å})(10^{-8}\ \text{cm/Å})} = 6.6 \times 10^{-17}\ \text{g-cm/sec}$$

For the electron

$$p = mu = (9.11 \times 10^{-28}\ \text{g})(1.0 \times 10^{8}\ \text{cm/sec})$$

$$p = 9.1 \times 10^{-20}\ \text{g-cm/sec}$$

Thus we see that the momentum of the γ-ray is 1400 times as large as that of the electron. Using a γ-ray to determine the position of an electron, therefore, is roughly equivalent to using a bowling ball to determine the position of a ping-pong ball.

In describing atomic structure, we want to be able to specify very precisely the energy of the electrons. To do this we need to know their momentums. The Heisenberg uncertainty principle does not prevent us from doing this if we sacrifice the knowledge of the position of the electron as it moves within the atom.

Although we may never be able to trace the path of the electron as it moves around in the atom, it is possible to predict the most probable place for finding it. The second postulate of the quantum theory provides the statistical basis for predicting the position of particles. Let us construct a very small box at a specified location in space. The second postulate of the quantum theory tells us how to calculate the probability of finding a particle in this box.

Probability gives us an expectation but not an absolute assurance for success. For example, the probability of drawing an ace of spades out of a deck of fifty-two cards is 1/52 = 0.019. The probability that a card randomly selected from a deck is an ace of spades is, therefore, rather low, but it could happen. On the other hand, the probability that a marble randomly selected from a bag containing 99 red marbles and one blue will be red is high (0.99), but there is still a chance that it will be blue. Probabilities range from zero, which means no chance for success, to unity, where success is certain.

Second postulate of the quantum theory: *The square of the wave function* (ψ^2) *evaluated at a specified location in space is the electron probability density at that location.*

The electron probability density is proportional to the probability of finding an electron at a specified location. A comparison of ψ with the wave function for electromagnetic radiation makes the postulate seem reasonable. We mentioned on page 345 that the square of the wave function for electromagnetic waves gives the energy density of the wave. The interpretation of ψ^2 as the probability density of the electron is analogous. We want to emphasize, however, that ψ^2 cannot be used to calculate the position of the particle, but only the probability of its being in a specified location.

The Third Postulate of the Quantum Theory— The Schrödinger Wave Equation

In a previous section we gave physical significance to the wave function ψ by relating it to the probability of finding the particle. The next postulate of the quantum theory enables us to determine the mathematical form of the wave function and to calculate observable properties of the system, in particular the energy.

In the early days of atomic theory, the electrons in an atom were thought to revolve around the atomic nucleus in orbits similar to the planetary orbits around the sun. From the uncertainty principle we know that we cannot restrict the position of the electron to an orbit. Instead, we describe the position in terms of the probability density (given by ψ^2).

Third postulate of the quantum theory: *The wave properties of a system are governed by a wave equation. The particular wave equation that applies to particles is the Schrödinger wave equation.*

For our immediate purposes we need to know only that solution of the **Schrödinger wave equation** yields values for ψ, the wave function, and E, the total energy. We will make use of the ψ and E information obtained from the wave equation, but leave the solution of the equation that provides ψ and E to more advanced courses.

For complicated systems it is impossible to solve the wave equation exactly. In such cases, approximate solutions and analogies to simpler systems become important.

In summary, the first postulate enables us to describe a system with a wave function. The second postulate gives some meaning to ψ by stating that ψ^2 is the electron probability density. The third postulate is the Schrödinger wave equation. Solution of this equation for any particular system gives ψ and E for that system. The equation has been successfully solved for only very simple systems. Our interest in the equation for the purposes of this chapter will be to describe the solutions obtained for the hydrogen atom.

13.2 Hydrogen Atom

The wave equation has been solved in detail for the hydrogen atom, and from that study much has been learned about the electronic states of all atoms. In this section we will discuss some of the results obtained from solving the Schrödinger equation for the hydrogen atom. We will consider such concepts as quantum numbers, orbitals, energy levels, and spectroscopy.

Quantum Numbers

The solution of the wave equation (for a system in which the motion of the particle is confined, such as the electron in the hydrogen atom) gives rise to three numbers, which are designated as the *n*, *l*, and *m* **quantum numbers.** They all have integral values and are related as follows:

> The quantum number n is a positive integer ($n = 1, 2, 3, \ldots$).
>
> The quantum number l is an integer between 0 and $n - 1$, inclusive.
>
> The quantum number m is any integer from $-l$ to $+l$, inclusive.

As an example, for $n = 3$, l can be either 0, +1, or +2, which are the integers from 0 to $n - 1$; and m can be -2, -1, 0, +1, or +2, which are the integers from $-l$ to $+l$.

Each set of quantum numbers represents a solution of the wave equation. The description of the electron probability obtained from the wave function, ψ (characterized by a given set of quantum numbers), is called an **orbital.** In Chapter 2 we used the common orbital notation of 1*s*, 2*s*, 3*d*, and so on. In that notation the numbers 1, 2, 3, . . . correspond to the values of the quantum number *n*. This number is called the principal quantum number.

The *l* quantum number corresponds to the *s*, *p*, *d*, and *f* notation that we introduced in Chapter 2. The relationship is

l	ORBITAL
0	*s*
1	*p*
2	*d*
3	*f*

The *m* quantum numbers identify three types of *p* (that later we will designate as p_x, p_y, and p_z), five types of *d*, and seven types of *f* orbitals.

The relationship described earlier among the quantum numbers limits the number of allowed states, as illustrated in the following example.

EXAMPLE 13.8 If the principal quantum number is 3, what values can the quantum numbers *l* and *m* have? How many total orbital states are possible for $n = 3$?

SOLUTION The possible values for *l* are the integers from 0 to $(n - 1)$; hence, for $n = 3$, l can be 0, +1, or +2.

The possible values for m are the integers from $-l$ to $+l$; hence, m can be any integer between -2 to $+2$; that is, -2, -1, 0, $+1$, or $+2$. The orbital states are as follows:

n	l	m	ORBITAL DESIGNATION
3	0	0	3*s* (one orbital)
3	1	−1	3*p* (three orbitals)
		0	
		+1	
3	2	−2	3*d* (five orbitals)
		−1	
		0	
		+1	
		+2	

Thus, for $n = 3$ there are three l values, and for the total of three l values there are $1 + 3 + 5 = 9\,m$ values. Hence, there are nine orbitals (one *s*, three *p*'s, and five *d*'s).

The relationships between the orbital designation and the quantum numbers are summarized in Table 13.1.

Energy

The quantum numbers determine the energy of the electrons in the atom. As a result of there being restrictions on the values of the quantum numbers, it fol-

TABLE 13.1

Relationship Between Quantum Number and Orbital Designation

n	l	m	ORBITAL DESIGNATION	NUMBER OF ORBITALS
1	0	0	1*s*	one
2	0	0	2*s*	one
	1	0, ±1	2*p*	three
3	0	0	3*s*	one
	1	0, ±1	3*p*	three
	2	0, ±1, ±2	3*d*	five
4	0	0	4*s*	one
	1	0, ±1	4*p*	three
	2	0, ±1, ±2	4*d*	five
	3	0, ±1, ±2, ±3	4*f*	seven

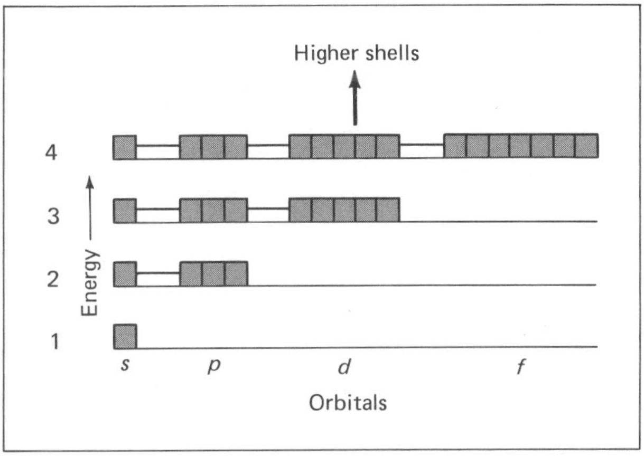

Figure 13.8
Energy levels of the hydrogen atom. The boxes represent orbitals.

lows that there are restrictions on the values of energy that an electron in an atom can have. This is the concept of **quantization of energy.** We sometimes liken quantization to stair steps. The electron in the atom has energy corresponding to the lowest step when it is in its **ground state** (lowest energy level, the normal condition). If an electron in a lower energy level absorbs energy, such as it does when excited in an arc or spark of a spectrograph, it makes a "quantum jump" from one stair step (the ground state) to one of the higher steps (an **excited state**), but it cannot stop between steps.

Solving the hydrogen-atom wave equation for energy gives

$$E_n = \frac{-2\pi^2 me^4}{n^2h^2} \qquad [13.10]$$

Solution of the wave equation for a free electron (one not confined within boundaries) does not lead to discrete energy levels; a free electron is not restricted in the energy it can possess.

where m and e are the mass and charge, respectively, of the electron, h is Planck's constant, and n is the principal quantum number. Notice that E for the H atom does not depend on the quantum numbers l and m. This means that in the hydrogen atom the $2s$ and $2p$ states have the same energy. Also, the $3s$, $3p$, and $3d$ states have the same energy, and the $4s$, $4p$, $4d$, and $4f$ states have the same energy. Orbitals with the same energy are said to be degenerate.

As we will see later, E does depend on the quantum number l in atoms more complex than H.

Figure 13.8 illustrates how the total energy of the electron in a hydrogen atom varies with the quantum number n. We see from Figure 13.8 that the energy of the electron is quantized. Discrete energy levels are available to the electron, and when the electron changes energy, it has to do so in such a way that it makes an abrupt change from one level to another. The energy levels get closer together as n increases, which was predicted in equation [13.10].

EXAMPLE 13.9 Calculate the energy of the electron in a hydrogen atom for one of the $3p$ states.

SOLUTION Equation [13.10] gives the energy

$$E_n = -\frac{2\pi^2 me^4}{n^2h^2}$$

Putting in the values of the constants gives

$1 \text{ esu} = 1 \frac{g^{\frac{1}{2}}cm^{\frac{3}{2}}}{sec}$ and is the unit of charge in the cgs system of units.

$$E_n = -\frac{2\pi^2(9.110 \times 10^{-28} \text{ g})(4.803 \times 10^{-10} \text{ esu})^4}{n^2(6.626 \times 10^{-27} \text{ erg-sec})^2}$$

$$E_n = -\frac{2.179 \times 10^{-11}}{n^2} \text{ ergs} \qquad [13.11]$$

For a $3p$ orbital, $n = 3$. Hence,

$$E_n = -\frac{2.179 \times 10^{-11}}{3^2} \text{ ergs} = -2.421 \times 10^{-12} \text{ ergs}$$

Electronic energies are often reported in electron volts (eV), and in these units

$$E_n = -(2.421 \times 10^{-12} \text{ ergs})\left(\frac{1 \text{ eV}}{1.6021 \times 10^{-12} \text{ erg}}\right) = -1.511 \text{ eV}$$

Energy difference, ΔE, is also easily calculated. For a change from n_1 to n_2,

$$\Delta E = E_2 - E_1 = \frac{2\pi^2 me^4}{h^2}\left(\frac{1}{n_1^2} - \frac{1}{n_2^2}\right) \qquad [13.12]$$

and as n_1 and n_2 get larger, ΔE gets smaller. When $n = 1$, the H atom is in its ground (stable) state. When $n = 2, 3, 4, \ldots$ the H atom is in an excited state. An interesting quantity of energy is that which is necessary to remove the electron completely from the atom. This energy, called the *ionization energy*, can be calculated from equation [13.12] by letting $n_1 = 1$ and $n_2 = \infty$. It has a value of 2.179×10^{-11} ergs for the hydrogen atom.

We will show later that as n increases, the average distance of the electron from the nucleus increases. When $n = \infty$, this average distance becomes infinitely large and the atom is ionized.

Atomic Spectroscopy for Hydrogen

Earlier, we described the production of line spectra. When an element such as hydrogen is heated to incandescence or excited by an electric spark, the atom emits electromagnetic radiation. If this radiation is dispersed (separated by wavelength) in a spectrograph (see Figure 13.6), a line spectrum is obtained (as shown in Figure 13.5).

The source of the spectrum is believed to be the energy emitted when an electron that has been promoted to a higher energy level falls back to a lower energy state. Since only quantized energy states are possible in the atom, only very specific wavelengths of radiation (corresponding to the position of the spectral lines) are emitted. For hydrogen, the wavelength can be readily calculated with the aid of equation [13.12].

EXAMPLE 13.10 Calculate the wavelength of light produced when the hydrogen atom in the $n = 2$ state changes to the $n = 1$ state.

SOLUTION The photon emitted must have an energy equal to the energy difference between the levels. This wavelength is calculated by combining equations [13.1], [13.4], and [13.12].

$$E_2 - E_1 = h\nu = \frac{hc}{\lambda}$$

$$E_2 - E_1 = -\frac{2\pi^2 me^4}{h^2}\left(\frac{1}{(2)^2} - \frac{1}{(1)^2}\right)$$

$$= -2.18 \times 10^{-11}(\tfrac{1}{4} - 1) = 1.63 \times 10^{-11} \text{ ergs}$$

$$E_2 - E_1 = \frac{hc}{\lambda} = 1.63 \times 10^{-11} \text{ ergs}$$

$$\lambda = \frac{(6.63 \times 10^{-27} \text{ erg-sec})(3.00 \times 10^{10} \text{ cm/sec})}{1.63 \times 10^{-11} \text{ erg}}$$

$$= 1.22 \times 10^{-5} \text{ cm}$$

Figure 13.9 summarizes the energy levels for atomic hydrogen and also gives the wavelength of the radiation emitted (or absorbed) when the electron

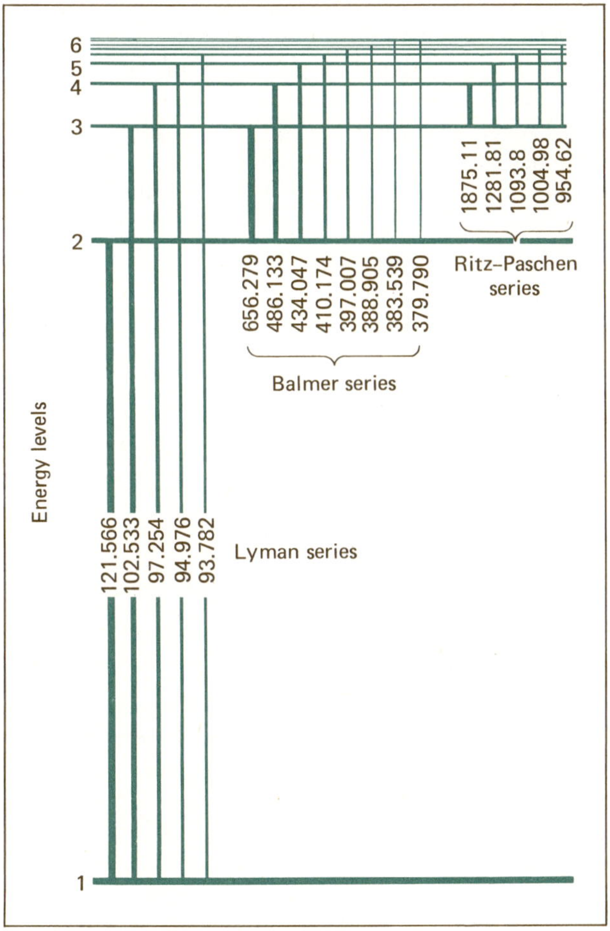

Figure 13.9
Energy levels of the hydrogen atom. The connecting lines indicate the transitions between energy levels; the wavelength of the spectral line resulting from the transition is given in nanometers.

changes energy levels. The emission spectrum of hydrogen results when excited atoms lose energy and go to a lower level. The series of lines that results when the final state is $n = 1$ is in the ultraviolet part of the spectrum and is known as the Lyman series. Visible radiation is emitted when the final state is $n = 2$; these spectral lines make up the Balmer series. When the final state is $n = 3$, infrared radiation results, and this is known as the Ritz-Paschen series.

The determination of wavelength is one of the most accurate of all experimental measurements. It is no accident that the metric unit of length is now defined in terms of the wavelength of one of the spectral lines of cadmium.

Spectroscopy provides one of the best verifications of the quantum theory. The wavelengths of the spectral lines calculated for hydrogen as shown in Example 13.10 agree with experimentally measured wavelengths as closely as they can be measured.

Orbital Geometry

The electron probability distribution in an atom is referred to as the **orbital geometry** and is predicted from ψ^2. We will first discuss a system of coordinates called **spherical polar coordinates,** used in describing the geometry of orbitals. It is convenient when working with wave equations to describe the location of a point in space by giving its spherical polar coordinates, a distance and two angles (r, θ, ϕ), instead of the Cartesian coordinates (x, y, z) with which you may be more familiar. The relationship between these two coordinate systems is shown in Figure 13.10. If point 0 in the figure is the center of the atom, and point P is the location of an electron, we could describe the location of the electron by giving its distance from the origin along the x, y, and z axes, or, alternatively, by giving the angles θ and ϕ together with the radial distance r. It is the latter system of polar coordinates that we find convenient for this problem.

Table 13.2 gives the wave functions, expressed in spherical polar coordinates, for the ground state (1*s*) and some of the low-lying excited states (2*s*, $2p_x$, . . .) for the H atom. We will use the wave functions to deduce the geometry of the orbitals.

Figure 13.10
Cartesian and spherical polar coordinates of the point P.

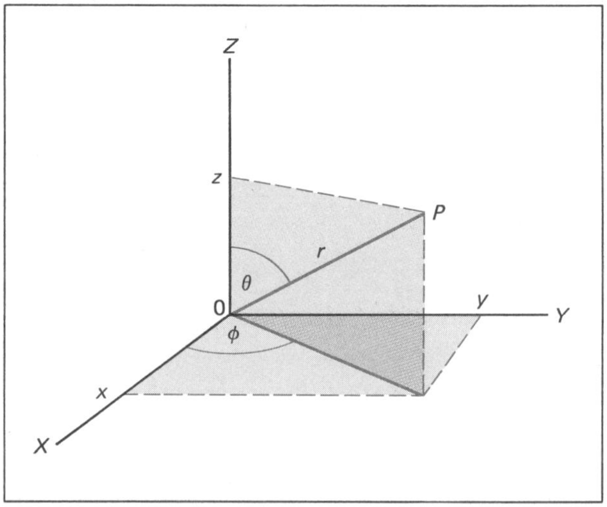

TABLE 13.2

Wave Functions for One-Electron Atoms

$$\psi(1s) = N_1 e^{-r/a_0}$$
$$\psi(2s) = N_2(2 - r/a_0)e^{-r/2a_0}$$
$$\psi(2p_z) = N_2(r/a_0)e^{-r/2a_0}\cos\theta$$
$$\psi(2p_x) = N_2(r/a_0)e^{-r/2a_0}\sin\theta\cos\phi$$
$$\psi(2p_y) = N_2(r/a_0)e^{-r/2a_0}\sin\theta\sin\phi$$
$$N_1 = \left(\frac{1}{\pi a_0^3}\right)^{\frac{1}{2}}; \quad N_2 = \frac{1}{4}\left(\frac{1}{2\pi a_0^3}\right)^{\frac{1}{2}}; \quad a_0 = \frac{h^2}{4\pi^2 me^2}$$

The *s* Orbitals

Consider first the 1*s* orbital. To determine its probability density function or orbital geometry, ψ^2(ls), we square its wave function as given in Table 13.2:

$$\psi^2(1s) = \frac{1}{\pi a_0^3} e^{-2r/a_0}$$

Examining this equation, we see that the electron probability density does not depend on the *direction* of a line extending out from the nucleus. The electron probability density at any given distance from the nucleus is the same regardless of which direction one takes. We know this because the angles θ and ϕ (spherical polar coordinates) do not appear in the wave function of any of the *s* orbitals. The result is that contours of equal probability density for an *s* orbital are spherical.

A plot of ψ^2(ls) against the distance *r* from the nucleus (see Figure 13.11) shows that the electron probability density is greatest at the nucleus and approaches zero probability as the distance from the nucleus increases.

Since ψ^2 is not zero for any finite value of *r*, the possibility exists that the electron can be found any distance from the nucleus. This is quite a different picture from the planetary model used in the early part of the century when electrons were believed to rotate in fixed orbits about the nucleus.

Part (b) of Figure 13.11 shows another way of visualizing the probability of locating an electron in a ls orbital. Here, the density of the dotted area (the number of dots per unit area) is used to represent the probability of finding an electron. The volume of the atom can be defined as that volume enclosing some arbitrarily decided percentage (for example, 99%) of the dots [part (c)].

From the information in Table 13.2, we can determine for a 2*s* orbital that

$$\psi^2(2s) = \frac{1}{32\pi a_0^3}\left(2 - \frac{r}{a_0}\right)^2 e^{-r/a_0}$$

The exponential term causes the probability to fall off more gradually with an increase in *r* in this case than does the exponential term in the equation for the

$$a_0 = \frac{h^2}{8\pi^2 me^2}$$

where *h* is Planck's constant, *m* is the mass of the electron, and *e* is the electronic charge. Putting in the values for the constants gives

$$a_0 = 0.5292 \times 10^{-8} \text{ cm}$$

Thus a_0 is a distance. It is called the Bohr radius since it is exactly the distance from the nucleus to the electron for a ground-state hydrogen atom in the old Bohr theory, in which the electron was assumed to be moving in a circular orbit around the nucleus. Knowing now about the uncertainty principle, we no longer accept the Bohr theory, which presumed that we could locate the precise position and momentum of the electron. In the wave-mechanical model, a_0 becomes the most probable electron-proton distance in the ground-state hydrogen atom.

The descriptions we give for energy and orbital geometry for the H atom can easily be extended to any system containing one positively charged particle interacting with a single negatively charged particle. Examples are He^+, Li^{2+}, Be^{3+}, ... All we need do is take into account the larger charge on the nucleus. In general, the energy is given by

$$E_n = \frac{-2\pi^2 mZ^2e^4}{n^2h^2}$$

where *Z* is the atomic number of the nucleus. ($Z = 1$ for H, 2 for He, 3 for Li, . . .). Similar adjustments must be made in the expressions for the wave functions. For example,

$$\psi(1s) = \left(\frac{Z^3}{\pi a_0^3}\right)^{\frac{1}{2}} e^{-Zr/a_0}$$

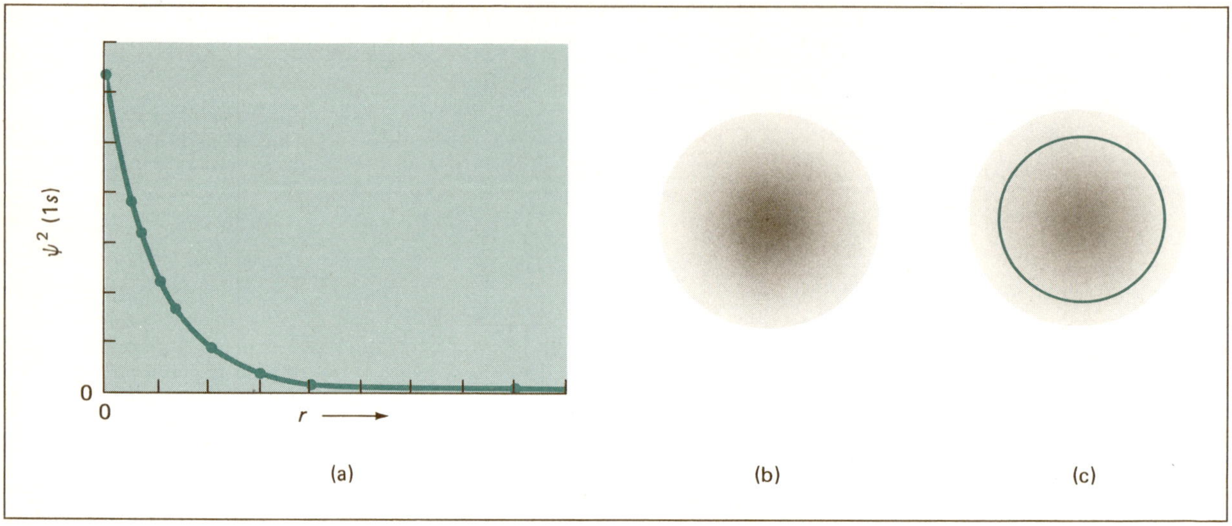

Figure 13.11
Probability density for a 1s electron at a distance r *from the nucleus.* Parts (b) and (c) represent cross sections through the three-dimensional atom. See text for explanation.

1*s* orbital. Hence, the electron in a 2*s* orbital is on the average farther away from the nucleus than is the 1*s* electron, and the 2*s* orbital is the larger orbital (in the sense in which we talked about size above). An interesting feature of the 2*s* orbital is that the wave function vanishes at $r = 2a_0$, that is, when the term $(2 - r/a_0)$ is zero. At this distance, ψ^2 will also become zero, and the probability density at $2a_0$ is zero. The picture shown in Figure 13.12 results. The set of points at which ψ^2 is zero is called a *node*. In this case, the node is the surface of a sphere at a distance $2a_0$ from the center of the atom and is known as a radial node.

Figure 13.12
*Two ways of showing the probability density of a 2*s *electron at a distance* r *from the nucleus.*

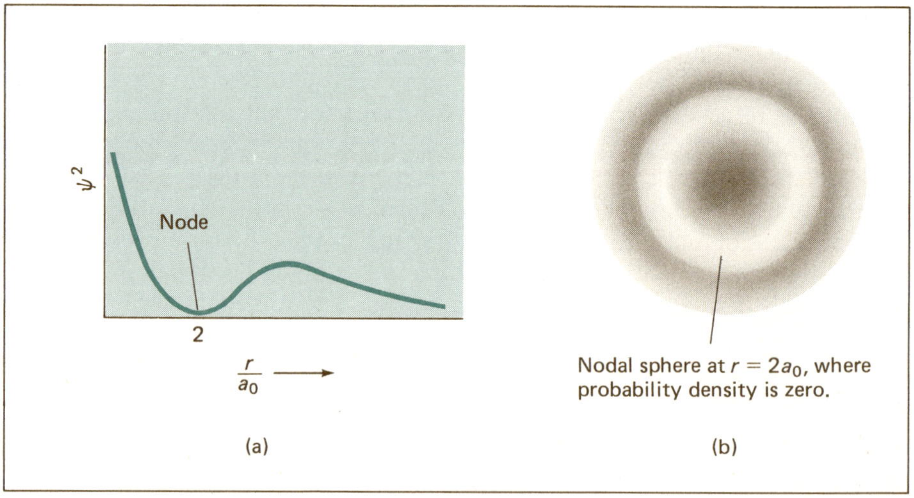

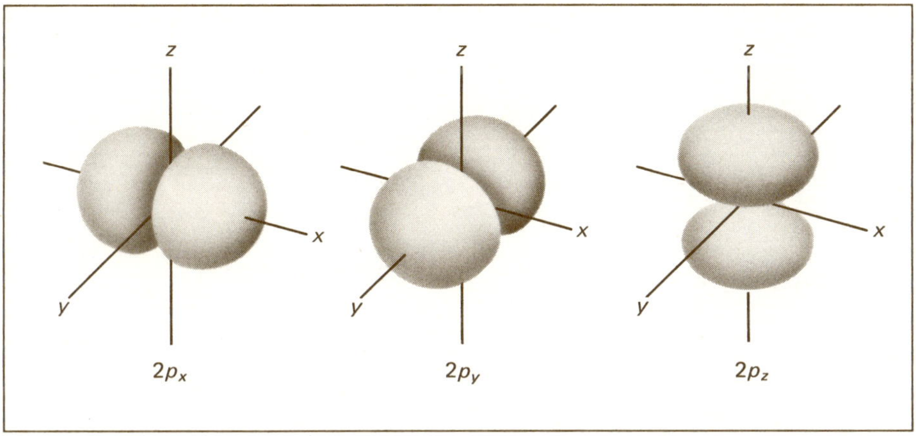

Figure 13.13
The 2p orbitals of the hydrogen atom.

If we examined $\psi(3s)$ and $\psi(4s)$, we would see that the 3*s* orbital has two radial nodes and the 4*s* orbital has three. In general, there are $n - 1$ radial nodes in the *s* orbitals. In each instance, ψ changes sign as it goes through the node.

The sizes of the *s* orbitals increase with increasing principal quantum numbers. On the average, the electron is farther from the nucleus in 4*s* than in 3*s*, and farther in 3*s* than in 2*s*.

The *p* Orbitals

Consider next the geometry of the *p* orbitals. We need take only a brief look at the 2*p* wave functions to see that ψ depends on the angles θ and ϕ, and thus ψ^2 will have directional properties. There are three 2*p* orbitals, designated $2p_x$, $2p_y$, and $2p_z$. A detailed analysis shows that the electron probability density is greatest along the *x*-axis of the $2p_x$ orbital, the *y*-axis of $2p_y$, and the *z*-axis of $2p_z$ (see Figure 13.13). The three orbitals differ only in their orientation relative to one another.

As with the 2*s* orbitals, the 2*p* orbitals contain a node—in this case a nodal plane. For the $2p_z$ orbital, the *xy* plane is a nodal plane. For the $2p_x$ orbital, it is the *yz* plane where $\psi^2 = 0$, and for $2p_y$, it is the *xz* plane. We call these nodes angular nodes since they occur in planes oriented at specified angles.

The *d* Orbitals

The five *d* orbitals also show angular dependence (see Figure 13.14). In three of these, d_{xy}, d_{xz}, and d_{yz}, the regions of high probability density are between the axes. The subscripts tell in which plane the orbitals have their greatest electron density. In the other two orbitals, the regions of high probability density are positioned along the axes. The $d_{x^2-y^2}$ orbital has its greatest density along the *x* and *y* axes, and the d_{z^2} orbital along the *z* axes.

The *d* orbitals each contain two angular nodes. For example, in d_{xy}, the *yz* and *xz* planes are nodal planes. In general, an orbital contains *l* angular nodes ($l = 1$ for *p* orbitals, and $l = 2$ for *d* orbitals) and $n - 1$ total nodes, and hence, $n - l - 1$ radial nodes.

Radial Probability Distribution

In addition to ψ^2, there is another probability function, known as the **radial probability distribution,** that is important. This function is quite different from the probability density discussed previously. Whereas ψ^2 is proportional to the probability of locating an electron at a particular location in space, the radial distribution function is related to the probability of locating an electron

The radial probability distribution is equal to the product of (1) the square of the *r*-dependent terms of the wave function and (2) the term $4\pi r^2$, which is the surface area

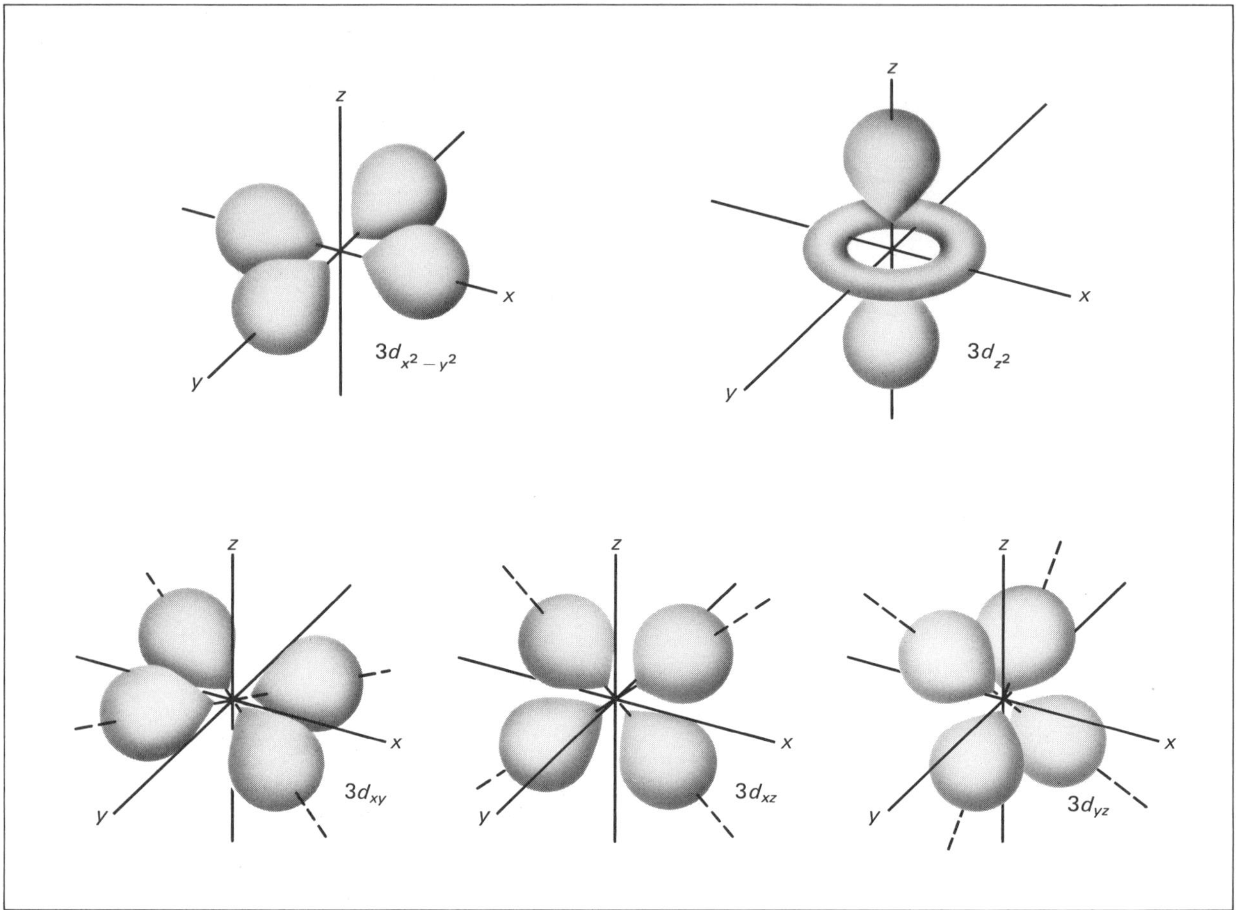

Figure 13.14
The 3d orbitals of the hydrogen atom.

of a sphere. The probability of finding an electron in a spherical shell r distance from the nucleus is $4\pi r^2 R^2(dr)$, where R is the radial dependent part of the wave function and $4\pi r^2(dr)$ is the volume of a very thin spherical shell of thickness dr. The shell must be thin since R^2 varies with r, and for a shell of finite thickness, R^2 on the outside of the shell will be different than on the inside.

The 1s orbital has a maximum radial probability at $r = a_0$.

anywhere in a thin spherical shell r distance from the nucleus. Figure 13.15 shows plots of the radial distribution function versus distance from the nucleus for several orbitals.

The maximum in the radial probability distribution gives the most probable distance from the nucleus of the electron. It is apparent from Figure 13.15 that a 2s electron is more likely to be found farther from the nucleus than a 1s electron, and that the most probable location of a 3s electron is yet farther removed from the nucleus. Note that the probability of finding an electron in a *shell* close to the nucleus is very low. This is because the volume of the shell is very small near the nucleus.

The relative abilities of s, p, and d electrons to closely approach the nucleus should be noted. This affects the chemistry of the elements, and is responsible for the difference in energy of s, p, d, and f orbitals in multielectron atoms, which we discuss next.

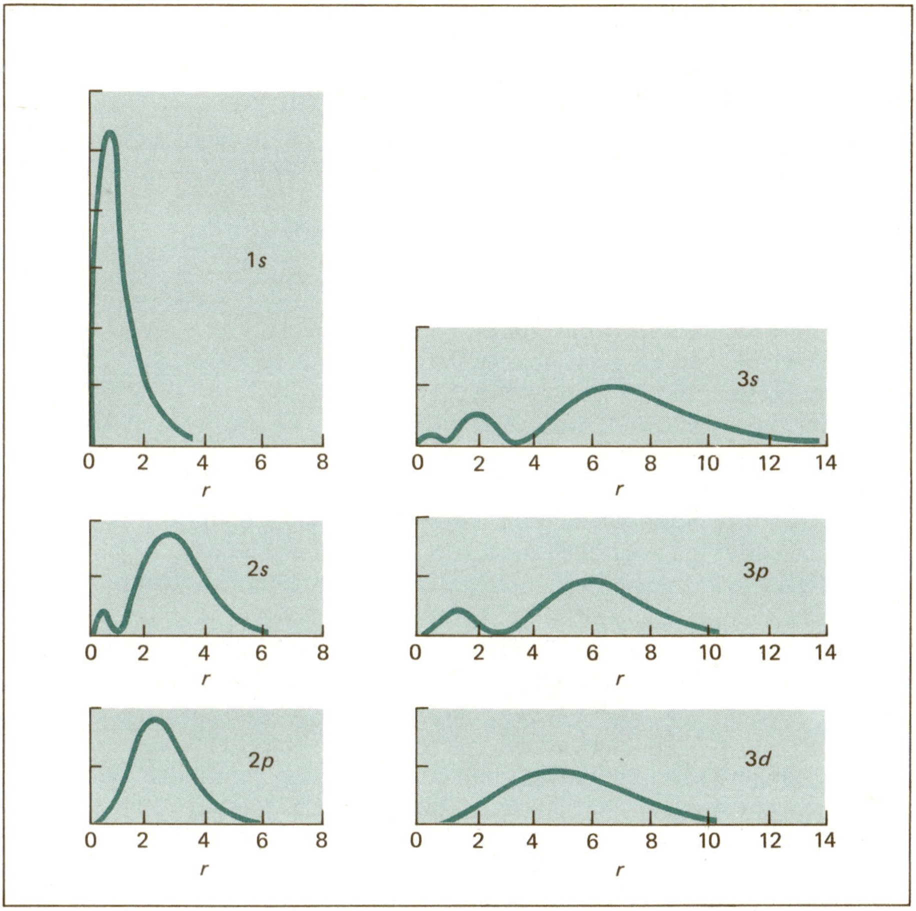

Figure 13.15
Radial probability density for low-lying orbitals of the hydrogen atom.

13.3 Multielectron Atoms

Orbital Description

In theory, we should be able to apply the wave description to any atomic system. However, rigorous solution of the Schrödinger equation for a system even as simple as the helium atom would be extremely difficult, and complete solutions of the equation for the other one hundred plus elements would seem to be beyond the realm of possibility.

It is important to realize that it is the difficulty of the mathematics and not a breakdown in the theory that prevents one from obtaining rigorous analytical expressions for heavier atoms. There are many indications that the quantum theory does apply to every atomic system. For example, line spectra are obtained for all the elements, indicating that the electrons are in quantized energy levels. Furthermore, although analytical expressions cannot be obtained for ψ and E, numerical calculations can be made. These calculations are highly complex and require a large high-speed computer. They have been made, however, and the resulting values for the energy levels are in agreement with those determined from the line spectra.

The experimental evidence suggests that all the elements have orbitals similar to those in hydrogen; for example, 1s, 2s, $2p_x$, ... One important difference between the orbital descriptions of the H atom and the multielectron atoms is that the orbitals that were degenerate in hydrogen (such as 2*s* and the 2*p*'s) have different energies in multielectron atoms. This occurs because some types of orbitals penetrate closer to the nucleus than do others. For example, as was shown in Figure 13.15, the 3*s* orbital, principally because of the two small peaks of high electron density, has high density close to the nucleus. The same is true, but to a somewhat lesser extent, of the 3*p* orbital. The 3*d* orbital, on the other hand, does not have high electron density peaks near the nucleus.

In a multielectron atom, such as $_{11}$Na, the 3*s* electron effectively penetrates close to the nucleus and feels the effect in varying degrees of the full +11 charge on the nucleus. If the valence electron in Na were excited to a 3*d* state, however, the nucleus would be almost completely shielded by the other 10 electrons, so that the 3*d* electron would feel the effect of only a +1 charge. A valence electron in a 3*p* orbital would be intermediate in its penetrating ability to the 3*s* and 3*d* electrons. Hence, the 3*s* orbital is at a lower energy level than that of the 3*p*, which in turn is lower than the 3*d*. These penetration and shielding effects are the explanation for the differences in energy of *s*, *p*, *d*, and *f* orbitals of the same principal quantum number. The same effect is operative in all multielectron atoms, and the result is a set of energy levels like those shown in Figure 13.16 for potassium.

Notice that screening does not destroy the degeneracy of orbitals with the same *n and l* values. The 2*p* orbitals are still degenerate, as are the 3*p*'s, 3*d*'s, 4*p*'s, and so on.

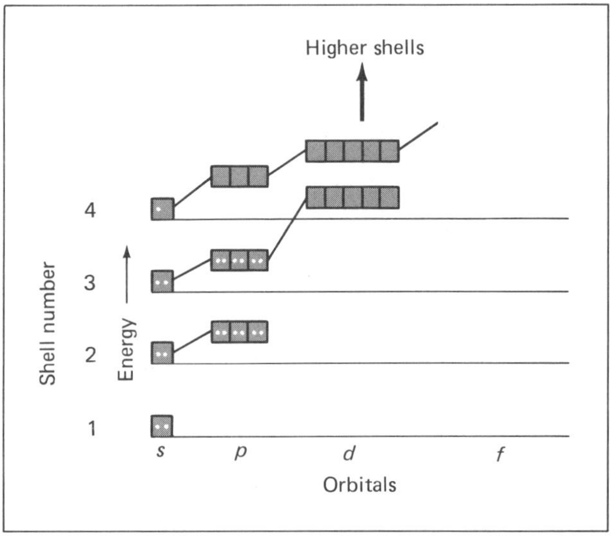

Figure 13.16
Schematic representation of the energy of the orbitals in the multielectron atom, $_{19}$K, showing the energy splitting between s, p, d, *and* f *states.*

Electronic Configuration

With an energy-level diagram such as that shown in Figure 13.16 and with the **Pauli exclusion principle,** we can predict the electronic structure of the elements. The Pauli exclusion principle states that *no two electrons in an atom can have the same set of quantum numbers.* The quantum numbers referred to are *n*, *l*, *m*, as well as a fourth quantum number designated by the symbol *s* and known as the **spin quantum number.** An electron in an atom behaves as if it were spinning. This motion generates a magnetic moment along the axis of rotation. The electron can spin in one of two opposite directions (equivalent to clockwise or counterclockwise); this spin then directs the magnetic moment in one of two opposite directions. If the electron is placed in a magnetic field, the spinning axis is aligned with, or against, the north pole of the magnet, depending on the direction of the magnetic moment. The spin quantum number makes the distinction between these two conditions. It has a value of $+1/2$ or $-1/2$.

We are now ready to use the **aufbau principle** to predict atomic structure. In this imaginary scheme, an electron and a proton (plus necessary neutrons) are added to an atom to form the atom with the next larger atomic number. This extra electron is added to an orbital with the lowest available energy in agreement with the Pauli exclusion principle. Let us start with hydrogen and see how it works. Hydrogen has a single electron, which we put in the 1*s* orbital (see Figure 13.17). We will represent this electronic configuration as

$$\text{H} \quad 1s$$

For helium we have a second electron, which can also go in the 1*s* orbital (see Figure 13.18). The two electrons have the quantum numbers *n*, *l*, *m*, and *s* of

Figure 13.17
In hydrogen, the single electron represented by the dot in the figure is placed in the 1*s* orbital.

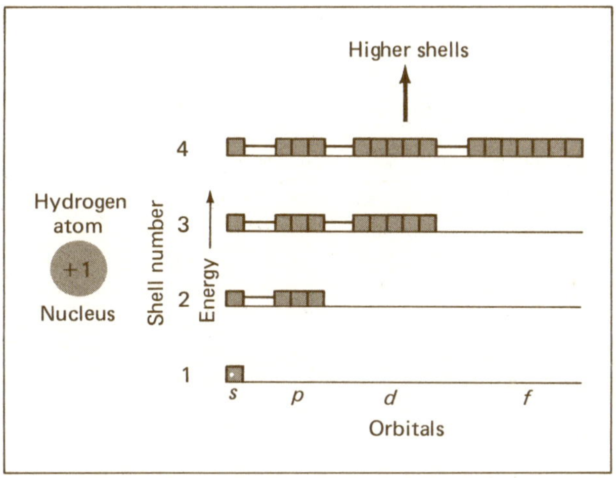

Figure 13.18
The two electrons in helium are placed in the 1*s* orbital (the dots represent the electron).

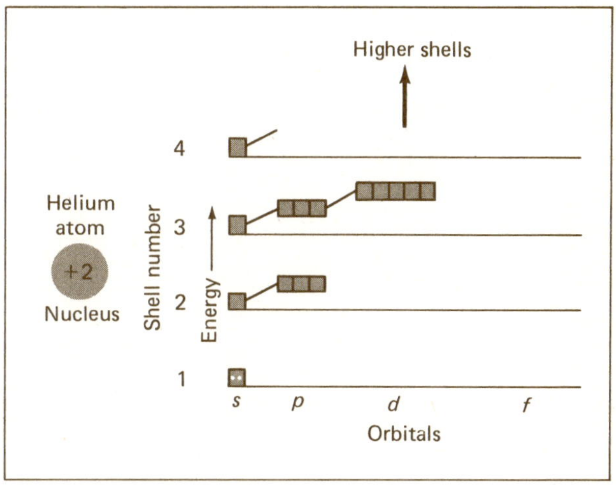

Figure 13.19
Occupied orbitals in Li.

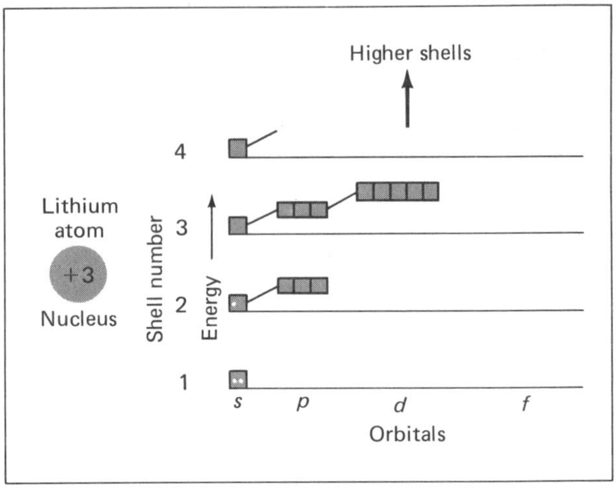

1, 0, 0, and 1/2, and 1, 0, 0, and −1/2, and the Pauli principle is not violated. We represent the electronic configuration as

$$\text{He} \quad 1s^2$$

The superscript "2" indicates that there are two electrons in the ls orbital.

Lithium is next. The extra electron (over helium) must go into the next higher energy orbital, the 2*s* (see Figure 13.19). The result is the electronic configuration

$$\text{Li} \quad 1s^2 2s$$

We can carry this scheme on to predict the configuration of all the elements. Some complications occur when the orbitals from different shells (different principle quantum numbers) overlap, but they are not difficult to handle.

The Pauli exclusion principle limits the population of each orbital to two electrons.

Note in Figure 13.16 that shielding effects for potassium have raised the energy of the 3*d* orbitals so they are higher than the 4*s* orbitals. Hence, potassium has the ground state configuration

$$\text{K} \quad 1s^2 2s^2 2p^6 3s^2 3p^6 4s$$

Through the aufbau principle, the orbitals are filled in the order 1*s*, 2*s*, 2*p*, 3*s*, 3*p*, 4*s*, 3*d*, ... Mnemonic schemes for remembering and predicting the order of filling were given in Chapter 2.

Note that in partially filled subshells, such as the 3*d* for titanium or the 3*p* for sulfur, the electrons are put in different orbitals. Thus, for

EXAMPLE 13.11 Predict the electronic configurations of the elements S($Z = 16$), Ar($Z = 18$), Mg($Z = 20$), and Ti($Z = 22$). (Z = atomic number.)

SOLUTION The electrons are added to the orbitals in order of increasing energy until the electrons are all included. The results are shown in Figure 13.20.

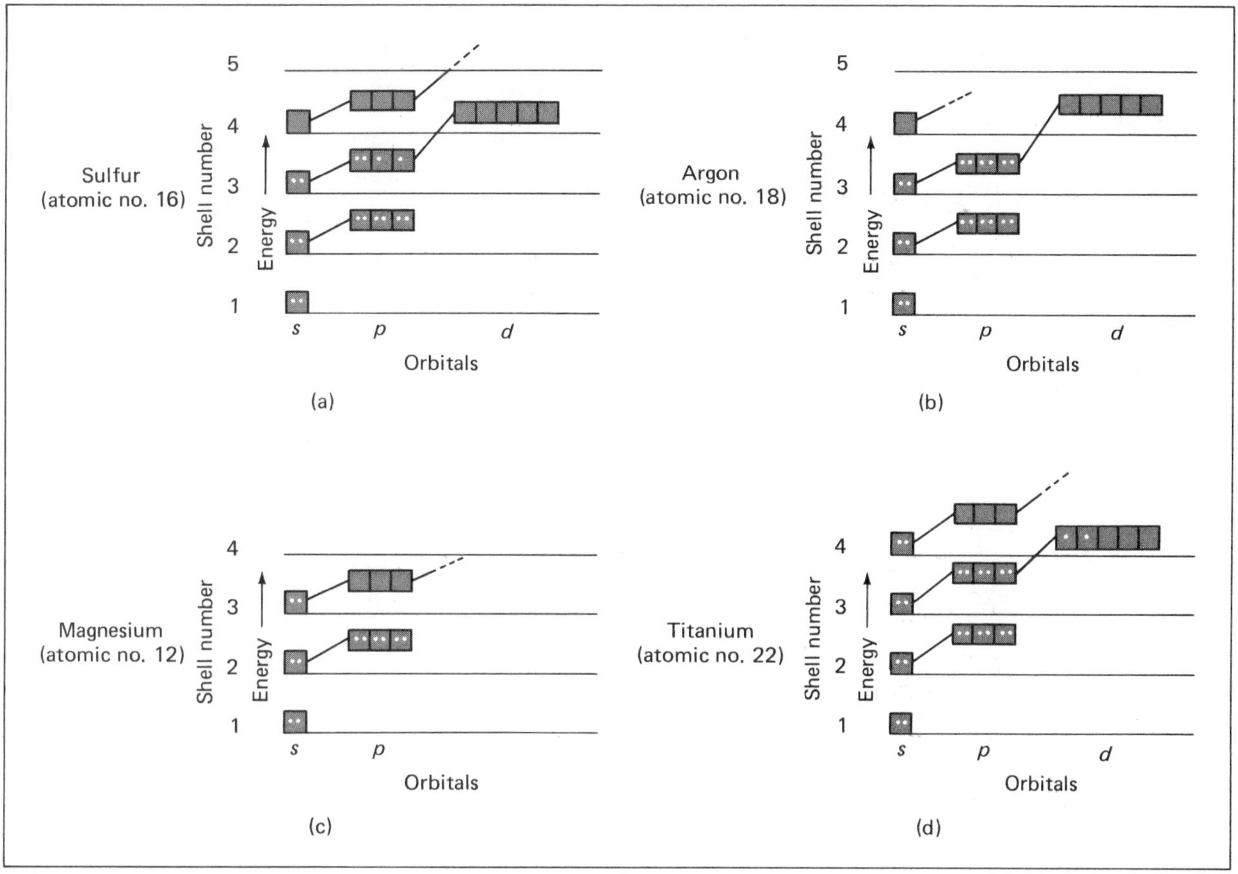

Figure 13.20
Filled orbitals in (a) S, (b) Ar, (c) Mg, and (d) Ti.

The configurations are:

S	$1s^2 2s^2 2p^6 3s^2 3p^4$
Ar	$1s^2 2s^2 2p^6 3s^2 3p^6$
Mg	$1s^2 2s^2 2p^6 3s^2$
Ti	$1s^2 2s^2 2p^6 3s^2 3p^6 3d^2 4s^2$

Usually we simplify the notation by writing the electronic configuration in terms of the preceding noble gas. For example, neon has the configuration $1s^2 2s^2 2p^6$. The above elements would be written

S	$[Ne]3s^2 3p^4$
Ar	$[Ne]3s^2 3p^6$
Mg	$[Ne]3s^2 3p^6$
Ti	$[Ar]3d^2 4s^2$

titanium we get

$3d$ 

instead of

$3d$ 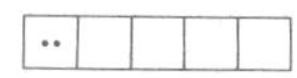

Electrons in separate orbitals can have the same spin, which is a lower energy condition. When there is a subshell of equivalent orbitals, the electrons go singly into each orbital before pairing.

TABLE 13.3

Electronic Configuration of Gaseous Atoms

ATOMIC NUMBER	ELEMENT	ELECTRONIC CONFIGURATION	ATOMIC NUMBER	ELEMENT	ELECTRONIC CONFIGURATION
1	H	$1s$	45	Rh	—$4d^8 5s$
2	He	$1s^2$	46	Pd	—$4d^{10}$
3	Li	[He]$2s$	47	Ag	—$4d^{10}5s$
4	Be	—$2s^2$	48	Cd	—$4d^{10}5s^2$
5	B	—$2s^2 2p$	49	In	—$4d^{10}5s^2 5p$
6	C	—$2s^2 2p^2$	50	Sn	—$4d^{10}5s^2 5p^2$
7	N	—$2s^2 2p^3$	51	Sb	—$4d^{10}5s^2 5p^3$
8	O	—$2s^2 2p^4$	52	Te	—$4d^{10}5s^2 5p^4$
9	F	—$2s^2 2p^5$	53	I	—$4d^{10}5s^2 5p^5$
10	Ne	—$2s^2 2p^6$	54	Xe	—$4d^{10}5s^2 5p^6$
11	Na	[Ne]$3s$	55	Cs	[Xe]$6s$
12	Mg	—$3s^2$	56	Ba	—$6s^2$
13	Al	—$3s^2 3p$	57	La	—$5d6s^2$
14	Si	—$3s^2 3p^2$	58	Ce	—$4f^2 6s^2$
15	P	—$3s^2 3p^3$	59	Pr	—$4f^3 6s^2$
16	S	—$3s^2 3p^4$	60	Nd	—$4f^4 6s^2$
17	Cl	—$3s^2 3p^5$	61	Pm	—$4f^5 6s^2$
18	Ar	—$3s^2 3p^6$	62	Sm	—$4f^6 6s^2$
19	K	[Ar]$4s$	63	Eu	—$4f^7 6s^2$
20	Ca	—$4s^2$	64	Gd	—$4f^7 5d6s^2$
21	Sc	—$3d4s^2$	65	Tb	—$4f^9 6s^2$
22	Ti	—$3d^2 4s^2$	66	Dy	—$4f^{10}6s^2$
23	V	—$3d^3 4s^2$	67	Ho	—$4f^{11}6s^2$
24	Cr	—$3d^5 4s$	68	Er	—$4f^{12}6s^2$
25	Mn	—$3d^5 4s^2$	69	Tm	—$4f^{13}6s^2$
26	Fe	—$3d^6 4s^2$	70	Yb	—$4f^{14}6s^2$
27	Co	—$3d^7 4s^2$	71	Lu	—$4f^{14}5d6s^2$
28	Ni	—$3d^8 4s^2$	72	Hf	—$4f^{14}5d^2 6s^2$
29	Cu	—$3d^{10}4s$	73	Ta	—$4f^{14}5d^3 6s^2$
30	Zn	—$3d^{10}4s^2$	74	W	—$4f^{14}5d^4 6s^2$
31	Ga	—$3d^{10}4s^2 4p$	75	Re	—$4f^{14}5d^5 6s^2$
32	Ge	—$3d^{10}4s^2 4p^2$	76	Os	—$4f^{14}5d^6 6s^2$
33	As	—$3d^{10}4s^2 4p^3$	77	Ir	—$4f^{14}5d^7 6s^2$
34	Se	—$3d^{10}4s^2 4p^4$	78	Pt	—$4f^{14}5d^9 6s$
35	Br	—$3d^{10}4s^2 4p^5$	79	Au	—$4f^{14}5d^{10}6s$
36	Kr	—$3d^{10}4s^2 4p^6$	80	Hg	[Au]$6s^2$
37	Rb	[Kr]$5s$	81	Tl	—$6s^2 6p$
38	Sr	—$5s^2$	82	Pb	—$6s^2 6p^2$
39	Y	—$4d5s^2$	83	Bi	—$6s^2 6p^3$
40	Zr	—$4d^2 5s^2$	84	Po	—$6s^2 6p^4$
41	Nb	—$4d^4 5s$	85	At	—$6s^2 6p^5$
42	Mo	—$4d^5 5s$	86	Rn	—$6s^2 6p^6$
43	Tc	—$4d^5 5s^2$	87	Fr	[Rn]$7s$
44	Ru	—$4d^7 5s$	88	Ra	—$7s^2$

TABLE 13.3 (*Cont.*)

Electronic Configuration of Gaseous Atoms

ATOMIC NUMBER	ELEMENT	ELECTRONIC CONFIGURATION	ATOMIC NUMBER	ELEMENT	ELECTRONIC CONFIGURATION
89	Ac	$—6d7s^2$	97	Bk	$—5f^97s^2$
90	Th	$—6d^27s^2$	98	Cf	$—5f^{10}7s^2$
91	Pa	$—5f^26d7s^2$	99	Es	$—5f^{11}7s^2$
92	U	$—5f^36d7s^2$	100	Fm	$—5f^{12}7s^2$
93	Np	$—5f^46d7s^2$	101	Md	$—5f^{13}7s^2$
94	Pu	$—5f^67s^2$	102	No	$—5f^{14}7s^2$
95	Am	$—5f^77s^2$	103	Lr	$—5f^{14}6d7s^2$
96	Cm	$—5f^76d7s^2$			

If you examine Table 13.3 in detail, you will notice some apparent exceptions to the predictions from the aufbau principle. For example, chromium has the configuration $[Ar]3d^54s$ rather than $[Ar]3d^44s^2$. There are five $3d$ orbitals. The $3d^5$ configuration puts one electron in each orbital and gives what is referred to as a "half-filled subshell." Half-filled subshells and filled subshells appear to be especially stable and are of lower energy. Hence, the last electron goes into the $3d$ orbital to give $3d^54s$ rather than into the $4s$ orbital to give $3d^44s^2$. A comparable example with a filled subshell is copper, where the configuration is $[Kr]3d^{10}4s^1$ rather than $[Kr]3d^94s^2$.

Table 13.3 summarizes the electronic configurations of all the elements. This table explains the periodic table of the elements and will be useful in later chapters where we consider periodic properties and chemical bonding.

13.4 A Brief Thirty-Year History

Our present concept of the wave-mechanical model of the atom developed over a period of time extending from about 1900 to 1930, thirty years of some of the most exciting and revolutionary thinking in the history of science. Let us review in chronological order some of these events.

The quantum theory of energy was born in 1900 when Max Planck (1858–1947) proposed that certain paradoxes in the classical physics of light could be resolved if one assumed that radiant energy exists in the form of small discreet quanta with the property $\epsilon = h\nu$. This proposal satisfactorily explained for the first time the nature of radiant energy from heated objects. In 1905, Albert Einstein (1879–1955) firmly established the validity of the quantum theory of light by using it to explain the photoelectric effect (emission of electrons from metallic surfaces irradiated by light).

The paradox referred to by physicists of that time as the "ultraviolet catastrophe" was resolved by the assumption of quantized energy. Classical physics predicted that radiant energy enclosed in a cube with reflecting mirrors for walls and containing a black body to absorb and re-emit radiation would result in the

In 1910 the concept of the nuclear atom was established from Rutherford's alpha-scattering experiment with gold foil (refer back to Chapter 2).

conversion of radiant energy of low frequencies to continually higher and higher frequencies. This "run away" of energy into the higher frequencies does not occur. If it did, it would be dangerous to sit in front of a fire in a fireplace. The pleasant infrared heat would become dangerous ultraviolet and x-radiation. The quantum theory resolved the paradox.

In 1913, Niels Bohr (1885–1962) proposed a planetary model for the atom in which electrons were believed to rotate about the nucleus in fixed orbits. The highly revolutionary feature in the Bohr model was the assumption that—contrary to classical behavior—an electron revolving around a nucleus would not spiral into the nucleus, but would be stable in certain specified orbits. Combining classical equations for (1) attraction of opposite charges and (2) centrifugal force with the nonclassical assumption that stable orbits could be obtained when the angular momentum of the electron was quantized, Bohr was able to calculate allowed orbits and energy levels in the H atom. The Bohr model appeared phenomenonally successful at first. The line spectrum of H contains several series of lines that could be accurately predicted from the differences in energies of the Bohr orbits. Furthermore, the size of the H atom predicted by the radius of the Bohr orbit was in agreement with independent methods for estimating the radius of the H atom.

However, the model failed when it was applied to atoms more complex than H. Attempts to correct the deficiencies in the Bohr model by the assumption of elliptical orbits were made, but within a few years after its proposal most of the rosy glow of optimism that scientists felt about the Bohr theory in 1913 had faded. Although the Bohr theory is now outdated, Bohr's contribution must not be underestimated. It was Niels Bohr who first introduced the radical and correct concept of quantized energy levels in the atom. This marked the beginning of the application of quantum theory to chemistry.

In 1924, the French physicist Louis deBroglie propounded the concept of the dual nature of matter. His $\lambda = h/p$ equation connected the wave and particle nature of matter. Three years after the deBroglie equation, C. Davisson and L. H. Germer obtained a diffraction pattern for a beam of electrons, demonstrating that the predictions of the deBroglie equation were correct—matter does indeed have wave character.

In 1926, Erwin Schrödinger (1887–1961) proposed the general wave equation that is the basis for our present wave-mechanical model of the atom.

In 1927, the uncertainty principle, $(\Delta x)(\Delta p) \geqq h/4\pi$, was described by Werner Heisenberg. In light of this principle, the Bohr model of the H atom with its electron traveling at a prescribed speed in a carefully defined orbit is now seen to be very naïve indeed.

Add to the above list Einstein's theory of relativity (special theory, 1905; general theory, 1915), and we have a most extraordinary thirty years of productivity for theoretical science. Considerable valuable progress has been made in the application of the quantum theory to chemistry in the years since 1930, but no accomplishments since are as revolutionary and far-reaching in their effects as those of 1900–1930.

Just preceding the turn of the century, natural science was in a somewhat dormant state. There are numerous stories of famous scientists advising young people to choose a career outside of natural science "since all the fundamental principles have already been discovered." This attitude, of course, made the advent of the quantum revolution all the more dramatic.

Summary

Electromagnetic radiation has a dual nature. Some of its properties are best explained on the basis of a wave model. The wavelength (λ) and frequency (ν) of the radiation are related through the velocity of light (c).

$$\nu = \frac{c}{\lambda}$$

Other properties are best explained by the photon model. The energy (ϵ) of these indivisible packets is related to the frequency of radiation through Planck's constant (h).

$$\epsilon = h\nu$$

The wave model and photon model are brought together in the de Broglie equation

$$p = \frac{h}{\lambda}$$

where p is the momentum of a photon of radiation with a wavelength of λ.

Not only does electromagnetic radiation have particle properties, but particles have wave properties. This wave property is the basis for the quantum-mechanical description of mass systems.

Quantum mechanics is based on a series of postulates:

(1) Each system has a wave function (ψ). The wave function describes fully the state of the system.
(2) The probability density of a particle in space is equal to the square of the wave function (ψ^2) evaluated at that location.
(3) The wave equation that applies to particles is the Schrödinger wave equation.

The first postulate recognizes the wave nature of matter. The wave description is useful only for very small, fast-moving particles (for example, an electron in an atom).

The second postulate allows one to calculate the *probability* of locating a particle in a specified place.

The Heisenberg uncertainty principle places limits on the accuracy with which we can know the position of a particle and its momentum simultaneously:

$$(\Delta p)(\Delta x) \geqq \frac{h}{4\pi}$$

The third postulate provides a way to get the wave function. Solving the Schrödinger equation for a system gives ψ and also the energy (E) of that system.

The wave equation has been solved exactly for the hydrogen atom. The solution gives rise to the three quantum numbers n, l, and m, which are restricted to specified integral values. This leads to the important concept of quantization of energy levels. The energy of the electron in an H atom is given by

$$E = -\frac{2\pi^2 me^4}{n^2h^2}$$

where n, the principal quantum number, can have values of 1, 2, 3, ...

The three quantum numbers are related as follows:

$$\begin{aligned} n &= 1, 2, 3, \ldots \infty \\ l &= 0, 1, 2, \ldots (n-1) \\ m &= 0, \pm 1, \pm 2, \ldots \pm l \end{aligned}$$

The quantum numbers describe the orbitals. The numerical value of the number n designates the number of the main energy shell; the value of l identifies the orbital as an s, p, d, or f type orbital; and the value of m identifies specific p, d, or f orbitals.

By examining the equations for ψ^2, we can deduce the geometry of the orbitals. Contours of equal probability density for s orbitals are seen to be spherical. ψ^2 for p and d orbitals has angular dependence.

The radial probability distribution is proportional to the probability of finding an electron anywhere in a thin spherical shell at distance r from the nucleus, information that is very useful in interpreting the chemistry of the elements.

The wave theory can, in principle, be applied to more complex systems than the H atom. In practice, however, one is prevented from making complete solutions in more complex systems by the difficulty of the mathematics. Relying on what was learned from the study of the H atom about orbitals, quantum numbers, and relative energy levels, we can interpret spectroscopic data to obtain a description of the electronic structure of complex atoms.

In atoms with more than one electron, an additional quantum number becomes important. It is the spin quantum number, which can have values of $+1/2$ or $-1/2$. The Pauli exclusion principle states that no two electrons in an atom can have the same set of quantum numbers. The effect of this rule is that there can be no more than two electrons in an orbital, and that these must have opposite spins. Knowledge of the electron configuration of the elements will be the basis for much of our discussion in subsequent chapters of the chemical properties of the elements.

Important Terms and Concepts

Quantum theory
Electromagnetic radiation
Wave model
Wavelength
Frequency
Wave function
Electromagnetic spectrum
Line spectrum
Photon
de Broglie equation
Postulates of the quantum theory
Heisenberg uncertainty principle
Schrödinger wave equation
Quantum numbers
Orbital
Quantization of energy
Energy of the hydrogen atom
Hydrogen atom wave functions
Ground state
Excited state
Orbital geometry
Radial probability distribution
Pauli exclusion principle
Spin quantum number
Aufbau principle
Electronic configuration of the elements

Exercises

13-1.
Compare the energy and frequency of the photons at $\lambda = 200$ nm (ultraviolet), 500 nm (visible), and 1000 nm (infrared).

13-2.
The high C the soprano hits in the "mad" scene of "Lucia di Lammermoor" has a frequency of 1024 sec^{-1}. What is the wavelength of this sound wave if the velocity of sound in air is 3.317×10^4 cm/sec?

13-3.
What is the wavelength of photons that possess an energy of 347 kjoules/mole, and in what part of the electromagnetic spectrum do we find this radiation?

13-4.
(a) Use enthalpy of formation data to calculate the heat evolved in the combustion of one mole of benzene (C_6H_6) to $CO_2(g)$ and $H_2O(l)$ at 298.15° K.
(b) How much mass is lost as a result of this combustion reaction?

13-5.
A helium atom moving with a velocity of 1.0×10^6 cm/sec is confined to bounce around in a small box. If we insist that the uncertainty in the velocity of this helium atom is not greater than 1.0×10^4 cm/sec (1% of the velocity), according to the uncertainty principle, how small (volume in Å^3) could we make the box and still be certain our helium atom is inside the box?

13-6.
Calculate the de Broglie wavelength of the following:

(a) A 22-caliber bullet ($m = 1.0$ g) moving with a velocity of 2.5×10^4 cm/sec.
(b) A thoroughbred race horse ($m = 1200$ lb) moving with a speed of 40 miles/hour.
(c) A neutron (atomic wt. = 1.008 g/mole) moving with a velocity of 1.0×10^6 cm/sec.

13-7.
A helium atom is moving with a velocity such that its wavelength is 1.0×10^{-3} Å. At what velocity must it be moving?

13-8.
Radioactive radium gives off alpha particles (helium nuclei) with an energy of 4.8 MeV (million electron volts). What is the de Broglie wavelength of one of these alpha particles?

13-9.
How fast is an electron moving if it has a velocity in cm/sec equal to its wavelength in cm?

13-10.
Complete the following table.

MINIMUM VALUE FOR PRINCIPAL QUANTUM NUMBER (n)	MAXIMUM VALUE FOR QUANTUM NUMBER l	MAXIMUM VALUE FOR QUANTUM NUMBER m	TYPE OF ORBITAL (s, p, d, etc.)
2			
	2		
		2	
			f

13-11.
What, if anything, is wrong with the following sets of n, l, m and s quantum numbers?

(a) $1, 1, 0, \frac{1}{2}$
(b) $3, 0, -1, \frac{1}{2}$
(c) $3, 2, 1, \frac{1}{2}$
(d) $2, -1, 1, \frac{1}{2}$
(e) $2, 1, -1, \frac{1}{2}$
(f) $1, 0, 0, 0$

13-12.
An electron in hydrogen has a principal quantum number of $n = 4$. List all the possible values for l and m that the electron can have.

13-13.
An electron is in a $4f$ orbital. What possible values for the quantum numbers n, l, m, and s can it have?

13-14.
The next set of orbitals after f would be g. These orbitals would be characterized by the quantum number $l = 4$.

(a) In which principal electronic level (that is, for what value of n) could the g levels first appear?
(b) How many different g levels are possible for a given n?
(c) Give the quantum numbers of all the electrons that would occupy the filled set of g orbitals for $n = 7$.

13-15.
Give all four quantum numbers for each electron in a neon atom ($Z = 10$) in its ground state.

13-16.
How much energy (kjoules/mole) is required to remove an electron from a hydrogen atom in the $4s$, $4p$, $4d$, and $4f$ states?

13-17.
(a) What is the difference in energy between the $1s$ and $2p$ orbitals in hydrogen?
(b) In the x-ray spectrum of copper, radiation with $\lambda = 1.54$ Å is emitted when the electron changes from the $2p$ to the $1s$ orbital. What is the difference in energy between the $1s$ and $2p$ orbitals in copper?

Problems

13-1.
(a) What is the wavelength of an electron moving with a velocity of 1.0×10^8 cm/sec?
(b) Repeat the calculation for a neutron, again with a velocity of 1.0×10^8 cm/sec (mass = 1.80×10^{-24} g).
(c) Repeat the calculation for a radon atom and decide which of these particles could be efficiently diffracted in an atomic crystal, where the spacings are about one angstrom.

13-2.
An electron microscope uses a "wave" of electrons to observe an object. The wavelength of the electrons used is much smaller than for visible radiation; hence, much smaller objects can be observed. To distinguish objects as small as 10 Å with good resolution, we would want the electrons to have a wavelength of about 0.1 Å. What accelerating voltage is necessary in the electron microscope to produce electrons with $\lambda = 0.1$ Å?

13-3.
It can be shown that for the one-electron atom, the average kinetic energy ($\bar{E}_k$) and the average potential energy ($\bar{E}_p$) are very simply related to the total energy E. The relations are

$$\bar{E}_p = 2E = \frac{-Z^2e^2}{n^2a_0}$$

$$\bar{E}_k = -E = \frac{Z^2e^2}{2n^2a_0}$$

(a) Calculate an average velocity of the electron in the $n = 2$ orbital for hydrogen.
(b) The average radius ($\bar{r}$) of the one-electron atom is

$$\bar{r} = \frac{n^2 a_0}{Z}$$

Calculate the average diameter of the hydrogen atom (in Å) for $n = 1, 2, 3$.
(c) Use the above equations and the de Broglie relation to show that the wavelength of the electron in the hydrogen atom ($Z = 1$) is given by $\lambda = 2\pi n a_0$. Then show that λ is one average circumference for $n = 1$, and that in general there are n wavelengths per circumference for the nth level.

13-4.

(a) Use equations from the chapter to derive the Rydberg equation for the wavelength of the radiation emitted when an excited hydrogen atom changes quantum states from n_2 to n_1:

$$\frac{1}{\lambda} = 1.0972 \times 10^5 \left(\frac{1}{n_1^2} - \frac{1}{n_2^2}\right) \text{cm}^{-1}$$

(b) The spectral lines obtained from hydrogen are grouped into series depending upon the part of the electromagnetic spectrum where the lines are found. These series are related to the quantum numbers n_1 and n_2 as follows:

$n_1 = 1, n_2 = 2, 3, 4 \ldots$ Lyman series, ultraviolet region
$n_1 = 2, n_2 = 3, 4, 5 \ldots$ Balmer series, visible region
$n_1 = 3, n_2 = 4, 5, 6 \ldots$ Ritz-Paschen series, infrared region

Calculate λ (in Å) for the first four spectral lines of the Balmer series, and identify the part of the visible spectrum where these lines would occur (red, yellow, green, etc.).

13-5.

The Rydberg equation in problem 13-4 can be used to calculate the wavelength of the radiation that would be absorbed to excite the electron in the hydrogen atom from a quantum state n_1 to state n_2. Use this equation to calculate ν and λ for the radiation that must be absorbed to make the transition

(a) $n_1 = 1 \rightarrow n_2 = 3$
(b) $n_1 = 2, l_1 = 0 \rightarrow n_2 = 3, l_2 = 1$
(c) $n_1 = 3, l_1 = 2 \rightarrow n_2 = 4, l_2 = 1$

13-6.

Suppose the spin quantum number could have the values $+\frac{1}{2}$, 0, $-\frac{1}{2}$. Assuming that both the rules governing the possible values for the other quantum numbers and the order in which the various sublevels are filled remain unchanged,

(a) what would be the electronic configuration of element number 10?
(b) what would be the atomic number of the element that would most closely resemble the element with atomic number 7 in physical and chemical properties?
(c) construct a periodic table of the first 40 elements using electronic configurations.

13-7.

The wavelengths of photons emitted during certain electronic transitions in the lithium atom are as follows:

TRANSITION	λ (nm)	TRANSITION	λ (nm)
$4f \rightarrow 3d$	1869.7	$4s \rightarrow 2p$	497.2
$2p \rightarrow 2s$	670.8	$4d \rightarrow 2p$	460.3
$3d \rightarrow 2p$	610.4	$3p \rightarrow 2s$	323.3

Calculate the energy difference in ergs between an electron in the $3p$ and the $3d$ orbitals of lithium.

13-8.

The ionization energy for sodium is 118.5 kcal/mole. Calculate the lowest possible frequency and corresponding wavelengths of light that can ionize a sodium atom.

13-9.

The energy required to remove the first, second, and third electrons from carbon and nitrogen atoms is 11.26, 24.38, and 47.87 electron volts for carbon and 14.54, 29.60, and 47.43 electron volts for nitrogen. Why is it easier to remove the first and second electrons from carbon but easier to remove the third electron from nitrogen?

13-10.

Horatio is an avid reader. One evening he curled up in his soft easy chair in front of the fireplace with a copy of "Mr. Tompkins in Wonderland," intent on an evening of interesting reading. Very quickly, however, the soft chair, warm fire, and Horatio's lack of scientific aptitude were too much, and he fell asleep. While asleep, he dreamed that Congress passed a law changing Planck's constant to 4π

joule-sec (instead of 6.63×10^{-27} erg-sec).† In this uncertain dream world, animals mysteriously escaped from locked cages in the zoo, bicycles jumped lanes of traffic, valuables disappeared from locked vaults, and a flashlight became a death-ray weapon.

Help Horatio and his imagination to describe this world by making the following calculations.

(a) Calculate the wavelength of a 2000-pound automobile traveling at 50 miles per hour, and decide if diffraction problems would be serious for automobiles passing through the Lincoln Tunnel into New York City.

(b) Calculate the uncertainty in the position of a baseball (mass = 142 g) thrown by a pitcher if the uncertainty in the velocity is one meter/sec. From this result predict how this new value for Planck's constant would affect such popular sports as baseball, basketball, and football.

(c) The electrons striking the inner face of a television picture tube have a velocity of approximately 6×10^9 cm/sec. Calculate the de Broglie wavelength for these electrons, and predict how television would be affected.

(d) The earth receives approximately 10^{35} photons/sec from the sun. Assume these photons have an average wavelength of 500 nm, and calculate (assuming $h = 4\pi$ joule-sec) the total momentum change per second (force) as these photons are absorbed by the earth. Compare this result to the momentum of the earth, and predict how the earth would be affected. The earth has a mass of 6×10^{27} grams, and it circles the sun once per year at an average distance of 9.3×10^7 miles.

† Legislative bodies have been known to try to pass laws as strange as this. A state legislature once attempted to pass a law requiring that π be 3.00000 in order that calculations of the circumference and area of a circle be simpler. It is fortunate that Congress does not know or care about the uncertainty principle for if h were set equal to 4π joule-sec by Congressional edict, then $(\Delta p)(\Delta x)$ would be unity, which would simplify calculations but also have some interesting results, as you will see in parts b–d of this problem.

14

The Chemical Bond

In the preceding chapter we used the quantum theory to explain the structure and energy of atomic systems. In this chapter we will apply the quantum theory to molecular systems. We will see that the theory successfully explains the stability of a chemical bond and correctly predicts the bonding and geometry of molecular systems. The general procedure is similar to that used in applying quantum theory to atomic systems. Solution of the Schrödinger wave equation for a molecular system, however, provides a description of *molecular* (rather than atomic) orbitals, and the electrons in the atoms of the molecule are no longer restricted to atomic orbitals; instead, they are distributed throughout the orbitals that belong to the molecule as a whole. This is the **molecular orbital theory,** which will be discussed below. Following this we will describe a less sophisticated but very useful theory of covalent bonding in which atomic orbitals maintain their identity in the molecule; this is the **valence bond theory.**

14.1 Molecular Orbital Theory

In attempting to apply quantum theory to molecules, we encounter the same problem that we had in applying quantum theory to atoms: The wave equation can be solved explicitly for only the simplest systems. Only approximate solutions and qualitative descriptions can be made for more complicated molecules. In this chapter we first describe the hydrogen molecule ion (H_2^+), a molecular system for which the wave equation *can* be solved explicitly. The results obtained for this system will help in our qualitative description later in the chapter of more complicated molecules.

The Hydrogen Molecule Ion

The H_2^+ ion is formed in the high-voltage glow discharge through hydrogen gas.

The hydrogen molecule ion is the simplest of all molecular systems. It is a stable but highly reactive molecule consisting of an electron and two protons.

The Schrödinger equation, written for the hydrogen molecule ion, can be solved explicitly for ψ and E. However, rather sophisticated mathematics is required to solve the equation, and the resulting mathematical expressions for ψ are complicated and difficult to interpret. Therefore, we will limit our discussion to a qualitative description of ψ^2 and E, which will help us to understand the bonding and stability of the molecule.

The Electron Density

Just as the square of the wave function for an atomic orbital gives a measure of the electron probability density (which we will abbreviate in the remaining discussion to *electron density*) in an atomic orbital, the square of the wave function for a molecular orbital is a measure of the electron density in a molecular orbital. A graphical representation of the electron density for the H_2^+ molecule is shown in Figure 14.1. In this figure electron density is plotted along a line that passes through the two atomic nuclei in the molecule. We see that the electron density is distributed symmetrically about the two nuclei and that the probability is greatest for finding the electron next to either nucleus; the

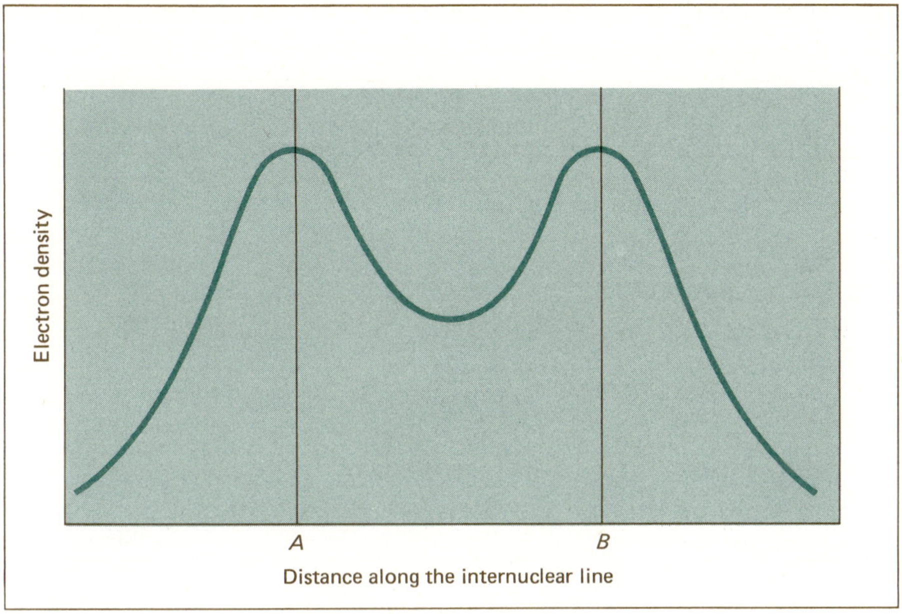

Figure 14.1
Electron density along the internuclear line for the stable diatomic molecule H_2^+. The nuclei are at *A* and *B*.

probability decreases as we move away from the nuclei. It is most important to note that the probability of finding an electron is greater between the two nuclei than it is on the far side of each nucleus. It is this increased electron density between the nuclei that holds them together. That is, the positively charged nuclei are held together by their mutual attraction for the negative charge that is between them. Hence, this orbital is called a **bonding orbital.**

Energy

We can add to our description of the bonding molecular orbital by considering the energy changes that occur as the bond is formed. Figure 14.2 shows how the energy changes as a proton (H^+) and a hydrogen atom (H) are brought together to form the hydrogen molecule ion

$$H + H^+ = H_2^+$$

The curve can be explained as follows: At a large separation, H^+ and H do not interact and the energy equals that of the isolated H^+ and H. As the particles are brought together, both attraction and repulsion occur: attraction as a result of the interaction of the electron of H and the proton of H^+ (unlike charges), and repulsion because of the proton-proton interaction (like charges). Initially the force of attraction exceeds that of repulsion. The result is a net force of attraction and a lowering in the energy. The energy continues to drop until the minimum is reached. This is the condition for greatest stability. As the particles are moved still closer together, the repulsion force becomes more pronounced until it eventually exceeds the force of attraction and the energy becomes positive.

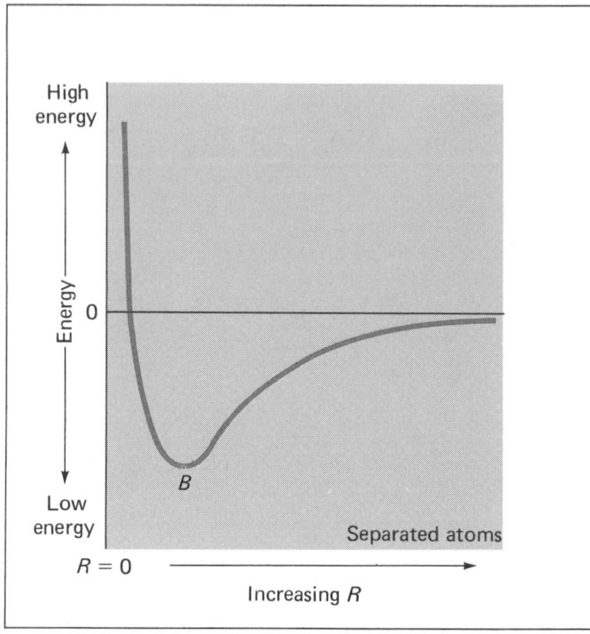

Figure 14.2
Energy of a system consisting of a hydrogen ion and a hydrogen atom as a function of internuclear separation.

As an analogy, a rock is most stable when it rolls to the bottom of a valley (state of lowest potential energy).

When H^+ and H are brought together, energy is released until the internuclear distance given by point *B* in Figure 14.2 is obtained. In H_2^+, this distance is 1.07 Å. It represents the stable internuclear separation or **bond length** in this molecule; either an increase or decrease in this bond length requires the absorption of energy. The energy required to break a bond is called the **bond energy.** In H_2^+, its value is 255.5 kjoules/mole. We use the bond energy as a measure of the stability of the bond, since the larger the bond energy, the larger the energy that must be added to break the bond.

The concepts we have discussed for H_2^+ can be qualitatively extended to more complicated molecules. For example, whenever a chemical bond forms, we expect an increase in electron density between the nuclei being bonded. We also expect a decrease in energy similar to that shown in Figure 14.2. Unfortunately, however, the methods that have been used for H_2^+ cannot be applied quantitatively to more complex molecules. Even for a molecule as simple as H_2, the Schrödinger equation is too difficult to solve exactly. However, there are methods for approximating the wave functions for such molecules.

LCAO Approximation

A simple, but effective, method for approximating the wave function for a molecular orbital is to appropriately combine atomic orbital wave functions. In one method, known as the **linear combination of atomic orbitals (LCAO) approximation,** the molecular wave function ψ is written mathematically as

$$\psi = \sum C_i \psi_i \qquad [14.1]$$

where ψ_i is the wave function for an atomic orbital and C_i is a constant. To illustrate the use of equation [14.1], we will apply the LCAO approximation method to the hydrogen molecule ion. In H_2^+, a molecular orbital is obtained by combining the 1*s* atomic orbitals from each hydrogen atom. Using equation [14.1], we write

$$\psi = C_1\psi_{1s_A} + C_2\psi_{1s_B} \qquad [14.2]$$

where ψ_{1s_A} is the atomic wave function for hydrogen atom A, ψ_{1s_B} is the atomic wave function for atom B, and C_1 and C_2 are constants.

The coefficients C_1 and C_2 in equation [14.2] are related. In H_2^+, atoms A and B are both H atoms and should contribute equally to a molecular orbital. In other words, the electron density should be the same around nucleus A and nucleus B. Mathematically, this equivalence is expressed by

$$C_1^2 = C_2^2 \qquad [14.3]$$

or

$$C_2 = \pm C_1 \qquad [14.4]$$

Remember that the square of the wave function gives the electron density. The equality of the squares of the coefficients C_1 and C_2 gives equal contributions of $\psi_{1s_A}^2$ and $\psi_{1s_B}^2$ to the total electron density.

The result of substituting equation [14.4] into [14.2] is that *two* wave functions and, hence, *two* orbitals are formed:

$$\psi = C_1(\psi_{1s_A} + \psi_{1s_B}) \qquad [14.5]$$

and

$$\psi = C_1(\psi_{1s_A} - \psi_{1s_B}) \qquad [14.6]$$

In other words, the combination of two atomic orbitals gives two molecular orbitals. This concept can be generalized to more complicated systems. In general, when the LCAO approximation is used to combine a number of atomic orbitals, an equal number of molecular orbitals are formed.

Let us see how equations [14.5] and [14.6] describe the molecular orbitals for H_2^+. Figure 14.3 is a graph of equation [14.5] along the axis of the molecule.

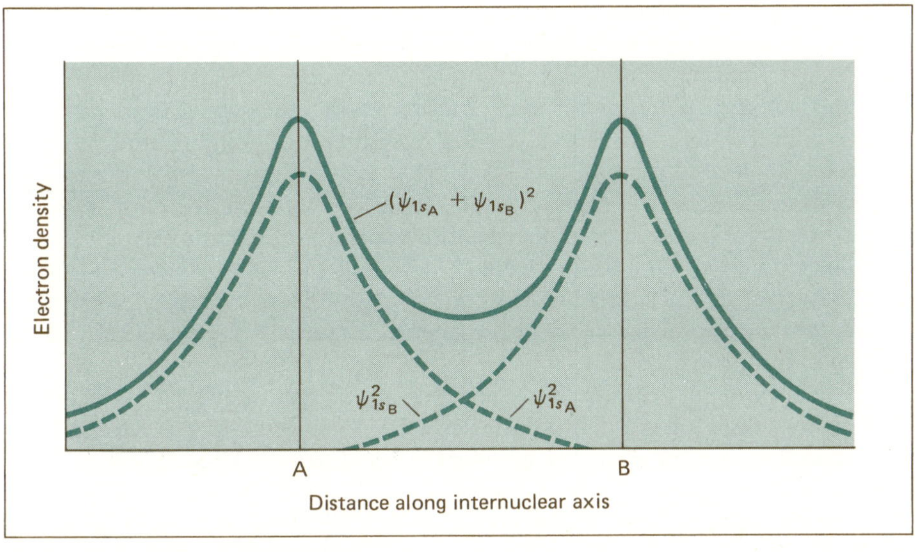

Figure 14.3
Electron densities of separate 1*s* atomic orbitals (dashed lines) and their linear combination into a molecular orbital (solid line).

The dotted lines are plots of $\psi^2_{1s_A}$ and $\psi^2_{1s_B}$ to show how these atomic wave functions fall off with distance from the nucleus. The solid line is the term $(\psi_{1s_A} + \psi_{1s_B})^2$ and represents the electron density of the molecular orbital. We see from the figure that ψ^2 is larger between than outside the two nuclei; hence, the probability of finding the electron is greater between than outside the nuclei.

The increased electron density between the nuclei makes this a bonding orbital. The argument for bonding is the same one we used in the previous section. The region of relatively high negative electron density holds the two positively charged nuclei together in a chemical bond. A pictorial representation of the bonding molecular orbital in H_2^+ is shown below:

The orbital has cylindrical symmetry about the axis of the molecule. That is, the electron probability density is symmetrical about the line connecting the two nuclei. If we could look down the axis of the molecule, the shape of the orbital would appear like this:

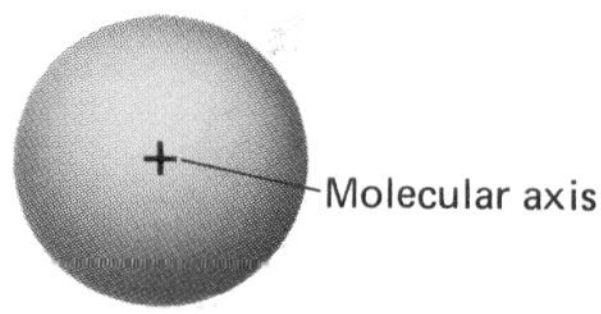

For a σ orbital, the electron density is symmetrical about the molecular axis.

Orbitals with this symmetry are called **σ orbitals.** The particular orbital we have described is designated as σ^b_{1s}. The 1*s* indicates that the σ molecular orbital is formed from the combination of 1*s* atomic orbitals. The *b* signifies that the orbital is a bonding orbital.

So far in the discussion we have considered only the orbital represented by equation [14.5], which, as we have seen, describes a bonding orbital. We now consider the second of the two orbitals that result from linear combinations of 1*s* orbitals, the orbital represented by the wave function in equation [14.6]. Figure 14.4 is a graph of ψ^2 along the molecular axis for that orbital.

Figure 14.4
The antibonding orbital of H_2^+.

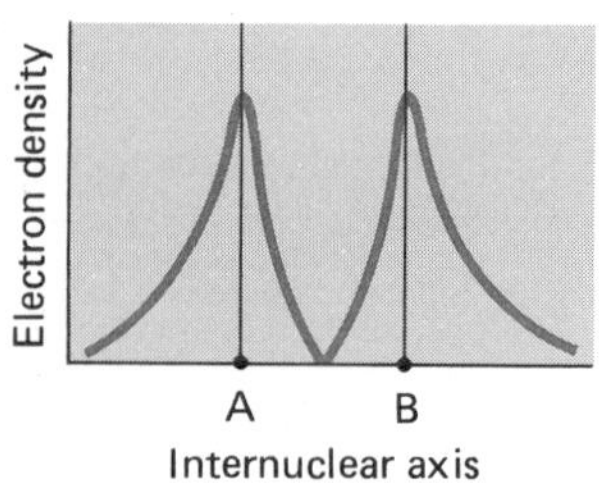

This time we see that the probability of finding the electron is greater outside than between the nuclei. In fact, a nodal plane (a plane with zero probability) is present equidistant between the two nuclei. Figure 14.5 is a pictorial representation of this orbital.

The orbital has cylindrical symmetry about the molecular axis, and it therefore is classed as a σ orbital. We designate this orbital as σ^*_{1s}. Again, the 1*s* indicates that the molecular orbital is formed from 1*s* atomic orbitals. The *(star) indicates that it is an **antibonding orbital.** When electrons are put into this orbital, they have more energy than they do in the atomic system. Figure 14.6

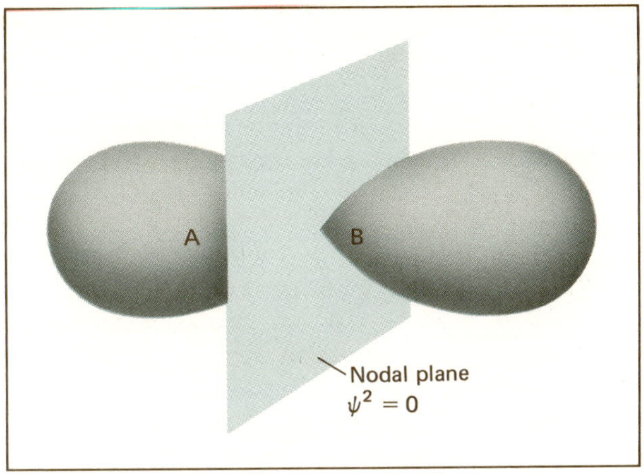

Figure 14.5
The antibonding orbital, σ^*_{1s}, has a nodal plane equidistant between the two nuclei.

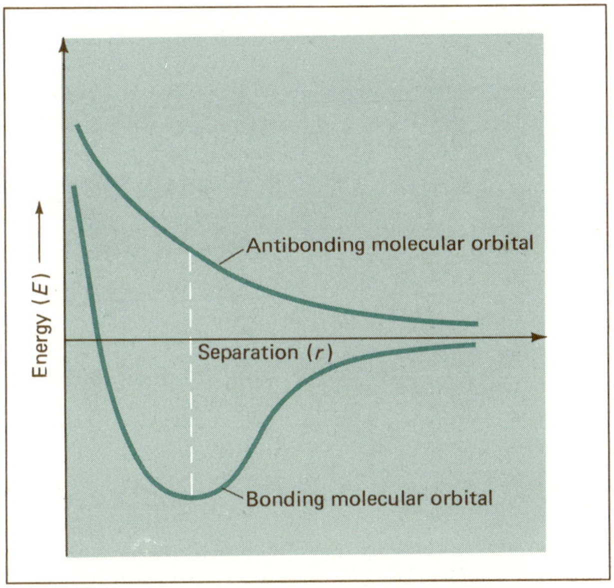

Figure 14.6
The total energy E *of the* H_2^+ *system as a function of the internuclear distance* r.

shows how the energy changes with internuclear separation as two hydrogen atoms are brought together to form σ^b_{1s} and σ^*_{1s} molecular orbitals. The presence of electrons in a σ^b_{1s} molecular orbital lowers the energy and results in a stable bond.

Electrons in a σ^*_{1s} molecular orbital increase the energy of the system, and no bond forms. We shall see later that putting electrons into a σ^* orbital cancels the bonding in a σ^b orbital. In other words, the σ^* orbital cancels chemical bonds, which accounts for the name *antibonding orbital.*

Molecular orbital theory can be extended to more complex molecules. Using the LCAO approximation to give a molecular orbital description of the bonding in simple homonuclear diatomic molecules (molecules composed of

Figure 14.7
1s atomic orbitals combine to form σ^b_{1s} and σ^*_{1s} molecular orbitals.

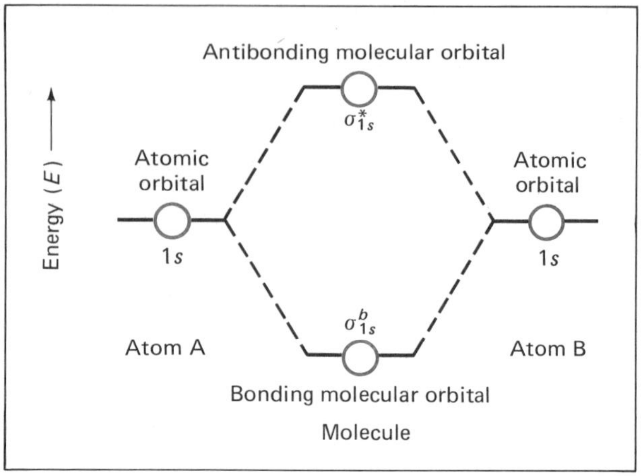

two identical atoms), we can proceed as follows:

(1) Combine the atomic orbitals to form molecular orbitals. An energy-level diagram illustrating this combination for 1s orbitals is shown in Figure 14.7.
(2) Place the electrons (maximum of two per orbital) in the lowest energy orbitals available. Notice that this is an application of the aufbau principle to molecules.

Let us now look at the molecular orbital descriptions that result for several of the simplest diatomic molecules.

H_2^+ has only one electron; it will occupy the σ^b_{1s} orbital as shown in Figure 14.8.

The line notation for the molecular orbital description shown in Figure 14.8 is $(\sigma^b_{1s})^1$, which says that there is one electron in the 1s bonding orbital.

The molecular orbital description of the H_2 molecule is shown in Figure 14.9. Its line notation is $(\sigma^b_{1s})^2$.

We arbitrarily define a **chemical bond** as two electrons in a bonding molecular orbital; hence, H_2 has one complete sigma bond, and we speak of a bond order of one. By the same definition, H_2^+ has only a one-half bond or a bond order of one-half, and the stability of that molecule would accordingly be expected to be less than that of H_2. Since *bond length* is influenced by how strongly the two atoms are bonded, one would also predict the bond length in H_2^+ to be greater than that in H_2.

He_2^+ and He_2 have a total of three and four electrons, respectively. The first two electrons go into a σ^b_{1s} molecular orbital, just as they do in H_2. The extra electrons, however, must go to the higher energy σ^*_{1s} molecular orbital (see Figure 14.10).

As mentioned earlier, electrons in antibonding orbitals (σ^*) cancel the

Figure 14.8
Molecular orbital description of H_2^+ (1 electron).

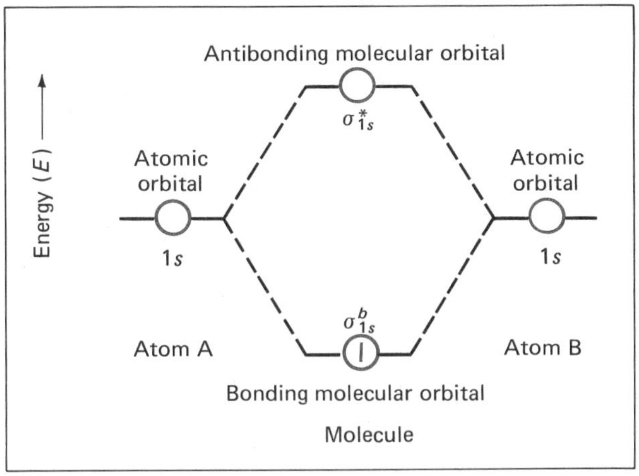

Figure 14.9
Occupied molecular orbitals in H_2. Both electrons occupy the σ_{1s} bonding orbital.

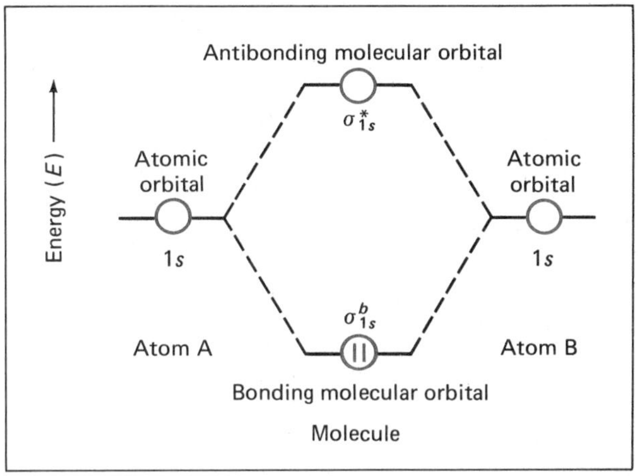

effect of electrons in the bonding orbitals (σ^b). In He_2^+, the one electron in σ^*_{1s} cancels the effect of one electron in the σ^b_{1s}. The result is the equivalent of a one-half bond. In He_2, the two electrons in σ^*_{1s} cancel the two electrons in σ^b_{1s}, and no bond is present. In summary, H_2^+ and He_2^+ should be stable but weakly bonded molecules (compared with H_2). H_2 should be stable, and He_2 should not exist as a molecule. The bond energies for H_2^+, H_2, He_2^+, and He_2 are compared in Table 14.1. The results are as we predicted.

Figure 14.10a
Occupied molecular orbitals in He_2^+.

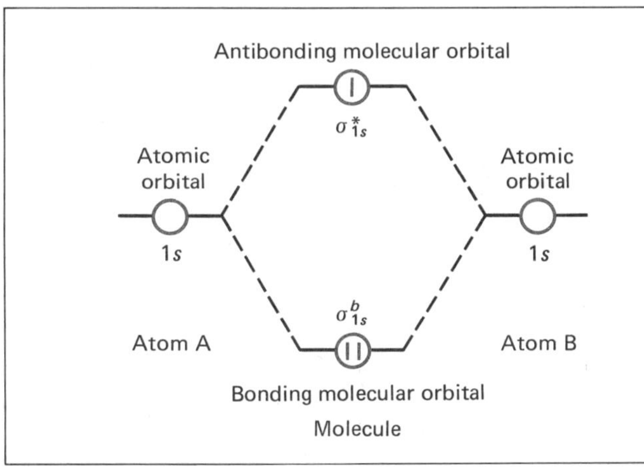

Figure 14.10b
Occupied molecular orbitals in He_2.

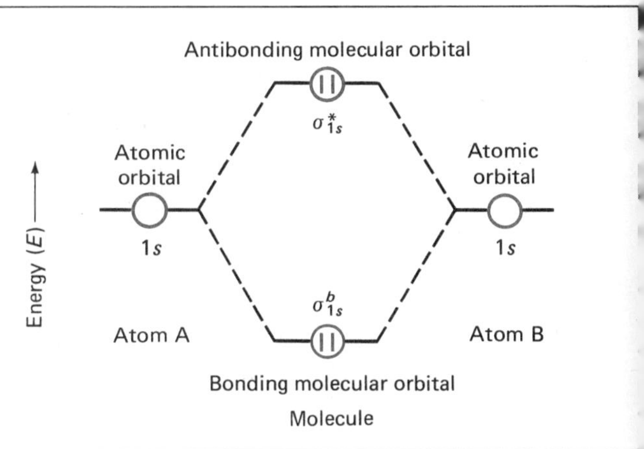

TABLE 14.1

Bond Energies and Bond Lengths in the Hydrogen and Helium Molecules

MOLECULE	BOND ENERGY (kjoules/mole)	BOND LENGTH (Å)
H_2^+	255.5	1.06
H_2	432.0	0.7412
He_2^+	—	1.080
He_2	Not stable	Not stable

EXAMPLE 14.1 Sketch a total energy versus interatomic distance diagram for (a) H_2^+, (b) H_2, and (c) He_2. Sketch carefully enough that the relative differences in energy and normal interatomic distances are apparent.

SOLUTION The diagram is shown in Figure 14.11. Note that the bond of H_2^+ is not as short as that of H_2, and that the bond energy of H_2^+ is only about half that of H_2. The very small attraction shown in He_2 is a result of van der Waals attraction only. This attraction is too weak to hold He atoms together in a diatomic molecule

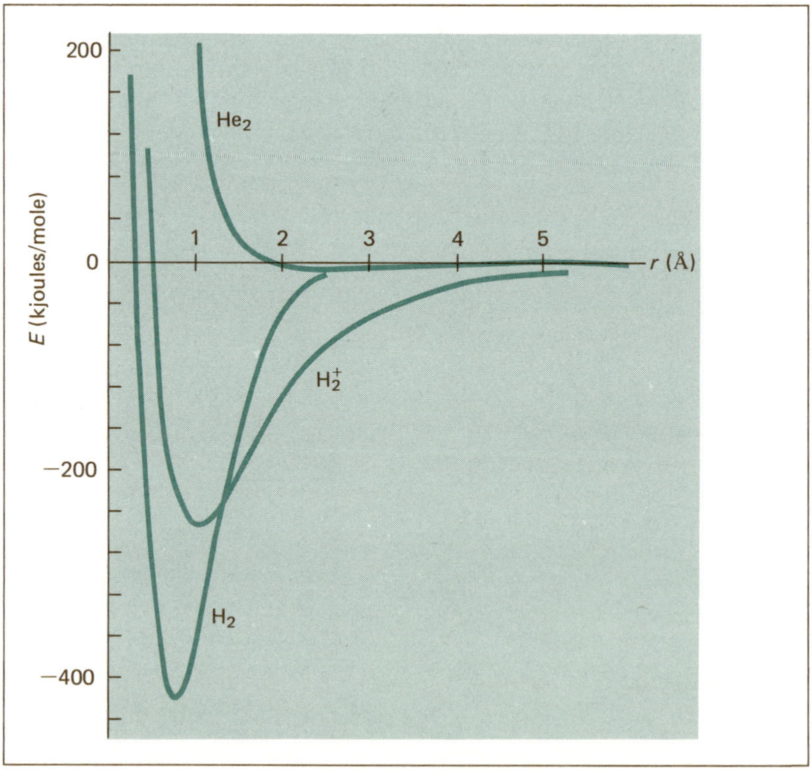

Figure 14.11 *Interatomic distance versus total energy for simple diatomic molecules.*

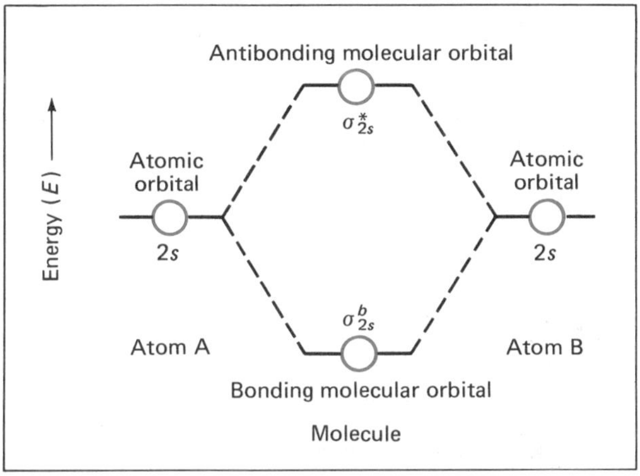

Figure 14.12
Combining the 2s atomic orbitals gives σ^b_{2s} and σ^*_{2s} molecular orbitals.

More Complex Homonuclear Diatomic Molecules

To describe the bonding in molecules containing more than four electrons, we need a more extensive energy-level diagram than that shown in Figure 14.10. We can obtain the information we need by combining the 2*s*, 2*p*, 3*s*, . . . atomic orbitals to form molecular orbitals in the same way we earlier combined the 1*s* atomic orbitals.

The 2*s* orbitals can be combined to give σ_{2s} molecular orbitals that are very similar to the σ_{1s} molecular orbitals. A σ^b_{2s} and a σ^*_{2s} molecular orbital are obtained by adding or subtracting the corresponding atomic wave functions. The σ^b_{2s} orbital is lower in energy than the 2*s* atomic orbitals, whereas the σ^*_{2s} is higher in energy. The result is pictured in Figure 14.12.

The 2*p* atomic orbitals can also be combined to give molecular orbitals. Let us arbitrarily designate the axis of the molecule as the *z* axis for the two atoms. In other words, the $2p_z$ atomic orbitals are along the molecular axis (see Figure 14.13). Notice that the two coordinate systems are mirror images with the positive *z* direction for both atoms between the nuclei.

Figure 14.13
Designation of coordinate axes in combining 2p *orbitals.*

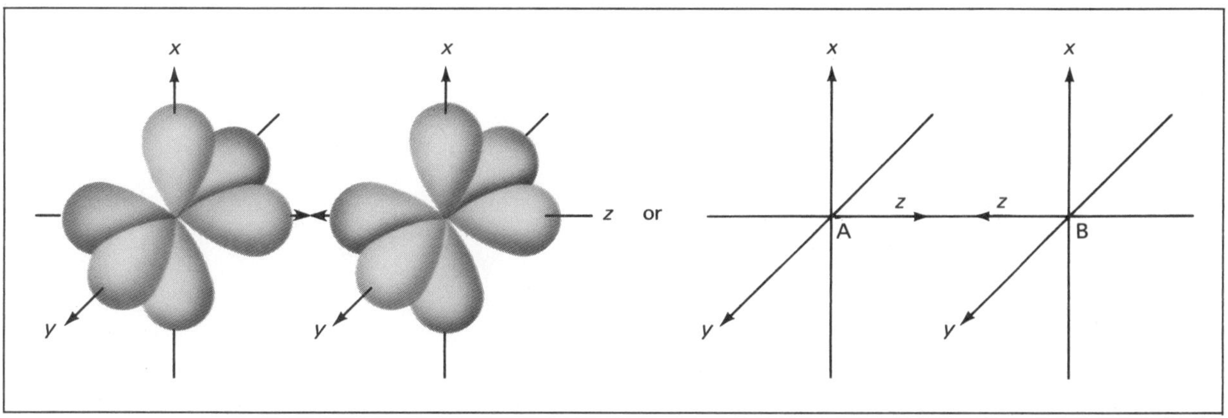

Figure 14.14
Representation of the σ_z^b molecular orbital.

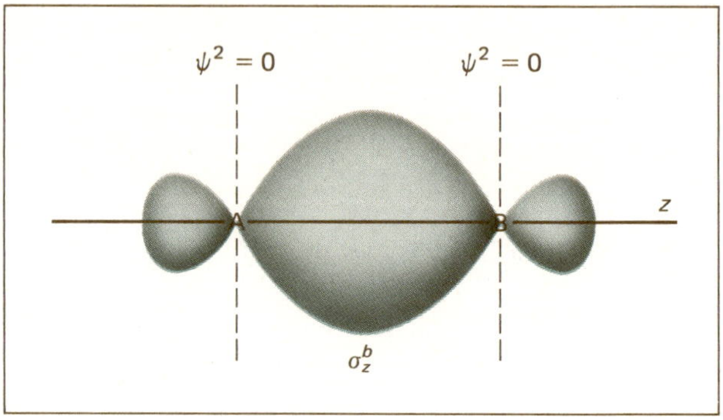

We can form molecular orbitals from the $2p$ atomic orbitals as follows:

(1) The sum of the $2p_z$ atomic orbitals gives a bonding molecular orbital. Figure 14.14 is a representation of this orbital. It is a σ orbital since it has cylindrical symmetry about the molecular axis. It is a bonding orbital since it has high electron density between the nuclei. This σ_z orbital differs from the σ_s orbitals in that the electron density is zero in a plane near and parallel to the xy plane of each atom (this plane is a nodal plane). We designate this orbital as the σ_z^b molecular orbital.

(2) Subtracting the $2p_z$ atomic orbitals gives an orbital designated as σ_z^*. It is an antibonding orbital with a nodal plane equidistant between the nuclei. Figure 14.15 is a pictorial representation of the σ_z^* orbital. Notice the cylindrical symmetry about the molecular axis, which is why it is designated as a σ orbital.

(3) When two $2p_x$ atomic orbitals are added together, the result is a molecular orbital different from those that we have discussed previously. This orbital does not have cylindrical symmetry along the molecular axis. In fact, the yz plane is a nodal plane. Figure 14.16 illustrates how the addition of the two $2p_x$ orbitals forms this molecular orbital. This orbital has a region of high electron density above and another below the yz

Figure 14.15
Representation of the σ_z^* molecular orbital.

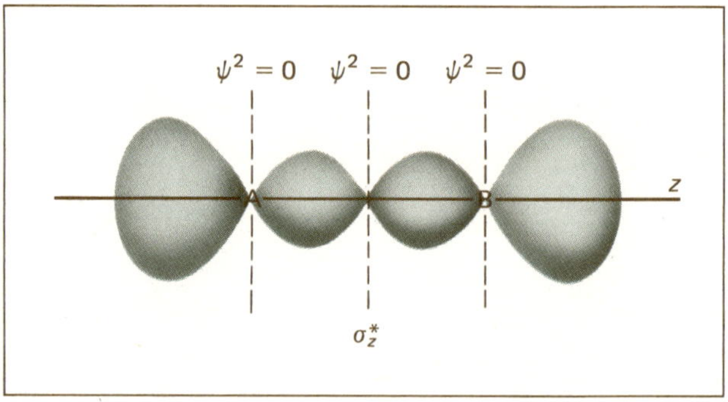

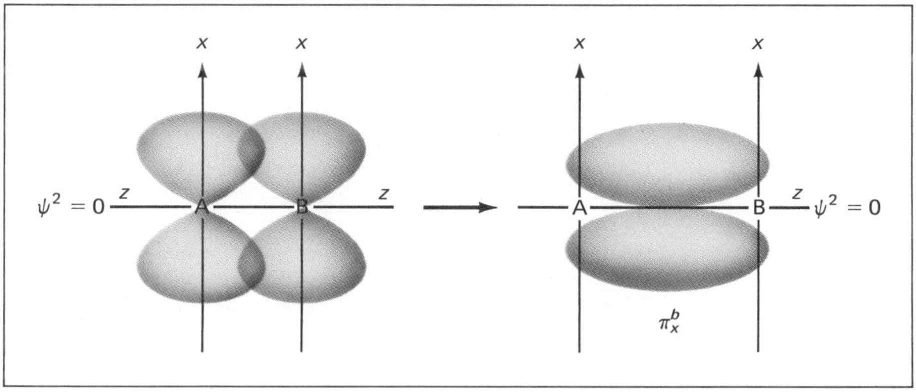

Figure 14.16
The overlap of $2p_x$ atomic orbitals gives the π_x^b molecular orbital.

nodal plane. If one could look down the axis of the molecule, the probability of finding the electron would look like this:

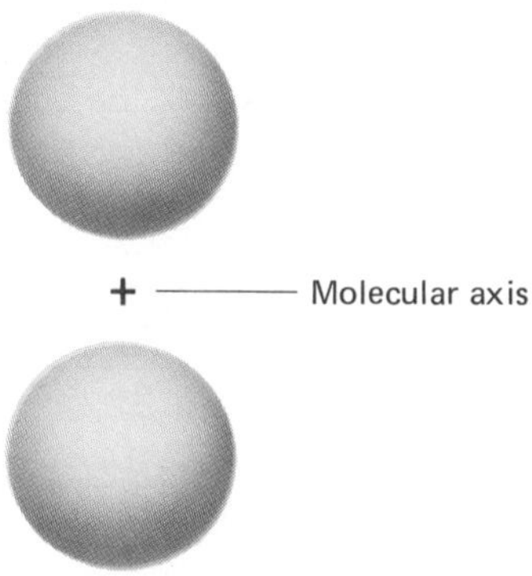

In contrast to the orbitals we described earlier, there is no cylindrical symmetry in this case. Orbitals without cylindrical symmetry are called **π orbitals.**

Notice that even though the electron density is zero along the molecular (z) axis, the density is nevertheless greater between, than outside, the nuclei. This is a bonding orbital, but it usually results in a weaker bond than a σ orbital. We designate this orbital as π_x. In a similar way, the $2p_y$ atomic orbitals can be combined to form a π_y molecular orbital. This orbital would have two regions of high density, one above and one below the xz plane.

(4) The $2p_x$ atomic orbitals can be subtracted to give a π_x^* molecular orbital. Figure 14.17 is a pictorial representation of this orbital. It shows that the orbital has a nodal plane midway between the nuclei (parallel to the xy plane). The yz plane is also a nodal plane. This orbital is an antibonding orbital since the electron density is greater outside than between the nuclei. A similar orbital is obtained by subtracting the $2p_y$ atomic orbitals. This orbital, designated as π_y^*, is identical to the π_x^* molecular orbital except that it is rotated by 90°.

Along the molecular axis the π^* orbitals appear much like the π^b orbitals.

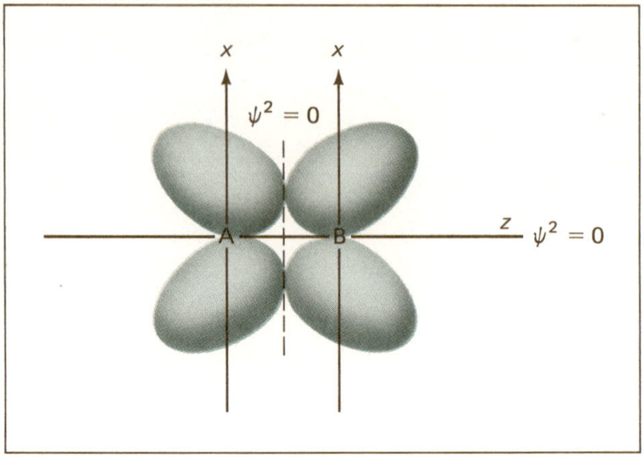

Figure 14.17
Representation of the π_x^* molecular orbital. It has a nodal plane between the nuclei.

In summary, combining the eight atomic orbitals of $n = 2$ ($2s$, $2p_x$, $2p_y$, and $2p_z$ from each atom) gives eight molecular orbitals. Four are bonding and four are antibonding. Figure 14.18 shows how these orbitals compare in energy. We will now use this energy-level diagram to describe the bonding in a number of homonuclear diatomic molecules.

EXAMPLE 14.2 Describe the bonding in Li_2.

The π_x and π_y orbitals have the same energy. In other words, they are degenerate orbitals (like the atomic p orbitals). The π_x^* and π_y^* orbitals are also degenerate. The π_x^b or π_y^b and the σ_z^b orbitals are very close in energy. There are conditions where the σ_z^b orbital is lower in energy than the π_x^b or π_y^b.

SOLUTION Two lithium atoms have a total of six electrons. Figure 14.19 shows how these six electrons are placed in the molecular orbitals. The filled

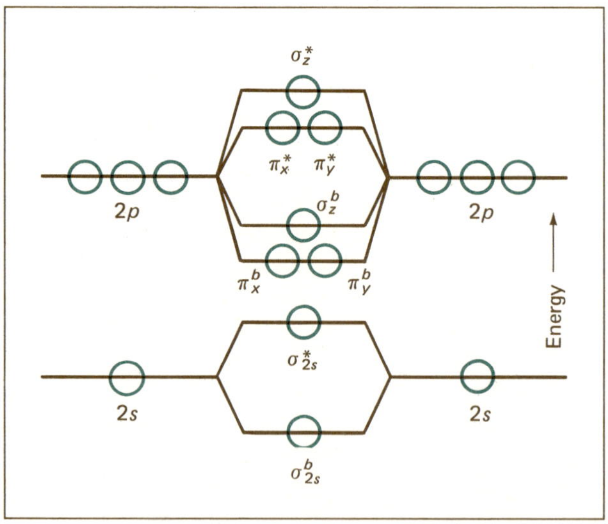

Figure 14.18
Comparison of the energy of the σ^b, σ^*, π^b, and π^* molecular orbitals.

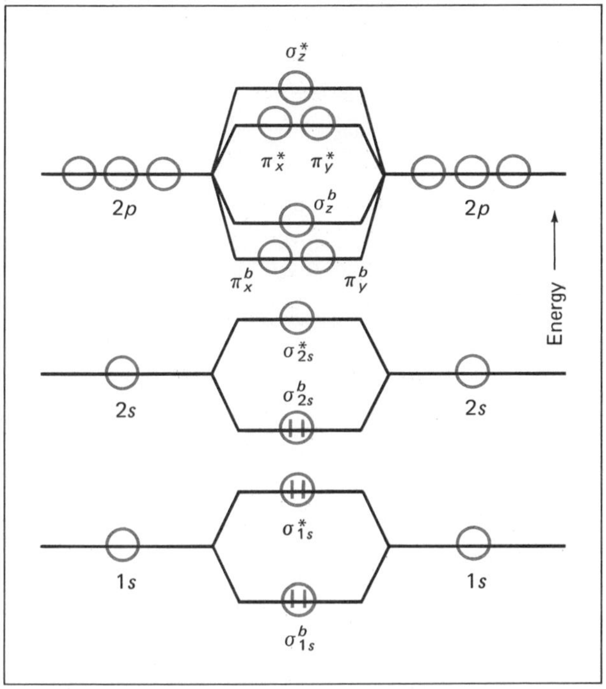

Figure 14.19
Occupied molecular orbitals in Li_2.

σ_{1s}^b and σ_{1s}^* orbitals cancel, leaving two electrons in the σ_{2s}^b orbital. Li_2 has a net of one σ bond.

It should be apparent that for molecules containing more than four electrons, the σ_{1s}^b and σ_{1s}^* are filled and cancel. We call them the **core electrons** since they do not enter into the bonding. They are centered around the two nuclei much as they would be in the atomic case. All we need to consider in the bonding are the electrons beyond these four, the **valence electrons.** The line notation for Li_2 is $KK(\sigma_{2s}^b)^2$, where KK represents filled $1s^b$ and $1s^*$ molecular orbitals (the core electrons).

EXAMPLE 14.3 Show from molecular orbital theory that Be will not exist as a diatomic molecule.

SOLUTION Two Be atoms have a total of eight electrons. Four of these electrons go into the σ_{1s}^b and σ_{1s}^* and cancel. The other four are the valence electrons; they go into the σ_{2s}^b and σ_{2s}^* orbitals (two in each). These two filled orbitals also cancel, leaving zero bonds.

Experimental data verify the molecular orbital prediction for Be. There is no experimental evidence for the existence of the Be_2 molecule.

EXAMPLE 14.4 Predict the occupied molecular orbitals and the bonding in B_2 and C_2.

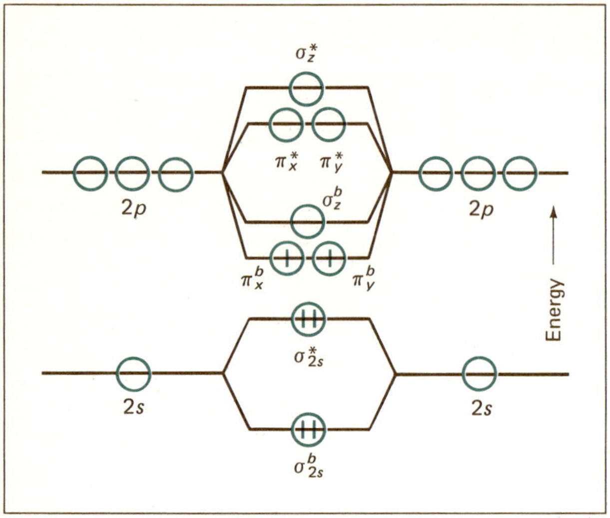

Figure 14.20a
Occupied molecular orbitals in B_2.

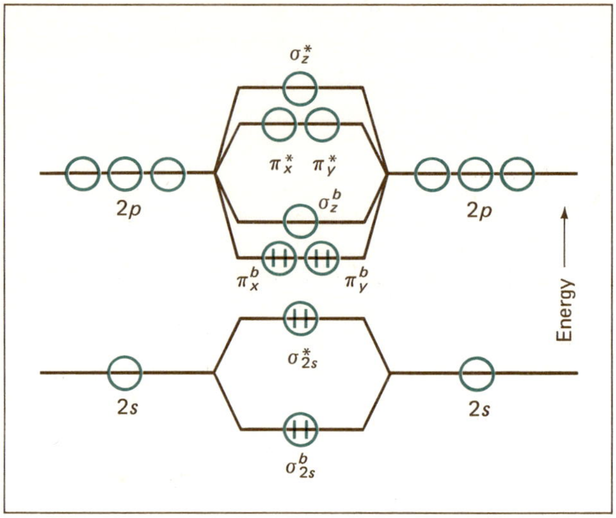

Figure 14.20b
Occupied molecular orbitals in C_2.

The electrons in B_2 are in separate orbitals and hence are unpaired, which results in paramagnetism (attraction into a magnetic field). The amount of paramagnetism observed in B_2 is consistent with it having two unpaired electrons. Molecular orbital theory very successfully predicts the presence of unpaired electrons and paramagnetism in molecules.

SOLUTION The electronic configurations for B and C are [He]$2s^22p$ and [He]$2s^22p^2$, respectively. Boron has 3 valence electrons, and carbon has 4; hence, 6 and 8 electrons are available for bonding in B_2 and C_2, respectively. These electrons are placed in the molecular orbitals as shown in Figure 14.20. The line notations are

$$B_2\colon\ KK(\sigma_{2s}^b)^2(\sigma_{2s}^*)^2(\pi_x^b)^1(\pi_y^b)^1$$
$$C_2\colon\ KK(\sigma_{2s}^b)^2(\sigma_{2s}^*)^2(\pi_x^b)^2(\pi_y^b)^2$$

For both B_2 and C_2, the σ_{2s}^b and σ_{2s}^* cancel. B_2 is left with one electron in each of the π_x^b and π_y^b orbitals. The result is two half-bonds or a net of one π bond. In C_2, the π_x^b and π_y^b orbitals each have two electrons for a net of two π bonds.

EXAMPLE 14.5 Predict the bonding in N_2.

SOLUTION The electronic configuration of nitrogen is [He]$2s^22p^3$. Hence, a total of ten electrons are present in the valence shells of two nitrogen atoms. These electrons are placed in the molecular orbitals to give one σ and two π bonds (see Figure 14.21); N_2 has a triple bond. The σ_z bond is represented in Figure 14.14. The two π bonds merge to give a doughnut-shaped electron cloud around the σ bond (see Figure 14.22).

With N_2, all the bonding orbitals are filled with electrons. When molecules with more electrons are formed, such as O_2 and F_2, electrons must go into the π_x^*, π_y^*, and σ_z^* antibonding orbitals, and the bonding in such molecules will

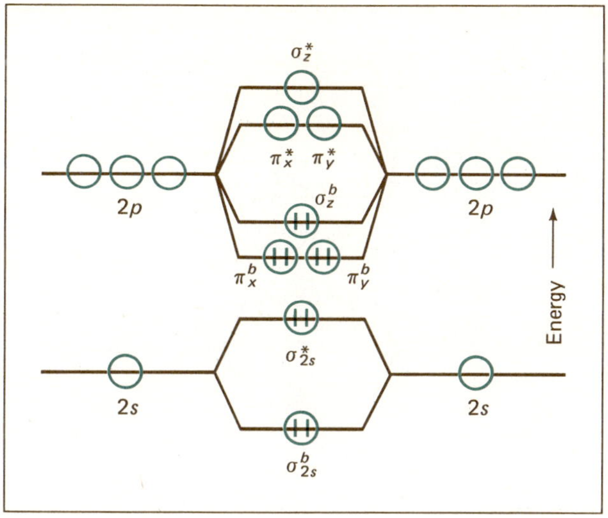

Figure 14.21
Occupied molecular orbitals in N_2.

be weaker than in N_2. It is easy to show that O_2 has a net of one σ and one π bond while F_2 has only a σ bond. The next element in the period, Ne, does not form a diatomic molecule, as all of its antibonding orbitals are completely filled.

O_2, like B_2, has unpaired electrons; hence, it is paramagnetic.

EXAMPLE 14.6 Compare the stability of the O_2 molecule with the O_2^+ and O_2^- ions.

SOLUTION Oxygen has the electronic configuration $[\text{He}]2s^22p^4$. O_2 thus has twelve electrons in the valence shell; O_2^+ would have eleven electrons; and O_2^-, thirteen electrons. These electrons fill the molecular orbitals as shown in Figures 14.23, 14.24, and 14.25. The net bonds are:

$$O_2^+: \quad 1\sigma + 1\tfrac{1}{2}\pi = 2\tfrac{1}{2} \text{ bonds}$$
$$O_2: \quad 1\sigma + 1\pi = 2 \text{ bonds}$$
$$O_2^-: \quad 1\sigma + \tfrac{1}{2}\pi = 1\tfrac{1}{2} \text{ bonds}$$

Figure 14.22
Representation of the π bonds in N_2.

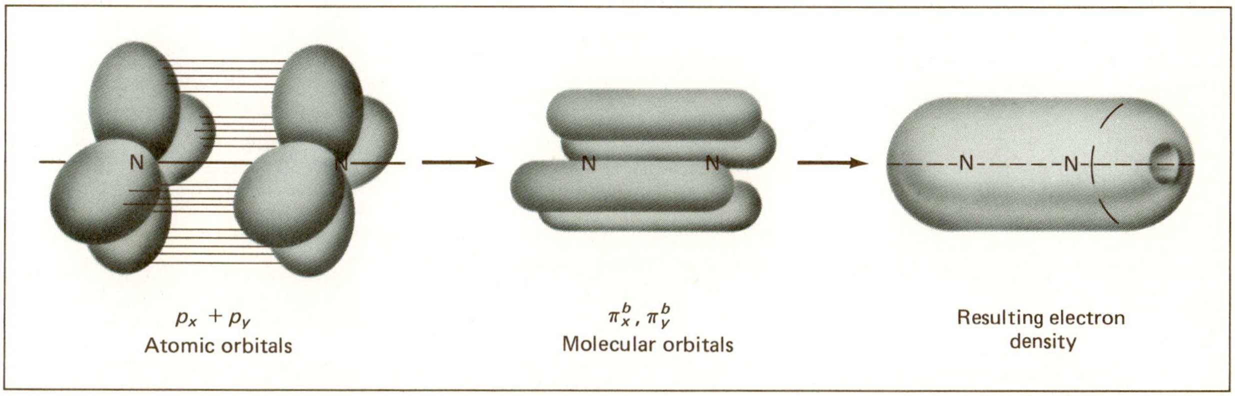

Figure 14.23
Occupied molecular orbitals in O_2^+.

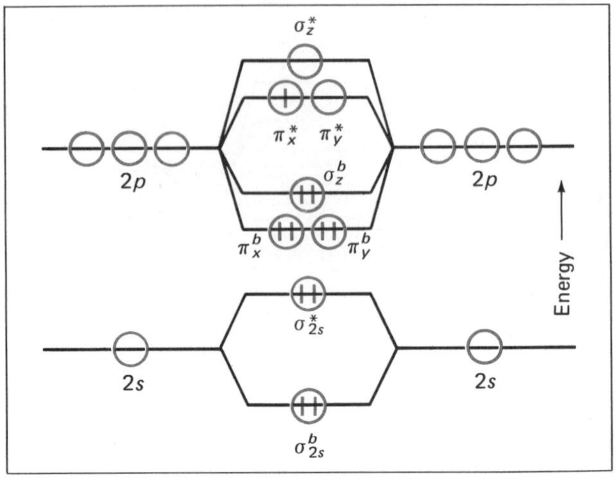

Figure 14.24
Occupied molecular orbitals in O_2.

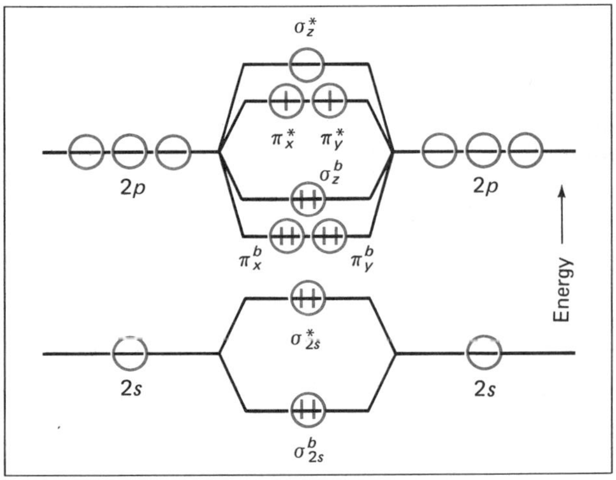

Figure 14.25
Occupied molecular orbitals in O_2^-.

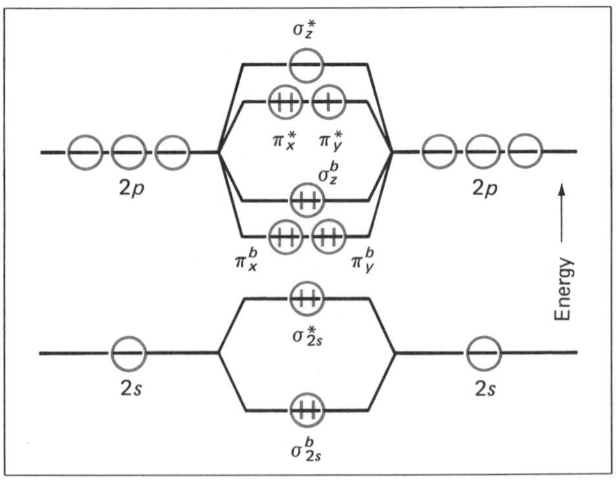

The most stable molecule is the one with the most bonds. Hence, in order of decreasing stability, we have $O_2^+ > O_2 > O_2^-$. Experimental data for the bond energy and bond length verify that O_2^+ is the most stable of the three molecules.

The description of homonuclear diatomic molecules does not end with neon. The $n = 3$ atomic orbitals can be combined to form σ_{3s}^b, σ_{3s}^*, σ_z^b, σ_z^*, π_x, π_y, . . . molecular orbitals, which are similar to those with $n = 2$. The process can be repeated for $n = 4$, $n = 5$, and so on, depending upon how many electrons need to be accommodated. Combining larger, less-concentrated atomic orbitals of $n = 3$, $n = 4$, $n = 5$, . . . produces weaker bonds with larger internuclear separations.

Heteronuclear Diatomic Molecules

The LCAO approximation can also be used to describe the bonding in heteronuclear diatomic molecules (a combination of two different kinds of atoms). The molecular orbital energy levels for a molecule AB in which both A and B have *s* and *p* valence atomic orbitals are shown in Figure 14.26.

The atomic orbitals of A and B combine to form σ^b, σ^*, σ_z, . . . molecular orbitals. These orbitals are very similar to those formed in a homonuclear diatomic molecule. The principle difference is that the atomic orbitals of A and B are not at the same energy. In Figure 14.26, B is more electronegative than A, and its orbitals are at a lower energy level. This difference in energy distorts the molecular orbital such that the electron density is larger around the nucleus whose atomic orbitals are closest in energy to the molecular orbital. In other words, with B more electronegative than A, the σ_s^b orbital has a higher electron density near nucleus B and the σ_s^* orbital has a higher electron density near nucleus A. The result is shown in Figure 14.27.

This unequal sharing of electronic charge makes the molecule polar. If the σ_s^b orbital in a molecule contains electrons while the σ_s^* orbital does not, the end of the molecule that contains the more electronegative atom will have a net negative charge while the other end will have a net positive charge. The molecular orbital description of a polar molecule such as CO is obtained as

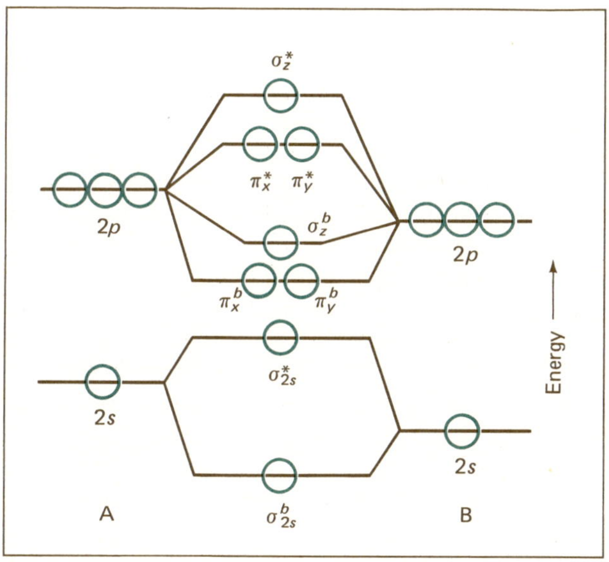

Figure 14.26
Molecular-orbital energy-level diagram for a molecule AB in which B is more electronegative than A.

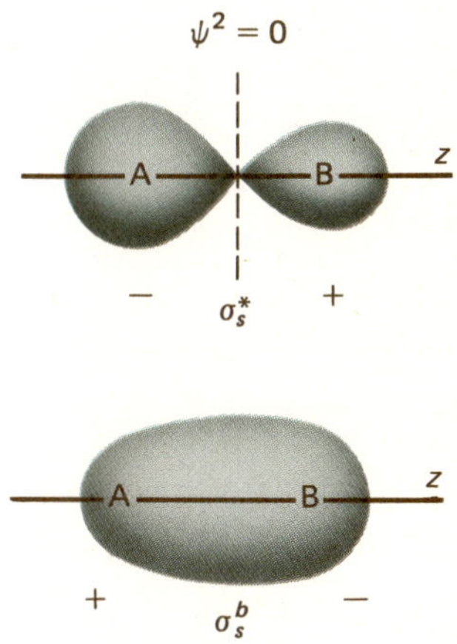

Figure 14.27
Unequal distribution of electronic charge results in a polar bond.

Figure 14.28
Occupied molecular orbitals in CO.

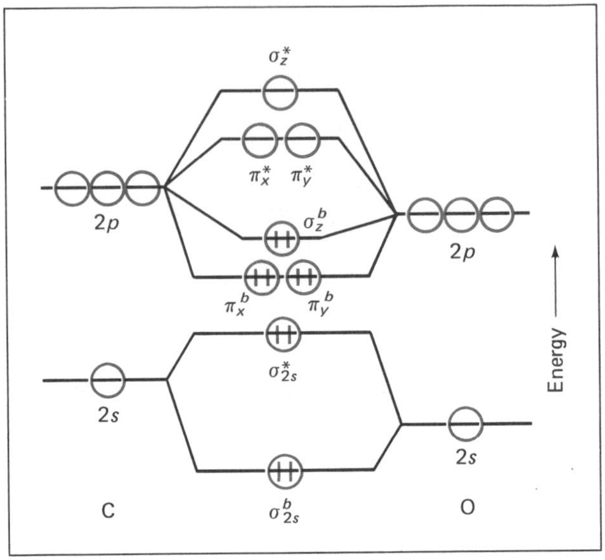

follows: The electronic configurations of carbon and oxygen are

$$\text{Carbon:}\quad [\text{He}]2s^22p^2$$
$$\text{Oxygen:}\quad [\text{He}]2s^22p^4$$

Thus, carbon has four valence electrons and oxygen has six. These ten electrons are put in the molecular orbitals as shown in Figure 14.28. (Since oxygen is the more electronegative element, the oxygen atomic orbitals are at a lower energy level than those of carbon.)

The σ_{2s}^b and the σ_{2s}^* orbitals cancel. This leaves filled π_x^b, π_y^b, and σ_z^b orbitals, making two π bonds and one σ bond. Since the oxygen atomic orbitals are lower in energy than those of carbon, we have a polar molecule with the oxygen end the more negative. In the extreme case of a very large electronegativity or energy difference, the two electrons in the orbital spend essentially all of their time around the more electronegative atom. In other words, when the bond forms, there is complete transfer of an electron from A to B. Atom B gets a complete unit of negative charge, and atom A is left with a unit of positive charge. This is the molecular orbital description of an *ionic bond*. We describe ionic bonding in much more detail in the next chapter.

Summarizing, we have seen that the LCAO approximation of molecular orbital theory is easily applied in describing the bonding in simple molecular systems. However, in more complicated systems in which the number of atoms, and hence the number of atomic orbitals, gets large, the molecular orbital description becomes difficult. We now describe the **valence bond theory,** which can be applied very simply to these more complicated systems with reasonably good results. We will see, however, that the valence bond theory is more qualitative than the molecular orbital theory, and one must be careful in its application since it does not correctly describe some of the details of chemical bonding.

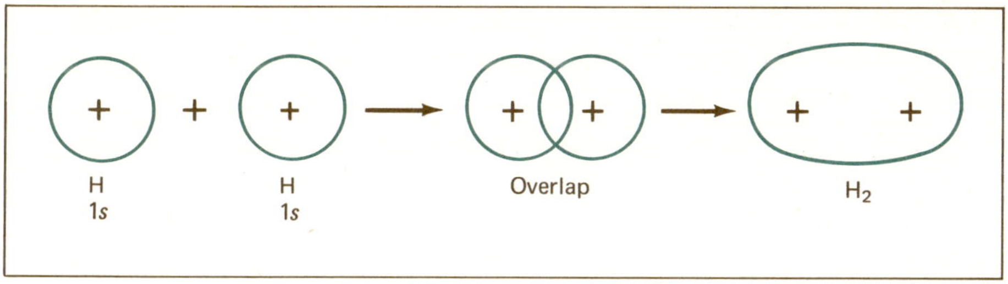

Figure 14.29
A chemical bond results when the 1*s* orbitals in hydrogen overlap.

14.2 The Valence Bond Theory

In the valence bond theory, unlike in the molecular orbital theory, the identities of the atomic orbitals are maintained. In this interpretation, a chemical bond forms when atomic orbitals interact. Consider, for example, the valence bond model of the hydrogen molecule. Each hydrogen atom has a single electron in a 1*s* orbital. If two H atoms are brought together, the spherical 1*s* orbitals are considered to overlap, as shown in Figure 14.29. Simplified line drawings are used to represent the orbitals.

The overlap of the orbitals results in a region of high electron density between the nuclei. For the same reasons as described earlier, this increase in electron density between the nuclei lowers the energy and holds the nuclei together.

The lowest energy is obtained when the electrons in the bond have opposite spin. In the valence bond theory, the electrons in the bond pair up with opposite spin.

The maximum number of electrons in a chemical bond is two. Overlapping the 1*s* orbitals of two helium atoms does not form a chemical bond. He_2 does not exist, since each 1*s* orbital has two electrons. Overlapping these orbitals would put four electrons in the bond. Li_2, however, is predicted to be a stable molecule. In this case, it is the 2*s* orbitals that overlap to form the bond. Be_2, like He_2, is predicted to be unstable since the 2*s* orbitals have two electrons.

Bonding, of course, is not restricted to the overlapping of *s* orbitals; *p*, *d*, *f*, . . . orbitals also participate in chemical bonding.

EXAMPLE 14.7 Describe the bonding in the Cl_2 molecule.

SOLUTION The electronic configuration of Cl is [Ne]$3s^2 3p^5$. One of the *p* orbitals from each chlorine atom has a single electron. These orbitals can overlap to form the bond. Simplified drawings of the orbitals are shown below.

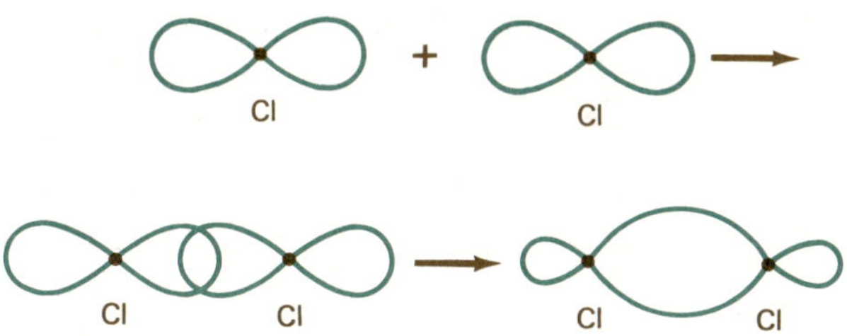

Just as with s orbitals, overlapping the p orbitals increases the electron density between the nuclei and thus holds the atoms together.

EXAMPLE 14.8 Use valence bond theory to describe the bonding in the HF molecule.

Fluorine is more electronegative than hydrogen. The electron density in the bond is larger near the fluorine. As a result, HF is a polar molecular ($\overset{\delta+}{H}$-$\overset{\delta-}{F}$).

SOLUTION Fluorine has the electronic configuration

$$[He]2s^22p^5$$

The p orbital, which contains one electron, overlaps the s orbital of hydrogen to form the bond.

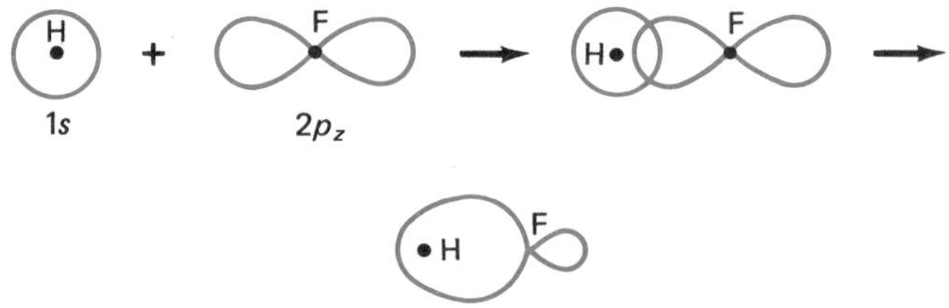

In H_2 or Li_2, the s orbitals overlap to form the region of highest electron density on the axis between the nuclei. The same is true for the p overlap in Cl_2 and the sp overlap in HF. The electron density has cylindrical symmetry in these overlapped orbitals; hence, bonds of this type are called **σ bonds.**

σ bonds are analogous to the σ^b molecular orbitals in molecular orbital theory. π bonds are analogous to the π molecular orbitals in molecular orbital theory.

Parallel p atomic orbitals can also overlap to form a chemical bond. This bond is called a **π bond.** With this bond, the probability of finding the electron along the molecular axis is zero. Consider O_2 as an example.

Oxygen has the atomic configuration $[He]2s^22p^4$. Two of the p atomic orbitals on each oxygen atom have a single electron. One orbital from each oxygen atom overlaps to form a σ bond.

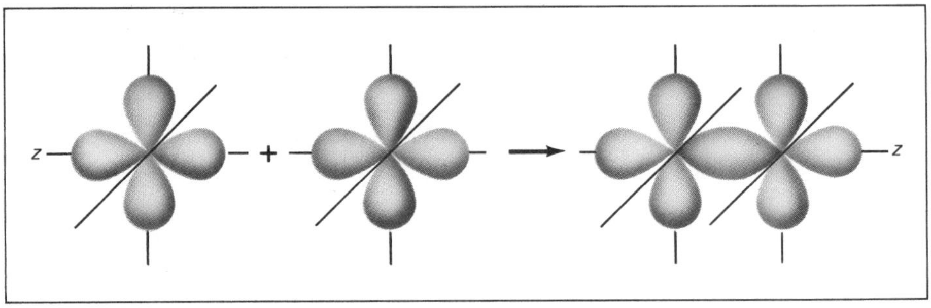

The parallel p orbitals then overlap to form a π bond.

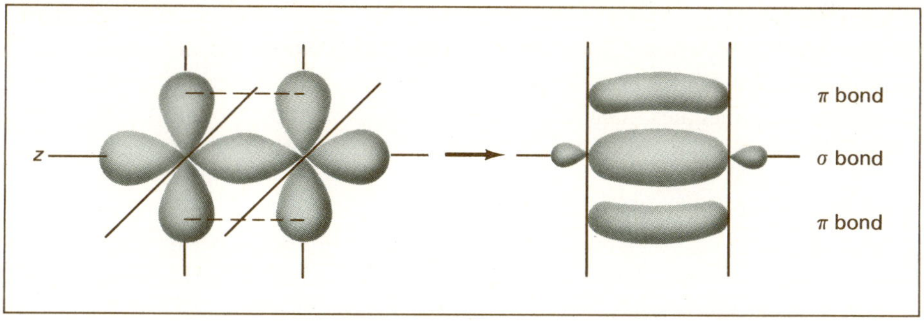

The π bond has two regions of high electron density: one above and one below a plane containing the molecular axis. For the π bond the probability of finding electrons in this axial plane is zero. In other words, the axial plane is a nodal plane.

When electrons are available, multiple π bonds can form. Consider N_2 as an example.

Using valence bond theory, we predict a double bond ($\sigma + \pi$) for O_2 with no unpaired electrons. You will recall that from molecular orbital theory we predicted a σ bond and a π bond with two unpaired electrons (see Example 14.6). O_2 is paramagnetic and does, therefore, have the unpaired electrons. Thus, molecular orbital theory gives a better description than does valence bond theory of the bonding in O_2.

EXAMPLE 14.9 Discuss the bonding in N_2 gas from the point of view of valence bond theory.

SOLUTION Nitrogen has the electronic configuration

$$[\text{He}]2s^22p^3$$

Each of the p orbitals has a single electron. Figure 14.30 illustrates how the orbitals overlap to form one σ and two π bonds.

Molecular orbital and valence bond theories agree in their descriptions of N_2. In each instance, one σ bond and two π bonds are formed.

Figure 14.30
Overlap of the σ and π orbitals in N_2 results in one σ and two π bonds. The π bonds are separated by 90°.

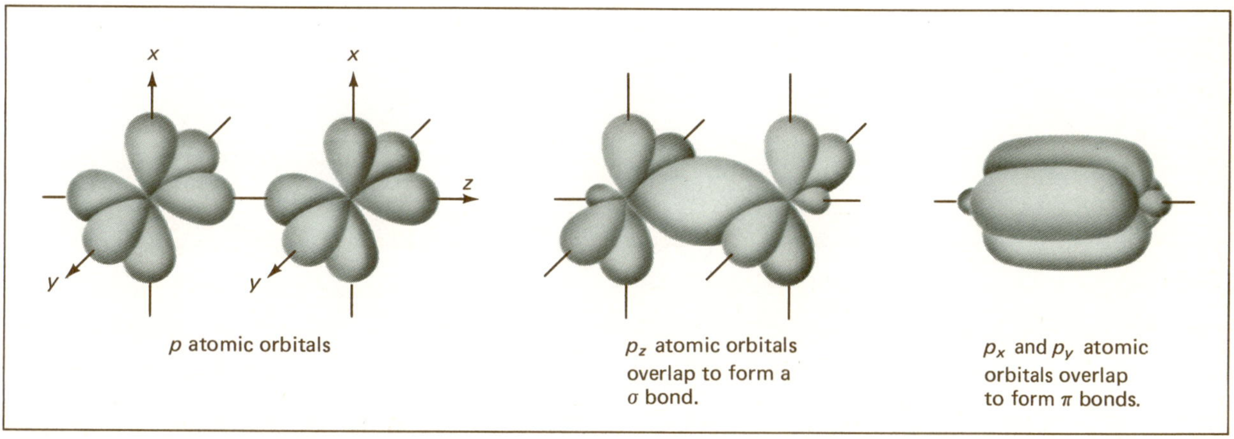

Hybridization

We are now ready to introduce the concept of the **hybrid atomic orbital,** a new kind of orbital that proves especially valuable to us in accounting for the correct geometry of molecules. There are several types of hybrid orbitals formed by combining *s*, *p*, and *d* orbitals in various combinations.

sp Hybridization

Let us use BeH_2 as an example. In this molecule, Be is a central atom bonded to two H atoms. Be has the electronic configuration [He]$2s^2$. To form two bonds, two orbitals must be available in Be, each with one electron. This can be done if we assume that one of the electrons in a 2*s* orbital is promoted to a 2*p* orbital, which we show as follows:

The expenditure of energy in this promotion step would be more than compensated for by the energy released in the bonding that follows.

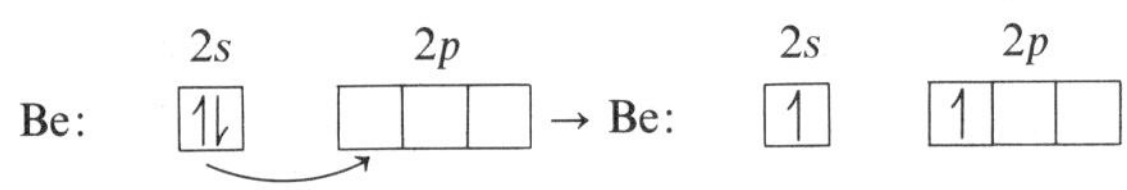

Each of the boxes represents an atomic orbital and the straight arrows represent the electrons. The two arrows in a box pointing in opposite directions represent paired electrons with opposite spins.

This excited state of Be could combine with hydrogen to form the molecule BeH_2. The 2*s* orbital could overlap with a H 1*s* orbital to form one σ bond, and the 2*p* orbital could overlap with the other H 1*s* orbital to form a second σ bond (see Figure 14.31). This representation would require that the two σ bonds be different and that the molecule be nonlinear. But experimental evidence shows that the two bonds are equivalent and that the molecule is linear. One way around the discrepancy is to assume that the two valence orbitals in Be (one *s* and one *p*) combine to form two new equivalent orbitals that are called ***sp* hybrid orbitals.** The shape of the *sp* hybrid is shown in simplified form in Figure 14.32a. The shape of the *sp* orbital shows characteristics of both the *s* and *p* orbitals from which it is formed. The large lobes of the two *sp* hybrid

An analogy to *sp* hybridization is the mixing of black and white paint to form gray paint.

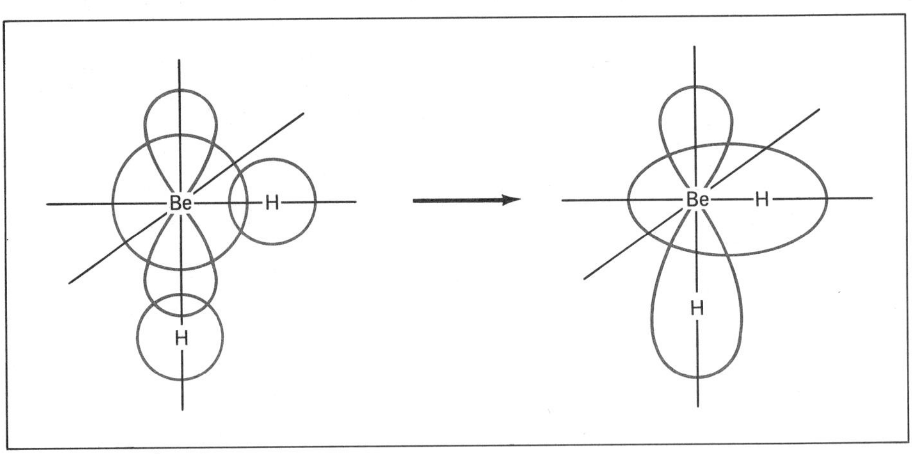

Figure 14.31
Overlap of Be and H atomic orbitals.

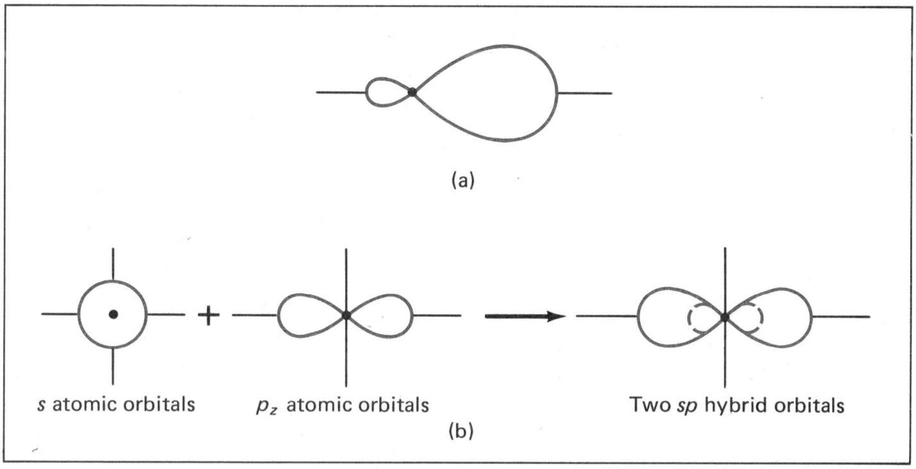

Figure 14.32
Representation of sp *orbitals.*

orbitals point in opposite directions, as shown in Figure 14.32b. This is important because it leads to a linear structure.

Making use of hybrid orbitals, we visualize the formation of a BeH_2 molecule from Be and H as follows:

(a) An electron in Be is promoted to the excited state.

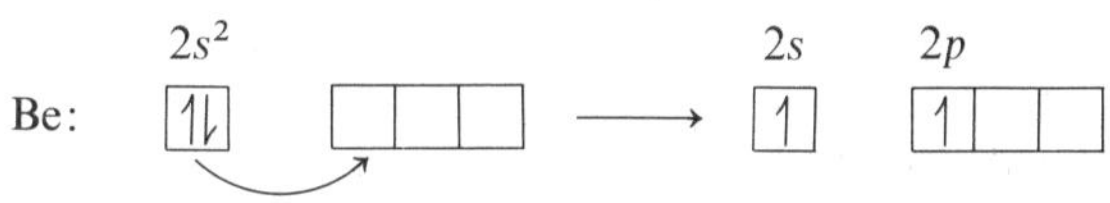

(b) Hybridization occurs to form two equivalent *sp* hybrid orbitals.

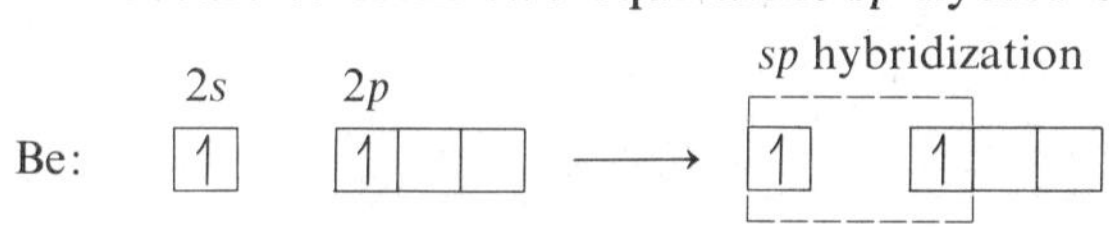

(c) The hybrid orbitals overlap with 1*s* orbitals from hydrogen to form a linear molecule with equivalent bonds, in full agreement with experimental fact:

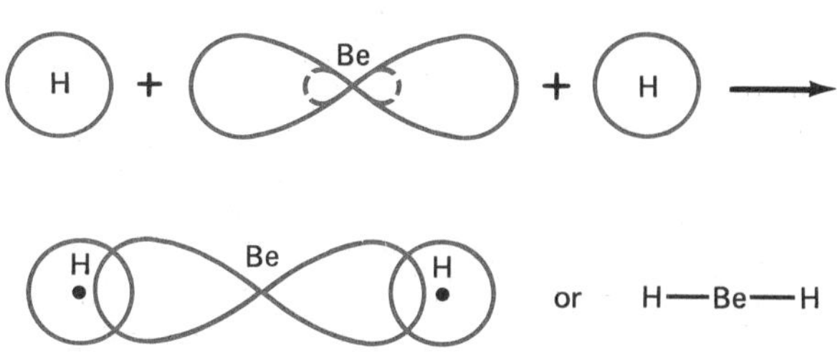

where the lines in the last representation are the bonds.

Another analogy is the mating of a horse and a donkey to produce a mule. Mules retain some of the characteristics of both the horse and the donkey, but they are a new, unique animal.

By no means do we imply that this is the actual path by which the BeH_2 molecule is formed. It is only a formal description of the steps we go through to describe the structure of the molecule.

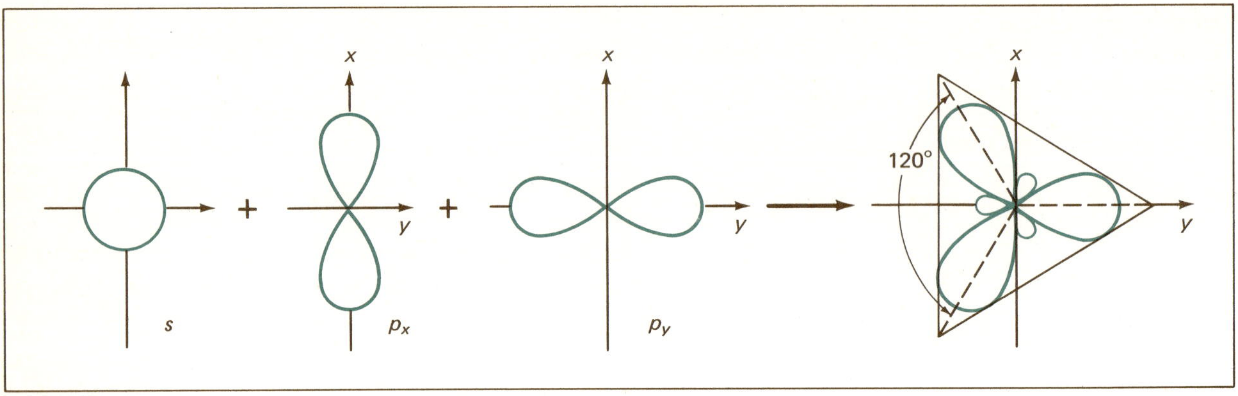

Figure 14.33
Combination of an *s* with two *p* orbitals gives three sp^2 hybrid orbitals. Note that the angle separating sp^2 hybrid orbitals is 120°. As with *sp* hybridization, the shape of the sp^2 orbitals has been simplified.

sp^2 Hybridization

A second type of hybrid orbital is called sp^2 since it forms from one *s* and two *p* orbitals. Three equivalent sp^2 orbitals result from hybridization of one *s* plus two *p* orbitals. The geometry of the three **sp^2 hybrid orbitals** taken together is a three-lobed structure with the lobes directed toward the corners of an equilateral triangle (see Figure 14.33).

sp^3 Hybridization

In a similar manner, an *s* orbital can be combined with three *p* orbitals to give four equivalent **sp^3 hybrid orbitals.** These orbitals are directed toward the corners of a tetrahedron.

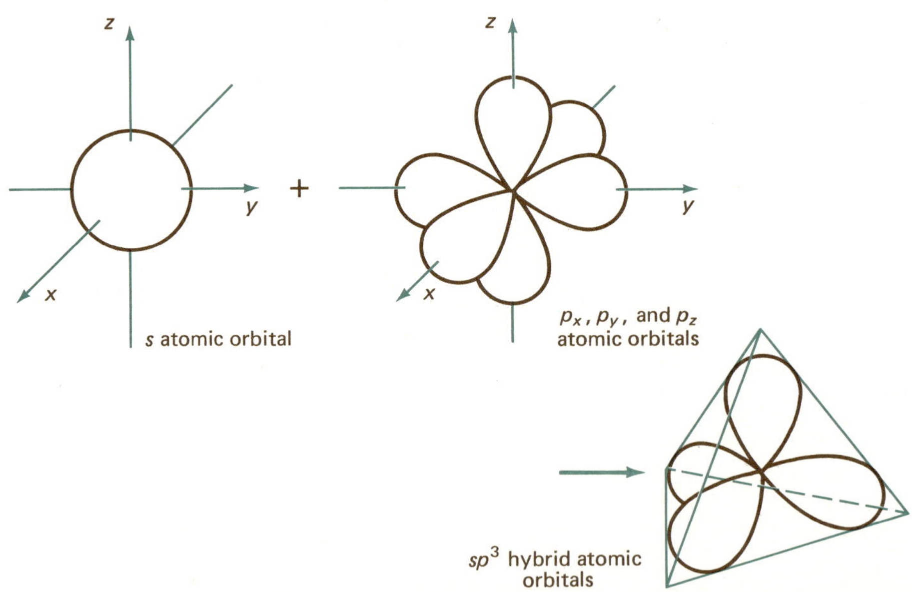

sp^3d Hybridization

Combining one s, three p, and one d orbital results in five ***sp^3d* hybrid orbitals.** The five hybrids are directed toward the corners of a trigonal bipyramid.

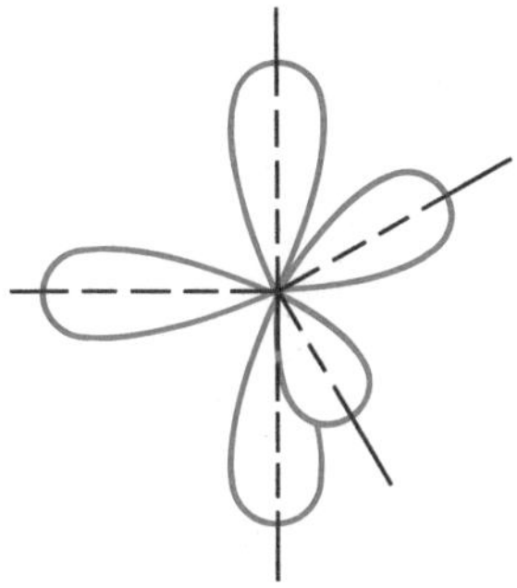

sp^3d hybrid orbitals

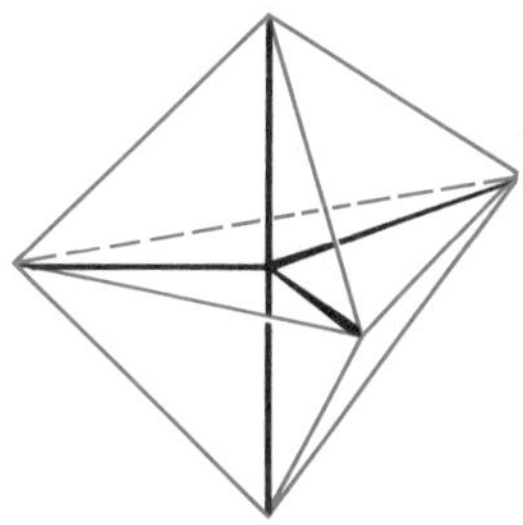

Trigonal bipyramid

sp^3d^2 Hybridization

The final hybridization we want to consider is sp^3d^2. When two d orbitals are combined with an s orbital and three p orbitals, six equivalent ***sp^3d^2* hybrid orbitals** are formed that are directed toward the corners of a regular octahedron.

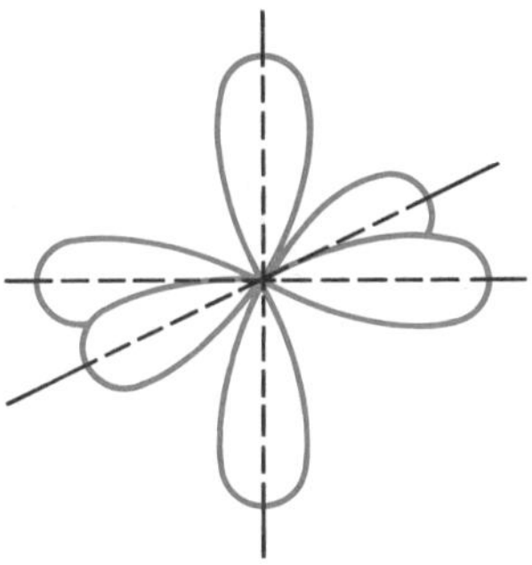

sp^3d^2 hybridization

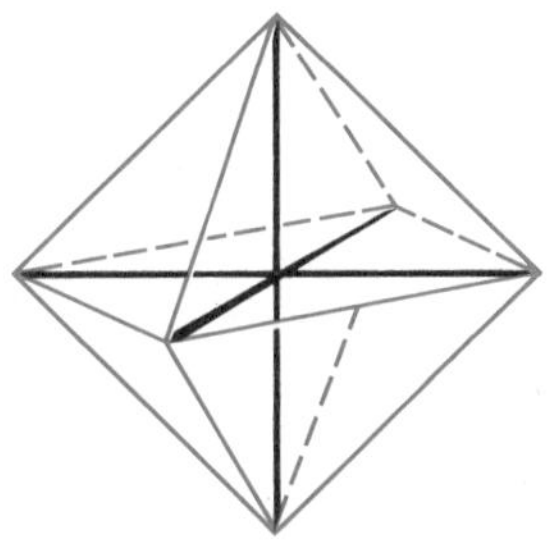

Regular octahedron

Table 14.2 summarizes the different types of hybridization.

TABLE 14.2

Summary of Types of Hybridization

HYBRIDIZATION	NO. OF BONDS	GEOMETRY	BOND ANGLES
sp	2	linear	180°
sp^2	3	triangular	120°
sp^3	4	tetrahedral	109°28′
sp^3d	5	trigonal bipyramidal	120°, 90°
sp^3d^2	6	octahedral	90°

A tetrahedron is a symmetrical four-sided, four-cornered figure.

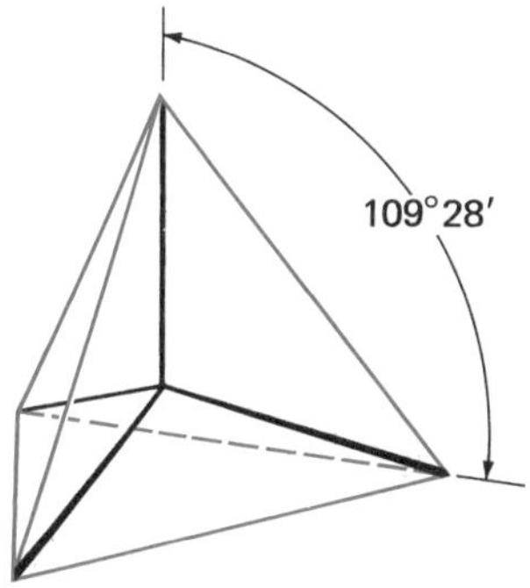

or

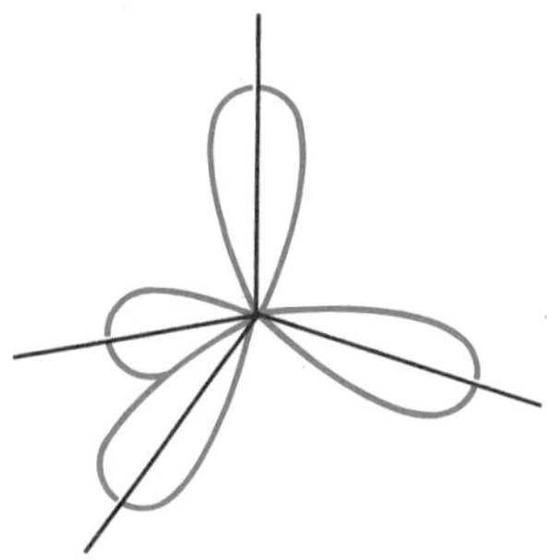

The bond angle in sp^3 hybridization is 109°28′. Tetrahedral bonding is often represented as follows:

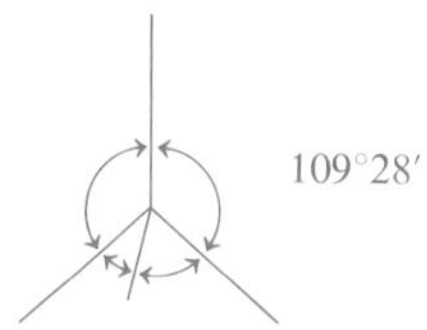

The top line is in the plane of the paper, the two long bottom lines come out of the paper, and the short line goes into the paper.

A trigonal bipyramid is a six-sided figure made by putting two three-sided pyramids base to base. This is our first example of a structure in which not all of the angles between hybrid orbitals (bond angles) are the same. The three positions in the base where the pyramids come together are not equivalent to the other two. The angles in the base are 120° while the others are 90°. We will often represent trigonal bipyramid bonding as follows:

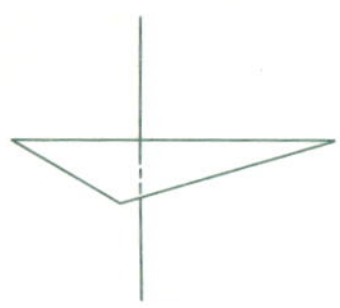

The triangle is perpendicular to the plane of the paper. The five hybrid orbitals are directed toward the corners of the triangle and along the vertical lines.

The octahedron is an eight-sided figure with six corners. It is made by putting together the bases of two square pyramids. All six corners in the octahedron are equivalent. The angle between hybrid orbitals is 90°. We will represent the octahedral structure as follows:

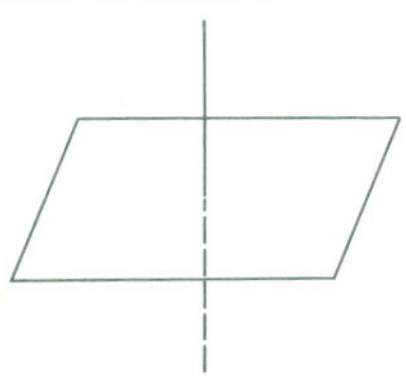

The square is perpendicular to the plane of the paper. The six orbitals are directed toward the corners of the square and along the vertical lines.

Applications of the Hybrid Orbital Concept

We are now ready to consider a number of examples illustrating the usefulness of the hybrid orbital concept. The steps necessary to obtain the bonding and structure of a molecule are as follows: (1) write the electron configuration of the central bonding atom; (2) determine the number of half-filled atomic orbitals needed for bonding; (3) promote electrons to obtain this number; and (4) then assume that sp, sp^2, sp^3, sp^3d, or sp^3d^2 hybridization occurs. An overlapping of orbitals from the peripheral atoms gives the bonding and the structure. The following examples illustrate applications of the theory.

EXAMPLE 14.10 Describe the bonding and geometry of the BF_3 molecule.

SOLUTION The boron is bonded to three fluorine atoms. The configuration of boron is

$$[\mathrm{He}]2s^22p$$

To form three bonds, assume that an electron is promoted from s to p, and then that three sp^2 hybrid orbitals are formed.

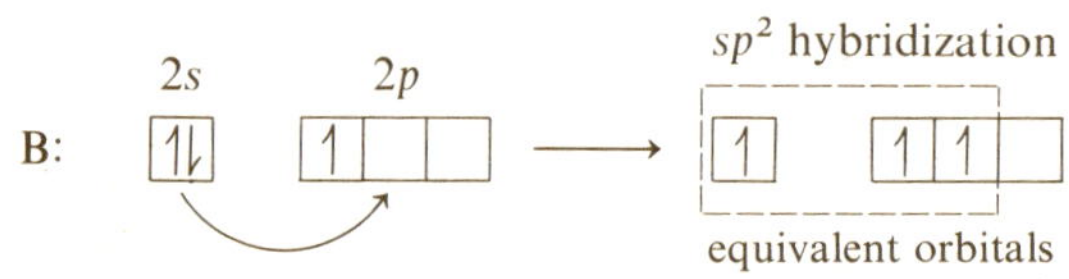

The configuration of fluorine is $[\mathrm{He}]2s^22p^5$. Overlap of a p orbital from each fluorine (the ones containing the single electron) with each of the hybrid orbitals of B gives the BF_3 molecule.

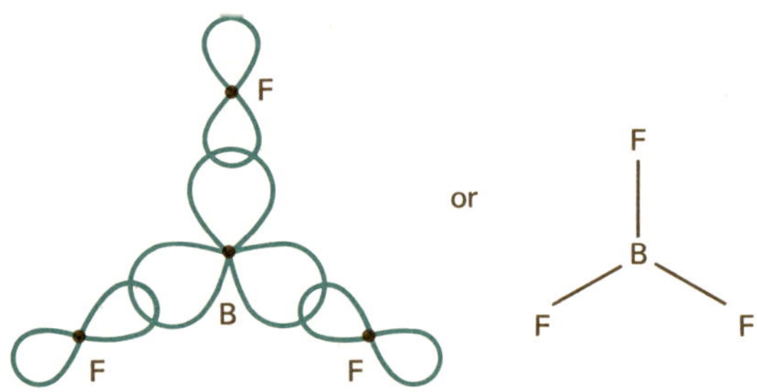

The result is a triangular molecule with three equivalent σ bonds, which is in agreement with experimental evidence.

We next consider an example of sp^3 hybridization, which leads to a tetrahedral structure.

EXAMPLE 14.11 Describe the bonding and geometry of the CH_4 molecule.

SOLUTION Carbon has the electronic configuration $[\mathrm{He}]2s^22p^2$. The four bonding orbitals required are obtained if an s electron is excited to a p state and sp^3 hybridization takes place.

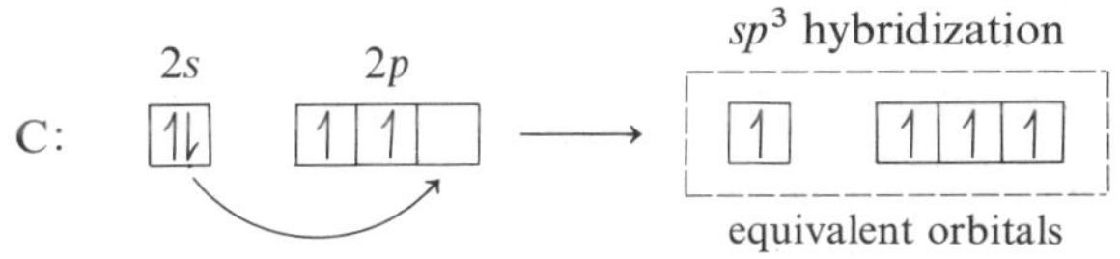

The 1*s* orbitals from hydrogen are overlapped with the sp^3 hybrid orbitals to form a tetrahedral CH_4 molecule.

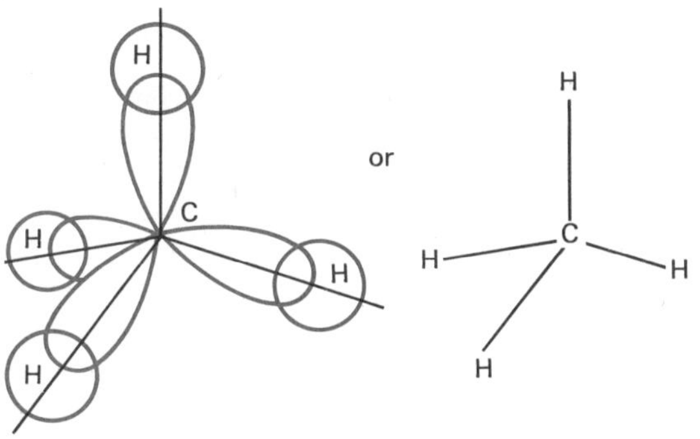

The Coordinate Covalent Bond

In the examples that we have discussed so far, each atom has supplied one electron to each bond. We now consider an example of **coordinate covalent bonding**—that is, one in which both electrons of the bond come from the same atom.

EXAMPLE 14.12 Describe the bonding and geometry of phosphoric acid, H_3PO_4.

SOLUTION The electronic configuration of phosphorus is $[Ne]3s^2 3p^3$, and the four bonding orbitals required are formed from sp^3 hybridization. A coordinate covalent bond is involved—that is, the phosphorus atom supplies both electrons in one of the bonds.

Note that one of the hybrid orbitals contains a pair of electrons.

Oxygen has the electron configuration $[He]2s^2 2p^4$, and two of the *p* orbitals contain a single electron. One of these orbitals overlaps the 1*s* orbital of hydrogen and the other overlaps one of the sp^3 hybrid orbitals from phosphorus:

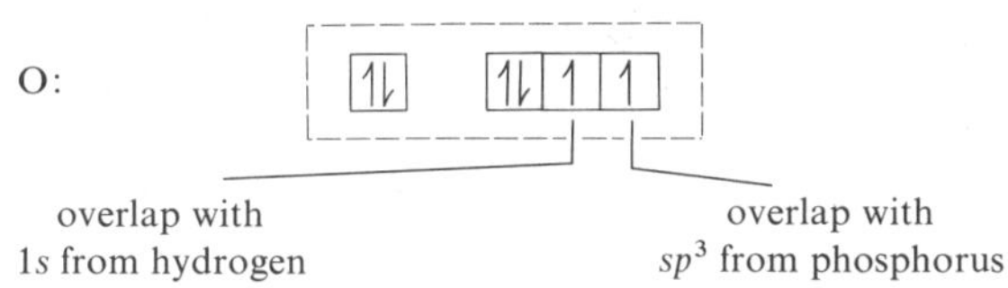

Three of the oxygen atoms are treated in this way. In the fourth, the electrons are paired to leave an empty orbital. This orbital overlaps the sp^3 hybrid orbital from the phosphorus atom that contains the pair of electrons. The result is a coordinate bond:

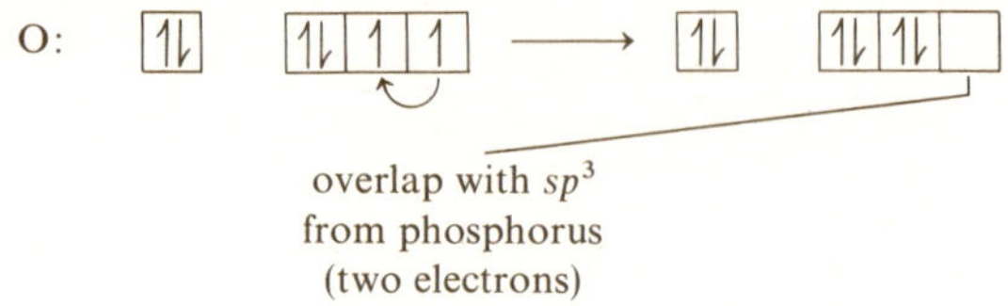

The geometrical configuration of phosphoric acid has phosphorus at the center of a tetrahedron with OH groups at three corners and a coordinately bonded oxygen at the fourth corner.

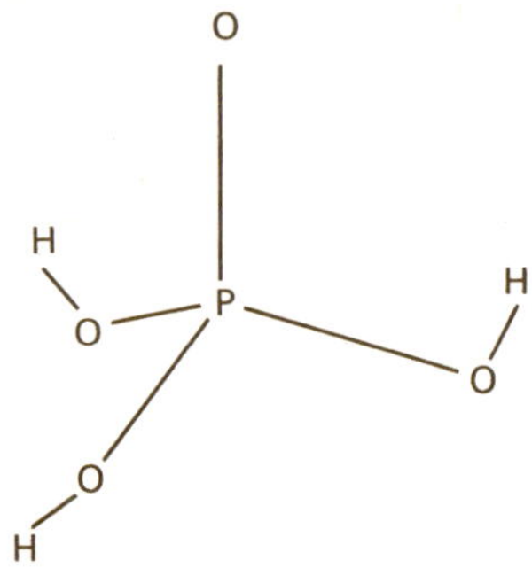

The tetrahedron will be slightly distorted since the corners are not all equivalent. However, when the acid is ionized to form the phosphate ion PO_4^{3-}, all corners will be equivalent and a regular tetrahedron results.

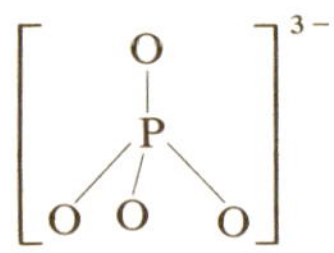

Coordinate covalent bonds are common in many compounds. In H_2SO_4 and SO_4^{2-}, for example, there are two coordinate covalent bonds. The bonding and structure are accounted for by assuming sp^3 hybridization of the valence orbitals of the central sulfur atom. Tetrahedral structure is predicted—distorted in H_2SO_4, and regular in SO_4^{2-}.

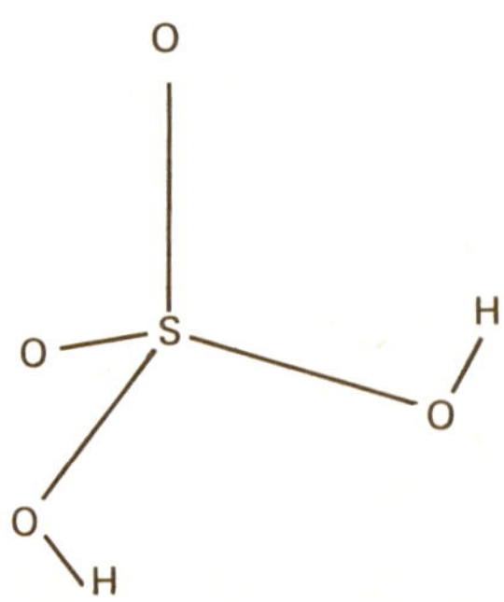

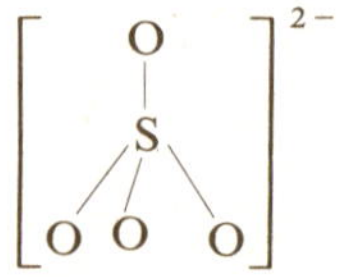

sulfate ion

Examples of Hybridization Involving *d* Orbitals

Formation of sp^3d and sp^3d^2 hybrid orbitals is likely when a central atom is bonded to five and six atoms, respectively. Examples of molecules with the required number of bonds are PF_5 and SF_6. Configurations of P and S are

$$\text{P:}\quad [\text{Ne}]3s^2 3p^3$$
$$\text{S:}\quad [\text{Ne}]3s^2 3p^4$$

To form the correct number of bonds, electrons are promoted and orbitals are hybridized:

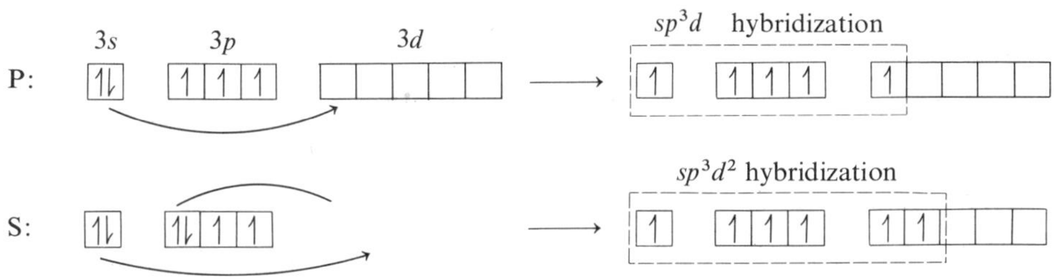

A fluorine *p* orbital containing one electron is then overlapped with each hybrid orbital to form trigonal bipyramidal PF_5 and octahedral SF_6.

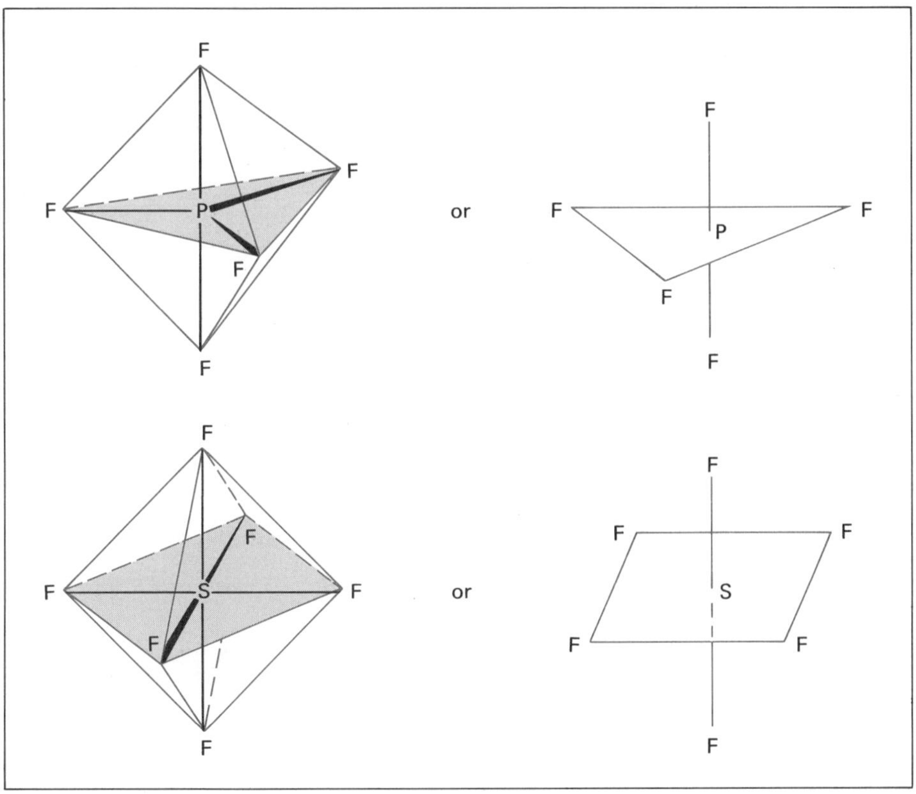

Examples of Bonding When Nonbonding Electrons Pairs Are Present

We must not conclude from the above examples that the number of bonds will always determine the geometry. To determine the molecular structure, one must always look at the electronic configuration. We will now consider several examples.

EXAMPLE 14.13 Describe the bonding and geometry of NH_3.

SOLUTION The electronic configuration of nitrogen is $[He]2s^22p^3$. To form the NH_3 molecule, we assume sp^3 hybridization for nitrogen:

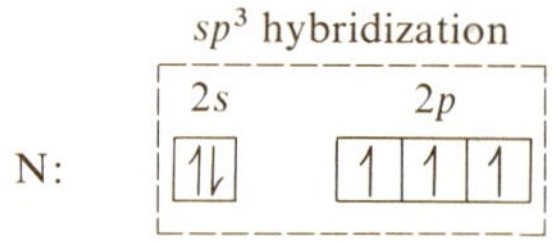

The three orbitals containing single electrons are then overlapped by the 1*s* orbitals from the three hydrogen atoms. The result is a tetrahedral electronic geometry with only a pair of electrons at one corner of the tetrahedron.

We need to distinguish between *electronic geometry*, which is the geometrical structure obtained with the nonbonded electrons making up one corner of the tetrahedron, and the *molecular geometry* determined only by the arrangement of the atoms. The molecular geometry of NH_3 is a triangular pyramid. The molecular geometry of NH_4^+, on the other hand, is the same as its electronic geometry, a tetrahedron.

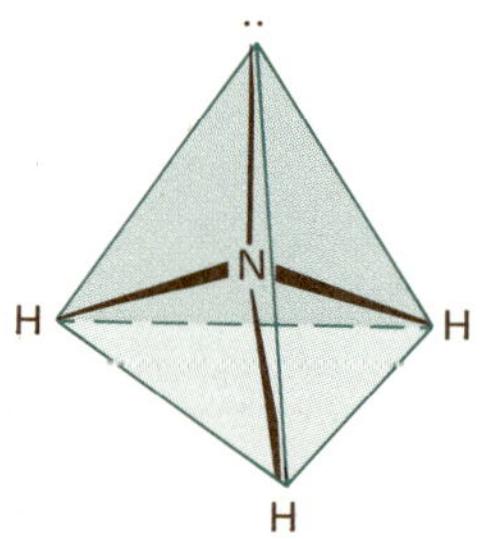

The H—N—H bond angle in NH_3 is 107.3°. It is not surprising that the bond angle is not exactly the tetrahedral value of 109°28′. The pair of electrons at the one corner is repelled by the electronic charge around the hydrogen atoms. As a result, the bond angle is less than the tetrahedral angle. It is not surprising that the bond angle in NH_4^+ ion is precisely the tetrahedral value. NH_4^+ is obtained by adding H^+ to NH_3.

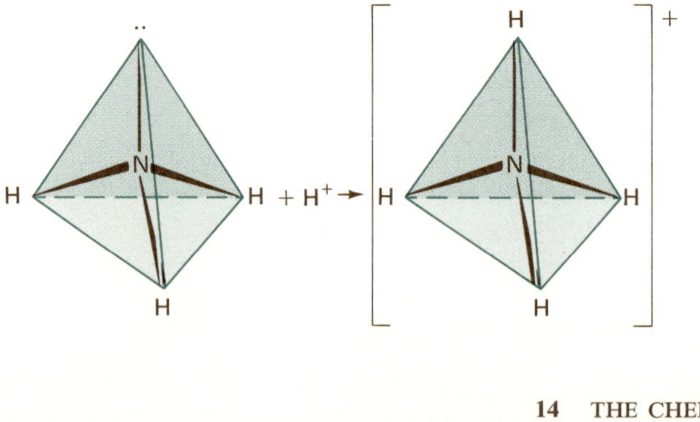

It is possible to explain the bonding in NH_3 and H_2O without hybridization. One could reason as follows: In NH_3, a hydrogen overlaps each of the *p* orbitals that contain one electron. The bond angle should be 90°, but repulsion between the hydrogen atoms increases the angle to 107.3°.

In a similar manner one could reason that hydrogen overlaps the two *p* orbitals in oxygen that contain only one electron to form H_2O. (The third *p* orbital contains a pair of electrons.) The bond angle

The symmetrical structure that results (all four corners now having a hydrogen atom) is tetrahedral.

The bonding in H_2O is also explained by sp^3 hybridization.

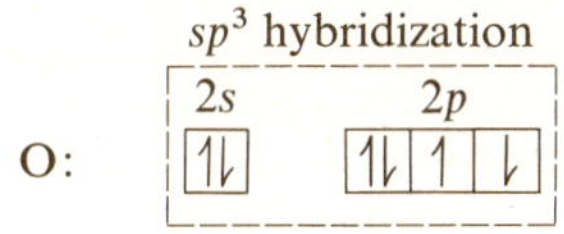

Overlap of 1*s* orbitals from hydrogen atoms with the two hybrid orbitals that contain only one electron gives H_2O. The result is a tetrahedral electronic structure with a pair of electrons at two corners.

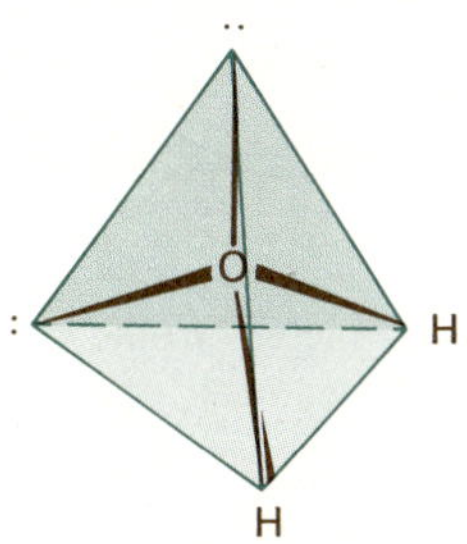

The H—O—H bond angle is 104.5°. The angle is once again less than the 109°28′ tetrahedral angle. The *molecular* structure of H_2O is a bent molecule:

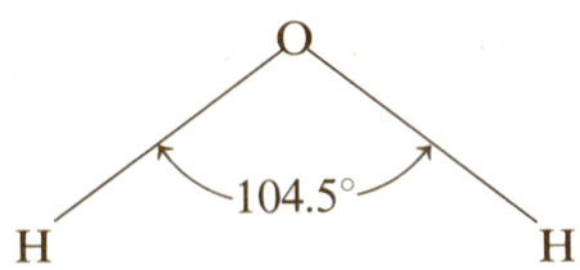

EXAMPLE 14.14 Describe the electronic geometry of SF_4 and IF_5.

SOLUTION The electronic configurations of S and I are

$$\text{S: } [\text{Ne}]3s^2 3p^4$$
$$\text{I: } [\text{Kr}]4d^{10} 5s^2 5p^5$$

To form four bonds with S and five bonds with I, we promote electrons to *d* orbitals and hybridize (the 4*d* orbitals of iodine are filled and do not participate in the bonding).

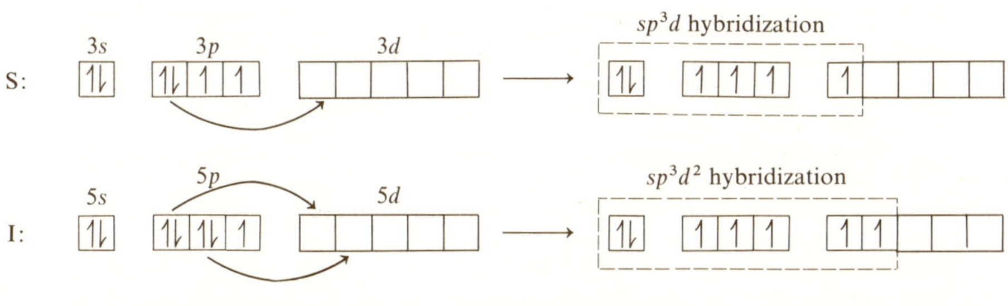

again should be 90°, but hydrogen repulsion again increases the angle, this time to 104.5°. For H_2O and NH_3, the explanation involving hybridization is superior.

The bond angles in both instances are much closer to 109°28′ than to 90°. In other words, the distortion from the tetrahedral angle is less than from the 90° angle of two *p* orbitals. Dipole moments are also closer to what one would predict if electron pairs are present at corners of a tetrahedron. It is interesting, however, that in the analogous PH_3 and H_2S molecules, the bond angle is close to 90°. The nonhybridization explanation appears better for these molecules.

With SF_4, two different structures are possible since the positions are not all equivalent.

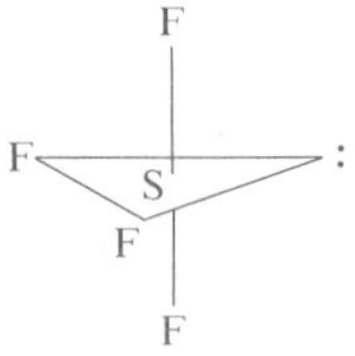

(1)

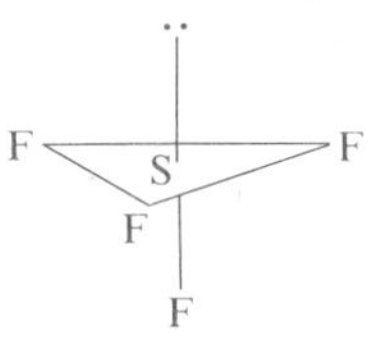

(2)

Experimental evidence supports structure (1) as the stable structure. A similar problem is not encountered with IF_5, since all octahedral positions are equivalent.

Fluorine p orbitals are then overlapped to form the bonds. The result is a trigonal bipyramid with a pair of electrons at one corner in SF_4 and an octahedron with a pair of electrons at one corner in IF_5.

sp^3d

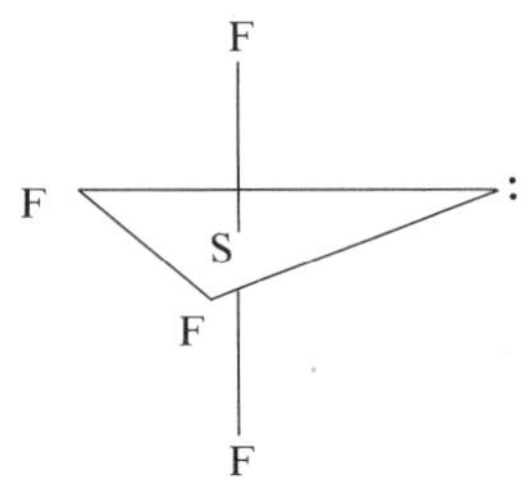

SF_4

sp^3d^2

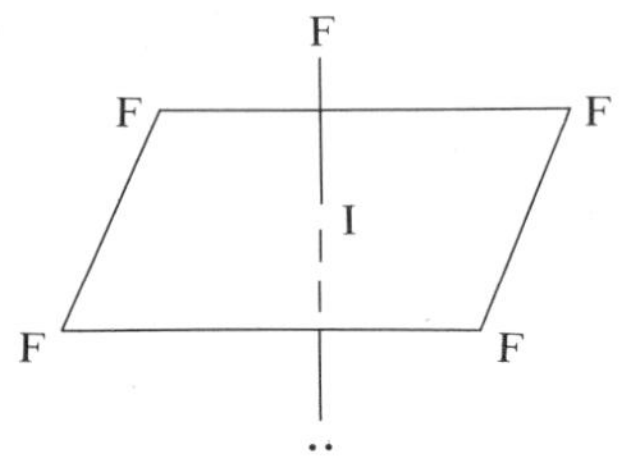

IF_5

EXAMPLE 14.15 Describe (1) the electronic geometry and (2) the molecular shape of XeF_4.

SOLUTION Xe has the electronic configuration $[Kr]4d^{10}5s^25p^6$. To bond Xe to four fluorine atoms, we promote two $5p$ electrons to $5d$ and hybridize.

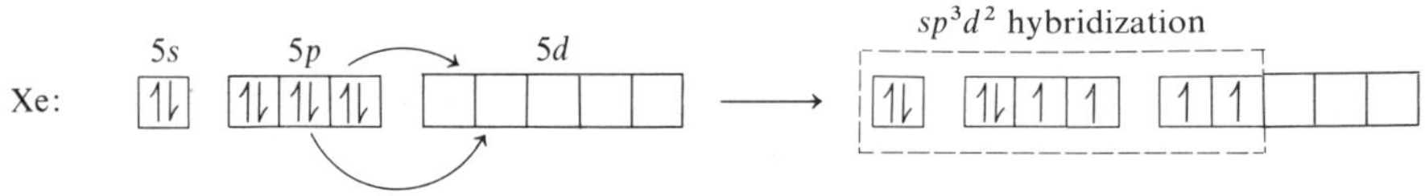

Again, two structures are possible:

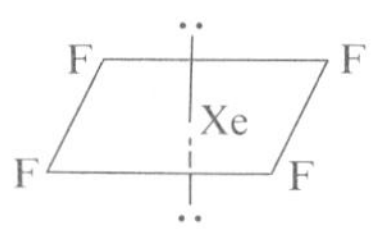

(1)

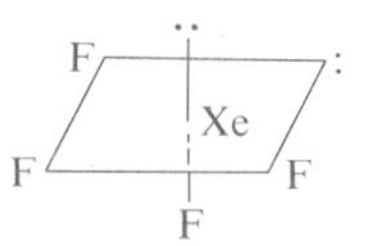

(2)

Structure (1) is more stable since it separates the pairs of electrons (which repel one another) by the greatest distance. XeF_4 may also be described as a square-planar molecule with a high electronic charge above and below the plane.

The hybrid orbitals are overlapped with p orbitals from fluorine. The result is an octahedron for the electronic geometry, with a pair of electrons at two corners.

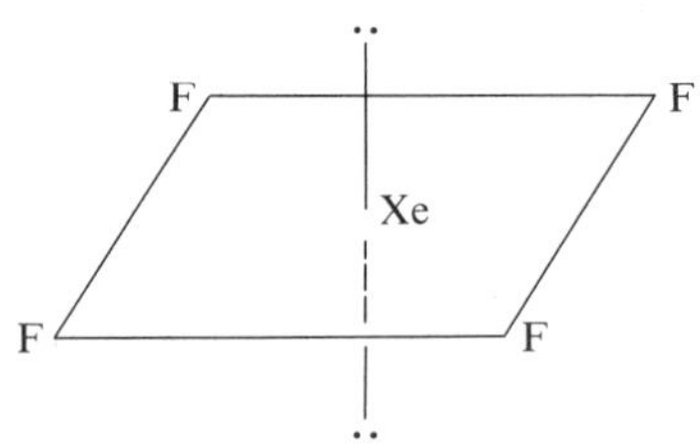

The molecular geometry, determined by the location of the five atomic nuclei, is square planar.

The Bonds of Carbon

In Example 14.11, we found that CH_4 is a tetrahedral molecule. Whenever carbon is bonded to four other groups, the structure is tetrahedral. For example:

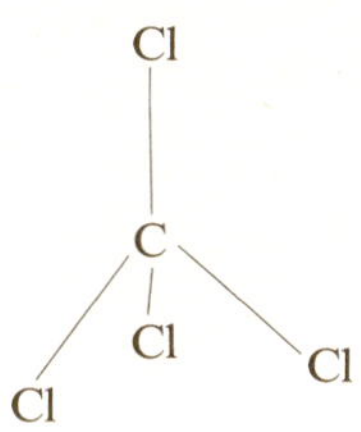

CCl_4
(carbon tetrachloride)
Tetrahedral carbon atom

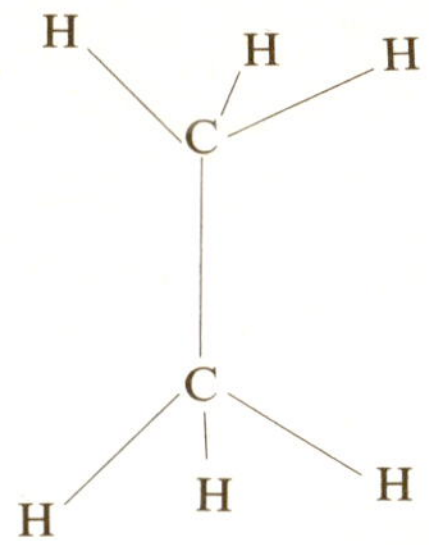

C_2H_6
(ethane)
Both carbon atoms in centers of tetrahedra

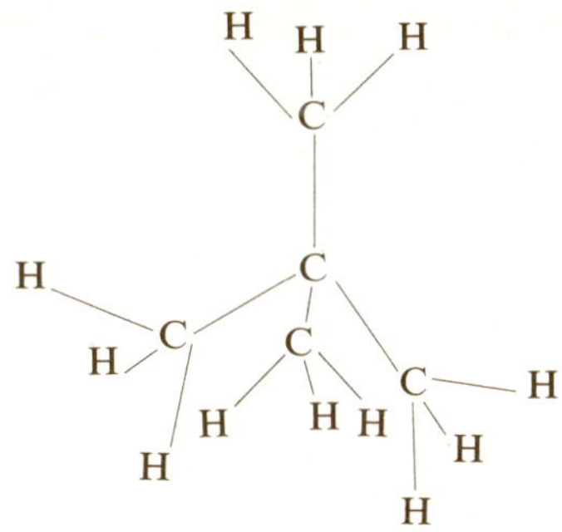

$C(CH_3)_4$
(neopentane)
All five carbon atoms in centers of tetrahedral structures

Double and triple bonds are common in carbon compounds. The description of the bonding in these instances involves π as well as σ bonds.

EXAMPLE 14.16 Describe the bonding in the C_2H_4 (ethylene) molecule.

SOLUTION For each carbon atom, we can assume that an electron is promoted and that hybridization to three sp^2 orbitals takes place.

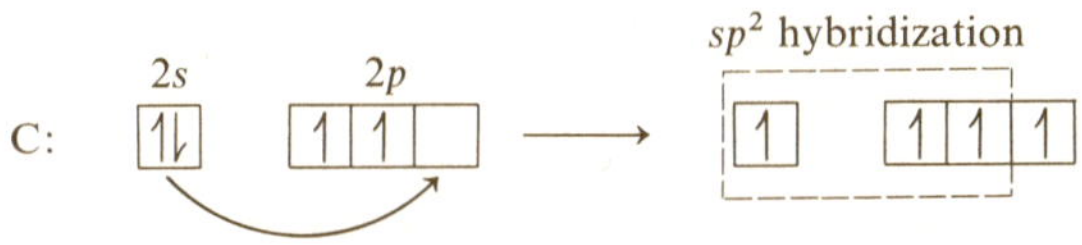

Hybrid orbitals from each carbon atom overlap to form a C—C sigma bond. The other two hybrid orbitals from each carbon atom then overlap a hydrogen 1s orbital to form σ bonds, and the σ bond structure is a flat, double triangle.

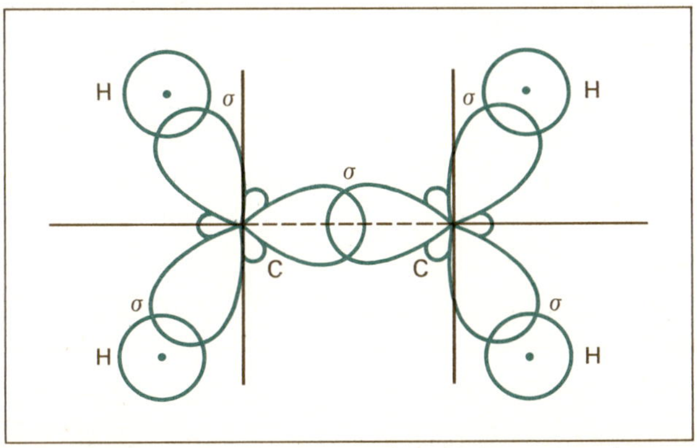

We still have available for bonding the p orbital from each carbon atom, which contains one electron. These p orbitals are directed perpendicularly to the triangle of the sp^2 hybrid orbitals.

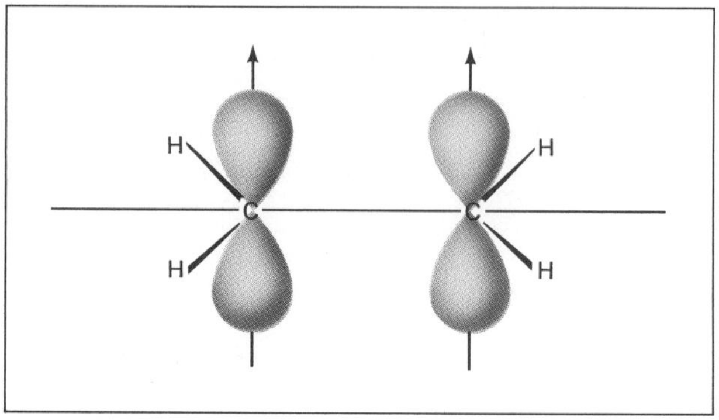

They can now overlap to form a π bond (above and below the plane of the triangles).

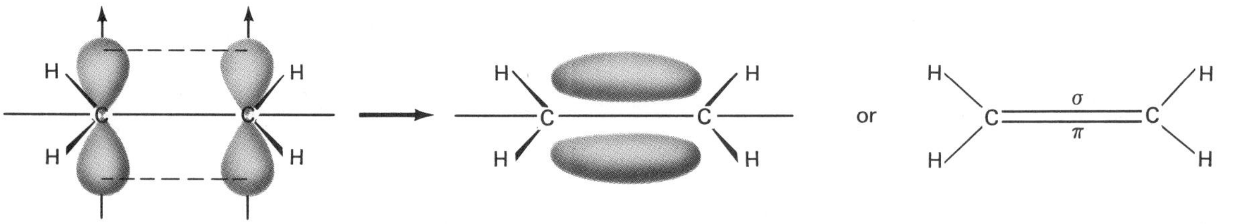

The C—C bond now is a double bond ($\sigma + \pi$).

In ethylene, carbon is bonded to three other atoms. The result is sp^2 hybridization with three σ bonds plus a π bond for each carbon atom. We will now consider an example in which carbon is bonded to two other atoms. The result is sp hybridization with two σ bonds and two π bonds.

EXAMPLE 14.17 Using bond hybridization theory, account for the bonding and geometry in acetylene (C_2H_2). Each carbon is bonded to a hydrogen and a carbon atom (H—C—C—H).

SOLUTION Promote an electron in each carbon and form *sp* orbitals.

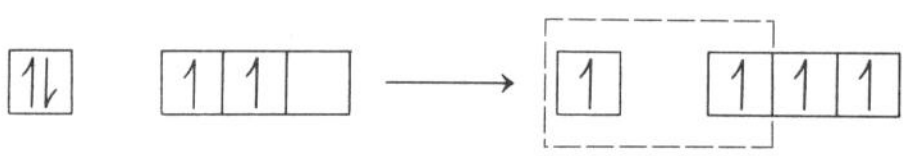

A hybrid orbital from each carbon atom can overlap to form a C—C σ bond. The other hybrid orbital (from each carbon) can then overlap a hydrogen 1*s* orbital. The result is a linear σ bond structure. Finally, the remaining *p* orbitals overlap to form two π bonds. Hence, the bonding between the two carbon atoms is similar to that in N_2 (see Figure 14.30).

Resonance

The valence bond theory, as we have described it so far, very effectively describes the bonding and geometry of a large number of molecules. However, for some molecules, the bonding cannot be described in such simple terms as specific bonds between adjacent atoms. In such systems the actual bonding does not correspond to any single structure that can be drawn, but rather to a structure intermediate between several formal structures. This description is called **resonance** and the molecule is said to be a **resonance hybrid** of the different formal structures.

The name resonance is somewhat misleading. The individual molecules do not change back and forth from one structure to another (resonate). All molecules are the same; each has a structure intermediate between the different resonance structures.

Consider the benzene (C_6H_6) molecule as an example. In benzene, the arrangement of carbon atoms and of hydrogen atoms is

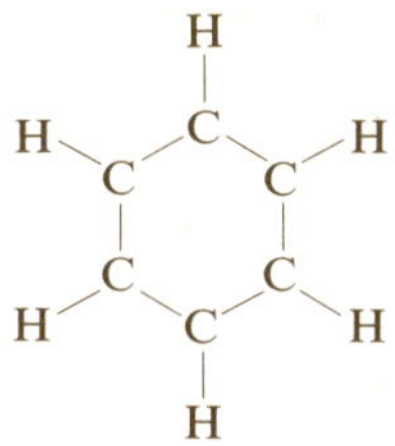

Since each carbon atom is bonded to three other atoms, in the valence bond description of the bonding in this molecule we assume that each carbon atom has sp^2 hybridization:

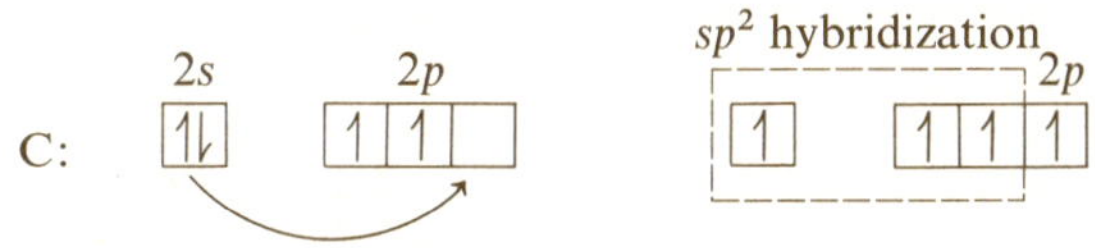

Overlap of the sp^2 hybrid orbitals of each carbon atom with similar orbitals on adjacent carbon atoms and with 1*s* orbitals from hydrogen gives the σ bond structure for the molecule.

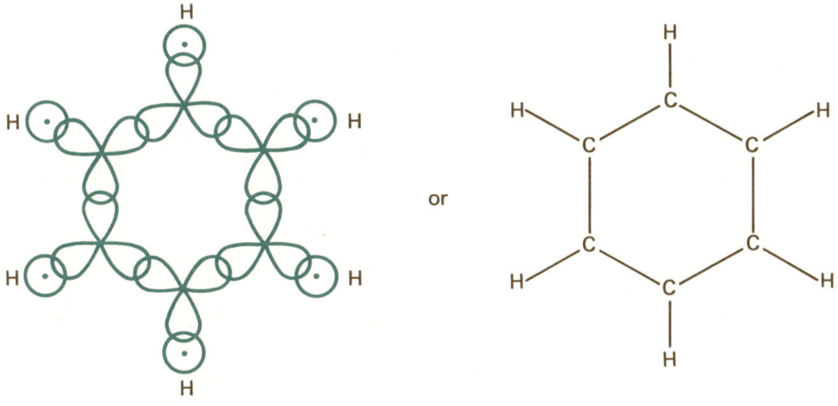

In the second representation the lines show the sigma bonds only. In addition, each carbon atom has a single electron in a p orbital perpendicular to the plane of the ring. These p orbitals can overlap to form π bonds with high electron density above and below the plane of the ring.

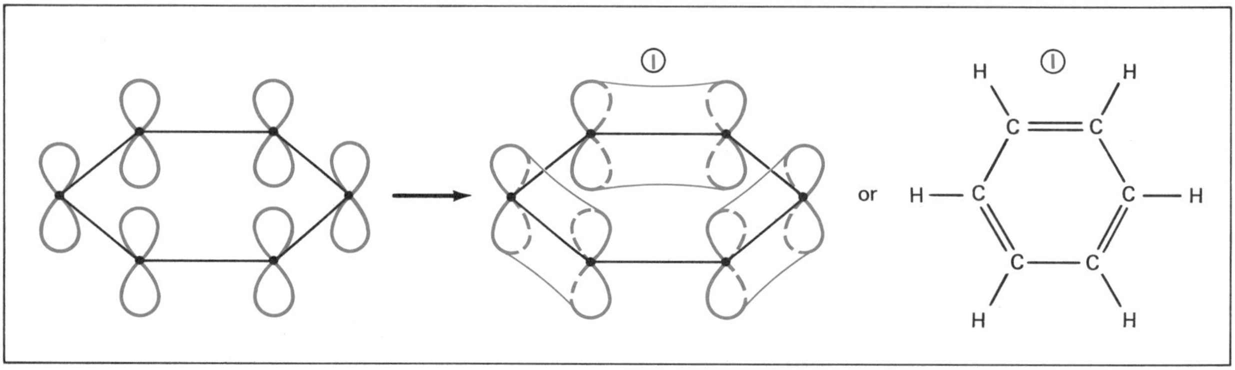

Thus far, the theory predicts a ring structure for benzene with alternating double and single bonds. However, the experimental evidence shows that this is not so. Structural determination indicates that all six carbon-carbon bonds are equivalent. To resolve this problem, we return to our description of the overlap of the p orbitals from the carbon atoms. An alternative way that these can overlap is as follows:

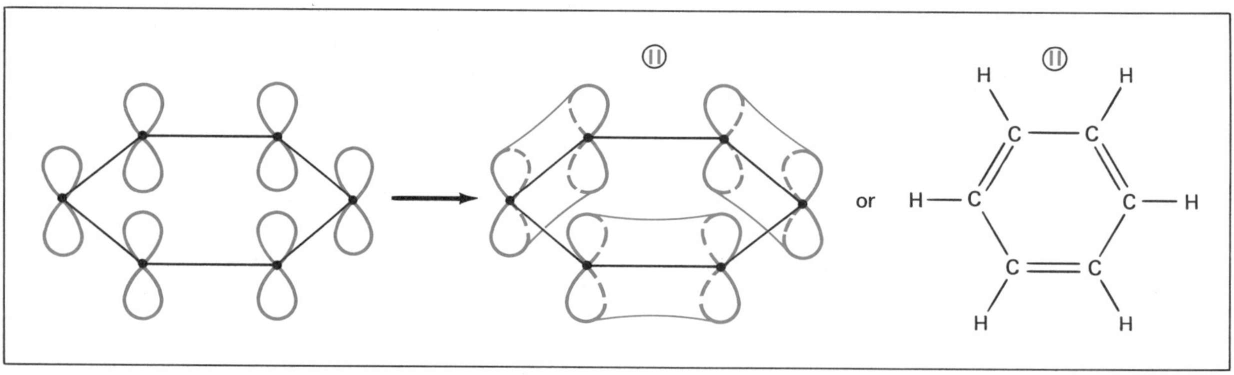

Structures I and II are equivalent. The actual structure is intermediate between the two. In other words, the actual structure is a resonance hybrid of structures I and II. We represent the structure of benzene as follows:

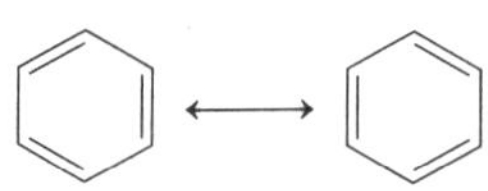

Kekulé structures

The double-ended arrow indicates that the actual structure is a resonance hybrid of the two structures. (We usually simplify the structural drawings of ring molecules, such as benzene, by not showing the hydrogen or carbon atoms.)

Many other molecules require a resonance description. The following are some commonly encountered examples:

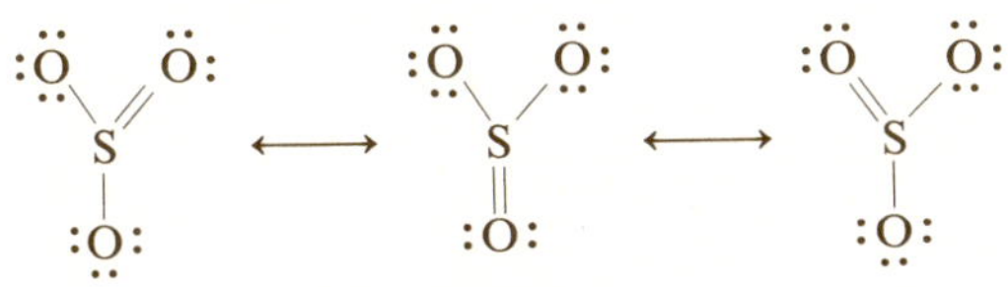

Sulfur trioxide

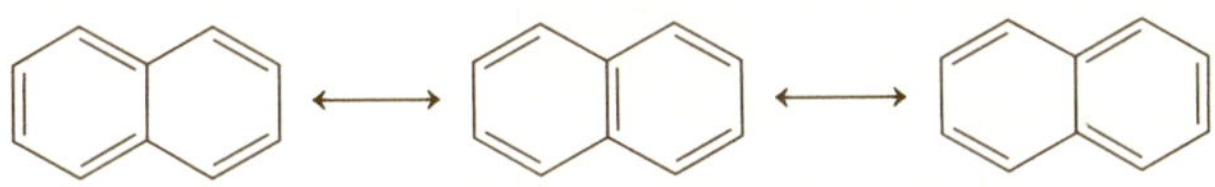

Naphthalene

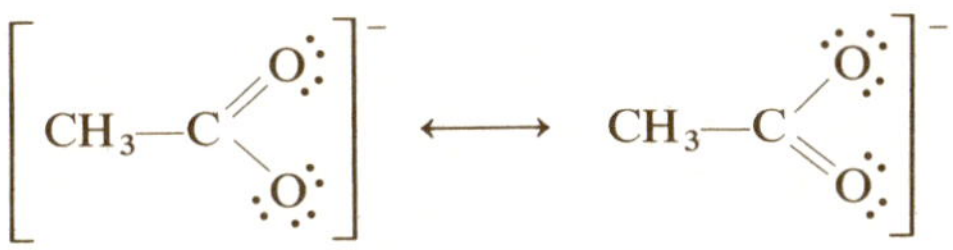

Acetate ion

Ozone

An interesting analogy to the resonance description is the medevial European traveler who returns from a trip to India and describes the rhinoceros as resembling a cross between a dragon and a unicorn. (A description of a real animal in terms of two familiar but imaginary animals.)

In the molecular orbital theory, electrons can be spread over the entire molecule in molecular orbitals, and a resonance description is not necessary. The molecular orbital representation for benzene is

The circle represents the six π electrons, which are uniformly distributed around the ring in molecular orbitals.

14.3 Dipole Moments and Geometry

With our understanding of molecular geometry, we are now ready to extend the discussions in Chapter 5 on polarity and dipole moments to more complicated molecules. It is important to recognize that molecules containing atoms with differing electronegativities may not necessarily be polar molecules. One of the simplest examples is carbon dioxide. This linear molecule is nonpolar even though oxygen is much more electronegative than carbon. It is the symmetry of the molecule that makes it nonpolar. In the drawings shown below, the electrical charges are those that would result if each bond were considered separately. In the case of CO_2, the positive charge is on the C atom, which is the center of the molecule, and the *center* of the negative charge is at the same place. Thus, there is no net displacement of charge in the molecule, and the molecule is nonpolar.

$$\underset{\delta^-}{O}=\underset{\delta^+}{C}=\underset{\delta^-}{O}$$

In like manner, a symmetrical triangular molecule is nonpolar. Consider BF_3 as an example.

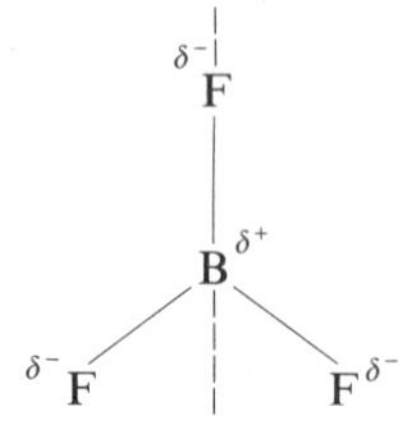

When the effects of the two fluorine atoms off the dotted line are added, they just balance the effect of the fluorine on the line, and no net separation of charge occurs. The same argument can be extended to tetrahedral, trigonal bipyramidal, and octahedral molecules. When the same kinds of atoms occupy all the exterior positions, no dipole moment is present. Examples are CF_4, PF_5, or SF_6.

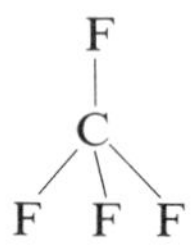

CF_4

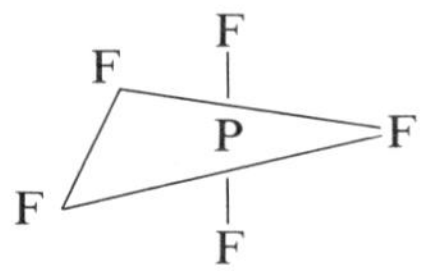

PF_5

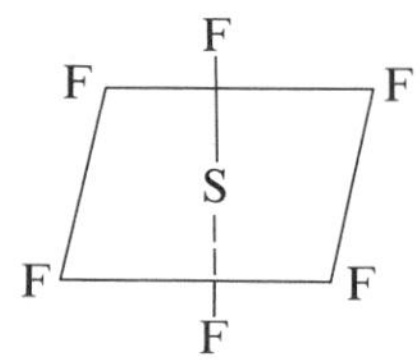

SF_6

When different kinds of atoms occupy the exterior positions, the molecule is polar, and the larger the electronegativity difference, the larger the dipole moment. For example, $CHCl_3$ is a polar molecule.

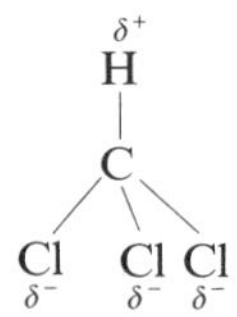

Summary

From the detailed solution of the Schrödinger wave equation for H_2^+ and from approximate methods for more complex systems, we are able to obtain wave functions and energies for molecular systems in a way very similar to that for atomic systems, as discussed in Chapter 13.

As with the atomic systems, however, detailed solutions of the wave equation have been made for only the most simple of the molecular systems. Hence, we are forced to use approximate theories to describe covalent bonding in molecules. One of these is a molecular orbital model, with the wave functions being estimated from a combination of those of the atomic orbitals. Description of molecular orbitals and quantized energy levels in the molecule are the result. Electronic configurations and, therefore, bonding in the molecule can be deduced.

The molecular orbital theory provides a very satisfactory description of covalent bonding for simple molecules. For example, relative bond distances, bond energies, and magnetic properties predicted from molecular orbital theory are in excellent agreement with the experimental information for a large number of diatomic molecules.

For complex molecules we resort to the less elegant, but highly useful, valence bond theory. In valence bond theory, the identities of the atomic orbitals are maintained, and bonding is described in terms of overlap of atomic orbitals.

When ordinary atomic orbitals (*s*, *p*, and *d*) are used in valence bond theory, erroneous predictions about the geometry of the molecules often occur. If, however, hybrid atomic orbitals (formed from *s*, *p*, and *d* orbitals) are used, valence bond theory proves to be very useful in accounting for both the nature of the bonding and the observed geometry of a large number of molecules. Each hybrid atomic orbital discussed and its geometry are as follows:

HYBRIDIZATION	NUMBER OF ORBITALS	ELECTRONIC GEOMETRY
sp	2	linear
sp^2	3	triangular
sp^3	4	tetrahedral
sp^3d	5	trigonal bipyramidal
sp^3d^2	6	octahedral

There are some molecules that can be described only as intermediate in their properties to two or more structures composed of formal bonds. This description of a molecule in terms of two or more structures (for example, the two Kekulé structures for C_6H_6) is called resonance.

Important Terms and Concepts

Molecular orbital theory
Valence bond theory
Bonding orbital
Bond length
Bond energy
Linear combination of atomic orbitals (LCAO) approximation
σ orbital
Antibonding orbital
Chemical bond
π orbital
Core electrons
Valence electrons
Valence bond theory
σ bond
π bond
Hybrid atomic orbital
sp, sp^2, sp^3, sp^3d, and sp^3d^2 hybrid orbitals
Coordinate covalent bonding
Resonance
Resonance hybrid

Exercises

14-1.
Use molecular orbital theory to compare the stability of the following molecules:

(a) Ar_2 and Ar_2^+
(b) Ar_2 and Ar_2^-
(c) Ar_2^+ and Ar_2^-

14-2.
Which of the following molecules will have unpaired electrons and hence be paramagnetic?

(a) B_2^+ (d) C_2^+
(b) B_2 (e) C_2
(c) B_2^- (f) C_2^-

14-3.
Use molecular orbital theory to describe the bonding and compare the stability of the following pairs of molecules:

(a) Cl_2 and Cl_2^- (d) Si_2 and Si_2^-
(b) S_2 and S_2^- (e) P_2 and Cl_2
(c) P_2 and P_2^-

14-4.
Predict from molecular orbital theory the relative bond energies of F_2^+, F_2, and F_2^-.

14-5.
Determine the bond order from molecular orbital theory for each of the following:

(a) H_2 (d) He_2^+
(b) H_2^+ (e) He_2
(c) HHe

14-6.
Use molecular orbital theory to predict for the CN molecule

(a) the number of π and σ bonds.
(b) whether CN or CN^+ would have the longer bond length.
(c) whether CN or CN^- would have the larger bond energy.

14-7.
(a) How many σ and π bonds are there in NO?
(b) Would NO or NO^- have the larger bond energy, and why?
(c) Would NO or NO^+ have the longer bond length, and why?
(d) Draw how you would expect the π^b and π^* orbitals to look around NO (make separate drawings for each type of orbital).

14-8.
Use molecular orbital theory to describe the bonding and compare the stability of the ClS, S_2, and PS molecules.

14-9.
Predict from molecular orbital theory which of the following molecules would have unpaired electrons, and, hence, be paramagnetic.

(a) NO
(b) CO
(c) PCl

14-10.
Use valence bond theory to describe the geometry of the hydrogen peroxide (H_2O_2) molecule.

14-11.
Use valence bond theory to predict the geometry of the following:

(a) NH_2OH (hydroxylamine)
(b) $NH_2—NH_2$ (hydrazine)

14-12.
Predict the hybridization in the following molecules:

(a) BeF_2
(b) BF_3
(c) CF_4

14-13.
Predict the hybridization and, hence, the geometry around the underlined atom for each of the following molecules.

(a) $\underline{Hg}(CH_3)_2$
(b) $\underline{Pb}(CH_3)_4$

14-14.
Predict the geometrical structure of each of the following molecules:

(a) CH_3NH_2 (methylamine)
(b) CH_3CN (acetonitrile)
(c) HCOOH (formic acid)
(d) CH_3OH (methyl alcohol)
(e) C_2Cl_4 (tetrachloroethylene)
(f) $(CH_3)_2CO$ (acetone)
(g) $(CH_3CH_2)_2O$ (diethylether)

14-15.
Predict the hybridization and, hence, the geometry around the underlined atom in each of the following:

(a) $\underline{C}OCl_2$
(b) $\underline{S}O_2F_2$
(c) $\underline{P}OCl_3$

14-16.
Use valence bond theory to predict the hybridization and, hence, the geometrical structure of the NCl_3 molecule.

14-17.
In carbon dioxide (CO_2), the carbon atom is between the oxygen atoms. Use valence bond theory to predict the bonding and the geometrical structure of the CO_2 molecule.

14-18.
Predict the hybridization and, hence, the geometrical arrangement around the underlined atoms in each of the following molecules:

(a) $\underline{Al}F_3$
(b) $\underline{S}F_2$
(c) $\underline{Cl}F_3$

14-19.
Use valence bond theory to describe the bonding and geometry of the KrF_2 and KrF_4 molecules.

14-20.
Use valence bond theory to predict the geometry of the following compounds. Assume covalent bonds form.

(a) $SnCl_2$
(b) $SnCl_4$
(c) $CdCl_2$
(d) $GaCl_3$

14-21.
Use valence bond theory to discuss the geometrical structure of each of the following molecules:

(a) PbF_4
(b) PoF_4
(c) RnF_4
(d) TeF_4

14-22.
Predict the hybridization and, hence, the molecular shape for each of the following ions:

(a) SO_4^{2-}
(b) PO_4^{3-}
(c) NO_3^-
(d) CO_3^{2-}

14-23.
Determine the hybridization of the underlined atoms and, hence, the geometrical structure of each of the following species:

(a) $\underline{Sn}(OH)_4$
(b) $\underline{Sb}(OH)_5$
(c) $\underline{Te}(OH)_6$
(b) $\underline{I}O(OH)_5$

14-24.
Predict which of the following molecules would have a dipole moment.

(a) CH_2Cl_2

(b) 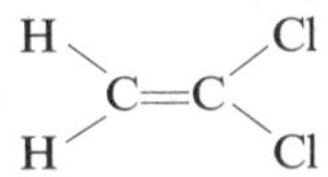

(c) 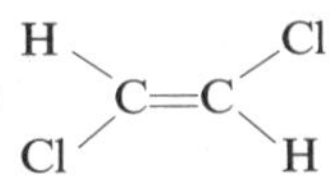

(d) 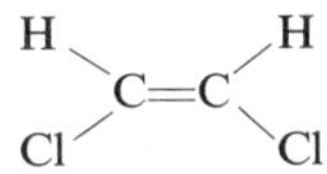

(e) $POCl_3$

(f) $H_2C{=}O$

14-25.
Use valence bond theory to predict which of the following molecules would have a dipole moment.

(a) OF_2
(b) SiF_4
(c) BeF_2
(d) IF_3
(e) SF_4

Problems

14-1.
The bond dissociation energies for some gaseous metal molecules as determined from spectroscopy are as follows:

MOLECULE	DISSOCIATION ENERGY (kjoules/mole)
Ag_2	160
Au_2	220
Zn_2	25
Cd_2	8

Use molecular orbital theory to explain why Ag_2 is so much more stable than Cd_2.

14-2.
Given the energies for the following reactions:

$Cl_2 \rightarrow Cl_2^+ + e^-$	1120 kjoules/mole
$Cl_2 \rightarrow Cl_2^{2+} + 2\,e^-$	3150 kjoules/mole
$Cl_2 \rightarrow 2\,Cl$	238 kjoules/mole
$Cl \rightarrow Cl^+ + e^-$	1260 kjoules/mole
$Cl^+ \rightarrow Cl^{2+} + e^-$	2300 kjoules/mole

(a) calculate the dissociation energies of Cl_2^+ and Cl_2^{2+} ($Cl_2^+ \rightarrow Cl + Cl^+$; $Cl_2^{2+} \rightarrow Cl + Cl^{2+}$).
(b) compare the dissociation energies of Cl_2, Cl_2^+, and Cl_2^{2+}, and discuss the results in terms of molecular orbital theory.

14-3.
Explain from molecular orbital theory why the electron affinity of O_2 is much smaller (14.6 kjoules/mole) than for CN (369 kjoules/mole).

14-4.
The oxygen molecule has an excited electronic state that is 431 kjoules/mole higher in energy than the ground state. This excited state differs from the ground state in that one of the bonding π electrons has been raised to a π^* antibonding orbital.

(a) Predict the bond order for O_2 in this excited state and then predict how the bond energy and bond length will differ from the ground state.
(b) What would be the wavelength of the photon that would be given off as the electron in this excited state falls back from the π^* to the π molecular orbital?

14-5.
The table given below summarizes the bond energies and bond lengths for the diatomic molecules Li_2, Na_2, and K_2.

	BOND ENERGY (kjoules/mole)	BOND LENGTH (Å)
Li_2	105	2.672
Na_2	72.4	3.078
K_2	49.4	3.923

Use molecular orbital theory to explain the systematic changes.

14-6.
The molecule borazole, $B_3N_3H_6$, has the following arrangement of atoms:

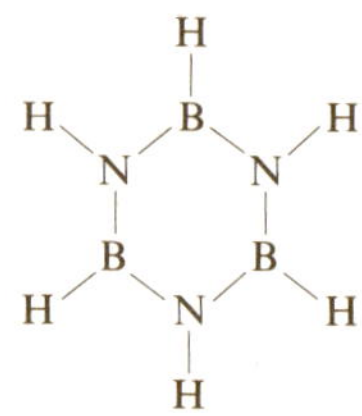

How does the bonding in this molecule compare with the bonding in benzene?

14-7.
Solid PCl_5 consists of equal numbers of PCl_4^+ and PCl_6^- ions. In the liquid and vapor states, however, the compound exists as PCl_5 molecules. Predict the hybridization and the geometrical structure for PCl_4^+, PCl_6^-, and PCl_5.

14-8.
Iodine heptafluoride, IF_7, is one of the very few substances in which seven atoms are bonded to a central atom.

(a) What hybridization would you predict for iodine in IF_7?
(b) By comparing IF_7 with such molecules as SnF_4, SbF_5, and TeF_6, can you suggest a likely structure for IF_7?
(c) Suggest a possible structure for XeF_6 by comparing it with IF_7.

14-9.
Draw important contributing resonance structures for the following molecules:

(a) 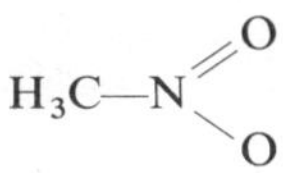 (nitromethane)

(b) 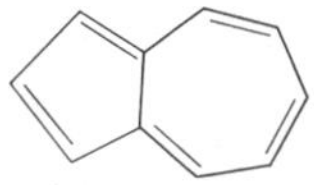 (azulene)

14-10.
In Chapter 24 we will give several examples of isomerism. Compounds are isomers if they have the same molecular formula but different arrangements of the atoms in the molecule. Examples of different kinds of isomerism are as follows:

(a) Ethyl alcohol and dimethyl ether are examples of structural isomers.

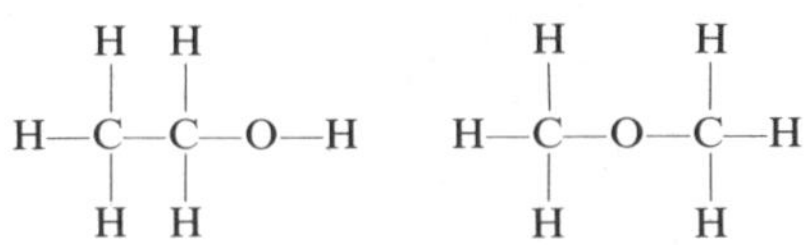

(b) Cis- and trans-dichloroethylene are examples of geometrical isomers.

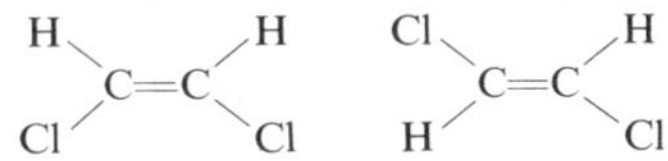

(c) *d*- and *l*-bromochlorofluoromethane are examples of optical isomers.

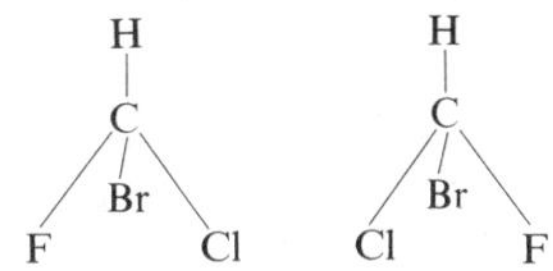

Construct three-dimensional models of the six molecules above, and show that in each of the three cases the molecules in a pair of isomers are indeed different. Try to ascertain what features characterize each of the three types of isomerism. Effective molecular models can be made by connecting different colors of gumdrops (representing the different kinds of atoms) with toothpicks.

15

The Solid State

In Chapter 6, we described a solid as a collection of molecules, atoms, or ions arranged in a geometrical array called a *lattice*. The lattice repeats itself in all three directions. As a result, the molecules are aligned in rows, columns, and planes. In this chapter we will consider the solid state in more detail. First we will describe the geometrical structure of some simple solids. Then we will see how the structure explains the macroscopic properties of the solid. Next, we will describe the experimental techniques that are used to determine the arrangement of atoms, ions, or molecules in a crystal; and finally, we will discuss the ionic bond.

15.1 The Crystal Lattice

A **lattice** is a regular geometrical arrangement of points or objects in space. Figure 15.1 is an example of a lattice. Atoms, molecules, or ions are at the points of the intersection of the lines. The lattice is characterized by the distance between successive points (a, b, c) along each of the three axes and by the angles (α, β, γ) between these axes.

The **unit cell** is useful in describing a lattice. The unit cell is shown in heavy outline in Figure 15.1. It is the smallest part of the lattice that retains all the geometrical properties of the lattice. One can see from the figure that the lattice is produced by successive displacements of the unit cell along each of the crystal axes.

When particles form a lattice, they usually pack together in such a way that the amount of unoccupied space is kept to a minimum. As a result, the crystal lattice for a particular solid depends greatly on the size and shape of the atoms, molecules, or ions that compose the solid. As an example, Figure 15.2 compares possible packing schemes for spherical atoms and for long molecules. Let us first describe the lattices formed by the packing of spherical atoms, after which we can describe the structure of more complicated systems.

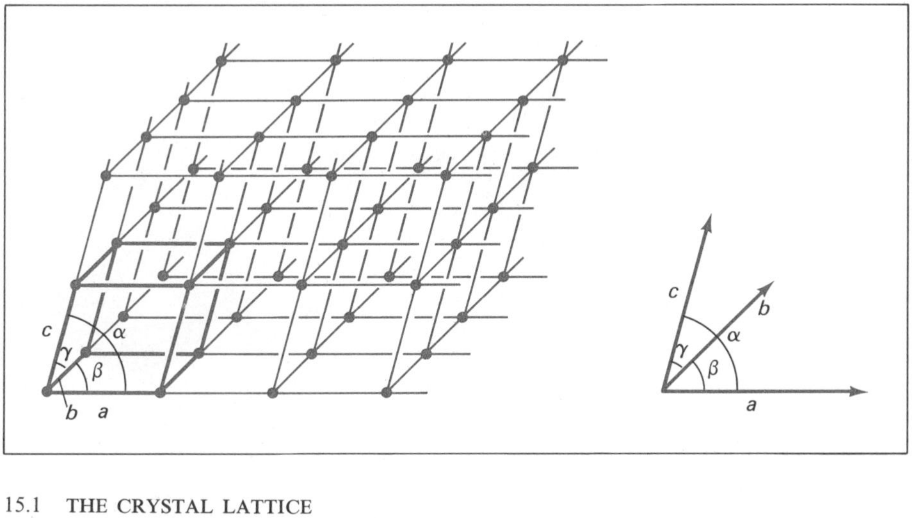

Figure 15.1
A lattice of points showing the unit cell in heavy outline. The three edges of the unit cell are of length a, b, and c. The angles are α, β, and γ.

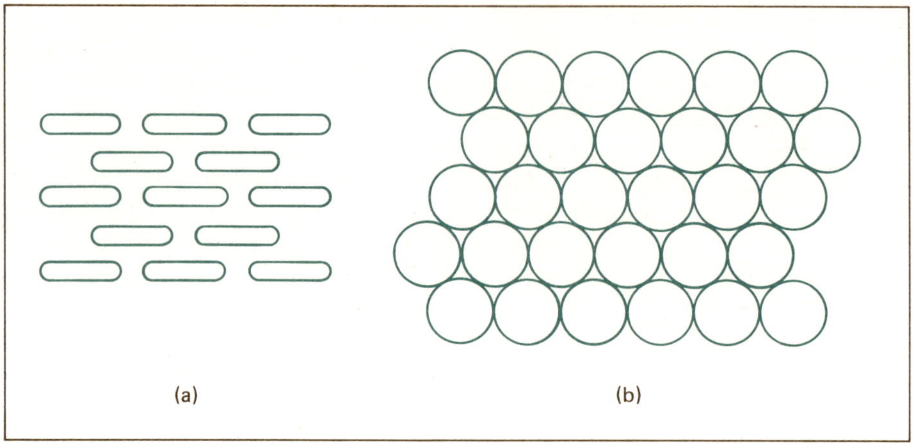

Figure 15.2
Possible packing schemes (two-dimensional) for (a) long molecules and (b) spherical atoms.

15.2 Atomic Solids

Simple Cubic Packing

The least complicated structure into which spherical particles (atoms) can be arranged is the **simple cubic lattice.** In this lattice the spheres are arranged in layers, as shown in Figure 15.3. Notice that the lattice points at the center of the spheres are at the corners of a square. The three-dimensional lattice results from

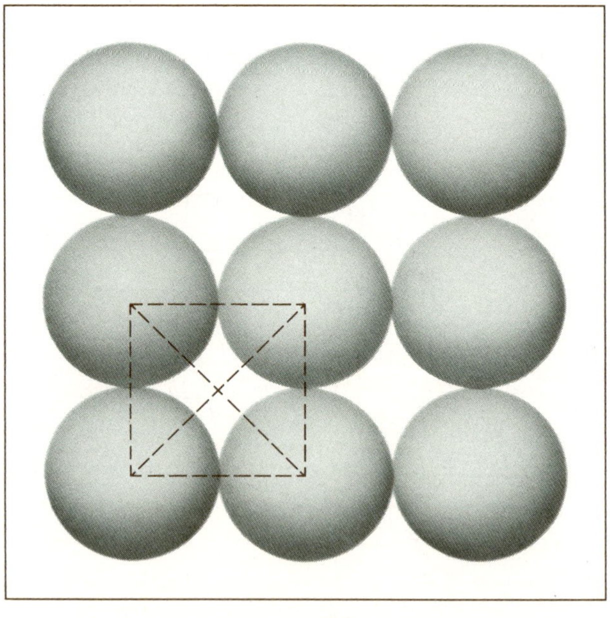

Figure 15.3
Layer of spheres in the simple cubic lattice.

Figure 15.4
Stacking of layers of spheres in the simple cubic lattice.

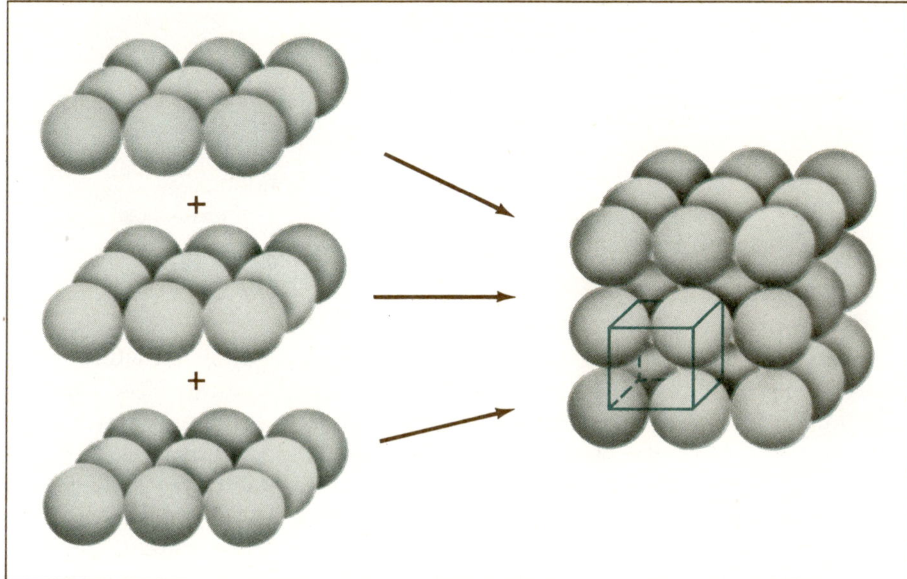

stacking of the layers directly on top of one another, as shown in Figure 15.4. The unit cell as outlined is a cube; hence, this lattice system is called *simple cubic.*

Figure 15.5a shows the simple cubic unit cell in more detail. Note that one-eighth of a sphere appears at each of the corners of the cube (see Figure 15.5b). Since there are eight corners, there is a total of one sphere per unit cell.

The corners of the unit cell are at the centers of the atoms.

The simple cubic structure is not an efficient way to pack spherical particles. Indeed, only 52% of the volume of the structure is occupied by the spheres and the remaining 48% is empty space.

Figure 15.5a
The simple cubic unit cell.

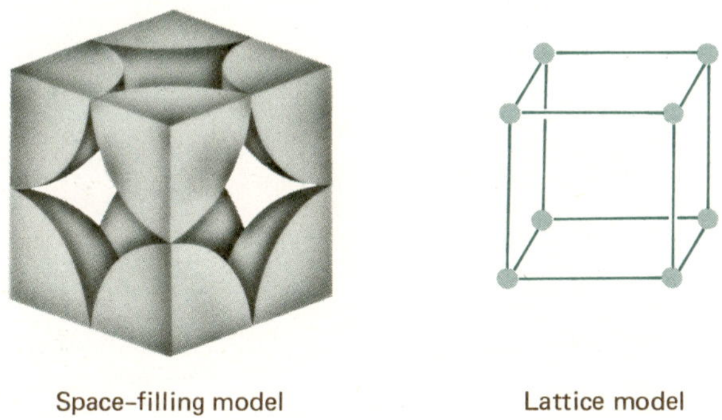

Figure 15.5b
Sharing of a corner sphere by adjacent unit cells. One-eighth of the sphere is in each cell.

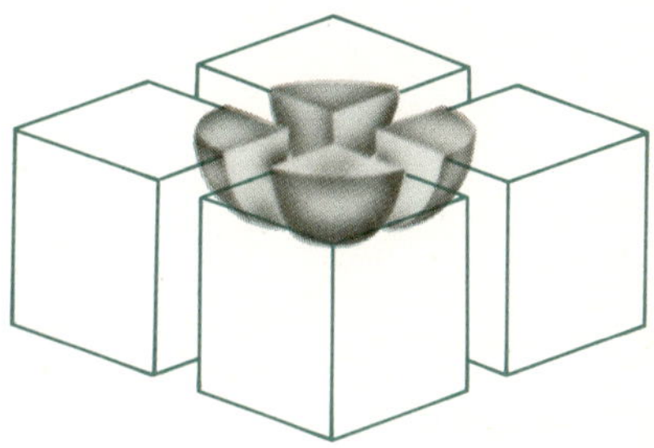

Figure 15.6
The length, a, of the edge of the simple cubic unit cell is 2r.

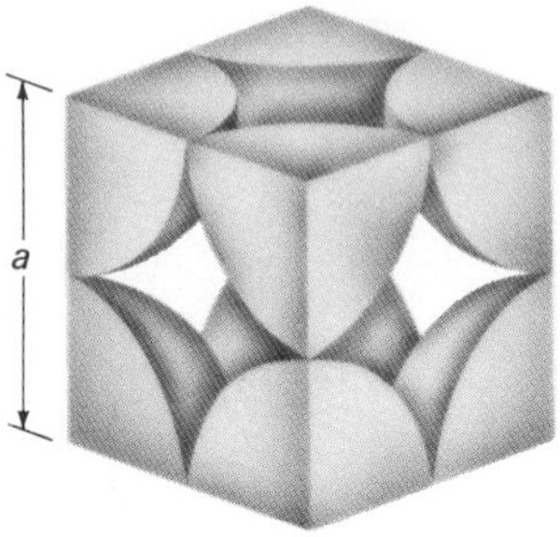

Figure 15.7
The body-centered cubic structure.

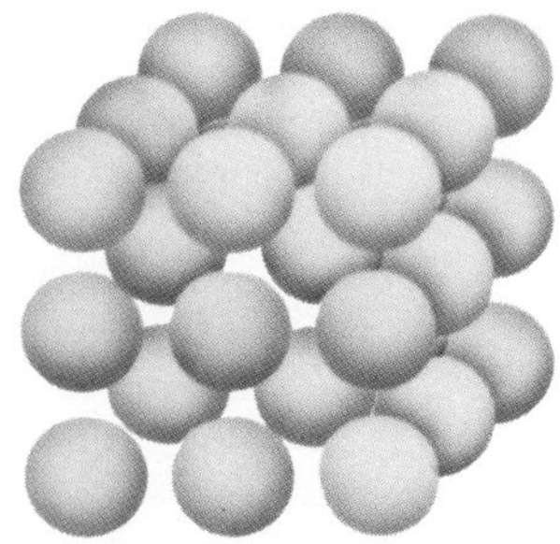

Body-Centered Cubic Packing

The very inefficient packing arrangement of the simple cubic structure is not observed in crystals. Attractive forces tend to pull the atoms close together and minimize the empty space. One arrangement with less empty space is the **body-centered cubic structure,** shown in Figure 15.7. (The unit cell is shown in Figure 15.8.) One way to describe the packing in this system is to imagine that an atom is placed in the large empty space in the center of the simple cubic structure. This hole, however, is not large enough to accommodate the atom without expansion of the cube; the corner atoms no longer touch along the cube edge. The body-centered cubic structure is a more efficient structure than is the simple cubic: 68% of the volume is occupied by the atoms in this structure.

The following example illustrates some of the kinds of information that can be calculated for a solid if the type of crystal structure is known.

EXAMPLE 15.1 Iron crystallizes in the body-centered cubic lattice. It has a density of 7.86 g/cm^3 and a mass of 55.847 g/mole. Calculate the dimensions of the unit cell and the radius of the atoms.

SOLUTION The volume per atom (including the empty space) is first calculated:

$$\frac{55.85 \text{ g/mole}}{(7.86 \text{ g/cm}^3)(6.022 \times 10^{23} \text{ atoms/mole})} = 1.18 \times 10^{-23} \text{ cm}^3\text{/atom}$$

Figure 15.8
The body-centered cubic unit cell.

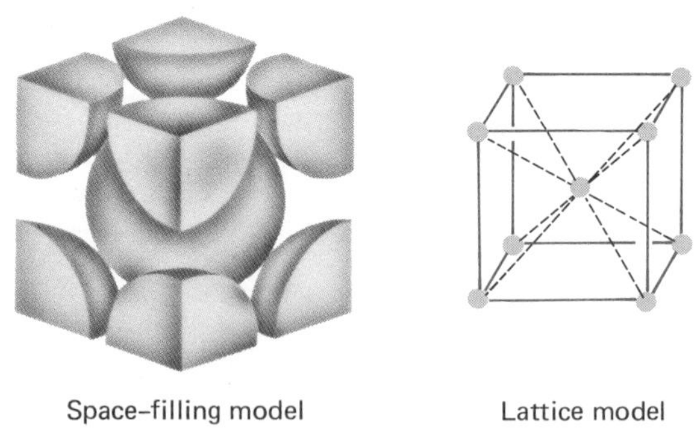

It is a simple problem in geometry to calculate the percentage of the structure occupied by the spheres. The length of the edge of the unit cell (Figure 15.6) is $2r$, where r is the radius of the spherical atom. Hence, the volume of the unit cell is

$$V = a^3 = (2r)^3 = 8r^3$$

The volume of the unit cell occupied by the spheres is just the volume of one sphere (there is a total of one atom in that unit cell) of radius r:

$$V_S = \frac{4}{3}\pi r^3$$

Therefore,

$$\% \text{ occupied} = \left(\frac{V_s}{V}\right)(100)$$

$$= \left(\frac{\frac{4}{3}\pi r^3}{8r^3}\right)(100)$$

$$= 52\%$$

Since the unit cell is the smallest part of the crystal that retains the geometric properties of the entire crystal, the calculation of the percent of occupied space for the unit cell applies as well to the crystal as a whole.

The body-centered cubic unit cell contains two atoms. Hence, the volume of the unit cell is given by

$$V = (1.18 \times 10^{-23}\ \text{cm}^3/\text{atom})(2\ \text{atoms/unit cell}) = 2.36 \times 10^{-23}\ \text{cm}^3/\text{unit cell}$$

The length of the cell edge a is then easily calculated from

$$a = \sqrt[3]{V} = \sqrt[3]{2.36 \times 10^{-23}\ \text{cm}^3} = 2.86 \times 10^{-8}\ \text{cm}$$

To get the radius r of the atom, we remember that in the body-centered cubic structure (see marginal note),

$$a = \frac{4r}{\sqrt{3}}$$

Hence,

$$r = \frac{\sqrt{3}a}{4} = \left(\frac{\sqrt{3}}{4}\right)(2.86 \times 10^{-8}\ \text{cm})$$

$$= (1.24 \times 10^{-8}\ \text{cm})\left(10^8\ \frac{\text{Å}}{\text{cm}}\right) = 1.24\ \text{Å}$$

Figure 15.9
The spheres touch along the diagonal of the cube in the body-centered cubic unit cell.

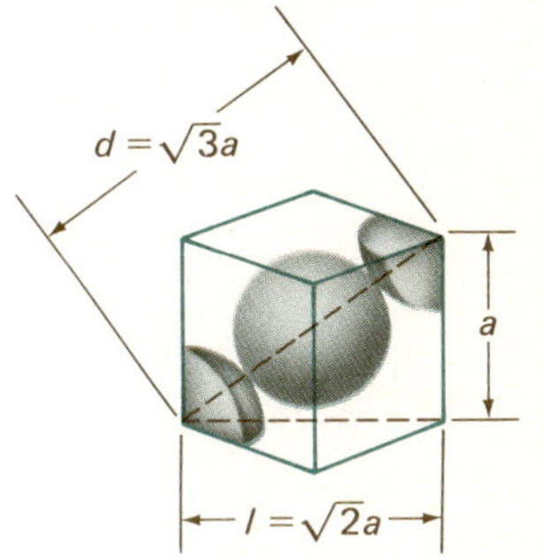

$\sqrt{3}a = 4r$
Body-centered unit cell

Figure 15.9 shows that the spheres touch along the diagonal of the cube in a unit cell of the body-centered cubic structure. The length of this diagonal is $4r$, where r is the radius of the spherical atom. The length of the diagonal l across the cube face is seen to be related to the length of an edge of the cube by

$$l^2 = a^2 + a^2 = 2a^2$$

From the dotted triangle,

$$d^2 = l^2 + a^2 = 3a^2$$

or

$$d = \sqrt{3}a$$

But earlier we saw that

$$d = 4r$$

Therefore,

$$a = \frac{4r}{\sqrt{3}}$$

and the volume of the cube is

$$V = a^3 = \left(\frac{4}{\sqrt{3}}\right)^3 r^3 = \frac{64r^3}{3\sqrt{3}}$$

Each body-centered cubic unit cell contains two atoms (a total of one from the eight

Face-Centered Cubic Packing

As the above example indicates, some atomic systems exist in the body-centered cubic structures. However, **face-centered cubic packing** is even more efficient. Figure 15.10 shows the arrangement of atoms in this lattice, and Figure 15.11 shows the unit cell. Notice that an atom is placed in the center of each face of the cube. The number of atoms per unit cell is four (one-eighth atom from each of eight corners plus one-half atom from each of six faces). It is not difficult to show that this unit cell contains only 26% empty space (see Exercise 15-1).

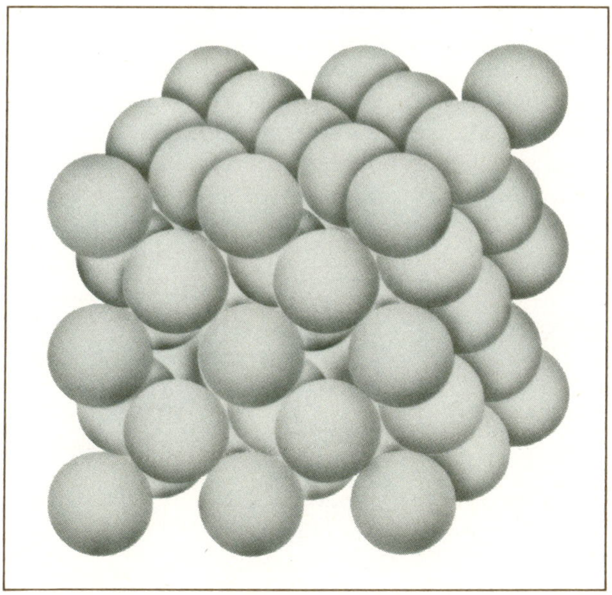

Figure 15.10
The face-centered cubic structure.

corners plus the one in the center). Hence, the volume occupied by the spherical atoms is

$$V_S = \left(\frac{4}{3}\pi r^3 \text{ per atom}\right) \times (2 \text{ atoms per unit cell})$$

$$V_S = \frac{8}{3}\pi r^3$$

The percent of the unit cell occupied by the two atoms is

$$\% \text{ occupied} = \left(\frac{\frac{8\pi}{3}r^3}{\frac{64}{3\sqrt{3}}r^3}\right)(100) = \left(\frac{\sqrt{3}\pi}{8}\right)(100) = 68\%$$

and the empty space is 32%.

There is no substitute for three-dimensional models in studying crystal lattices. Students are encouraged to construct models by stacking such spherical objects as marbles or ping-pong balls.

Within each cubic system, the coordination number is the same for every atom. Note that all the atoms in a given system are equivalent. For example, the atom at the center of the unit cell of body-centered cubic packing has the same environment as those at the corners. The same is true for the atoms in the center of the faces and at the corners of the face-centered cubic structure. In both cases, it is possible to reverse the role of the corner and center atoms by shifting the boundaries of the unit cell (see Figures 15.7 and 15.10).

Figure 15.11
The face-centered cubic unit cell.

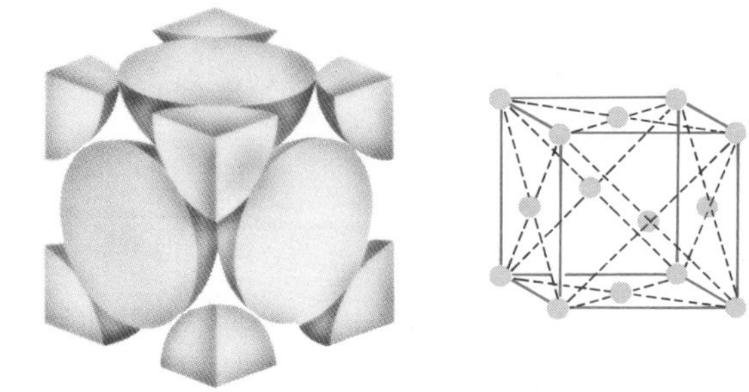

Table 15.1 is a comparison of the properties of the three cubic systems. Included in the table is a comparison of the **coordination number** for each atom. The coordination number gives the number of "nearest neighbor" atoms in contact with each atom. It can be seen from Figures 15.4, 15.7, and 15.10 that the coordination numbers are 6, 8, and 12 for the three cubic systems. Notice that in these atomic systems, the coordination number is also a measure of the efficiency of packing (the larger the coordination number, the more efficient the packing).

Twenty-six percent is the minimum empty space for packing equal-size spheres. Face-centered cubic packing is often called **cubic close-packed** to emphasize that this arrangement gives the closest possible packing for equal-sized spherical atoms.

Closed-Packed Structures

The cubic close-packed structure is only one of two packing arrangements that result in this maximum packing efficiency. The second is a hexagonal (six-sided) structure known appropriately as **hexagonal close-packed.** The cubic and hexagonal close-packed structures can be produced by stacking spheres in close-packed layers. The relationship between the stacked layers determines whether the cubic or hexagonal structure is obtained. To see this relationship, consider stacking spheres as follows: First, form a layer as shown in Figure 15.12a. Notice that alternate rows are displaced half a diameter to give the

TABLE 15.1

Comparison of the Cubic Lattice Structures

SYSTEM	ATOM/UNIT CELL	% EMPTY SPACE	COORDINATION NUMBER
Simple cubic	1	48	6
Body-centered cubic	2	32	8
Face-centered cubic	4	26	12

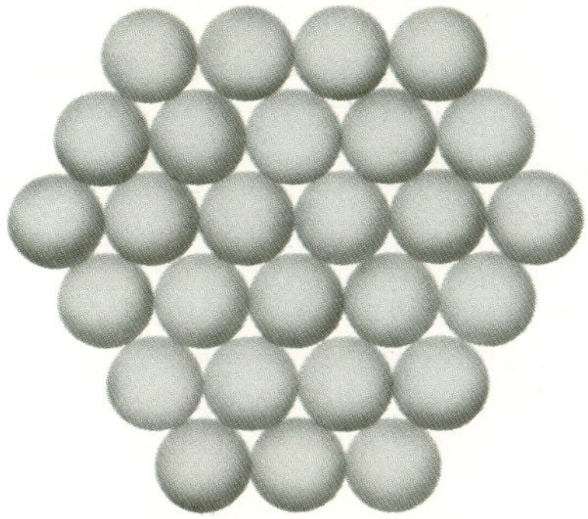

(a)

Figure 15.12a
The packing of spheres for the first layer of a close-packed structure.

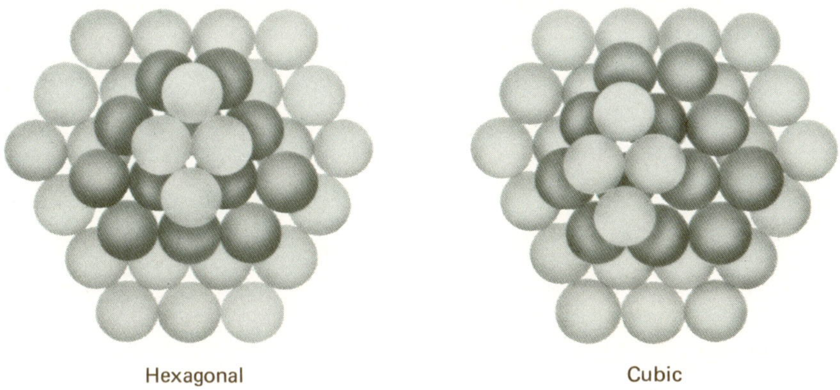

(b)

Figure 15.12b
In either the hexagonal or the cubic structure, the second layer (dark spheres) is placed in a close-packed arrangement over holes in the first layer. In the third layer, however, the placement is different for the two structures. In the hexagonal structure the spheres of the third layer are directly above the spheres of the first layer, while in the cubic structure the spheres of the third layer are above holes of the first layer.

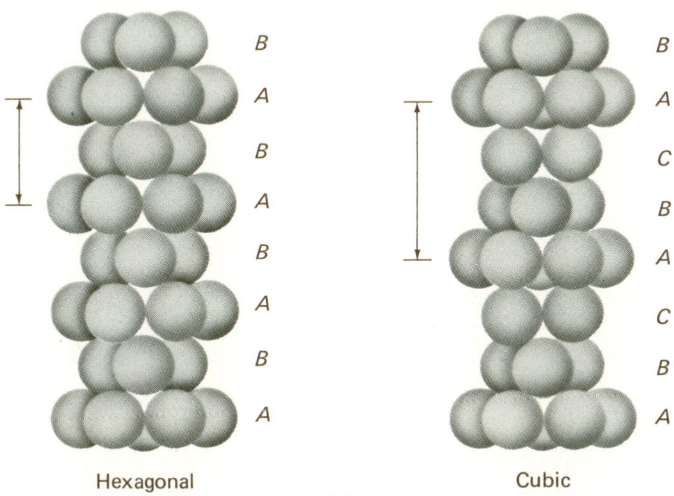

(c)

Figure 15.12c
Side views of the hexagonal and cubic close-packed structures. Notice that the repeating pattern is *ABABAB* for the hexagonal structure and *ABCABCABC* for the cubic structure.

closest possible packing. This first layer is identical in cubic and hexagonal systems.

Next, stack a second layer on top of the first (Figure 15.12b). Place the spheres of the second layer over the holes between spheres of the first layer in order to achieve the closest packing between layers. The placement of this second layer is the same in the cubic and hexagonal systems.

Now add the third layer. Again, the spheres of this layer are placed over the holes between spheres in the second layer. When we examine the first two layers, however, we see that there are two nonequivalent positions for stacking this third layer (Figure 15.12b): (1) over the holes of the *first* layer, or (2) over the

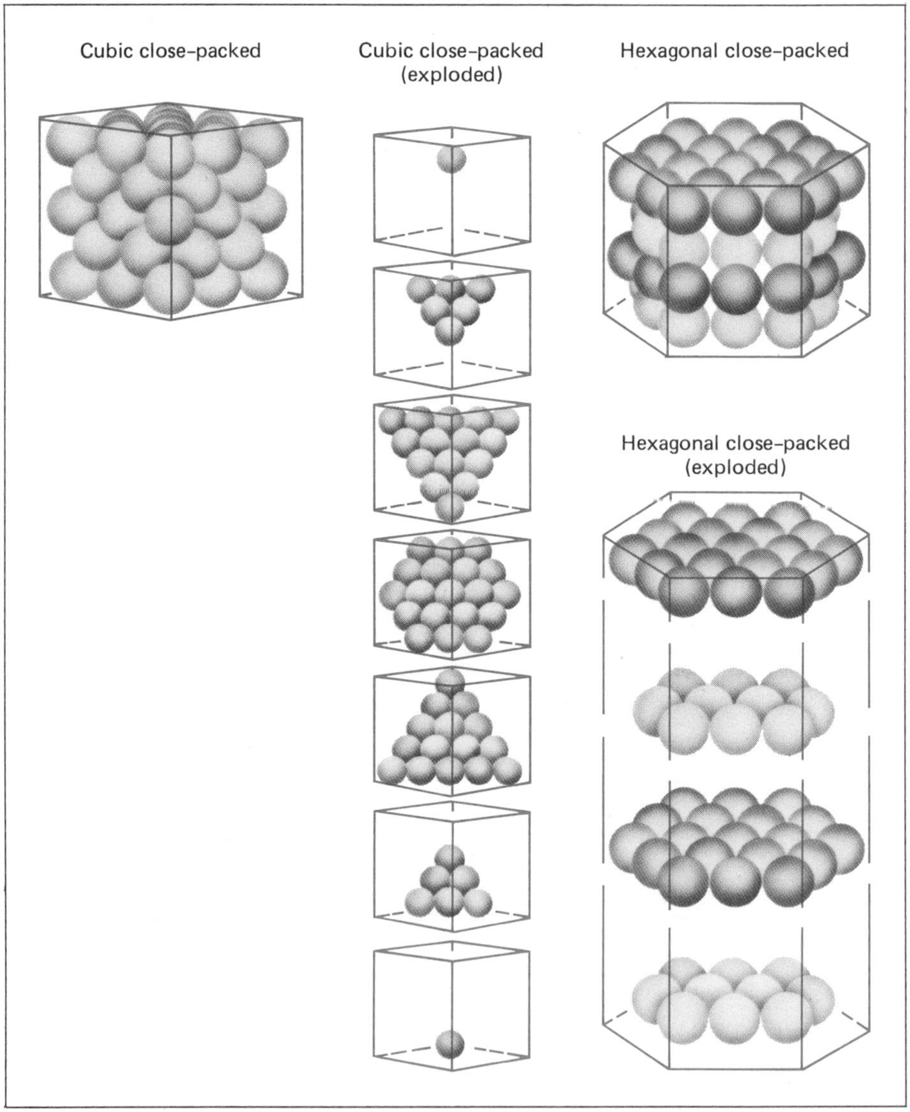

Figure 15.13
Comparison of the packing in the hexagonal and cubic closed-packed structures. The stacked layers in the cubic close-packed structure run diagonally through the cube.

TABLE 15.2

Crystal Structures of the Metallic Elements

Li	Be	B									
I	II	—									
Na	Mg	Al									
I	II	III									
K	Ca	Sc	Ti	V	Cr	Mn	Fe	Co	Ni	Cu	Zn
I	III	II	II	I	I	I	I	II	III	III	II
Rb	Sr	Y	Zr	Nb	Mo	Tc	Ru	Rh	Pd	Ag	Cd
I	III	II	II	I	I	II	II	III	III	III	II
Cs	Ba	La	Hf	Ta	W	Re	Os	Ir	Pt	Au	Hg
I	I	II	II	I	I	II	II	III	III	III	—

I Body-centered cubic
II Hexagonal close-packed
III Cubic close-packed

spheres of the first layer. If the first option is followed and repeated, the cubic close-packed structure results. If the second option is followed and repeated, the hexagonal structure results (Figure 15.12c). Figure 15.13 compares the packing in the cubic and hexagonal systems, and Figure 15.14 shows the hexagonal close-packed unit cell. As stated earlier, the cell is 26% empty space, and the coordination number is 12, the same as in the cubic close-packed system.

Later in this chapter, we will describe the experimental methods used to determine crystal structure. Table 15.2 gives the stable crystal structures at room temperature of many of the metallic elements. Many substances undergo phase transitions between solid phases; the crystal changes from one of the structures to another at a transition temperature.

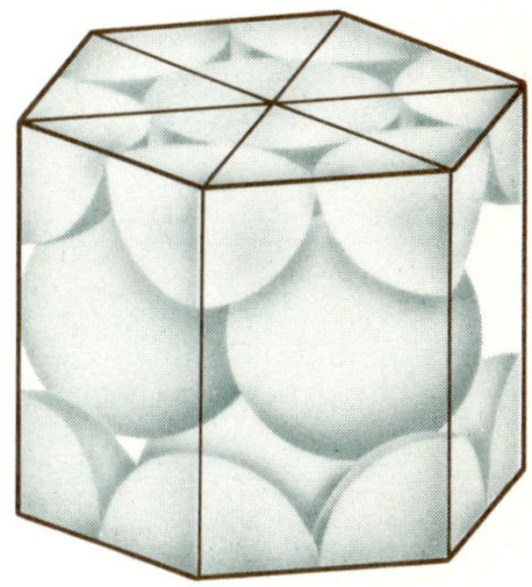

Figure 15.14
The hexagonal close-packed unit cell.

15.3 Binary Solids

The discussion so far has been limited to systems in which identical spherical atoms occupy all the sites in the crystal lattice. We now extend the discussion to include *binary solid compounds*, in which more than one kind of spherical atom or ion occupy lattice sites.

The crystal structure of many binary compounds of the type AB, AB_2, and A_2B are related in a simple manner to the cubic arrangements discussed earlier. Often, *the A atoms can be pictured as spheres that form a cubic structure while the B atoms occupy the interstices or holes between the spheres.*

The Cubic Hole

A cubic hole is the hole at the center of the simple cubic unit cell. The coordination number about an atom placed in the hole is eight. An important characteristic of the hole is the **radius ratio** R:

$$R = \frac{r_h}{r}$$

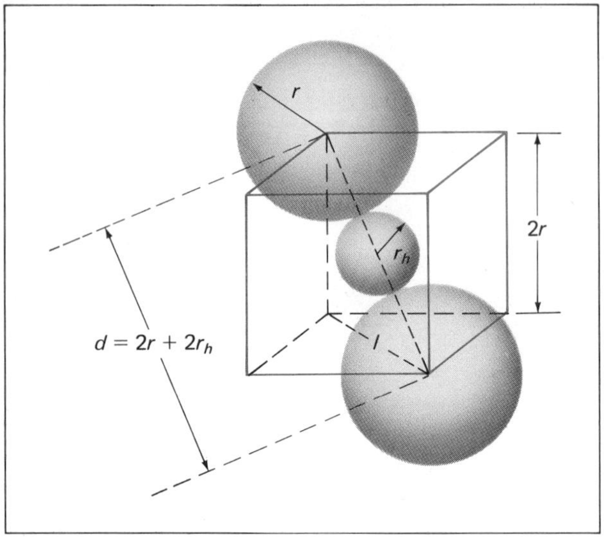

Figure 15.15
The atom that just fills the cubic hole touches the corner atoms along the cube diagonal d.

The radius ratio is calculated as follows: In Figure 15.15, an atom that just fills the cubic hole touches the corner atoms along the cube diagonal d. Hence,

$$d = 2r + 2r_h \quad [15.1]$$

The length of the cube edge is $2r$, since the atoms touch along this edge. The cube diagonal d can be related to r by the following steps:

$$l^2 = (2r)^2 + (2r)^2 = 8r^2$$

where l is the length of the diagonal in the cube face. From the dotted triangle,

$$l^2 + (2r)^2 = d^2$$

Hence,

$$d^2 = 8r^2 + 4r^2 = 12r^2$$

and

$$d = 2\sqrt{3}r \quad [15.2]$$

Combining equations [15.1] and [15.2] gives

$$2r + 2r_h = 2\sqrt{3}r$$

Solving for r_h and dividing by r gives

$$R = \frac{r_h}{r} = 0.732 \quad [15.3]$$

That is, an atom with radius $r_h = 0.732r$ will just fill the cubic hole in a simple cubic lattice of atoms of radius r.

where r_h is the radius of a spherical atom that will just fill the hole and r is the radius of the atoms at the corners of the cube. For a simple cubic structure the radius ratio is 0.732.

It is also useful to know the number of holes per unit cell. Each simple cubic unit cell has one hole at its center. Since these are the only cubic holes in the lattice, there is one cubic hole per unit cell. Hence, there are equal numbers of atoms and cubic holes in the simple cubic lattice.

The Octahedral Hole

In the face-centered cubic lattice there are several types of holes. The first of these to be described is the **octahedral hole.** This hole is surrounded by six atoms whose centers are at the corners of a regular octahedron (see Figure 15.16). Figure 15.17 shows a layer of a face-centered cubic lattice. The centers of the octahedral holes are shown as X's in the figure. To show that these are octahedral holes, we remember that the center of a regular octahedron falls on an equatorial plane formed by the center of four spheres. This plane is defined by points 1, 2, 3, and 4 in Figure 15.16. The next layers of spheres in either direction are displaced until they are above or below the holes. The atoms in these layers form the vertexes of the octahedron above and below the equatorial plane.

For the octahedral hole, all the spheres at the corners of the octahedron would contact a sphere of proper size placed in the hole; hence, the coordination number is six. The radius ratio for atoms, the smaller of which would just fill

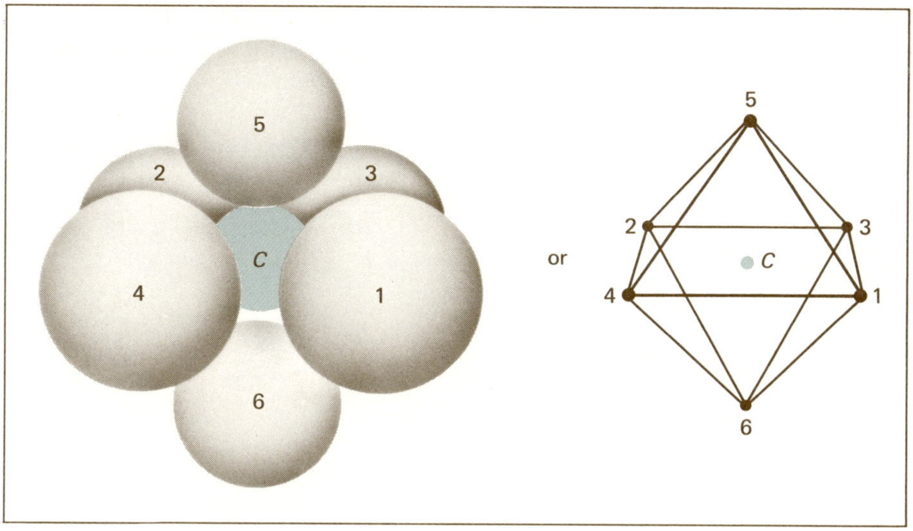

Figure 15.16
The sphere at *C* occupies the octahedral hole.

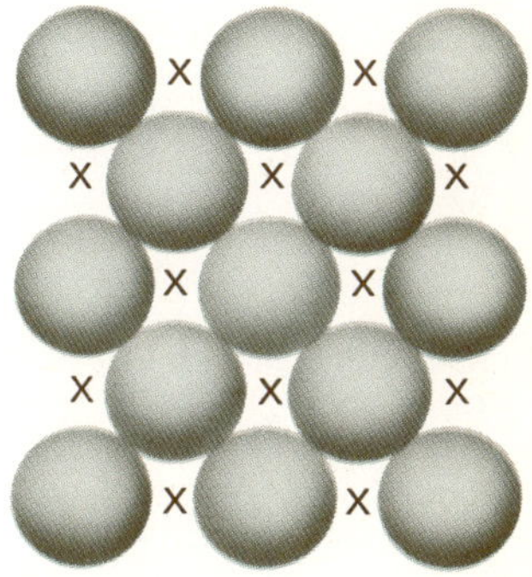

Figure 15.17
Location of octahedral holes in the face-centered cubic lattice.

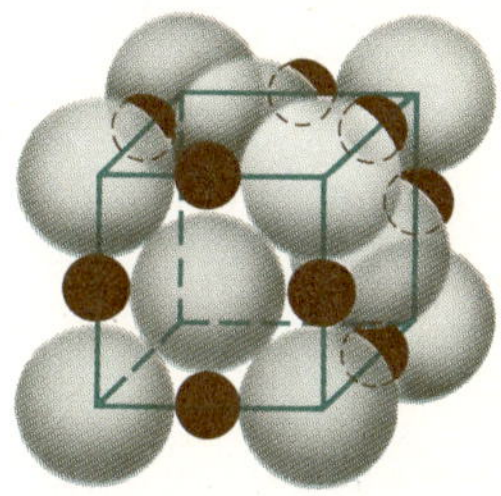

Figure 15.18
Position of the octahedral holes (shaded dark) in the face-centered cubic lattice. One additional hole is at the center of the cube.

an octahedral hole, is 0.414; this is readily calculated from the geometry of the lattice. The number of holes per unit cell can be determined by examining the unit cell of the face-centered cubic lattice. From Figure 15.18 we see that there are octahedral holes along each edge of the face-centered cubic unit cell. However, only one-fourth of each of these holes is within the unit cell. Since there are twelve edges to the unit cell, the holes on the edge contribute 12 edges × 1/4 hole/edge = 3 holes. There is also one hole in the center of the cell for a total of four holes/unit cell. Since there are also four atoms per unit cell, there are equal numbers of octahedral holes and atoms in the unit cell.

The Tetrahedral Hole

One of the basic structural units of the close-packed lattice is one sphere resting upon three others, as shown in Figure 15.19. The centers of these four spheres are at the corners of a regular tetrahedron. The space at the center of this tetrahedron is called a **tetrahedral hole.** The coordination number is four since an atom placed in the hole would contact all four of the spheres at the corners of the tetrahedron.

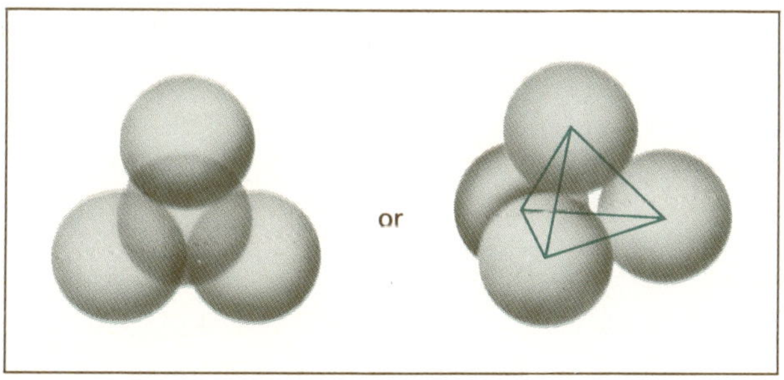

Figure 15.19
The tetrahedral hole is at the center of the tetrahedron made by placing one sphere on top of three others.

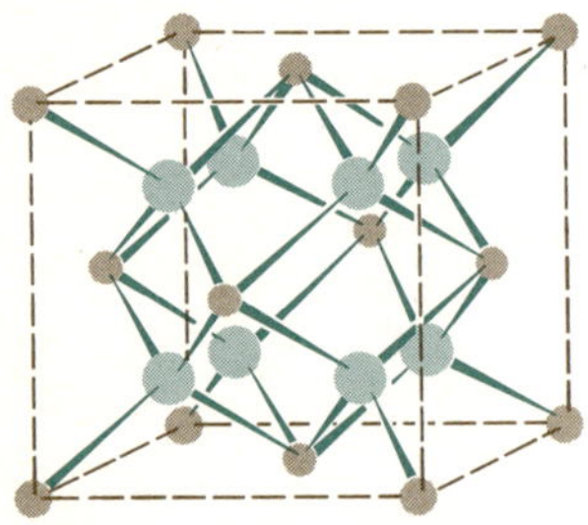

Figure 15.20
Positions of the tetrahedral holes (shaded dark) in the face-centered cubic lattice. There are eight holes in each unit cell.

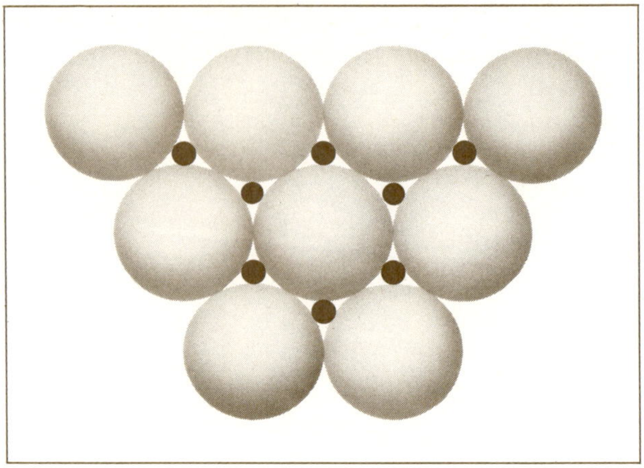

Figure 15.21
Position of the triangular hole (shaded dark) in the close-packed lattice.

It can be shown that the radius ratio is 0.225 for the tetrahedral hole. In the close-packed structure, each sphere is in contact with three others in the layer above it, and with three more in the layer below. As a result, there are two tetrahedral holes for each sphere. In the face-centered cubic unit cell there are four atoms and, hence, eight tetrahedral holes. The positions of these holes are shown in Figure 15.20.

Do not confuse triangular and tetrahedral holes. The tetrahedral holes at the center of a tetrahedron are displaced *above* the plane of the triangle that makes up the base of the tetrahedron. The triangular holes are *within* this triangular base.

The Triangular Hole

Whenever three atoms are together in a layer in the face-centered cubic structure, their centers are at the corners of a triangle. The hole at the center of this triangle is called a **triangular hole** (see Figure 15.21). It is obvious that the coordination number is three for triangular holes. The radius ratio is calculated to be 0.155.

The Linear Hole

In some crystal systems, an atom of element B is placed directly between two atoms of element A in a lattice. The so-called **linear hole,** when occupied, has a coordination number of two (see Figure 15.22). It has a radius ratio of zero. The different kinds of holes available in the simple cubic or face-centered lattices are summarized in Table 15.3.

TABLE 15.3

Summary of Holes

TYPE OF HOLE	COORDINATION NUMBER	RADIUS RATIO (TO JUST FILL HOLE)
Cubic	8	.732
Octahedral	6	.414
Tetrahedral	4	.225
Triangular	3	.155
Linear	2	0

As mentioned earlier, the crystal structure of many binary compounds can be described as the placing of one kind of atom or ion in the holes in the lattice of the second atom or ion. This description is especially useful for ionic compounds, in which ions of one charge occupy the holes in the lattice of ions of the opposite charge. By remembering that ions of like charge repel and those of opposite charge attract, one can (with few exceptions) predict the type of hole occupied. The rules that help us in the prediction follow:

(1) The ion that goes in the hole must be at least large enough to fill the hole.
(2) Each ion tends to be surrounded by the largest possible number of ions of the opposite charge.

Figure 15.22
The linear hole (shaded dark) has a coordination number of two.

The first rule follows from the repulsion of ions of like charge. The ion in the hole must be large enough to fill the hole and keep the ions that surround it from touching (and repelling) one another. The second rule follows from the attraction between ions of opposite charge. The more contacts between these oppositely charged ions, the lower the energy of the system.

The two rules listed above tend to limit the radius-ratio range within which a certain crystal structure is preferred. For example, the radius of the ion in the octahedral hole must be at least 0.414 times the radius of the other ion to fill the hole. However, if the ion is 0.732 times as large (or larger), the cubic hole can be occupied and is favored since the coordination number is larger. The result is that octahedral holes are filled when the ratio of the radii of the ions is between 0.414 and 0.732. The same type of argument can be used to assign ion radius-ratio ranges to the other types of holes. The result is as shown in Table 15.4.

TABLE 15.4

Relationship Between the Radius Ratio and the Type of Lattice Hole Filled

RATIO OF ATOMIC RADII

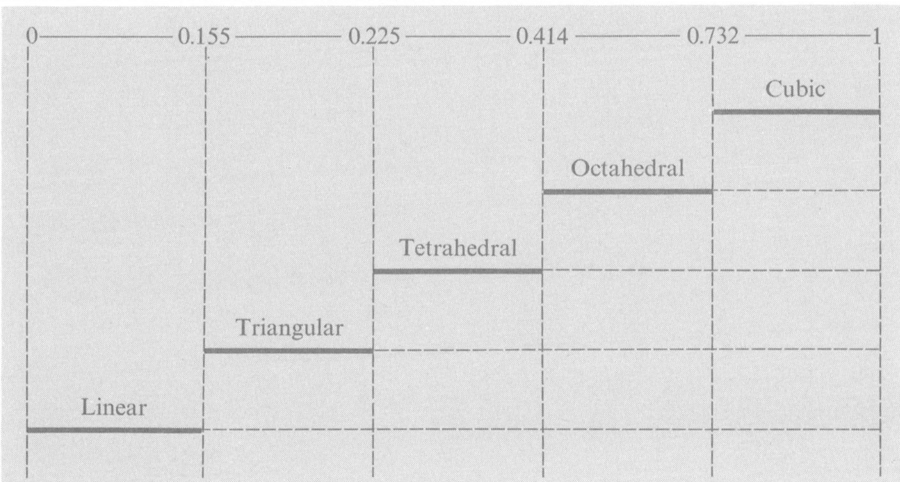

The solid line gives the ion radius-ratio range expected for each kind of hole. The dotted line gives an extended range where exceptions sometimes occur. For example, if the radius ratio of a substance is 0.5, the table shows that octahedral holes would probably be filled, but tetrahedral, triangular, or linear holes could be used instead. Cubic holes are not possible. The exceptions in the regions indicated by the dotted lines are violations of the second rule. Having the maximum number of contacts does not seem to be as important as keeping ions of the same charge widely separated from one another. This wide separation occurs when a larger ion is placed in the smaller hole.

Table 15.5 gives the ionic radii for a number of common ions. Using these radii and the ion radius-ratio ranges from Table 15.4, one can predict and describe the bonding of many binary ionic compounds. We will first describe the crystal structure of the substances that are used as representative examples of the types of lattices for AB, A_2B, and AB_2 compounds. At the same time, we will see how well the radius ratio predicts the crystal structure.

TABLE 15.5

Ionic Radii

ION	RADIUS (Å)	ION	RADIUS (Å)
H^-	2.08	Sr^{2+}	1.10
F^-	1.33	Ba^{2+}	1.29
Cl^-	1.81	Ra^{2+}	1.40
Br^-	1.96	Zn^{2+}	0.74
I^-	2.19	Cd^{2+}	0.97
O^{2-}	1.40	Hg^{2+}	1.10
S^{2-}	1.84	Sn^{2+}	1.10
Se^{2-}	1.98	Pb^{2+}	1.32
Te^{2-}	2.21	Mn^{2+}	0.82
Li^+	0.68	Fe^{2+}	0.76
Na^+	0.98	Co^{2+}	0.74
K^+	1.33	Ni^{2+}	0.68
Rb^+	1.48	Cu^{2+}	0.72
Cs^+	1.67	Bi^{3+}	0.20
Cu^+	0.96	Al^{3+}	0.45
Ag^+	1.26	Sc^{3+}	0.69
Au^+	1.37	Y^{3+}	0.93
Tl^+	1.40	La^{3+}	1.15
NH_4^+	1.48	Ga^{3+}	0.60
Be^{2+}	0.30	In^{3+}	0.81
Mg^{2+}	0.65	Tl^{3+}	0.95
Ca^{2+}	0.94	Fe^{3+}	0.63
		Cr^{3+}	0.61

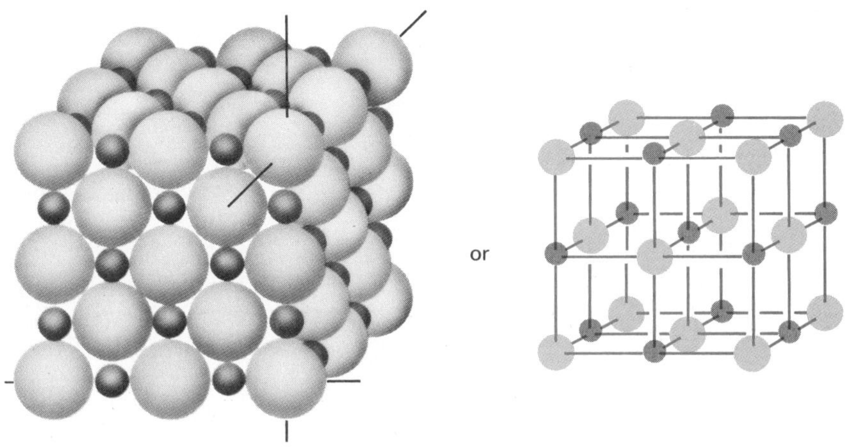

Figure 15.23
The NaCl structure. The large spheres represent Cl^- ions and the small spheres Na^+ ions.

It is easier to see the positions in the lattice model than in the space-filling model. When using the lattice model, one must keep in mind that the ions do fill the space such that ions of opposite charge are touching.

Many of these crystal structures were first cataloged by mineralogists and are referred to by the name of the mineral, such as rock salt, zinc blende, and fluorite. Chemists have retained these mineralogical names for the structures.

The rock salt (NaCl) structure. The experimentally determined crystal structure of NaCl is shown in Figure 15.23. The Cl^- ions are in a face-centered cubic lattice with the Na^+ ions in the octahedral holes. All the octahedral holes are filled since there are the same numbers of lattice positions and holes in the face-centered cubic lattice.

Note the equivalence of the Na^+ and Cl^- sites. The Na^+ ions are also in a face-centered cubic lattice with Cl^- ions in octahedral holes. The NaCl lattice is sometimes described as interpenetrating face-centered cubic lattices of Cl^- and Na^+.

The radius ratio for NaCl is (see Table 15.5)

$$\frac{r_{Na^+}}{r_{Cl^-}} = \frac{0.98}{1.81} = 0.54$$

The radius ratio does predict correctly that octahedral holes are occupied.

The CsCl structure. Cubic holes are filled in CsCl. Figure 15.24 shows the Cl^- ions in the cubic holes in the Cs^+ simple cubic lattice. The converse is also true: The Cs^+ are in cubic holes in the Cl^- simple cubic lattice. Another description is that CsCl consists of interpenetrating Cs^+ and Cl^- simple cubic lattices. The radius ratio is

$$\frac{r_{Cs^+}}{r_{Cl^-}} = \frac{1.67}{1.81} = 0.92$$

and the structure is as predicted from the radius ratio.

Figure 15.24
The CsCl structure. The Cl^- ions are shaded dark.

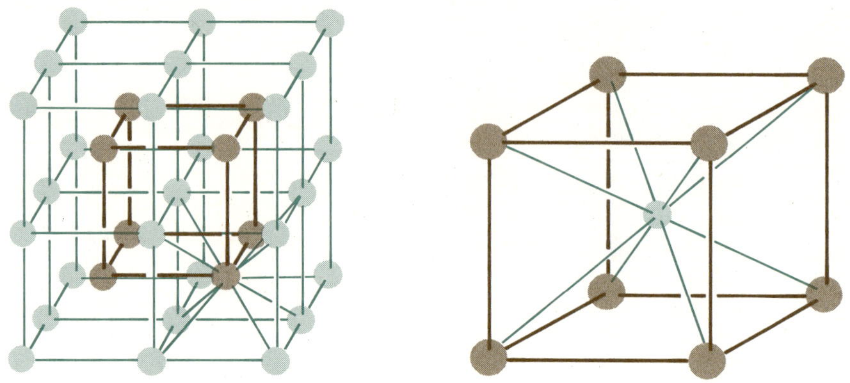

Zinc sulfide also crystallizes in a hexagonal structure, with the Zn^{2+} ions in tetrahedral holes. This crystal structure is known as wurtzite.

The zinc blende (cubic ZnS) structure. Cubic ZnS has the structure shown in Figure 15.25. The zinc ions fill tetrahedral holes in the sulfide ion face-centered cubic lattice. As with the examples described previously, the reverse is also true: The sulfide ions fill tetrahedral holes in a Zn^{2+} face-centered cubic lattice. Only half the tetrahedral holes are filled since there are twice the number of tetrahedral holes as there are lattice sites in the face-centered cubic lattice. In the zinc blende structure, alternate tetrahedral holes are filled. The radius ratio for ZnS is 0.40 (see Table 15.5). Again, the radius ratio correctly predicts the type of holes to be occupied.

Figure 15.25
The zinc blende (cubic ZnS) lattice. The small spheres are the Zn^{2+} ions.

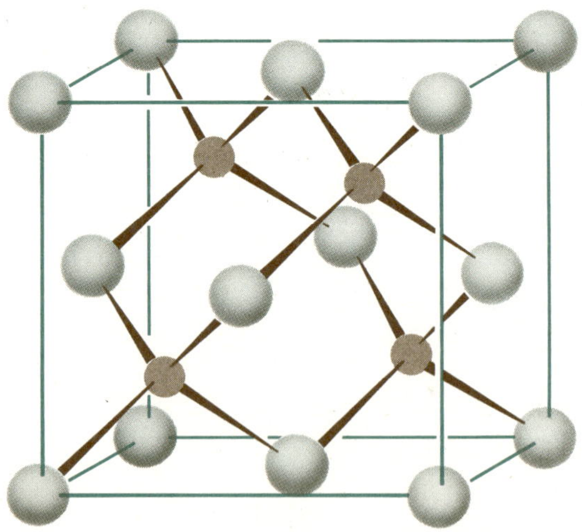

If we keep in mind the exceptions that can occur, we can use the radius ratio to predict the probable structures of 1:1 ionic compounds.

EXAMPLE 15.2 Predict the crystal structures of AgCl and AgI from the sizes of the ions.

SOLUTION From Table 15.5 the radius ratios are

$$\text{AgCl:} \qquad R = \frac{r_{Ag^+}}{r_{Cl^-}} = \frac{1.26}{1.81} = 0.70$$

$$\text{AgI:} \qquad R = \frac{r_{Ag^+}}{r_{I^-}} = \frac{1.26}{2.19} = 0.58$$

Referring to Table 15.4 we predict that neither AgCl nor AgI can have the cesium chloride structure. Both crystals have ratios of ionic radii that make the sodium chloride structure probable, and that is our prediction. We do recognize, however, that structures are also possible that lead to a greater separation of ions of like charge. In fact, AgCl does have the sodium chloride structure while AgI has tetrahedral coordination. This example shows how careful use of the ratio of ionic radii can point to probable crystal structures, but that exceptions must be expected. One should remember that the rule that favors a maximum coordination number is sometimes violated when the violation causes a larger separation of ions with like charges.

Table 15.6 lists crystal structures for several 1:1 ionic solids.

TABLE 15.6

Crystal Structure Types for 1:1 Ionic Solids

SODIUM CHLORIDE		CESIUM CHLORIDE	ZINC BLENDE OR WURTZITE
LiF	CaTe	CsCl	AgI
LiCl	KF	CsBr	ZnO
LiBr	KCl	CsI	ZnS
LiI	KBr		ZnSe
NaF	KI		ZnTe
NaCl	RbF		CdS
NaBr	RbCl		CdSe
NaI	RbBr		CdTe
AgF	RbI		HgS
AgCl	MgO		HgSe
AgBr	MgS		HgTe
CaO	MgSe		CuBr
CaS	CdO		
CaSe			

Crystal structures that are more complex can be obtained when nonspherical molecules or ions occupy lattice points (for example, S_8, C_6H_6, and $NaNO_3$) and when there is an ion stoichiometry other than 1:1 (for example, CaF_2, TiO_2, and Al_2O_3).

As an example, the mineral fluorite (CaF_2) is cubic with a structure in many ways similar to zinc blende. The Ca^{2+} ions are in a face-centered cubic lattice with all the tetrahedral holes filled with F^- ions (instead of with half the tetrahedral holes filled, as in ZnS). Since there are twice as many tetrahedral holes as lattice points in the face-centered cubic structure, the ratio is just right for the 1:2 compound.

15.4 Other Crystal Systems

We will now consider a few other common substances whose crystal systems are very interesting.

Carbon

Carbon exists in two crystal structures: diamond and graphite. Figure 15.26 shows the crystal structure of diamond. The structure is identical to zinc blende except that only one kind of atom (carbon) occupies all the sites. Half the carbon atoms occupy alternate tetrahedral holes in the face-centered cubic lattice of the other half. All carbon atom positions are equivalent, however, since those atoms that we have relegated to tetrahedral holes are also in a face-centered cubic lattice, with the atoms we originally placed at the corners now in the tetrahedral holes. We can think of the structure as two interpenetrating face-centered cubic lattices.

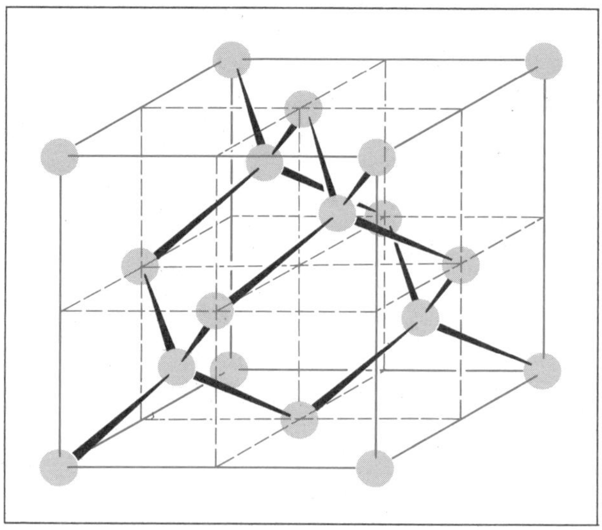

Figure 15.26
The diamond lattice.

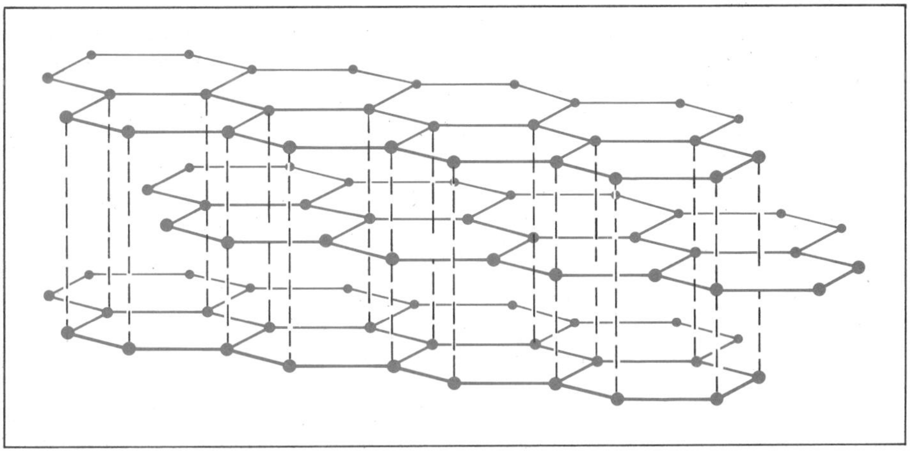

Figure 15.27
The graphite lattice. The atoms in alternate planes are directly under one another.

The coordination number of each carbon atom is four. The diamond crystal can be thought of as one large covalently bonded molecule. In the valence bond description of the bonding, the orbitals of the carbon atom are treated as four tetrahedral hybrid (sp^3) orbitals. Overlap of these orbitals with the same type of orbitals on adjacent atoms results in four covalent bonds for each carbon.

The structure of graphite is shown in Figure 15.27. The carbon atoms within each layer are bonded together by strong covalent bonds. In the valence bond description of the bonding, one *s* and two 2*p* orbitals of each atom are considered to have formed three sp^2 hybrid orbitals:

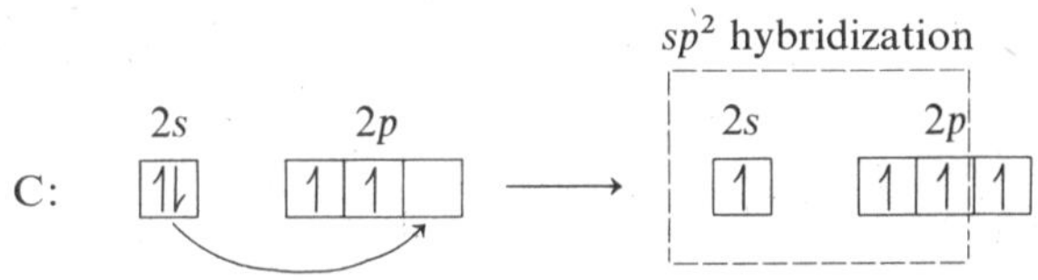

These overlap with similar orbitals on adjacent atoms, forming the hexagonal rings. The electrons in the *p* orbitals also overlap to bond the layers together. However, this bond is much weaker. Within each plane the carbon atoms are tightly held, but between the planes the attraction is much smaller. The distance between atoms in a plane is 1.34 Å, but the distance between the planes is 3.41 Å.

Notice the similarity between the bonding in graphite and in benzene (Chapter 14).

Water (Ice)

Figure 15.28 shows the lattice arrangement of the oxygen atoms in ice. Figure 15.29 shows the position of the hydrogen atoms. They occupy linear holes between pairs of oxygen atoms. Figure 15.30 shows the tetrahedral arrangement around one of the oxygen atoms. An interesting feature of the structure is that the hydrogen is not equidistant between the two oxygen atoms it separates. Each oxygen atom contacts four hydrogen atoms. Two of these hydrogen atoms are covalently bonded to the oxygen. This strong bond results

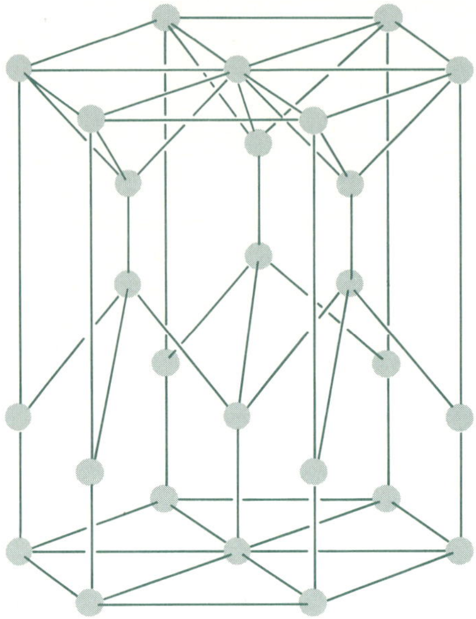

Figure 15.28
The lattice arrangement of oxygen in ice.

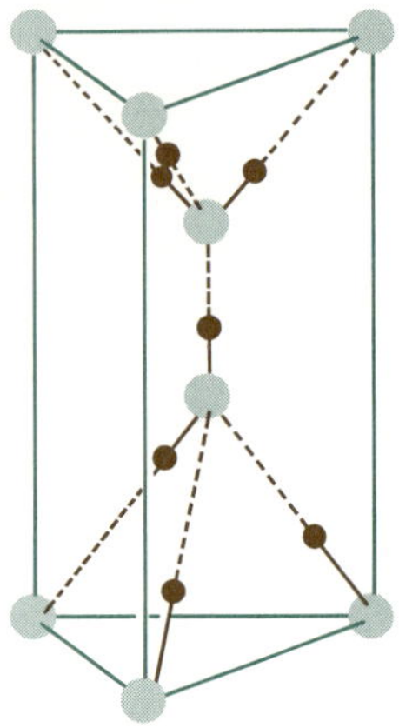

Figure 15.29
The hydrogen atoms (small spheres) are in linear holes between the oxygen atoms (large spheres) in ice.

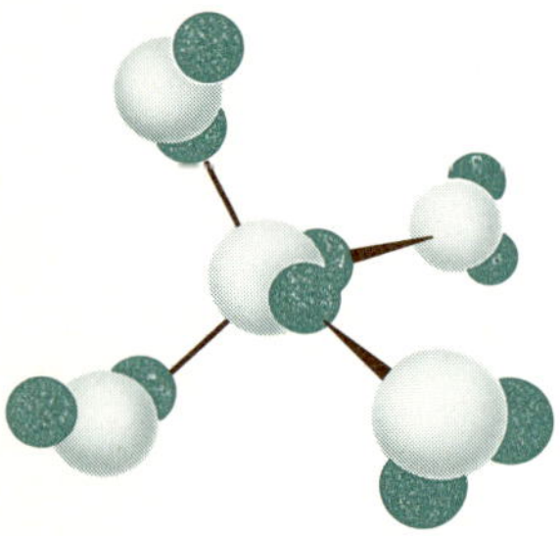

Figure 15.30
Tetrahedral arrangement around the oxygen in a water molecule.

in a short O—H internuclear separation. The other two hydrogens are bonded by the much-weaker hydrogen bonds. The O—H separation in the hydrogen bond is much larger than in the covalent bond.

CO_2 (Dry Ice)

As a final example, Figure 15.31 shows the crystal structure of solid CO_2. The face-centered cubic structure with a CO_2 molecule at each of the lattice sites is evident in this system.

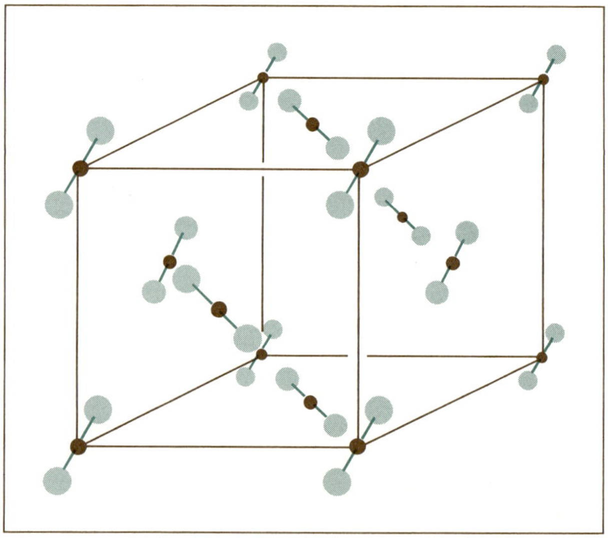

Figure 15.31
The lattice structure for dry ice (CO_2). The dark spheres are carbon and the light spheres are oxygen.

15.5 Physical Properties of Crystals

The physical properties of crystals are closely related to their structures. The lattice description of crystals is very helpful in explaining the observed physical properties. For example, crystals are anisotropic; that is, their properties are different in different directions. Electrical conductivity, refractive index, hardness, and shear strength may be quite different depending upon the direction of measurement. This anisotropy is a consequence of an ordered crystal structure. In graphite, for example, the shear strength varies greatly with the direction.

Many substances are polycrystalline; that is, they consist of a large number of very small crystals. These substances do not appear anisotropic. The many small crystals are randomly aligned and any directional effects average out.

Anisotropy does not occur in liquids or gases in which the arrangement of the particles is random and disordered. As a result of this molecular chaos, all directions are equivalent.

The geometrical shape of a crystal is often a magnification of the lattice structure. For example, naturally occurring hexagonal quartz crystals are a reflection of the hexagonal structure on the atomic level. The mineral asbestos occurs in long fibers, reflecting the long fiberlike structure of the individual asbestos molecules. Finally, sodium chloride crystals have 90° angles, indicative of the cubic structure of the lattice.

Crystals possess the property of cleavage. When fractured, they break along flat planes, and the angles between these planes are always the same. Cleavage is easily explained from the lattice model when one remembers that the particles are arranged in planes, and the angles between the planes are fixed. When the crystal is broken, separation occurs along these planes.

To describe other properties of the solid, it is useful to distinguish between ionic, molecular, covalent, and metallic solids.

In ionic solids, positive and negative ions occupy the lattice positions. Strong electrostatic attractive forces hold the ions together.

Examples of ionic crystals are NaCl, CsCl, and CaF_2 (see Figures 15.23 and 15.24).

Examples of molecular crystals are CO_2 (see Figure 15.31), ice (see Figure 15.28), I_2, and Ar.

Examples of covalent crystals are diamond and graphite (see Figures 15.26 and 15.27), SiO_2, and SiC.

In molecular crystals, the repeating unit in the lattice is a neutral atom or molecule. Usually, weak van der Waals or dipole forces hold the molecules together.

In covalent solids, atoms occupying the sites in the lattice are held together with strong covalent bonds. The covalent crystal is in effect a single giant molecule with all the atoms covalently bonded together.

In the metallic crystal, metal atoms occupy the lattice sites. The properties of the metallic state—such as the silvery luster and reflectivity of light, the high electrical and heat conductivity, and the ease with which metals can be drawn into wire, hammered into sheets, and bent—are best explained by assuming that the valence electrons are not localized on each atom, but are free to move throughout the crystal. In a simplified picture, we may think of a metal as a collection of positive ions occupying the lattice sites and immersed in a "sea" of mobile electrons (see Figure 15.32).

Figure 15.32
In the metallic crystal, positive ions are immersed in a "sea" of mobile valence electrons.

This model of the metallic crystal, while useful, is an oversimplification, since metals differ so widely in their physical properties.

An important feature in explaining many properties of crystals is the strength of the interaction between the particles. For example, covalent crystals are very hard since the deformation or cutting of the crystal breaks strong covalent bonds. The electrostatic attractions in ionic crystals are also strong, with the result that ionic crystals are not easily deformed but are very brittle. The explanation for this is that it is difficult to move one plane of ions relative to another with opposite charge. Once movement starts, however, ions with the same charge come in contact, repulsions occur, and the crystal breaks apart.

The diamond cutter obtains the octahedral structure of a jewelry diamond by cleaving the gem along the octahedral planes of the cubic diamond structure (similar to that shown in Figure 15.26). For a large diamond, the cutter will spend long hours locating these planes in the crystal. He then strikes the diamond in a way that will cause it to break along one of the desired planes. Occasionally a cutter makes an expensive miscalculation, shattering the crystal.

Molecular crystals are usually soft. Deformation of these crystals requires only that molecules held together by weak interactions be separated.

Table 15.7 compares the energy required to separate the particles in the different kinds of solids. The high energy required to separate the ions in ionic crystals and break the covalent bonds in covalent solids is consistent with strong bonds. Little energy is required to overcome the weak attractive forces in molecular crystals. The rather high energy required to remove atoms from metallic crystals is related to the need to collect electrons from the "sea of electrons" and remove them as part of a metal atom.

The thermal properties of the different types of crystals are summarized in Table 15.8. The high heat of fusion and vaporization in ionic and covalent substances reflect the large amount of energy required to break down the crystal (for melting) and separate the atoms or molecules (in vaporization).

TABLE 15.7

Comparison of the Energy Required to Separate the Particles in Ionic, Molecular, Covalent, and Metallic Crystals

IONIC CRYSTALS	ENERGY REQUIRED TO SEPARATE IONS (kjoules/mole)
LiF	1032
NaCl	779
ZnO	4030
MOLECULAR CRYSTALS	**ENERGY REQUIRED TO SEPARATE MOLECULES (kjoules/mole)**
Ar	8
Cl_2	20
CO_2	25
CH_4	8
COVALENT CRYSTALS	**ENERGY REQUIRED TO SEPARATE ATOMS (kjoules/mole)**
C(diamond)	711
Si	439
SiO_2	1180
METALLIC CRYSTALS	**ENERGY REQUIRED TO SEPARATE ATOMS (kjoules/mole)**
Li	159
Al	322
Fe	414
W	837

TABLE 15.8

Comparison of the Thermal Properties of the Different Types of Solids

TYPE OF PROPERTY	IONIC	MOLECULAR	COVALENT	METALLIC
Melting point	High	Low	High	Low to high
Enthalpy of fusion	High	Low	High	Moderate
Boiling point	High	Low	Very high	Low to high
Enthalpy of vaporization	Very high	Low	Very high	High

15.6 Determination of Crystal Structure

The lattice arrangement of atoms in a crystalline solid is determined principally from x-ray diffraction studies. A beam of x-rays is directed onto a crystal as shown in Figure 15.34. The crystal is then slowly rotated until those orientations that produce maximum intensity of the emerging x-ray beam are located. The emerging x-ray waves will be in phase (and therefore produce a beam of maximum intensity) when the angle of incidence (measured relative to a repeating plane of atoms) is equal to the angle of reflection. This is the condition shown in Figure 15.34. Under these conditions, the emerging x-rays reinforce each other, and we say that the crystal has produced diffraction of the x-ray beam.

The relationship between the angle where diffraction occurs (θ), the wavelength of the x-rays (λ), and the distance between repeating planes of atoms (d) is given by

$$n\lambda = 2d \sin \theta \qquad [15.4]$$

in which $n = 1, 2, 3, \ldots$.

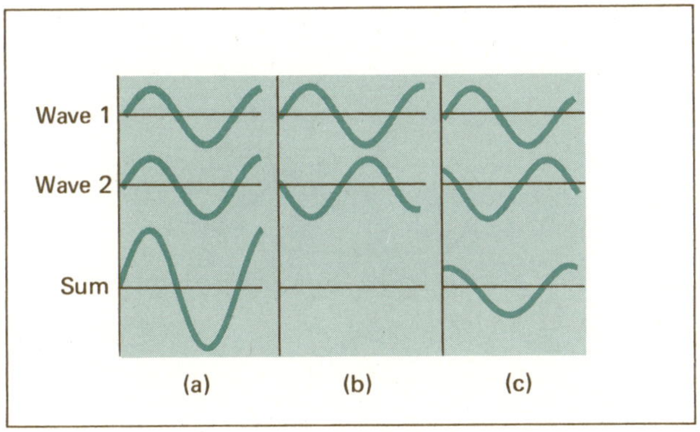

As seen in Figure 15.33, when two electromagnetic waves are in phase, their amplitudes (intensities) add to give a wave with an amplitude equal to the sum of the two. When the two waves are exactly out of phase, they cancel.

Figure 15.33
Electromagnetic waves reinforce one another when they are in phase (a), cancel when out of phase (b), and partially cancel when partly out of phase (c).

Figure 15.34
Diffraction from successive planes of atoms. Diffracted waves are in phase when $n\lambda = 2d \sin\theta$.

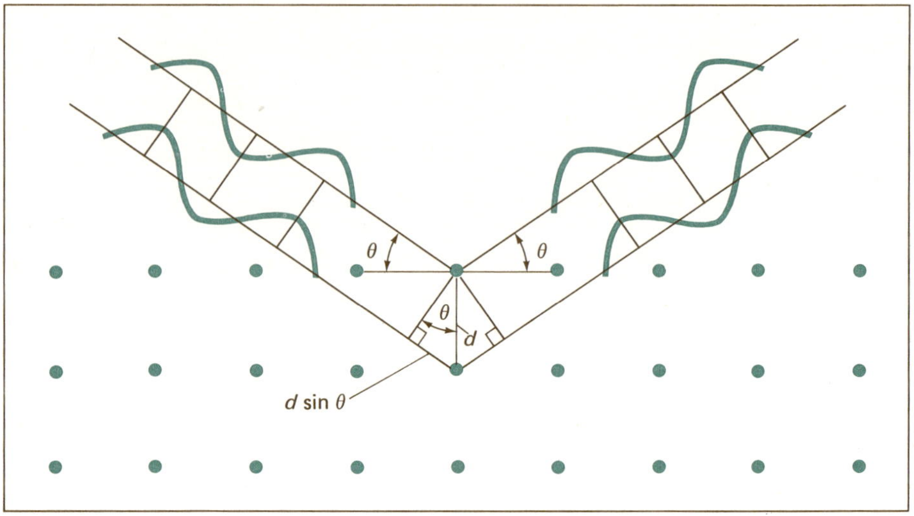

Figure 15.35
Representation of some of the planes of lattice points for the crystal lattice shown in Figure 15.34. Diffraction occurs from such planes as these.

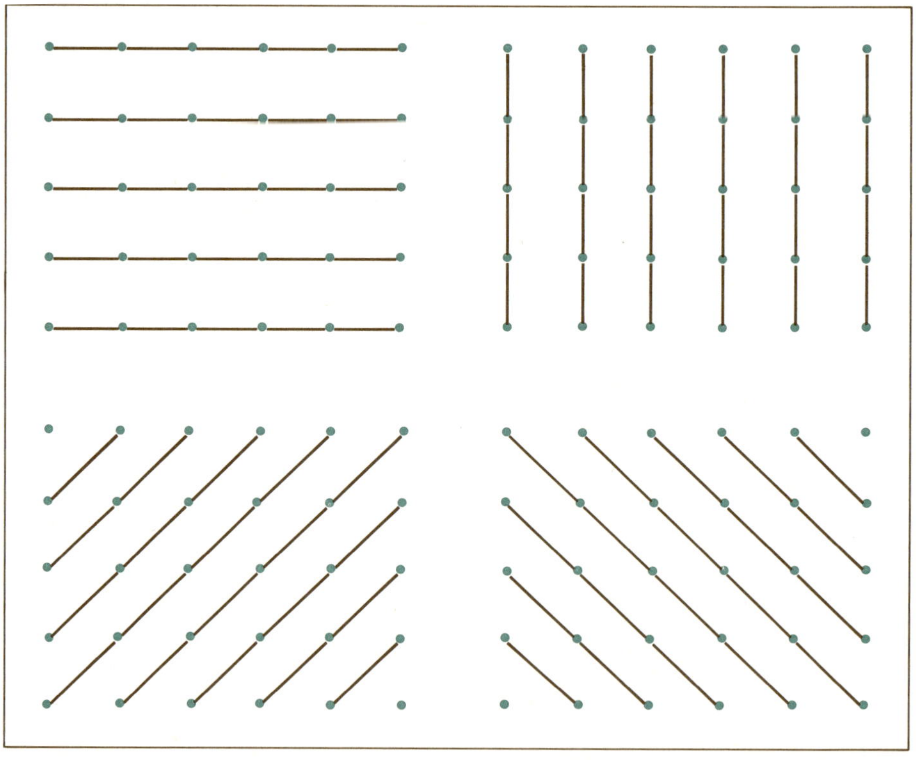

Equation [15.4] follows from the reasoning that the x-rays being reflected from atoms in parallel planes at different depths in the crystal (refer to Figure 15.34) will be in phase if the difference in the length of the path they travel is a whole-number multiple of their wavelength. The difference in the distances traveled by the two waves in Figure 15.34 is seen to be twice the distance $d \sin \theta$. Since the waves enter the crystal in phase, the reflected waves will emerge in phase if θ is such that $2d \sin \theta$ is equal to λ (or some whole-number multiple of λ).

Equation 15.4 is called the Bragg equation in honor of W. H. and W. L. Bragg, a father and son team who received the Nobel Prize for their contributions to x-ray diffraction studies with crystals.

Figure 15.35 shows other sets of planes that are present in a lattice of the kind illustrated in Figure 15.34. Diffraction can also occur from these planes, but at different angles, since the spacing d is different. By rotating the crystal to find the angles at which diffraction occurs, the crystallographer can use equation [15.4] to determine all the distances between planes. Having the information about the distances between all possible parallel planes in a crystal, the crystallographer can locate the positions of the atoms by assuming that the atoms will be at the intersections of the planes. In complex crystals, the process of deducing the structure can be very difficult. Modern high-capacity computers have proved to be a most valuable tool for the crystallographer.

15.7 The Ionic Bond

Ionic bonds form between atoms having large electronegativity differences. Metal-nonmetal compounds are usually best described as ionically bonded since most metals have low electronegativities and nonmetals high electronegativities. The best examples are compounds of the alkali metals (Li, Na, K, Rb, Cs) and the alkaline earth metals (Mg, Ca, Sr, Ba) with such nonmetals as the halogens (F, Cl, Br, I), oxygen, and sulfur. In these compounds the ions occupy the lattice points of ionic crystals.

Metals near the right side of the periodic table are less electropositive. The result is that compounds such as $HgCl_2$, $ZnCl_2$, $AlCl_3$, $SnCl_4$, and $TiCl_4$ are best described as covalently bonded.

In Chapter 2 we described the ionic bond as the limiting case of a very polar covalent bond in which the bonding electrons spend all their time around one of the nuclei. The result is a complete separation of charge, with one of the bonding atoms positively charged and the other negatively charged. Quantum mechanics can be used to describe the energy and stability of the ionic bond. That is, approximate wave functions can be written to describe the electron density, and the approximate bond energy can be calculated. However, it is much simpler and in many ways very satisfactory to think of the bond as the result of a simple electrostatic attraction between positively and negatively charged particles.

Let us calculate the energy change that results from forming a gaseous NaCl molecule from gaseous Na and Cl atoms. This energy is known as the **bond energy** for NaCl and is a measure of the stability of the bond. What we want to do is obtain ΔE for the reaction

$$Na(g) + Cl(g) = NaCl(g) \quad \Delta E_1 \qquad [15.5]$$

This energy change can be calculated as the sum of several steps:

	$Na(g) = Na^+(g) + e^-$	ΔE_2 = ionization energy	[15.6]
	$e^- + Cl(g) = Cl^-$	ΔE_3 = negative of electron affinity	[15.7]
	$Na^+(g) + Cl^-(g) = NaCl(g)$	ΔE_4 = energy released when ions are brought together	[15.8]
Sum:	$Na(g) + Cl(g) = NaCl(g)$	$\Delta E_1 = \Delta E_2 + \Delta E_3 + \Delta E_4$	

The sum of ΔE_2 and ΔE_3 is 146.5 kjoules/mole; energy must be added to the gaseous atoms to form gaseous ions, even though the latter have completed octets. It is evident that something more than the gaining of noble-gas configurations is required to make sodium and chlorine react.

The numerical values of ΔE_2, ΔE_3, and ΔE_4 are obtained in the following way: ΔE_2 is the ionization energy for sodium, and is 495.4 kjoules/mole. ΔE_3 is the negative of the electron affinity of chlorine, and is -348.9 kjoules/mole. ΔE_4 is more difficult to obtain. It is the potential energy change when the two isolated gaseous ions are brought together. Figure 15.36 shows how the total potential energy varies as a function of the distance separating the oppositely charged ions, and also shows the repulsion and attraction energies, which add to give the total. The total potential energy, U_{total}, reaches a minimum at the equilibrium bond distance r_0 in the molecule. The potential energy of attraction can be calculated from electrostatic theory, and for this example it is equal to 583.7 kjoules. The energy of repulsion is much more difficult to obtain, but it is seen in Figure 15.36 to be small relative to the energy of attraction at r_0; thus it can (for a good approximation) be neglected to give

$$\Delta E_4 = -583.7 \text{ kjoules/mole}$$

and

$$\Delta E_1 = \Delta E_2 + \Delta E_3 + \Delta E_4 = -437.2 \text{ kjoules/mole}$$

Figure 15.36
The total potential energy change as charged ions are brought together is the sum of a repulsion term and an attraction term.

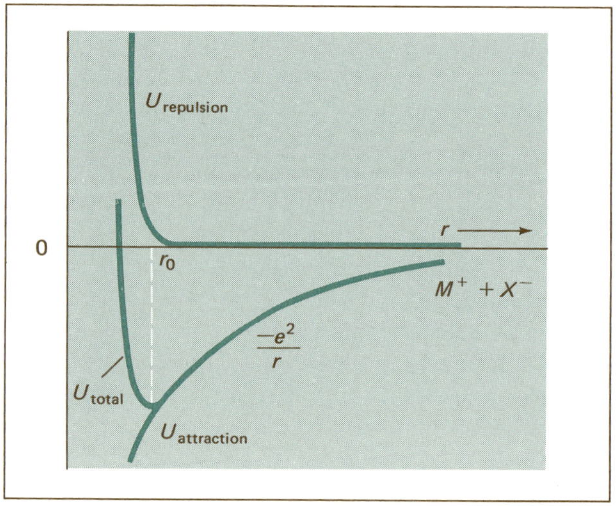

This result indicates that the reaction (equation [15.5]) is energetically favorable; that is, energy is released in the formation of the bond. The reason for the stability of the bond is clear: The energy required to transfer the electron from the Na atom to the Cl atom is more than compensated for by the potential energy decrease that results from the electrostatic attraction of the Na^+ and Cl^- ions.

Another energy change that is of great importance is that for the process

$$Na^+(g) + Cl^-(g) = NaCl(s)$$

The energy change for this reaction is called the **lattice energy.** It is the energy change associated with the formation of the crystal from its components, in this case gaseous Na^+ and Cl^- ions. A large negative lattice energy indicates a stable crystal. It can be calculated from theoretical considerations or from experimental data through use of a thermodynamic cycle known as the **Born-Haber cycle.** Good agreement is obtained by the two approaches. We will show the use of the Born-Haber cycle.

In this cyclic process, the elements are converted to the solid crystal by a path that first involves converting the elements to gaseous atoms. The atoms are then ionized and condensed to the solid crystal. The sum of all the steps gives the energy of formation of the solid.

The Born-Haber cycle has the following steps when applied to NaCl:

Remember that energy is a state function; the change is the same regardless of the path.

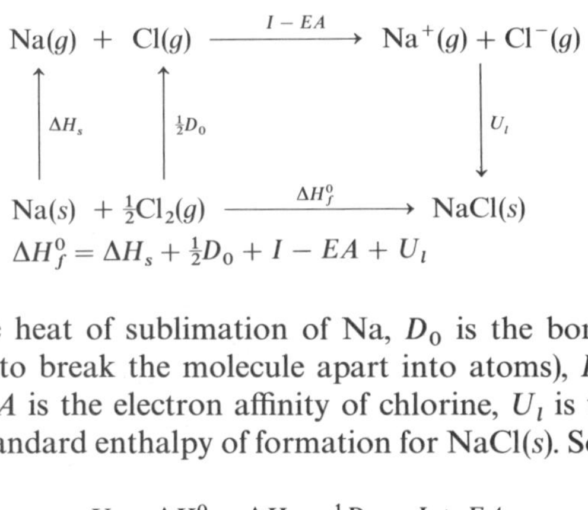

$$\Delta H_f^0 = \Delta H_s + \tfrac{1}{2}D_0 + I - EA + U_l$$

where ΔH_s is the heat of sublimation of Na, D_0 is the bond energy for Cl_2 (energy required to break the molecule apart into atoms), I is the ionization energy for Na, EA is the electron affinity of chlorine, U_l is the lattice energy, and ΔH_f^0 is the standard enthalpy of formation for NaCl(s). Solving for U_l gives

In the Born-Haber cycle we use the more readily available ΔH_s and ΔH_f^0 in place of ΔE_s and ΔE_f^0. For these processes the difference between change in energy and change in enthalpy is very small.

$$U_l = \Delta H_f^0 - \Delta H_s - \tfrac{1}{2}D_0 - I + EA$$

For Na, Cl_2 and NaCl at 25° C,

$$\Delta H_f^0 = -410.9 \text{ kjoules/mole}$$
$$\Delta H_s = 108.4 \text{ kjoules/mole}$$
$$D_0 = 241.8 \text{ kjoules/mole}$$
$$I = 495.4 \text{ kjoules/mole}$$
$$EA = 348.9 \text{ kjoules/mole}$$

TABLE 15.9

Comparison of the Lattice Energies of Alkali Metal Halides

SALT	LATTICE ENERGY (kjoules/mole)	SALT	LATTICE ENERGY (kjoules/mole)
NaF	−916	LiCl	−849
NaCl	−787	NaCl	−787
NaBr	−732	KCl	−715
NaI	−628	RbCl	−682
		CsCl	−669

Hence,

$$\begin{aligned} U_l &= -410.9 - 108.4 - \tfrac{1}{2}(241.8) - 495.4 + 348.9 \\ &= -786.7 \text{ kjoules/mole} \end{aligned}$$

This large negative value shows that there is much energy released when the lattice is formed; thus, the lattice is very stable.

Similar calculations can be made for other ionic compounds. Table 15.9 compares the lattice energies obtained from the Born-Haber cycle for a number of alkali metal halides. The trends in the table show the effect of ionic size. The size of the halide ion increases as we go down the left column, and the size of the alkali metal ion increases as we go down the right column. In both cases, the lattice energy decreases. This is as expected, since U_l varies inversely with r_0, the internuclear separation in the lattice.

Summary

A crystalline solid is a collection of molecules, atoms, or ions arranged in a repeating geometrical pattern called a crystal lattice. The unit cell is the smallest segment of the lattice that contains all the geometrical properties of the lattice.

In this chapter we discussed the structure, coordination number, and packing efficiency of the unit cells of the cubic systems: simple cubic, body-centered cubic, and face-centered cubic. Coordination number and packing efficiency increase as one goes from the simple to body-centered to face-centered cubic structures.

When spherical particles of the same size, such as the atoms of a metal or a noble gas, crystallize in the most compact structure possible, either the cubic close-packed (face-centered cubic unit cell) or the hexagonal close-packed (hexagonal unit cell) structure results. Many metals crystallize into one or the other of these close-packed structures. Other metals form body-centered cubic structures.

Three types of holes are common in crystal lattices: tetrahedral and octahedral holes in close-packed structures and cubic holes in simple cubic lattices. The structure of many ionic compounds can be viewed as lattices of anions with cations in the lattice holes, or the reverse structure. The rock salt (NaCl) type lattice, for example, is a cubic close-packed lattice of anions with the cations occupying the octahedral holes of the lattice. The zinc blende (ZnS) type lattice is a cubic close-packed lattice of anions with the cations occupying every other tetrahedral hole. Another representative ionic structure is that of the CsCl type lattice. In this structure, Cs^+ ions form a simple cubic lattice, and the Cl^- ions occupy the cubic holes.

The radius ratio of the ions plays an important role in determining the type of ionic lattice that results. When the radius ratio of the ions is large, lattices with large holes, such as the cubic lattice, are formed. When the ion radius ratio is small, as in ZnS, tetrahedral holes are favored. Octahedral holes are intermediate in size to cubic and tetrahedral holes.

Crystals are conveniently classified as ionic, molecular, covalent, or metallic. In ionic crystals, ions are at the lattice points. Strong electrostatic forces hold the crystal together. Ionic crystals tend to be brittle, hard, and high-melting. In molecular crystals, molecules occupy the lattice points. The bonds holding the molecules together are weak intermolecular forces, such as van der Waals or dipole forces. Such crystals tend to be soft and low-melting. In covalent crystals, atoms occupying the lattice points are held together by strong covalent bonds. In metallic crystals the lattice sites are viewed as positive ions surrounded by mobile valence electrons that are not associated with a specific atom. This model accounts for the high electrical and thermal conductivity of metals. It does not account, however, for the wide variability in melting points, boiling points, and hardness of the metals.

The ionic bond can be thought of as an extreme case of a polar covalent bond in which there is complete separation of charge with one of the bonding atoms positively charged and the other negatively charged. The bond energy and lattice energy are two important quantities that give a measure of the stability of ionic bonds and crystal lattices, respectively.

Important Terms and Concepts

Lattice
Unit cell
Simple cubic packing
Body-centered cubic packing
Face-centered cubic packing
Coordination number
Cubic close-packed
Hexagonal close-packed
Cubic hole
Radius ratio
Octahedral hole
Tetrahedral hole
Triangular hole
Linear hole
Rock salt (NaCl) structure
CsCl structure
Zinc blende (cubic ZnS) structure
Ionic bond
Bond energy
Lattice energy
Born-Haber cycle

Exercises

15-1.
Assuming equal rigid spheres in contact with each other, calculate the percent of empty space if the spheres are arranged in the face-centered cubic lattice.

15-2.
Sr metal crystallizes in a face-centered cubic crystal structure. If the unit cell is 6.10 Å on an edge, calculate the density of Sr metal in g/cm^3.

15-3.
Aluminum crystallizes in a face-centered cubic unit cell. The density is 2.70 g/cm^3.

(a) Calculate the length of an edge of the Al unit cell.
(b) Calculate the atomic diameter of the Al atom in the solid metal.

15-4.
Potassium crystallizes in a body-centered cubic lattice with the length of the cell edge equal to 5.33 Å. Calculate the dimensions of a cube that would contain one mole of K.

15-5.
Tantalum crystallizes in a cubic lattice, and the edge of the unit cell is 3.30 Å in length. If the density of Ta is 16.6 g/cm^3, what type of cubic unit cell does Ta form?

15-6.
The atomic radius of the iron atom is 1.24 Å. Iron occurs in both body-centered cubic and the face-centered cubic structures. Calculate the volume change in cm^3/mole when iron changes from face-centered cubic to body-centered cubic.

15-7.
Show that the radius ratio for a triangular hole is 0.155.

15-8.
For the following ionic compounds, predict from the radius ratio the expected crystal structure.

(a) BeO (c) BaO
(b) NaI (d) RbF

15-9.
Use radius ratios to predict the structures of the oxides and sulfides of Ni^{2+}, Co^{2+}, and Fe^{2+}

15-10.
From the table of ionic radii, predict the coordination number of the cation in each of the following compounds.

(a) BaSe
(b) BeSe

15-11.
Carbon, hydrogen, and nitrogen form compounds with a variety of metals. These are called interstitial compounds because the added atoms fit into holes in the cubic close-packed metallic structure. Given the following table of atomic radii

(a) predict whether hydrogen will occupy tetrahedral or octahedral holes in the titanium lattice.
(b) predict the type of holes carbon will occupy in a Zr lattice when the two react.

ELEMENT	ATOMIC RADII (Å)
C	0.77
H	0.37
Hf	1.44
N	0.70
Ti	1.32
Zr	1.45

15-12.
What is the simplest formula of a solid in which A atoms occupy only the corners of a face-centered cubic arrangement and B atoms occupy only the center of the faces?

15-13.
A particular solid crystallizes in a cubic unit cell with tungsten atoms located at each cube corner, oxygen atoms at the centers of each cube edge, and sodium atoms at the center of the cube.

(a) What is the simplest formula for this compound?
(b) The length of the unit cell edge is 3.86 Å. Calculate the density of the solid.

15-14.
A spinel is a mineral found in nature containing oxide ions and two types of metal ions. In one such spinel, the oxide

ions are in a face-centered cubic lattice. One-eighth of the tetrahedral holes are occupied by Zn^{2+} ions and one-half of the octahedral holes by Al^{3+} ions. What is the formula of this spinel?

15-15.

(a) How many ions of each type are there in the unit cell of cubic zinc blende, ZnS?
(b) The density of cubic ZnS is 4.10 g/cm^3. What is the length of the edge of the unit cell in ZnS?

15-16.

The density of solid NaCl is 2.164 g/cm^3. Assume that the radius of the chloride ion is exactly twice the radius of the sodium ion and calculate

(a) the radii (in cm) of the chloride and sodium ions.
(b) the percent of empty space in the NaCl crystal.

15-17.

LiF has the rock salt (NaCl) crystal structure. What is the volume in Å^3 of the LiF unit cell?

15-18.

At what angles would first-order ($n = 1$) and second-order ($n = 2$) reflections be observed for diffraction from a set of crystal planes with a spacing of 3.03 Å if the x-rays used have a wavelength of 0.710 Å?

15-19.

When x-rays with $\lambda = 1.54$ Å are diffracted from a crystal, a first-order reflection is found at an angle of 10°0′. What wavelength of x-rays would show this same reflection at an angle of 20°20′?

15-20.

Calculate the lattice energy for MgO from the following data:

$O(g) + 2e^- \rightarrow O^{2-}(g)$	$\Delta H = 736$ kjoules
$Mg(s) \rightarrow Mg(g)$	$\Delta H = 149$ kjoules
$Mg(g) \rightarrow Mg^{2+}(g) + 2\,e^-$	$\Delta H = 2190$ kjoules
$Mg(s) + \frac{1}{2}\,O_2(g) \rightarrow MgO(s)$	$\Delta H = -602$ kjoules
$O_2(g) = 2\,O(g)$	$\Delta H = 494$ kjoules

Problems

15-1.

Manganese metal changes crystal structure from face-centered cubic to body-centered cubic at 1130° C and one atm pressure.

(a) Would you expect this transition to be accompanied by an increase or decrease in density?
(b) Use thermodynamics to predict the effect on the transition temperature of increasing the pressure.

15-2.

It has been determined from x-ray diffraction measurements that lithium crystallizes in a body-centered cubic lattice with a unit cell 3.50 Å on an edge. The atomic weight of Li is 6.94 g/mole and the density is 0.534 g/cm^3. Calculate Avogadro's number from these data.

15-3.

Titanium monoxide, TiO, has the NaCl structure. However, the crystal is imperfect in that some of the Ti^{2+} sites and an equal number of O^{2-} sites are vacant. Given that the density of TiO is 4.93 g/cm^3 and the length of the unit cell edge (as determined from x-ray diffraction) is 4.235 Å, calculate the fraction of the sites that are vacant.

15-4.

Using only the following data for NaCl, calculate Avogadro's number.

$d = 2.165$ g/cm^3	(density)
$r_0 = 2.819$ Å	(internuclear distance)
$M = 58.448$ g/mole	(molecular weight)

15-5.

Xenon difluoride, XeF_2, crystallizes in a body-centered tetragonal lattice with a XeF_2 molecule at each lattice point. The base of the tetragonal unit cell is a square 4.32 Å on a side, and the height of the unit cell is 6.99 Å. What is the density of the solid XeF_2? (In counting the number of molecules per unit cell, assume that the molecule at the center is completely within the unit cell and that the molecules at the corners are equally shared by all the unit cells meeting at the corner.)

15-6.

When x-rays of wavelength $\lambda = 2.743$ Å are directed upon the face of a body-centered cubic crystal of barium metal, the smallest angle of incidence that produces a maximum reflected beam is $\theta = 32°20'$. Calculate for Ba

(a) the distance between the crystal planes.
(b) the atomic radius of Ba.

15-7.

Diffraction with x-rays of 0.586 Å wavelength occurs in the NaCl crystal from planes of ions parallel to the face of the face-centered cubic lattice at $\theta = 5°58'$.

(a) Calculate the separation between the planes.
(b) What is the smallest Na-Cl internuclear separation?
(c) What is the smallest Cl-Cl internuclear separation?

15-8.

Given the following data, calculate the bond energy for gaseous HF and HCl assuming they are completely ionic compounds. That is, calculate ΔE for the process

$$H(g) + F(g) = H^+F^-(g)$$
$$H(g) + Cl(g) = H^+Cl^-(g)$$

	FIRST IONIZATION POTENTIAL (kcal/mole)	ELECTRON AFFINITY (kcal/mole)	IONIC RADIUS
F	401.8	79.5	1.36 Å (F^-)
Cl	300.0	83.3	1.81 Å (Cl^-)
H	313.6	17.4	~0 (H^+)

15-9.

Horatio spent one winter semester sorting oranges. He had worn out his welcome at the university just because of two small fires and one medium-sized explosion in the chemistry laboratory. The operator of the fruit packing plant recognized qualities in Horatio outside the normal range, so taught him to repack the boxes as they came from the orchard. Horatio took esthetic delight in the orderly columns and rows of oranges as he followed his supervisor's instructions. He was, however, confused by a constant surplus of oranges; the box would never hold all the oranges that it had contained when it came from the field.

Had Horatio spent more time studying chemistry and less time building fires, he could have understood the matter. The unscrupulous operator had taught him to repack the oranges in a body-centered cubic arrangement, while the pickers let them go into their natural arrangement. Since the close-packed structure is the lowest energy structure, oranges poured into a box will be approximately in such an arrangement.

Assume that the oranges were changed from a close-packed to a body-centered cubic arrangement and calculate the dishonest gain made per box by the operator. Assume that the cost is $10.00 a box.

16

Periodic Properties

In Chapter 2 we gave a very brief introduction to atomic structure, chemical bonding, and some periodic properties of the elements. Now that we have explored atomic structure (Chapter 13) and the chemical bond (Chapter 14) in greater detail, we can return to the subject of periodic properties. This will serve as our introduction to the vast and interesting subject of *descriptive chemistry*. The properties of the oxides of the elements display periodic relationships to a remarkable extent; therefore, we will find it useful later in the chapter to consider in detail the reactions of oxides in aqueous solutions. One should not expect that all the detailed chemistry of the elements can be explained by the periodic relationships, but the beginning student can take comfort from the knowledge that here there is an excellent simplifying and unifying framework upon which to support many facts of chemical reactions.

16.1 Physical Properties

Review Chapter 2 for the different groupings in the periodic table: alkali metals, transition elements, and so forth.

Correlations can be made between periodic table position and physical properties of the elements. Figures 16.1 and 16.2 give the melting and boiling temperatures for the elements. Note the very regular trends in such groups as the alkali metals, halogens, and noble gases. The changes are not quite so regular for other families, such as the carbon family and the nitrogen family, but trends are still evident. Notice, too, the very high melting and boiling temperatures for the transition elements in Groups IIIB to VIII and the lower melting and boiling temperatures for the other metals.

The metals in Groups IIIB to VIII have partially filled *d* orbitals. The overlap of these orbitals forms chemical bonds that must be broken in melting and boiling; high temperatures are required for this to happen.

Figure 16.3 compares the densities of the elements. You can see the general trend toward higher density as one moves down the columns of the table. Again the transition elements are the most impressive; the elements osmium, iridium, and platinum have the highest densities.

In the nineteenth century there was great excitement among chemists as the periodic table was being developed. From the properties of adjacent elements chemists predicted the properties of unknown elements where gaps appeared in the table. The remarkable accuracy of these predictions is illustrated by comparing Mendeleev's predictions about the unknown element "ekasilicon" with the actual properties of germanium, as it was called after its discovery.

	PREDICTED PROPERTIES OF EKASILICON	ACTUAL PROPERTIES OF GERMANIUM
Atomic Wt.	72	72.5
Density of Element	5.5 g/ml	5.46 g/ml
Oxide	EsO_2, density = 4.7 g/ml	GeO_2, density = 4.76 g/ml
Chloride	$EsCl_4$, liquid with boiling point slightly below 100° C and density = 1.9 g/ml	$GeCl_4$, liquid with boiling point at 86° C and density = 1.887 g/ml
Fluoride	EsF_4, not gaseous	GeF_4, white solid

IA	IIA	IIIB	IVB	VB	VIB	VIIB	VIII			IB	IIB	IIIA	IVA	VA	VIA	VIIA	Noble Gases
-259 H																	He
181 Li	1283 Be											2027 B	C	-210 N	-219 O	-219 F	-249 Ne
98 Na	656 Mg											659 Al	1410 Si	44 P	119 S	-101 Cl	-189 Ar
63 K	850 Ca	1400 Sc	1677 Ti	1917 V	1903 Cr	1244 Mn	1539 Fe	1495 Co	1455 Ni	1083 Cu	420 Zn	30 Ga	937 Ge	As	217 Se	-7 Br	-157 Kr
39 Rb	770 Sr	1500 Y	1852 Zr	2497 Nb	2617 Mo	2127 Tc	2400 Ru	1966 Rh	1550 Pd	961 Ag	321 Cd	156 In	232 Sn	630 Sb	450 Te	114 I	-112 Xe
29 Cs	710 Ba	920 La	2222 Hf	2997 Ta	3377 W	3180 Re	2700 Os	2454 Ir	1770 Pt	1063 Au	-39 Hg	304 Tl	328 Pb	272 Bi	254 Po	At	-71 Rn

Figure 16.1
Melting points of the elements (°C).

IA	IIA	IIIB	IVB	VB	VIB	VIIB	VIII			IB	IIB	IIIA	IVA	VA	VIA	VIIA	Noble Gases
-253 H																	-269 He
1331 Li	2477 Be											3927 B	3900† C	-196 N	-183 O	-188 F	-246 Ne
890 Na	1127 Mg											2447 Al	2677 Si	280 P	445 S	-34 Cl	-186 Ar
766 K	1477 Ca	2477 Sc	3277 Ti	3377 V	2627 Cr	2067 Mn	2887 Fe	2887 Co	2887 Ni	2582 Cu	908 Zn	2237 Ga	2827 Ge	613† As	685 Se	58 Br	-153 Kr
701 Rb	1367 Sr	3227 Y	4377 Zr	4927 Nb	4627 Mo	4627 Tc	3700 Ru	3700 Rh	3100 Pd	2177 Ag	765 Cd	2047 In	2687 Sn	1637 Sb	987 Te	183 I	-108 Xe
685 Cs	1677 Ba	3367 La	5227 Hf	5427 Ta	5527 W	5627 Re	4200 Os	4100 Ir	3900 Pt	2707 Au	357 Hg	1467 Tl	1751 Pb	1559 Bi	962 Po	377 At	-62 Rn

† Sublimation Point

Figure 16.2
Boiling points of the element (°C).

IA	IIA	IIIB	IVB	VB	VIB	VIIB	VIII			IB	IIB	IIIA	IVA	VA	VIA	VIIA	Noble Gases
H																	He
0.53 Li	1.85 Be											2.54 B	2.25 C	N	O	F	Ne
0.97 Na	1.74 Mg											2.70 Al	2.40 Si	1.82 P	2.07 S	Cl	Ar
0.86 K	1.54 Ca	2.5 Sc	4.5 Ti	5.96 V	6.92 Cr	7.2 Mn	7.86 Fe	8.9 Co	8.90 Ni	8.92 Cu	7.14 Zn	5.90 Ga	5.36 Ge	5.72 As	4.80 Se	3.12 Br	Kr
1.53 Rb	2.6 Sr	4.47 Y	6.40 Zr	8.57 Nb	10.2 Mo	11.5 Tc	12.4 Ru	12.4 Rh	12.0 Pd	10.5 Ag	8.65 Cd	7.30 In	7.31 Sn	6.68 Sb	6.24 Te	4.97 I	Xe
1.90 Cs	3.51 Ba	6.17 La	12.1 Hf	16.6 Ta	19.3 W	21.40 Re	22.7 Os	22.6 Ir	21.5 Pt	19.3 Au	13.5 Hg	11.85 Tl	11.34 Pb	9.8 Bi	(9.2) Po	At	Rn

Figure 16.3
Densities of elements (g/cm^3).

Many other properties, such as hardness, electrical conductivity, thermal conductivity, and chemical reactivity, can also be correlated. But we will restrict the remainder of our discussion in this chapter to four properties of special interest: atomic size, ionization potential, oxidation states, and chemical properties of the oxides of the elements.

16.2 Sizes of Atoms

There is an obvious problem associated with determining the volume or size of an atom. Where in the electronic cloud does the atom end? We have seen (in Chapter 13) that the shape and size of the electronic orbitals are given by probabilities and not by some definite limits.

Figure 16.4 is a plot of atomic volume versus atomic number. The volumes in this figure are based on calculations made from the densities and atomic weights of the elements. That is, from the molar volume, we calculate the average volume occupied by an atom (see Problem 16-2). Even with such a crude estimate as this, which ignores problems of crystal structure and temperature effects upon the volume, one obtains an impressive relationship between "volume" and position in the periodic table. The relationship can be explained in terms of the electronic configurations. The alkali metals, having just gained an electron with a new principal quantum number, have the enlarged size that results from the larger radial probability of that electron. The minimums that occur in the plot between alkali metals result from the competition between two opposing interactions. Each electron that is added increases the repulsive forces between electrons, and the effect of this tends to be an enlargement of the atomic size.

Figure 16.4
The relationship of atomic volume and atomic number.

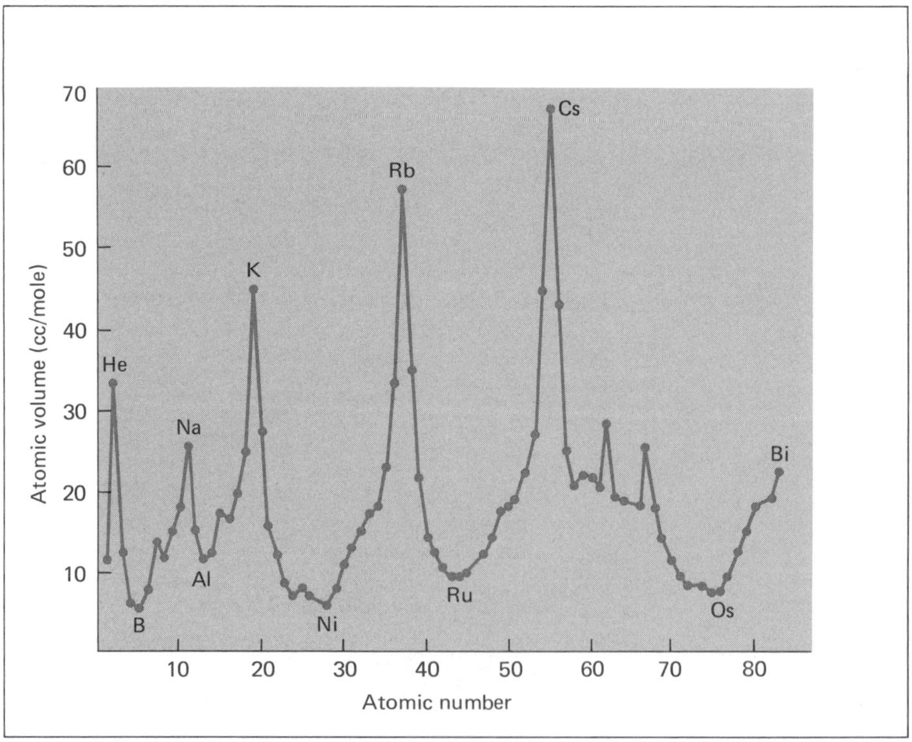

At the same time, the nuclear charge increases with atomic number, which increases the attractive force and tends to decrease the atomic size.

The sizes of atoms in the third period illustrate the result of the increase in electron repulsion and the increased nuclear attraction. The nuclear attraction has the greater effect in the change from sodium to magnesium to aluminum, and the atomic volume decreases. Then the repulsion of electrons becomes dominant, and silicon has a slightly larger volume than aluminum. Increased volume with atomic number characterizes the remainder of the period, except when the first pairing occurs in a *p* orbital. You should notice that this same discontinuity occurs in the second period.

There are better ways to estimate the size of an atom than by the density-atomic weight method used for calculating the values in Figure 16.4. The use of x-ray diffraction to determine interatomic distances in a crystal has been used effectively for metallic elements. Figure 16.5 gives atomic radii of metals that were determined by this method. The data in that table clearly show the initial decrease in atomic size with atomic number in each period and the subsequent increase toward the end of the period. Particularly striking is the decrease in size in the sixth period between lanthanum and hafnium. This decrease is called the *lanthanide contraction* and results from the addition of fourteen 4*f* electrons to the lanthanum metals after the 5*s*, 5*p*, and 6*s* electrons are in the atoms. The 4*f* electrons lie deep within the atoms and have relatively little effect upon the size, while the added protons in the nucleus tend to shrink the atom by their attractive forces on the electron cloud. You should notice that some sixth-period elements have radii the same, or virtually the same, as the fifth-period elements immediately above them. That this occurs even though there are electrons having higher principal quantum numbers in the sixth period is a clear indication of the effect of the lanthanide contraction.

You will see in subsequent chapters that there are aspects of the chemistry of the elements just beyond the lanthanides that are explained in terms of the lanthanide contraction.

In Figure 16.6, the radii of several ions are plotted against atomic number, with solid lines connecting ions that are isoelectronic. This plot shows clearly the effect of nuclear charge upon the size of the ion. Each isoelectronic set includes ions with the same electronic configuration. Thus, Al^{3+} and O^{2-} have the same electronic configuration but strikingly different radii. The change from 8 to 13 protons in the nucleus has a large effect upon the ionic size because of the greatly increased attractive forces on the electronic cloud.

The electronic configurations of the neutral O and Al atoms are very different.

$${}_{8}O;\ 1s^22s^22p^4$$
$${}_{13}Al;\ 1s^22s^22p^63s^23p^1$$

but the configurations of the ions are identical:

$${}_{8}O^{2-};\ 1s^22s^22p^6$$
$${}_{13}Al^{3+};\ 1s^22s^22p^6$$

Figure 16.5
Atomic radii of metallic elements expressed in angstroms.

Li 1.33	**Be** 0.89											
Na 1.57	**Mg** 1.36											**Al** 1.25
K 2.03	**Ca** 1.74	**Sc** 1.44	**Ti** 1.32	**V** 1.22	**Cr** 1.17	**Mn** 1.17	**Fe** 1.16	**Co** 1.16	**Ni** 1.15	**Cu** 1.17	**Zn** 1.25	**Ga** 1.24
Rb 2.16	**Sr** 1.91	**Y** —	**Zr** 1.45	**Nb** 1.34	**Mo** 1.29	**Tc** —	**Ru** 1.24	**Rh** 1.25	**Pd** 1.25	**Ag** 1.34	**Cd** 1.41	**In** 1.50
Cs 2.35	**Ba** 1.98	**La** 1.69	**Hf** 1.44	**Ta** 1.34	**W** 1.30	**Re** 1.28	**Os** 1.26	**Ir** 1.26	**Pt** 1.29	**Au** 1.34	**Hg** 1.44	**Tl** 1.55

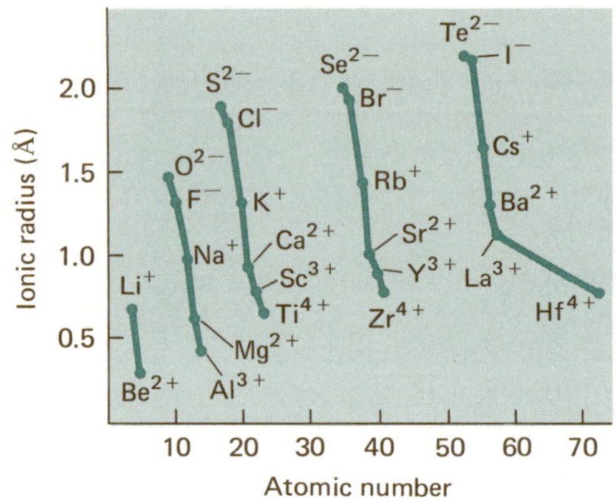

Figure 16.6
Ionic radius as a function of atomic number. The solid lines connect isoelectronic ions.

16.3 Ionization Energy

The ionization energy is the energy required to remove an electron from a gaseous atom or ion.

Figure 16.7 is a duplicate of Figure 2.13 on page 31. We reproduce it here for your convenience.

Figure 16.7, a plot of ionization energy versus atomic number, shows the striking periodicity of the minimum energy required to remove an electron from a gaseous atom to form the positive ion. One can recognize the relative

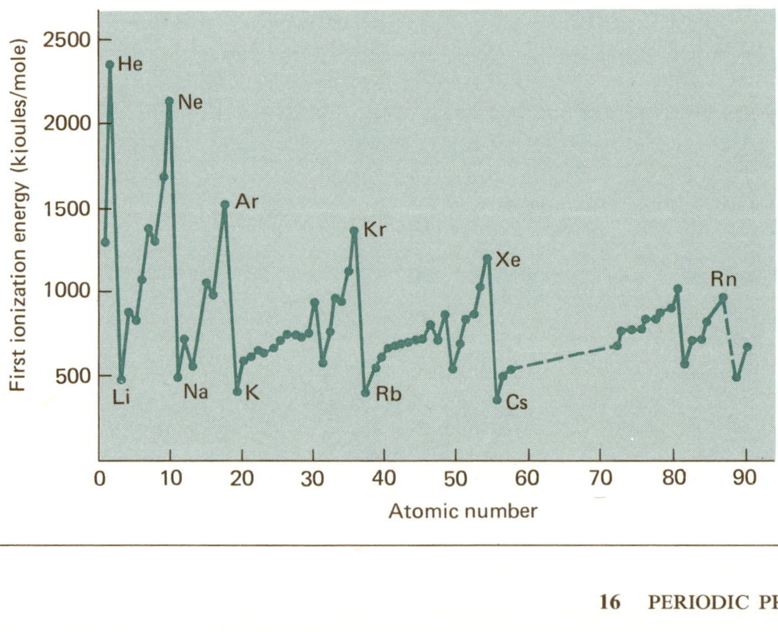

Figure 16.7
Ionization energies of the elements as a function of atomic number.

ease of removing an electron from an alkali metal ion and notice the high energy required to ionize a noble gas. But the plot contains much more detailed information than just that trend. Let us review the relationships of ionization energy and atomic number.

Not only do the alkali metals have relatively low ionization energies, but they also show a distinct trend within the family. In any pair of alkali metals, less energy is required to ionize the member that has the higher atomic number. A similar but even more striking trend is shown by the noble gases. Even though the energy required to ionize radon is high, every noble gas that has a lower atomic number requires more energy to be ionized (see Table 16.1.)

In Figure 16.7, the peaks and valleys are most easily compared, but it can be seen that, in general, the same kind of trend occurs in each family from high toward lower ionization energy as atomic number increases. This trend is readily explained in terms of the atomic structures and the sizes of the atoms. In a given family, the outermost or valence electron structures are similar except that as one goes down the family, the average distance of these electrons from the nucleus increases. This average distance of the valence electrons increases, in general, as the principal quantum number increases, and these valence electrons are the ones removed first in ionization. Larger atoms require less energy to cause ionization than do smaller atoms. Hence, potassium is ionized with less energy than is lithium, and krypton has a higher ionization energy than xenon.

The large decrease in ionization energy that is observed in going from a noble gas to the adjacent alkali metal can be explained in terms of the "screening" or "shielding" effect that *s* and *p* electrons in filled sublevels have upon an *s* electron with a higher principal quantum number. The maximum radial probability density of the electrons in the filled sublevels is at a smaller radius than is that of the outer *s* electron of the alkali metal. Because of the presence of these inner electrons, the outer electron is effectively attracted by a decreased fraction of the nuclear charge. Owing to this screening effect, the outer *s* electron of an alkali metal atom is effectively attracted by approximately the equivalent

TABLE 16.1

Atomic Radii and Ionization Energies of the Noble Gases

	ATOMIC RADIUS (Å)	IONIZATION ENERGY (kjoules/mole)
He	0.93	2371
Ne	1.12	2080
Ar	1.54	1520
Kr	1.69	1351
Xe	1.90	1170
Rn		1037

of a single positive charge, and the electron requires little energy to be removed. In the noble-gas atom, the outermost *p* electrons have little screening effect upon one another because they all have the same radial probability. The attractive force between the nucleus and the electron that is removed is much larger in the noble gases than in the alkali metals, and this is reflected in the large value of the ionization energy.

Let us now consider the elements of the second period, their electronic configurations, and their relative ionization energies. We have seen that the 2*s* electron in lithium is screened from the +3 charge on the nucleus by the two 1*s* electrons. The ionization energy is low, since the 2*s* electron is effectively attracted by a charge of only +1 and the radial probability of the 2*s* electron is relatively large. Beryllium has two 2*s* electrons and a nuclear charge of +4. The 1*s* electrons again screen the 2*s* electrons from the nucleus, but since the 2*s* electrons have little screening effect upon each other (both having the same radial probability), each is attracted by an effective charge of approximately +2. The ionization energy is significantly higher for Be than for Li. In the boron atom the ionization energy is slightly lower than that for beryllium. This can be explained in terms of screening effects when one reviews the radial probabilities of *s* and *p* electrons. Since the 2*s* electrons have a greater probability of being closer to the nucleus than has a 2*p* electron, the latter is more effectively screened from the nuclear charge and hence is slightly more easily removed, even though there is an increased nuclear charge.

The sequence from boron to carbon to nitrogen is straightforward. Carbon has two 2*p* electrons and nitrogen has three. These electrons are able to occupy separate orbitals, thus avoiding the extra energy required to pair electrons. The screening effect stays relatively constant, but the nuclear charge increases, making the ionization energy increase through the sequence. The slight drop in ionization energy from nitrogen to oxygen is related to the extra energy required to add the fourth 2*p* electron, since this has to pair with another 2*p* electron. The plot in Figure 16.7 shows that the repulsion between electrons that makes their pairing difficult more than offsets the effect of the increased nuclear charge, and oxygen has a slightly lower ionization energy than has nitrogen. The addition of the two remaining 2*p* electrons is accompanied by increases in ionization energy, showing that the increased nuclear charge is having a more pronounced effect than is the electron repulsion in the orbitals.

If you remember that in the *aufbau* buildup of boron, carbon, and nitrogen, the electrons are placed successively in p_x, p_y, and p_z orbitals, which are directed perpendicularly to one another, it will be easy for you to visualize why these electrons are relatively ineffective in screening the nuclear charges from one another.

Subsequent periods show similar relationships. The third period is a close duplication of the second in the relative ionization energies. The fourth, fifth, and sixth periods are similar but with a smaller overall spread of values and with many ionization energies very close to one another as *d* and *f* electrons are added.

Table 16.2 gives numerical values of first, second, third, and fourth ionization energies. While the second ionization energy is always larger than the first, and the third is larger than the second, you should notice that there are some other interesting and predictable kinds of variation. For example, the difference between first and second ionization energies is larger for an alkali metal than for its adjacent noble gas. The alkali metal atom, upon losing an electron,

TABLE 16.2

Ionization Energies of Gaseous Atoms (10^6 joules/mole)

ATOMIC NUMBER	ELEMENT	I_1	I_2	I_3	I_4
1	H	1.312			
2	He	2.371	5.248		
3	Li	0.520	7.295	11.81	
4	Be	0.899	1.756	14.842	21.00
5	B	0.807	2.425	3.658	25.02
6	C	1.086	2.352	4.618	6.222
7	N	1.402	2.855	4.576	7.48
8	O	1.313	3.387	5.295	7.47
9	F	1.681	3.372	6.044	8.40
10	Ne	2.080	3.953	6.0	9.4
11	Na	0.496	4.562	6.910	9.54
12	Mg	0.737	1.450	7.729	10.54
13	Al	0.577	1.816	2.744	11.58
14	Si	0.786	1.576	3.230	4.35
15	P	1.011	1.902	2.911	4.96
16	S	0.999	2.25	3.38	4.56
17	Cl	1.250	2.296	3.820	5.16
18	Ar	1.520	2.665	3.88	5.77
19	K	0.419	3.05	4.4	5.88
20	Ca	0.589	1.145	4.91	6.5
21	Sc	0.631	1.235	2.388	7.13
22	Ti	0.658	1.309	2.650	4.17
23	V	0.650	1.413	2.828	4.6
24	Cr	0.652	1.591	2.986	4.8
25	Mn	0.717	1.509	3.250	—
26	Fe	0.759	1.561	2.956	—
27	Co	0.758	1.645	3.230	—
28	Ni	0.736	1.752	3.430	—
29	Cu	0.745	1.957	3.553	—
30	Zn	0.906	1.733	3.830	—
31	Ga	0.579	1.979	2.962	6.19
32	Ge	0.762	1.537	3.300	4.41
33	As	0.946	1.794	2.734	4.83
34	Se	0.941	2.05	3.1	4.1
35	Br	1.139	2.08	3.56	4.56
36	Kr	1.351	2.350	3.56	—

achieves the low-energy noble-gas configuration. That first electron had effective screening from the nuclear charge by the filled *s* and *p* orbitals. The second electron to be removed now must leave the electrostatic field of a nucleus having a positive charge larger than the negative charge of all the electrons; moreover, this *p* electron is not shielded from the nuclear charge by the electrons. Hence, the difference between first and second ionization energies should be larger for an alkali metal than for a noble gas. For similar reasons the third ionization energy for an alkaline earth metal should be significantly larger than the second, since again the change is from the removal of a well-screened *s* electron to a relatively unscreened *p* electron. The fourth ionization energy for Group III elements is much larger than the third for the same reason.

Table 16.2 also allows us to see the effect of nuclear charge upon ionization energy because we can compare the energies required to remove electrons from isoelectronic species. Ne, Na^+, and Mg^{2+} all have ten electrons, but the nuclear charges are +10, +11, and +12, respectively. The ionization energies are 2080, 4562, and 7729 kjoules/mole. The differences in nuclear charge (and hence in the net charge of the initial particle) result in an almost fourfold spread in ionization energy of these species, all of which have the same electronic configuration.

16.4 Oxidation States

We saw in Chapter 4 that maximum-positive and maximum-negative oxidation states can be predicted from the position of an element in the periodic table. We made these predictions on the basis that electrons are added to or removed from the atom to achieve the low-energy noble-gas configuration. There is an impressively simple relationship between oxidation state and position in the periodic table, as can be seen in Figure 16.8. While we have discussed oxidation states in detail earlier, we present the relationship here in graphic form to stress the periodicity of oxidation numbers. For the representative elements, note

Figure 16.8
Common oxidation states of the elements as a function of atomic number.

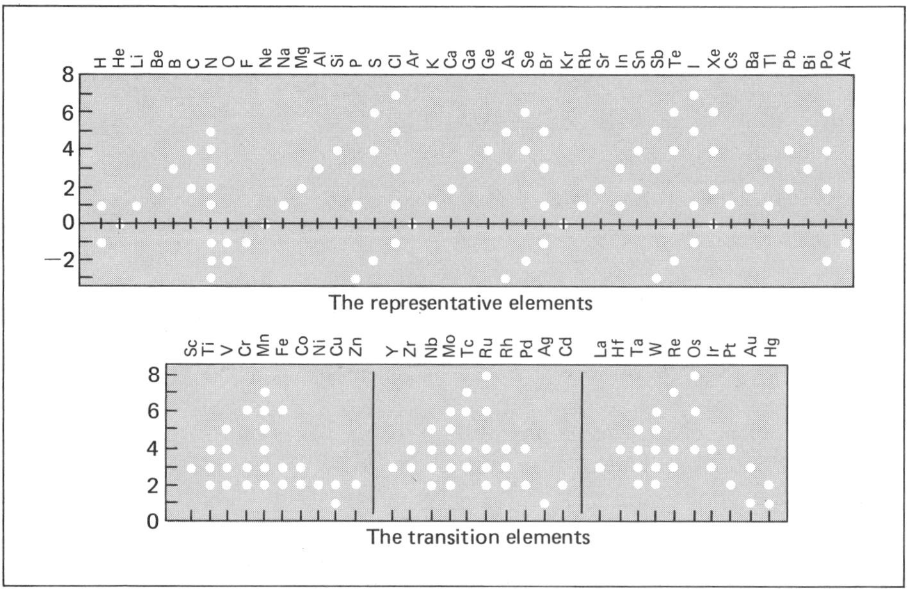

how within a family the oxidation numbers repeat with increasing atomic number. Also, alternating even and odd oxidation states are common with neighboring elements (nitrogen, with all oxidation states represented from +5 to −3, is an exception).

For the transition elements, the multitude of positive oxidation states for many of these metals is especially significant. Ruthenium and osmium are interesting as examples of elements with the +8 oxidation state.

The compounds that show the +8 oxidation state are OsO_4, RuO_4, OsF_8, and RuF_8.

16.5 Acidic and Basic Oxides

It is well to remember that the periodic classification of the elements had its origin in relatively simple observations of the properties and reactions of the elements and their compounds. More than in other kinds of compounds, we see in the oxides of the elements a relatively complete range of periodic properties. Oxides of nearly all the elements are known; the high prevalence of naturally occurring oxides stems from both the chemical stability of many oxides and the great abundance of oxygen on the earth. Study of the oxides provides an excellent framework upon which to support an understanding of descriptive chemistry—the properties and reactions of the elements and their compounds.

Classification of Oxides

Oxides are classified as *basic*, *acidic*, or *amphoteric*, depending upon their chemical behavior. Dual definitions are required for both basic and acidic oxides to cover soluble and insoluble substances. This requirement of a dual definition is not nearly so awkward as it appears at first encounter. As you study and learn the definitions, remember that the oxides are binary compounds of oxygen and a second element; we are not discussing all oxygen-containing compounds. Thus X_mO_n is of interest to us in this discussion, but $X_mY_nO_p$ is not. Cr_2O_3 is an oxide, Na_2CrO_4 is not; SO_2 is an oxide, SO_4^{2-} is not. Moreover, the considerations of solubility, hydrogen ion concentration, and hydroxide ion concentration all pertain to aqueous solutions; the acidic, basic, and amphoteric properties of the oxides are expressed in terms of behavior in solutions in which water is the solvent.

Basic Oxides

A basic oxide is either (1) a slightly soluble substance whose solubility increases in acidic solution and decreases in alkaline solution or (2) a substance that dissolves in water to give an alkaline solution. Ferric oxide, Fe_2O_3, has very low solubility in water or in alkaline solution. An increase in the hydrogen ion concentration causes a significant increase in the solubility of the oxide. Therefore, ferric oxide is a basic oxide because it meets the requirements of the first definition given above. Sodium oxide dissolves readily in water to give an alkaline solution. Therefore, sodium oxide is a basic oxide because it meets the requirements of the second definition. These two oxides, Fe_2O_3 and Na_2O, are typical of basic oxides.

Study the following reactions carefully to verify your understanding of why each set represents the behavior of a basic oxide:

$$Fe_2O_3(s) + H_2O = \text{no reaction}^{\dagger}$$
$$Fe_2O_3(s) + OH^- = \text{no reaction}$$
$$Fe_2O_3(s) + 6\,H^+ = 2\,Fe^{3+} + 3\,H_2O$$

$$Ag_2O(s) + H_2O = \text{no reaction}^{\dagger}$$
$$Ag_2O(s) + OH^- = \text{no reaction}$$
$$Ag_2O(s) + 2\,H^+ = 2\,Ag^+ + H_2O$$

$$MgO(s) + H_2O = \text{no reaction}^{\dagger}$$
$$MgO(s) + OH^- = \text{no reaction}$$
$$MgO(s) + 2\,H^+ = Mg^{2+} + H_2O$$

$$Na_2O(s) + H_2O = 2\,Na^+ + 2\,OH^-$$

$$BaO(s) + H_2O = Ba^{2+} + 2\,OH^-$$

$$Li_2O(s) + H_2O = 2\,Li^+ + 2\,OH^-$$

From the above information you should be able to clearly identify each of the oxides listed as a basic oxide. Statements of the following kind should similarly lead you to identify basic oxides:

> Nickel oxide is insoluble in water and in sodium hydroxide solution but dissolves in perchloric acid solution.
>
> CuO does not dissolve appreciably in potassium hydroxide solution. It dissolves in sulfuric acid solution.
>
> Rubidium oxide dissolves readily in water giving a solution with pH greater than 7.

Acidic Oxides

An acidic oxide is either (1) a slightly soluble substance whose solubility increases in alkaline solution and decreases in acidic solution or (2) a substance that dissolves in water to give an acidic solution. Silicon dioxide, SiO_2, has very low solubility in water or in acidic solution but is much more soluble in alkaline solution. Sulfur dioxide, SO_2, dissolves in water to give an acidic solution. Hence, silicon dioxide and sulfur dioxide are typical of the acidic oxides.

Equilibrium and *rate* must not become confused here. The rate of dissolving of SiO_2 in alkaline solution is ordinarily very low unless the SiO_2 has been freshly precipitated. At equilibrium, however, relatively large quantities of the oxide dissolve in a strongly alkaline solution.

† In water there is some conversion of these basic oxides to hydroxides. For example:

$$Fe_2O_3(s) + 3\,H_2O = 2\,Fe(OH)_3(s)$$

Since the hydroxides are insoluble, such substances as Fe_2O_3, Ag_2O, and MgO satisfy the requirements of basic oxides.

The following equations and statements all identify acidic oxides:

$$SiO_2(s) + H_2O = \text{no reaction}$$
$$SiO_2(s) + H^+ = \text{no reaction}$$
$$SiO_2(s) + 2\ OH^- = SiO_3^{2-} + H_2O$$
$$SO_2(g) + H_2O = H_2SO_3 = H^+ + HSO_3^-$$
$$CO_2(g) + H_2O = H_2CO_3 = H^+ + HCO_3^-$$
$$As_2O_5(s) + 3\ H_2O = 2\ H_3AsO_4 = 2\ H^+ + 2\ H_2AsO_4^-$$

Nitrogen dioxide dissolves in water to give a solution having pH less than 7.

Cl_2O_7 dissolves in water to form an acidic solution.

Amphoteric Oxides

An amphoteric oxide has properties that are both acidic and basic. It has a low solubility in water and increased solubility in both acidic and basic solutions. Aluminum oxide (Al_2O_3) is only slightly soluble in water but can be dissolved by the addition of either a strong acid or a strong base. The reactions are

$$Al_2O_3 + 6\ H^+ = 2\ Al^{3+} + 3\ H_2O$$

and

$$Al_2O_3 + 2\ OH^- + 3\ H_2O = 2\ Al(OH)_4^- \quad [16.1]$$

It is sometimes more convenient to write reactions such as [16.1] in terms of the hydroxides (or hydrated oxides) in the following way:

$$Al(OH)_3(s) + 3\ H^+ = Al^{3+} + 3\ H_2O$$
$$Al(OH)_3(s) + OH^- = Al(OH)_4^-$$

Once one recognizes the rather simple formal relationship between hydroxides and hydrated oxides, then the use of hydroxides in place of oxides in these reactions will be seen as an appropriate substitution. Consider solid aluminum oxide with water:

$$Al_2O_3(s) + 3\ H_2O = 2\ Al(OH)_3(s)$$

Formally, the solid may be considered as $Al_2O_3{\cdot}3\ H_2O$, $Al(OH)_3$, or wet Al_2O_3. In any reaction aluminum(III)† is present in the solid, and an abundance of water is available. For the sake of simplification, we will write the oxides as hydroxides when showing the amphoteric reactions.

In Chapter 20 we will consider many complex ions including hydroxy complexes such as $Al(OH)_4^-$. There we will observe that most complex ions formed by the representative elements and OH^- have four hydroxides in tetrahedral positions about the central metal ion. This tetrahedral configuration results from sp^3 hybridization of the atomic orbitals, as we saw in Chapter 14. Similarly, species such as $Zn(OH)_4^{2-}$, $Be(OH)_4^{2-}$, and $Cr(OH)_4^-$ are tetrahedral complexes. In each of these complex ions the central metal ion is bonded to four anions called *ligands* or *coordinating groups*. Later we will encounter ligands that are neutral molecules; in this chapter we will consider only one ligand, the hydroxide ion.

† A Roman numeral is used to indicate the oxidation state of the element.

Other examples of amphoteric oxides are given in the following equations and statements. Each oxide has low solubility in water but has increased solubility in both acidic and alkaline solutions.

$$Cr(OH)_3(s) + H_2O = \text{no reaction}$$
$$Cr(OH)_3(s) + 3\,H^+ = Cr^{3+} + 3\,H_2O$$
$$Cr(OH)_3(s) + OH^- = Cr(OH)_4^-$$

$$Be(OH)_2(s) + H_2O = \text{no reaction}$$
$$Be(OH)_2(s) + 2\,H^+ = Be^{2+} + 2\,H_2O$$
$$Be(OH)_2(s) + 2\,OH^- = Be(OH)_4^{2-}$$

$$Zn(OH)_2(s) + H_2O = \text{no reaction}$$
$$Zn(OH)_2(s) + 2\,H^+ = Zn^{2+} + 2\,H_2O$$
$$Zn(OH)_2(s) + 2\,OH^- = Zn(OH)_4^{2-}$$

Lead (II) oxide is insoluble in water but dissolves in a sodium hydroxide solution and in a nitric acid solution.

Characteristics That Affect the Nature of Oxides

We will now examine some of the characteristics of elements that cause them to form oxides having acidic, basic, or amphoteric properties. First, we will look at the effect of the ionic or covalent character of the oxides, next at the effect of oxidation number, and finally at the effect of atomic size.

Ionic and Covalent Oxides

By considering the nature of the bonds in various oxides we can see why some of these oxides are basic, why others are acidic, and why still others have intermediate properties. Oxygen, as we have already noted, is a very electronegative element, second only to fluorine. Consequently, the bond formed between an oxygen atom and a metal atom that has low electronegativity will be an ionic bond. Na_2O, Cs_2O, BaO, and MnO are all *ionic oxides*, displaying the properties of ionic substances. They have high melting points and high boiling points, and, to the extent that they dissolve in water, they form solutions that conduct electricity. The conducting species in such solutions are not the metal ion and oxide ion, however. The second acid constant for water is so much smaller than the first that the oxide ion does not exist in appreciable concentration in aqueous solution. The two acid constants for water are

$$H_2O = H^+ + OH^-; \qquad K_1 = 10^{-14}$$
$$OH^- = H^+ + O^{2-}; \qquad K_2 < 10^{-36}$$

The equilibrium constant for the reaction of oxide ion and water (obtained by subtracting the second reaction from the first) is greater than 10^{22}:

$$O^{2-} + H_2O = 2\,OH^-; \qquad K = K_1/K_2$$

Therefore, the addition of a soluble ionic oxide to water results in the formation of the hydroxide ion, and the conductivity is due to the metal ion and the hydroxide ion. Stated another way, when an ionic oxide dissolves in water, the concentration of OH^- increases and the solution becomes more alkaline.

Some ionic oxides are soluble and some have very low solubilities. You should be careful to avoid confusion between *solubility* and *extent of ionization.* In this section, we are concerned with the distinction between ionic and covalent oxides, but we are not primarily concerned with how soluble a given oxide is.

Covalent oxides are formed between oxygen and nonmetallic elements that have electronegativity values near that of oxygen. For example, when Cl_2O is dissolved in water, there is a very different reaction from that obtained with Na_2O. The sodium oxide gives Na^+ and OH^- in solution, but the reaction of Cl_2O is

$$Cl_2O + H_2O = 2\,ClOH$$

Here also the oxygen atom acquires a proton, but there is no significant ionization to give hydroxide ions. Instead, the Cl-O covalent bond remains, and there is distortion of the electron density about the central oxygen atom toward the chlorine atom (which is more electronegative than hydrogen), with the result that the O-H bond is weakened. The slight ionization that occurs is

$$ClOH = ClO^- + H^+$$

showing that ClOH has acidic and not basic properties. Hence, the formula for the substance is more commonly written HClO.

It is clear, then, that the electronegativity of the atom that is combined with oxygen to form an oxide contributes greatly to the character of the oxide. One expects a trend across the periodic table, with the electropositive elements on the left forming basic (ionic) oxides and the electronegative elements on the right forming acid (covalent) oxides. This trend is clearly observed in the behavior of the alkali metal oxides, the alkaline earth oxides, and the nonmetal oxides. The alkali metal oxides are very basic, the alkaline earth oxides only slightly less so, while the nonmetal oxides are very acidic. Intermediate elements show a general but somewhat erratic trend in the properties of their oxides. As one would expect, this is the region where amphoteric oxides are found. However, we must consider the effects of oxidation state and atomic size to have a clear understanding of trends in acidic and basic properties.

Effect of Oxidation State

Most elements in Group III and higher groups have more than one oxidation state, and for a given element the oxides formed from these oxidation states often differ markedly from one another in properties. Thus, manganese(II) forms a basic oxide while manganese (VII) forms an acidic oxide; chromium (II) oxide is basic, chromium (III) oxide is amphoteric, and chromium (VI) oxide is acidic. This transition toward more acidic properties with higher positive oxidation number is to be expected from the increased positive charge on the atom bonded to oxygen. This charge distorts the electron cloud away from hydrogen and weakens the hydrogen-oxygen bond.

Consider, as an example, the two acids formed when the oxides of sulfur are added to water:

$$SO_2 + H_2O = H{-}O{-}\overset{\displaystyle O}{\overset{|}{S}}{-}O{-}H$$

$$SO_3 + H_2O = H{-}O{-}\underset{\displaystyle O}{\underset{|}{\overset{\displaystyle O}{\overset{|}{S}}}}{-}O{-}H$$

The extra oxygen in the second acid (where sulfur is in the higher oxidation state) causes a greater displacement of charge from the H-O bonds and weakens them more. The first acid, sulfurous acid, has a K_a of 1.3×10^{-2}:

$$H_2SO_3 = H^+ + HSO_3^-; \qquad K_a = 1.3 \times 10^{-2}$$

while the second acid, sulfuric acid, is completely ionized:

$$H_2SO_4 = H^+ + HSO_4^- \qquad \text{(complete dissociation)}$$

As a second example, consider the oxygen acids that can be thought of as being formed by hydration of chlorine oxides. The acid constants of these oxygen acids are as follows:

CHLORINE OXIDE	OXIDATION STATE	OXYGEN ACID		K_a
Cl_2O	+1	$H{-}O{-}Cl$	($HClO$)	10^{-8}
$(Cl_2O_3)^{\dagger}$	+3	$H{-}O{-}Cl{-}O$	($HClO_2$)	10^{-2}
$(Cl_2O_5)^{\dagger}$	+5	$H{-}O{-}\overset{\displaystyle O}{\overset{\vert}{Cl}}{-}O$	($HClO_3$)	completely ionized
Cl_2O_7	+7	$H{-}O{-}\underset{\displaystyle O}{\underset{\vert}{\overset{\displaystyle O}{\overset{\vert}{Cl}}}}{-}O$	($HClO_4$)	completely ionized

† Existence of the two oxides in parentheses is uncertain. All the known oxides of chlorine are explosive; therefore it is not surprising that some chlorine oxides are not found as products of reactions.

Both $HClO_3$ and $HClO_4$ are strong acids in water solution; that is, they are completely ionized. The two acids can be differentiated in other solvents where, as expected, $HClO_4$ is the stronger acid.

Even the +1 chlorine oxide Cl_2O is weakly acidic, but the acidic character increases with oxidation state until the +7 oxide Cl_2O_7 forms the very strong perchloric acid.

Effect of Atomic Size

Within a vertical column of the periodic table the oxides that are formed by the elements in a given oxidation state show a trend toward more basic (less acidic) behavior as the atomic number increases. This trend illustrates the effect of atomic size, since the oxidation number remains constant and the outermost electrons are in corresponding configurations, differing only in the value of *n*, the principal quantum number. The smaller atom at the top of a column has the greater concentration of charge and has, consequently, the greater attraction for the electrons of the oxygen atom. The result is that the hydrogen-oxygen bond of the hydrated oxide is weakened more when the central atom is small. Conversely, a large atom from the same column will form an oxide that is more basic (or less acidic), since the positive charge of the larger central atom will be more diffuse, causing less weakening of the hydrogen-oxygen bond in the hydrated oxide. The Group VIA elements illustrate this trend. The first dissociation constants of the acids formed from sulfur, selenium, and tellerium in the +4 oxidation state have the following values:

	$K_a(1)$
H_2SO_3	$\sim 10^{-2}$
H_2SeO_3	$\sim 10^{-3}$
H_2TeO_3	$\sim 10^{-6}$

A similar progression is seen in the acid constants of the hypohalous acids:

	K_a
HClO	$\sim 10^{-8}$
HBrO	$\sim 10^{-9}$
HIO	$\sim 10^{-11}$

In these two examples, the acids, which can be viewed as hydrated oxides, become progressively weaker as atomic size increases, verifying our prediction that the oxides tend to become less acidic with increasing size of the central atom.

The alkaline earth group shows a comparable trend. The oxides of the alkaline earths have a wide range of solubilities in water. The solubility is lowest for beryllium oxide and increases with the atomic weight of the alkaline earth. The solubility products for the hydroxides (hydrated oxides) given in Table 16.3 illustrate that trend. All the oxides are very soluble in acidic solution, but only beryllium oxide is more soluble in alkaline solution than in water. That is, beryllium oxide is an amphoteric oxide, having both acidic and basic properties, while magnesium, calcium, strontium, and barium oxides are all basic oxides.

TABLE 16.3

Properties of the Alkaline Earth Elements

	Be	Mg	Ca	Sr	Ba
Atomic radii (Å)	0.89	1.36	1.74	1.91	1.98
Ionic radii M^{2+} (Å)	0.30	0.65	0.94	1.10	1.29
Ionization constant $\frac{[M^{2+}][OH^-]}{[M(OH)^+]}$	10^{-8}	10^{-2}	0.031	—	0.23
Solubility products $[M^{2+}][OH^-]^2$	10^{-26}	10^{-11}	10^{-6}	10^{-4}	10^{-3}

Beryllium, the smallest of the alkaline earths, forms a more acidic (less basic) oxide than do larger members of the family. What of the oxides formed by the subsequent elements of the group? We have mentioned that all have only basic properties, but Table 16.3 shows a trend in the solubilities in water that clouds the issue. The clearest picture of strengths of the bases formed is given by the second ionization constant of the bases, $M(OH)_2$:

$$M(OH)^+ = M^{2+} + OH^-$$

$$K_2 = \frac{[M^{2+}][OH^-]}{[MOH^+]}$$

Values for this constant, also shown in Table 16.3, display the expected increase in base strength with atomic size.

Periodic Trends in the Properties of Oxides

We can summarize the periodic trends in acid-base properties of the oxides as follows: Among the representative elements the change is along a diagonal line from very basic oxides in the lower left corner of the periodic table to very acidic oxides in the upper right corner. Thus, all the alkali metals form basic oxides that are very soluble in water and completely dissociated. The equation

$$M_2O + H_2O = 2\,M^+ + 2\,OH^-$$

represents the reaction that occurs between one of these oxides and water.

Of the alkaline earth oxides, BeO is amphoteric but the other alkaline earth oxides are bases that have increasing water solubility as one goes down the family from MgO to BaO.

At the other extreme, the many soluble oxides of the nonmetals prove to be the anhydrides of acids. Table 16.4 lists several of these anhydrides and their corresponding acids.

TABLE 16.4

Some Anhydrides and Their Corresponding Acids

ANHYDRIDE	ACID
CO_2	H_2CO_3
N_2O_5	HNO_3
N_2O_3	HNO_2
SO_2	H_2SO_3
SO_3	H_2SO_4
P_4O_{10}	H_3PO_4
P_4O_6	H_3PO_3
As_4O_{10}	H_3AsO_4
As_4O_6	H_3AsO_3
B_2O_3	H_3BO_3

The representative elements in the middle region of the periodic table tend to form amphoteric oxides. You will see that the members of the list in Table 16.5 all fall in that middle region, although the correlation with position in the periodic table is imperfect. The oxidation state of transition elements plays a more important part in determining acid-base behavior than does position in the periodic table. The change is from basic to amphoteric to acidic oxide as the oxidation number increases. Common elements that form amphoteric oxides and their oxidation states are given in Table 16.5. Higher oxidation states of the elements listed are acidic while lower oxidation states of these elements and the oxidation states of most other common transition metal oxides are basic.

TABLE 16.5

Common Elements Forming Amphoteric Oxides

(Notice that the oxidation state must be specified since the acidic-basic character depends on this.)

Be(II)	V(IV), V(V)
Al(III)	Cr(III)
Ga(III)	Cu(II)†
Sn(II), Sn(IV)	Zn(II)
Pb(II)	

† Copper(II) oxide has only slight amphoterism. Its properties are those of a basic oxide, except that in concentrated alkaline solution a very small quantity of the oxide dissolves, giving the pale blue color of the cuprate ion, $Cu(OH)_4^{2-}$.

Separations Based on Acid-Base Behavior of Oxides

We will not take space here to discuss in detail the difficulties of carrying out such a separation in the laboratory. In this example one would need to take steps to get the gelatinous ferric hydroxide (hydrated ferric oxide) into a more condensed and compact form; otherwise a large fraction of the aluminum would be carried along with the voluminous precipitate.

Separations of the amphoteric oxides can, in principle, be readily made from either the insoluble basic oxides or the insoluble acidic oxides. Consider, for example, an acidic solution of Fe^{3+} and Al^{3+}. Ferric ion forms a basic oxide; upon neutralization of the solution a precipitate of $Fe(OH)_3$ will form that will not dissolve even if the solution is made very alkaline. Aluminum (III) forms an amphoteric oxide; the precipitate of $Al(OH)_3$ that forms when the solution is neutralized is redissolved as $Al(OH)_4^-$ when excess hydroxide ion is added. The two species can be separated, therefore, by making the solution strongly alkaline. In this mixture the precipitate is the ferric hydroxide, and the aluminum remains in solution as $Al(OH)_4^-$. Filtration or centrifugation can be used to separate the solid and liquid phases.

EXAMPLE 16.1 A solution is prepared from $Zn(NO_3)_2$ and $AgNO_3$. Suggest a method for separating the Ag^+ from the Zn^{2+} by use of the differences in behavior of their hydroxides.

SOLUTION Both cations form insoluble hydroxides. $Zn(OH)_2$, however, is amphoteric and can be put back into solution with an excess of base. AgOH, on the other hand, becomes more insoluble as the concentration of OH^- is increased. Therefore, the separation can be accomplished if the solution is made strongly alkaline, thus precipitating AgOH and forming $Zn(OH)_4^{2-}$.

$$Ag^+ + OH^- = AgOH(s)$$
$$Zn^{2+} + 2\ OH^- = Zn(OH)_2(s)$$
$$Zn(OH)_2(s) + 2\ OH^- = Zn(OH)_4^{2-}$$

Summary

The physical properties of many elements show trends closely related to position in the periodic table. Atomic volume, melting point, boiling point, density, ionization energy, and maximum oxidation state can all be correlated with the periodic table.

A basic oxide either (a) dissolves in water to form a basic solution or (b) is more soluble in acid solution than in neutral or basic solution.

An acidic oxide either (a) dissolves in water to form an acidic solution or (b) is more soluble in basic solution than in neutral or acidic solution.

An amphoteric oxide has properties that are both basic and acidic; it is more soluble in acidic solution and in basic solution than it is in water.

Elements form acidic or basic oxides depending upon factors of electronegativity, atomic size, and oxidation number:

> The bond between oxygen and a metal atom that has very low electronegativity will be ionic, and the oxide will be a basic oxide. Conversely, the bond between oxygen and an atom having high electronegativity will be covalent, and the oxide will be an acidic oxide.

Within a family the element with the largest atomic size will form the most basic oxide.

As the oxidation number of an element increases, the oxides that it forms become more acidic.

Since electronegativity, atomic size, and oxidation number are all affected by position in the periodic table, the nature of the oxide formed by an element can be correlated with its position in the periodic table. The elements at the upper right form the most acidic oxides while those at the lower left form very basic oxides. Amphoteric oxides are formed by elements in intermediate positions.

Exercises

16-1.
Predict which member of each of the following pairs of atoms is larger, and explain your reasoning: (a) Ge or Br; (b) P or Ar; (c) F or Cl; (d) K or Se.

16-2.
Predict, with explanation, which member of each of the following pairs of atoms is larger: (a) Fe or Ga; (b) Sr or Te; (c) Rb or Cs; (d) Cd or Hg.

16-3.
Predict, with explanation, which member of each of the following pairs of species is larger: (a) Na or Na^+; (b) F or F^-; (c) Ca or Ca^{2+}; (d) Ca^{2+} or K^+.

16-4.
Predict, with explanation, which member of each of the following pairs of species is larger: (a) I^- or Xe; (b) I^- or Be^{2+}; (c) Br^- or Se^{2-}; (d) Ca^{2+} or Sr^{2+}.

16-5.
Predict, with explanation, which member of each of the following pairs of atoms has the higher ionization energy: (a) Na or K; (b) Na or Mg; (c) F or Ne; (d) Na or F.

16-6.
Predict, with explanation, which member of each of the following pairs of atoms has the higher ionization energy: (a) Ar or Xe; (b) Cl or I; (c) Li or I; (d) Cs or Ba.

16-7.
Predict, with explanation, which member of each of the following pairs of species has the higher ionization energy: (a) Li or Li^+; (b) Li^+ or He; (c) He^+ or F; (d) F or F^-.

16-8.
Predict, with explanation, which member of each of the following pairs of species has the higher ionization energy: (a) Sr^{2+} or Rb^+; (b) Rb^+ or Na^+; (c) I^- or F^-; (d) Cs^{2+} or Xe^+.

16-9.
List three species (atoms or simple ions) that are isoelectronic with (a) Ar; (b) Sr^{2+}; (c) I^-; (d) K^+.

16-10.
Refer only to a periodic table to answer this exercise. List five oxides that you expect to be basic. List five oxides that you expect to be amphoteric. List five oxides that you expect to be acidic. Explain your choices.

16-11.
Predict, with explanation, which of each of the following pairs of oxides is more acidic: (a) SO_2 or TeO_2; (b) SiO_2 or PbO_2; (c) Al_2O_3 or Tl_2O_3; (d) As_2O_3 or Bi_2O_3.

16-12.
You are given two white oxides, MgO and ZnO. Describe tests by which you can identify which sample is which oxide.

16-13.
A student asked her roommates, both chemistry majors, for help in restoring the interior of her aluminum saucepan to its original appearance. One suggested that she use vinegar, because it is acidic. The other suggested that she use a cleanser with ammonia, because it is alkaline. Discuss the merits of the suggestions.

Problems

16-1.

Suppose you were taking chemistry prior to the discovery of strontium, element 38. Consider its position in the periodic table, and predict the following properties of Sr by comparing it with the surrounding elements. Explain the reasoning behind your choices.

(a) Melting point and boiling point of the element.
(b) Density of the element.
(c) First ionization potential.
(d) Atomic radius.
(e) Chemical formula of its most common oxide.
(f) Solubility of its oxide in water.
(g) Formula of the metal sulfate.

16-2.

The density of potassium is 0.86 g/cm^3. Calculate

(a) the volume occupied by a mole of potassium in cm^3.
(b) the average volume occupied by a potassium atom in $Å^3$. Compare your answer with the value given in Figure 16.4.

16-3.

Copper crystallizes in a face-centered cubic lattice with a density of 8.93 g/cm^3. Calculate the atomic radius of a copper atom.

16-4.

Lithium crystallizes in a body-centered cubic lattice with a density of 0.534 g/cm^3. Calculate the atomic radius of the lithium atom.

16-5.

Potassium chloride, KCl, crystallizes in the rock salt (NaCl) structure (see Chapter 15 for a description of this structure). The density of KCl is 1.989 g/cm^3 at 18° C.

(a) Calculate the separation between positive and negative ions in the lattice.
(b) For KCl, the radius ratio is $r_{K^+}/r_{Cl^-} = 0.735$. Calculate the ionic radii of K^+ and Cl^-.
(c) The K^+ and Cl^- ions are isoelectronic. Explain the differences in size.

17

Precipitation Reactions

A large number of important ionic reactions can be classified as *precipitation reactions*, those in which ions combine to form an insoluble salt. A precipitation reaction can be expected to occur in an ionic solution if any combination of the ions present will result in an insoluble substance. For example, a mixture of $AgNO_3$ and NaCl solutions reacts to form AgCl(*s*). The net ionic reaction is

$$Ag^+ + Cl^- = AgCl(s)$$

The extent to which the reaction occurs is directly related to the solubility of the precipitate. The equilibrium-constant expression for the reaction above, $1/[Ag^+][Cl^-]$, is seen to be simply the reciprocal of the K_{sp} of AgCl. Knowing that the K_{sp} of AgCl is 1.8×10^{-10}, we can predict that $AgNO_3$ and NaCl solutions will react extensively, since the value of $K_{\text{precipitation}}$ is so large:

$$K_{\text{precipitation}} = \frac{1}{1.8 \times 10^{-10}} = 5.6 \times 10^9$$

Thus, we can predict the extent to which a precipitation reaction will occur from a knowledge of solubility data.

17.1 Solubility Rules

As we shall see later in the chapter, the extent to which we can predict solubilities from theoretical models is quite limited. For this reason we resort to formulating empirical rules that generalize the observed solubility properties of several classes of salts. Following are some of these useful rules of solubility:

Solubility is a relative term. For our purposes, we will call a salt soluble if at least a 0.1 molar solution of it can be prepared.

(1) *Alkali metal salts* and *ammonium salts* are generally very soluble. Examples are $LiNO_3$, NaCl, K_2SO_4, Rb_3PO_4, $CsNO_3$, and NH_4Br.
(2) *Nitrates*, *chlorates*, and *acetates* are generally very soluble. Examples are $AgNO_3$, $Ca(ClO_3)_2$, and $(CH_3COO)_2Mg$.
(3) With the exceptions of the Ag(I), Hg(I), and Pb(II) compounds, the *chloride* salts are very soluble. Even $PbCl_2$ is quite soluble in warm water. The same rule applies to *bromides* and *iodides*, with the Hg(II) compounds added to the list of exceptions.
(4) Most metallic *oxides* and *hydroxides* are insoluble. The notable exceptions are those of the alkali metals, barium, and Tl(I).
(5) Most of the common metallic *sulfides* are insoluble, with the exceptions of the Groups IA and IIA sulfides, ammonium sulfide, and aluminum sulfide.
(6) The following *sulfates* are insoluble: $SrSO_4$, $BaSO_4$, Hg_2SO_4, $HgSO_4$, $PbSO_4$ and Ag_2SO_4. Most other sulfates are soluble.
(7) *Carbonates* and *phosphates*, with the exceptions of the compounds formed with IA metals and NH_4^+, are insoluble.

EXAMPLE 17.1 Write the formula of the precipitate (if any) that forms when equal quantities of 0.1 *M* solutions of the following salts are mixed.

(a) Na_2SO_4 and $Pb(NO_3)_2$
(b) KBr and $AgNO_3$
(c) $Mg(NO_3)_2$ and NaOH
(d) $NaNO_3$ and $CuSO_4$
(e) Rb_3PO_4 and $Al(NO_3)_3$
(f) $(NH_4)_2S$ and $HgCl_2$
(g) $CuSO_4$ and $Ba(OH)_2$

SOLUTION

(a) $PbSO_4(s)$ (Rule 6)
(b) $AgBr(s)$ (Rule 3)
(c) $Mg(OH)_2(s)$ (Rule 4)
(d) No reaction (Rules 1 and 2)
(e) $AlPO_4(s)$ (Rule 7)
(f) $HgS(s)$ (Rule 5)
(g) $Cu(OH)_2(s)$ and $BaSO_4(s)$ (Rules 4 and 6)

Example 17.1 should convince you that by remembering and applying a relatively few simple rules, you can predict the results of a large number of chemical reactions. The rules, however, do not explain the reasons for the solubility trends. This is the subject we consider next.

17.2 Factors That Determine Solubility

Figure 17.1 illustrates the process that is believed to take place when an ionic solid dissolves in water. In the first step, the polar water molecules orient themselves around the ions exposed on the surface of the crystal and compete with

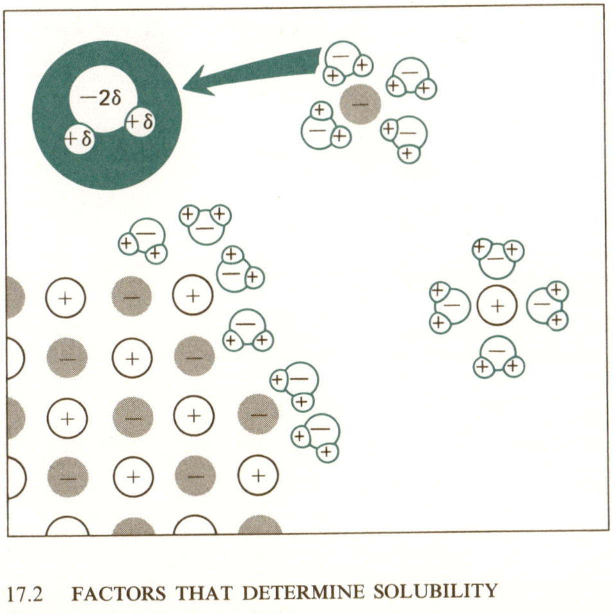

Figure 17.1
Ionic solid dissolving in water.

Figure 17.2
Ion-dipole interaction.

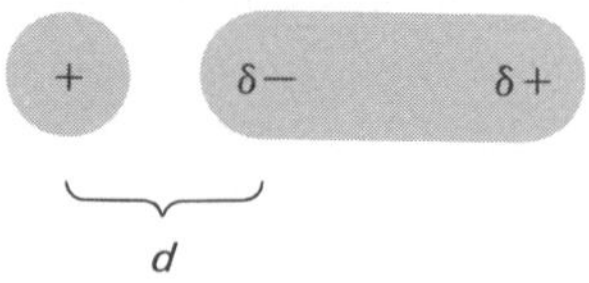

The real criterion for whether dissolution of a crystal will take place is the same as it is for any process: $\Delta G < 0$ (constant T and P). When we say that the energy of hydration has to exceed the ionic lattice energy, we are using energy data as a rough approximation of the less easily available free energies.

Thus small ions with large charge have large hydration energies. The very small H^+ (a single proton) produces a strong electrostatic field and would, therefore, cause water to bond to it strongly. The result is the H_3O^+ ion.

the crystal for the ions. If the *energy of hydration* (the energy released when the dry ion reacts with the polar water molecules) exceeds the *lattice energy* (the net energy of attraction of the ions for each other in the crystal) the solid goes into solution. Solubility is thus seen to be a function of (a) hydration energy and (b) lattice energy. We will consider each effect separately.

Hydration of ions involves an electrostatic attraction between the polar water molecules and the ion. Coulomb's law applies, which is

$$\textit{force} \text{ of attraction} = k\frac{(z^+)(z^-)}{d^2} \qquad [17.1]$$

If we are considering the hydration of a cation, (z^+) is the charge of the ion, (z^-) is the partial charge on the negative end of the water dipole (see Figure 17.2), d is the distance of closest approach of the *centers* of charge of the ion and molecule (again refer to Figure 17.2), and k is a proportionality constant.

The energy of attraction of the cation for the negative end of the polar molecule is

$$\textit{energy} \text{ of attraction} = \frac{k(z^+)(z^-)}{d} \qquad [17.2]$$

Thus, the energy of hydration is directly proportional to the charge on the ion and inversely proportional to the radius of the ion. For example, Al^{3+} ion, because of both its small size and large charge, would be expected to have a high negative energy of hydration. On the other hand, Cs^+, with only a single charge and a relatively large size, has a small negative hydration energy. Hydration of anions is generally less significant than hydration of cations because of the relatively larger size of most anions; for example, the energy of hydration of I^-, ClO_4^-, or NO_3^- would be considerably less negative than that of Li^+, Na^+ or K^+.

Now, let us consider the breakdown of the crystal. As we saw in Chapter 15, the ionic lattice energy is principally determined by the charge and size of the ions in the crystal. The ionic lattice energy represents the sum of the attractions between ions of unlike charge and the repulsions between ions of like charge.

In Chapter 15 we saw that in crystal lattices, where the anion is large relative to the cation, anion-anion repulsions can be important in determining the lattice energy. This is because the small cations allow close anion-to-anion contact as shown schematically in Figure 17.3.

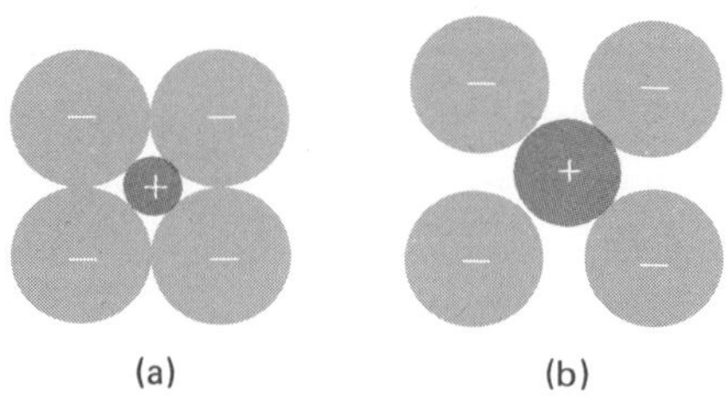

Figure 17.3
In (a), the cation is not large enough to keep the anions separated, and repulsion is high. In (b), the larger cation separates the anions, reducing the repulsion.

Coulomb's law applies to both effects, and quite obviously (since the crystal exists) the energy of attraction always exceeds the energy of repulsion. In those compounds where the ions are small and highly charged, the lattice energy is large.

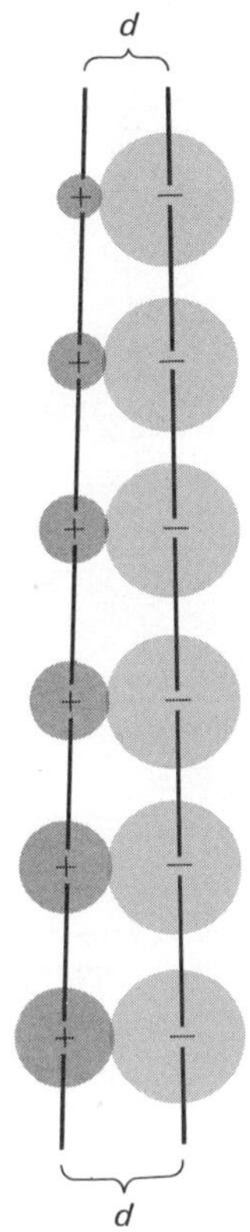

Figure 17.4
When the anion is large, the distance of closest approach of the centers of charge is not influenced significantly by the size of the small cations.

Both factors that affect solubility—(1) the energy of hydration and (2) the lattice energy—depend on the same variables—(1) the size of the ions and (2) their charges. Unfortunately (for predicting solubilities) the energy of hydration and lattice energy are opposing effects. If, for example, the ions are small and highly charged, both the energy of hydration and the lattice energy will have large negative values. However, a large negative value for the energy of hydration favors dissolution of the crystal, while a large negative value of the lattice energy creates resistance to the dissolution of the crystal. With our present state of understanding we are not always able to predict which effect will be predominant. This is why we have to settle for empirical solubility rules. Understanding the effects of hydration and lattice energies does, however, help us to see the rationale for the rules and allows us to make some qualitative predictions about solubilities.

A case in point is found in the relative solubilities of the Group IIA sulfates and hydroxides. As shown in Table 17.1, the solubility of the sulfates decreases from the top to the bottom of the IIA family; whereas the opposite effect is observed for the hydroxides.

The explanation for the behavior of the sulfates is that the sulfate ion is so large compared with the cations that the attraction between ions is not greatly affected by the variations in the size of the IIA cations. A schematic representation of this case is shown in Figure 17.4.

With the lattice energies about constant, the hydration of the cation becomes the important factor in determining the solubility. Since the hydration energy is greatest at the top of the family, where the cation is the smallest, the hydration effect contributes to an increased solubility in the compounds at the top of the family. And since SO_4^{2-} is large compared with Mg^{2+}, anion-to-anion repulsion should contribute to the weakening of the crystal and assist solubility in $MgSO_4$. As one goes down the family and the size of the cation increases, this repulsion should become less important. Hence, the hydration effect determines the solubility, and both favor increased solubility as one proceeds from the bottom to the top of the family. One would expect similar

TABLE 17.1

Trends in Solubility-Product Constants for IIA Hydroxides and Sulfates

	HYDROXIDES	SULFATES
Mg	1×10^{-11}	very soluble
Ca	1×10^{-6}	6×10^{-5}
Sr	3×10^{-4}	5×10^{-7}
Ba	5×10^{-3}	1×10^{-10}

behavior for the chromates, arsenates, and other such compounds containing large anions.

The case for the hydroxides is just the reverse of that for the large anions. Because the hydroxide ion is small, the size of the cation does contribute significantly to the distance of closest approach. A schematic representation is shown in Figure 17.5.

Since we know that the measured solubility decreases as one goes from the bottom to the top of the family, we conclude that for this case the strength of the lattice, which increases as one goes from the bottom to the top of the family, is the predominant factor in determining the solubility. However, one has to know the trend in solubilities of these compounds in order to reach that conclusion.

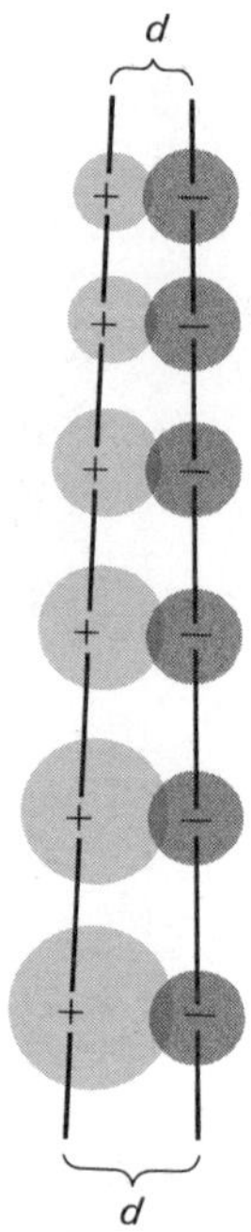

Figure 17.5
Where the anion is small, the size of the cation does significantly affect the distance of closest approach and, hence, the lattice energy of the crystal.

EXAMPLE 17.2 Based on what you know about the size of ions, predict whether $LiClO_4$ or $CsClO_4$ is the more soluble.

SOLUTION Because of the relatively large size of ClO_4^-, the energy of attraction in the lattice will not be appreciably different in the two compounds; hence, the hydration effect will determine the relative solubilities. Since the smaller Li^+ will release more energy during hydration than will Cs^+, $LiClO_4$ should be more soluble. (The measured solubilities are $LiClO_4$, 60.0 g/100 g H_2O; $CsClO_4$, 2.00 g/100 g H_2O at 25° C.)

EXAMPLE 17.3 Rationalize the known fact that the oxides and sulfides of Mn, Fe, Co, Ni, Cu, Zn, and Hg are highly insoluble whereas their chlorides are quite soluble.

SOLUTION The difference in solubilities of the two groups appears to be controlled by their lattice energies. The doubly charged oxide and sulfide ions would result in a much stronger lattice than would the singly charged chloride ion. The difference in hydration energies of the sulfides and oxides as compared with the chlorides would appear to be less significant. It is generally true that the hydration effect of anions, which are usually large ions, is of less importance in determining solubility than that of cations.

17.3 Effect of Temperature on Solubility

Qualitatively, the effect of temperature on solubility is quite clear. As predicted by LeChatelier's principle, the solubility of a solute increases with an increase in temperature if the solution process absorbs heat. Conversely, solubility decreases with an increase of temperature if the solution process is exothermic. The former is the more common occurrence when the solute is a solid, although, as can be seen in Table 17.2, examples of solids showing a decrease in solubility with an increase in temperature are known.

solution we have the following equilibria applying:

$$(1) \qquad AgF(s) = Ag^+ + F^-; \qquad K = K_{sp(AgF)}$$

$$(2) \quad F^- + H^+ = HF; \qquad K = \frac{1}{K_{a(HF)}}$$

If we add the two reactions, we obtain

$$AgF(s) + H^+ = Ag^+ + HF; \; K = \frac{K_{sp}}{K_a}$$

The equilibrium constant for this reaction is obtained by multiplying the equilibrium constants for reactions (1) and (2). From that expression we can see that the weaker the acid formed (HF in this example), the farther to the right the reaction proceeds and the more the solid dissolves. In the case of AgCl, a reaction analogous to (2) above is not possible since HCl is a fully dissociated acid. If we were to attempt to write a reaction analogous to (2) we would get only that

$$H^+ + Cl^- = H^+ + Cl^-$$

which although true is neither very productive nor very interesting.

EXAMPLE 17.5 Calculate the solubility of CH_3COOAg in (a) pure water, and (b) a solution in which the final concentration of H^+ is 0.01 M. $K_{sp(CH_3COOAg)} = 4.0 \times 10^{-3}$; $K_{a(CH_3COOH)} = 1.8 \times 10^{-5}$.

SOLUTION

(a) The equilibrium that applies is

$$CH_3COOAg = CH_3COO^- + Ag^+$$

Let x = solubility = $[CH_3COO^-]$.

$$K = [CH_3COO^-][Ag^+] = x^2 = 4.0 \times 10^{-3}$$
$$x = 6.3 \times 10^{-2}\ M$$

(b) With the acid present, there are two sources of Ag^+, one from the solubility equation in part (a) and a second source from the reaction with the acid:

$$CH_3COOAg(s) + H^+ = CH_3COOH + Ag^+$$

The equilibrium constant expression for this reaction is

$$K = \frac{[CH_3COOH][Ag^+]}{[H^+]}$$

We illustrate in this example an alternative way to combine two equilibrium constants to get the K for a third. Compare this with the procedure used above in the silver fluoride discussion and in Example 17.6.

By multiplying numerator and denominator by $[CH_3COO^-]$ we obtain

$$K = \frac{[CH_3COOH][Ag^+][CH_3COO^-]}{[H^+][CH_3COO^-]} = \frac{K_{sp}}{K_a} = \frac{4.0 \times 10^{-3}}{1.8 \times 10^{-5}} = 220$$

Comparing this value with the constant for the reaction in part (a), 4.0×10^{-3}, we conclude that of the two reactions the second is the more important in dissolving $CH_3COOAg(s)$. As an approximation, then, we neglect the reaction in part (a) and make the calculation from the acid solubility reaction alone. Letting $[CH_3COOH] = [Ag^+] = x$, we have in a 0.010 M H^+ solution (equilibrium concentration)

$$K = \frac{x^2}{.010} = 220$$

$$x^2 = 2.2$$

$$x = 1.5\ M$$

which is 24 times the solubility obtained in pure water.

17.6 Solubility and Complex Ion Formation

The formation of a complex ion can have a dramatic effect on solubility in much the same way that the formation of a weak acid does. For example, $AgCl(s)$ will dissolve in an excess of Cl^- because of the formation of the stable complex $AgCl_2^-$. The equilibrium that applies is

$$AgCl(s) + Cl^- = AgCl_2^-$$

$AgCl(s)$ also dissolves in an ammonia solution because of the reaction

$$AgCl(s) + 2\ NH_3 = Ag(NH_3)_2^+ + Cl^-$$

Knowing the formation constant of the complex ion makes it an easy matter to calculate the solubility.

EXAMPLE 17.6 Calculate the solubility of $AgCl(s)$ in (a) pure water, and (b) a 1.0 M NH_3 solution. $K_{sp(AgCl)} = 1.8 \times 10^{-10}$; $K_{d(Ag(NH_3)_2^+)} = 6 \times 10^{-8}$.

SOLUTION

(a)

$$AgCl(s) = Ag^+ + Cl^-$$

Let $x = [Ag^+] = [Cl^-] =$ solubility.

$$K = [Ag^+][Cl^-] = x^2 = 1.8 \times 10^{-10}$$

$$x = 1.3 \times 10^{-5}\ M$$

(b) Consider the reaction

$$AgCl + 2\,NH_3 = Ag(NH_3)_2^+ + Cl^-$$

Its equilibrium constant can be calculated from the data given in the problem by combining the following reactions:

$$AgCl = Ag^+ + Cl^-; \qquad K_{sp} = 1.8 \times 10^{-10}$$

$$Ag^+ + 2\,NH_3 = Ag(NH_3)_2^+; \qquad \frac{1}{K_d} = \frac{1}{6 \times 10^{-8}}$$

$$AgCl(s) + 2\,NH_3 = Ag(NH_3)_2^+ + Cl^-; \qquad K = \frac{K_{sp}}{K_d} = 3 \times 10^{-3}$$

The equilibrium constant for this reaction shows it to be more significant than the reaction in part (a) in dissolving AgCl(*s*). Hence, we will neglect the small effect of part (a). Let x = solubility (via complex ion formation). We can then write

$$K = \frac{[Ag(NH_3)_2^+][Cl^-]}{[NH_3]^2} = \frac{x^2}{(1.0 - 2x)^2} = 3 \times 10^{-3}$$

Solving for x by means of the quadratic formula gives

$$x = 0.05\ M$$

which is 4000 times the solubility in pure water.

The subject of complex ion formation is discussed further in Chapter 20.

17.7 Selective Precipitation

By careful control of the concentration of a precipitating agent in a solution, it is often possible to make a quantitative separation of two substances by selectively precipitating one of them. This may be accomplished even though both ions form very insoluble compounds with the precipitant. As an example, consider a solution that is 0.0010 M in both Mg^{2+} and Mn^{2+}. Both ions form very insoluble hydroxides: $Mg(OH)_2(K_{sp} = 10^{-11})$, and $Mn(OH)_2(K_{sp} = 10^{-14})$. However, if we limit the concentration of OH^- to a value just less than that required to precipitate $Mg(OH)_2$, we will find that enough $Mn(OH)_2$ will have precipitated to decrease the Mn^{2+} left in solution to a very low value.

We will first calculate the maximum concentration of OH^- that can be present without causing any $Mg(OH)_2$ to precipitate. If $Mg(OH)_2$ is not to be present, the product $[Mg^{2+}][OH^-]^2$ must not exceed the K_{sp} for $Mg(OH)_2$. Hence, in 0.0010 M Mg^{2+} solution we must have the condition that

$$(0.0010)[OH^-]^2 < 10^{-11}$$

$$[OH^-]^2 < \frac{10^{-11}}{10^{-3}} = 10^{-8}$$

$$[OH^-] < 10^{-4}$$

Thus, if we keep the concentration of OH^- below 10^{-4} *M*, solid $Mg(OH)_2$ cannot be present. Now let us calculate the concentration of Mn^{2+} left in the solution when the concentration of $[OH^-]$ is just under 10^{-4} *M*. We have

$$[Mn^{2+}][OH^-]^2 = 10^{-14}$$

Let $[Mn^{2+}] = x$ and $[OH^-] = 10^{-4}$:

$$(x)(10^{-4})^2 = 10^{-14}$$

$$x = \frac{10^{-14}}{10^{-8}} = 10^{-6}\ M$$

In practice one would not be likely to achieve such a clear separation. Oftentimes impurities are occluded in the fast-growing crystals of the precipitating substance. Sometimes, where the size and charge of the ionic impurities are similar to those of the ion being precipitated, the impurity will be incorporated into the crystal structure of the precipitating compound. The term *co-precipitation* is used to describe all such phenomena.

Thus, the $[Mn^{2+}]$ is decreased to nearly 10^{-6} *M* before any $Mg(OH)_2$ will precipitate. Removing all but 10^{-6} *M* Mn^{2+} from the solution is a good separation.

A convenient way to carefully control the $[OH^-]$ is by means of an NH_3-NH_4^+ buffer, for which the equilibrium reaction is

$$NH_3 + H_2O = NH_4^+ + OH^-$$

$$K = \frac{[NH_4^+][OH^-]}{[NH_3]} = 1.8 \times 10^{-5}$$

Solving for the ratio of $[NH_4^+]/[NH_3]$ when $[OH^-] = 1.0 \times 10^{-4}$ gives

$$\frac{[NH_4^+]}{[NH_3]} = \frac{1.8 \times 10^{-5}}{1.0 \times 10^{-4}} = 0.18$$

The buffer could be conveniently prepared by adding 0.18 mole of NH_4Cl for each mole of NH_3 in the solution. Thus, if NH_4^+ and NH_3 are added to the Mn^{2+} solution to buffer it at $[OH^-] = 10^{-4}$, $Mn(OH)_2$ will precipitate until $[Mn^{2+}] = 10^{-6}$, and Mg^{2+} will be left in solution.

17.8 Additional Examples of Equilibrium Calculations Involving Slightly Soluble Compounds

The best way to gain mastery of almost any subject in chemistry is to have the experience of working many varieties of problems that test your knowledge of the principles involved. The examples that follow are intended to demonstrate methods for solving somewhat complex solubility problems.

EXAMPLE 17.7 At 100° C, one liter of water dissolves 0.0211 g of silver chloride. A 0.2923 g sample of sodium chloride was dissolved in 100 ml of water, and to this was added 100 ml of 0.100 *M* $AgNO_3$. The mixture was heated to boiling and filtered while still at 100° C. Calculate the weight of silver chloride lost with the filtrate.

SOLUTION The loss of silver chloride is the result of its solubility in the hot solution. We will approach the problem with a step-by-step procedure:

(1) Calculate K_{sp} for AgCl at 100° C from its solubility in pure water.
(2) Calculate the concentration of the ion (Ag^+ or Cl^-) in the solution that is in excess as a result of mixing the NaCl and $AgNO_3$.
(3) Calculate the solubility of AgCl in the solution.
(4) Calculate the weight of AgCl that dissolves in the 200 ml.

(1) K_{sp}. Since 0.0211 g of AgCl dissolves in a liter, the concentration of either Ag^+ or Cl^- is given by

$$\left(\frac{0.0211 \text{ g}}{\text{liter}}\right)\left(\frac{1 \text{ mole}}{143 \text{ g}}\right) = 1.48 \times 10^{-4} \frac{\text{mole}}{\text{liter}}$$

Hence, $K_{sp} = [Ag^+][Cl^-] = (1.48 \times 10^{-4})^2 = 2.18 \times 10^{-8}$.

(2) *The concentration of Ag^+ ion in excess.* The mixture was made by adding

$$(0.100 \text{ liter})\left(0.100 \frac{\text{mole}}{\text{liter}}\right) = 1.00 \times 10^{-2} \text{ mole } Ag^+$$

to

$$(0.2923 \text{ g})\left(\frac{1 \text{ mole}}{58.45 \text{ g}}\right) = 5.00 \times 10^{-3} \text{ mole } Cl^-$$

Therefore, 5.00×10^{-3} mole AgCl precipitates, and half the Ag^+ remains in solution:

$$[Ag^+] = (5.00 \times 10^{-3} \text{ mole})\left(\frac{1}{0.200 \text{ liter}}\right) = 2.5 \times 10^{-2}\ M$$

(3) *The solubility of AgCl in this solution.* Now that we know that Ag^+ is in excess, we can readily calculate $[Cl^-]$ from $[Ag^+]$ and K_{sp}. $[Cl^-]$ would be zero if no AgCl dissolved; every mole of Cl^- in solution is the result of a mole of AgCl dissolving:

$$[Cl^-] = \frac{K_{sp}}{[Ag^+]} = \frac{2.18 \times 10^{-8}}{2.50 \times 10^{-2}} = 8.7 \times 10^{-7}\ M$$

Therefore, in 200 ml,

$$\left(8.7 \times 10^{-7} \frac{\text{mole}}{\text{liter}}\right)(0.200 \text{ liter}) = 1.7 \times 10^{-7} \text{ mole of AgCl dissolves}$$

(4) *The weight of AgCl dissolved* is only

$$(1.7 \times 10^{-7} \text{ mole})\left(143 \frac{\text{g}}{\text{mole}}\right) = 2.4 \times 10^{-5} \text{ g}$$

It is evident that even though the solubility product of AgCl at 100° C is 100 times its values at 25° C, the loss of precipitate in 200 ml of 2.5×10^{-2} *M* Ag^+ is very small.

This example illustrates how a problem that appears complex can often be quite straightforward if approached in small steps.

EXAMPLE 17.8 A 0.1936 g sample that was a mixture of pure KCl and pure KBr was dissolved in water, and the halides were precipitated with $AgNO_3$. The precipitate weighed 0.3311 g. Calculate the weight of KCl in the original sample.

SOLUTION Both AgCl and AgBr are so insoluble (Appendix F) that they can be assumed to be completely precipitated. The original sample contained only the pure salts KCl and KBr, while the precipitate contained only AgCl and AgBr. We can attack the problem by making use of the simple relationship that each mole of KCl leads to a mole of AgCl and each mole of KBr leads to a mole of AgBr.

Let x = moles KCl and y = moles KBr. Therefore, x = moles AgCl and y = moles AgBr. We can write two simultaneous equations that express the weights of sample and precipitate in terms of x and y:

$$(x \text{ moles KCl})\left(\frac{74.55 \text{ g KCl}}{\text{mole KCl}}\right) + (y \text{ moles KBr})\left(\frac{119.0 \text{ g KBr}}{\text{mole KBr}}\right) = 0.1936 \text{ g}$$

$$(x \text{ moles AgCl})\left(\frac{143.3 \text{ g AgCl}}{\text{mole AgCl}}\right) + (y \text{ moles AgBr})\left(\frac{187.8 \text{ g AgBr}}{\text{mole AgBr}}\right) = 0.3311 \text{ g}$$

Simultaneous solution of these two equations gives

$$x = 9.97 \times 10^{-4} \text{ mole KCl}$$

which is 7.43×10^{-2} g KCl.

EXAMPLE 17.9 Calculate the weight of $BaSO_4$ that can be converted to $BaCO_3$ by treatment of the $BaSO_4$ with 100 ml of 1.0 *M* Na_2CO_3. This is an example of a metathesis of one slightly soluble substance into another. This reaction is useful for putting $BaSO_4$ into solution. $BaSO_4$ is first converted to $BaCO_3$ and the latter is readily dissolved in acidic solution because of the formation and evolution of CO_2:

$$BaCO_3 + 2\,H^+ = Ba^{2+} + CO_2(g) + H_2O$$

SOLUTION The solubility products of the barium compounds show that $BaCO_3$ is *more* soluble than $BaSO_4$:

$$BaCO_3\text{:} \quad K_{sp} = 7.0 \times 10^{-9}$$
$$BaSO_4\text{:} \quad K_{sp} = 1.0 \times 10^{-10}$$

The equilibrium constant for conversion of $BaSO_4$ to $BaCO_3$ is therefore smaller than unity, and the conversion is not very favorable:

$$BaSO_4(s) + CO_3^{2-} = BaCO_3(s) + SO_4^{2-}$$

$$K = \frac{[SO_4^{2-}]}{[CO_3^{2-}]} = \frac{[SO_4^{2-}][Ba^{2+}]}{[CO_3^{2-}][Ba^{2+}]} = \frac{1.0 \times 10^{-10}}{7.0 \times 10^{-9}} = 0.014$$

To determine the quantity of $BaSO_4$ that will be metathesized, let us assume that an excess of $BaSO_4$ is present so that some remains at equilibrium. Let $x = [SO_4^{2-}]$ at equilibrium; then $1.0 - x = [CO_3^{2-}]$. Hence,

$$\frac{x}{1.0 - x} = 0.014$$

and

$$x = 0.014 = [SO_4^{2-}]$$

The weight of $BaSO_4$ converted, since each mole of SO_4^{2-} in solution is the result of a mole of $BaSO_4$ having reacted, is

$$\left(0.014\,\frac{\text{mole}}{\text{liter}}\right)(0.100\text{ liter})\left(233\,\frac{\text{g}}{\text{mole}}\right) = 0.32\text{ g}$$

The calculations above must be seen only as rough approximations, for it would be inaccurate to assume that such high molar concentrations can be used for activities.

Even though the conversion is unfavorable, a significant weight of $BaSO_4$ is changed to $BaCO_3$; increasing the concentration of CO_3^{2-} would cause an increased conversion.

EXAMPLE 17.10 Copper and manganese can be separated by precipitation of the copper as CuS from 1.0 M H^+ solution saturated with $H_2S(g)$ at 1 atm pressure. Under these conditions, manganese (II) is not precipitated as the sulfide. Make calculations to show that this separation should be effective. The following data are useful:

MnS:	$K_{sp} = 10^{-22}$
CuS:	$K_{sp} = 10^{-42}$
H_2S:	$K_1 = 1.1 \times 10^{-7}$
	$K_2 = 1 \times 10^{-13}$

$$\text{Henry's law constant for } H_2S = 10\,\frac{\text{atm}}{\text{mole/liter}}$$

SOLUTION From the Henry's law constant and the pressure of H_2S, we can calculate the concentration of H_2S in solution:

$$P = k[H_2S]$$

$$[H_2S] = \frac{1.0\text{ atm}}{\dfrac{10\text{ atm}}{\text{mole/liter}}} = 0.1\ M$$

The two-step ionization of H_2S is

$$\begin{array}{cccc} H_2S & = H^+ & + HS^-; & K_1 = 1.1 \times 10^{-7} \\ (0.10 - x) & (x + y + 1.0) & (x - y) & \end{array}$$

$$\begin{array}{cccc} HS^- & = H^+ & + S^{2-}; & K_2 = 1 \times 10^{-13} \\ (x - y) & (x + y + 1.0) & y & \end{array}$$

in which x is the $[H^+]$ produced in the first ionization, y is the $[H^+]$ from the second ionization, and 1.0 is the $[H^+]$ added as a common ion.

Recognizing that x is small compared with 0.10, that both x and y are small compared with 1.0, and that y is small compared with x, we have

$$\begin{array}{cccc} H_2S = & H^+ + & HS^-; & K_1 = 1.1 \times 10^{-7} \\ 0.10 & 1.0 & x & \end{array}$$

$$\begin{array}{cccc} HS^- = & H^+ + & S^{2-}; & K_2 = 1 \times 10^{-13} \\ x & 1.0 & y & \end{array}$$

Solving for x from the equilibrium-constant expression for the first equation above, and substituting its value into the equilibrium expression for the second equation gives

$$y = [S^{2-}] = 1 \times 10^{-21}$$

we can now calculate the concentrations of Cu^{2+} and Mn^{2+} that would be at equilibrium with that concentration of S^{2-} ion:

$$[Cu^{2+}] = \frac{10^{-42}}{1 \times 10^{-21}} = 10^{-21}$$

$$[Mn^{2+}] = \frac{10^{-22}}{1 \times 10^{-21}} = 10^{-1}$$

From this, we see that $[Cu^{2+}]$ is decreased to a very low value, while $[Mn^{2+}]$ can be as high as 10^{-1} *M* before any MnS would precipitate. Hence, it should be possible to separate Cu^{2+} and Mn^{2+} by precipitating the former as sulfide under the conditions used in this example.

Summary

Precipitation reactions, those in which ions combine to form insoluble ionic solids, represent an important class of inorganic chemical reactions. We can predict the extent to which a precipitation reaction will occur from a knowledge of solubility data; hence, it is important to become familiar with the rules for solubility, given early in the chapter.

Solubility is shown to be a function of (1) hydration energy and (2) lattice energy. Both factors depend on the same variables: (1) the sizes of the ions and (2) their charges. Small, highly charged ions have high (negative) energies of hydration, and that favors high solubility. On the other hand, small ions of high charge also produce high (negative) lattice energies, which promote insolubility. With our present state of understanding we are not always able to predict with certainty which of these opposing effects will be predominant, but understanding the effects of hydration and lattice energies is often helpful to rationalize observed results and to make at least qualitative predictions about solubilities.

The effect of temperature on solubility can be predicted from knowledge of the enthalpies of solution. If the dissolving process is endothermic (ΔH is positive), an increase in temperature increases solubility. If the process is exothermic, raising the temperature decreases the solubility.

Other factors affecting solubility are (a) common ions (which were also discussed in Chapter 10), (b) acidity, and (c) complex ion formation. Acidity will increase the solubility of any ionic salt in which the anion of the salt is derived from a weak acid. Because of the high stability of many complex ions, formation of a complex ion usually has a very dramatic effect on solubility. Quantitative calculations can be made of the effects of both acidity and complex ions.

By careful control of the concentration of a reactant, it is often possible to quantitatively separate substances by selectively precipitating one of them. The application of chemical equilibrium in this section should be valuable in reinforcing your understanding of that subject.

Exercises

17-1.
Write the equations for any precipitation reaction that occurs when equal volumes of a 0.1 *M* solution of the following salts are mixed. (The equations should be written showing ionic species as ions.)

(a) Na_2SO_4 and KNO_3
(b) $Ba(NO_3)_2$ and Li_2SO_4
(c) $AgNO_3$ and $NaClO_3$
(d) KI and $AgNO_3$
(e) $Hg_2(NO_3)_2$ and NaCl
(f) Na_2S and KI
(g) $Fe(NO_3)_2$ and Na_2S
(h) $AgNO_3$ and NaOH

17-2.
(a) Making use of the relative solubilities of the hydroxides, suggest a means for identifying whether a white solid is $BaCl_2$ or $MgCl_2$.
(b) Making use of the relative solubilities of the sulfates, suggest a means for identifying whether a white solid is $BaCl_2$ or $MgCl_2$.

17-3.
Explain why Fe_2O_3 is much more soluble in nitric acid solution than in water, while the solubility of AgCl is very low in both water and nitric acid solution.

17-4.
A white solid is either $BaCO_3$ or $BaSO_4$. Suggest a simple solubility test for determining which substance is present.

17-5.
A solid sample is either $AgNO_3$ or $NaNO_3$. Suggest a simple solubility test for determining which substance is present.

17-6.
A substance is either $Ca(NO_3)_2$ or $AgNO_3$. Suggest a simple solubility test for determining which substance is present.

17-7.
A substance is either NaCl or $NaNO_3$. Suggest a simple solubility test for determining which substance is present.

17-8.
A black solid is either Ag_2S or PbS. Suggest a simple solubility test for determining which substance is present. (Remember that H_2S is both a weak acid and a gas that can be removed by boiling the solution.)

Problems

(K_{sp} *values are given in Appendix F.*)

17-1.
Calculate the weight of $BaSO_4$ that will dissolve at equilibrium in 1.5 liters of water at 25° C.

17-2.
Calculate the solubility in moles/liter of each of the following substances:

(a) AgCl
(b) Ag_2CrO_4
(c) $PbCl_2$

17-3.
From the solubilities given here, calculate the solubility product for each substance:

(a) $BaSO_4$ 1.0×10^{-5} moles/liter
(b) Ag_2CrO_4 8×10^{-5} moles/liter
(c) Ag_3PO_4 1.6×10^{-5} moles/liter

17-4.
The solubility of $Ce(IO_3)_3$ is found to be 110 mg in 100 ml. Calculate the weight of $Ce(IO_3)_3$ that will dissolve in 500 ml of 0.010 *M* IO_3^- solution.

17-5.
The silver ion in a solution is precipitated by the addition of excess S^{2-}. When the mixture reaches equilibrium, the concentration of S^{2-} is 0.10 *M*. Calculate the equilibrium concentration of Ag^+.

17-6.
A solution contains 10.0 mmoles of $Pb(NO_3)_2$. What percentage of the total lead is precipitated as $PbCl_2$ if 10.0 ml of 0.80 *M* HCl is added and the mixture diluted to 100 ml?

17-7.
Calculate the equilibrium concentrations of Ag^+ and Cl^- that result when to 50.00 ml of a 0.1000 *M* Cl^- solution there are added (a) 49.90 ml, (b) 50.00 ml, and (c) 50.10 ml of a 0.1000 *M* Ag^+ solution.

17-8.
What weight of $FeCl_3$ can be added to 500 ml of solution at pH 3.0 before $Fe(OH)_3$ precipitates? (K_{sp} of $Fe(OH)_3$ is 10^{-36}.)

17-9.
A 100 ml solution is 0.10 *M* in Br^- and 0.10 *M* in I^-. One desires to separate these halides by precipitating the I^- as AgI(*s*), leaving the Br^- in solution. Calculate the weight of I^- left in solution if Ag^+ is added until the solution is just saturated with AgBr. That is, no AgBr has precipitated, but no more Ag^+ can be added without causing AgBr to precipitate. What percent of the I^- is left in solution?

17-10.
A 200 ml solution is 0.10 *M* Cl^-, 0.10 *M* Br^-, and 0.10 *M* I^-. To this is added a 50 ml portion of 1.0 *M* Ag^+ solution. Calculate the equilibrium concentrations of Cl^-, Br^-, I^- and Ag^+.

17-11.
The solubility of AgCl is 1.92×10^{-3} g/liter at 25° C. $\mathscr{E}^0$ for $Ag^+ + e^- = Ag(s)$ is +0.799 V. From these data and the atomic weights calculate $\mathscr{E}^0$ for $AgCl(s) + e^- = Ag(s) + Cl^-$.

17-12.
A mixture was prepared from 1.0 ml of 1.0 *M* Cl^-, 1.0 ml 0.50 *M* Ag^+, and water, to make the total volume 10 ml. Then 10 ml of 0.040 *M* I^- was added, and the mixture was allowed to come to equilibrium. Calculate the weight and composition of the precipitate and the concentrations of each of the ions in solution.

17-13.
The relative solubilities of AgCl and AgBr in 1 *M* NH_3 solution are sometimes used to distinguish the two silver salts. Calculate the solubility in 1 *M* $NH_3(aq)$ of each salt in grams per 100 ml. Discuss the possibility of distinguishing AgCl and AgBr by a solubility test in ammonia solution.

17-14.
$AgNO_3$ solution is carefully added to a solution that is originally 0.10 *M* in Cl^- and 0.10 *M* in CrO_4^{2-}. Calculate the concentration of Cl^- present in the solution when the first crystal of $Ag_2CrO_4(s)$ forms. $K_{sp(AgCl)} = 1.8 \times 10^{-10}$, and $K_{sp(Ag_2CrO_4)} = 2.0 \times 10^{-12}$.

This calculation illustrates an application of selective precipitation in analytical chemistry. In the analysis of Cl^- by titration with Ag^+, CrO_4^{2-} is added to the sample as an end-point indicator. Both AgCl and Ag_2CrO_4 are insoluble compounds, but the solubility of Ag_2CrO_4 is high enough compared with that of AgCl that no Ag_2CrO_4 precipitates until the precipitation of Cl^- is almost complete. The first appearance of Ag_2CrO_4, detectable because of its red color (AgCl is white), signals that the Cl^- concentration has been decreased (as you saw in your calculation) to a negligibly small concentration, and the titration of Cl^- is considered at that point to be complete.

18

The Representative Elements: Groups IA–IVA

In this chapter and the one following we discuss the descriptive chemistry of the representative, or A-Group, elements. These are the elements in which the inner shells of electrons are complete, and the *differentiating electron* is either an *s* or a *p* valence electron. The differentiating electron is the last electron added in the *aufbau* process; that is, it is the electron that makes the difference in the electronic composition of an element and that of the immediately preceding element in the periodic table. The differentiating electron in the representative groups is an *s* or *p* electron, while the differentiating electron in the transition, or B-Group, elements is either a *d* or an *f* electron. Discussion of the transition elements is reserved for Chapter 21.

You have already been exposed to some of the reactions that will be discussed in the next few chapters. They have been used as examples of precipitation, oxidation-reduction, and acid-base reactions. We now look at them again, together with other reactions, from the point of view of family relations. We will be interested in noting characteristics of families, gradations of properties within families, and relationships among families. These group relationships are of inestimable value in mastering the large body of descriptive chemistry.

18.1 Group IA: The Alkali Metals

The members of Group IA, the alkali metals, are lithium (Li), sodium (Na), potassium (K), rubidium (Rb), cesium (Cs), and francium (Fr). They are the most homogeneous family in terms of physical and chemical properties. Physically, they are characterized by softness, low melting temperature, high reflectivity, low density, and high electrical and thermal conductivity. Chemically, they are highly reactive reducing agents. They react readily with the nonmetals, especially with oxygen and the halogens, to form salts. They react vigorously with water and must, therefore, be stored under conditions in which both water and oxygen are excluded. In their compounds, the Group IA metals exist only in the +1 oxidation state as singly charged positive ions. This makes their chemistry relatively simple.

TABLE 18.1

Electronic Configuration of Group IA Elements

ELEMENT	CONFIGURATION
$_{3}Li$	He core + $2s^1$
$_{11}Na$	Ne core + $3s^1$
$_{19}K$	Ar core + $4s^1$
$_{37}Rb$	Kr core + $5s^1$
$_{55}Cs$	Xe core + $6s^1$
$_{87}Fr$	Rn core + $7s^1$

TABLE 18.2

Properties of the Group IA Elements (Alkali Metals)

	Li	Na	K	Rb	Cs
Atomic number	3	11	19	37	55
Atomic radius (Å)	1.33	1.57	2.03	2.16	2.35
Ionic radius (Å)	0.68	0.98	1.33	1.48	1.67
Density (g/cm^3)	0.53	0.97	0.86	1.53	1.90
Melting point (°K)	454	371	336	312	302
Boiling point (°K)	1604	1163	1039	974	958
$\mathscr{E}^\circ$ (M^+, M) (volts)	−3.045	−2.714	−2.925	−2.925	−2.923
Ionization energy (I) (kjoules/mole)	520	496	419	402	377
Ionization energy (II) (kjoules/mole)	7295	4562	3050	2653	2423
$\Delta H_{\text{hydration}}$ (of M^+) (kjoules/mole)	−506	−397	−318	−289	−259
ΔH_{fus} (kjoules/mole)	3.02	2.60	2.32	2.34	2.10
ΔH_{vap} (kjoules/mole)	135	89.0	77.4	69.2	65.9

Properties of the Group IA Elements

Table 18.1 gives the electronic configurations of the Group IA elements. The single *s* electron in the valence shell of each atom is responsible for the similarities in chemical properties of the elements. They react by giving up the *s* electron to form a singly charged positive ion with a noble-gas configuration. The physical properties of high reflectivity, high electrical conductivity, and high thermal conductivity are also associated with this loosely held electron. Table 18.2 is a listing of several important properties of the Group IA elements.

The standard single-electrode reduction potentials (Table 18.2), corresponding to the half-reactions

$$M^+ + e^- = M$$

are all near to −3 volts, showing that the reverse reaction (the oxidation of the metal)

$$M = M^+ + e^-$$

has a very strong tendency to occur. In other words, the alkali metals are very strong reducing agents.

Because of the decrease in ionization energy from top to bottom of the Group IA elements, one might at first expect that the corresponding oxidation potentials would increase. The ionization energy, however, while important, is only one of the factors that affect the oxidation potential. As in our discussion of the Born-Haber cycle (Chapter 15), consider the oxidation half-reaction to take place in the following steps:

(1)	$M(s) = M(g)$	(sublimation)
(2)	$M(g) = M^+(g) + e^-$	(ionization)
(3)	$M^+(g) + aq = M^+(aq)$	(hydration)
	$M(s) = M^+(aq) + e^-$	(oxidation half-reaction)

After learning that ionization energies decrease from top to bottom of a representative family, one might be surprised to learn that the Li^+, Li couple has a slightly more negative reduction potential than the Cs^+, Cs couple. The explanation lies in the relatively small size of the Li^+ ion, which results in a relatively high negative heat of hydration.

The energies required in steps (1) and (2) both decrease as we go from top to bottom of Group IA. The energy supplied by step (3), however, also decreases from top to bottom of the group. Apparently, the decrease in the energy requirements of the sublimation and ionization steps are quite evenly offset by the hydration step. The net effect is that the free energy for the overall process is approximately the same for each element in the group.

Note that the experimental energies of hydration listed in Table 18.2 are in line with what one would expect from the relative sizes of the ions; that is, the smaller the ion (charge held constant), the greater is the energy released in the hydration process.

Chemical Reactions of Group IA Elements

Sodium and potassium are relatively abundant in the earth's crust, between 2 to 3% each. Lithium, rubidium, and cesium are relatively rare, of the order of .01 to .001%. Francium is detectable in nature, but only in minute trace amounts. The little information we have about francium comes from the study of the small quantities of the element produced in specific nuclear reactions. None of the alkali metals exists in the elemental form in nature; they are far too reactive for that. Some of the common naturally occurring compounds of sodium and potassium are NaCl, $Na_2SO_4 \cdot 10\,H_2O$, $Na_2B_4O_7 \cdot 10\,H_2O$, KCl, $KCl \cdot MgCl_2 \cdot 6\,H_2O$, $NaNO_3$, and a large variety of silicate minerals such as the feldspars ($KAlSi_3O_8$, for one example). Sodium chloride makes up 2.8% of sea water.

As can be seen in Table 18.2, the melting and boiling temperatures of the alkali metals decrease regularly from top to bottom of the group. The melting temperatures are among the lowest of the metals. For example, compare these melting points with those of the IB elements: Cu—1356° K, Ag—1234° K, and Au—1336° K.

The Group IA elements are also among the softest metals. Softness increases from top to bottom of the family. For example, lithium can be cut with a knife with difficulty; cesium, with ease. The freshly cut metals are silvery white in appearance (high reflectivity), but the surface corrodes almost immediately in air because of reactions with water vapor and oxygen.

The density of the IA elements is seen to increase with atomic weight. Note how low the densities are for metals: The first three are less dense than water. By way of comparison, the densities of the IB metals are Cu—8.92 g/cm^3, Ag—10.5 g/cm^3, and Au—19.3 g/cm^3.

All the alkali metals have high thermal conductivity. Because of this property, sodium and mixtures of sodium and potassium find use as coolants in the valves of some internal combustion engines and in the heat exchange systems of nuclear reactors.

Only a cursory examination of the large energy difference between first and second ionization energies (Table 18.2) is needed to see why the alkali metals form +1 ions exclusively. Loss of the lone outermost *s* electron to form the +1 ion leaves a stable noble-gas configuration. Hence, it would not be expected that additional electrons would be lost. The decrease in ionization energy from top to bottom of the family can be accounted for by the increase in size of the atom as one goes down the group.

The high oxidation potentials of the alkali metals make electrochemical reduction of the molten salts the most feasible method of producing the metals. Electrolysis of molten NaCl, with inert electrodes, is an important industrial process. The products are sodium metal and chlorine gas.

As mentioned previously, all the alkali metals react vigorously with water. Sodium is used in the equation below as an example:

$$2\,Na + 2\,H_2O = 2\,Na^+ + 2\,OH^- + H_2$$

Often the heat of the reaction will ignite the mixture of H_2 (from the reaction) and O_2 (from the air) to produce a violent explosion. Analogous reactions can be written for the other alkali metals.

All the elements of Group IA combine directly with oxygen to form solid oxides. Lithium, however, is the only one that forms the normal oxide by direct combination. Sodium produces a peroxide; potassium, rubidium, and cesium give superoxides:

$$\begin{aligned} 4\,Li + O_2 &= 2\,Li_2O && \text{(normal oxide)} \\ 2\,Na + O_2 &= Na_2O_2 && \text{(peroxide)} \\ \left.\begin{aligned} K + O_2 &= KO_2 \\ Rb + O_2 &= RbO_2 \\ Cs + O_2 &= CsO_2 \end{aligned}\right\} & && \text{(superoxides)} \end{aligned}$$

The oxidation number of the alkali metal in all of these compounds is +1. Peroxide and superoxide ions are discussed in Chapter 19.

Normal oxides of Na, K, Rb, and Cs can be obtained by heating their solid hydroxides or nitrates with the free metal. Using sodium as an example, the reactions are

$$2\,NaOH(s) + 2\,Na(l) = 2\,Na_2O(s) + H_2(g)$$
$$2\,NaNO_3(s) + 10\,Na(l) = 6\,Na_2O(s) + N_2(g)$$

The normal oxides of the alkali metals dissolve readily in water to form strong bases. A typical reaction is

$$Na_2O(s) + H_2O = 2\,Na^+ + 2\,OH^-$$

The reaction can be thought of as a hydrolysis of the oxide ion:

$$O^{2-} + HOH = 2\,OH^-; \qquad K > 10^{22}$$

which, for all practical purposes, is complete.

As would be expected from their positions in the periodic table, the Group IA metals react vigorously with the Group VIIA nonmetals to form salts. Typical reactions are

$$2\,Na(s) + Cl_2(g) = 2\,NaCl(s); \qquad \Delta G^0 = -768.2 \text{ kjoules}$$
$$2\,K(s) + Br_2(l) = 2\,KBr(s); \qquad \Delta G^0 = -758.4 \text{ kjoules}$$

The strong tendency for reaction between members of the Group IA and Group VIIA elements is evident from the large decrease in free energy attending those reactions. One way to gain an appreciation for the magnitude of these free-energy decreases is to calculate the extent to which the reverse reactions take place—that is, calculate the extent to which the salt decomposes to the elements. This calculation is made for NaCl in Example 18.1. The result shows how impressively stable the alkali metal halides are relative to the elements from which they form.

The weak acid formed in the hydrolysis of O^{2-} ion is OH^-, a weak acid for a certainty! The equilibrium constant for the ionization of OH^-,

$$OH^- = O^{2-} + H^+$$

is estimated to be $< 10^{-36}$. Hence,

$$K_h = \frac{K_w}{K_{a(OH^-)}} > \frac{10^{-14}}{10^{-36}} > 10^{22}$$

At elevated temperatures, Group IA metals (as well as some of the more active Group IIA metals) react with molecular hydrogen to form hydrides, ionic compounds in which hydrogen has a -1 charge. An example is

$$2\,Na(l) + H_2(g) = 2\,NaH(s)$$

The ionic hydrides are strong reducing agents as can be seen in their reaction with water:

$$NaH(s) + H_2O = Na^+ + OH^- + H_2(g).$$

EXAMPLE 18.1 Calculate the partial pressure of $Cl_2(g)$ in equilibrium with NaCl(s) and Na(s) at 25° C. The value for $\Delta G^0_{f(NaCl)}$ at this temperature is -384 kjoules/mole.

$$Na(s) + \frac{1}{2}Cl_2(g) = NaCl(s); \qquad \Delta G^0 = -384 \text{ kjoules}$$

SOLUTION The reverse of the formation reaction is

$$NaCl(s) = Na(s) + \frac{1}{2}Cl_2(g); \qquad \Delta G^0 = 384 \text{ kjoules}$$

Since the activities of Na(s) and NaCl(s) are unity (pure solids), the equilibrium constant for the decomposition reaction is

$$K = P_{Cl_2}^{\frac{1}{2}}$$

The numerical value of K is found from the ΔG^0 of the reaction through the relation

$$\Delta G^0 = -RT \ln K$$

$$\ln K = -\frac{\Delta G^0}{RT} = \frac{-384{,}000 \text{ joules}}{\left(8.31 \frac{\text{joules}}{°K}\right)(298° \text{ K})}$$

$$K = 4.5 \times 10^{-68}$$

Hence,

$$P_{Cl_2}^{\frac{1}{2}} = 4.5 \times 10^{-68}$$

and

$$P_{Cl_2} = 2.0 \times 10^{-135} \text{ atm}$$

This represents less than one molecule of chlorine in the estimated volume of the universe! The other alkali metal halides exhibit comparable stability.

Because of the low ionization potentials of the alkali metals and the high electron affinities of the halogens, the bonding in alkali metal halides approaches the purely ionic bond. As we will see in Example 18.2, however, even these "purely ionic" bonds have the beginnings of covalent character.

EXAMPLE 18.2 The bond distance in gaseous KCl is 2.67 Å. The unit charge is 4.80×10^{-10} esu. Calculate the dipole moment for a gaseous KCl ion pair and compare it with the experimental value of 10.0 debyes.

SOLUTION If you need to refresh your memory on the subject of dipole moments, re-read the discussion in Chapter 6.

$$\begin{aligned} \mu &= \delta l \\ \mu(\text{calc}) &= (4.80 \times 10^{-10} \text{ esu})(2.67 \times 10^{-8} \text{ cm}) \\ &= 12.8 \times 10^{-18} \text{ esu-cm} \\ &= 12.8 \text{ debyes} \end{aligned}$$

The difference between this value, calculated on the assumption of spherical ions, and the observed value of 10.0 debyes is a measure of the extent to which the shape of the anion is distorted because of the attraction for its electrons by the positively charged cation. These polarization effects are most pronounced when a small positive ion acts on a large negative ion. The polarization effect represents the beginning of covalent character in the bond.

18.2 Group IIA: The Alkaline Earths

The Group IIA elements are beryllium (Be), magnesium (Mg), calcium (Ca), strontium (Sr), barium (Ba), and radium (Ra). None of these is found in the elemental form in nature because of their high reactivity. Like the Group IA metals, they are strong reducing agents and react with a variety of nonmetals. Calcium and magnesium are the most abundant elements of the group. Calcium composes 2.0%, and magnesium 3.5%, of the earth's crust. The others are present in amounts less than 0.05%; the percent of radium is on the order of only 10^{-10}%.

TABLE 18.3

Electronic Configuration of Group IIA Elements

ELEMENT	CONFIGURATION
$_{4}Be$	He core + $2s^2$
$_{12}Mg$	Ne core + $3s^2$
$_{20}Ca$	Ar core + $4s^2$
$_{38}Sr$	Kr core + $5s^2$
$_{56}Ba$	Xe core + $6s^2$
$_{88}Ra$	Rn core + $7s^2$

Properties of the Alkaline Earths

Table 18.3 gives the electronic configurations of the Group IIA elements. The Group IIA elements are seen to have inert-gas cores and a pair of *s* electrons. It is this similarity in electronic configuration that gives the members of the family similar chemical properties. In their compounds, these elements exhibit only the +2 oxidation state, which arises principally by the atom donating two electrons to a nonmetal and forming a +2 ion. Table 18.4 lists some of the important properties of the Group IIA elements. The alkaline earths, as a group, are seen to have higher melting and boiling points than the alkali metals. The Group IIA elements are appreciably harder and more dense than their IA counterparts; however, the IIA elements are still rather low-density elements and, for the most part, are too soft to have structural applications.

Magnesium alloyed with aluminum, zinc, and manganese, however, forms moderately strong lightweight material that has applications wherever weight is an important consideration, as in the aircraft industry.

TABLE 18.4

Properties of the Group IIA Elements (Alkaline Earths)

	Be	Mg	Ca	Sr	Ba
Atomic number	4	12	20	38	56
Atomic radius (Å)	0.89	1.36	1.74	1.91	1.98
Density (g/cm^3)	1.85	1.74	1.54	2.6	3.51
Ionic radius (Å)	0.30	0.65	0.94	1.10	1.29
Melting point (°K)	1556	929	1123	1043	983
Boiling point (°K)	2750	1400	1750	1640	1950
$\mathscr{E}^\circ$ (M^{2+}, M) (volts)	−1.85	−2.37	−2.87	−2.89	−2.90
Ionization energy (I) (kjoules/mole)	899	737	589	548	502
Ionization energy (II) (kjoules/mole)	1756	1450	1145	1063	967
Ionization energy (III) (kjoules/mole)	14842	7729	4910		
$\Delta H_{hydration}$ (of M^{2+}) (kjoules/mole)	−2384	−1925	−1653	−1485	−1276
ΔH_{fus} (kjoules/mole)	11.7	8.95	8.66	9.20	7.66
ΔH_{vap} (kjoules/mole)	295	129	150	139	151

As can be seen in Table 18.4, the size of both the atoms and the ions increases from top to bottom of the family as would be expected since the electrons are found, on the average, farther from the nucleus as the principal quantum number increases.

If one compares the ionic radius of a IIA element with that of the IA element of the same period (Mg^{2+} with Na^+, Ca^{2+} with K^+, Sr^{2+} with Rb^+, and Ba^{2+} with Cs^+), one sees a marked decrease in ionic size in going from IA to IIA. The difference in the nuclear charge appears to be the explanation. The ions being compared are isoelectronic, and the higher-charged nucleus of the IIA ions would be expected to draw the electrons closer to the nucleus.

The smaller size and larger charge of the IIA ions result in a dramatically greater negative energy of hydration of IIA over IA ions. It is the large amount of energy released in the hydration of +2 ions that is responsible for the alkaline earths forming +2 instead of +1 ions. It can be shown, for example, that $Ca^+(aq)$ is unstable relative to $Ca^{2+}(aq)$. Even though the +2 ion requires additional expenditure of energy to bring about the removal of a second electron, the much-larger quantity of energy released in the hydration of the +2 ion over that of the +1 ion drives the ionization on to the +2 stage. The prohibitively high energy required for dislodging a third electron in a IIA metal (a process that involves breaking into a noble-gas core) limits the charge on the hydrated ion to +2.

The $\mathscr{E}^0$ data of Table 18.4 show that the alkaline earths are, like the alkali metals, strong reducing agents. As a group, however, the IIA metals are seen to be somewhat less active than the IA metals. This is particularly true of Be, the first member of the IIA group. One will note that Be is immediately next to the line that divides the metals from nonmetals and does show considerably less metallic character than do the other members of Group IIA.

Chemical Reactions of Group IIA Elements

The elements of Group IIA are all sufficiently active reductants that they are not found in nature in the elemental state. Beryllium is found principally in the form of the complex alumino-silicate mineral beryl, $Be_3Al_2(SiO_3)_6$. Magnesium is present in a number of silicates and in the minerals dolomite, $MgCO_3 \cdot CaCO_3$, and magnesite, $MgCO_3$. Magnesium is also found in sea water, being the second most-abundant metal ion in the sea. Calcium is found abundantly as the carbonate, $CaCO_3$, in the form of limestone, marble, and chalk. Egg shells, oyster shells, and coral are principally $CaCO_3$. Barium and strontium are present in smaller quantities and principally as sulfates.

Oxides

The IIA elements combine directly with oxygen to form normal oxides. Using magnesium as an example, we have

$$2\,Mg + O_2 = 2\,MgO$$
$$\Delta H^0 = -1204 \text{ kjoules}$$
$$\Delta G^0 = -1139 \text{ kjoules}$$

Magnesium ribbon or foil burns brightly in the oxygen of the air. It can be ignited with the flame of a match. Magnesium powder has been used in flares, and the foil at one time was commonly used in photographic flashbulbs.

The large negative value for ΔG^0 shows the strong driving force of the reaction, and the high negative ΔH^0 shows the large amount of heat evolved in the reaction. The thermodynamic data for the other IIA oxides are comparable. With the exception of BaO, which is moderately soluble, the oxides of the Group IIA metals are all quite insoluble in water. They all dissolve in acids. Magnesium oxide is a typical example:

$$MgO + 2\,H^+ = Mg^{2+} + H_2O$$

Beryllium oxide is different from the oxides of the other IIA elements in its chemical properties; it is an amphoteric oxide. Being amphoteric, it reacts not only with acids, as do all IIA oxides, but also with strong bases. The tendency toward acidic properties in the oxide is a consequence of the beginning semblance of nonmetallic character in the element. The reaction of BeO with a strong base is

$$BeO + 2\,OH^- + H_2O = Be(OH)_4^{2-}$$

In Chapter 20, we will discuss in detail the composition, nomenclature, and structure of complex ions, such as beryllate ion.

The ion $Be(OH)_4^{2-}$ is commonly referred to as the beryllate ion. Its structure is tetrahedral: four hydroxide groups arranged at the corners of a tetrahedron with Be^{2+} in the center.

Carbonates

Like the oxides, the Group IIA carbonates are also insoluble in water. They dissolve readily in dilute acid by reacting to liberate carbon dioxide gas. The reaction of calcium carbonate with acid is used as a representative example:

$$CaCO_3(s) + 2\,H^+ = Ca^{2+} + H_2O(l) + CO_2(g)$$

Unlike the Group IA carbonates, which are relatively stable toward heat, the carbonates of Group IIA can be decomposed to the oxide and carbon dioxide when heated from roughly 500° C for $MgCO_3$ to 1400° C for $BaCO_3$. Using the reaction of $CaCO_3$ as an example, we have

These decomposition temperatures are the ones at which the decomposition pressure—in this case the CO_2 pressure—reaches one atmosphere.

$$CaCO_3(s) = CaO(s) + CO_2(g)$$

The thermodynamic data for the thermal decomposition of Group IIA carbonates are interesting. Some of these are recorded in Table 18.5.

From the ΔH^0 values, we see that the amount of heat required to decompose the carbonates increases from top to bottom of the IIA group. This trend can be explained in terms of lattice energies. Because of the large size of the carbonate ion, the lattice energies of the IIA carbonates (the reactants in Table 18.5) are roughly the same. On the other hand, the lattice energies of the IIA oxides (the solid products of the decomposition) will decrease from top to bottom of the family. The net result is that the heat required to decompose the carbonates increases from top to bottom of the IIA family.

TABLE 18.5

Standard Enthalpies and Free Energies of the Thermal Decomposition Reactions of Group IIA Carbonates

REACTION	$\Delta H^0_{298^\circ K}$ (kjoules)	$\Delta G^0_{298^\circ K}$ (kjoules)
$MgCO_3 = MgO + CO_2$	117	67
$CaCO_3 = CaO + CO_2$	176	130
$SrCO_3 = SrO + CO_2$	238	188
$BaCO_3 = BaO + CO_2$	268	218

The ΔG^0 data, which are a measure of the tendency for the decomposition to take place, show that the stability of the carbonates increases from top to bottom of the family.

EXAMPLE 18.3 Even though Group IIA carbonates can be decomposed at elevated temperatures, they are very stable at room temperature. To see how stable they are, calculate the decomposition pressure of $CaCO_3$ at 25° C.

SOLUTION

$$CaCO_3(s) = CaO(s) + CO_2(g)$$

From Table 17.5, ΔG^0 for the decomposition reaction above is 130 kjoules. Since $\Delta G^0 = -RT \ln K$,

$$\ln K = \frac{-\Delta G^0}{RT} = \frac{-130{,}000 \text{ joules}}{\left(8.31 \dfrac{\text{joules}}{^\circ\text{K}}\right)(298^\circ \text{ K})}$$

$$K = 1.6 \times 10^{-23}$$

$$K = P_{CO_2}$$

$$P_{CO_2} = P_{\text{decomposition}} = 1.6 \times 10^{-23} \text{ atm}$$

Stated another way, the pressure of CO_2 gas that is in equilibrium with CaO(s) and $CaCO_3$(s) is only 1.6×10^{-23} atmospheres at 25° C.

The dissociation pressure varies with temperature. At 825° C the equilibrium pressure of CO_2 is 1 atm.

18.3 Group IIIA: The Boron Family

The members of Group IIIA are boron (B), aluminum (Al), gallium (Ga), indium (In), and thallium (Tl). There is far more variation in properties in this family than in the IA and IIA groups. We will see that in going from top to bottom of the family, we encounter changes from nonmetallic to semimetallic to definitely metallic properties; from acidic to amphoteric to basic oxides; and from principally covalent to principally ionic halides.

TABLE 18.6

Electronic Configuration of Group IIIA Elements

ELEMENT	CONFIGURATION
$_5B$	He core + $2s^22p^1$
$_{13}Al$	Ne core + $3s^23p^1$
$_{31}Ga$	Ar core + $3d^{10}4s^24p^1$
$_{49}In$	Kr core + $4d^{10}5s^25p^1$
$_{81}Tl$	Xe core + $4f^{14}5d^{10}6s^26p^1$

Properties of the IIIA Elements

Table 18.6 gives the electronic configuration of the IIIA elements. The maximum oxidation state of IIIA elements is seen to be +3 (noble-gas or pseudo-noble-gas configuration) and is the usual oxidation state for IIIA elements except for Tl, which is commonly found in the +1 state.

Table 18.7 summarizes some of the important properties exhibited by the elements of Group IIIA. Of particular interest in Table 18.7 are the high melting point of boron and the relatively low melting point (just above room temperature) of gallium; the significant increase in reduction potentials from top

TABLE 18.7

Properties of Group IIIA Elements

	B	Al	Ga	In	Tl
Atomic number	5	13	31	49	81
Atomic radius (Å)	0.80	1.25	1.24	1.50	1.55
Ionic radius (Å)	—	0.45	0.60	0.81	0.95
Density (g/cm^3)	2.54	2.70	5.90	7.30	11.85
Melting point (°K)	2300	932	303	429	577
Boiling point (°K)	4200	2720	2510	2320	1740
$\mathscr{E}^0$ (M^{3+}, M) (volts)	—	−1.66	−0.53	−0.34	0.72
Ionization energy (I) (kjoules/mole)	807	577	579	556	590
Ionization energy (II) (kjoules/mole)	2425	1816	1979	1820	1971
Ionization energy (III) (kjoules/mole)	3658	2744	2962	2703	2874
$\Delta H_{hydration}$ (of M^{3+}) (kjoules/mole)	—	−4690	−4703	−4159	−4117
ΔH_{fus} (kjoules/mole)	22.2	10.7	5.60	3.26	4.27
ΔH_{vap} (kjoules/mole)	538.9	294	256	226	162

to bottom of the group; the high (relative to the other members of the group) ionization energy of the nonmetal of the group (boron); and the large increase in density from top to bottom of the family.

Boron

Boron, the lightest element of Group IIIA, is classified as a nonmetal. Typical of nonmetals, and unlike the heavier elements of the family, boron's electrical conductivity is low and *increases* with a rise in temperature, its ionization energy is high, and the bonding in its compounds is covalent. A comparison of the melting points of boron halides with aluminum and gallium halides gives reinforcement to the statement above concerning covalent bonding in boron compounds. For example, BF_3 melts at $-127°$ C and boils at $-101°$ C, which strongly suggests that BF_3 (solid, liquid, or gas) exists as discrete BF_3 molecules, not boron and fluoride ions. By comparison, the more ionically bonded AlF_3 and GaF_3 have normal sublimation points of 1291° C and 800° C, respectively.

The high-charge density that would result from the formation of the B^{3+} ion precludes its existence. B^{3+} would be a very small ion (it is isoelectronic with He and has a higher nuclear charge). Such a small ion with a $+3$ charge would have a very strong polarizing effect on neighboring anions; that is, B^{3+} would attract the electrons of neighboring anions. The result would be a sharing of the electron density, which is covalence.

The boron halides act as Lewis acids; that is, they will accept a share of a pair of electrons furnished by a Lewis base. Examples are

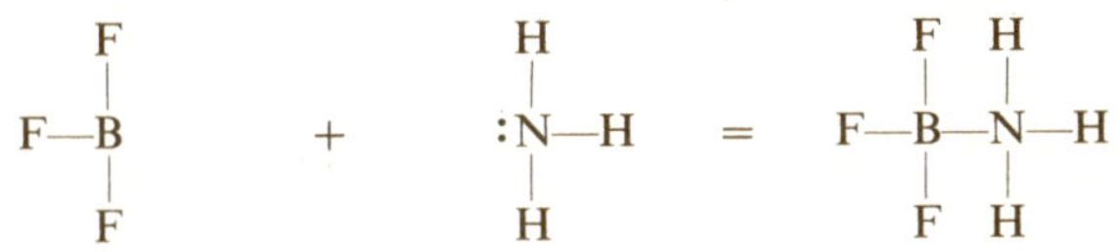

Lewis acid (electron-pair acceptor) | Lewis base (electron-pair donor)

and

$$BF_3 + F^- = BF_4^-$$

Boric Acid and Borates

B_2O_3 is, as would be expected from the nonmetallic character of boron, an acidic oxide. It reacts with water to form $B(OH)_3$, which, because it ionizes as an acid, is usually written as H_3BO_3. Either formula is referred to as boric acid. Boric acid is a very weak, *monoprotic* acid. Its ionization reaction can be written as

$$B(OH)_3 + H_2O = B(OH)_4^- + H^+$$

The reaction may be looked upon as a Lewis acid-base reaction, with water acting as the Lewis base.

There are two categories of boranes; one with the formula B_nH_{n+4}, and the other, B_nH_{n+6}. Boranes containing as many as ten boron atoms to the molecule have been synthesized.

Boranes

Boron forms a series of compounds with hydrogen called *boranes*. The simplest of these is B_2H_6, diborane. The lower molecular weight boranes are highly volatile, are spontaneously inflammable in moist air, and are hydrolyzed readily to boric acid. Their reaction with oxygen liberates large quantities of heat as shown below by the large negative ΔH of combustion of diborane:

$$B_2H_6 + 3\ O_2 = B_2O_3 + 3\ H_2O; \qquad \Delta H^0 = -2017\ \text{kjoules}$$

Hence, the boranes appear to have potential as rocket fuels, although they currently are not being used as such.

Probably the most interesting property of the boranes is the unusual bonding in the molecule. In ethane, C_2H_6, which in terms of the number of atoms is analogous to B_2H_6, there are fourteen valence electrons from which seven classical covalent bonds form, as can be seen in the Lewis structure below:

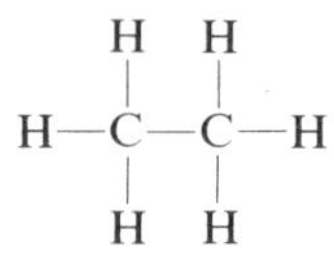

In diborane, however, there are only twelve valence electrons to bond the eight atoms, and the explanation of the bonding becomes more difficult.

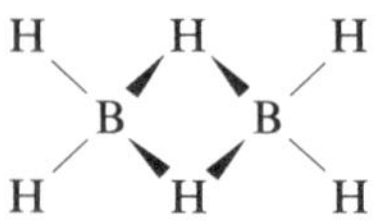

Figure 18.1
The structure of B_2H_6. The "bridge" hydrogens are in a plane that is at right angles to the plane in which the other six atoms lie.

The structure of diborane supplies a clue (see Figure 18.1). The two boron atoms in Figure 18.1 and the four hydrogen atoms on the ends of the molecule all lie in the plane of the page. The other two hydrogen atoms are positioned one behind and the other in front of the page, forming "bridges" between the boron atoms. Each boron atom is connected to four hydrogen atoms, but the B-H bonds are not all the same. The B-H bond distance in the "bridges" is 1.33 Å. The other B-H distances are 1.19 Å. Figure 18.2 is another representation of the geometry of B_2H_6.

The shorter bonds are thought to be normal covalent bonds, but the H-B-H bridges are explained in terms of three-center, electron-pair bonds; that is,

Figure 18.2
Geometry of the B_2H_6 molecule.

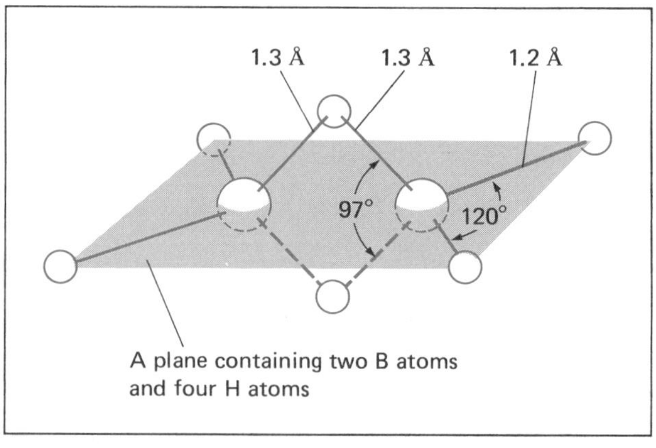

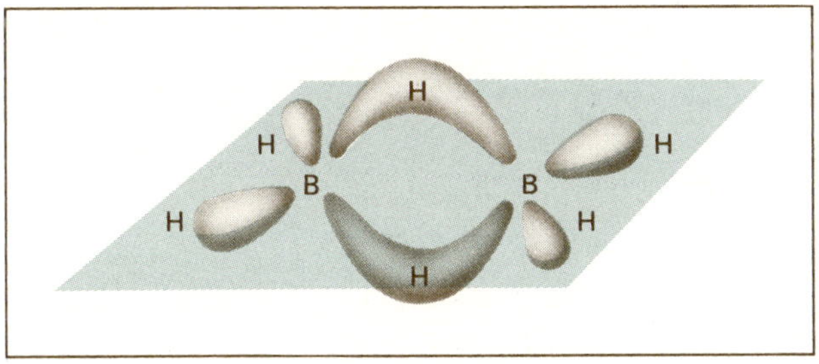

Figure 18.3
Molecular orbital representation of B_2H_6 showing the two three-center, two-electron "bridge" bonds as well as the four more-conventional two-center bonds.

each bridge is a single molecular orbital constructed from a hybrid atomic orbital from each of the two boron atoms and the single 1*s* orbital of hydrogen. The molecular orbital representation of B_2H_6 is shown in Figure 18.3. We will see shortly that similar three-center bonds are present in the aluminum halides.

Nitrides

When boron is heated with elemental nitrogen, a white solid compound of empirical formula BN results. Boron nitride is of interest for several reasons, one of which is its similarity in structure to graphite (see Figure 18.4). At high pressures, boron nitride is converted to a more dense, very hard (approaching that of diamond) allotropic modification called borazon. Borazon has the crystal structure of diamond, as seen in Figure 18.5.

The striking similarity in structure of the boron nitrides and the carbon allotropes is to be expected. Carbon (element 6) is between boron (element 5)

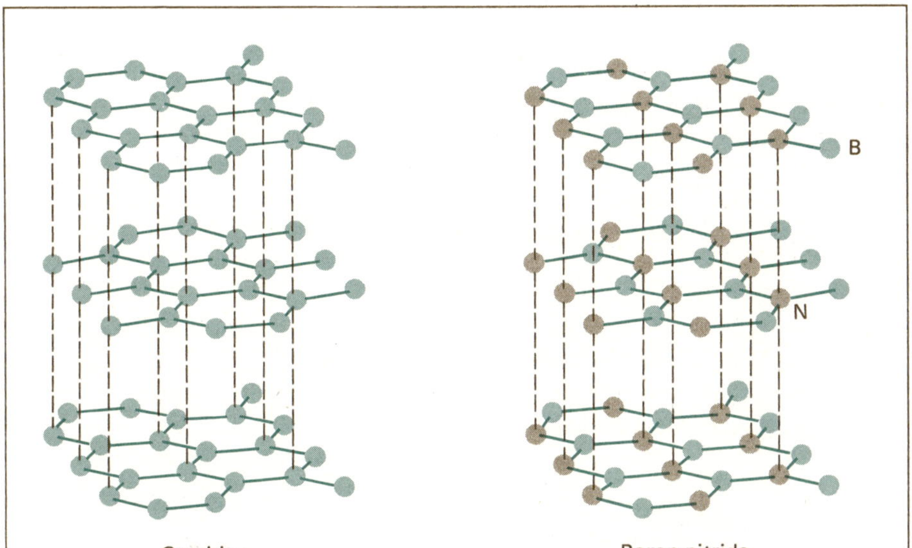

Figure 18.4
Boron nitride has the graphite structure.

Figure 18.5
Borazon has a structure analogous to diamond.

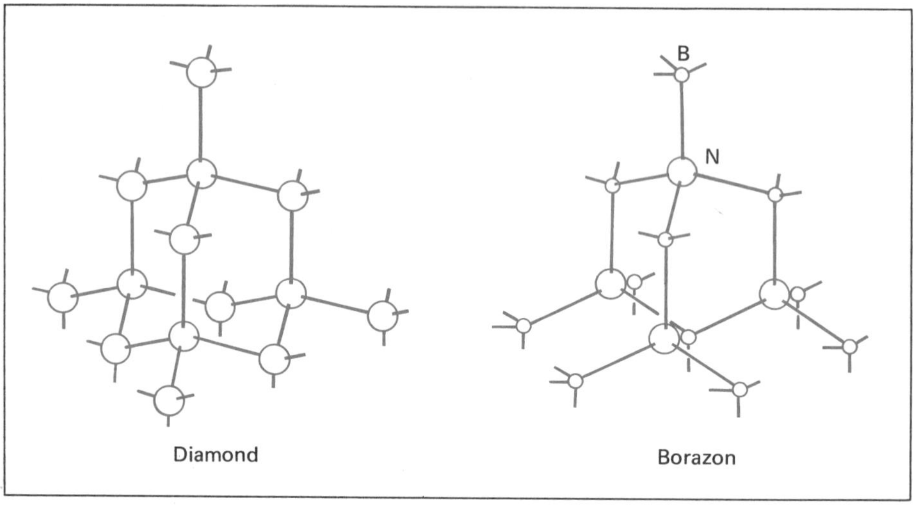

and nitrogen (element 7) in the second period; hence, a B atom plus an N atom have the same number of valence electrons (eight) as two C atoms. Stated another way, the B-N bond is isoelectronic with the C-C bond. Both involve the same type of molecular orbitals and have the same number of electrons to occupy the orbitals.

Aluminum and Its Compounds

Aluminum is the third most-abundant element in the earth's crust. Only oxygen and silicon are present in greater quantities. Since both the latter are nonmetals, aluminum is the *most*-abundant *metal.* Aluminum is an active metal as evidenced by its high reduction potential (Table 18.7) and is not found, therefore, in the elemental form in nature. It is found most commonly combined in the form of complex alumino-silicates such as the feldspar minerals ($KAlSi_3O_8$ as an example) and as the mineral bauxite, a hydrated aluminum oxide ($Al_2O_3 \cdot x\,H_2O$). The latter is the ore from which aluminum is obtained commercially.

Elemental Aluminum

An early method for obtaining aluminum metal was reduction of Al_2O_3 by Na:

$$6\,Na + Al_2O_3 = 2\,Al + 3\,Na_2O$$

The process is thermodynamically favorable, but both inconvenient and expensive.

Even though aluminum compounds are abundant in nature, for years no economical way could be found to obtain the metal from its compounds. Electrolytic reduction of molten alumina (Al_2O_3) would appear to be the obvious solution to the problem, but alumina melts at 2050° C. It would not only be expensive to work at such high temperatures, but difficult to find a container for the reaction, since most steels melt well below 2000° C. In fact, alumina is the substance often used as the container for high-temperature studies of other compounds.

In 1886, two young men, Paul Heroult of France and Charles Martin Hall of the United States, working independently, conceived the idea of dissolving the alumina in a lower-melting aluminum compound, the mineral cryolite,

$AlF_3 \cdot 3\ NaF$. Cryolite melts at around 1000° C; alumina dissolves well in it; and the electrolysis is highly successful. It was on this process that the Aluminum Company of America (Alcoa) was founded and has prospered.

Aluminum has properties that make it highly desirable in structural and electrical applications. It has low density, high electrical conductivity, and when alloyed with small amounts of other metals, such as copper, a tensile strength comparable to that of steel. In addition, aluminum is decorative and corrosion resistant. These two latter properties need some explanation.

The standard reduction potential of aluminum (−1.66 volts) suggests that the metal will oxidize easily and, therefore, corrode readily. This does happen to the surface of aluminum, but the resulting layer of Al_2O_3 that forms on the surface provides a tough, transparent coating, which leaves the metal with its original metallic luster and yet provides protection against further oxidation.

Amalgamation (alloying with mercury) prevents the formation of the Al_2O_3 coating. Under these conditions, aluminum becomes an active reductant.

Al_2O_3

The standard enthalpy and free energy of formation of Al_2O_3 are very large negative values:

$$2\ Al + \frac{3}{2}O_2 = Al_2O_3; \qquad \Delta H^0 = -1670 \text{ kjoules} \quad \Delta G^0 = -1576 \text{ kjoules}$$

Thus, the reaction for formation of Al_2O_3 has a strong driving force (large decrease in free energy) and is accompanied by evolution of much heat (large negative ΔH). The free energy of formation of Al_2O_3 is so negative that aluminum will reduce most metallic oxides to the free metal. An example is

$$2\ Al + Fe_2O_3 = Al_2O_3 + 2\ Fe; \qquad \Delta G^0 = -540 \text{ kjoules} \quad \Delta H^0 = -845 \text{ kjoules}$$

which is called the *thermite process* and is used in welding large pieces of steel. In this procedure, a mixture of Al and Fe_2O_3 is poured around the two pieces of steel where the weld is desired. The large quantity of heat evolved when the mixture is ignited softens the iron and produces the weld at the points of contact of the two pieces of steel. The thermite mixture is usually ignited by using a piece of magnesium ribbon for a fuse. As described earlier, magnesium ribbon, which burns with a very hot flame, can be easily ignited.

There are several crystalline forms of Al_2O_3. The most common one is α-alumina (corundum). It is mechanically strong, thermally stable, and chemically quite inert. It is frequently used as electrical insulation. Pure, crystalline α-alumina is colorless and highly transparent. Traces of impurities, however, produce beautifully colored gems. For example, ruby is α-alumina containing small amounts of Cr^{3+}. Blue sapphire is α-alumina contaminated slightly by Fe^{2+}, Fe^{3+}, and Ti^{3+}.

In the crystal structure of α-alumina, the oxide ions are arranged in the hexagonal close-packed lattice, and aluminum ions occupy interstitial sites in such a way that each oxygen ion is surrounded by four aluminum ions.

Al_2O_3 and $Al(OH)_3$ are amphoteric compounds. They are very insoluble in water, but dissolve in strong acids and strong bases. The reactions with acids are

$$Al_2O_3 + 6\,H^+ = 2\,Al^{3+} + 3\,H_2O$$
$$Al(OH)_3 + 3\,H^+ = Al^{3+} + 3\,H_2O$$

and with bases:

$Al(OH)_4^-$, commonly called aluminate ion, is a complex ion. The bonding, structure, and scientific name for such complex ions will be discussed in Chapter 20.

$$Al_2O_3 + 3\,H_2O + 2\,OH^- = 2\,Al(OH)_4^-$$
$$Al(OH)_3 + OH^- = Al(OH)_4^-$$

Use is made of the amphoteric properties of Al_2O_3 in removing Fe_2O_3 impurities from bauxite ($Al_2O_3 \cdot x\ H_2O$) ores. The ore is treated with a hot solution of base, which dissolves the Al_2O_3 as $Al(OH)_4^-$ but does not dissolve the basic oxide Fe_2O_3. The Fe_2O_3 can then be removed by filtration.

As would be expected from its relatively large negative reduction potential, aluminum metal reacts with dilute acids to release H_2 gas:

$$2\,Al(s) + 6\,H^+ = 2\,Al^{3+} + 3\,H_2(g)$$

Aluminum also reacts with a strong base to release H_2 gas. The reaction is

$$2\,Al(s) + 2\,OH^- + 6\,H_2O = 2\,Al(OH)_4^- + 3\,H_2(g)$$

It is the high stability of the complex ion $Al(OH)_4^-$ that is responsible for the reaction of aluminum with a base. Reaction both with strong acids and strong bases to produce hydrogen is typical of those elements whose oxides are amphoteric.

Hydrolysis of Al^{3+}

Al^{3+} is a small ion (a radius of 0.45 Å) and has a high charge (+3). The result of a large charge being concentrated on a small ion is an intense electrical field surrounding the ion. We speak of this condition as high *charge-density*. Cations of high charge-density are capable of polarizing (by their attraction for electrons) neighboring anions and molecules. Al^{3+} acts on water molecules in that way. The water molecule in its normal state is a polar molecule. In the presence of the electrical field of Al^{3+}, the polarization of the water molecule is enhanced to the extent that significant ionization takes place. The reaction can be thought of as the hydrolysis of a "dry" Al^{3+} ion, as shown below:

$$Al^{3+} + HOH = AlOH^{2+} + H^+ \qquad [18.1]$$

or as the ionization of a hydrated $Al(H_2O)_6^{3+}$ ion:

$$Al(H_2O)_6^{3+} = Al(H_2O)_5OH^{2+} + H^+ \qquad [18.2]$$

The net effect in either way of writing the reaction is the splitting of a water molecule to release a H^+. The equilibrium constant (for either reaction) is 1.0×10^{-5}, which shows that a water solution of Al^{3+} has acidity comparable to that of acetic acid ($K_a = 1.8 \times 10^{-5}$).

All cations can be expected to hydrolyze to some extent, the degree of hydrolysis being related to the size and charge of the cation. As you might expect, however, the hydrolysis of a large monovalent cation, such as Rb^+, is negligible. Among the representative elements, the highly "acidic" cations are Be^{2+}, Al^{3+}, Ga^{3+}, and Sn^{2+}.

Aluminum Halides

Aluminum fluoride, AlF_3, is a typical ionic solid in that it shows low volatility (normal sublimation temperature of 1291° C). Aluminum chloride, on the other hand, sublimes at only 180° C, suggesting a lattice composed of discrete molecules. Studies of the vaporized compound show that aluminum chloride gas consists not of $AlCl_3$, but Al_2Cl_6 molecules. Analogous formulas apply also to the bromide and iodide.

It has been shown that the structures of Al_2Cl_6, Al_2Br_6, and Al_2I_6 have a bridge feature comparable to that in B_2H_6, the bridged halogens being tied to two aluminum atoms by three-center, two-electron bonds. A representation of the structure for Al_2Cl_6 is shown below:

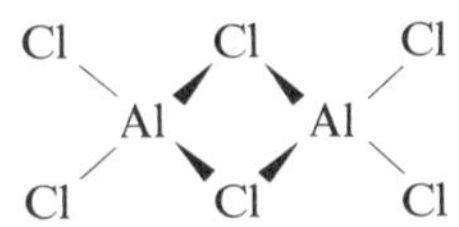

The geometry of the molecule is such that the bridge halogens lie in a plane that is at right angles to the plane containing the other six atoms. See Figures 18.2 and 18.3 for an analogous structure.

Gallium and Its Compounds

The chemistry of gallium is very similar to that of aluminum. It reacts with both acids and bases to produce H_2 gas:

$$2\,Ga + 6\,H^+ = 2\,Ga^{3+} + 3\,H_2$$
$$2\,Ga + 2\,OH^- + 6\,H_2O = 2\,Ga(OH)_4^- + 3\,H_2$$

As might be expected from the position of gallium in the periodic table and from the reactions just given, the oxide and hydroxide of gallium are amphoteric. The reactions with acids are

$$Ga_2O_3 + 6\,H^+ = 2\,Ga^{3+} + 3\,H_2O$$
$$Ga(OH)_3 + 3\,H^+ = Ga^{3+} + 3\,H_2O$$

Equations 18.1 and 18.2 represent only the first and most important step in the hydrolysis of a trivalent cation. A second step in the hydrolysis of Al^{3+} would be written as

$$AlOH^{2+} + HOH = Al(OH)_2^+ + H^+$$

or

$$Al(H_2O)_5OH^{2+} = Al(H_2O)_4(OH)_2^+ + H^+$$

The third step would be

$$Al(OH)_2^+ + HOH = Al(OH)_3(s) + H^+$$

or

$$Al(H_2O)_4(OH)_2^+ + HOH = Al(OH)_3(s) + 4\,H_2O + H^+$$

Since the second and third steps involve the hydrolysis of larger, lower-charged cations than the first, the first step is by far the more important.

Adding OH^- to a water solution of Al^{3+}, however, drives each of the three steps to the right (by removal of H^+), and the final precipitation of $Al(OH)_3$ can be thought of as passing through the three steps of hydrolysis.

The reactions with base (entirely analogous to those with Al_2O_3 and $Al(OH)_3$) are

$$Ga_2O_3 + 2\,OH^- + 3\,H_2O = 2\,Ga(OH)_4^-$$
$$Ga(OH)_3 + OH^- = Ga(OH)_4^-$$

Ga^{3+} ion is even more extensively hydrolyzed than Al^{3+}. The equilibrium constant for

$$Ga(H_2O)_6^{3+} = Ga(H_2O)_5OH^{2+} + H^+$$

is 2.5×10^{-3}, two orders of magnitude greater than the ionization constant of acetic acid!

Indium and Thallium

We have seen that the first member of Group IIIA, boron, is a nonmetal, and its oxide is typically acidic. As one proceeds down the group, the metallic character increases. All the other elements of this family are classified as metals, although, as noted earlier, the second and third members, Al and Ga, maintain enough nonmetallic character that their oxides are amphoteric. The last two elements, In and Tl, are typical metals, and their oxides are basic. They will dissolve in strong acids but not in bases.

Both In and Tl are scarce elements and have few commercial uses. One application worthy of mention, however, is the use of indium to make mirrors for scientific optical instruments. It is a soft metal but will take a high polish that results in mirrors of excellent quality.

Both In and Tl form compounds in the +1 as well as +3 oxidation states. The +3 state is more common for In, and the +1 for Tl. Except that TlCl is quite insoluble, Tl(I) compounds resemble closely the compounds of the Group IA elements.

Reference was made earlier to the increase in reduction potential (decrease in reducing ability) from aluminum to thallium. Differences in hydration energies of the ions because of differences in size of the ions appear to be the controlling factor in this trend. Differences in sublimation and ionization energies are dwarfed by the differences in hydration energies of these ions.

18.4 Group IVA: The Carbon Family

The elements of Group IVA are carbon (C), silicon (Si), germanium (Ge), tin (Sn), and lead (Pb). While they have similar electronic configurations, as shown in Table 18.8, and all exhibit a *maximum* oxidation state of +4, they differ widely in other properties. As in Group IIIA, gradations in metal-nonmetal properties range from the distinctly nonmetal (C) at the top of the family to the metal (Pb) at the bottom. Smooth gradations from nonmetallic to metallic

TABLE 18.8

Electronic Configuration of Group IVA Elements

ELEMENT	ELECTRONIC CONFIGURATION
$_{6}C$	He core + $2s^22p^2$
$_{14}Si$	Ne core + $3s^23p^2$
$_{32}Ge$	Ar core + $3d^{10}4s^24p^2$
$_{50}Sn$	Kr core + $4d^{10}5s^25p^2$
$_{82}Pb$	Xe core + $4f^{14}5d^{10}6s^26p^2$

character, acidic to basic properties of the oxides, covalent to ionic bonding in their compounds, and prevalence of the +2 over the +4 oxidation state are evident as one moves down the family from top to bottom.

Table 18.9 summarizes other properties of the elements of this family. Among other trends, note the gradation and wide range in melting points. Carbon is an atomic solid, with the units of the lattice (atoms) covalently bonded in the graphite or the diamond array. Tin and lead, on the other hand, are metallic crystals, as evidenced by their high electrical and thermal conductivity. The wide variation in density between the nonmetals carbon and silicon and the metallic elements of the group is apparent. Note that lead is five times more dense than graphite.

TABLE 18.9

Properties of the Group IVA Elements

	C	Si	Ge	Sn	Pb
Atomic number	6	14	32	50	82
Atomic radius (Å)	0.77	1.17	1.22	1.41	1.54
Ionic radius (M^{2+}) (Å)	—	—	—	1.10	1.32
Density (g/cm^3)	2.25†	2.40	5.36	7.31	11.34
Melting point (°K)	3500	1683	1210	505	601
Boiling point (°K)	4200	2950	3100	2960	2024
$\mathscr{E}^0_r(M^{2+}, M)$ (volts)	—	—	—	−0.14	−0.13
Ionization energy I (kjoules/mole)	1086	786	762	707	715
ΔH_{fus} (kjoules/mole)	—	46.4	31.8	7.20	4.77
ΔH_{vap} (kjoules/mole)	—	—	334	290	179

† Graphite.

Carbon and Its Compounds

Graphite and Diamond

Carbon exists in two allotropic forms, graphite and diamond. Their crystal structures were described on pages 444–445 of Chapter 15, and sketched in Figures 15.26 and 15.27. Of the two forms, graphite is the more stable at 1 atm pressure and 25° C, as can be seen from the value of the standard free energy for the reaction

$$\text{C(graphite)} = \text{C(diamond)}; \qquad \Delta G^0_{298.15^\circ\text{K}} = 2866 \text{ joules}$$

$$d = 2.25\ \frac{\text{g}}{\text{cm}^3} \qquad d = 3.51\ \frac{\text{g}}{\text{cm}^3}$$

Since diamond is the more dense form, one would predict from the principle of LeChatelier that pressure would favor the conversion from graphite to diamond. Converting graphite into diamonds is an even better increase in value than was the old alchemists' dream of changing lead into gold. The first verified synthesis of diamond from graphite was accomplished by H. Tracy Hall at General Electric Corp. in 1954. Commercial diamonds for industrial use (cutting tools, not gemstones) are now made by several companies. High temperatures as well as high pressures are used.

Oxides of Carbon

When carbon or any carbonaceous material (for example, gasoline, coal, wood) is burned in a limited amount of oxygen, carbon monoxide, CO, results:

$$\text{C}(s) + \frac{1}{2}\text{O}_2(g) = \text{CO}(g); \qquad \begin{aligned} \Delta H^0 &= -110.5 \text{ kjoules} \\ \Delta G^0 &= -137.3 \text{ kjoules} \end{aligned}$$

It is readily apparent from the thermodynamic data that CO forms with the evolution of considerable heat ($\Delta H^0 = -110.5$ kjoules) and is a very stable compound ($\Delta G^0 = -137.3$ kjoules).

At room temperature and one atmosphere pressure, CO is a colorless, odorless gas with practically no solubility in water. It is highly toxic, a property that results from its ability to combine with hemoglobin of the blood and block its role in transporting oxygen throughout the body.

Carbon monoxide is isoelectronic with the diatomic nitrogen molecule. As can be seen below, CO and N_2 also have similar structures; as might be expected, they resemble each other in many of their *physical* properties.

$$:\text{C}\equiv\text{O}: \qquad :\text{N}\equiv\text{N}:$$

At high temperatures, CO is an effective reducing agent and is employed as such in many metallurgical processes. Examples of reduction of metal oxides with CO are

$$\text{Fe}_2\text{O}_3(s) + 3\ \text{CO}(g) = 2\ \text{Fe}(l) + 3\ \text{CO}_2(g)$$
$$\text{CuO}(s) + \text{CO}(g) = \text{Cu}(l) + \text{CO}_2(g)$$

Carbon monoxide can be burned in air to form the dioxide:

$$CO + \frac{1}{2}O_2 = CO_2; \qquad \begin{aligned} \Delta H^0 &= -283 \text{ kjoules} \\ \Delta G^0 &= -257 \text{ kjoules} \end{aligned}$$

This reaction is seen to be highly exothermic, and the high negative free-energy change shows that the reaction goes far to the right. In other words, CO_2 is very stable relative to CO plus O_2.

Carbon dioxide is a colorless, odorless gas with enough solubility in water that a 0.34 *M* solution can be formed when the CO_2 pressure is 1 atm and the temperature is 25° C. (See the discussion of Henry's law in Chapter 8). The molecule is linear, and the electron dot formula is

$$\ddot{\underset{\cdot\cdot}{O}}{=}C{=}\ddot{\underset{\cdot\cdot}{O}}$$

CO_2 is the end product of oxidation of carbon and carbon-containing compounds, whether the oxidation takes place in a flame or in the slower oxidation of sugars in the body.

Carbonic Acid and Carbonates

Carbon dioxide is the anhydride of carbonic acid, H_2CO_3. The reaction is usually written as

$$CO_2(g) + H_2O = H_2CO_3 \qquad [18.3]$$

Carbonic acid is a weak, diprotic acid, whose first ionization,

$$H_2CO_3 = H^+ + HCO_3^- \qquad [18.4]$$

has a K_1 value of 3×10^{-7}. Its second ionization,

$$HCO_3^- = H^+ + CO_3^{2-}$$

is, as is typical, four to five orders of magnitude weaker:

$$K_2 = 6 \times 10^{-11}$$

In a saturated solution of CO_2 in water, the dissolved CO_2 is almost entirely in the form of $CO_2(aq)$ rather than the form of H_2CO_3 molecules. Hence, it would be entirely appropriate to write equation [18.3] as

$$CO_2(g) + H_2O = CO_2(aq)$$

and equation [18.4] as

$$CO_2(aq) + H_2O = H^+ + HCO_3^- \qquad [18.5]$$

Because $CO_2(aq)$ and H_2CO_3 are in equilibrium, the equilibrium constant for [18.4] and [18.5] is the same number, 3×10^{-7}.

EXAMPLE 18.4 Calculate the pH of a buffer solution that is 0.10 *M* in the weak acid H_2CO_3 and also 0.10 *M* in its salt $NaHCO_3$.

SOLUTION One way of solving the problem is to treat it as an ionization equilibrium:

$$H_2CO_3 = H^+ + HCO_3^-; \qquad K_1 = 3 \times 10^{-7}$$

We will neglect the much-smaller second ionization. If we let x equal the moles/liter of H_2CO_3 that ionize to H^+ and HCO_3^-, we will have the following concentrations at equilibrium:

$$[H_2CO_3] = 0.10 - x$$
$$[H^+] = x$$
$$[HCO_3^-] = x + 0.10$$

Substituting into the equilibrium-constant expression gives

$$\frac{[H^+][HCO_3^-]}{[H_2CO_3]} = \frac{(x)(x + 0.10)}{(0.10 - x)} = 3 \times 10^{-7}$$

Neglecting x relative to 0.10, we have

$$\frac{0.10x}{0.10} = 3 \times 10^{-7}$$
$$x = 3 \times 10^{-7}$$
$$\text{pH} = -\log(3 \times 10^{-7}) = 6.5$$

A second solution to the same problem is to approach it in terms of the hydrolysis equilibrium, which also applies.

$$HCO_3^- + H_2O = H_2CO_3 + OH^-$$
$$K_h = \frac{K_w}{K_1} = \frac{1.0 \times 10^{-14}}{3 \times 10^{-7}} = 3 \times 10^{-8}$$

Letting y equal the moles/liter of HCO_3^- that hydrolyze, we have at equilibrium

$$K_h = \frac{[H_2CO_3][OH^-]}{[HCO_3^-]} = \frac{(y + 0.10)(y)}{(0.10 - y)} \cong \frac{0.10y}{0.10} = 3 \times 10^{-8}$$
$$y = [OH^-] = 3 \times 10^{-8}$$
$$\text{pOH} = -\log[OH^-] = -\log(3 \times 10^{-8}) = 7.5$$
$$\text{pH} = 14.0 - \text{pOH} = 6.5$$

The answer is the same, of course, since the same species are involved in both sets of equilibria.

The H_2CO_3-HCO_3^- equilibrium is a buffer system that plays an important role in nature. It is, for example, one of the important buffer systems in soil.

Neutralization of carbonic acid produces two sets of salts, the hydrogen carbonates and the normal carbonates:

$$\begin{array}{l} OH^- + H_2CO_3 = HCO_3^- + H_2O \\ \underline{HCO_3^- + OH^- \quad = CO_3^{2-} + H_2O} \\ H_2CO_3 + 2\,OH^- = CO_3^{2-} + 2\,H_2O \end{array}$$

Both HCO_3^- and CO_3^{2-} hydrolyze, the latter fairly extensively:

$$HCO_3^- + HOH = H_2CO_3 + OH^-$$

$$K_{h(HCO_3^-)} = \frac{K_w}{K_{a(H_2CO_3)}} = 3 \times 10^{-8}$$

$$CO_3^{2-} + HOH = HCO_3^- + OH^-$$

$$K_{h(CO_3^{2-})} = \frac{K_w}{K_{a(HCO_3^-)}} = 2 \times 10^{-4}$$

Carbonates and hydrogen carbonates, in general, react with *strong* acids to liberate $CO_2(g)$. Typical examples are

$$CaCO_3(s) + 2\ H^+ = Ca^{2+} + H_2O + CO_2(g)$$
$$NaHCO_3(s) + H^+ = Na^+ + H_2O + CO_2(g)$$

When carbonates are treated with carbonic acid, hydrogen carbonate ion results:

$$CaCO_3(s) + H_2CO_3 = Ca^{2+} + 2\ HCO_3^-$$

The dissolving of carbonate minerals by CO_2-laden water as it passes through and over the earth is the most usual method by which such cations as Ca^{2+}, Mg^{2+}, and Fe^{2+} get into water supplies to produce "hard" water. Ca^{2+}, Mg^{2+}, and Fe^{2+} mainly, but other divalent and trivalent cations as well, are objectionable in water because (1) they form insoluble precipitates with soap (Chapter 24), and (2) when hard water is heated, the soluble calcium, magnesium, and iron hydrogen carbonates expel CO_2 to form deposits of insoluble normal carbonates (on the sides of boiler pipes, for example). A typical reaction is

$$Ca^{2+} + 2\ HCO_3^- = CaCO_3(s) + H_2O + CO_2(g)$$

Further heating of $CaCO_3$ decomposes it to the oxide:

$$CaCO_3(s) = CaO(s) + CO_2(g) \qquad [18.6]$$

Reactions analogous to equation [18.6] take place with most carbonates, including Li_2CO_3, but excepting the other alkali metal carbonates. The latter are thermally very stable compounds.

A series of interesting reactions apply to the manufacture and "curing" of masonry mortar: (1) An impure $CaCO_3$ (limestone) is heated to form CaO (commercially known as *lime*); (2) water is added to produce a slurry of $Ca(OH)_2$ ("slaked lime"); (3) sand is added, and the resulting mortar is placed between the bricks; (4) CO_2 from the air reacts with the slaked lime to produce an impure calcium carbonate (limestone); and the circle is complete. We start with

There are various ways of softening water; that is, removing the polyvalent cations. One way is to add alkali metal carbonates to precipitate the Ca^{2+}, Mg^{2+}, and Fe^{2+} as carbonates. Alkali metal phosphates and borates can be used in the same way. In effect, these methods replace the polyvalent cations in the solution with the innocuous monovalent alkali metal ions. (Alkali metal compounds, including alkali-metal ion soaps, are all soluble.)

The most usual method of replacing polyvalent ions in solution with monovalent ions is by use of a cation-exchange resin. Various synthetic resins have the property of *ad*sorbing (attaching to the surface of) ions. The procedure consists of saturating the surface of a cation-exchange resin with Na^+ by passing concentrated NaCl solution over a column of the resin, washing out the excess NaCl solution, and then passing the hard water through the column. Because polyvalent ions are adsorbed more tenaciously than monovalent ions, the former replace sodium ions on the resin:

$$\text{Resin—(Na}^+)_x + \frac{x}{2}\,\text{Ca}^{2+} =$$

$$\text{Resin—(Ca}^{2+})_{\frac{x}{2}} + x\,\text{Na}^+$$

The sodium ion resin is regenerated by treating the column periodically with concentrated NaCl solution, which reverses the reaction above.

limestone and end with limestone. The reactions are

$$CaCO_3(s) = CaO(s) + CO_2(g)$$

$$CaO(s) + H_2O = Ca(OH)_2(s)$$

$$Ca(OH)_2(s) + CO_2(g) = CaCO_3(s) + H_2O$$

Several Million Other Carbon Compounds

Carbon forms more compounds by far than all of the other elements together. To date, well over two million carbon compounds have been identified and characterized. The number of *possible* carbon compounds is astronomical; yet, only a few elements are involved. Carbon, hydrogen, oxygen, nitrogen, phosphorus, sulfur, and chlorine are the main elements of this large body of carbon-containing compounds, the study of which is called *organic chemistry*. An introduction to this vast and intriguing field is provided in Chapter 24.

Silicon

Silicon, composing roughly 25% of the earth's crust, is second only to oxygen in abundance. It is found in the form of SiO_2 (silica), of which quartz is the most familiar mineral, and in numerous complex silicate minerals such as the feldspars, micas, and clays. The element silicon is classified as a nonmetal. It has the luster of a metal but otherwise shows no marked metallic properties. It is chemically quite inert except at high temperatures. At high temperatures it will react with oxygen to form SiO_2 and with the halogens to form tetrahalides ($SiCl_4$, for example).

SiO_2 is an acidic oxide. It is insoluble in water and acid but dissolves slowly in strong base solutions. The rapid way to dissolve SiO_2 is to fuse the SiO_2 with the oxides or carbonates of alkali metals. A series of soluble complex silicates results.

Silicon Tetrahalides

At high temperatures, silicon will combine with the halogens to form silicon tetrahalides. Silicon tetrachloride is an example. It is a covalent, low-boiling liquid that hydrolyzes readily:

$$SiCl_4(l) + 2\,H_2O(l) = SiO_2(s) + 4\,HCl(g)$$

Silanes and Silicones

Silicon forms a short series of hydrides, called *silanes*, the simplest member of which is SiH_4. The silanes are analogous in structure to the hydrocarbon series that begins with CH_4, as shown in Figure 18.6. Note the bond between silicon atoms. Unlike the C-C linkage, Si-Si is quite unstable, and the silanes ignite spontaneously and often explosively in air. The more silicon atoms in the molecule, the less stable is the silane; Si_6H_{14} (five Si-Si linkages) is the highest molecular weight silicon hydride that has been prepared.

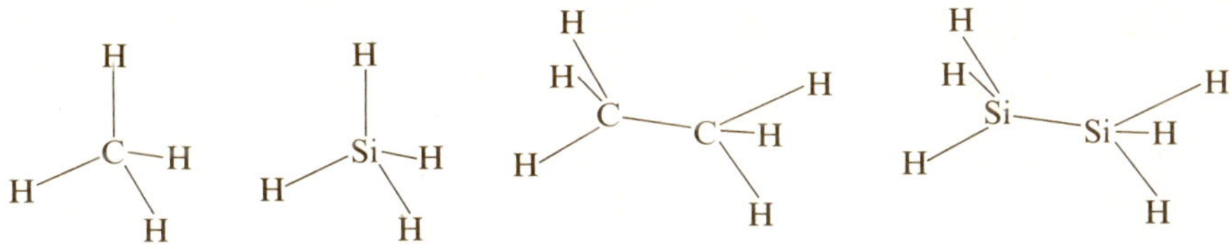

CH$_4$ SiH$_4$ C$_2$H$_6$ Si$_2$H$_6$

Figure 18.6
The silanes are analogous in structure to the carbon hydrides.

When treated with HCl(g), SiH_4 forms a series of compounds called chlorosilanes, SiH_3Cl being an example. It is also possible to synthesize such substituted chlorosilanes as $Si(CH_3)_3Cl$, $Si(CH_3)_2Cl_2$, and $Si(CH_3)Cl_3$. These and other substituted chlorosilanes are commercially valuable compounds since they hydrolyze to form the high-molecular-weight polymers known as *silicones*. Silicones are characterized by silicon-to-oxygen-to-silicon . . . bonds. This linkage, unlike Si-Si, is very stable, and polymers of very high molecular weight result. When the starting material is $Si(CH_3)_2Cl_2$, a silicon chain results, a part of which is shown below:

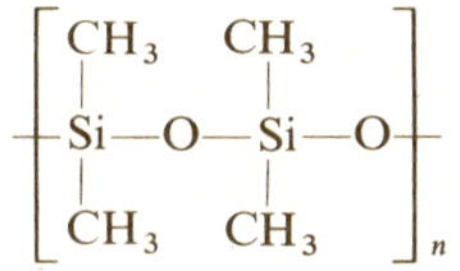

Hydrolysis of $Si(CH_3)Cl_3$ gives a two-dimensional cross-linked chain:

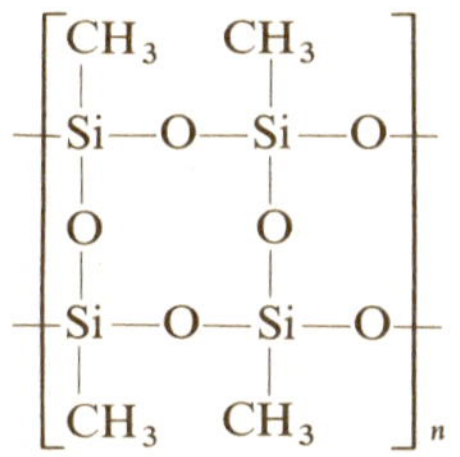

Depending upon the amount of cross-linking, the physical properties vary all the way from light fluids to plastic solids. All the silicones are water repellent, heat resistant, and chemically quite inert, making them useful as protective coatings. They have the desirable quality of undergoing only a very slight change in viscosity when subjected to a rather large change in temperature, making them valuable as lubricants, especially under conditions of extreme variations in temperature.

You are probably familiar with the properties of one of the heavily cross-linked silicones marketed as a toy called "silly-putty." When subjected to a slow sustained force, it undergoes plastic flow and can be molded. When given a sharp, quick blow, however, it is elastic and bounces.

Germanium and Its Compounds

Germanium (Ge), though generally classified as a metal (or, at least, a semimetal) has many properties that are more similar to the nonmetal silicon than to tin and lead, the more metallic elements of Group IVA. The oxide of germanium, GeO_2, being a weakly acidic oxide like SiO_2, is an example of the similarity. The most stable compounds of germanium are those in which the germanium is in the +4 oxidation state. In contrast to tin and lead, +2 compounds of germanium are not common. Examples of common germanium compounds are $GeCl_4$, GeO_2, GeS_2, and GeH_4 (quite unstable). The element germanium is light gray in color, quite resistant to most chemicals at room temperature, and a semiconductor.

Tin and Its Compounds

Tin is one of the metals that has been known from ancient times. Bronze, an alloy of tin and copper, has been used in statuary and ornaments for many centuries. The important ore of tin is SnO_2, the mineral cassiterite. It is found in only a few places on the earth: England, Malaysia, and Bolivia being the main sources. The metallurgy of tin is relatively simple. The crushed ore is concentrated by washing away the lighter rock, and reduction is accomplished with coke. The reaction is

$$SnO_2 + 2\,C = Sn + 2\,CO$$

There are three allotropes of solid tin: (1) "white tin," stable between 13° C and 161° C, (2) "gray tin," an amorphous powder stable below 13° C, and (3) "brittle tin," which shatters when struck with a sharp blow, stable above 161° C. Before the phenomenon was fully understood, the conversion (in cold weather) of "white tin" to the amorphous powder that is "gray tin" was known as "tin disease."

The metal is purified electrolytically. Its principal use is in coating sheet metal for use in the manufacture of "tin" cans. Tin is a constituent of several important alloys, among which are soft solder (Sn and Pb), bronze (Sn and Cu), and "white metal" (Sn, Pb, Sb, and Cu). Door handles on automobiles and other light castings are commonly made of "white metal."

In its compounds, tin exhibits both +2 and +4 oxidation states. Typical reactions involving the +2 state compounds are

$$Sn + 2\,H^+ = Sn^{2+} + H_2 \qquad [18.7]$$

$$Sn^{2+} + H_2O = SnOH^+ + H^+; \quad K = 10^{-2} \qquad [18.8]$$

$$SnO + 2\,H^+ = Sn^{2+} + H_2O \qquad [18.9]$$

$$SnO + OH^- + H_2O = Sn(OH)_3^- \qquad [18.10]$$

Equation [18.7] shows that tin acts as a reducing agent toward H^+, but barely ($\mathscr{E}^0_{red} = -0.13$ volts). To achieve this standard reduction potential, one must have a specially prepared tin surface, since tin, like aluminum, forms a thin protective film of dense oxide (in this case, SnO_2) on its surface. Equations [18.9] and [18.10] show that SnO is amphoteric. Analogous reactions could be written for $Sn(OH)_2$. Equation [18.8] shows that Sn^{2+} is one of the *highly* "acidic" cations, producing H^+ from hydrolysis to about the same extent that HSO_4^- ($K_a = 1.2 \times 10^{-2}$) does by ionization.

Tin(II) ion is frequently used as a reducing agent in the laboratory, being easily oxidized to the +4 state. An example of such a reaction that is useful in analytical chemistry is

$$Sn^{2+} + 2\,Hg^{2+} + 8\,Cl^- = SnCl_6^{2-} + Hg_2Cl_2(s) \qquad [18.11]$$

Equation [18.11] not only is an example of the reducing ability of Sn(II) but demonstrates another property that both Sn(II) and Sn(IV) have, that of their propensity for forming complexes with halide ions, as in $SnCl_6^{2-}$.

When tin is made to react with an oxidizing acid, such as nitric acid, tin in the +4 state results:

$$Sn(s) + 4\,H^+ + 4\,NO_3^- = SnO_2(s) + 2\,H_2O + 4\,NO_2(g)$$

Sn(IV) compounds also result from the direct combination of tin with oxygen and with halogens:

$$Sn + O_2 = SnO_2$$

$$Sn + 2\,Cl_2 = SnCl_4$$

It is not surprising that it is the +4 state rather than the +2 that results from such reactions, since oxygen and chlorine are strong oxidizing agents. $SnCl_4$ is a covalent, low-boiling-point liquid. It hydrolyzes completely when added to water in much the same way that $SiCl_4$ does:

$$SnCl_4(l) + 2\,H_2O(l) = SnO_2(s) + 4\,HCl(g)$$

Solutions of stannous ion exposed to the air oxidize to the +4 state:

$$Sn^{2+} + \frac{1}{2}O_2 + 2\,H^+ + 6\,Cl^- = SnCl_6^{2-} + H_2O$$

Sn(II) solutions can be stabilized by adding a few pieces of tin metal. The elemental tin reduces any Sn(IV) that might form.

$$Sn + SnCl_6^{2-} = 2\,Sn^{2+} + 6\,Cl^-$$

Lead and Its Compounds

Lead is more abundant than tin, and, unlike tin, the deposits of lead ores are not concentrated in just a few places on the earth. The chief ore of lead contains the mineral galena, PbS; the metallurgy is relatively straightforward, consisting of a "roasting" process to convert the sulfide to the oxide, followed by reduction with coke:

$$2\,PbS + 3\,O_2 = 2\,PbO + 2\,SO_2$$

$$PbO + C = Pb + CO$$

Purification is usually by electrolysis. Silver and gold are common impurities. The refining process is carried out not only to get a purer grade of lead, but also to recover the silver and gold.

The +2 oxidation state is the most common for lead. Typical reactions involving compounds in the +2 state are

$$Pb + 2\,H^+ = Pb^{2+} + H_2; \qquad \mathscr{E}^0 = 0.126 \text{ volt} \qquad [18.12]$$

$$PbO + 2\,H^+ = Pb^{2+} + H_2O \qquad [18.13]$$

$$PbO + OH^- + H_2O = Pb(OH)_3^- \qquad [18.14]$$

$$Pb^{2+} + HOH = PbOH^+ + H^+; \qquad K = 10^{-8} \qquad [18.15]$$

Equation [18.12] shows that lead and tin are very similar in their standard reduction potentials, only sightly less negative than $H_2(g)$. Equations [18.13] and [18.14] illustrate the amphoteric character of PbO. Analogous reactions can be written for $Pb(OH)_2$. Equation [18.15] shows the hydrolysis of Pb^{2+} to be six orders of magnitude less than that of Sn^{2+}. The larger size of Pb^{2+} apparently makes it considerably less effective in polarizing and eventually ionizing water molecules.

Many of the lead salts are very insoluble. A few important ones are listed below:

$$PbSO_4(s) = Pb^{2+} + SO_4^{2-}; \qquad K_{sp} = 2 \times 10^{-8}$$
$$PbCO_3(s) = Pb^{2+} + CO_3^{2-}; \qquad K_{sp} = 1.5 \times 10^{-13}$$
$$PbCrO_4(s) = Pb^{2+} + CrO_4^{2-}; \qquad K_{sp} = 8 \times 10^{-14}$$
$$PbS(s) = Pb^{2+} + S^{2-}; \qquad K_{sp} = 10^{-28}$$
$$PbCl_2(s) = Pb^{2+} + 2\,Cl^{-}; \qquad K_{sp} = 2 \times 10^{-4}$$

Lead (II) chloride is one of the few chlorides that are not soluble.

You may recall from Chapter 17 that the other insoluble chlorides are AgCl and Hg_2Cl_2. TlCl is also relatively insoluble.

The oxides of lead are interesting. There are four of them, one in the +2 oxidation state, one in the +4, and two "mixed" oxides, in which both Pb(II) and Pb(IV) are present. Lead (II) oxide, PbO, commonly called litharge, is a yellow, crystalline, amphoteric oxide. The +4 oxide, PbO_2, is black in color, and it is a strong oxidizing agent:

$$PbO_2 + 4\,H^+ + 2\,e^- = Pb^{2+} + 2\,H_2O; \qquad \mathscr{E}^0 = 1.46 \text{ volts}$$

We have already mentioned in Chapter 11 its application in storage batteries. The two "mixed" oxides have the formulas Pb_2O_3 and Pb_3O_4. Writing Pb_2O_3 as $PbO{\cdot}PbO_2$ and Pb_3O_4 as $2\,PbO{\cdot}PbO_2$, we see that in the former the ratio of Pb(II) to Pb(IV) in the lattice is 1:1; in Pb_3O_4, it is 2:1.

Do not misconstrue the term "mixed" oxide to mean a mixture of two oxides. Pb_2O_3 and Pb_3O_4 are true, stoichiometric compounds, each with its own crystal structure. The "mixed" notation refers to the presence of lead in two oxidation states within the same compound. We will encounter another example of a "mixed" oxide in Chapter 21 when we discuss the oxides of iron.

Lead compounds, like those of most of the heavy metals and some of the lighter ones, too (for example, beryllium compounds), are poisonous. Before the toxic properties of lead compounds were recognized, they were common components of paint. Many people became ill from lead poisoning by simply living in a house where specks of paint were flaking off the woodwork and walls and getting into the food and water. Children were especially vulnerable because of their habit of chewing on toys, many of which were painted with lead paints. Lead compounds are no longer used in house paints.

Lead cooking utensils were popular among the wealthy ruling class of early Rome because they "sweetened" the food. Corrosion of the lead cooking ware by reaction with the acids of the food apparently produced sweet-tasting lead salts. Lead acetate, $Pb(C_2H_3O_2)_2$, was once commonly called "sugar of lead." It has been seriously suggested that slow lead poisoning from the use of lead utensils among the ruling class of Rome was a contributing factor to the decline of the Roman Empire. One symptom of slow lead poisoning is extreme lethargy.

Summary

The electronic configurations of the representative elements are characterized by having either a noble-gas or pseudo-noble-gas core plus one valence electron in Group IA, two valence electrons in Group IIA, and so on up to seven valence electrons in the halogen family.

Similarities in properties of members of the same group are directly related to their having the same number of valence electrons. Gradations within groups are principally a function of the regular gradations in size of the atoms.

Alkali Metals

The Group IA elements are a silvery-white, soft, low-density, highly conducting (both thermal and electrical) group of chemically active metals. They are very active reducing agents, reacting readily with nonmetals to form ionic salts. They react vigorously with water to release hydrogen gas. In their compounds they exhibit only the +1 oxidation state. Both physical and chemical properties are associated with the single loosely held valence electrons.

As a group, the IA atoms are large and increase regularly in size as one goes from top to bottom of the group. This effect shows up in a general decrease in ionization energy of the elements from top to bottom. Francium is the most metallic element in the periodic table.

Alkaline Earths

The Group IIA elements, all of which are metals, have noble-gas cores plus two valence electrons. In their compounds they exhibit only the +2 oxidation state. As a group they are harder, more dense, higher melting, and less homogeneous in their properties than Group IA elements. They also have higher ionization energies and are somewhat less active as reducing agents than Group IA elements. Group IIA elements, however, are still very active metals—calcium, barium, and strontium having reduction potentials that compare favorably with those of the IA elements.

The least typical member of Group IIA is beryllium, the first element in the family. The transition to nonmetallic character is beginning to appear in beryllium, as evidenced by the amphoteric nature of its oxide.

Group IIIA

The Group IIIA elements have three valence electrons, and the usual oxidation state for these elements is +3, Tl(I) being an exception. There is far more variation in properties in this group than in Groups IA and IIA. In going from top to bottom of the IIIA family we encounter changes all the way from nonmetallic to definitely metallic properties; from acidic to amphoteric to basic oxides; and from principally covalent to principally ionic halides.

Aluminum is by far the most abundant and most industrially important element of the group. Its oxide and hydroxide are amphoteric. The relatively small, highly charged Al^{3+} ion hydrolyzes appreciably to produce an acidic solution.

Group IVA

The members of the carbon family have four valence electrons and exhibit a *maximum* oxidation number of four in their compounds. Their properties range all the way from distinctly nonmetallic at the top of the family to metallic at the bottom. Smooth gradations from nonmetallic to metallic character, acidic to basic properties of the oxides, covalent to ionic bonding in their compounds, and preference for the +2 over the +4 oxidation state occur from top to bottom of the family.

Carbon is the "backbone" of the organic world; silicon, of the inorganic. Tin and lead are metals that have been known and used since ancient times. As reducing agents, they are only slightly more active than H_2. Their +4 halides are principally covalent compounds. Their +2 oxides and hydroxides are amphoteric.

Exercises

18-1.
Complete the following table.

NAME	FORMULA	NAME	FORMULA
	RbO_2	Stannous chloride	
	SnO		HCO_3^-
	SnO_2	Magnesium carbonate	
Barium peroxide			H_3BO_3
Calcium chloride		Aluminum chloride	

18-2.
Complete the following equations for the reaction of the alkali metals with oxygen.

(a) $Li + O_2 =$
(b) $Na + O_2 =$
(c) $K + O_2 =$
(d) $Rb + O_2 =$
(e) $Cs + O_2 =$

18-3.
Complete and balance the following chemical reactions involving the alkaline and alkaline earth metals.

(a) $K + H_2O =$
(b) $Na + Cl_2 =$
(c) $CaCO_3 \xrightarrow{\text{heat}}$
(d) $BaO + H_2SO_4 =$
(e) $Ba^{2+} + H_2SO_4 =$
(f) $Ca^{2+} + CO_2 + H_2O =$
(g) $CaO + H_2O =$
(h) $MgCO_3 + H^+ =$
(i) $Ba + H_2O =$

18-4.
Complete and balance the following chemical reactions involving the Group IIIA elements.

(a) $Al + OH^- =$
(b) $B_2O_3 + OH^- =$
(c) $BF_3 + :NF_3 =$
(d) $B_2H_6 + O_2 =$
(e) $Al + Fe_2O_3 \xrightarrow{\text{heat}}$
(f) $Al + Cl_2 =$

18-5.
Complete and balance the following chemical reactions involving the Group IVA elements.

(a) $Pb + H^+ =$
(b) $Pb^{2+} + H_2S =$
(c) $Sn^{2+} + Fe =$
(d) $Sn + OH^- =$
(e) $Si_2H_6 + O_2 =$
(f) $Sn(OH)_2 + OH^- =$
(g) $PbO + OH^- + H_2O =$

18-6.
Write the chemical reactions that occur when an aqueous hydrochloric acid solution is mixed with the following oxides. (*Hint:* Amphoteric and basic oxides neutralize acid.)

(a) Na_2O
(b) BeO
(c) CaO
(d) B_2O_3
(e) Al_2O_3
(f) In_2O_3
(g) CO_2
(h) SnO
(i) PbO

18-7.
Write the chemical reactions that occur when an aqueous sodium hydroxide solution is mixed with the following oxides. (*Hint:* Acidic and amphoteric oxides neutralize base.)

(a) Na_2O
(b) BeO
(c) CaO
(d) B_2O_3
(e) Al_2O_3
(f) Ga_2O_3
(g) CO_2
(h) SnO
(i) PbO

Problems

18-1.
Write the chemical reactions that occur in the following processes:

(a) A mercuric chloride solution is mixed with a stannous chloride solution.
(b) Water saturated with carbon dioxide is run over solid limestone ($CaCO_3$).
(c) Tin tetrachloride is added to water.
(d) Lithium oxide is added to water.
(e) Magnesium metal is burned in oxygen.

18-2.
Outline a method for separating the following mixtures:

(a) Na^+ and Mg^{2+}
(b) SiO_2 and PbO
(c) CaO and Al_2O_3
(d) $BaSO_4$ and $MgSO_4$
(e) BeO and MgO

18-3.
Write the chemical reactions that could be used to make the following conversions:

(a) $Na \rightarrow Na_2O$
(b) $SnO \rightarrow SnO_2$
(c) $CO \rightarrow CO_2$
(d) $MgCl_2 \rightarrow MgCO_3$
(e) $KCl \rightarrow K$
(f) $CaCO_3 \rightarrow Ca(OH)_2$

18-4.
Calculate the pH of the following solutions:

(a) 0.10 *M* KCl. (HCl is a strong acid.)
(b) 0.10 *M* H_3BO_3. (See Appendix E for K_a.)
(c) 0.10 *M* $Ga(NO_3)_3$. [K_a for $Ga(H_2O)_6^{3+}$ is 2.5×10^{-3}.]

18-5.
Calculate the equilibrium constant at 25° C for the reaction

$$CO_2(g) = CO(g) + \frac{1}{2}O_2(g)$$

Then calculate the number of CO molecules in equilibrium with a mole of $CO_2(g)$ at 1 atm pressure and 25° C. See Appendix B for additional data. (Assume that only CO_2 was initially present.)

18-6.
Calculate the equilibrium constant for the reaction

$$CaCO_3(s) + CO_2 + H_2O = Ca^{2+} + 2\,HCO_3^-$$

from the K_{sp} for $CaCO_3$ and K_a for CO_2 (see Appendixes E and F). Then calculate $[Ca^{2+}]$ in a solution that is 0.010 *M* in CO_2 and in equilibrium with $CaCO_3$.

19

The Representative Elements: Groups VA–VIIIA

In this chapter we complete the discussion of the representative elements that was begun in Chapter 18. As in that chapter, we will be interested not only in individual elements, but also in noting characteristics of groups and gradations of properties within and among families.

We are now getting deep into the nonmetal area of the periodic table. Only three of the elements to be discussed in this chapter are classified as metals.

19.1 Group VA: The Nitrogen Family

Group VA consists of nitrogen (N), phosphorus (P), arsenic (As), antimony (Sb), and bismuth (Bi). The first two elements of the group, N and P, are typical nonmetals. The third element, As, exhibits principally a nonmetallic character in the compounds that it forms, but it has some metallic properties in its elemental form. It is classified as a metalloid. The last two members of the family, Sb and Bi, while still exhibiting some nonmetallic characteristics in the higher oxidation states, have definite metallic character in the elemental and low-oxidation states.

Table 19.1 gives the electronic configuration of the elements of Group VA. The valence electrons for this family are thus two *s* and three *p* electrons (s^2p^3), which suggests that the characteristic oxidation states will be -3, $+3$, and $+5$ (see page 68 in Chapter 4). The two most nonmetallic elements, N and P, can *gain* three electrons to form nitride (N^{3-}) and phosphide (P^{3-}) ions. The two most metallic, Sb and Bi, can *lose* three electrons to form Sb^{3+} and Bi^{3+} ions. The $+3$ and -3 states are also achieved by *sharing* three electrons with other elements, $+3$ if the other element is more electronegative, and -3 if it is less electronegative than the Group VA element. The $+5$ oxidation state results from sharing all five electrons with a more electronegative element.

Table 19.2 lists some other properties of the elements of Group VA. Especially note the trend in electronegativity and size of atom. These are the properties that are largely responsible for variations in the chemistry of the elements within a family. Other trends and variations in Group VA elements will become apparent when we consider their compounds.

TABLE 19.1

Electronic Configuration of Group VA Elements

ELEMENT	CONFIGURATION
$_{7}N$	He core + $2s^22p^3$
$_{15}P$	Ne core + $3s^23p^3$
$_{33}As$	Ar core + $3d^{10}4s^24p^3$
$_{51}Sb$	Kr core + $4d^{10}5s^25p^3$
$_{83}Bi$	Xe core + $4f^{14}5d^{10}6s^26p^3$

TABLE 19.2

Properties of Group VA Elements

	N	P	As	Sb	Bi
Atomic number	7	15	33	51	83
Atomic radius (Å)	0.74	1.10	1.21	1.41	1.52
Electronegativity	3.0	2.1	2.0	1.9	1.9
Ionization energy (kjoules/mole)	1402	1011	946	833	703
Melting point (° K)	63	317	sublimes	903	545
Boiling point (° K)	77	553	886[†]	1910	1832
ΔH_{fus} (kjoules/mole)	0.720	0.630	—	19.83	10.90
ΔH_{vap} (kjoules/mole)	5.58	12.38	—	67.8	151.5

[†] Normal sublimation point.

Nitrogen and Its Compounds

Nitrogen (N_2), the first element of Group VA, is a colorless, odorless diatomic gas. It is the most abundant element in the air, composing approximately 79 mole percent of the atmosphere. The element nitrogen is a major constituent of protein and hence is one of the essential elements for both plant and animal life. Very large quantities of nitrogen are used each year in fertilizers, principally as ammonium sulfate [$(NH_4)_2SO_4$], ammonium nitrate (NH_4NO_3), and the organic compound urea

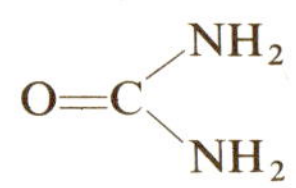

Urea is popular as a lawn fertilizer. Since it is not ionic, it does not draw water out of plant tissue by osmosis (causing lawn "burn") to the same extent that inorganic fertilizers do. Plants do not use urea directly, but over a period of time the urea releases its nitrogen in the form of NH_4^+ by a hydrolysis reaction. The reaction that takes place slowly in the soil is

$$CO(NH_2)_2 + 2\,H_2O = 2\,NH_4^+ + CO_3^{2-}$$

A second major use of nitrogen compounds is in the manufacturing of explosives. Trinitrotoluene (TNT) and glyceryl trinitrate (nitroglycerin) are two well-known nitrogen explosives.

The chemistry of nitrogen is especially interesting because of the wide variety of oxidation states it can assume in its compounds. As can be seen in Table 19.3, all nine integral oxidation states from -3 to $+5$ are represented. The descriptive chemistry that will be presented in this course will be restricted principally to the compounds listed in this table.

Elemental Nitrogen and Nitrogen Fixation

Molecular nitrogen is relatively inert. You may recall from molecular orbital theory that its bond order is three, and its bond dissociation energy is correspondingly high (941 kjoules/mole), which accounts for its stable character. Hence, while nitrogen gas is abundant in the atmosphere, only a small quantity of it is found combined with other elements in the earth's crust. Some mineral deposits of sodium nitrate ($NaNO_3$) are known (the extensive deposits in Chile,

TABLE 19.3

Representative Examples of the Principal Oxidation States of Nitrogen

OXIDATION STATE	COMPOUND OR ION
+5	N_2O_5 (dinitrogen pentoxide) HNO_3 (nitric acid) NO_3^- (nitrate ion)
+4	NO_2 (nitrogen dioxide) N_2O_4 (dinitrogen tetroxide)
+3	HNO_2 (nitrous acid) NO_2^- (nitrite ion)
+2	NO (nitric oxide)
+1	N_2O (dinitrogen monoxide or nitrous oxide)
0	N_2 (molecular nitrogen)
−1	NH_2OH (hydroxylamine)
−2	N_2H_4 (hydrazine)
−3	NH_3 (ammonia) NH_4^+ (ammonium ion) N^{3-} (nitride ion)

for example), but the main source of nitrogen compounds is industrial "fixation" of the N_2 gas in the atmosphere.

The very successful commercial process for "fixing" nitrogen into a compound was developed during World War I in Germany by Fritz Haber. The Haber process is based on the reaction

$$N_2 + 3\,H_2 = 2\,NH_3 \qquad [19.1]$$

This is not only a very simple reaction, but in spite of the high dissociation energy of N_2, the reaction has favorable thermodynamics. For reaction [19.1] at 25°C,

$$\Delta H^0 = -92{,}380 \text{ joules}$$
$$\Delta G^0 = -33{,}280 \text{ joules}$$
$$K = 6.77 \times 10^5$$

The large decrease in free energy accompanying the reaction results in the large equilibrium constant. At equilibrium, the yield of NH_3 would thus be favorable. The kinetics of the reaction, however, at 25° C are very unfavorable. At this temperature, the reaction is immeasurably slow. To increase the rate of the reaction, the temperature is raised and a suitable catalyst is added. However, since the reaction is exothermic (ΔH^0 is negative), raising the temperature forces the point of equilibrium to the left, and the yield of the reaction is decreased. To regain the yield lost from increasing the temperature, the reaction is carried out at high pressures. Since the product occupies a smaller volume (2 moles of gas) than the reactants (4 moles of gas), pressure moves the point of equilibrium to the right. The Haber process is an excellent example of the way thermodynamic and kinetic effects can sometimes be balanced against each other. The usual operating conditions are temperatures in the range of 450–600° C and pressures of 200–600 atmospheres. These conditions result in a satisfactory yield in a reasonable length of time.

Ammonia and Its Salts

Ammonia is a colorless gas with a characteristic pungent odor. It boils at −33.4° C, which, as we saw on pages 139–140 in Chapter 6, is abnormally high when compared with the boiling points of the other hydrides of Group VA. The abnormal behavior is ascribed to the existence of hydrogen bonding in ammonia.

Ammonia is very soluble in cold water, one volume of 20° C water dissolving 700 volumes of NH_3 gas at 1 atmosphere pressure. The aqueous ammonia, $NH_3(aq)$, thus formed partially dissociates to ammonium ion (NH_4^+) and hydroxide ion. Thus, aqueous ammonia is a weak base. The reaction is conveniently written in the Brønsted-Lowry form:

$$NH_3 + H_2O = NH_4^+ + OH^-; \qquad K_b = 1.8 \times 10^{-5} \qquad [19.2]$$

Aqueous ammonia reacts with acid in the same way that other bases do to form salts and water. With a strong acid, the reaction is

$$NH_3 + H^+ = NH_4^+ \qquad [19.3]$$

With a weak acid, such as acetic acid, the neutralization reaction would be written

$$NH_3 + CH_3COOH = CH_3COO^- + NH_4^+ \qquad [19.4]$$

Since aqueous ammonia is a weak base, ammonium salts will hydrolyze to form acidic solutions as shown in the equation

$$NH_4^+ + H_2O = NH_3 + H_3O^+ \qquad [19.5]$$

$$K_h = \frac{K_w}{K_b} = \frac{1.0 \times 10^{-14}}{1.8 \times 10^{-5}} = 5.6 \times 10^{-10} \qquad [19.6]$$

The small value for K_h shows that the extent of the hydrolysis reaction is slight.

EXAMPLE 19.1 Calculate the pH of 0.10 *M* NH_4Cl.

SOLUTION We need to consider only the hydrolysis of NH_4^+; Cl^-, coming from the strong acid HCl, does not hydrolyze.

$$\underset{0.10 - x}{NH_4^+} + H_2O = \underset{x}{NH_3} + \underset{x}{H_3O^+}$$

where x is the concentration of NH_3 produced.

$$\frac{[NH_3][H_3O^+]}{[NH_4^+]} = \frac{(x)(x)}{0.10 - x} = \frac{K_w}{K_b} = 5.6 \times 10^{-10}$$

Solving for x gives

$$x = 7.5 \times 10^{-6} = [H_3O^+]$$

and

$$pH = 5.1$$

If an ammonium salt is treated with a strong base and warmed, the odor of ammonia gas can be readily detected. This procedure provides a common and quick, if not completely reliable, identification of an ammonium salt. The reaction can be thought of as a Brønsted-Lowry acid-base reaction:

$$NH_4^+ + OH^- = H_2O + NH_3(g) \qquad [19.7]$$

Reaction [19.7] can be used for a laboratory preparation of ammonia. When this is done, either very concentrated solutions of strong base and ammonium salts are used, or an analogous reaction can be carried out in the solid state with an ammonium salt. An example is

$$CaO(s) + 2\ NH_4Cl(s) = CaCl_2(s) + 2\ NH_3(g) + H_2O \qquad [19.8]$$

All ammonium salts are unstable at high temperatures. Some examples are

$$NH_4Cl(s) = NH_3(g) + HCl(g) \qquad [19.9]$$

$$(NH_4)_2CO_3(s) = 2\ NH_3(g) + H_2O(g) + CO_2(g) \qquad [19.10]$$

The white film that forms on hydrochloric acid and aqueous ammonia reagent bottles in the laboratory is caused by the reverse reaction of [19.9]. Ammonia gas, in equilibrium with the aqueous ammonia solutions, diffuses to the hydrochloric acid bottles and reacts to form the white

Before leaving the −3 oxidation state of nitrogen, we should mention that nitrogen will combine directly with several very active metals to form ionic compounds in which the nitrogen is in the −3 oxidation state. Examples are Li_3N and Ca_3N_2. The nitride ion (N^{3-}) does not exist in water solution. With its high charge and small size, one would expect it to hydrolyze even more than ammonium chloride solid; hydrogen chloride gas from the hydrochloric acid solutions finds its way to the aqueous ammonia bottles to engage in the same reaction.

oxide ion (O^{2-}). The reaction with water of an ionic nitride is practically complete. The reaction of lithium nitride with water is shown below:

$$Li_3N(s) + 3\,H_2O = 3\,Li^+ + 3\,OH^- + NH_3 \qquad [19.11]$$

Oxides of Nitrogen

NITROUS OXIDE (+1 OXIDATION STATE) Nitrous oxide (N_2O) is a colorless, fairly unreactive gas. It can be prepared by the thermal decomposition of ammonium nitrate. The reaction is carried out below 200° C.

$$NH_4NO_3 = N_2O + 2\,H_2O \qquad [19.12]$$

Caution! Heating NH_4NO_3 to high temperatures can result in an explosion. The N_2O formed in reaction [19.12] decomposes to N_2 and O_2, with the release of 82 kjoules of heat per mole of N_2O decomposed. The great stability of N_2 with its bond energy of 941 kjoules (second only to the C-O bond in carbon monoxide) accounts for the fact that many nitrogen-containing compounds are explosive; that is, they decompose to form N_2 gas and simultaneously liberate large amounts of energy. The explosion is the rapid production and expansion (from heating) of large volumes of gases.

Nitrous oxide is isoelectronic with carbon dioxide and has physical properties somewhat like those of CO_2. For example, the normal boiling point of N_2O is −88° C, which compares well with the normal sublimation point of CO_2 at −78° C. Both molecules have linear structures. N_2O, however, has a slight dielectric constant (0.17 D), whereas CO_2, being symmetrical (Chapter 18), has no dipole moment. The bond lengths of N_2O are shown below:

$$\underset{1.12\,\text{Å}}{N\text{——}N}\underset{1.19\,\text{Å}}{\text{——}O}$$

It may be considered to be a resonance hybrid of the structures

$$:N{\equiv}N{-}\ddot{\underset{..}{O}}: \quad \text{and} \quad :\ddot{N}{=}N{=}\ddot{O}:$$

Nitrous oxide has an interesting physiological effect. It is sometimes used by dentists as a mild anaesthetic. However, if not properly administered, it can cause a mild hysteria in the patient; hence, its common name, "laughing gas."

NITRIC OXIDE Nitric oxide, NO, is a colorless, fairly reactive gas that can be formed, among other ways, by the direct combination of N_2 with O_2 at high temperatures:

$$N_2 + O_2 = 2\,NO \qquad [19.13]$$
$$\Delta H^0_{298} = 181 \text{ kjoules}$$
$$\Delta G^0_{298} = 173 \text{ kjoules}$$

The reaction at room temperature is seen to be thermodynamically unfavorable (ΔG^0_{298} is positive), but since the reaction is endothermic (ΔH^0 is positive), high temperatures favor the formation of the product. NO, as well as other oxides of nitrogen, forms in arcs and sparks taking place in mixtures of N_2 and O_2—for example, lightning discharges in the atmosphere and the combustion reactions in automobile engines.

Nitric oxide reacts readily with atmospheric oxygen to form nitrogen dioxide (NO_2) at ordinary room temperatures. The reaction is

$$2\,NO + O_2 = 2\,NO_2$$

NITROGEN DIOXIDE (NO_2) AND DINITROGEN TETROXIDE (N_2O_4) Nitrogen dioxide (NO_2) is a dark red-brown, highly corrosive gas in which the nitrogen is in the +4 oxidation state. It, together with NO, is a common reduction product in reactions involving +5 nitrogen compounds. Several examples are discussed in the section on nitric acid. NO_2, with its 17 valence electrons, is an "odd" molecule, which accounts for its paramagnetism and probably its color. Nitrogen dioxide dimerizes to form the almost colorless dinitrogen tetroxide, which has no unpaired electrons. The reaction is

$$\underset{\text{dark red-brown}}{2\,NO_2} = \underset{\text{colorless}}{N_2O_4} \qquad [19.14]$$

$$\Delta H^0_{298} = -58.0 \text{ kjoules}$$
$$\Delta G^0_{298} = 5.40 \text{ kjoules}$$

It is evident from the enthalpy change and the stoichiometry of the reaction that the formation of N_2O_4 is favored at low temperatures and high pressures.

The standard free energy of formation of all the oxides of nitrogen are positive at 25° C. Hence, they are all intrinsically unstable with respect to N_2 and O_2 at ordinary laboratory temperatures. They exist only because their decomposition reactions are very slow.

DINITROGEN PENTOXIDE (N_2O_5) Dinitrogen pentoxide (+5 oxidation state for nitrogen) is a white solid that can be prepared by dehydration of pure HNO_3. It sublimes slightly above room temperature, and its vapor decomposes at elevated temperatures. The decomposition reaction is

$$2\,N_2O_5 = 4\,NO_2 + O_2 \qquad [19.15]$$

The kinetics of equation [19.15] have been extensively studied. See page 323 in Chapter 12 for the proposed mechanism of the reaction.

Dinitrogen pentoxide is the most acidic of the nitrogen oxides. This is predictable since the nitrogen is in its highest oxidation state (see pages 473–474 in Chapter 16). Dinitrogen pentoxide is the acid anhydride of nitric acid:

The anhydride of an acid is the oxide that dissolves in water to give the acid.

$$N_2O_5 + H_2O = 2\,H^+ + 2\,NO_3^- \qquad [19.16]$$

Nitric Acid and Nitrates

Nitric acid is prepared commercially by the oxidation of ammonia. The first step in the process is a high-temperature catalytic oxidation of ammonia with oxygen to nitric oxide. The nitric oxide is then readily oxidized to nitrogen dioxide by atmospheric oxygen at room temperature. The third step consists of bubbling the NO_2 into water. The nitrogen dioxide disproportionates to form nitric oxide and nitric acid. A summary of these steps, known as the *Ostwald process*, is given below:

$$(1)\quad 4\,NH_3 + 5\,O_2 \underset{500\text{–}1000^\circ\,C}{\overset{\text{catalyst}}{=}} 4\,NO + 6\,H_2O \qquad [19.17]$$

$$(2)\quad 2\,NO + O_2 = 2\,NO_2 \qquad [19.18]$$

$$(3)\quad 3\,NO_2 + H_2O = 2\,H^+ + 2\,NO_3^- + NO \qquad [19.19]$$

The nitric oxide produced in the third step can be fed back into the process at step (2).

Nitric acid is a strong acid. In water solution it is considered to be fully ionized; hence, we usually write its formula in the ionized form as in equation [19.19]. Nitric acid has all the typical properties of an acid, such as reaction with (a) a base, (b) a carbonate, (c) a metal oxide, and (d) an active metal. Examples are

$$H^+ + OH^- = H_2O$$
$$H^+ + NH_3 = NH_4^+$$
$$2\,H^+ + CO_3^{2-} = H_2O + CO_2(g)$$
$$H^+ + HCO_3^- = H_2O + CO_2(g)$$
$$2\,H^+ + MgO = Mg^{2+} + H_2O$$
$$2\,H^+ + Zn = Zn^{2+} + H_2(g)$$

In addition, nitric acid (that is, the nitrate ion in acidic solution) is a moderately strong oxidizing agent, as shown by its standard reduction potentials:

	$\mathscr{E}^0$
$NO_3^- + 2\,H^+ + e^- = NO_2(g) + H_2O$	0.79
$NO_3^- + 4\,H^+ + 3\,e^- = NO(g) + 2\,H_2O$	0.96
$NO_3^- + 10\,H^+ + 8\,e^- = NH_4^+ + 3\,H_2O$	0.87

The positive voltages show that there is a driving force for the reactions; that is, NO_3^- in acid solution has a fairly strong tendency to be reduced to NO_2, NO, and NH_4^+. Other reduction states are possible, but the three listed above are the most commonly encountered.

Typical examples of reactions that involve nitric acid as an oxidizing agent are discussed below. Copper, silver, and mercury, which are below hydrogen in oxidation potential and, therefore, are not oxidized by H^+ (see pages 292–293 in Chapter 11), *are* oxidized by nitric acid. A variety of reduction products results, depending upon the reducing agent and the conditions. With concentrated nitric acid, the principal reduction product is nitrogen dioxide:

$$Cu(s) + 4\,H^+ + 2\,NO_3^- = Cu^{2+} + 2\,NO_2(g) + 2\,H_2O$$
$$Ag(s) + 2\,H^+ + NO_3^- = Ag^+ + NO_2(g) + H_2O$$
$$Hg(l) + 4\,H^+ + 2\,NO_3^- = Hg^{2+} + 2\,NO_2(g) + 2\,H_2O$$

In dilute nitric acid, the principal reduction product for reaction with Cu, Ag, and Hg is NO. As an example, consider the reaction with Cu:

$$3\,Cu(s) + 8\,H^+ + 2\,NO_3^- = 3\,Cu^{2+} + 2\,NO(g) + 4\,H_2O$$

Dilute nitric acid with an active metal, such as Zn or Al, is reduced all the way to the -3 state:

$$4\,Zn(s) + 10\,H^+ + NO_3^- = 4\,Zn^{2+} + NH_4^+ + 3\,H_2O$$

A convenient way of summarizing electrode-potential data for an element and its compounds is the reduction-potential diagram, an example of which is shown below (in acid solution).

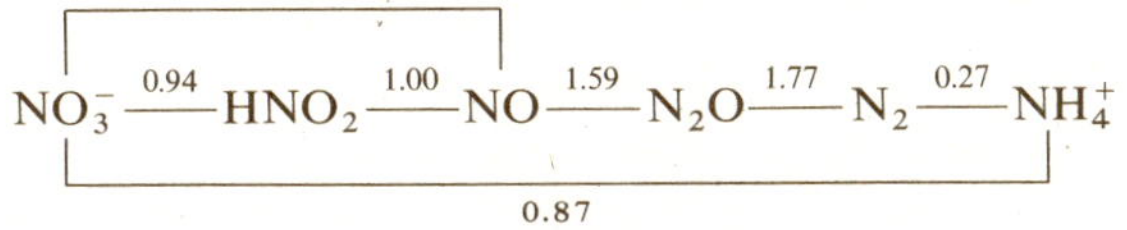

A line connecting two species in the diagram gives the reduction potential for the half-reaction between the two species. For example, one sees at a glance that the reduction of NO_3^- to NO has a standard potential of 0.96 volts, and that oxidation of NH_4^+ to N_2 has an $\mathscr{E}^0$ of -0.27 volts.

To obtain $\mathscr{E}^0$ data for a reaction between two species not directly connected by a line (for example, N_2O to NH_4^+), it would be necessary to go through the intermediate species, N_2. To do this, one would need to add the ΔG^{0}'s for the intermediate reactions (N_2O to N_2, and N_2 to NH_4^+) and calculate $\mathscr{E}^0$ for the desired reaction from the sum of the ΔG^{0}'s (see page 289 in Chapter 11).

The reduction-potential diagram for nitrogen compounds and ions in basic solution is

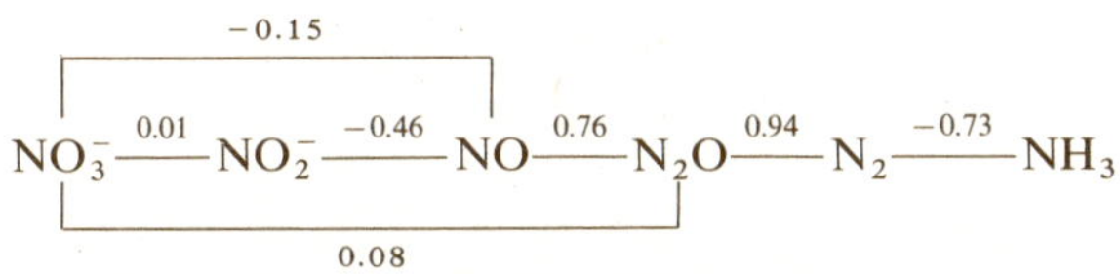

The nitrate ion is seen to be a much weaker oxidizing agent in basic solution: For example, $\mathscr{E}^0$ for NO_3^- to NO in basic solution is -0.15. This is to be compared with $\mathscr{E}^0 = 0.96$ in acid solution. The greater oxidizing ability of NO_3^- in acid solution is easily accounted for by examining the balanced half-reactions for reduction of NO_3^- to NO. In acid solution, the balanced reaction is

$$NO_3^- + 4\,H^+ + 3\,e^- = NO + 2\,H_2O$$

Since H^+ appears on the left-hand side of the equation, the higher the concentration of H^+, the farther to the right the point of equilibrium of the reaction is shifted. In a basic solution, the H^+ concentration is very low, and the driving force of the reaction to the right is diminished.

Nitrous Acid and Nitrites

Nitrous acid (HNO_2) can be formed by adding N_2O_3 (or an equivalent mixture of NO and NO_2) to cold water. Pure HNO_2 is too unstable to be isolated, but it does exist in water solution; however, even in dilute solutions it undergoes measurable disproportionation to nitric acid and nitric oxide. The instability of nitrous acid is predictable from $\mathscr{E}^0$ data. Consider the standard reduction-potential diagram given below, which is part of the more complete diagram given in the previous section:

$$NO_3^- \overset{0.94}{\text{——}} HNO_2 \overset{1.00}{\text{——}} NO$$

We see that HNO_2 can be oxidized to NO_3^- (acid solution) with a standard potential of -0.94 V, and it can be reduced to NO with a potential of 1.00 V.

Since the sum of these two potentials (one written as an oxidation, the other as a reduction) is positive (+0.06 V), disproportionation is favorable. The balanced reactions are

Standard reduction-potential diagrams are convenient for predicting disproportionation reactions. Any species intermediate in oxidation number will disproportionate if the $\mathscr{E}^0$ on its right is more positive than the $\mathscr{E}^0$ on its left.

	$\mathscr{E}^0$
$2HNO_2 + 2\,H^+ + 2\,e^- = 2\,NO + 2\,H_2O$	1.00
$HNO_2 + H_2O = NO_3^- + 3\,H^+ + 2\,e^-$	−0.94
$3\,HNO_2 = NO_3^- + H^+ + 2\,NO + H_2O$	+0.06

The salts of nitrous acid are called nitrites. The nitrite ion in basic solution is considerably more stable than nitrous acid. This fact is also predictable from the standard potential diagram for nitrogen compounds in basic solution. A part of the more complete diagram given in the previous section is shown below:

$$NO_3^- \xrightarrow{0.01} NO_2^- \xrightarrow{-0.46} NO$$

Since the sum of the potentials (one written as an oxidation, the other as a reduction) is $(-0.01) + (-0.46) = -0.47$, disproportionation of the +3 nitrogen does not take place in basic solution. The thermal stability of NO_2^- is demonstrated by the fact that alkali metal nitrites can be heated to their melting points without significant decomposition.

As is evident from their intermediate position in the reduction-potential diagrams, nitrous acid and nitrites may act as either oxidizing or reducing agents. Nitrous acid is a moderately strong oxidizing agent, its reduction potential to nitric oxide being 1.00 volt. As a reducing agent, nitrous acid is relatively weak ($\mathscr{E}^0$ for HNO_2—NO_3^- is −0.94 V). Oxidizing agents as strong as MnO_4^-, $Cr_2O_7^{2-}$, and Cl_2 (all in acid solution), however, are capable of oxidizing HNO_2 to NO_3^-.

Phosphorus and Its Compounds

Phosphorus, the second member of Group VA, makes up 0.14% of the earth's crust and is the twelfth most abundant element. It is chemically too active to be found in nature in the elemental state, the only member of Group VA that is not. It occurs principally in the form of hydrated calcium phosphate [$Ca_3(PO_4)_2 \cdot H_2O$] and related minerals high in $Ca_3(PO_4)_2$. Phosphorus, like nitrogen, is essential for both plant and animal life. Large quantities of phosphorus, mostly in the form of calcium dihydrogen phosphate, are used as fertilizers. $Ca_3(PO_4)_2$ is the main constituent of teeth and bones, and organic phosphorus compounds play a vital role in many biochemical processes.

Other phosphate minerals are apatite [$Ca_3(PO_4)_2 \cdot CaF_2$] and phosphorite [$3Ca_3(PO_4)_2 \cdot Ca(OH)_2$]. The enamel of teeth is an apatite-like mineral.

Table 19.4 lists representative compounds of the principal oxidation states of phosphorus.

Elemental Phosphorus

Elemental phosphorus is obtained from $Ca_3(PO_4)_2$ by treatment with silica and coke at high temperatures in electrical furnaces. The reactions are

$$2\,Ca_3(PO_4)_2 + 6\,SiO_2 = 6\,CaSiO_3 + P_4O_{10}$$

TABLE 19.4

Representative Examples of the Principal Oxidation States of Phosphorus

OXIDATION STATE	COMPOUND OR ION
+5	P_4O_{10} (phosphorus pentoxide) H_3PO_4 (phosphoric or orthophosphoric acid) $H_2PO_4^-$ (dihydrogen phosphate ion) HPO_4^{2-} (monohydrogen phosphate ion) PO_4^{3-} (phosphate ion or orthophosphate ion)
+3	P_4O_6 (phosphorus trioxide) H_3PO_3 (phosphorous acid) PCl_3 (phosphorus trichloride)
0	P_4 (elemental phosphorus)
−3	PH_3 (phosphine) PH_4^+ (phosphonium ion) P^{3-}(phosphide ion)

Figure 19.1
Structure of the P_4 molecule. A phosphorus atom occupies each corner of a tetrahedron.

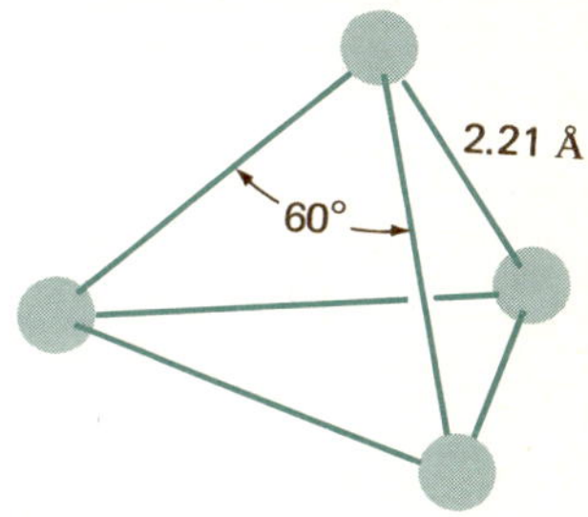

and

$$P_4O_{10} + 10\ C = P_4 + 10\ CO$$

The $CaSiO_3$ collects at the bottom of the furnace as a molten slag. The P_4 and CO gases leave at the top of the furnace, where the phosphorus is then condensed to solid white phosphorus under water.

There are at least three allotropic forms of phosphorus. The white form produced in the process described above consists of discrete tetratomic molecules, P_4, whose structure is shown in Figure 19.1. White phosphorus is soft, low-melting (44.1° C), and highly reactive. It is insoluble in water and is usually stored under water to prevent its spontaneous ignition in atmospheric oxygen.

Red phosphorus, a second allotrope, is obtained by heating the white form to about 250° C in the absence of oxygen. The properties of the red form are quite different from those of the white. The red form is a polymeric material in which the phosphorus atoms are joined in a three-dimensional network, the details of which are not well established. It does not melt till 600° C and is stable toward atmospheric oxygen at room temperatures. It undergoes the same reactions as the white form, but only at considerably higher temperatures.

Exposure to light also causes a slow conversion of white to red phosphorus. White phosphorus that has been exposed to light for a period of time usually has a yellow tinge to it probably caused by small amounts of red phosphorus present as an impurity.

A third, less frequently encountered, allotrope called black phosphorus can be produced by subjecting the white form to high pressures. It is the least reactive of the three phosphorus allotropes.

Phosphine

Phosphorus forms phosphides by direct combination with the active Group IA and IIA metals in much the same way that nitrogen does. The ionic phosphides react with water to form phosphine (PH_3), again analogous to the reactions of nitrides with water to form NH_3:

$$Ca_3P_2(s) + 6\,H_2O = 2\,PH_3(g) + 3\,Ca(OH)_2(s)$$

Phosphine is less soluble in water than ammonia and is a very much weaker base. Like ammonia, phosphine combines with gaseous HCl to form an ionic solid containing the phosphonium ion, PH_4^+:

$$PH_3(g) + HCl(g) = PH_4Cl(s)$$

PH_4Cl, however, when added to water, hydrolyzes almost completely to give PH_3 and hydrochloric acid:

$$PH_4Cl(s) + H_2O = PH_3(g) + H_3O^+ + Cl^-$$

Phosphoric Acid and Phosphates

Compounds in which the phosphorus is in the +5 oxidation state are by far the most important of the phosphorus compounds. When phosphorus is burned in an excess of oxygen, the principal product is P_4O_{10}, called phosphorus pentoxide, a name that goes back to a time when its molecular formula was thought to be P_2O_5. The reaction with abundant oxygen is

$$P_4 + 5\,O_2 = P_4O_{10}$$

With limited oxygen, P_4O_6 is formed. Figure 19.2 shows the structures of these phosphorus oxides.

Phosphorus pentoxide is the acid anhydride of orthophosphoric acid (more commonly referred to as *phosphoric acid*):

$$P_4O_{10} + 6\,H_2O = 4\,H_3PO_4$$

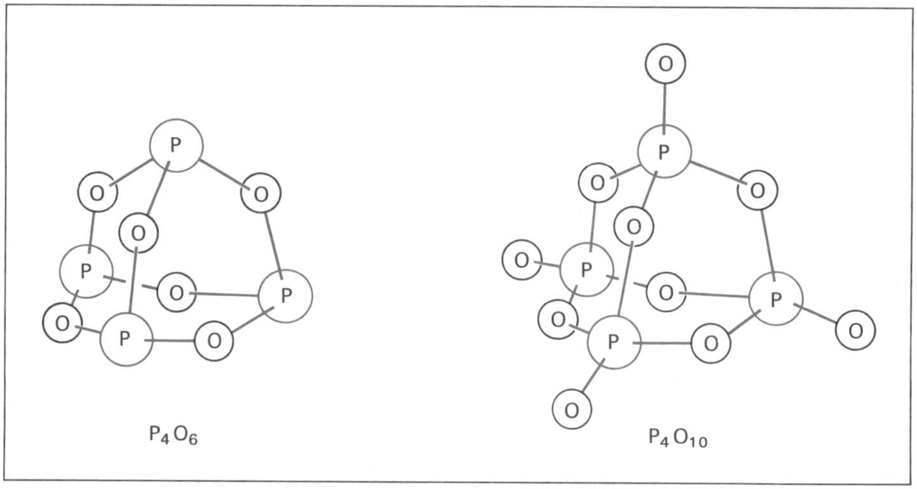

Figure 19.2 *Structures of* P_4O_6 *and* P_4O_{10}. P_4O_6 can be thought of as derived from P_4 by placing an oxygen atom between each pair of phosphorus atoms. In P_4O_{10} an additional oxygen is placed on each phosphorus atom.

Phosphoric acid is a weak, triprotic acid, whose electron-dot formula is

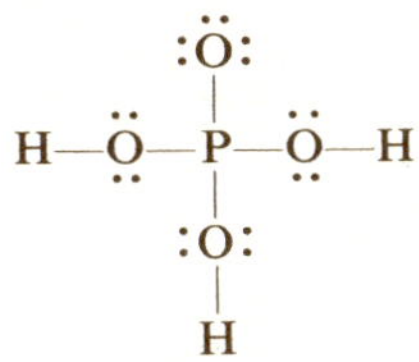

Note that there are three covalent and one coordinate covalent bonds, and all of the ionizable hydrogens are connected to the central phosphorus atom through an oxygen atom. The structure of H_3PO_4 is tetrahedral, slightly distorted because three of the corners contain an H atom and the fourth does not.

The equilibrium constants for the three successive ionization reactions of phosphoric acid are

	K_a	
$H_3PO_4 = H^+ + H_2PO_4^-$	7.5×10^{-3}	[19.20a]
$H_2PO_4^- = H^+ + HPO_4^{2-}$	6.8×10^{-8}	[19.20b]
$HPO_4^{2-} = H^+ + PO_4^{3-}$	4×10^{-13}	[19.20c]

Thus, H_3PO_4, $H_2PO_4^-$, and HPO_4^{2-} are all seen to be weak acids. It is predictable that a second ionization, equation [19.20b], should be weaker than a first, equation [19.20a], and a third reaction, equation [19.20c], weaker than a second. In the first ionization, a proton is breaking away from a neutral molecule, but in the second and third ionizations, a proton (+ charge) has to break away from an already negatively charged ion. A difference of a factor of 10^5 between the K_a's for two successive ionizations in polyprotic acids is fairly common. This rule is seen to be roughly followed in the ionization data given above.

Since phosphoric acid is triprotic, three series of salts result from its neutralization. Examples and their names are

NaH_2PO_4	sodium dihydrogen phosphate
Na_2HPO_4	disodium monohydrogen phosphate
Na_3PO_4	trisodium phosphate (normal sodium phosphate)

Salts such as NaH_2PO_4 and Na_2HPO_4, which still contain ionizable hydrogen, are appropriately called acid-salts. Mixtures of the acid phosphates and normal phosphate are commonly used as buffer systems. They are commercially available in high purity; therefore, solutions of the desired concentrations are easily prepared by direct weighing. The approximately 10^5 factor difference between K_a's of successive ionizations allows one to prepare effective buffers over a wide pH range.

EXAMPLE 19.2 Describe how to prepare a phosphate buffer of (a) pH = 7.0 and (b) pH = 12.0, using the sodium salts of phosphoric acid.

SOLUTION

(a) The flattest portion of a titration curve (see page 257 in Chapter 10) of a weak acid occurs at the point of half-neutralization. Near this point the change in pH with amount of base added is smallest; thus, the buffering effect is greatest. At the point of half-neutralization, $[HA] \cong [A^-]$ and, as seen from the argument below, $K_a = [H^+]$:

$$HA = H^+ + A^-$$

$$K_a = \frac{[H^+][A^-]}{[HA]}$$

$$[A^-] = [HA]$$

$$K_a = [H^+]$$

Hence, for best buffering capacity, choose an acid whose K_a is near the hydrogen ion concentration of the buffer desired. For preparing a buffer of pH = 7.00, the acid salt $H_2PO_4^-$ ($K_a = 6.8 \times 10^{-8}$) is excellent.

$$H_2PO_4^- = H^+ + HPO_4^{2-}$$

$$\frac{[H^+][HPO_4^{2-}]}{[H_2PO_4^-]} = K_a$$

$$\frac{[HPO_4^{2-}]}{[H_2PO_4^-]} = \frac{K_a}{[H^+]} = \frac{6.8 \times 10^{-8}}{1.0 \times 10^{-7}} = 0.68$$

Therefore, mix Na_2HPO_4 and NaH_2PO_4 in the ratio of 0.68 moles of the former to 1 mole of the latter, and add water.

(b) Choose HPO_4^{2-} ($K_a = 4 \times 10^{-13}$) for the acid.

$$HPO_4^{2-} = H^+ + PO_4^{3-}$$

$$\frac{[PO_4^{3-}]}{[HPO_4^{2-}]} = \frac{K_a}{[H^+]} = \frac{4 \times 10^{-13}}{1.0 \times 10^{-12}} = 0.40$$

Therefore, mix Na_3PO_4 with Na_2HPO_4 in the ratio of 0.40 moles of the former to one of the latter, and add water.

As might be predicted, dihydrogenphosphates (singly charged, large anions) are fairly soluble compounds. Monohydrogen, and especially normal phosphates, with their multiple-charged anions are, on the other hand, quite insoluble. The "rock phosphate" (phosphate ore) found in nature is principally Ca_3PO_4 and is too insoluble to be useful as a fertilizer. It is easily converted, however, to the soluble $Ca(H_2PO_4)_2$ by treatment with sulfuric acid:

$$Ca_3(PO_4)_2(s) + 2\,H^+ + 2\,HSO_4^- = Ca^{2+} + 2\,H_2PO_4^- + 2\,CaSO_4(s)$$

The oxidation-reduction characteristics of phosphorus compounds are summarized in the following standard reduction-potential diagrams:

$$H_3PO_4 \xrightarrow{-0.28} H_3PO_3 \xrightarrow{-0.50} P_4 \xrightarrow{-0.06} PH_3 \qquad \text{(acidic solution)}$$

$$PO_4^{3-} \xrightarrow{-1.12} HPO_3^{2-} \xrightarrow{-1.73} P_4 \xrightarrow{-0.89} PH_3 \qquad \text{(basic solution)}$$

It is readily apparent that the phosphorus compounds are poor oxidizing agents (large negative reduction potentials). This is especially true in basic solution. It is also apparent that those phosphorus compounds that are not already in the highest (+5) oxidation state are easily oxidized (large positive oxidation potentials), which makes them good reducing agents.

Phosphorous Acid and Related Compounds

Phosphorus trioxide (P_4O_6) and phosphorus trihalides (PX_3) can be prepared by direct combination of the elements. Representative equations are

$$P_4 + 3\,O_2 = P_4O_6$$
$$P_4 + 6\,Cl_2 = 4\,PCl_3$$

In an excess of O_2 or chlorine, the +5 compounds P_4O_{10} and PCl_5 result.

Reaction of either PX_3 or P_4O_6 with water produces phosphorous acid:

$$PX_3 + 3\,H_2O = H_3PO_3 + 3\,HX(g)$$
$$P_4O_6 + 6\,H_2O = 4\,H_3PO_3$$

It may come as a surprise to learn that H_3PO_3 is a *diprotic* acid; that is, only two of the hydrogen atoms are bonded to oxygen atoms and capable of ionizing. The third hydrogen atom is bonded directly to the central phosphorus atom. The electron-dot formula is

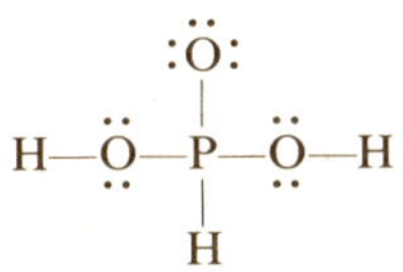

There is an oxygen acid of +1 phosphorus, H_3PO_2 (hypophosphorous acid), that has only one ionizable H. To account for its being monoprotic, we assign to it the following electron-dot configuration:

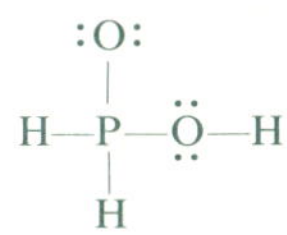

Its structure is a distorted tetrahedron. The ionization constants for phosphorous acid are

$$H_3PO_3 = H^+ + H_2PO_3^- \qquad K_1 = 5 \times 10^{-2}$$
$$H_2PO_3^- = H^+ + HPO_3^{2-}; \qquad K_2 = 7 \times 10^{-7}$$

Both ionizations are seen to be weak. The salts of phosphorous acid are called phosphites.

We have previously observed that phosphorous acid and the phosphites are poor oxidizing agents but good reducing agents. An example of phosphorous acid acting as a reducing agent is its reaction with permanganate ion.

$$5\,H_3PO_3 + 2\,MnO_4^- + 6\,H^+ = 5\,H_3PO_4 + 2\,Mn^{2+} + 3\,H_2O$$

Arsenic and Its Compounds

Elemental Arsenic

Arsenic is not an abundant element, occurring only to the extent of 5×10^{-4} percent in the earth's crust. Its ores are principally in the form of sulfides. The sulfide is converted to the oxide by heating in air, and then reduced to the element with coke:

$$2\ As_2S_3 + 9\ O_2 = As_4O_6 + 6\ SO_2$$
$$As_4O_6 + 6\ C = As_4 + 6\ CO$$

Arsenic is intermediate between the two nonmetals, nitrogen and phosphorus, and the two metals, antimony and bismuth, in Group VA. In many of its compounds, arsenic is similar to phosphorus. In its elemental form its partial metallic character becomes evident. The more stable of the two known allotropes of arsenic is referred to as the metallic or gray form. It has characteristic metallic luster and is a fair conductor of heat and electricity.

In the gas phase, arsenic exists as discrete tetratomic molecules, As_4, with the arsenic atoms arranged at the four corners of a tetrahedron as in P_4.

Compounds of Arsenic

MINUS 3 OXIDATION STATE The principal oxidation states of arsenic are, as in phosphorus, -3, $+3$, and $+5$. Representing the -3 oxidation states are the arsenides. As with nitrogen and phosphorus, these form by direct combination of arsenic with Group IA and IIA metals. The bonding, however, is more covalent. No ion analogous to the nitride (N^{3-}) is present in the arsenides.

The arsenides react with water in a manner similar to that of the nitrides and phosphides:

$$Na_3As(s) + 3\ H_2O = 3\ Na^+ + 3\ OH^- + AsH_3(g)$$

The gaseous hydride AsH_3 is known as arsine. As would be predicted, its properties are even less basic than are phosphine's.

PLUS 3 OXIDATION STATE Representative of the $+3$ oxidation state are the trihalides (AsX_3), arsenious oxide (As_4O_6), and arsenious acid (H_3AsO_3). Arsenic and the halogens combine directly to form trihalides. The trihalides decompose readily in water to yield arsenious acid. The reactions, using the chloride as the example, are

The only pentahalide of arsenic that exists is the fluoride, AsF_5. In the synthesis of AsF_3, an excess of halogen is avoided to prevent formation of some AsF_5.

$$As_4 + 6\ Cl_2 = 4\ AsCl_3$$
$$AsCl_3 + 3\ H_2O = H_3AsO_3 + 3\ HCl(g)$$

The reactions are seen to be analogous to those for phosphorus. Arsenious acid is an extremely weak acid:

$$H_3AsO_3 = H^+ + H_2AsO_3^-; \qquad K_a = 6 \times 10^{-10}$$

The neutralization of arsenious acid with a strong base can be written as

$$H_3AsO_3 + OH^- = H_2O + H_2AsO_3^-$$

Arsenious acid, like phosphorous acid, is diprotic. The second acid constant is 3×10^{-14}.

Arsenious acid and arsenites are reducing agents, as shown in the standard reduction-potential diagram below:

$$H_3AsO_4 \xrightarrow{0.559} H_3AsO_3 \xrightarrow{0.25} As \xrightarrow{-0.60} AsH_3 \qquad \text{(acidic solution)}$$

$$AsO_4^{3-} \xrightarrow{-0.67} H_2AsO_3^- \xrightarrow{-0.68} As \xrightarrow{-1.43} AsH_3 \qquad \text{(basic solution)}$$

Arsenious acid is commonly used in analytical chemistry to determine the concentration of an oxidizing agent. The usual procedure is to weigh out a sample of arsenious oxide (which can be obtained in high purity and is convenient to weigh), dissolve it in a strong base, acidify, and titrate with the oxidizing agent. The following series of reactions are those that apply when permanganate ion solution is standardized with arsenious oxide:

$$As_2O_3 + 2\,OH^- + H_2O = 2\,H_2AsO_3^-$$
$$H_2AsO_3^- + H^+ = H_3AsO_3$$
$$5\,H_3AsO_3 + 2\,MnO_4^- + 6\,H^+ = 5\,H_3AsO_4 + 2\,Mn^{2+} + 3\,H_2O$$

PLUS 5 OXIDATION STATE—ARSENIC ACID Arsenic acid, H_3AsO_4, is the most important of the +5 oxidation-state compounds of arsenic. It is a triprotic, weak acid that resembles phosphoric acid in its acidic properties. Step-by-step neutralization of arsenic acid with a base results in three salts: dihydrogen-, monohydrogen-, and normal arsenates, analogous to the phosphates. The ionization constants of H_3AsO_4 are relatively close to those of H_3PO_4:

$$H_3AsO_4 = H^+ + H_2AsO_4^-; \qquad K_1 = 5.6 \times 10^{-3}$$
$$H_2AsO_4^- = H^+ + HAsO_4^{2-}; \qquad K_2 = 1.7 \times 10^{-7}$$
$$HAsO_4^{2-} = H^+ + AsO_4^{3-}; \qquad K_3 = 3 \times 10^{-12}$$

One important way in which phosphoric and arsenic acid differ, however, is in their oxidizing ability. As can be seen in the reduction-potential diagrams that were given earlier in this chapter, arsenic acid is a moderately strong oxidizing agent ($\mathscr{E}^0$ for arsenic acid to arsenious acid = 0.56), whereas phosphoric acid is a very poor oxidizing agent ($\mathscr{E}^0$ for phosphoric acid to phosphorous acid = −0.28).

A most important difference in the properties of the salts of the two acids is that while phosphates are necessary for life, arsenates are highly poisonous.

Antimony (Sb), Bismuth (Bi), and Their Compounds

Antimony and bismuth show definite metallic properties not only in their elemental states, but also in the properties of the compounds that they form. For example, their sesquioxides, Sb_4O_6 and Bi_2O_3, dissolve in strong acids to form

+3 cations, Sb^{3+} and Bi^{3+}. As might be expected from their high charge, however, the cations hydrolyze extensively. Antimonyl (SbO^+) and bismuthyl (BiO^+) ions result. These can be looked upon as being formed by the loss of a molecule of water from the product of a second-stage hydrolysis of the trivalent ions. Using Bi^{3+} as an example, we have

$$Bi^{3+} + H_2O = BiOH^{2+} + H^+$$
$$BiOH^{2+} + H_2O = Bi(OH)_2^+ + H^+$$
$$Bi(OH)_2^+ = BiO^+ + H_2O$$

The oxide of +3 antimony (Sb_4O_6) reacts with strong bases as well as strong acids, thus qualifying as an amphoteric oxide. Its reaction with an acid can be written as

$$Sb_4O_6 + 4\,H^+ = 4\,SbO^+ + 2\,H_2O$$

and its reaction with a base as

$$Sb_4O_6 + 4\,OH^- = 4\,SbO_2^- + 2\,H_2O$$

The antimonite ion, SbO_2^-, in solution is probably more correctly written as $Sb(OH)_4^-$, analogous to $Al(OH)_4^-$. Loss of two molecules of water from $Sb(OH)_4^-$ gives SbO_2^-, which is the formula found for the anion in dry solid salts. Arsenious oxide also shows amphoteric properties, but its acidic character is much more pronounced than its basic character. In antimonous oxide, the basic character is dominant. The sesquioxide of bismuth (Bi_2O_3) is properly classified as a basic oxide; it does not react with strong bases. These gradations from acidic to basic character in the oxides are, of course, in line with the gradations in the nonmetal to metal character of the elements as one moves down the family of elements.

The fact that Bi_2O_3 does not react with a base while Bi_2O_5 does is another example (see page 473 in Chapter 16) of the acidic character of an oxide increasing with an increase in oxidation number of the metal.

The pentoxides of antimony and bismuth (Sb_2O_5 and Bi_2O_5) exist, but corresponding acids analogous to phosphoric and arsenic acids do not. However, antimonates [$Sb(OH)_6^-$] and bismuthates (BiO_3^-) can be prepared by treating the corresponding pentoxides with strong bases. Antimonates are fair oxidizing agents, and bismuthates are powerful ones, stronger than permanganate ion and about the same strength as cerium (IV) ion. In acid solution, the +5 bismuth-bismuthyl couple has a standard reduction potential of 1.6 V. By comparison, the permanganate ion-manganous ion couple is 1.51 V.

It is apparent that +5 bismuth in acid solution will oxidize manganous ion to permanganate. This fact is the basis for a qualitative test for manganous ion. If a purple color (MnO_4^-) appears after $NaBiO_3$ is added to an acidified solution of the ion being tested, manganous ion is presumed to be present.

In general, the stability of the +5 oxidation state of the Group VI elements decreases from top to bottom of the family. For example, +5 nitrogen compounds (nitric acid and the nitrates) are more stable than the +3 nitrogen compounds (nitrous acid and the nitrites); however, as one moves down through P, As, and Sb, a gradual reversal takes place, and it is the +3 bismuth compounds that are stable, and the +5 bismuth compounds that are unstable. This is what one would predict knowing that the metallic character increases from top to bottom. Plus 5 and higher oxidation states of metals frequently are strong oxidizing agents: for example, V(V), Cr(VI), and Mn(VII).

19.2 Group VIA: The Oxygen Family

The Group VIA elements are oxygen (O), sulfur (S), selenium (Se), tellurium (Te), and polonium (Po). As a group they are more nonmetallic than Group VA. For example, oxygen, the first member of the group, is second only to fluorine in electronegativity. As in the other A-Group elements, the metallic character increases from top to bottom of the group, and polonium, the last member of the family, is a metal.

The electronic configurations of Group VIA elements are given in Table 19.5. From this table the common oxidation states for elements of this family would be predicted to be -2, $+4$, and $+6$. The -2 state is achieved either by accepting two electrons from a metal to form a doubly charged negative ion, or by sharing electrons with a less electronegative nonmetal or with hydrogen. The $+4$ and $+6$ states are obtained by sharing either four or all six valence electrons with a more electronegative element, usually oxygen.

Several physical properties of the Group VIA elements are listed in Table 19.6.

TABLE 19.5

Electronic Configuration of Group VIA Elements

ELEMENT	CONFIGURATION
$_{8}O$	He core + $2s^22p^4$
$_{16}S$	Ne core + $3s^23p^4$
$_{34}Se$	Ar core + $3d^{10}4s^24p^4$
$_{52}Te$	Kr core + $4d^{10}5s^25p^4$
$_{84}Po$	Xe core + $4f^{14}5d^{10}6s^26p^4$

TABLE 19.6

Properties of Group VIA Elements

	O	S	Se	Te	Po
Atomic number	8	16	34	52	84
Ionization energy (kjoules/mole)	1313	999	941	870	812
Atomic radius (Å)	0.74	1.04	1.17	1.37	1.64
Melting point (°K)	54	392	490	723	527
Boiling point (°K)	90	718	958	1260	1235
Electronegativity	3.5	2.5	2.4	2.1	2.0
ΔH_{fus} (kjoules/mole)	0.445	1.23	5.44	17.5	12.6
ΔH_{vap} (kjoules/mole)	6.82	10.5	26.3	50.6	60.2

Oxygen and Its Compounds

Table 19.7 lists representative compounds and ions of the common oxidation states of oxygen.

Silicon is the second most abundant element (approximately 25 percent by weight); aluminum is third (approximately 8.5 percent); iron is fourth (approximately 5 percent). Calcium, magnesium, sodium, and potassium each constitute 1–2 percent of the earth's crust.

Oxygen constitutes 21 mole percent of the earth's atmosphere and approximately 50 percent by weight of the earth's crust. Thus, it is by far the most abundant element on earth.

Oxygen is a very active element, combining with almost every other element, including some of the noble gases. The stable molecular form of oxygen, O_2, is a colorless, odorless, diatomic gas. Its electronic configuration has been referred to earlier (Chapter 14). Magnetic measurements show the molecule to have two unpaired electrons, and its bond energy is of the magnitude expected of a double bond. You will recall that we were unable to write an electron-dot formula for O_2 that would be consistent with these facts, but that the molecular orbital theory accounts for its properties very well.

The molecular orbital descriptions for O_2, as well as for the two molecule ions, O_2^- and O_2^{2-}, that we will discuss later, are given below:

FORMULA	MOLECULAR ORBITAL DESCRIPTION	BOND ORDER	BOND LENGTH
O_2	KK, $(\sigma_{2s}^b)^2(\sigma_{2s}^*)^2(\sigma_z^b)^2(\pi_x^b)^2(\pi_y^b)^2(\pi_x^*)^1(\pi_y^*)^1$	2	1.21
O_2^-	KK, $(\sigma_{2s}^b)^2(\sigma_{2s}^*)^2(\sigma_z^b)^2(\pi_x^b)^2(\pi_y^b)^2(\pi_x^*)^2(\pi_y^*)^1$	1.5	1.28
O_2^{2-}	KK, $(\sigma_{2s}^b)^2(\sigma_{2s}^*)^2(\sigma_z^b)^2(\pi_x^b)^2(\pi_y^b)^2(\pi_x^*)^2(\pi_y^*)^2$	1	1.49

Note the two unpaired electrons in the π_x^* and π_y^* antibonding orbitals that are responsible for the paramagnetism of O_2. When liquefied, oxygen has a

TABLE 19.7

Representative Examples of the Principal Oxidation States of Oxygen

OXIDATION STATE	COMPOUND OR ION
0	O_2 (molecular oxygen) O_3 (ozone)
$-\frac{1}{2}$	O_2^- (superoxide ion)
−1	H_2O_2 (hydrogen peroxide) HO_2^- (hydrogen peroxide ion) O_2^{2-} (peroxide ion)
−2	H_2O (water) OH^- (hydroxide ion) O^{2-} (oxide ion)

pale-blue color, which is related to the energy associated with the excitation of the unpaired electrons.

There is an allotrope of molecular oxygen that is triatomic (O_3), called ozone. It is the less stable of the two molecular forms of oxygen, and it is found in the lower atmosphere in only minute amounts (parts per billion concentration range, usually). Unlike O_2, it has a characteristic odor that can be detected around sparking electrical motors or around some office copy machines. Ozone can be formed by passing an electrical discharge through O_2 gas:

$$3\,O_2 = 2\,O_3 \qquad [19.21]$$
$$\Delta H^\circ = 284 \text{ kjoules}$$
$$\Delta G^\circ = 327 \text{ kjoules}$$

The thermodynamic data point up the instability of O_3 relative to O_2.

Ozone is a nonlinear molecule with the bond lengths and bond angle shown below:

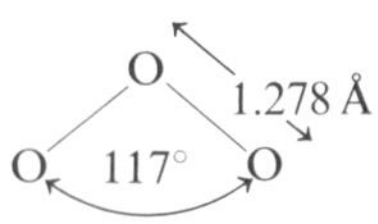

There is no detectable difference in the lengths of the two O-O bonds. To write a satisfactory electron-dot formula for O_3 thus requires writing the two resonance structures shown below:

Each bond is thus seen to be an average of a single and a double bond. The experimentally measured bond length and energy are consistent with 1.5 bond character.

Ozone is a very powerful oxidizing agent as shown by its high reduction potential:

$$O_3 + 2\,H^+ + 2\,e^- = O_2 + H_2O; \qquad \mathscr{E}^0 = 2.07 \text{ V}$$

By way of comparison, the O_2-H_2O couple is only 1.23 V. It is obvious that high concentrations of ozone would be deleterious to delicate plant and animal membranes.

Oxides

Oxygen is an active nonmetal. As shown by its electronegativity of 3.5, it is the second most electronegative element in the periodic table (second only to fluorine). It behaves in a typical nonmetal fashion, forming principally ionic

Reference has been made previously to the existence of an ozone layer in the upper atmosphere, and to the role it plays in filtering out harmful ultraviolet radiation from the sun (see pages 115–116 in Chapter 5). In view of the unfavorable thermodynamics for the formation of ozone from oxygen, the existence of a stable ozone layer needs some explanation. As you may suspect, the answer is found in the kinetics of the reactions involved. At altitudes of 100–150 km, oxygen molecules are dissociated to oxygen atoms by the high-energy ultraviolet radiation of the sun:

$$O_2 + \text{u.v. radiation} = 2\,O$$

The atomic oxygen formed is quite stable at the high altitudes because of the low concentration of "third-bodies" (other molecules, such as N_2 or O_2). In order to recombine, oxygen atoms require a third-body to carry away the large energy associated with recombination. Thus, oxygen atoms are free to diffuse to lower altitudes until they reach a concentration of third-body molecules that is high enough to allow reaction of atomic oxygen. By about 50 km, the concentration of molecular oxygen and other third-body molecules is high enough for the following reaction to take place with appreciable speed:

$$O + O_2 + M = O_3 + M$$

Below about 20 km, the rate of reaction falls off rapidly because the supply of oxygen atoms falls off. Hence, production of O_3 is limited above 50 km by the low concentration of third-body molecules, and below about 20 km by the low concentration of oxygen atoms. The result is an ozone *layer*. The ultimate source of the energy required to convert O_2 to O_3 is the ultraviolet radiation from the sun.

The normal oxides of sodium, barium, potassium, rubidium, and cesium (Na_2O, BaO, K_2O, Rb_2O, and Cs_2O) also exist, but these five are not made by direct combination of the elements.

The normal oxides can be obtained by heating either the nitrate or hydroxide with the free metal in an oxygen-free atmosphere. For example,

$$2\ KNO_3 + 10\ K = 6\ K_2O + N_2$$

compounds with metals, and principally covalent compounds with the nonmetals.

The -2 oxidation state (normal oxides) is by far the most common for oxygen. It is only in compounds in which oxygen bonds to itself (peroxides and superoxides) or to the more electronegative element, fluorine, that oxygen has an oxidation number other than -2. We have discussed the acid-base properties of the normal oxides in Chapter 16, and have referred to various other properties of the normal oxides in connection with the descriptive chemistry of other elements; hence, we will not repeat that discussion here.

We should mention, however, two special types of oxides: the *superoxide* (oxidation state $-1/2$) and the *peroxide* (oxidation state -1). Superoxides form when the heavier alkali metals react directly with oxygen. An example is

$$K + O_2 = KO_2$$

The oxygen in the superoxides (KO_2, RbO_2, CsO_2) is in the $-1/2$ oxidation state. The molecular orbital description given above for the molecule ion O_2^- shows how the existence of stable compounds with oxygen in the $-1/2$ oxidation state is possible.

When sodium or barium combines directly with oxygen, a peroxide forms:

$$2\ Na + O_2 = Na_2O_2$$

$$Ba + O_2 = BaO_2$$

In these compounds, the oxygen is in the -1 oxidation state.

Hydrogen Peroxide

Adding an acid (H_2SO_4 is used in the commercial process) to BaO_2 produces liquid hydrogen peroxide (H_2O_2) that can be distilled from the mixture

$$BaO_2(s) + H^+ + HSO_4^- = BaSO_4(s) + H_2O_2$$

Pure H_2O_2 is quite similar in some of its *physical* properties to H_2O. It is a colorless liquid, somewhat more viscous and dense than water, with a boiling point of 150° C and a freezing point of -0.89° C. As in water, hydrogen bonds form between molecules, and it is this characteristic that results in the similarity in the freezing and boiling points of the two compounds.

The electron-dot formula (not the geometry) of H_2O_2 is

$$H—\ddot{\underset{\cdot\cdot}{O}}—\ddot{\underset{\cdot\cdot}{O}}—H$$

Its geometry is such that the bond angles are 102°. The four atoms are not coplanar; one hydrogen atom projects at a right angle from the plane of the other three atoms.

Hydrogen peroxide is very much *unlike* water in its chemical properties. Hydrogen peroxide is a very strong oxiding agent ($\mathscr{E}^0_{\text{reduction}} = 1.77$ V) and also a fair reducing agent ($\mathscr{E}^0_{\text{oxidation}} = -0.68$ V). Its redox characteristics are summarized in the following reduction-potential diagram (acid solution):

$$O_2 \overset{0.68}{\text{———}} H_2O_2 \overset{1.77}{\text{———}} H_2O$$

The oxidation state of oxygen in hydrogen peroxide is -1. When acting as an oxidizing agent, the oxygen is reduced to -2. Examples of hydrogen peroxide acting as an oxiding agent ($H_2O_2 \rightarrow H_2O$) are

$$H_2O_2 + H_2S = 2\,H_2O + S \qquad [19.22]$$

$$H_2O_2 + 2\,I^- + 2\,H^+ = I_2 + 2\,H_2O \qquad [19.23]$$

$$PbS + 4\,H_2O_2 = PbSO_4 + 4\,H_2O \qquad [19.24]$$

Equation [19.24] is the basis for one of the steps in the restoration process of old paintings. The works of the old masters are dark (before they have been restored) because basic lead carbonate [$Pb(OH)_2 \cdot PbCO_3$] was used in the paints. Over the years, the small quantity of H_2S in the atmosphere has changed the lead compound from white basic carbonate to black lead sulfide (PbS). Treatment with hydrogen peroxide restores the original color to the lead compound since lead sulfate is white. There are many examples of hidden features in old paintings being uncovered by use of hydrogen peroxide.

When acting as a reducing agent, the oxygen in hydrogen peroxide is oxidized to the zero oxidation state. An example of hydrogen peroxide acting as a reductant ($H_2O_2 \rightarrow O_2$) is

$$2\,MnO_4^- + 5\,H_2O_2 + 6\,H^+ = 2\,Mn^{2+} + 8\,H_2O + 5\,O_2$$

As would be predicted from the reduction-potential diagram above, H_2O_2 is unstable. It disproportionates into $O_2 + H_2O$:

$$2\,H_2O_2 = 2\,H_2O + O_2$$

This reaction is induced by light; hence, hydrogen peroxide is stored in dark-colored bottles.

Hydrogen peroxide acts as a weak monoprotic acid in water solution. For the dissociation reaction

$$H_2O_2 = H^+ + HO_2^-$$

$K_a = 2.4 \times 10^{-12}$ at 25° C. Even though the extent of ionization is small, the principal species in basic solution is the HO_2^- ion. Oxidation-reduction reactions involving hydrogen peroxide in basic solution should be written in terms of the HO_2^- ion.

The use of H_2O_2 as a hair bleach is based on its ability to oxidize the pigments of hair.

Sulfur and Its Compounds

Elemental Sulfur

Sulfur, the second element of Group VIA is a bright-yellow solid (melting point 115° C), insoluble in water, but very soluble in carbon disulfide (CS_2). Sulfur occurs in nature in the elemental form, as sulfides and as sulfates. The main source of elemental sulfur in ancient times was the open-pit mines of Sicily. Here sulfur (brimstone) was melted away from the rock by heating the mixture in large open kettles located at the bottom of the pit. The molten sulfur frequently caught fire, and the flames and pungent fumes of SO_2 constantly billowed up from the pit. It is probable that Dante received his inspiration for the hell-fire and brimstone of the *Divine Comedy* from these open-pit operations in Sicily.

Most of the world's supply of sulfur now comes from extensive sulfur deposits in Texas and Louisiana. These deposits are buried under several hundred feet of rock and quicksand. They are, however, economically brought to the surface by a clever process developed by the American engineer Herman Frasch. Three concentric pipes are drilled down into the deposit (Figure 19.3). Hot water (170° C) under pressure (100 lbs/in^2) is forced down the outer pipe to melt the sulfur, and then compressed air is sent down the center pipe to force the molten sulfur up to the surface in the middle pipe.

While still molten, the sulfur is transported to a large flat area (the size of a football field). Here it is contained by metal forms until it cools and solidifies. A huge block of solid sulfur sitting on the surface of the earth is the result. This block is then blasted apart with dynamite, scooped up with power shovels, and shipped by boat and train to all parts of the world. The purity of the product is a surprising 99.5%. No further refining is needed for most industrial uses.

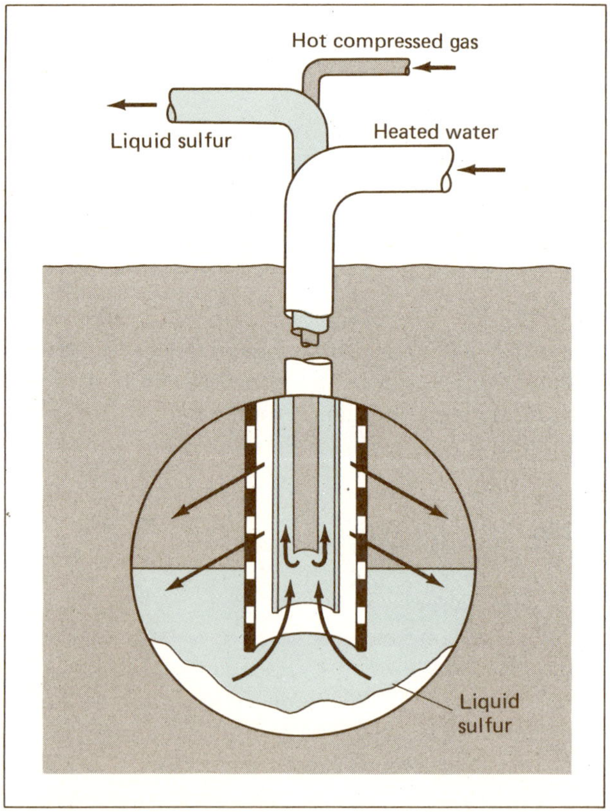

Figure 19.3
Frasch process for mining sulfur.

Another source of sulfur, and one that is becoming increasingly more important, is sulfur-containing fossil fuels. Sulfur is being removed from high-sulfur-content coal and petroleum in order to prevent excessive air pollution. The burning of the coal that contains sulfur produces SO_2 gas.

There are two crystalline forms of sulfur stable at atmospheric pressure: rhombic sulfur (stable below 95.6° C) and monoclinic sulfur (stable above 95.6° C). The molecules that occupy the points on the lattices of both forms are discrete S_8 molecules with the puckered-ring structure shown below:

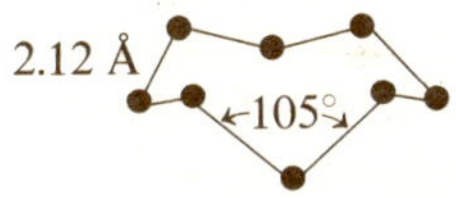

There is sound evidence that the S_8 rings break when sulfur melts (115° C), and that short chains of four to six sulfur atoms are present in the light yellow, mobile liquid that results. If heated above approximately 160° C, however, the liquid becomes increasingly viscous and dark in color. This effect is thought to be caused by the formation of long chains of sulfur atoms.

The ability of one atom of an element to bond to another atom of the same element (*catenation*) is characteristic of sulfur. Sulfur is second only to carbon in displaying this tendency.

In its chemical properties, sulfur is a typical nonmetal that forms ionic compounds (oxidation state of −2) with the metals, and covalent compounds of oxidation states +2, +2.5, +4, and +6 with the nonmetals. Table 19.8 lists representative compounds and ions in the various oxidation states.

TABLE 19.8

Representative Examples of the Principal Oxidation States of Sulfur

OXIDATION STATE	COMPOUND OR ION
+6	SO_3 (sulfur trioxide) H_2SO_4 (sulfuric acid) HSO_4^- (hydrogen sulfate ion) SO_4^{2-} (sulfate ion)
+4	SO_2 (sulfur dioxide) $SO_2(aq)$ or H_2SO_3 (sulfurous acid) HSO_3^- (hydrogen sulfite ion) SO_3^{2-} (sulfite ion)
+2.5	$S_4O_6^{2-}$ (tetrathionate ion)
+2	$S_2O_3^{2-}$ (thiosulfate ion)
0	S_8 (elemental sulfur)
−2	H_2S (hydrogen sulfide) HS^- (hydrogen sulfide ion) S^{2-} (sulfide ion)

Sulfides and Hydrogen Sulfide

The metal sulfides (S in oxidation state -2) are formed by direct combination of the elements:

$$16\ \text{Na} + \text{S}_8 = 8\ \text{Na}_2\text{S}$$
$$8\ \text{Fe} + \text{S}_8 = 8\ \text{FeS}$$

Many metal sulfides are quite insoluble, as can be seen from the following list of solubility product constants:

	K_{sp}
CdS	10^{-28}
CuS	10^{-42}
CoS	10^{-27}
HgS	9×10^{-52}
NiS	10^{-19}
FeS	10^{-22}
PbS	10^{-28}
MnS	10^{-16}
Ag_2S	6×10^{-52}
ZnS	10^{-25}

The soluble sulfides, as we shall see shortly, hydrolyze extensively and are, therefore, not found in nature. The insoluble sulfides in nature are important since they make up the principal ores of such important elements as silver, lead, copper, arsenic, antimony, cobalt, nickel, zinc, and cadmium.

Most of the metal sulfides are black in color: for example, Ag_2S, Bi_2S_3, PbS, Cu_2S, CuS, FeS, CoS, NiS, SnS, and HgS. Exceptions are CdS (yellow), ZnS (white), MnS (pink), As_2S_3 (yellow), and Sb_2S_3 (orange).

When a strong acid is added to a metal sulfide, hydrogen sulfide (H_2S) gas is generated. An example is

$$\text{FeS}(s) + 2\ \text{H}^+ = \text{H}_2\text{S}(g) + \text{Fe}^{2+}$$

Hydrogen sulfide, commonly called "rotten-egg gas" because of its odor, is a highly poisonous gas. Fortunately, its odor is so strong and disagreeable that it is easily detected even in very dilute concentrations.

The Henry's law constant for the solubility of H_2S gas in water is approximately 10 atm/(mole/liter) (refer to Chapter 8); hence, it is possible with one atmosphere pressure of H_2S gas above the solution to prepare 0.1 M solution of H_2S. In aqueous solution H_2S is a very weak diprotic acid, as shown by its small ionization constants:

$$\text{H}_2\text{S} = \text{H}^+ + \text{HS}^-; \qquad K_{a(1)} = 1.1 \times 10^{-7} \qquad [19.25]$$
$$\text{HS}^- = \text{H}^+ + \text{S}^{2-}; \qquad K_{a(2)} = 1 \times 10^{-13} \qquad [19.26]$$

Since H_2S and HS^- are such weak acids, it is to be expected that the soluble sulfides hydrolyze extensively:

$$S^{2-} + H_2O = HS^- + OH^- \qquad [19.27]$$

$$K_{h(1)} = \frac{K_w}{K_{a(2)}} = \frac{1.0 \times 10^{-14}}{1 \times 10^{-13}} = 0.1$$

$$HS^- + H_2O = H_2S + OH^- \qquad [19.28]$$

$$K_{h(2)} = \frac{K_w}{K_{a(1)}} = \frac{1.0 \times 10^{-14}}{1.1 \times 10^{-7}} = 9.1 \times 10^{-8}$$

Although the hydrolysis constant for sulfide ion is seen to be relatively large ($K_h = 0.1$), hydrolysis of sulfide ion is not nearly so complete as is that of the oxide ion. The hydrolysis of the latter, you will recall, is almost total. The HS^- ion ($K_h = 9.1 \times 10^{-8}$) is only slightly hydrolyzed.

EXAMPLE 19.3 Calculate the sulfide ion concentration present at equilibrium in a solution prepared by dissolving 0.1 mole of Na_2S in enough water to make 1 liter of solution.

SOLUTION The equilibria that apply are those represented by equations [19.27] and [19.28]. For this calculation, we need not consider the second hydrolysis (equation [19.28]). Its effect ($K_h = 9.1 \times 10^{-8}$) is negligible compared with the first hydrolysis ($K_h = 0.1$).

Let x = moles/liter of S^{2-} that reacts.

$$\underset{0.1 - x}{S^{2-}} + H_2O = \underset{x}{HS^-} + \underset{x}{OH^-}$$

At equilibrium;

$$\frac{x^2}{0.1 - x} = 0.1$$

Solving this gives

$$x = 0.06 \text{ mole/liter}$$

$$[S^{2-}] = 0.04$$

Hence, the sulfide ion under these conditions is roughly $\left(\frac{0.06}{0.1}\right) 100 = 60\%$ hydrolyzed.

The extent of hydrolysis of oxide and sulfide ion are compared with some other commonly encountered anions in Table 19.9.

TABLE 19.9

Hydrolysis of Anions

ANION	EXTENT OF HYDROLYSIS
O^{2-}	Almost completely hydrolyzed
S^{2-}	Extensively hydrolyzed ($K_h = 0.1$)
PO_4^{3-} AsO_4^{3-} CO_3^{2-}	Moderately hydrolyzed ($K_h = 10^{-2}$–10^{-4})
CN^- HS^- ClO^- HCO_3^- HPO_4^{2-} SO_3^{2-}	Slightly hydrolyzed ($K_h = 10^{-6}$–10^{-9})
F^- NO_2^- $H_2PO_4^-$ SO_4^{2-} HSO_3^-	Weakly hydrolyzed ($K_h = 10^{-11}$–10^{-13})
Cl^- Br^- I^- NO_3^- HSO_4^- ClO_4^-	Not hydrolyzed

Sulfur Dioxide and Sulfites

Sulfur burns in air to form sulfur dioxide gas (SO_2) as shown in the reaction

$$S_8(s) + 8\ O_2(g) = 8\ SO_2(g)$$

Another source of sulfur dioxide is from heating sulfide minerals in air. This "roasting" process is a first step in the metallurgy of most sulfide ores. An example is

$$2\ ZnS(s) + 3\ O_2(g) = 2\ ZnO(s) + 2\ SO_2(g)$$

Sulfur dioxide is a colorless gas with the characteristic odor associated with burning sulfur. The fumes are moderately irritating to the sensitive membranes of the nose and throat. It is an easily liquefied gas, condensing to a liquid at only $-10°$ C at one atmosphere pressure.

As predicted from molecular orbital and valence bond theories, the structure of SO_2 is nonlinear. The two S—O bonds are equivalent; therefore, two resonance structures are required to describe the molecule by the electron-dot notation:

One predicts from molecular orbital theory that triatomic molecules of more than 16 valence electrons will be nonlinear.

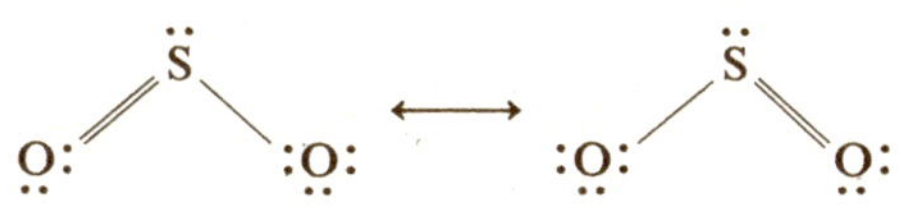

Sulfur dioxide dissolves in water to produce a weakly acidic solution that is often called sulfurous acid and given the formula H_2SO_3. Pure H_2SO_3, however, has never been isolated, and the weakly acidic solution called sulfurous acid is probably more accurately described by the formula $SO_2(aq)$. The acidity arises from the reactions

$$SO_2 + H_2O = H^+ + HSO_3^-; \qquad K_1 = 1.3 \times 10^{-2}$$

and

$$HSO_3^- = H^+ + SO_3^{2-}; \qquad K_2 = 5 \times 10^{-6}$$

The species HSO_3^- is called the hydrogen sulfite ion, and SO_3^{2-}, sulfite ion. Since $SO_2(aq)$ is diprotic, neutralization with a base produces both acid sulfites and normal sulfites:

$$SO_2(aq) + OH^- = HSO_3^-$$
$$HSO_3^- + OH^- = SO_3^{2-} + H_2O$$

Since $SO_2(aq)$ and HSO_3^- are weak acids, the sulfites and hydrogen sulfites hydrolyze to some extent. The reactions are

$$SO_3^{2-} + HOH = HSO_3^- + OH^- \qquad [19.29]$$
$$K_{h(1)} = 2.0 \times 10^{-9}$$
$$HSO_3^- + HOH = SO_2(aq) + OH^- \qquad [19.30]$$
$$K_{h(2)} = 7.7 \times 10^{-13}$$

EXAMPLE 19.4 Calculate the pH of (a) 0.10 M Na_2SO_3 solution; (b) 0.10 M $NaHSO_3$ solution

SOLUTION

(a) The hydrolysis of the SO_3^{2-} will produce a basic solution. The reactions are equations [19.29] and [19.30], given above. We will neglect the small effect of the second hydrolysis and make our calculation from equation [19.29]. Let x equal the moles/liter of SO_3^{2-} that reacts.

$$\begin{array}{llll} SO_3^{2-} & + H_2O = & HSO_3^- & + OH^- \\ 0.10 - x & & x & x \end{array}$$

$$\frac{x^2}{0.10 - x} = 2.0 \times 10^{-9} \cong \frac{x^2}{0.10}$$

$$x = 1.4 \times 10^{-5} = [OH^-]$$

$$pOH = -\log [OH^-] = 4.85$$

$$pH = 14.0 - pOH = 9.15$$

(b) The HSO_3^- ion can both ionize (which will make an acidic solution) and hydrolyze (which will produce a basic solution). Both processes take place, but one fortunately (for our calculation) does so to a much greater extent than the other. The two processes and their respective equilibrium constants are

$$HSO_3^- = H^+ + SO_3^{2-}; \qquad K_a = 5.0 \times 10^{-6}$$

$$HSO_3^- + H_2O = H_2SO_3 + OH^-; \qquad K_h = 7.7 \times 10^{-13}$$

The more important reaction (by a factor of about 10^7) is seen to be the acid ionization reaction. Neglecting the hydrolysis reaction and letting x equal the concentration of HSO_3^- that ionizes, we have

$$\begin{array}{llll} HSO_3^- & = H^+ & + SO_3^{2-}; & K_a = 5.0 \times 10^{-6} \\ 0.10 - x & x & x & \end{array}$$

$$\frac{x^2}{0.10 - x} = 5.0 \times 10^{-6} \cong \frac{x^2}{0.10}$$

$$x = [H^+] = 7.1 \times 10^{-4}$$

$$pH = -\log [H^+] = 3.15$$

An acid solution of sulfur dioxide is a mild reducing agent, and the sulfites in basic solution are strong reducing agents. The oxidized product is S in the +6 oxidation state (HSO_4^- or SO_4^{2-}). These facts can be deduced from the standard reduction-potential diagrams given below:

Acidic solution

$$HSO_4^- \xrightarrow{0.17} SO_2 \xrightarrow{0.40} S_2O_3^{2-} \xrightarrow{0.50} S \xrightarrow{0.14} H_2S$$

$$SO_2 \xrightarrow{0.51} S_4O_6^{2-} \xrightarrow{0.08} S_2O_3^{2-}$$

Basic solution

$$SO_4^{2-} \xrightarrow{-0.93} SO_3^{2-} \xrightarrow{-0.58} S_2O_3^{2-} \xrightarrow{-0.74} S \xrightarrow{-0.48} S^{2-}$$

$$SO_3^{2-} \xrightarrow{-0.61} S^{2-}$$

Sulfur Trioxide, Sulfuric Acid, and Sulfates

There are two oxides of sulfur, SO_2 and SO_3. When sulfur is burned in air, the product is the dioxide:

$$S + O_2 = SO_2; \qquad \Delta G^0 = -300.4 \text{ kjoules}$$

The high negative value of ΔG^0 shows that the reaction proceeds far to the right; that is, SO_2 is very stable relative to S and O_2. However, SO_2 will react further with oxygen to form the still more thermodynamically stable sulfur trioxide:

$$SO_2 + \frac{1}{2}O_2 = SO_3; \qquad \Delta G^0 = -70.0 \text{ kjoules}$$

The latter reaction, while thermodynamically favorable, takes place so slowly that the principal product from burning sulfur in oxygen is SO_2 rather than SO_3. Vanadium pentoxide, however, will catalyze the oxidation of SO_2 to SO_3, and this is one step in the procedure for manufacturing sulfuric acid. Since SO_3 is the anhydride of H_2SO_4, adding SO_3 to water will produce the acid:

$$SO_3(g) + H_2O = H^+ + HSO_4^-$$

In the commercial production of sulfuric acid it is found to be easier to bring the water and SO_3 into contact by first dissolving the SO_3 in pure $H_2SO_4(l)$ to form *pyrosulfuric* or "*fuming sulfuric*" acid ($H_2S_2O_7$ or $H_2SO_4 \cdot SO_3$).

$$H_2SO_4 + SO_3 = H_2S_2O_7$$

Water can then be added to the pyrosulfuric acid to produce sulfuric acid:

$$H_2S_2O_7(l) + H_2O = 2\ H_2SO_4(l)$$

Concentrated sulfuric acid forms stable hydrates and is thus a good dehydrating agent. It not only removes water that is physically attached to a substance, but in some of its reactions it removes the elements of water from another molecule. For example, concentrated sulfuric acid chars sugar (and clothing and your flesh) in a reaction of the type

$$C_6H_{12}O_6 \text{ (a sugar)} \overset{H_2SO_4}{=} 6\ C + 6\ H_2O$$

Hot, concentrated sulfuric acid is a moderately strong oxidizing agent. This fact is not readily apparent from the standard reduction-potential diagram given earlier. The diagram shows that sulfuric acid is a rather poor oxidizing agent, but we need to remember that the $\mathscr{E}^0$ values in the diagram refer to the standard state concentration of one-molar solutions. In these dilute solutions, sulfuric acid does have little oxidizing ability, but as the concentration of the reactants H^+ and HSO_4^- is increased, the reduction half-reaction is driven to the right, and hot, concentrated sulfuric acid is a fairly strong oxidizing agent.

The electron-dot formula of SO_3 is

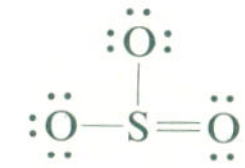

but since the molecule is known to be a planar structure in which all three sulfur-to-oxygen bonds are the same length and all three O-S-O angles are equivalent, the molecule is best represented as a hybrid of the following three resonance structures:

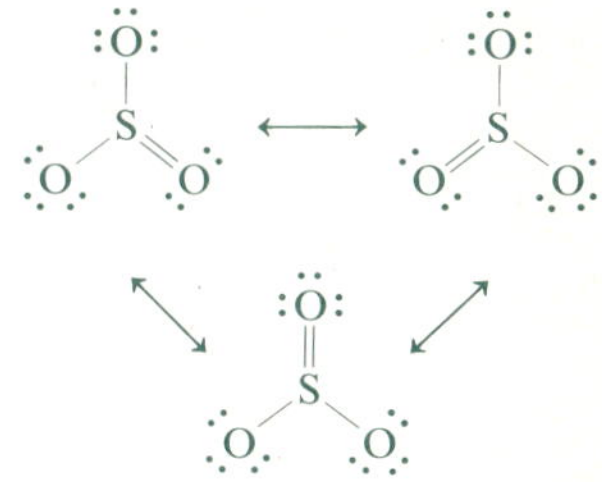

Sulfuric acid is used in so many industrial processes that the quantity of sulfuric acid being manufactured can be used as a fairly reliable index of the economic health of a nation. When the production of H_2SO_4 is high, it is safe to assume that the economy in general is doing well.

For example, it will oxidize copper. (Recall that H^+ alone will not react with copper.)

$$Cu + 3H^+ + \underset{\text{(hot, concentrated)}}{HSO_4^-} = SO_2 + Cu^{2+} + 2\,H_2O$$

Sulfuric acid is a diprotic acid whose first ionization is complete; therefore, aqueous solutions of sulfuric acid are written as $H^+ + HSO_4^-$. The ionization constant of HSO_4^- is 1.2×10^{-2}, which makes it one of the stronger weak acids. Hence, HSO_4^- is not hydrolyzed, and SO_4^{2-} is only very weakly hydrolyzed. Because sulfuric acid is diprotic, two types of salts are possible from the neutralization of the acid: acid sulfates, such as $NaHSO_4$, and normal sulfates, such as K_2SO_4.

The structure of SO_3 is trigonal. This might be expected since SO_3 is isoelectronic (valence electrons) with BF_3, CO_3^{2-}, and NO_3^-, other well-established trigonal structures. Sulfate ion, like PO_4^{3-} and AsO_4^{3-}, is tetrahedral.

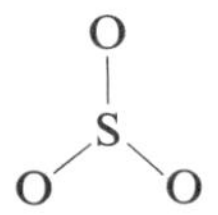
Sulfur trioxide

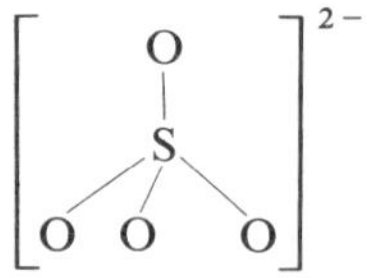
Sulfate ion

Thiosulfates and Tetrathionates

Sulfur will dissolve in basic solutions of sulfite ion to produce thiosulfate ion ($S_2O_3^{2-}$):

$$SO_3^{2-} + S = S_2O_3^{2-}$$

As can be seen from the reduction-potential diagrams, $S_2O_3^{2-}$ is stable in basic solution:

$$SO_3^{2-} \xrightarrow{-0.58} S_2O_3^{2-} \xrightarrow{-0.74} S$$

but disproportionates to sulfur and SO_2 in acid solution:

$$SO_2 \xrightarrow{0.40} S_2O_3^{2-} \xrightarrow{0.50} S$$

$\mathcal{E}^0$ for disproportionation in acid solution is $+0.50$ V $-$ 0.40 V $= +0.10$ V.

The prefix "thio" in thiosulfate means that sulfur has replaced oxygen. Thiosulfate ion is thus sulfate ion with one oxygen atom replaced by a sulfur atom. The electron-dot formula for thiosulfate is

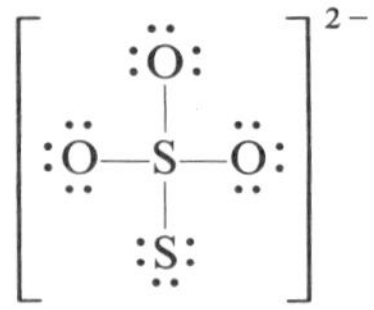

which should be compared with that of sulfate ion,

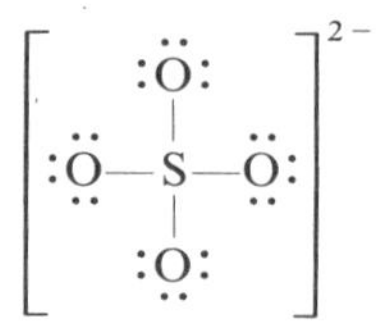

Thiosulfate ion, like sulfate ion, is tetrahedral. In tetrathionate ion two tetrahedra are joined through sulfur atoms.

Thiosulfate ion finds application in analytical chemistry as a reducing agent. In its analytical use, it is generally oxidized to the tetrathionate ion ($S_4O_6^{2-}$), in which the sulfur atoms are in the +2.5 oxidation state. An example is the reaction of $S_2O_3^{2-}$ with I_2:

$$2\,S_2O_3^{2-} + I_2 = 2\,I^- + S_4O_6^{2-}$$

The electron-dot configuration of $S_4O_6^{2-}$ is

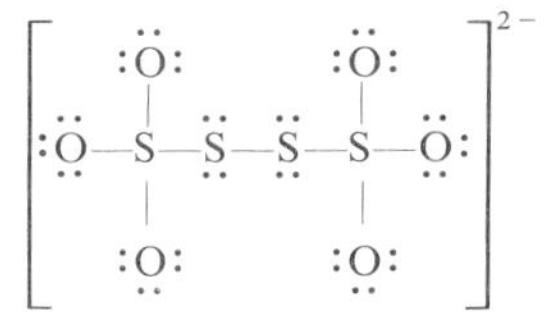

The Compounds of Selenium and Tellurium

The type of compounds formed by selenium and tellurium resemble those of sulfur, as can be seen in Table 19.10.

Observed differences in analogous compounds of S, Se, and Te can usually be accounted for by the difference in metallic character of the elements. As one moves down the family from S to Se,

(1) the hydrides become less stable ($\Delta G^0_{f(H_2S)} = -33.1$ kjoules/mole; $\Delta G^0_{f(H_2Se)} = 71.1$ kjoules/mole; and $\Delta G^0_{f(H_2Te)} = 138$ kjoules/mole).
(2) the extent of ionization of the aqueous hydrides increases (the first acid constants are 1.1×10^{-7} for H_2S; 2×10^{-4} for H_2Se; and 2.3×10^{-3} for H_2Te).
(3) the oxides become more ionic (SO_2 is a gas, SeO_2 is a solid with partial ionic character, and TeO_2 is considered to have a principally ionic lattice).
(4) the strengths of the +6-oxidation-state acids decrease. [The first ionization of H_2SO_4 is strong. $Te(OH)_6$, which can be thought of as $H_2TeO_4 \cdot 2H_2O$, is very weak ($K_{a(1)} \cong 10^{-7}$).]

The structure of $Te(OH)_6$ is octahedral with the tellurium atom occupying the central octahedral hole. The large size of the tellurium atom is no doubt responsible for its preference for the octahedral structure rather than the tetrahedral structure of H_2SO_4 and H_2SeO_4.

Polonium

As would be expected from its position in the family, polonium has typical metallic character. It, like many other heavy metals, is radioactive. Although it occurs in nature, it is present only in very small quantities in the earth's crust. It was discovered by Marie and Pierre Curie, and named for Madame Curie's native land, Poland.

TABLE 19.10

Some Hydrides, Oxides, Oxygen Acids, and Oxyanions of S, Se, and Te

OXIDATION STATE	S	Se	Te
−2	H_2S	H_2Se	H_2Te
+4	SO_2	SeO_2	TeO_2
	$SO_2(aq)$ or H_2SO_3	H_2SeO_3	—
	SO_3^{2-}	SeO_3^{2-}	TeO_3^{2-}
+6	H_2SO_4	H_2SeO_4	$Te(OH)_6$

19.3 Group VIIA: Halogens and Their Compounds

The Group VIIA elements are fluorine (F), chlorine (Cl), bromine (Br), iodine (I), and astatine (At). All members of this family can be classified as nonmetals and, therefore, can be conveniently discussed together. There are, of course, wide variations in the extent of nonmetallic character in the family.

The electronic configuration of the Group VIIA elements are given in Table 19.11. From this table we would predict the commonly encountered oxidation states to be -1, $+5$, and $+7$. Other important oxidation states for Group VIIA elements are $+1$ and $+3$. Fluorine, being the most electronegative element, has only the negative oxidation state. All the other halogens exist in positive oxidation states as well as the -1 state.

Table 19.12 lists some of the physical properties of the Group VIIA elements. Astatine, the last member of the group, is not included. It does not occur in nature, and all of its isotopes are radioactive of fairly short half-lives. Hence, quantitative data on its properties have been difficult to obtain.

TABLE 19.11

The Electronic Configurations of the Halogens

ELEMENT	CONFIGURATION
$_{9}F$	He core + $2s^22p^5$
$_{17}Cl$	Ne core + $3s^23p^5$
$_{35}Br$	Ar core + $3d^{10}4s^24p^5$
$_{53}I$	Kr core + $4d^{10}5s^25p^5$
$_{85}At$	Xe core + $4f^{14}5d^{10}6s^26p^5$

TABLE 19.12

Properties of the Halogens

	F	Cl	Br	I
Atomic number	9	17	35	53
Ionization energy (kjoules/mole)	1681	1250	1139	1008
Atomic radius (Å)	0.72	0.99	1.14	1.33
Ionic radius (X^-)(Å)	1.33	1.81	1.96	2.19
Melting point (° K)	54	172	266	387
Boiling point (° K)	85	239	331	456
$\mathscr{E}^0$ ($X_2 + 2e^- = 2X^-$)(V)	2.87	1.36	1.10	0.54
Electronegativity	4.0	3.0	2.8	2.5
ΔH_{fus} (kjoules/mole)	1.56	6.41	10.5	15.7
ΔH_{vap} (kjoules/mole)	6.31	20.4	30.0	41.7

Elements of Group VIIA

Each of the halogens exists as discrete diatomic molecules in all phases. Hence, we write their molecular formulas as F_2, Cl_2, Br_2, and I_2 whether gaseous, liquid, or solid. The gradations in physical state and color as one proceeds from top to bottom of the family are shown below:

F_2	pale-yellow gas
Cl_2	light-green gas
Br_2	dark-red liquid
I_2	gray-black solid

I_2 vapor is purple, as is I_2 dissolved in CCl_4.

None of the halogens is found in the elemental form in nature. They are all highly reactive nonmetals. They react with a wide variety of both metals (as implied by the name *halogen*, which means *salt former*), and nonmetals. All of the halogens are oxidizing agents, fluorine being an exceptionally powerful one $\mathscr{E}^0_{\text{reduction}} = 3.06$ V in acid solution). The strength as an oxidizing agent decreases as one goes from top to bottom of the family:

	$\mathscr{E}^0$ (V)
$F_2 + 2\,H^+ + 2\,e^- = 2\,HF\,(aq)$	3.06
$Cl_2 + 2\,e^- = 2\,Cl^-$	1.36
$Br_2 + 2\,e^- = 2\,Br^-$	1.07
$I_2 + 2\,e^- = 2\,I^-$	0.54

For comparison purposes,

$H_2O_2 + 2\,H^+ + 2\,e^- = 2\,H_2O$	1.77
$O_2 + 4\,H^+ + 4\,e^- = 2\,H_2O$	1.23
$NO_3^- + 4\,H^+ + 3\,e^- = NO + 2\,H_2O$	0.96

Thus, both F_2 and Cl_2 are stronger oxidizing agents than O_2, and all but I_2 are stronger oxidizing agents than nitric acid.

Fluorine gas is such a strong oxidizing agent that its preparation from the fluoride ion is by means of electrolysis only. There is no common chemical oxidizing agent capable of oxidizing fluoride ion. In the commercial process, fused KHF_2 (potassium hydrogen fluoride) is electrolyzed between inert electrodes. The products are $F_2(g)$ at the anode and $H_2(g)$ at the cathode. Chlorine gas is also prepared commercially by electrolysis. Bromine is commercially prepared from the bromide ion in sea water by oxidation with chlorine:

$$Cl_2 + 2\,Br^- = Br_2 + 2\,Cl^-$$

Iodine occurs in nature as iodates (+5 oxidation state). These are conveniently reduced to elemental iodine with hydrogen sulfites. The reaction is

$$2\,IO_3^- + 5\,HSO_3^- + 2\,H^+ = 5\,HSO_4^- + H_2O + I_2(s)$$

Elemental iodine can, of course, also be obtained from the iodide ion by oxidation. Either Cl_2 or Br_2 would serve as an oxidant:

$$Cl_2 + 2\,I^- = I_2 + 2\,Cl^-$$
$$Br_2 + 2\,I^- = I_2 + 2\,Br^-$$

Oxidation State −1: Metal Halides and Hydrogen Halides

Metal Halides

Binary halides of most metals are capable of being prepared by direct combination of the elements. As an example, consider

$$2\,Na + I_2 = 2\,NaI$$
$$Mg + Br_2 = MgBr_2$$
$$Sn + 2\,Cl_2 = SnCl_4$$

With the Groups IA and IIA metals, the metal halides that are formed are highly ionic in character. With less active metals and with metals in high oxidation states (such as $SnCl_4$), the halide takes on covalent character. The melting points of the halides reflect qualitatively the extent of ionic character in the halide. Table 19.13 shows melting points of the Groups IA–VA halides.

Some rather interesting relations are apparent from this table. Groups IA and IIA halides have the relatively high melting points characteristic of ionic compounds. Beryllium, being right next to the dividing line between metals and nonmetals, has less metallic character than the other members of Group IIA, and the increased covalency in its compounds is shown by the lower melting point of its chloride.

The first member of Group IIIA, boron, is on the nonmetal side of the dividing line, and the increased covalency of BCl_3 over that of $BeCl_2$ is very pronounced. Even Al_2Cl_6 and $GaCl_3$ show significant covalent character. The increased metallic character of the next two elements down in the family, In and Tl, is enough to raise the melting point up into the range of the principally ionic compounds.

The pronounced effect of the oxidation number of the metal can be seen by comparing $SnCl_2$ with $SnCl_4$, $PbCl_2$ with $PbCl_4$, and $SbCl_3$ with $SbCl_5$. It is readily apparent that the higher oxidation states lead to greater covalent character in the compound.

Hydrogen Halides

The hydrogen halides are colorless, covalent gaseous molecules that can be prepared by direct combination of the elements; that is,

$$X_2 + H_2 = 2\,HX$$

where X represents any one of the halogens. As might be expected, the speed and vigor of the reaction becomes progressively milder from fluorine to iodine. The reaction with fluorine is violent under any conditions; the reaction with chlorine is explosive in the presence of sunlight or at elevated temperatures;

TABLE 19.13

Melting Points of the Groups IA–VA Halides

	HALIDE	MELTING POINT (° C)
Group IA	LiCl	614
	NaCl	808
	KCl	776
	RbCl	715
	CsCl	646
Group IIA	$BeCl_2$	440
	$MgCl_2$	708
	$CaCl_2$	772
	$SrCl_2$	873
	$BaCl_2$	963
Group IIIA	BCl_3	−107
	Al_2Cl_6	190†
	$GaCl_3$	78
	$InCl_3$	586
	TlCl	430
Group IVA	CCl_4	−23
	$SiCl_4$	−70
	$GeCl_4$	−50
	$SnCl_2$	249
	$SnCl_4$	−30
	$PbCl_2$	501
	$PbCl_4$	−15
Group VA	NCl_3	−40
	PCl_3	−112
	$AsCl_3$	−9
	$SbCl_3$	73
	$SbCl_5$	3

† 2.5 atms; sublimes at lower pressures

the reaction with bromine is slow at room temperatures, but proceeds at moderate rates when heated; and the reaction with iodine requires a catalyst and heat. The thermodynamic data for the reactions (forming two moles of product) at 25° C are given below:

	ΔH_f^0 (kjoules)	ΔG_f^0 (kjoules)
HF	−538	−542
HCl	−184.6	−190.5
HBr	−72.4	−106.4
HI	51.8	2.6

Looking at the ΔG_f^0 data, we see that the stability of the hydrogen halides relative to the elements of which they are composed increases from HI to HF. HF is very stable toward decomposition into H_2 and F_2, whereas HI is seen to be unstable under standard state conditions and 25° C.

From the ΔH_f^0 data, we see that the reactions for formation of the hydrogen halides are all exothermic except for HI. Formation of HF is accompanied by an evolution of 538 kjoules of heat (constant pressure) per mole of HF formed, whereas HI requires the absorption of 51.8 kjoules.

Another method for preparing the hydrogen halides is to treat metal halides with a strong acid of low volatility (high boiling point) and drive out the very volatile hydrogen halides by heating. Sulfuric acid fits the requirements for the acid, and the following equations are typical reactions for producing HF and HCl:

$$CaF_2(s) + H^+ + HSO_4^- = 2\,HF(g) + CaSO_4(s)$$
$$NaCl(s) + H^+ + HSO_4^- = HCl(g) + NaHSO_4(s) \qquad [19.31]$$

$NaHSO_4$ is very soluble in water but only slightly soluble in concentrated sulfuric acid; hence, in equation [19.31] it is properly written $NaHSO_4(s)$.

HBr and HI cannot be produced by analogous reactions because hot, concentrated sulfuric acid is capable of oxidizing the bromide and iodide ions. The reaction for Br^- is given below. An analogous equation can be written for I^-.

$$2\,Br^- + 3\,H^+ + HSO_4^- = Br_2 + SO_2 + 2\,H_2O$$

Phosphoric acid is a high boiling point, nonoxidizing acid that can be used to produce HI and HBr from their salts. An example is

$$NaI(s) + H_3PO_4 = NaH_2PO_4(s) + HI(g)$$

A third method for producing the hydrogen halides is the hydrolysis of phosphorus trihalides:

$$PX_3 + 3\,H_2O = H_3PO_3 + 3\,HX(g)$$

The hydrogen halides of chlorine, bromine, and iodine are simple, discrete molecules of the formulas HCl, HBr, and HI. Hydrogen fluoride, on the other hand, does not exist as simple molecules but as hydrogen-bonded aggregates of as many as five monomeric units (gas phase). The structure of these polymeric species is shown below:

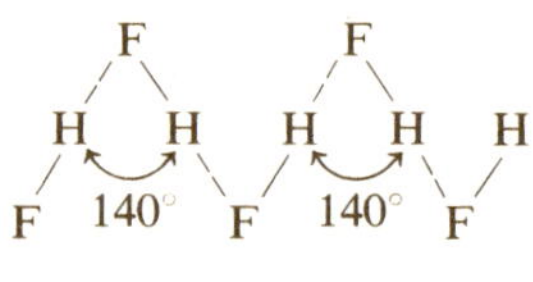

The hydrogen bonding in HF has a pronounced effect on the physical properties of the compound. Table 19.14 allows one to compare the abnormally high (for a molecular weight corresponding to HF) melting and boiling points, and heats of fusion and vaporization, for hydrogen fluoride with those of the other

TABLE 19.14

Physical Properties of Hydrogen Halides

PROPERTY	HF	HCl	HBr	HI
Melting point (°C)	−83.8	−114.2	−86.9	−50.8
Boiling point (°C)	19.5	−85.0	−66.7	−35.3
ΔH^0_{fus} (kjoules/mole)	4.577	1.99	2.41	2.83
ΔH^0_{vap} (kjoules/mole)	6.73	16.2	17.6	19.8

hydrogen halides. (See Chapter 6 for a plot of boiling points of the Group IV, V, VI, and VII hydrides.) This ability of hydrogen fluoride to undergo self-association through intermolecular hydrogen bonding is a direct result of the high electronegativity of the fluoride ion. The same tendency has been noted in H_2O and in NH_3.

The hydrogen halides are very soluble in water, in which they dissociate to form acids:

$$HX + H_2O = H_3O^+ + X^-$$

or

$$HX = H^+ + X^-$$

The aqueous solutions of the hydrogen halides are called hydrofluoric, hydrochloric, hydrobromic, and hydroiodic acids. The latter three are strong acids, almost completely dissociated. Hydrofluoric acid, however, is only partially ionized ($K_a = 7.4 \times 10^{-4}$). The very high strength of the hydrogen-fluorine bond contributes to the lack of complete ionization of hydrofluoric acid.

In concentrated hydrofluoric acid solutions, hydrogen fluoride is believed to be present principally as the hydrogen-bonded dimer, H_2F_2, which would have a structure of

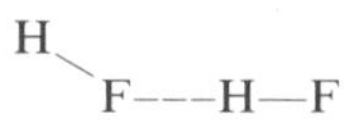

There is evidence that H_2F_2 ionizes in water to form HF_2^- anions:

$$H_2F_2 + H_2O = H_3O^+ + HF_2^-$$

Salts of the formula KHF_2 are obtained by neutralizing concentrated hydrofluoric acid solutions.

One additional property of hydrofluoric acid that is not observed with the other hydrohalic acids is its reaction with silica and silicates. Hydrofluoric acid cannot be contained in glass. It etches glass within a few minutes of contact, and in time dissolves the glass completely. With an excess of hydrofluoric acid, the complex ion hexafluorosilicate (IV) forms. The reaction of hydrofluoric acid on silica is used as a representative example:

$$SiO_2 + 6\,HF = SiF_6^{2-} + 2\,H^+ + 2\,H_2O$$

Oxygen Acids of the Halogens

Fluorine, being the most electronegative element, cannot exist in any positive oxidation state. The other halogens, however, are less electronegative than oxygen and can, therefore, assume positive oxidation states in the oxides and oxygen acids. Table 19.15 lists the oxygen acids of the halogens by formula and by name.

We will not discuss the oxides of the halogens in detail. There is a large variety of them. The oxides of chlorine are well known for their instability. Cl_2O_7 is the most stable of the chlorine oxides, and it explodes when heated or subjected to thermal shock. Cl_2O_7 is the anhydride of perchloric acid, $HClO_4$. I_2O_5, the anhydride of iodic acid, is a stable oxide. The formulas of some of the other known oxides are Cl_2O, ClO_2, Cl_2O_4, Cl_2O_6, Br_2O, BrO_2, BrO_3, I_2O_4, and I_2O_7.

In naming oxygen acids such as nitrous and nitric, sulfurous and sulfuric, or phosphorous and phosphoric, where there are only two oxidation states involved, the acid with the central atom (S, N, P) in the higher oxidation state (H_2SO_4, HNO_3, and H_3PO_4) takes the *ic* ending, and the lower oxidation state, the *ous* ending. The salts of the *ic* acids take an *ate* suffix, and the salts of the *ous* acids, an *ite* ending. For example:

HNO_3 (nitr*ic* acid); $NaNO_3$ (sodium nitr*ate*)
HNO_2 (nitr*ous* acid); $NaNO_2$ (sodium nitr*ite*)

H_2SO_4 (sulfur*ic* acid); Na_2SO_4 (sodium sulf*ate*)
H_2SO_3 (sulfur*ous* acid); Na_2SO_3 (sodium sulf*ite*)

H_3PO_4 (phosphor*ic* acid); Na_3PO_4 (sodium phosph*ate*)
H_3PO_3 (phosphor*ous* acid); Na_2HPO_3 (sodium phosph*ite*)

With chlorine, however, there are four oxygen acids (oxidation states +1, +3, +5, and +7). Here the *ic* and *ous* endings are given to the middle two (*ous* to the +3 and *ic* to the +5). The +1 acid then receives the prefix *hypo* as well as the suffix *ous*, and the +7 gets the prefix *per* in addition to the suffix *ic*. Hence,

$HClO$ (*hypo*chlor*ous* acid); ClO^- (*hypo*chlor*ite* ion)
$HClO_2$ (chlor*ous* acid); ClO_2^- (chlor*ite* ion)
$HClO_3$ (chlor*ic* acid); ClO_3^- (chlor*ate* ion)
$HClO_4$ (*per*chlor*ic* acid); ClO_4^- (*per*chlor*ate* ion)

TABLE 19.15

Oxygen Acids of the Halogens

OXIDATION NUMBER OF HALOGEN	CHLORINE	BROMINE	IODINE	NAME: ACID	NAME: SALT
+7	$HClO_4$	——†	HIO_4, H_5IO_6	per——ic	per——ate
+5	$HClO_3$	$HBrO_3$	HIO_3	——ic	——ate
+3	$HClO_2$	——	——	——ous	——ite
+1	$HClO$	$HBrO$	HIO	hypo——ous	hypo——ite

† Stable perbromates such as $KBrO_4$ have been prepared in recent years.

$HClO$	hypochlorous acid
$HBrO$	hypobromous acid
HIO	hypoiodous acid
$HClO_3$	chloric acid
$HBrO_3$	bromic acid
HIO_3	iodic acid

It is instructive to write the electron-dot formulas for the oxyhalogen acids. The chlorine acids are used as examples.

hypochlorous acid

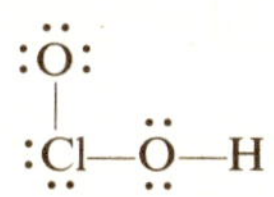

chlorous acid

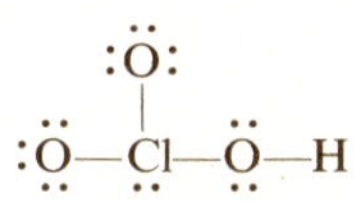

chloric acid

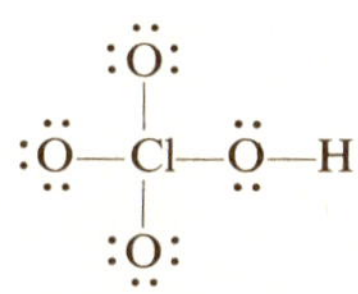

perchloric acid

In order to follow the octet rule, one sees that three of the bonds in $HClO_4$, two in $HClO_3$, and one in $HClO_2$ are required to be coordinate covalent bonds. Bond-hybridization theory suggests sp^3 hybridization of the chlorine orbitals. In the case of $HClO_4$, where four oxygen atoms are bonded to the chlorine, the structure of the molecule is a tetrahedron (slightly distorted because of the influence of the hydrogen atom). The structure of ClO_4^- is a perfect tetrahedron.

Periodic acid exists in two forms:

HIO_4	metaperiodic
H_5IO_6	paraperiodic

The structure of the former is analogous to perchloric acid. Iodine in HIO_4 is the central atom of a tetrahedral structure. The paraperiodic acid structure is the octahedral (sp^3d^2) structure shown below:

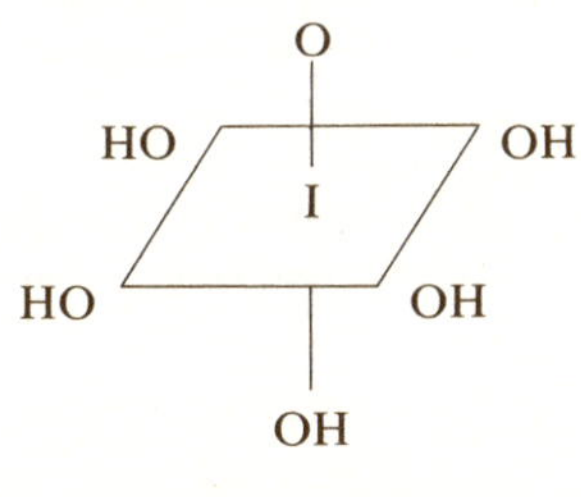

One finds a similar situation occurring in Group VA. Nitrate ion is a trigonal structure, but phosphorus and arsenic are large enough to occupy a tetrahedral position. Hence, the formula for the +5 oxyanion of nitrogen is NO_3^-, but the formulas for the +5 oxyanions of phosphorus and arsenic are PO_4^{3-} and AsO_4^{3-}.

When one wishes to add H^+ to a system and produce a minimal effect from the anion that has to accompany the H^+, dilute perchloric acid is often used. The large size and small charge of the perchlorate ion gives it practically no tendency to complex with cations or, for the same reasons, to form insoluble compounds. Theoretically, perchlorate ion is a strong oxidizing agent even in dilute solution; however, in practice, because of the slow kinetics associated with its reactions with inorganic compounds, it is quite unreactive in dilute solutions at room temperature.

It is the large size of the iodine atom that allows it to assume the octahedral structure. The other halogens are apparently too small for compounds analogous to paraperiodic acid to form. Paraperiodic acid is the predominant species in strongly acidic solutions. It is monoprotic and might be looked upon as $HIO_4 \cdot 2H_2O$. When periodic acid (in any form) is neutralized with a base, the periodate ion IO_4^- results.

The extent of ionization of the oxygen acids of the halogens shows a gradual change from the +7 acids, which are almost completely dissociated, to the +1 acids, which are only weakly ionized. The variation in acid strength can be explained as the result of an inductive effect of the very electronegative oxygen atoms. In perchloric acid, three oxygens attract the electrons around the central atom. This effect in turn is transmitted to the oxygen bonded to the hydrogen. With the increased polarity thus given to the O-H bond, it is relatively easy for a water molecule to complete the job of ionizing the hydrogen. The electron-dot formula below illustrates the shift in the electron density of the molecule that occurs as a result of the affinity of the oxygen atoms for electrons:

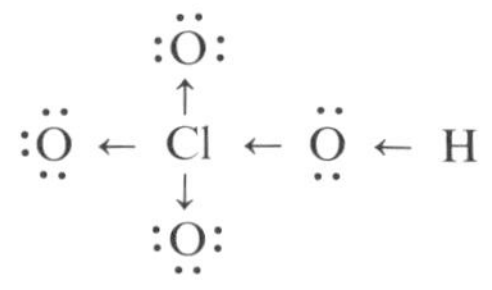

The fewer oxygens present, the weaker the inductive effect becomes, and the weaker the acid; hence, listed in decreasing order of acid strength we have $HClO_4$, $HClO_3$, $HClO_2$, and $HClO$.

The same effect is found in H_2SO_4 compared with H_2SO_3, and in HNO_3 compared with HNO_2. The higher oxidation-state acid in each pair is the stronger acid.

The redox properties of the oxygen acids and oxyanions of chlorine are summarized below:

Acidic solution

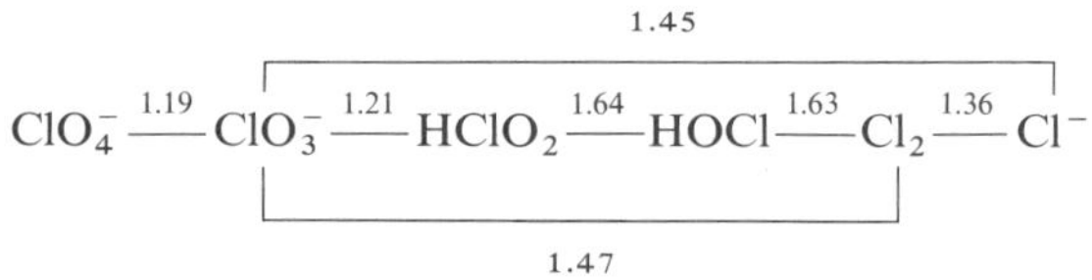

Basic solution

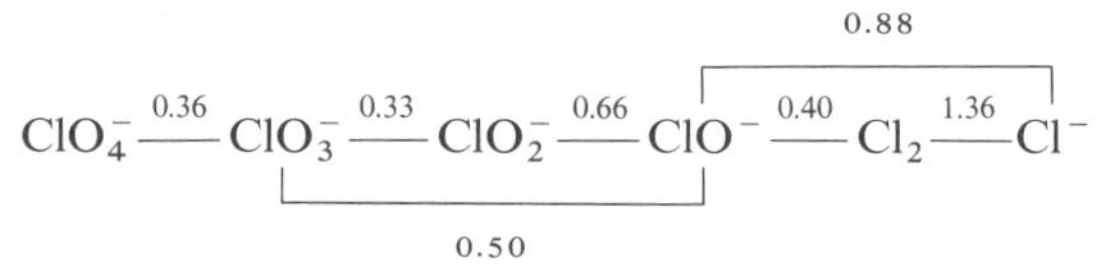

All of the oxygen acids of chlorine and their oxyanions are seen to be oxidizing agents, and in acid solution, they are *powerful* oxidizing agents. The possibility for several disproportionation reactions is evident from the diagrams above. There are two disproportionation reactions in basic solution that are especially important:

$$Cl_2 + 2\,OH^- = Cl^- + ClO^- + H_2O; \qquad \mathscr{E}^0 = 0.96\ V \qquad [19.32]$$

$$3\,ClO^- = ClO_3^- + 2\,Cl^-; \qquad \mathscr{E}^0 = 0.39\ V \qquad [19.33]$$

The first reaction is the basis for a method of obtaining hypochlorites, and the second is used to prepare chlorates.

Chlorine also disproportionates in acid solution, but to a much more limited extent. As a matter of fact, $\mathscr{E}^0$ for the reaction

$$Cl_2(g) + H_2O = Cl^- + HOCl + H^+ \qquad [19.34]$$

is -0.27 V, which means that the reaction would go to the left if both reactants and products were in their standard state concentrations (1 atm for the Cl_2 gas and 1 *M* concentration for H^+, Cl^-, and HOCl). Decreasing the H^+ concentration of the solution by adding a base drives the reaction to the right, and we end up with the reaction shown in equation [19.32].

The redox information for bromine is

Acidic solution

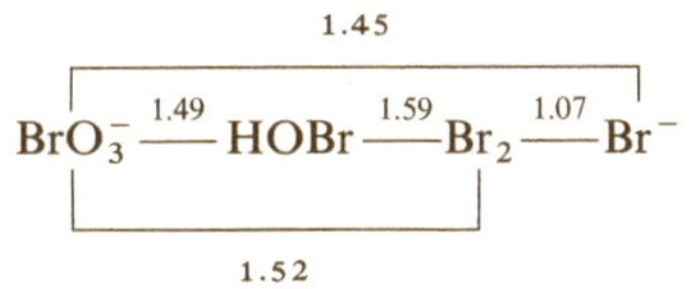

Basic solution

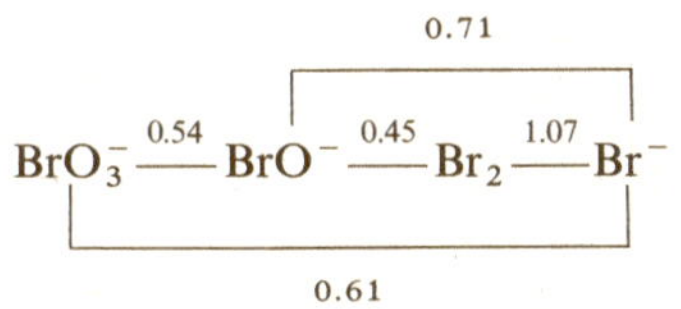

As with chlorine, the oxyacids and oxyanions of bromine are seen to be strong oxidizing agents, especially in acid solution. Disproportionation reactions analogous to those just discussed for chlorine are also seen to be possible.

The redox story for iodine is similar:

Acidic solution

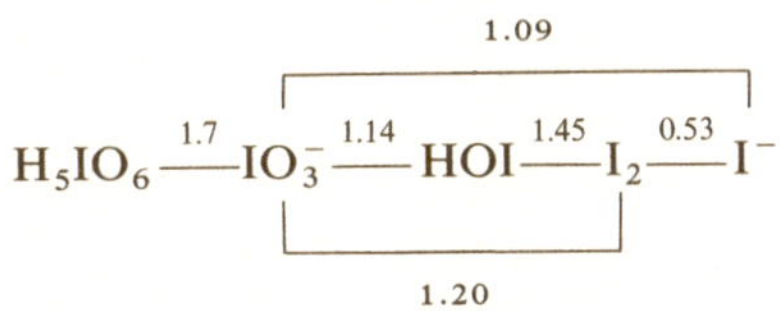

Basic solution

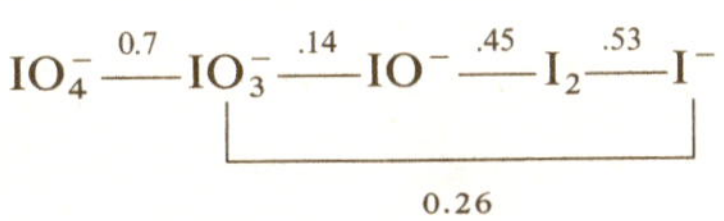

Again we see that the oxyacids and oxyanions are all strong oxidizing agents, especially in acid solution. Disproportionations analogous to those discussed with chlorine and bromine also occur with iodine. For example, in basic solution we have

$$I_2 + 2\,OH^- = I^- + IO^- + H_2O; \qquad \mathscr{E}^0 = 0.08\text{ V}$$

and

$$3\,IO^- = 2\,I^- + IO_3^-; \qquad \mathscr{E}^0 = 0.35\text{ V}$$

If we multiply the first reaction by three and add it to the second, we get

$$3\,I_2 + 6\,OH^- = 5\,I^- + IO_3^- + 3\,H_2O; \qquad \mathscr{E}^0 = 0.43\text{ V}$$

The large positive $\mathscr{E}^0$ shows that this reaction goes far to the right. The analogous reaction in acid solution, however, has an $\mathscr{E}^0$ of -0.67, which tells us that the reverse reaction takes place in acid solution (standard state condition):

$$IO_3^- + 5\,I^- + 6\,H^+ = 3\,I_2 + 3\,H_2O; \qquad \mathscr{E}^0 = 0.67\text{ V} \qquad [19.35]$$

This reaction (the reverse of disproportionation) will then have an $\mathscr{E}^0$ of $+0.67$ V, which corresponds to an equilibrium constant of 3.5×10^{56}. Reaction [19.35] thus proceeds far to the right. It is so complete, in fact, that it finds application in the quantitative analysis of iodide, iodate, and hydrogen ions.

The analogous reaction for the oxidation of bromide ion with bromate ion also takes place, but to a somewhat lesser extent:

$$BrO_3^- + 5\,Br^- + 6\,H^+ = 3\,Br_2 + 3\,H_2O; \qquad \mathscr{E}^0 = 0.45\text{ V}$$
$$K = 1.08 \times 10^{38}$$

Perchloric acid is only one of many unstable oxyhalogen compounds. We discussed earlier in this chapter the instability of the halogen oxides. Salts, such as the chlorates and perchlorates, also decompose readily when heated. In a standard laboratory preparation of O_2 gas, potassium chlorate is heated to remove the oxygen by decomposition:

$$2\,KClO_3(s) \xrightarrow{\text{heat}} 2\,KCl(s) + 3\,O_2(g)$$

(MnO_2 is often used as a catalyst to increase the rate of decomposition—see Chapter 12.)

Even the reaction between chloride and chlorate ions in *acid* solution goes far to the right:

$$\mathscr{E}^0 = 0.11\text{ V}$$
$$K = 1.98 \times 10^9$$

Before leaving the discussion of the oxyacids and oxyanions of the halogens, we want to caution you about the violence of the reactions of these strong oxidants when mixed with carbon or organic materials. Potassium chlorate, for example, is a component of black powder. Pure perchloric acid decomposes explosively when heated. For that reason, it is usually stored and used in no greater concentration than about 67% $HClO_4$ in water. Several serious industrial accidents have resulted from organic matter inadvertently coming into contact with hot perchloric acid solutions.

19.4 Group VIIIA: The Noble Gases

Helium was discovered on the sun before it was known to be on the earth. The discovery of

Before 1962, the Group VIIIA elements—He (helium), Ne (neon), Ar (argon), Kr (krypton), Xe (xenon), and Rn (radon)—were appropriately known as the inert gases. Until that time, no reactions of these gases with other elements were known, and overconfidence in the octet rule (Group VIIIA elements already have a full octet) probably discouraged attempts to produce compounds involving them.

In 1962, however, N. Bartlett produced a reaction between xenon and PtF_6 that stimulated considerable interest in noble-gas compounds. Several are now known, and their syntheses are not excessively difficult to accomplish. For example, XeF_2, XeF_4, and XeF_6 are prepared by direct combination of the elements. Xenon trioxide (XeO_3) is prepared by complete hydrolysis of XeF_6:

$$XeF_6 + 3\ H_2O = XeO_3 + 6\ HF$$

Bond-hybridization theory accounts well for the bonding and observed structures of the xenon fluorides (see Chapter 14).

Xenon trioxide has been prepared in quantity, and its properties studied. It shows promise of being a very useful oxidizing agent in synthetic inorganic chemistry. When dry it explodes violently, but in solution it behaves well. It is both a powerful and a convenient oxidizing agent. The reduction half-reaction is

$$XeO_3 + 6\ H^+ + 6\ e^- = Xe + 3\ H_2O; \qquad \mathscr{E}^0 = 2.12\ V.$$

The description of XeO_3 as a *powerful* oxidant is justified by the high $\mathscr{E}^0$ value, and its description as *convenient* comes from the fact that its only reduction product is a noble gas that can be allowed to escape from the reaction vessel. Contamination of the system with byproducts of the reaction is thus avoided.

The noble gases have some interesting and, in the case of He, most unusual physical properties; they also have important laboratory and commercial uses. Because of its low density and inertness, helium is a highly desirable gas for "lighter-than-air" aircraft. Its lifting power is smaller than that of H_2, but it has the advantage of not being inflammable. Liquid helium is used as a very-low-temperature refrigerant. Helium boils at a lower temperature than any other substance, approximately 4° K, and by "pumping" on the liquid helium (that is, evacuating the gases above it so as to increase the rate of evaporation), temperatures as low as 1° K can be achieved.

Argon, the most abundant of the noble gases in the atmosphere (nearly 1% by volume) is used in arc welding to provide a nonoxidizing atmosphere. Neon is well-known for its use in advertising lighting, and a modern use for Ne and He is in chemical lasers.

the helium spectrum in the sun's chromosphere was made in 1868. In 1895, Ramsay identified the inert gas that Hillebrand had discovered in 1890 (from heating uranium minerals) as helium.

Helium is produced naturally on the earth during the decay of radioactive elements such as uranium. Because of its low molecular weight, it escapes fairly rapidly from the gravitational field of the earth; hence, its concentration in the atmosphere is very low, 0.00052 percent by volume. It is interesting to note that helium is the only non-radioactive element on the earth that did not come from the sun. All helium on the earth apparently was "born" on the earth. The commercial source of helium is natural-gas deposits in the United States and Canada. The concentration of helium in these natural gases is in the range of 0.1 to 2 percent.

Liquid helium occurs in two forms, He-I and He-II, with a pronounced transition point evident as one form changes into the other. The "high-temperature" form, He-I, stable above 1.774° K is a normal liquid. He-II, however, behaves unlike any other known substance. It has ten times the heat capacity of He-I and a conductivity for heat of about 3×10^6 times that of He-I, and it behaves as if it had zero viscosity.

Summary

Nitrogen Family

The Group VA elements, with five valence electrons, have a maximum oxidation number of +5 in their compounds. Variable oxidation states are more common in this family than in the IA–IVA Groups, discussed in Chapter 18.

Because of its interesting chemistry, the necessity of its compounds in living systems, and the industrial value of its compounds, nitrogen has to be ranked as one of the most important of the elements.

Nitrogen can be found in its compounds in all nine of the integral oxidation states from -3 to $+5$. Two of the most important of its compounds are NH_3 and HNO_3. Ammonia in water solution is a weak base; it reacts with H^+ to form NH_4^+. HNO_3 is a strong acid and a strong oxidizing agent; it reacts with OH^- to form NO_3^-. In acting as an oxidizing agent it is reduced to NO_2, NO, N_2, and even NH_4^+, depending on the strength of the reducing agent and other conditions of the reaction.

Direct combination of N_2 and H_2 to form NH_3 is the commercially important process for "fixing" the relatively unreactive N_2 gas of the atmosphere into useful compounds. Devising optimum conditions of temperature, pressure, and catalysts for this process and the subsequent oxidation of NH_3 to HNO_3 is a noteworthy achievement of science and engineering, and it is an excellent example of the application of thermodynamics and chemical kinetics.

Because the compounds of phosphorus, like those of nitrogen, are essential to life, phosphorus is also a highly important element. Probably the most useful compounds of phosphorus are H_3PO_4 and its calcium salts, the latter having inestimable value to agriculture as commercial fertilizers. H_3PO_4 is a weak acid in all three of its ionizations. Unlike HNO_3, it has practically no oxidizing capabilities.

Oxygen Family

Group VIA contains the important nonmetals, oxygen and sulfur. Oxygen is by far the most abundant element in the earth's crust, making up approximately 50 percent of the mass of the crust. It is the second-most electronegative element, and it forms compounds with almost all other elements, both metals and nonmetals. By far its most common oxidation number is -2. H_2O_2 is an important compound in which the oxidation number is not -2. In the peroxides oxygen is linked to oxygen, resulting in a -1 oxidation state. Ozone, O_3, is an allotrope of the more common diatomic form of molecular oxygen, O_2. A layer of ozone exists in the upper atmosphere and plays an important role in preventing excessive ultraviolet radiation from reaching the surface of the earth.

The most important oxidation states of sulfur are $+6$ (SO_3 and H_2SO_4 and its salts), $+4$ (SO_2 and H_2SO_3 and its salts), and -2 (H_2S and its salts).

Halogen Family

The elements of Group VIIA are fascinating because of the large number and variety of compounds they form. As their name implies, the halogens are the "salt formers." They are active nonmetals that react with metals to form principally ionic salts in which the halide ion has a -1 charge. The halogens react with hydrogen to form gaseous hydrogen halides, which when dissolved in water give the important hydrohalic acids, hydrochloric acid being an example. The halogens form a wide variety of oxygen acids. These are all strong oxidizing agents. Several interesting disproportionation reactions take place among them.

Noble Gases

The elements of Group VIIIA are all monatomic gases at room temperature. As might be expected from the fully completed *s* and *p* energy levels, their chemistry is quite limited. We no longer, however, look upon them as being as inert as they were once thought to be. They have several important industrial uses, based mainly on their low boiling points and relatively inert character.

Exercises

19-1.
Complete the following table.

NAME	FORMULA	NAME	FORMULA
Sodium nitrite			$Mg_3(PO_4)_2$
	H_3PO_3		H_3AsO_4
	KO_2	Phosphoric acid	
Sodium thiosulfate		Nitric acid	
	PH_3		$H_2S_2O_7$

19-2.
Complete the following table, which gives the names of the halogen acids.

HALOGEN OXIDATION STATE	CHLORINE		BROMINE		IODINE	
	NAME OF ACID	FORMULA	NAME OF ACID	FORMULA	NAME OF ACID	FORMULA
−1	Hydrochloric					
+1				HOBr		
+3						
+5						
+7		$HClO_4$				H_5IO_6

19-3.
Complete and balance the following equations, which represent reactions of nitrogen.

(a) $HNO_3 + Cu =$
(b) $NH_4^+ + CaO =$
(c) $NH_4NO_3 \overset{\text{heat}}{=}$
(d) $NO + O_2 =$
(e) $NO_2 + H_2O =$
(f) $N_2O_5 + H_2O =$
(g) $HNO_2 =$
(h) $NH_3 + O_2 \underset{\text{catalyst}}{\overset{\text{Pt}}{=}}$
(i) $HNO_2 + Cr_2O_7^{2-} = Cr^{3+} +$ (acid solution)

19-4.
Complete and balance the following equations, which represent reactions of phosphorus and arsenic.

(a) $H_3AsO_3 + MnO_4^- =$ (acid solution)
(b) $P_4O_6 + H_2O =$
(c) $AsCl_3 + H_2O =$
(d) $P_4O_{10} + OH^- =$
(e) $Ca_3(PO_4)_2 + H_2SO_4(aq) =$
(f) $P_4 + O_2 =$
(g) $Ca_3P_2 + H_2O =$

19-5.
Complete and balance the following equations, which represent chemical reactions of sulfur.

(a) $S + O_2 =$
(b) $H_2SO_4 + NaCl \overset{\text{heat}}{=}$
(c) $H_2SO_4 + NaI \overset{\text{heat}}{=}$
(d) $SO_3 + H_2O =$
(e) $H_2SO_4 + SO_3 =$
(f) $S + SO_3^{2-} =$
(g) $SO_2 + OH^- =$
(h) $S_2O_3^{2-} + I_2 =$

19-6.
Complete and balance the following equations, which represent chemical reactions of the halogens.

(a) $Cl_2 + Br^- =$
(b) $I^- + IO_3^- =$ (acid solution)
(c) $NaI + H_3PO_4 \overset{\text{heat}}{=}$
(d) $I_2 + OH^- =$
(e) $Sn + Cl_2 =$
(f) $IO_3^- + H_2SO_3 =$ (acid solution)
(g) $H_2 + F_2 =$
(h) $PCl_3 + H_2O =$
(i) $HBr(g) + H_2O =$
(j) $HOCl + H_2SO_3 =$

19-7.
Complete and balance the following equations for reactions of hydrogen peroxide.

(a) $H_2O_2 + MnO_4^- = Mn^{2+} +$ (acid solution)
(b) $H_2O_2 + Cr_2O_7^{2-} = Cr^{3+} +$ (acid solution)
(c) $H_2O_2 + H_2O_2 =$
(d) $H_2O_2 + H_2SO_3 =$ (acid solution)
(e) $H_2O_2 + OH^- =$

Problems

19-1.
Write the chemical reactions that could be used to make the following conversions:

(a) $O_2 \rightarrow H_2O_2$
(b) $S_8 \rightarrow S_2O_3^{2-}$
(c) $N_2 \rightarrow HNO_3$
(d) $P_4 \rightarrow H_3PO_4$
(e) $S_8 \rightarrow H_2S$
(f) $Ca_3(PO_4)_2 \rightarrow P_4$
(g) $Cl_2 \rightarrow ClO_3^-$

19-2.
Write the chemical reaction that occurs in each of the following processes:

(a) Lithium nitride is added to water.
(b) Phosphorus is burned in a limited supply of oxygen.
(c) Zinc sulfide is roasted (heated in a stream of air).
(d) Silver metal is dissolved in concentrated nitric acid.
(e) Hydrofluoric acid is stored in a quartz (SiO_2) container.
(f) Hydrochloric acid is added to a solution of sodium thiosulfate.
(g) Solid potassium chlorate is heated.
(h) Chlorine gas is bubbled through a hot aqueous solution of sodium hydroxide.

19-3.
The reaction for the explosion of ammonium nitrate is

$$2\,NH_4NO_3(s) = 2\,N_2(g) + 4\,H_2O(g) + O_2(g)$$

Calculate ΔG^0_{298} for this reaction to show that it is spontaneous, and calculate ΔH^0_{298} to show the large amounts of energy released during the explosion. (For NH_4NO_3, $\Delta G_f^0 = -43.98$ kcal/mole and $\Delta H_f^0 = -87.37$ kcal/mole at 298° K. Other data are given in Appendix B.)

19-4.

A mixture of NH_3 and NH_4^+ makes a good buffer solution. Calculate the ratio of the concentrations of these two species that would make a buffer solution at a pH of eight.

19-5.

Calculate the pH of the following phosphate solutions:

(a) 0.10 M H_3PO_4
(b) 0.10 M $H_2PO_4^-$
(c) 0.10 M HPO_4^{2-}
(d) 0.10 M PO_4^{3-}
(e) A solution 0.10 M in NaH_2PO_4 and 0.050 M in Na_2HPO_4

(Additional data can be found in Appendix E.)

20

Complex Ions

In its most general definition, a complex ion is any ion composed of two or more elements. This broad definition would thus encompass such ions as CO_3^{2-}, NO_3^-, PO_4^{3-}, and SO_4^{2-}. In this chapter, however, we will concern ourselves with the more restricted group of complex ions in which *a metal ion is bonded to one or more anions or neutral molecules.* An example is $Ag(NH_3)_2^+$; it consists of a central Ag^+ bonded to two NH_3 molecules. The result is a linear $[H_3N—Ag—NH_3]^+$ structure. The ions or molecules that bond to the central metal ion are called *coordinating groups* or *ligands.* Examples of common ligands are

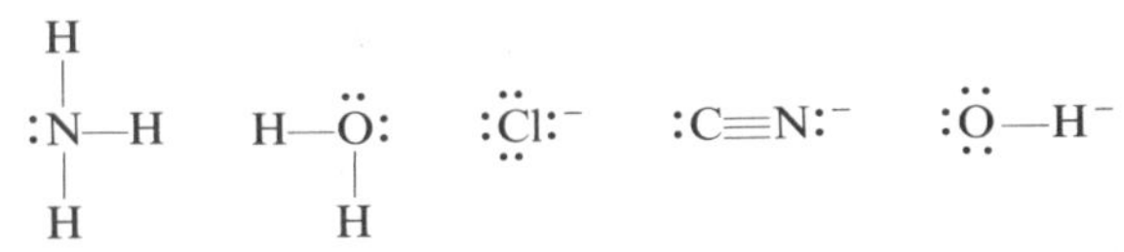

The characteristic that they all have in common is a pair of electrons that can be donated to form a coordinate covalent bond. The characteristic that the central metal ions have in common is that they have empty valence orbitals, as seen in the examples below:

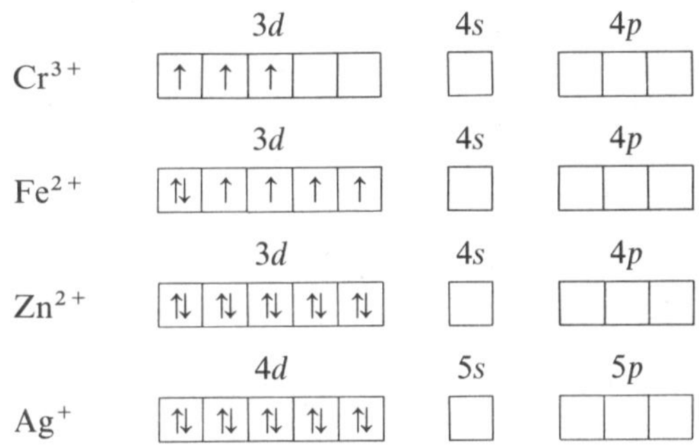

Complex-ion formation might thus be looked upon as a donor-acceptor reaction, with the bonding between the central atom and the ligand being coordinate covalent linkages. We need to be cautious, however, about oversimplification, for as we shall see later (pages 593–597), this useful model does have some deficiencies.

Complex ions are often referred to as **coordination complexes** for this reason.

20.1 Composition of Some Representative Complex Ions

It will be helpful for you to have an overall view of the composition of the common complex ions before learning about the detailed chemistry of any one of them. In this section, therefore, we have grouped the complexes in such a way as to emphasize similarities in composition. This will help you to master the many new formulas that you will encounter in this chapter.

Ammonia Complexes

As we saw earlier (Chapter 17), metallic ions do not exist in water solution as simple ions but as hydrated species. For example, $Cu^{2+}(aq)$ is not a simple ion but a complex in which the ligand groups are water molecules. Hence, we have the aquo complex of probable formula $Cu(H_2O)_4^{2+}$. A water solution of Cu^{2+} is pale blue in color because of the visible-light absorption properties of $Cu(H_2O)_4^{2+}$.

The blue color is the result of the complementary color being absorbed from white light, leaving the blue to be transmitted and reflected. It is interesting that anhydrous $CuSO_4$ is a white powder. Hence it is not Cu^{2+} but $Cu(H_2O)_4^{2+}$ that gives the characteristic blue color to solutions of copper salts.

When NH_3 is added to the solution, however, a definite change in color to a *deep* blue takes place. The NH_3 molecules have replaced water as the ligand, and the very stable $Cu(NH_3)_4^{2+}$ complex ion, which is dark blue in color, has been formed. The reaction is

$$\underset{\text{(light blue)}}{Cu(H_2O)_4^{2+}} + 4\,NH_3 = \underset{\text{(dark blue)}}{Cu(NH_3)_4^{2+}} + 4\,H_2O$$

Several cations form stable complexes with ammonia. Some **ammonia complexes** that are commonly encountered are listed below, grouped according to the charge of the central metal ion. Note that the number of ligands coordinated with the metal ion is in most cases twice the charge of the metal ion. $Ni(NH_3)_6^{2+}$ and $Co(NH_3)_6^{2+}$ are two of the exceptions. However, the rule works often enough that it is an aid in remembering the formulas.

This list and those that follow emphasize the similarities in composition of complex ions. Reactions involving many of these ions are discussed later in this chapter.

+1	+2	+3
$Cu(NH_3)_2^+$	$Cu(NH_3)_4^{2+}$	$Cr(NH_3)_6^{3+}$
$Ag(NH_3)_2^+$	$Zn(NH_3)_4^{2+}$	$Co(NH_3)_6^{3+}$
$Au(NH_3)_2^+$	$Cd(NH_3)_4^{2+}$	
	$Ni(NH_3)_4^{2+}$	
	$Ni(NH_3)_6^{2+}$	
	$Co(NH_3)_6^{2+}$	

It is apparent that the charge on the central atom is at least one important factor that determines the number of ligands that the metal ion can hold. If one thinks in terms of a donor-acceptor model, it seems reasonable to expect that a large positive charge on an acceptor ion would allow it to accept several sets of electron pairs, and hence to coordinate with several ligand molecules.

Complexes with Cyanide Ion

Cyanide ion serves as a ligand to produce another important class of complex ions. Some of the common **cyanide ion complexes** are listed below, again according to charge on the central metal ion. Note the parallelism in formula with those of the ammonia complexes. The ligands here, however, are anions instead of molecules, and hence the net charge of the complex ion will, of course, be different from that of the ammonia complexes.

+1	+2	+3
$Cu(CN)_2^-$	$Zn(CN)_4^{2-}$	$Fe(CN)_6^{3-}$
$Ag(CN)_2^-$	$Cd(CN)_4^{2-}$	$Co(CN)_6^{3-}$
$Au(CN)_2^-$	$Hg(CN)_4^{2-}$	$Au(CN)_4^-$
	$Fe(CN)_6^{4-}$	
	$Mn(CN)_6^{4-}$	

Some of the cyanide complexes are exceptionally stable. $Ag(CN)_2^-$, for example, has a very low dissociation constant:

$$Ag(CN)_2^- = Ag^+ + 2\,CN^-; \qquad K_d = 1 \times 10^{-21}$$

Other very stable complex cyanides are $Fe(CN)_6^{4-}$, $Fe(CN)_6^{3-}$, and $Co(CN)_6^{3-}$. The silver cyanide complex and these three are only slightly decomposed to HCN and the metal ion by a strong acid even though HCN is an extremely weak acid that would strongly promote the decomposition. The calculation is made below for $Ag(CN)_2^-$.

EXAMPLE 20.1 Calculate the ratio of $[Ag^+]$ to $[Ag(CN)_2^-]$ in a solution of $Ag(CN)_2^-$ that is equimolar in H^+ and $HCN(aq)$.

SOLUTION

$$Ag(CN)_2^- + 2\,H^+ = 2\,HCN(aq) + Ag^+$$

$$K = \frac{[HCN(aq)]^2[Ag^+]}{[H^+]^2[Ag(CN)_2^-]} = \frac{K_d}{K_a^2} = \frac{1 \times 10^{-21}}{(6 \times 10^{-10})^2} = 3 \times 10^{-3}$$

where K_d is the dissociation constant for $Ag(CN)_2^-$ and K_a is the acid dissociation constant for HCN. Since we are to have equimolar concentrations of H^+ and $HCN(aq)$,

$$[H^+] = [HCN(aq)]$$

and

$$\frac{[Ag^+]}{[Ag(CN)_2^-]} = 3 \times 10^{-3}$$

The small ratio of Ag^+ to complex ion attests to the stability of the latter.

Chloride Ion Complexes

The halides serve as ligands in many complexes. Some of the common stable **chloride ion complexes,** grouped according to the charge on the central ion, are

+1	+2	+3	+4
$AgCl_2^-$	$CuCl_4^{2-}$	$AuCl_4^-$	$SnCl_6^{2-}$
$AuCl_2^-$	$HgCl_4^{2-}$		$PtCl_6^{2-}$
	$CdCl_4^{2-}$		

Reactions involving many of these, as well as other complex ions, are discussed later in this chapter.

Hydroxide Ion Complexes

The **hydroxide ion complexes** play important roles in analytical separations of elements. Some of the more important of these complexes are given on page 586.

+2	+3	+4
$Be(OH)_4^{2-}$	$Al(OH)_4^-$	$Sn(OH)_6^{2-}$
$Zn(OH)_4^{2-}$	$Cr(OH)_4^-$	
$Pb(OH)_3^-$		
$Sn(OH)_3^-$		

The metal ions that form the complexes with hydroxide ions are seen to be the ones whose oxides and hydroxides are amphoteric (Chapter 16).

20.2 Nomenclature of Complex Ions

To the uninitiated, the names of the coordinate complexes may seem *complex* indeed. When one first reads hexaamminechromium (III) hexacyanoferrate (II), for example, one may well wonder how difficult it will be to master such language. You are now to be let in on the few simple rules that determine the names of the coordinate complexes. You will see that the system produces names that describe simply and clearly the composition of the compound.

Complex ions are named by giving in the following order (1) the number of ligands, (2) the name of the ligand, (3) the name of the central metal ion, and (4) the oxidation state of the metal, shown as a Roman numeral in parentheses.

The number of ligands is designated by a prefix:

2	di-	5	penta-
3	tri-	6	hexa-
4	tetra-		

Special names are given to the following molecular ligands:

H_2O	aquo
NH_3	ammine
CO	carbonyl
NO	nitrosyl

The names of all other molecular ligands are just those of the compounds. Anionic ligands take -o endings. Examples are

Cl^-	chloro
CN^-	cyano
OH^-	hydroxo
NO_2^-	nitro
SO_4^{2-}	sulfato
$S_2O_3^{2-}$	thiosulfato

If the complex ion is a cation, the name of the central atom is the name of the metal unchanged. Hence, $Ag(NH_3)_2^+$ is diamminesilver(I) ion, and $Cr(NH_3)_6^{3+}$ is hexaamminechromium(III) ion.

If the complex ion is an anion, the ending "-ate" is used with the central atom. Again the oxidation number is shown by a Roman numeral. Examples are

Note that the name of the complex is one word, and that it is neither capitalized nor hyphenated.

$PtCl_6^{2-}$	hexachloroplatinate (IV) ion
$Cu(CN)_4^{2-}$	tetracyanocuprate (II) ion
$AgCl_2^-$	dichloroargentate (I) ion
$Fe(CN)_6^{4-}$	hexacyanoferrate (II) ion
$Fe(CN)_6^{3-}$	hexacyanoferrate (III) ion
$Au(CN)_4^-$	tetracyanoaurate (III) ion
$Zn(OH)_4^{2-}$	tetrahydroxozincate ion

The names cuprate, argentate, ferrate, and aurate are from the Latin names of the elements.

When both neutral and anionic ligands are in the same complex ion, one names the neutral (molecular) ligand first. Hence, $Cr(NH_3)_4Cl_2^+$ is tetraamminedichlorochromium(III) ion. If the complex compound is a salt, the cation (either simple or complex) is named first. For example,

Since all good chemists know that zinc ion exists in only the one oxidation state (+2), the Roman numeral designation can be omitted from the name of $Zn(OH)_4^{2-}$.

$Cu(NH_3)_4Cl_2$	tetraamminecopper (II) chloride
K_2PtCl_6	potassium hexachloroplatinate (IV)
$[Cr(NH_3)_6]_4[Fe(CN)_6]_3$	hexaamminechromium (III) hexacyanoferrate (II)

The system of nomenclature just described is that recommended by the International Union of Pure and Applied Chemistry (IUPAC). Sometimes common names are also used, especially where they are clearly more convenient and there is little chance for misunderstanding. For example, ferrocyanide (ferro- from ferrous) ion is used for hexacyanoferrate(II) ion, ferricyanide (ferri- from ferric) ion for hexacyanoferrate(III) ion, and simply zincate ion and aluminate ion for $Zn(OH)_4^{2-}$ and $Al(OH)_4^-$, respectively.

Again, there is no need to specify the oxidation state of zinc and aluminum since each exists in only one oxidation state in its compounds.

20.3 Structure of Complex Ions

The study of the structures of complex ions has been a fascinating subject since the first discovery of such ions in the nineteenth century. We will find in this section that their structures conform to the same basic patterns that we have previously encountered in simpler molecules.

As you can see from the listings of complex ions previously given, the common coordination numbers in complex ions are two, four, and six. The coordination number for Cu(I), Ag(I), and Au(I) is two; these are linear complexes as shown in Figure 20.1a. Four-fold coordination with tetrahedral structure (Figure 20.1b) is common in the complexes of the representative elements and the d^{10} metal ions Zn(II), Cd(II), and Hg(II). The majority of the transition elements that have coordination numbers of four have the square-planar structure shown in Figure 20.1c. Six-fold coordination with octahedral structure,

Two important exceptions to the 2,4,6 rule for coordination numbers are $Sn(OH)_3^-$ and $Pb(OH)_3^-$. Bond orbital-hybridization theory gives a good explanation for these exceptions (see Exercise 20-10).

Figure 20.1
Structures of complex ions.

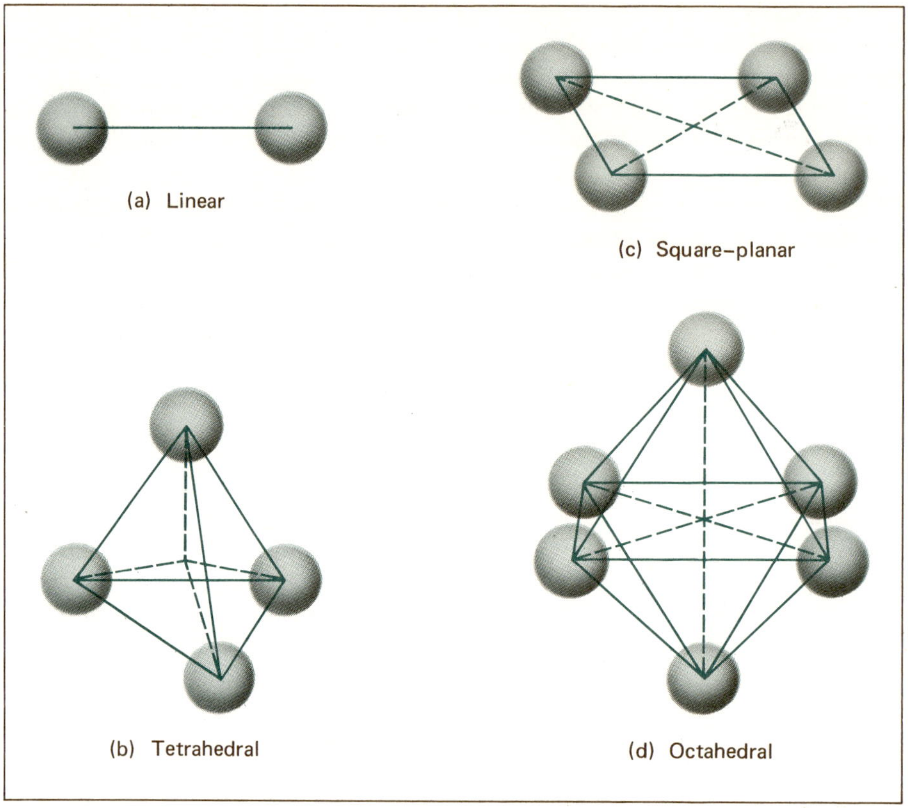

however, is the most commonly encountered structure in complex ions (see Figure 20.1d).

Structural Isomerism

From the point of view of structure, the chlorocomplexes of chromium are especially interesting. There are three distinct compounds with the formula $Cr(H_2O)_6Cl_3$. One is violet in color, one light green, and one dark green. They not only have different colors, but their chemical behavior is quite different. One mole of the violet compound dissolved in water reacts rapidly with three moles of Ag^+ to precipitate three moles of AgCl. One mole of the light-green compound reacts with only two moles of Ag^+, and the dark-green compound reacts with only one mole of Ag^+. Apparently the three chlorine atoms are not bonded the same way in all three compounds. Assuming that the chlorine that is precipitated is the anion of a salt and not a ligand coordinated to the metal ion, we can deduce the three structures shown in Figure 20.2.

Still further substantiation for the proposed structures comes from electrical conductivity data, which indicate four ions in the first compound, three in the second, and two in the third.

Additional evidence for these structures is obtained by subjecting the three salts to mild dehydration. The light- and dark-green compounds lose one and two moles of water, respectively, the loosely held water of crystallization. The water bonded as a ligand is not readily removed.

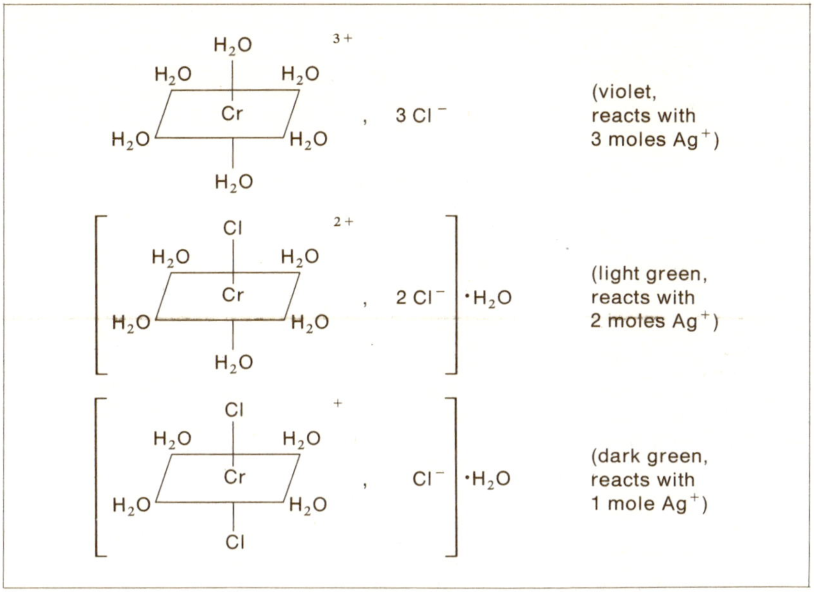

Figure 20.2
Three structures of $Cr(H_2O)_6Cl_3$.

The three separate structures with the same formula provide an example of **structural isomerism.** Compounds with the same formula but different structures are called *structural isomers.* Another example among the complex ions is seen in the two compounds $[Pt(NH_3)_4Cl_2]Br_2$ and $[Pt(NH_3)_4Br_2]Cl_2$. In the first, the chlorine atoms are part of the coordination sphere; in the second, they are the simple anions of the salt.

Structural isomerism is very common in organic chemistry. For example, there are two isomers of C_4H_{10}, three of C_5H_{12}, five of C_6H_{14}, and 62,491,178,805,831 possible structures for $C_{40}H_{82}$. (See Chapter 24).

Geometrical Isomerism

A second type of isomerism is illustrated by the two square-planar compounds diamminedichloroplatinate(II) shown in Figure 20.3. These two distinct compounds, with different physical and chemical properties, have the same formula, $Pt(NH_3)_2Cl_2$. They also have the same structural features, each having two Pt-NH_3 and two Pt-Cl linkages; hence, they are not structural isomers. *Their difference is in the geometrical arrangement of the atoms.* This is an example of

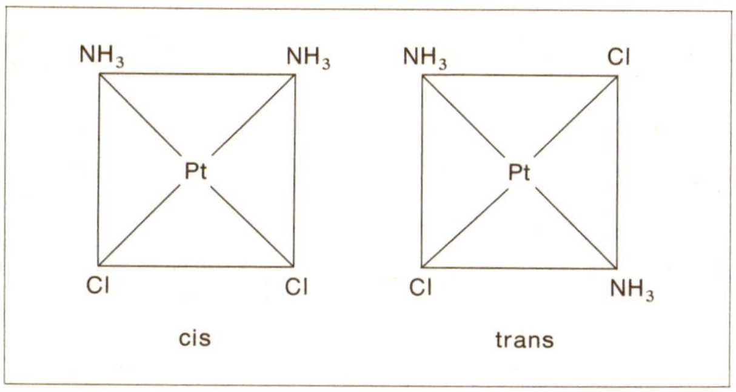

Figure 20.3
Geometrical isomers of the square-planar complexes of $Pt(NH_3)_2Cl_2$.

Figure 20.4
Geometrical isomers of $Co(NH_3)_4Cl_2$.

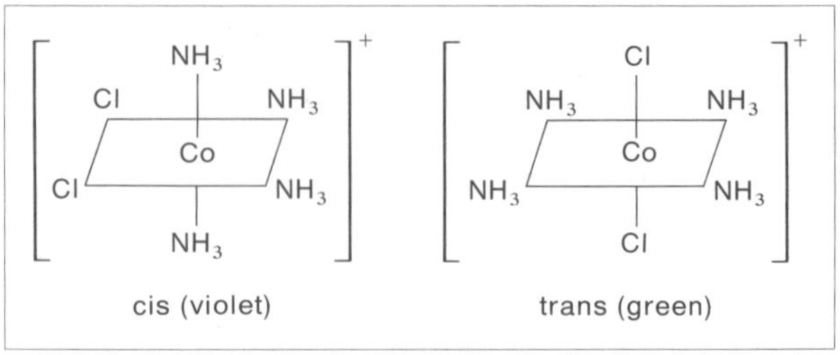

geometrical isomerism. The isomer designated as *cis* has two chloride-ion ligands next to each other (at the adjacent corners of the square), whereas the *trans* (across) isomer has the like groups located at the corners connected by a diagonal of the square.

Geometrical isomerism is possible also in octahedral structures, as shown in Figure 20.4.

Geometrical isomers are not possible for tetrahedral structures because all ligand-ligand distances are equivalent in this structure (see Figure 20.5), and there can be no "next to" and "across from" positions.

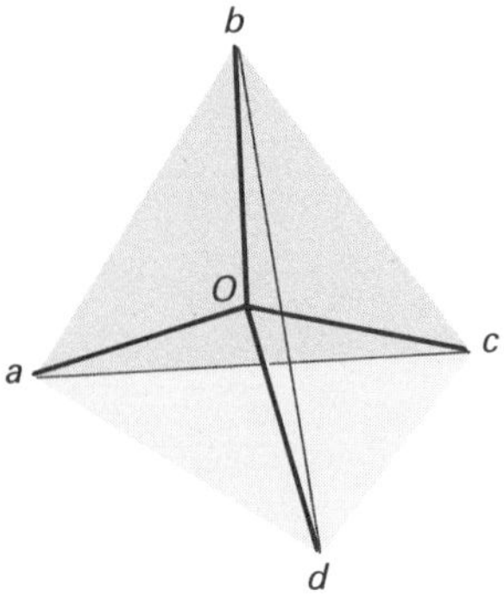

Figure 20.5
No geometrical isomerism is possible in the tetrahedral structure, since $ab = ad = ac = bc = bd = cd$ and $\angle aob = \angle boc = \angle cod = \angle aod$.

Before modern methods for structural determination were available, the existence of isomers was the basis for assigning the structure of complex compounds having four ligands. If no geometrical isomers could be found, the compound was considered to be tetrahedral; if isomers were found, then the structure was considered square-planar. By similar reasoning, the octahedral structure for complexes with six ligands was deduced by Alfred von Werner (1866–1919), who was responsible for much of the early work on inorganic complexes. These early successes in deducing structures of compounds are considered to be among the great triumphs of deductive reasoning. We will have more to say about this subject when we discuss the structure of carbon compounds (Chapter 24).

20.4 Bonding Theories for Inorganic Complexes

The nature of the bonding in complex ions is still an active research topic. In this section we will discuss two bonding theories for complex formation, each of which contributes something to the understanding of the subject, but neither of which is completely adequate.

Valence-Bond Theory

The simplest of the bonding theories is the **valence-bond theory** that was mentioned at the first of the chapter. In this model each ligand donates a pair of electrons to the metal ion to form a coordinate-covalent metal-ligand bond. The electrons donated by the ligand are accepted in empty hybridized orbitals of the metal ion. The types of hybridization that account for the observed structures are

sp	linear
sp^3	tetrahedral
dsp^2	square-planar
d^2sp^3	octahedral

The application of the theory can best be seen by a number of examples:

(a) Diamminesilver(I) ion is a linear complex. The electronic configuration of Ag^+ is

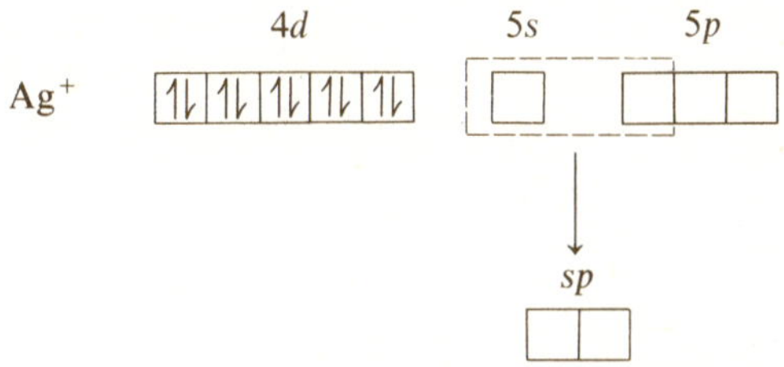

In order to bond with two ligands, two vacant orbitals are required. Hybridization of the empty 5*s* orbital and one of the empty 5*p* orbitals provides the two vacant *sp* orbitals, each of which can accept a pair of electrons from a $:NH_3$ ligand. The *sp* hybridization is consistent with the observed linear structure $[H_3N—Ag—NH_3]^+$.

(b) Tetracyanozincate(II) ion is tetrahedral. The electronic configuration of Zn^{2+} is

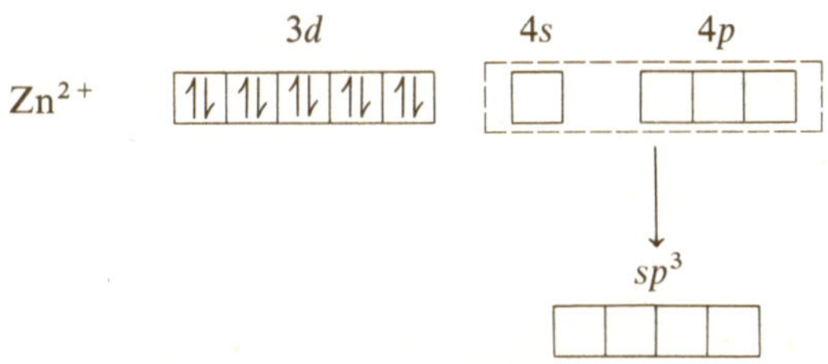

The empty 4*s* and three 4*p* orbitals hybridize to give four tetrahedrally oriented sp^3 orbitals, which accept electron pairs from four cyanide ions, $:C{\equiv}N:^-$. The predicted (and observed) structure is tetrahedral.

(c) Tetracyanonickelate(II) ion is representative of the square-planar complexes. Ni^{2+} has the configuration

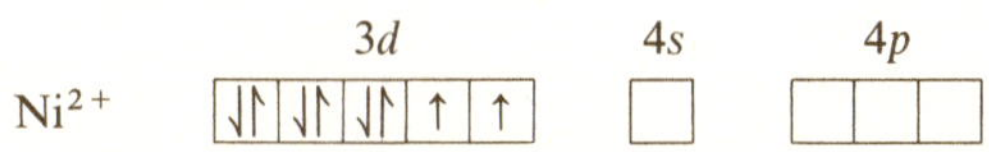

The first step in bond formation would be the pairing of the two single electrons to give a vacant *d* orbital. The vacant 3*d*, 4*s*, and two 4*p* orbitals

could then hybridize as shown below:

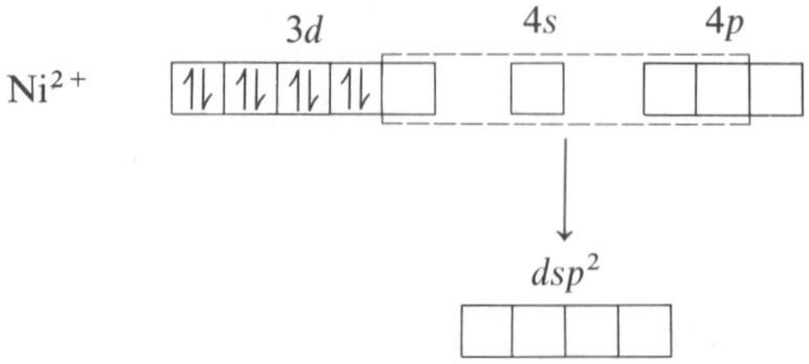

An electron pair from each of four $:NH_3$ molecules could then fill the dsp^2 orbitals. The hybridization postulated is consistent with the observed square-planar structure.

(d) $Cr(NH_3)_6^{3+}$, $Fe(CN)_6^{3-}$, and $Co(NH_3)_6^{3+}$ are typical of the octahedral complexes. The electronic configuration and postulated hybridization for Cr^{3+} in $Cr(NH_3)_6^{3+}$ are

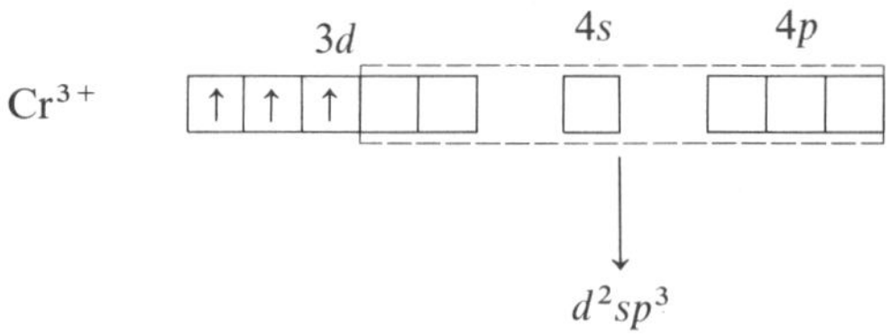

(e) The electronic configuration for the ferric ion is

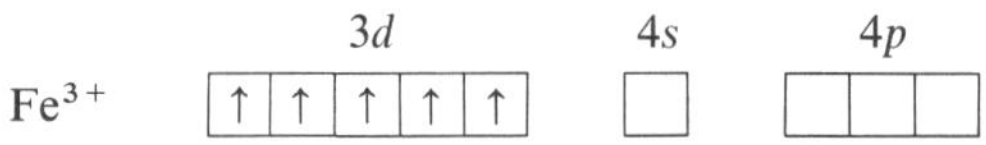

We can imagine the necessary orbitals becoming available by the pairing of four of the single d electrons followed by d^2sp^3 hybridization as shown below:

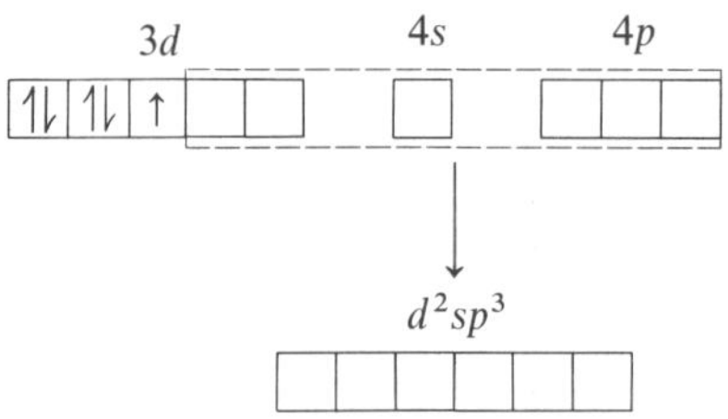

(f) The formation of $Co(NH_3)_6^{2+}$ as interpreted by valence-bond theory would appear as

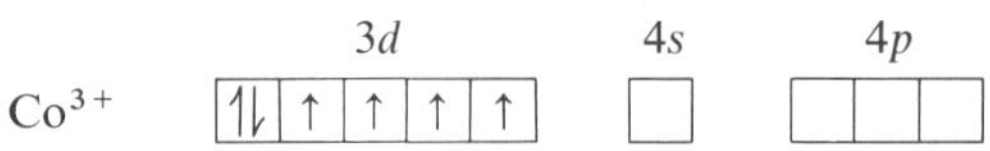

Pairing of the single d electrons followed by d^2sp^3 hybridization gives

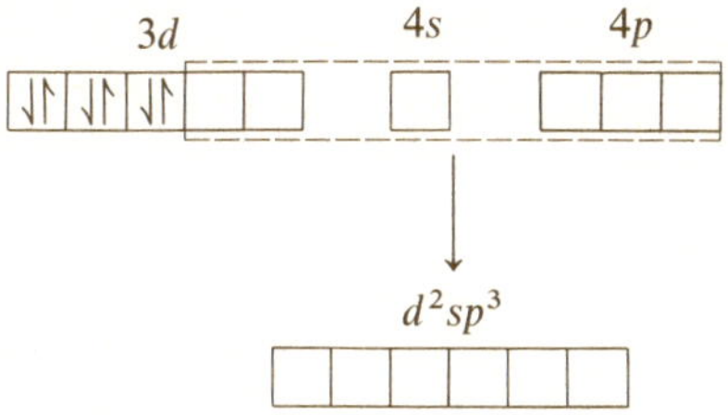

which would be consistent with the observed octahedral structure.

This delightfully simple model, however, has deficiencies when it comes to explaining the observed magnetic properties of certain complexes. For example, according to valence-bond theory all Co(III) complexes would have no unpaired electrons, and hence they would be diamagnetic. While this is true of the NH_3 complex, the fluoro complex, CoF_6^{3-}, is paramagnetic. The magnetic moment of CoF_6^{3-}, in fact, shows the presence of four unshared electrons. In a like manner, magnetic measurements on $Fe(CN)_6^{3-}$ show the presence of one unpaired electron as predicted by valence-bond theory, but similar measurements on FeF_6^{3-} show the presence of five unpaired electrons. Modifications in the theory have been suggested, but these do not appear to be entirely satisfactory.

Crystal-Field Theory

A second, and more modern, theory of bonding in complex ions is referred to as **crystal-field theory.** Bonding in the crystal-field theory is assumed to be purely electrostatic, caused by the attraction of the positively charged metal ion for either the negative charge of an anionic ligand or the negative end of a polar-molecular ligand. Since definite covalent character of metal-ligand bonds can be demonstrated, the model in its simplest form makes no pretense of giving an accurate description of the metal-ligand bond. On the other hand, the theory does account nicely for the magnetic and light-absorption properties of complex ions.

Crystal-field theory is often called ligand-field theory. The latter name, however, is sometimes reserved for an outgrowth of crystal-field theory that employs molecular orbitals.

Crystal-field theory is concerned with the effect that a ligand has on the d orbitals of the transition-metal ion. In the isolated state, the five d orbitals of a transition-metal ion have exactly the same energy. Although the energies are equal, the spatial orientation of the d orbitals are quite different, as is shown in Figure 20.6a. It is seen that when the lobes of the $d_{x^2-y^2}$ and d_{z^2} orbitals are placed *on the axes* of a Cartesian coordinate system, the other three—d_{xy}, d_{xz}, and d_{yz}—lie *between* the axes. The two orbitals that are directed along the axes are called d-gamma (or e_g orbitals) and the three between the axes are called d-epsilon (or t_{2g} orbitals).

We speak of the states in which energy levels are the same as *degenerate states.* Since in this case all five are equal, we have five-fold degeneracy.

Once a complex between ligand and metal ion forms, the metal ion is no longer isolated. The positively charged metal ion now is surrounded by negative charges either from the anionic ligand or the polar-molecular ligand. This electrostatic interaction affects the energies of the d orbitals. The extent of

Although crystal-field theory can be applied to tetrahedral and square-planar complexes, we will restrict our discussion here to the more commonly encountered octahedral complexes of the transition metals.

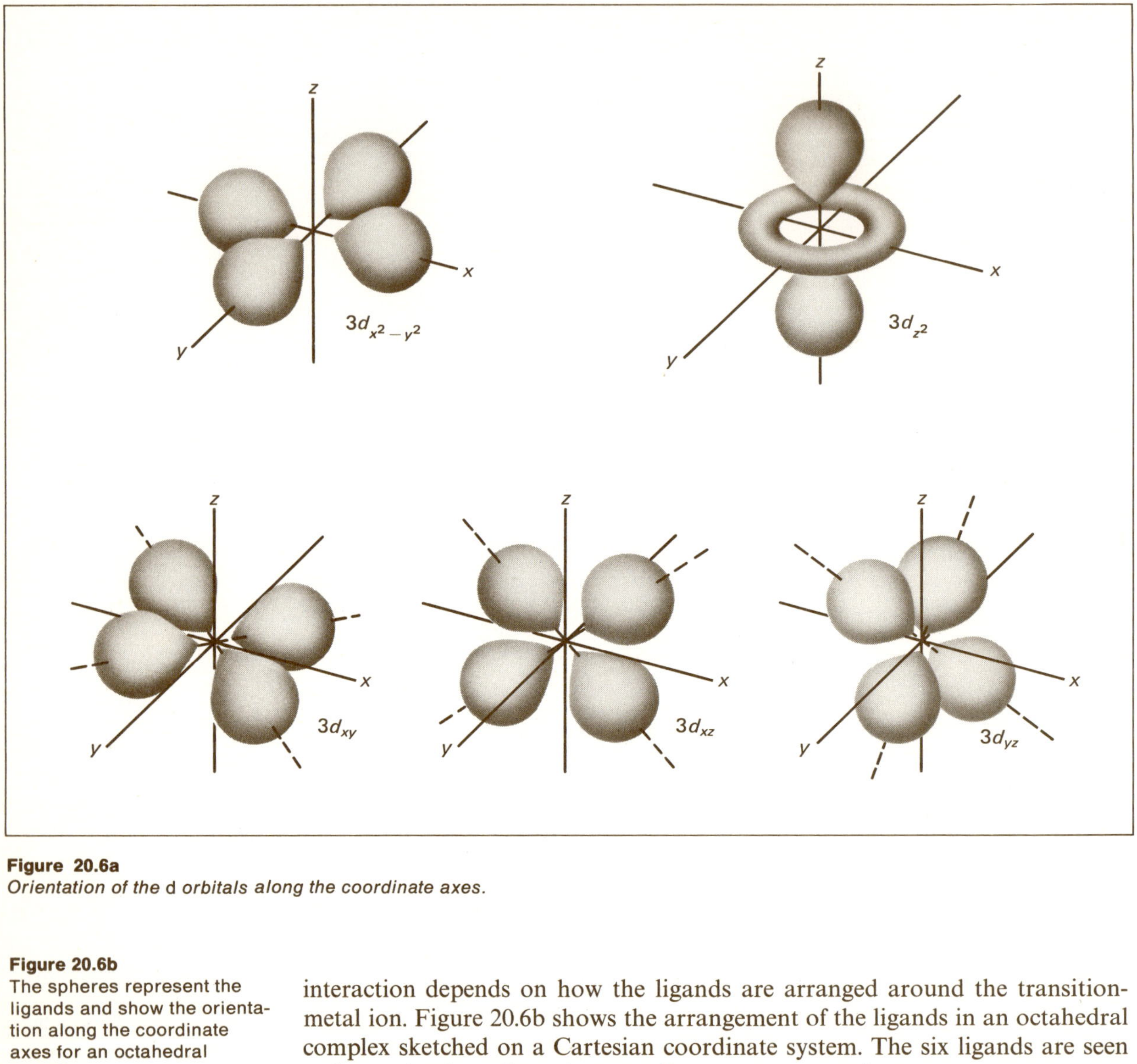

Figure 20.6a
Orientation of the d *orbitals along the coordinate axes.*

Figure 20.6b
The spheres represent the ligands and show the orientation along the coordinate axes for an octahedral complex.

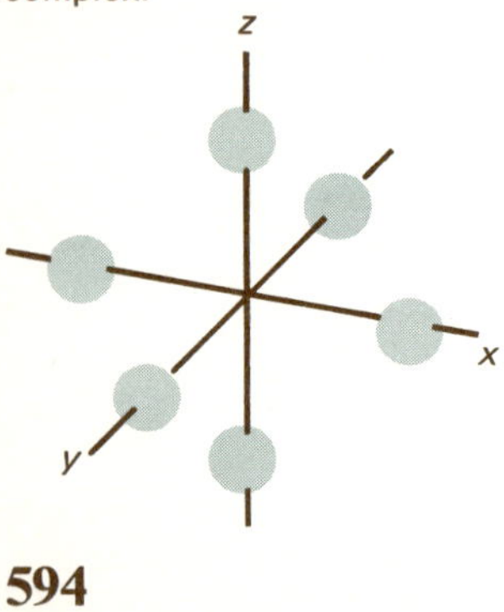

interaction depends on how the ligands are arranged around the transition-metal ion. Figure 20.6b shows the arrangement of the ligands in an octahedral complex sketched on a Cartesian coordinate system. The six ligands are seen to lie on the three Cartesian coordinate axes *x*, *y*, and *z*.

As the ligand approaches the metal ion, there is attraction between the unlike charges, but there is also repulsion between the electrons in the *d* orbitals and those of the ligands. Since the orientation of the ligands is to be along the coordinate axes, those orbitals that are also directed along the axes (the two e_g orbitals) will interact strongly with the ligand field. The other three *d* orbitals lying between the axes (the t_{2g} orbitals) interact less strongly with the ligand. The result is a splitting in the energy of the *d* orbitals into a pair of higher-energy orbitals (doublet) and a lower-energy triplet as shown in Figure 20.7. The extent of crystal-field splitting is designated by Δ_0. Its magnitude depends

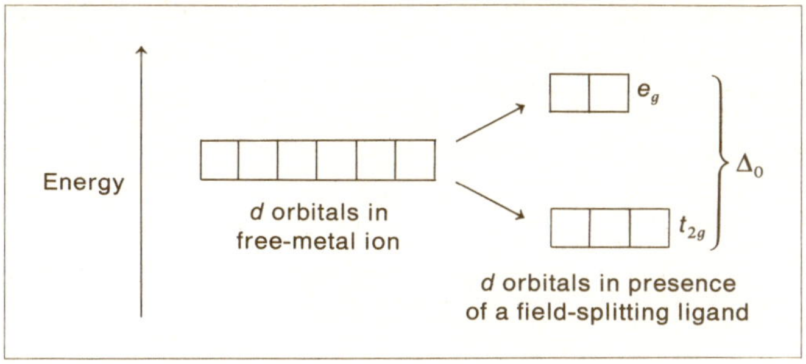

Figure 20.7

on the nature of the ligand. For example, cyanide ion produces a large separation, whereas Cl^- has a much smaller effect.

An experimental value for Δ_0 can be obtained from the light-absorption spectra of complexes. The complex ions of transition metals are almost always highly colored, which means that they absorb light in the visible range. A photon of light can be absorbed if its energy is just right to raise an electron from a lower-energy t_{2g} orbital to the higher-energy e_g orbital, as shown in Figure 20.7. The amount of energy necessary to do this corresponds to wavelengths in the range of visible light. This transition from a t_{2g} to an e_g state is thus responsible for the colors that are so evident in transition-metal complex ions. A simple case is that of $[Ti(H_2O)_6]^{3+}$. Ti^{3+} is referred to as a d^1 ion, because it has one d electron, as shown below.

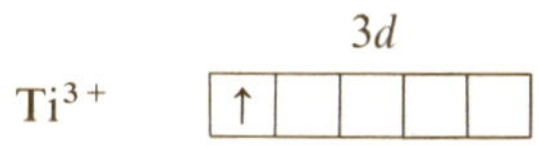

Crystal-field splitting that exists in the aquo complex gives

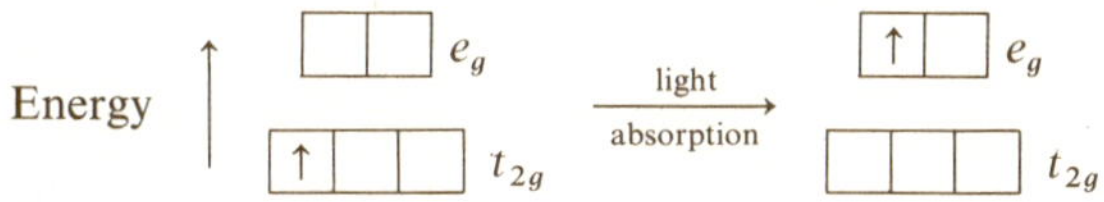

Excitation of the single t_{2g} electron to the e_g state is responsible for the purple color observed in $Ti(H_2O)_6^{3+}$.

Since different ligands produce different Δ_0 values, the color of the complex of a metal ion changes as the ligand is changed. Measurements of the absorption spectra make it possible to rank the ligands in order of their ability to induce a crystal-field split in the d orbitals of any metal ion. This series, known as the spectrochemical series, in order of increasing value of Δ_0 is

$$I^- < Br^- < Cl^- < F^- < C_2O_4^{2-} < H_2O < NH_3 < NO_2^- < CN^-$$

Magnetic Properties

In contrast to the valence-bond theory, the crystal-field theory very capably explains the observed magnetic properties of the complex. As we have seen, Ti^{3+} is a d^1 ion, and hence the titanium complexes, because of this unpaired electron, will be paramagnetic. Since the three t_{2g} orbitals are degenerate, the two d electrons in a d^2 ion will occupy separate t_{2g} orbitals, as shown below; hence, complexes containing central d^2 ions will also be paramagnetic:

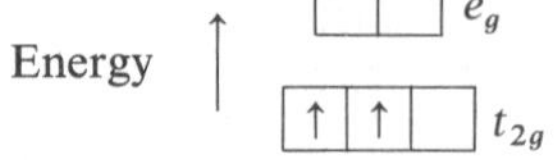

The same is true of d^3 metal ions:

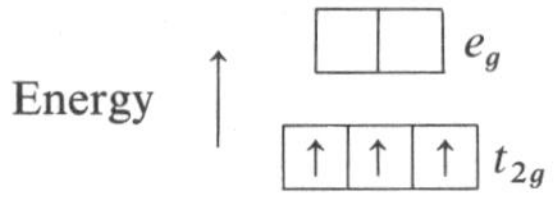

The d^4 central ions, however, have two possibilities for the location of the fourth electron: (1) It can pair with an electron in the t_{2g} state, or (2) it can go to a higher-energy e_g orbital. The two possibilities are illustrated in Figure 20.8. Both alternatives require an expenditure of energy. If Δ_0 is small enough, less energy will be required to locate the electron in an e_g orbital than to pair it with an electron in a t_{2g} orbital. This is the state represented in Figure 20.8a. On the other hand, if Δ_0 is large, the energetically favorable state will be that shown in Figure 20.8b, where the fourth electron pairs with a t_{2g} electron. The magnetic properties of the state shown in part (a) will be those that correspond to four unpaired electrons. This is the **high-spin state** for a d^4 ion. Part (b) shows the **low-spin state,** which has a magnetic susceptibility corresponding to two unpaired electrons.

The possibility for both high-spin and low-spin states exists in the d^4 through d^7 ions. Diagrams for the arrangement of electrons for d^5, d^6, and d^7 ions are given in Figure 20.9.

In the discussion of valence-bond theory we mentioned that $Co(NH_3)_6^{3+}$ was diamagnetic, but that CoF_6^{3-} was paramagnetic with a magnetic susceptibility that corresponds to four unpaired electrons. We can now explain this observation very well using the crystal-field theory. It is apparent from the spectrochemical series that NH_3 causes a large field-splitting and that F^- causes only small separations of the energies of e_g and t_{2g} orbitals. Hence, in $Co(NH_3)_6^{3+}$ we have the low-spin electronic configuration of a d^6 ion, $(t_{2g})^6(e_g)^0$. With CoF_6^{3-}, however, Δ_0 is low enough that the high-spin state, $(t_{2g})^4(e_g)^2$, results. On the

Figure 20.8
Representation of high- and low-spin states for a d^4 ion.

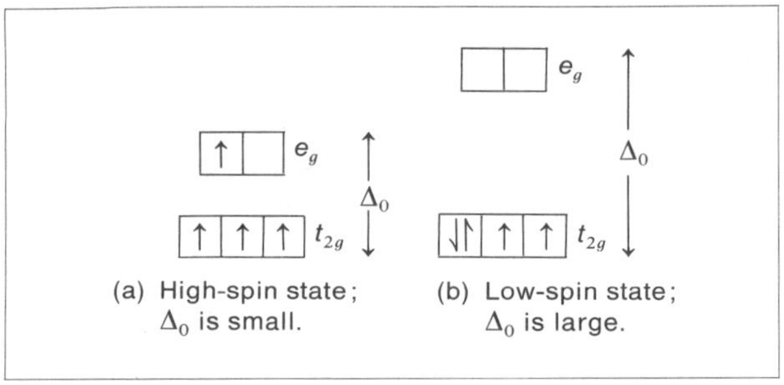

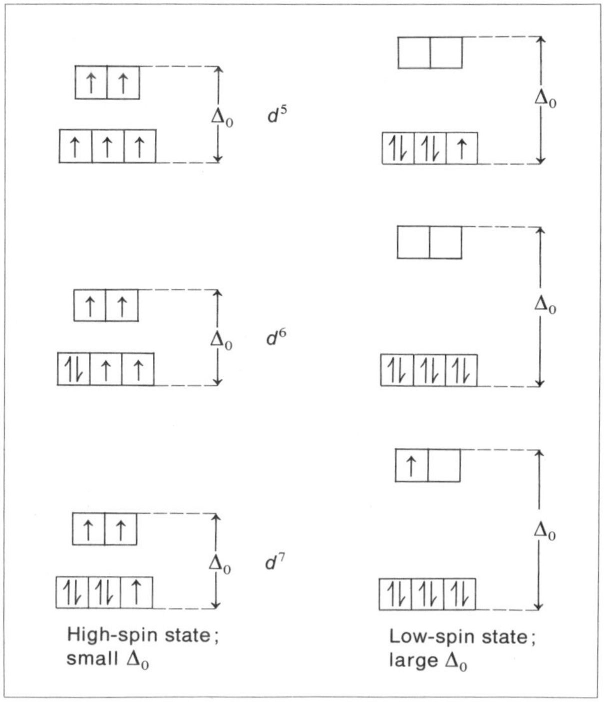

Figure 20.9
Representation of high- and low-spin states for d^5, d^6, and d^7 ion.

basis of crystal-field theory, we therefore can predict paired electrons and diamagnetism for $Co(NH_3)_6^{3+}$ and four unpaired electrons with their corresponding paramagnetism for CoF_6^{3-}, which is in complete agreement with the observed magnetic properties.

EXAMPLE 20.2 Using crystal-field theory, predict the electronic configuration and magnetic properties of (a) $Fe(CN)_6^{3-}$ and (b) FeF_6^{3-}.

SOLUTION Fe(III) is a d^5 ion. In an Fe(III) complex, two alternatives exist:

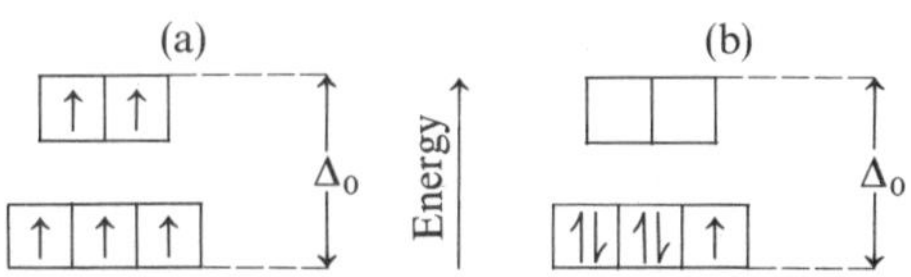

From the spectrochemical series, we see that CN^- has a large effect and F^- a small effect on field splitting. Where Δ_0 is large, as in the cyanide complex, we would expect the low-spin state to occur; we therefore can predict one unpaired electron for $Fe(CN)_6^{3-}$. The fluoro complex, however, with its low Δ_0 value, would be expected to assume the high-spin state of five unpaired electrons. Both complexes are, therefore, paramagnetic: $Fe(CN)_6^{3-}$ having a magnetic susceptibility corresponding to one electron, and FeF_6^{3-} having the magnetic properties associated with five unpaired electrons.

20.5 Chelates

Unidentate comes from the Latin for "single-toothed."

A ligand that coordinates with the metallic ion at only one place on the ligand is called a *unidentate ligand*; NH_3, CN^-, H_2O, and Cl^- are examples. A ligand with two coordinating groups is a **bidentate ligand.** A common example is ethylenediammine (abbreviated as en):

$$H_2\ddot{N}—CH_2—CH_2—\ddot{N}H_2$$

One molecule can span the distance between two points on an octahedral complex, as shown below; coordination takes place at both nitrogen atoms.

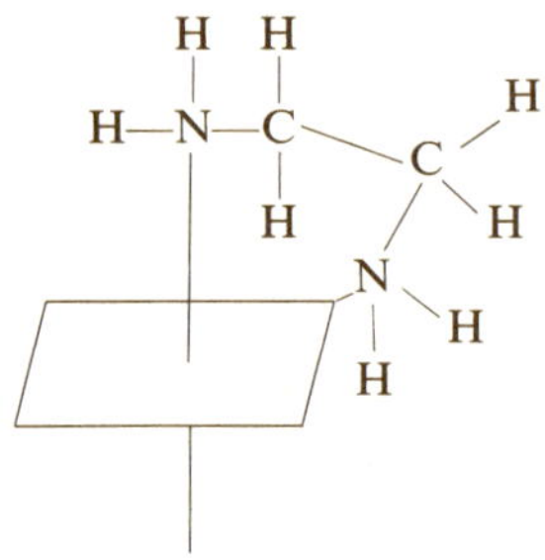

$[Co(en)_3]^{3+}$ has three bidentate ligands. Its structure is as follows:

The loops represent the carbon chain in the ethylenediammine. Only the nitrogen atoms are shown.

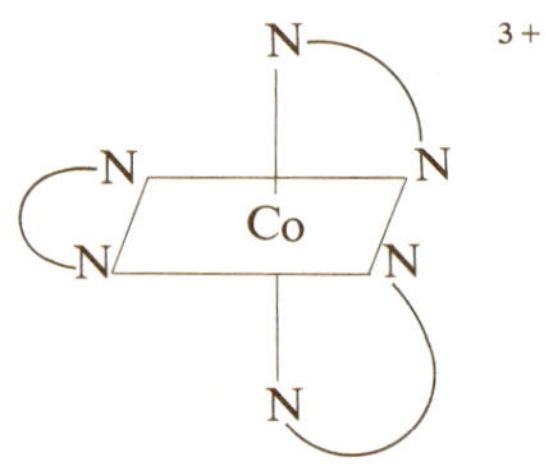

Chelate is from the Greek *chele*, meaning "claw."

The pincerlike structure obtained with a bidentate ligand has led to the name of **chelates** for compounds in which the metal ion is enclosed by a ligand attached at two or more coordinating positions. Ligands with up to six coordinating positions are known. A widely used chelating agent of this type is ethylenediammine tetraacetate ion (EDTA), whose structure is

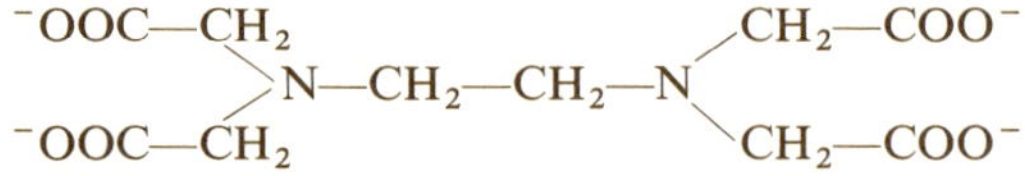

Coordination takes place at the two nitrogen and four terminal oxygen groups. One EDTA ion is thus capable of wrapping itself around all six coordinating positions in an octahedral structure. The structure of the complex of EDTA

with Fe(III) is shown in abbreviated form below:

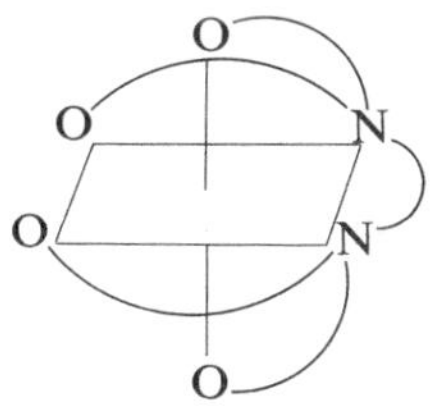

The loops represent the carbon chains in the EDTA ion. Only the oxygen and nitrogen atoms are shown.

The EDTA complex with Fe(III) has proved to be effective in combating iron chlorosis in plants grown in Fe^{3+}-deficient soils. Soils in the western United States often have very high total iron content, but very little soluble Fe^{3+} available to the plant. The high pH of these soils keep the iron tied up in insoluble hydrous oxides:

$$2\,Fe^{3+} + 6\,OH^- = 2\,Fe(OH)_3(s) = Fe_2O_3 \cdot xH_2O(s)$$

Iron chlorosis is characterized by a yellowing of the leaves of plants. Typically, the veins remain green and stand out in bold contrast to the rest of the leaf. Peach trees, roses, and many ornamental shrubs are very sensitive to iron deficiency.

Addition of soluble Fe^{3+} salts to the soil is not helpful since the added Fe^{3+} is immediately precipitated in the highly buffered, basic medium. Addition of soluble Fe^{2+} salts is also not helpful because they are oxidized so readily and completely to $Fe(OH)_3$ in basic solution:

$$4\,Fe^{2+} + O_2 + 8\,OH^- + 2\,H_2O = 4\,Fe(OH)_3(s)$$

The Fe(III)-EDTA complex, however, is sufficiently stable that it can exist in solution in moderate concentrations of OH^-; it is, therefore, available for absorption by the plant.

There are several classes of chelates found in nature that are vital to our existence. **Hemoglobin,** the red-colored component of our blood; **myoglobin,** the substance that stores oxygen in the muscles; and **chlorophyll,** the green-colored substance in plants, are examples. The multidentate ligands in all these complexes belong to the same family of compounds, the **porphyrins.**

The porphyrins can be considered to be derivatives of the simpler compound porphin, which is shown in Figure 20.10. Porphin forms a square-planar complex with several cations, including Mg^{2+}, Fe^{2+}, Cu^{2+}, and Zn^{2+}. The

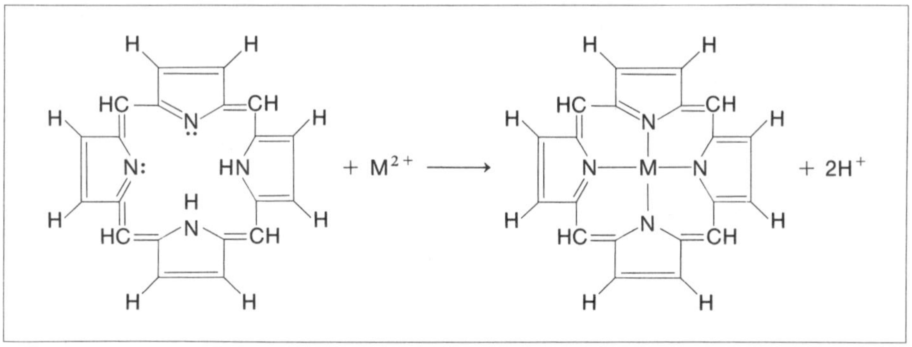

Figure 20.10
Porphin reacts with a +2 metal ion to form a square-planar tetradentate complex.

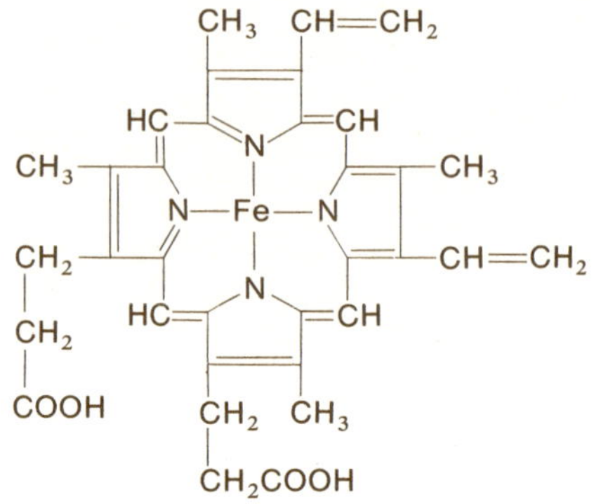

Figure 20.11
Heme, the Fe^{2+}-porphyrin complex of hemoglobin.
A fifth bond from the octahedral Fe^{2+} complex is attached to protein, and the sixth position can be occupied by O_2.

porphyrins have side groups substituted for the outermost hydrogens of the porphin molecules.

The Fe^{2+} porphyrin complex shown in Figure 20.11 is the *heme* part of hemoglobin. The hemoglobin molecule consists of four such heme groups attached to a large (molecular weight $\cong$65,000) protein unit. Hemoglobin serves the vital function in the body of transporting oxygen from the lungs to all the tissues in the body. This is accomplished through the formation of a complex of oxygen with the Fe^{2+} in hemoglobin. The Fe^{2+} in heme appears to be in the center of an octahedral (rather than a square-planar) complex. Four of the bonds from iron go to the porphyrin nitrogens; a fifth is attached to the protein; and, in the oxygenated porphyrin, O_2 occupies the sixth position. When not oxygenated, the sixth bond is also attached to the protein. The following equilibrium applies:

$$\text{Hb} + O_2 = \text{HbO}_2$$

where Hb is used as an abbreviation for hemoglobin. The high concentration of O_2 in the lungs favors the formation of oxyhemoglobin as the blood passes through the capillaries of the lungs; however, when the oxyhemoglobin reaches oxygen-deficient tissue in the rest of the body, the point of equilibrium is shifted to the left, and the oxygen is released. Thus, hemoglobin serves as a transporting agent of O_2, picking it up in the lungs and delivering it to the muscles.

The oxygen is stored in the muscle by myoglobin, another porphyrin and protein complex of Fe^{2+}. Of the six positions around the Fe^{2+}, four are taken up by nitrogens from the porphyrin rings, the fifth is attached to a nitrogen of a complex protein, and the sixth is available to accept an O_2 molecule. Thus, the functional part of myoglobin is similar to hemoglobin; the protein portions of the two molecules, however, are different.

Muscle requires oxygen in order to oxidize sugar in the energy-producing reaction

$$C_6H_{12}O_6 + 6\,O_2 = 6\,CO_2 + 6\,H_2O; \qquad \Delta H = -2800 \text{ kjoules} \qquad [20.1]$$

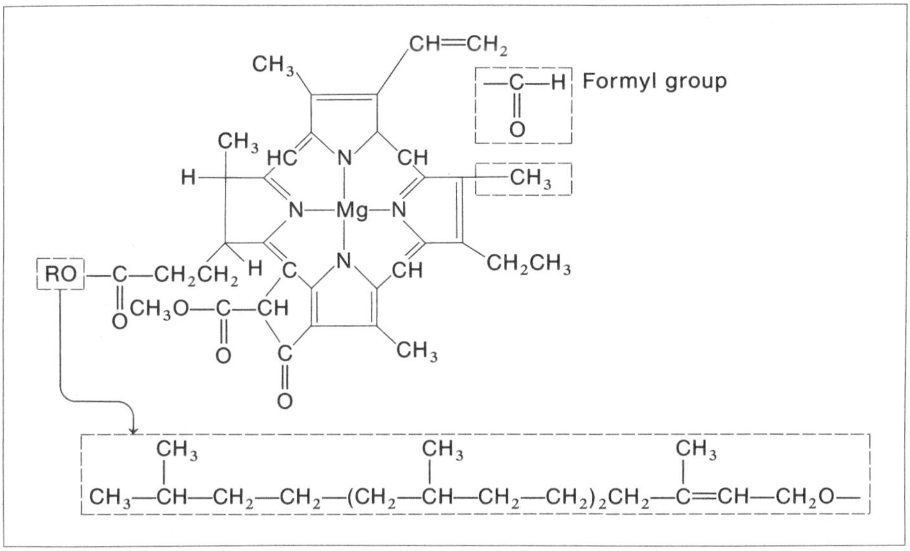

Figure 20.12 *Chlorophyll* a*—a* *Mg^{2+}-porphyrin complex.* Chlorophyll *b* differs from *a* by having a formyl group side-chain in place of the methyl group (upper right-hand side).

The reverse of reaction [20.1] is the process of photosynthesis. The source of the energy for this highly endothermic reaction is sunlight, with chlorophyll serving as the light absorber. Chlorophyll is a square-planar Mg^{2+} porphyrin (see Figure 20.12). Chlorophyll is one further example of the ability of metal complexes to absorb light in the visible range.

There is increasing evidence that chelates are responsible for the absorption and functioning of other metal ions in plant and animal nutrition. It may be that chromium, which is needed in only minute amounts in the body but which plays a role in transporting sugar through cell membranes, may function by a mechanism involving complex ions.

Before leaving the subject of biochemical complexes, we wish to mention an undesirable complex of hemoglobin, the carbon monoxide complex. Carbon monoxide forms a complex with hemoglobin in much the same way that O_2 does; that is, it occupies the position of the sixth ligand for the Fe^{2+}. The stability of HbCO, however, is many times greater than HbO_2, as shown by the equilibrium constant of the reaction

$$HbO_2 + CO(g) = HbCO + O_2(g)$$

$$\frac{[HbCO]P_{O_2}}{[HbO_2]P_{CO}} = 210 \quad \text{(normal body temperature)} \qquad [20.2]$$

At equal pressures of O_2 and CO the concentration of HbCO is seen to be 210 times that of HbO_2. Another way of looking at the equilibrium is to calculate the ratio of CO to O_2 that would exist if one had equal concentrations of HbCO and HbO_2. Setting [HbCO] = [HbO_2] in equation [20.2] gives

$$\frac{P_{O_2}}{P_{CO}} = 210 = \frac{n_{O_2}}{n_{CO}}$$

Hence, we see that a ratio of only one molecule of CO to 210 of O_2 would make one-half the hemoglobin unavailable for transporting O_2. One begins to understand why the CO level of the air, though numerically small, can be critical. On crowded freeways, in city traffic, or in long tunnels it is not uncommon for the CO content of the air to reach 0.005%, a dangerous level.

20.6 Applications of Complex-Ion Chemistry

Inorganic complexes are prominent in the reactions that are used in analytical chemistry. They function in two ways: (1) The formation of a complex ion can oftentimes be used to separate one or a group of cations from a mixture, and (2) the distinctive colors of many complexes can be used as a qualitative identification of the element. A few characteristic and useful applications are discussed below:

Silver

Ag^+, Hg_2^{2+}, and Pb^{2+} are the common cations that form insoluble chlorides; hence, these three metal ions can be separated from a mixture by precipitating them as chlorides and then separating the precipitate from the solution by filtering and washing. The silver can then be separated from the other two by treating the precipitate with $NH_3(aq)$. The AgCl will dissolve; the reaction is

$$AgCl(s) + 2\ NH_3(aq) = Ag(NH_3)_2^+ + Cl^-$$

To confirm the presence of Ag^+, one can acidify the filtrate, which decomposes the ammine complex and allows the AgCl(*s*) to be regenerated. The reaction is

$$Ag(NH_3)_2^+ + 2\ H^+ + Cl^- = AgCl(s) + 2\ NH_4^+$$

In the precipitation of the chlorides, a large excess of Cl^- is avoided; otherwise there is the possibility of dissolving the AgCl(*s*) as a chloro complex:

$$AgCl(s) + Cl^- = AgCl_2^-$$

It is also for this reason that nitric acid rather than hydrochloric acid is usually used to acidify the $Ag(NH_3)_2^+$ filtrate in the confirmatory reaction, since enough chloride ion to form AgCl is already present in the solution, and one wants to guard against a too-large excess.

An interesting series of reactions involving insoluble silver compounds and complex ions is given below. This series of reactions illustrates very well the relative stabilities of a number of the commonly encountered silver ion complexes and insoluble compounds.

$$Ag^+ + Cl^- = AgCl(s)$$
$$AgCl(s) + 2\ NH_3(aq) = Ag(NH_3)_2^+ + Cl^-$$
$$Ag(NH_3)_2^+ + Br^- = AgBr(s) + 2\ NH_3$$
$$AgBr(s) + 2\ S_2O_3^{2-} = Ag(S_2O_3)_2^{3-} + Br^-$$
$$Ag(S_2O_3)_2^{3-} + I^- = AgI(s) + 2\ S_2O_3^{2-}$$
$$AgI(s) + 2\ CN^- = Ag(CN)_2^- + I^-$$
$$2\ Ag(CN)_2^- + S^{2-} = Ag_2S(s) + 4\ CN^-$$

The equilibrium constants for these reactions are all sufficiently large that it is practical in turn to dissolve AgCl with $NH_3(aq)$, precipitate with Br^-, dissolve with $S_2O_3^{2-}$, precipitate with I^-, dissolve with CN^-, and, finally, precipitate with S^{2-}. The order of stability is thus

$$AgCl < Ag(NH_3)_2^+ < AgBr < Ag(S_2O_3)_2^{3-} < AgI < Ag(CN)_2^- < Ag_2S$$

Copper and Nickel

Copper sulfate pentahydrate, $CuSO_4 \cdot 5\ H_2O$, is a blue crystalline compound. X-ray diffraction studies show that four of the water molecules in the pentahydrate are coordinated with the Cu^{2+}, and the fifth is water of crystallization. Hence, the formula would be more appropriately written $[Cu(H_2O)_4]SO_4 \cdot H_2O$. The blue color apparently comes from the aquo complex ion of copper (II), since upon complete dehydration of the salt, the blue color disappears, leaving a white powder of formula $CuSO_4(s)$. If a small quantity of NH_3 solution is added to an aqueous solution of copper (II) (light blue), one can obtain the light-blue precipitate of $Cu(OH)_2$. The reactions are

Heating $CuSO_4 \cdot 5\ H_2O$ will first drive out the one molecule that is attached as water of crystallization. Further dehydration will remove the water in the complex ion.

$$NH_3 + H_2O = NH_4^+ + OH^-$$
$$Cu(H_2O)_4^{2+} + 2\ OH^- = Cu(OH)_2(s) + 4\ H_2O$$

Upon the addition of larger amounts of NH_3, however, the precipitate dissolves and the deep blue color of the amine complex appears:

$$Cu(OH)_2(s) + 4\ NH_3(aq) = Cu(NH_3)_4^{2+} + 2\ OH^-$$

The distinctive dark-blue color of tetraamminecopper (II) ion is an aid in identifying Cu^{2+}.

If HCl is now added carefully, one can reach the point where the ammine complex is destroyed but the solution is still sufficiently basic for $Cu(OH)_2(s)$ to reappear:

$$Cu(NH_3)_4^{2+} + 4\ H^+ = Cu^{2+} + 4\ NH_4^+$$
$$Cu^{2+} + 2\ OH^- = Cu(OH)_2(s)$$

Addition of further acid dissolves the $Cu(OH)_2$:

$$Cu(OH)_2(s) + 2\ H^+ + 2\ H_2O = Cu(H_2O)_4^{2+}$$

With enough HCl in excess, one obtains as a final product the yellow-green-colored tetrachlorocuprate (II) ion:

$$Cu(H_2O)_4^{2+} + 4\ Cl^- = CuCl_4^{2-} + 4\ H_2O$$

A series of various shades and hues of yellow-green colors are observed as the HCl is added, indicating that the copper complex makes a step-by-step transition from the aquo to the chloro complex. In other words, species such as $Cu(H_2O)_3Cl^+$, $Cu(H_2O)_2Cl_2$, and $Cu(H_2O)Cl_3^-$, each with a slightly different color, are evident.

The complexes of Ni(II) ion are similar to those of Cu(II). Ni^{2+} in aqueous solution is green in color. Addition of NH_3 solution produces first a green precipitate, $Ni(OH)_2$. Further addition of NH_3 produces the light-blue $Ni(NH_3)_4^{2+}$, and still higher concentrations of NH_3 give the violet-colored $Ni(NH_3)_6^{2+}$. Like copper (II), Ni(II) forms a series of light-green–yellow chloro complexes.

Be^{2+}, Al^{3+}, Cr^{3+}, Zn^{2+}, Sn^{2+}, and Pb^{2+}

Separations based on acid-base properties are very useful and were discussed more fully in Chapter 16. The oxides and hydroxides of Be^{2+}, Al^{3+}, Cr^{3+}, Zn^{2+}, Sn^{2+}, and Pb^{2+} are the common amphoteric compounds. The reactions of these compounds with excess strong base occur because of the formation of the hydroxo complexes $Be(OH)_4^{2-}$, $Al(OH)_4^-$, $Cr(OH)_4^-$, $Zn(OH)_4^{2-}$, $Sn(OH)_3^-$, and $Pb(OH)_3^-$. A typical reaction is

$$Zn(OH)_2(s) + 2\ OH^- = Zn(OH)_4^{2-}$$

EXAMPLE 20.3 Devise a scheme to separate a mixture of Cu^{2+}, Al^{3+}, and Fe^{3+}.

SOLUTION Add NH_3 solution. Initially, the hydroxides of all three ions—$Cu(OH)_2$(*s*, light blue), $Al(OH)_3$(*s*, white), and $Fe(OH)_3$(*s*, red-brown)—will precipitate. As the concentration of NH_3 in the solution is increased, however, the $Cu(OH)_2(s)$ will go into solution as $Cu(NH_3)_4^{2+}$, and the solid mixture of $Fe(OH)_3$ and $Al(OH)_3$ can then be removed by filtration or centrifugation. The precipitate of $Al(OH)_3(s)$ and $Fe(OH)_3(s)$ can be separated by treating with NaOH. $Al(OH)_3$, being amphoteric, will dissolve by forming $Al(OH)_4^-$; the $Fe(OH)_3(s)$ will remain in the solid state. A diagram for the separation is given below:

<table>
<tr><td colspan="3">Cu^{2+}, Al^{3+}, Fe^{3+}
(Add NH_3 solution.)</td></tr>
<tr><td colspan="2">Precipitate: $Fe(OH)_3(s)$, $Al(OH)_3(s)$
(Add excess NaOH.)</td><td rowspan="2">Solution: $Cu(NH_3)_4^{2+}$</td></tr>
<tr><td>Precipitate: $Fe(OH)_3(s)$</td><td>Solution: $Al(OH)_4^-$</td></tr>
</table>

Iron

Several highly colored complexes of iron exist and are useful in identifying both ferrous and ferric ions. The deep-red solution that results when thiocyanate ion, CNS^-, is added to Fe^{3+} is a sensitive test for Fe^{3+}. Complexes that range in composition from $Fe(H_2O)_5CNS^{2+}$ at low CNS^- concentrations to $Fe(CNS)_6^{3-}$ at high concentrations are responsible for the color.

Neither Fe(II) nor Fe(III) forms ammine or hydroxo compounds, but both form chloro and cyano complexes. The chloro compounds form readily upon addition of chloride ion; the cyano compounds require special methods of preparation. The ferrocyanide ion can be used as a test for the presence of ferric ion, and the ferricyanide ion can be used as a test for ferrous ion. In both cases, a positive test is the formation of a deep-blue color (Prussian blue).

Gold and Platinum

The only way to have gold and platinum in solution in appreciable quantities is as complex ions. Neither of these noble metals is attacked by nitric acid, but both are oxidized and go into solution when treated with *aqua regia*, a mixture of concentrated HNO_3 and concentrated HCl. The reactions are

$$Au(s) + 6\,H^+ + 3\,NO_3^- + 4\,Cl^- = AuCl_4^- + 3\,NO_2(g) + 3\,H_2O$$

and

$$Pt(s) + 8\,H^+ + 4\,NO_3^- + 6\,Cl^- = PtCl_6^{2-} + 4\,NO_2(g) + 4\,H_2O$$

The stability of the chloro complexes that form gives the extra drive to make the oxidation take place. This can be more readily seen by writing the reactions in a two-step process. For the gold reaction we have

$$Au + 6\,H^+ + 3\,NO_3^- = Au^{3+} + 3\,NO_2 + 3\,H_2O$$
$$Au^{3+} + 4\,Cl^- = AuCl_4^-$$

The free-energy decrease of the second equation is sufficient to overcome the positive ΔG of the first, giving a negative ΔG (in the presence of high concentrations of NO_3^- and Cl^-) for the net reaction; hence, the oxidation of Au in the presence of a high concentration of Cl^- is a natural reaction.

Another example of the formation of a complex ion aiding an oxidation is found in the formation of the very stable $Au(CN)_4^-$ ion. Gold can be oxidized by $O_2(g)$ in the presence of cyanide ion. This is the basis of the cyanide leaching process that is used to extract elemental gold from its ore. The reaction is

Warning: Cyanides are highly poisonous. In the presence of an acid, HCN gas will form. Treating KCN pellets with an acid is the procedure that has been used to generate HCN gas for executions in federal penitentiaries.

$$4\,Au(s) + 3\,O_2(g) + 16\,CN^- + 6\,H_2O = 4\,Au(CN)_4^- + 12\,OH^-$$

Again, the oxidation of Au by O_2 is possible only because of the large decrease in free energy accompanying the formation of $Au(CN)_4^-$. The ore is finely ground and treated with an alkaline cyanide solution, and air is bubbled through the mixture. After the gold is dissolved, it is reclaimed in the solid form by treating the leachate with Zn. The reaction is

$$2\,Au(CN)_4^- + 3\,Zn(s) + 4\,CN^- = 3\,Zn(CN)_4^{2-} + 2\,Au(s)$$

Cobalt and Chromium: Nonlabile Complexes

Complexes can be classified as *labile* or *nonlabile* depending on whether substitution reactions involving replacement of ligands occur rapidly (labile) or only slowly (nonlabile or inert). Most of the commonly encountered complexes fall in the first category, exchange reactions occurring almost instantaneously. Complexes of Cr(III) and Co(III) are notable exceptions. For example $Co(NH_3)_6^{3+}$, while thermodynamically unstable in acid solution, can exist in dilute acids for weeks because the rate of the reaction

$$Co(NH_3)_6^{3+} + 6\,H^+ + 6\,H_2O = Co(H_2O)_6^{3+} + 6\,NH_4^+$$

is very slow.

Because of their nonlabile character, formation of $Cr(NH_3)_6^{3+}$ and $Co(NH_3)_6^{3+}$ requires special procedures. Simply adding NH_3 to the aquo complexes does not work because of the unfavorable kinetics. If a soluble Cr(III) salt is heated in concentrated NH_3 solution, there is slow formation of the pale-violet $Cr(NH_3)_6^{3+}$. $Co(NH_3)_6^{3+}$ is conveniently prepared from $Co(NH_3)_6^{2+}$ by air oxidation. The free-energy change of this reaction very strongly favors the oxidation, and since no exchange of ligands is involved, there is no problem with the kinetics.

Sometimes the formation of a complex ion is the only way to obtain a particular oxidation state of a metal in aqueous solution. This is true for Co(III). If we examine the oxidation-reduction potentials, we see that Co(III) is a strong oxidant, capable of oxidizing water. The data at 25° C are

	$\mathscr{E}^0$(V)
$4\,e^- + 4\,Co(H_2O)_6^{3+} = 4\,Co(H_2O)_6^{2+}$	1.82
$2\,H_2O = O_2 + 4\,H^+ + 4\,e^-$	−1.23
$4\,Co(H_2O)_6^{3+} + 2\,H_2O = 4\,Co(H_2O)_6^{2+} + O_2 + 4\,H^+$	+0.59

The positive voltage shows that the oxidation takes place. Hence, Co^{3+} is unstable in aqueous solution. The $\mathscr{E}^0$ data for $Co(NH_3)_6^{3+}$, however, shows it to be very stable in water:

	$\mathscr{E}^0$(V)
$4\,Co(NH_3)_6^{3+} + 4\,e^- = 4\,Co(NH_3)_6^{2+}$	0.11
$2\,H_2O = O_2 + 4\,H^+ + 4\,e^-$	−1.23
$4\,Co(NH_3)_6^{3+} + 2\,H_2O = O_2 + 4\,H^+ + 4\,Co(NH_3)_6^{2+}$	−1.12

The large negative voltage rules out the oxidation-reduction reaction.

Copper (I) ion in aqueous solution is another example of an ion in an unstable state that can be stabilized by the formation of a complex. Cu(I) ion in water solution disproportionates as shown by the reaction

$$2\,Cu^{+} = Cu(s) + Cu^{2+}; \qquad \begin{array}{l} \mathscr{E}^0 = +0.184\text{ V} \\ \Delta G^0 = -35{,}500\text{ joules} \end{array}$$

The Cu(I) complex ions $Cu(NH_3)_2^+$, $CuCl_2^-$, and $Cu(CN)_2^-$, however, are quite stable in water solution.

Summary

Complex ions represent an important class of substances. The central metal ion in the complex is typically a transitional metal, such as Fe, Co, Ni, Cu, Ag, Cr, Au, Hg, or Zn; or a representative metal that forms an amphoteric hydroxide, such as Be, Al, Sn, and Pb. Typical ligands are NH_3, H_2O, Cl^-, I^-, CN^-, and OH^-. Depending on the composition of the complex ion and the charge on the ligand, a complex ion can be either a cation or an anion. Lists of common complex ions, arranged so as to facilitate your becoming easily acquainted with them, are given in the text, together with a simple, logical system of nomenclature.

Common coordination numbers in complex ions are 2, 4, and 6. Two ligands per metal ion result in a linear complex. Four-fold coordination is either a square-planar structure (common among the transition-metal complex ions) or a tetrahedral structure (the usual case for a representative-metal complex). Six-fold coordination with octahedral structure is by far the most common structure for complex ions. Both structural and geometrical (cis-trans) isomerism is possible in some complex ions, and this adds to their variety and interest.

Two main bonding theories for complex ions are discussed: (1) valence-bond theory and (2) crystal-field theory. Each has advantages and deficiencies, and we make use of both theories. It is assumed in the valence-bond theory that the ligands donate pairs of electrons to hybridized orbitals of the metal ion to form a coordinate covalent metal-ligand bond. This model accounts very well for the observed shapes of complex ions, but it does not account for the observed magnetic properties of many of the complexes. The crystal-field theory, on the other hand, does accurately account for magnetic and light-absorption properties of complex ions, but it makes no pretense at giving an accurate description of the metal-ligand bond.

Multidentate ligands (chelates) are common in nature and vital to our existence. Examples are hemoglobin, myoglobin, and chlorophyll.

Complex ions play an important role in analytical chemistry, functioning as a means of separating elements from one another and of identifying elements through distinctive colors of their complex ions. The selected group of highly important reactions involving complex ions that are discussed near the end of the chapter should be carefully studied.

Important Terms and Concepts

Coordination complex	Crystal-field theory
Ammonia complexes	High-spin state
Cyanide ion complexes	Low-spin state
Chloride ion complexes	Bidentate ligand
Hydroxide ion complexes	Chelates
Nomenclature of complex ions	Hemoglobin
Structural isomerism	Myoglobin
Geometrical isomerism	Chlorophyll
Valence-bond theory	Porphyrins

Exercises

20-1.
Write formulas for each of the following:

(a) tetracyanonickelate (II) ion
(b) pentaaquochlorochromium (II) chloride
(c) potassium amminepentachloroplatinate (IV)
(d) tetraamminecopper (II) tetrachloroplatinate (II)
(e) tetracarbonylnickel (0) (carbonyl is CO)
(f) hexafluoroferrate (III) ion

20-2.
Name each of the following compounds (including isomers where applicable):

(a) $[Ni(NH_3)_6]Cl_2$
(b) $Pd(CN)_6^{2-}$
(c) $Ag(NH_3)_2^+$
(d) $Na_3[Co(CN)_5Br]$
(e) $K[Cr(NH_3)_2Cl_4]$
(f) $[Cr(NH_3)_6][Co(CN)_6]$

20-3.
When HCl is added to a solution of $Cr(NO_3)_3$, the color changes from violet to green. Give a possible explanation in terms of complex ions.

20-4.
Propose a formula and a structure for a coordination compound with the following composition: 20.73% Cr, 11.17% N, 56.53% Cl, 9.16% Na, and 2.41% H.

20-5.
Draw the structural formulas of the two known compounds with the formula $Co(NH_3)_5ClBr$.

20-6.
Draw the octahedral structures for the six possible isomers of $CrCl_3(H_2O)_6$ (not all of them are known).

20-7.
Which of the following ions would you expect to be colorless when hydrated?

(a) Ti^{3+}
(b) V^{4+}
(c) Sc^{3+}
(d) Cu^+
(e) Zr^{4+}

20-8.
The complex $Ni(CN)_4^{2-}$ is square-planar, and the complex $NiCl_4^{2-}$ is tetrahedral. Use the valence-bond theory to predict the number of unpaired electrons in each.

20-9.
Draw the geometric structures for each of the following diamagnetic ions:

(a) $AgCl_2^-$
(b) $ZnCl_4^{2-}$
(c) $PdCl_4^{2-}$
(d) $RhCl_6^{3-}$
(e) $SnCl_3^-$

20-10.
Use the valence-bond theory to predict the geometry of the $Pb(OH)_3^-$ and the $Sn(OH)_3^-$ complex ions. In each example, an electron pair occupies one of the possible ligand positions.

20-11.
Which of the following pairs would you expect to absorb the highest energy visible radiation?

(a) $CrCl_6^{3-}$ or $Cr(NH_3)_6^{3+}$
(b) $Ni(H_2O)_6^{2+}$ or $Ni(NH_3)_6^{2+}$

20-1.

A compound has the following composition: 46.2% Pt, 33.6% Cl, 16.6% N, and 3.6% H. From the electrical conductance of an aqueous solution of the compound, one predicts a 3:1 electrolyte. Addition of a silver nitrate solution to this solution precipitates 3 moles of Cl^- as AgCl for each formula weight of compound. Write the chemical formula for the compound, show its probable geometry, and name it.

20-2.

Two compounds have the same empirical formula, $Co(en)_2NO_2Cl_2$. When Ag^+ is added to a solution of the first, one mole of AgCl precipitates per formula weight of compound. When Ag^+ is added to a solution of the second, no precipitate forms. Both compounds are 1:1 electrolytes. Draw the possible geometrical structures of each compound. (En represents ethylenediamine, a bidentate ligand.)

20-3.

How many unpaired electrons are present in octahedral complexes of the ions Ni^{2+}, Ni^{3+}, Mn^{2+}, Mn^{3+}, V^{2+}, and V^{3+} with (a) ligands that induce a large Δ_0 and (b) ligands that induce a small Δ_0?

21

The Transition Elements

The transition elements are those in which the differentiating electron is either a *d* or an *f* electron. These elements occupy the center block and the two rows at the bottom of the expanded periodic table shown in Figure 21.1. They are all metals, exhibit variable oxidation states, have a propensity for complex-ion formation, frequently form colored compounds, and generally engage in a wide variety of chemical reactions. It is the partially filled *d* and *f* energy levels that are responsible for many of their interesting properties.

Table 21.1 shows the electronic configurations of the elements of the first transition series (the 3*d* elements). It is seen that Sc is a d^1 element (one *d* electron); Ti, a d^2; V, a d^3; but the next element, Cr, is a d^5. There is apparently enough extra stability in a half-filled *d* energy level that chromium can promote one of its 4*s* electrons to the higher 3*d* level and obtain a more stable state than if it were a d^4 element. In the next element, Mn, the "borrowed" 4*s* electron is replaced, and Mn is also a d^5 element. A similar adjustment occurs in Cu. Copper promotes an *s* electron to the *d* level in order to achieve the stability associated with a full 3*d* level. The energy expended in promoting the 4*s* electron to the 3*d* level is apparently more than compensated for by the extra stability associated with either a *half-filled* 3*d* level as in chromium or a *completely filled* 3*d* level as in copper.

The electronic configurations of the ions of this series should be commented on. It is the *s*, not the *d*, electrons that are lost first in the ionization process. Hence, Cr(III) ion, as an example, has the configuration $[Ar]3d^3$; Fe(II) ion, $[Ar]3d^6$; and Fe(III) ion, $[Ar]3d^5$. It may surprise you, at first, to learn that although the *d* electrons were the last to be added in these elements in the *aufbau* process, they are not the first to leave via ionization. Remember, that in the *aufbau* process you are imagining the addition of a proton plus an electron as you move from one element to the next, whereas the ionization process is entirely different. In the latter, the nuclear charge is left undisturbed while an electron is removed. Thus, ionization is not to be construed as the reverse of the *aufbau* process. The two are not comparable.

It is instructive to examine the transition elements both in terms of rows and of families. In several respects, similarities in properties of members of the series are more apparent than they are in the families of these elements. Similarities are especially pronounced in the first row of transition elements. Table 21.2 lists several of the properties of this first transition series. Note especially the similarities in ionic size, melting points, boiling points, and ionization energies of the elements in the series.

Relative to the elements of Groups IA, IIA, and IIIA, the elements of the first transition series are seen to have high melting points, high boiling points, and high densities. The somewhat exceptional transition elements are those of Group IIB. Since the valence *d* orbitals of these elements are completely filled, one would expect them to behave more like the representative elements, as they in fact do. The stronger bonding in the other transition metals, evidenced by their higher melting points (and higher heats of fusion), shows that the *d* electrons are involved in the metallic bonding of transition elements.

Notice in Table 21.2 the general (but small) increase in ionization energies as the atomic number increases. The discrepancies, such as the rather large

1*s*	1 H																			2 He
2*s*	3 Li	4 Be													2*p*	5 B	6 C	7 N	8 O	9 F	10 Ne
3*s*	11 Na	12 Mg													3*p*	13 Al	14 Si	15 P	16 S	17 Cl	18 Ar
4*s*	19 K	20 Ca	3*d*	21 Sc	22 Ti	23 V	24 Cr	25 Mn	26 Fe	27 Co	28 Ni	29 Cu	30 Zn	4*p*	31 Ga	32 Ge	33 As	34 Se	35 Br	36 Kr
5*s*	37 Rb	38 Sr	4*d*	39 Y	40 Zr	41 Nb	42 Mo	43 Tc	44 Ru	45 Rh	46 Pd	47 Ag	48 Cd	5*p*	49 In	50 Sn	51 Sb	52 Te	53 I	54 Xe
6*s*	55 Cs	56 Ba	5*d*	57– La	72 Hf	73 Ta	74 W	75 Re	76 Os	77 Ir	78 Pt	79 Au	80 Hg	6*p*	81 Tl	82 Pb	83 Bi	84 Po	85 At	86 Rn
7*s*	87 Fr	88 Ra	6*d*	89– Ac	104	105	106							7*p*						

4*f*	58 Ce	59 Pr	60 Nd	61 Pm	62 Sm	63 Eu	64 Gd	65 Tb	66 Dy	67 Ho	68 Er	69 Tm	70 Yb	71 Lu
5*f*	90 Th	91 Pa	92 U	93 Np	94 Pu	95 Am	96 Cm	97 Bk	98 Cf	99 Es	100 Fm	101 Md	102 No	103 Lr

Figure 21.1
The periodic table, showing the separation into the s-, p-, d-, *and* f-*blocks*. The transition elements consist of those elements in the central block (3*d*, 4*d*, 5*d*, 6*d*) and the two "inner transition" rows at the bottom of the table (4*f*, 5*f*). Those elements with the differentiating electron in the 4*f* energy level are called the lanthanides, and those with 5*f* differentiating electrons, the actinides.

Table 21.1

Electronic Configurations of the First Series of Transition Elements

ELEMENT	ELECTRONIC CONFIGURATION
$_{21}$Sc (scandium)	Ar core, $4s^2 3d^1$
$_{22}$Ti (titanium)	Ar core, $4s^2 3d^2$
$_{23}$V (vanadium)	Ar core, $4s^2 3d^3$
$_{24}$Cr (chromium)	Ar core, $4s^1 3d^5$
$_{25}$Mn (manganese)	Ar core, $4s^2 3d^5$
$_{26}$Fe (iron)	Ar core, $4s^2 3d^6$
$_{27}$Co (cobalt)	Ar core, $4s^2 3d^7$
$_{28}$Ni (nickel)	Ar core, $4s^2 3d^8$
$_{29}$Cu (copper)	Ar core, $4s^1 3d^{10}$
$_{30}$Zn (zinc)	Ar core, $4s^2 3d^{10}$

decrease in the third ionization energy in going from $_{24}$Mn to $_{25}$Fe, can usually be accounted for by considering the added stability enjoyed by an element with either half-filled (d^5) or completely filled (d^{10}) valence orbitals. In the Mn-Fe example, an Fe atom after losing three electrons achieves the stable d^5 state, whereas to obtain Mn(III) the electrons removed break up a half-filled level.

The radii of the atoms of the first transition series are worth special comment. They decrease as the atomic number increases until they reach $_{28}$Ni, after which there is a slight increase. The decrease in size brought about by the effect of increased nuclear charge as one goes from left to right in the series is apparently quite evenly balanced by the expansion in size brought about by additional electrons. Note, however, that the overall changes in size from one end of the series to the other are relatively small.

Note the generally high melting points of the transition elements relative to those of the representative metals. Compare, for example, the elements of the first transition series with K, Ca, and Ga, which are in the same period and whose melting points are 336° K, 1123° K, and 303° K, respectively.

The ionic radii of the divalent ions of the first series of transition elements are seen to be slightly smaller than that of Ca^{2+}, and the radii of the +3 ions are slightly larger than Ga^{3+}. One might expect, therefore, that both the solubility and the basicity of the oxides of the first transition series would be similar to those of CaO for the +2 oxides and Ga_2O_3 for the +3 oxides. This prediction is verified experimentally.

As seen in Table 21.2, all of the first-series transition elements except copper will reduce H^+ in the standard state. The rate at which these metals react with H^+ is slowed down with several of the metals because of protective oxide coatings on the surface of the metals. The impervious layer of Cr_2O_3 that forms on chromium surfaces is a case in point. In spite of its high oxidation potential, chromium can be used as a protective coating on steel because the chromium is in turn protected by the relatively inert oxide layer.

The melting point of Zn (693° K) is much lower than any of the other metals of the first transition series. The other members of Group IIB, cadmium and mercury, also have anomalous melting points. As implied previously, electrons in filled *d* orbitals do not contribute to metallic bonding; hence, in their melting points and heats of fusion, the IIB elements resemble representative elements.

TABLE 21.2

Properties of the Elements of the First Transition Series

	Sc	Ti	V	Cr	Mn	Fe	Co	Ni	Cu	Zn
Ionization energy (kjoules)										
I_1	631	658	650	652	717	759	758	736	745	906
I_2	1235	1309	1413	1591	1509	1561	1645	1752	1957	1733
I_3	2388	2650	2828	2986	3250	2956	3230	3430	3553	3820
Atomic radius (Å)	1.44	1.32	1.22	1.17	1.17	1.16	1.16	1.15	1.17	1.25
Melting point (°K)	1673	1950	2190	2176	1517	1812	1768	1728	1356	693
Boiling point (°K)	2750	3550	3650	2900	2340	3160	3160	3160	2855	1181
Density (g/cm^3)	2.5	4.5	5.96	6.92	7.2	7.86	8.9	8.90	8.92	7.14
Ionic radii (Å)										
M^{2+}	—	0.72	0.65	—	0.82	0.76	0.72	0.68	0.72	0.74
M^{3+}	0.69	0.61	0.62	0.61	0.62	0.63	0.56	—	—	—
$\Delta H_{\text{hydration}}$, (kjoules/mole)										
M^{2+}	—	1870	1900	1920	1862	1950	2080	2120	2120	2050
M^{3+}	3960	4300	4410	4620	4590	4490	4710	—	—	—
Reduction potential (V)										
$M^{2+} + 2e^- = M$	—	−1.63	−1.2	−0.91	−1.18	−0.44	−0.28	−0.25	+0.34	−0.76
$M^{3+} + 3e^- = M$	−2.1	−1.21	−0.88	−0.74	−0.28	−0.04	0.42	—	—	—

21.1 Scandium Group

Group IIIB, the scandium family, is composed of scandium (Sc), yttrium (Y), lanthanum (La) and the fourteen elements (numbers 58–71) collectively known as the lanthanides, and actinium and the elements 90–103, known as the actinides. The discussion of the actinides, which is principally nuclear chemistry, is deferred until Chapter 23.

Table 21.3 lists some of the properties of the Group IIIB elements (including the lanthanides, but not the actinides). Scandium, yttrium, and lanthanum exhibit only the +3 oxidation state in their compounds (loss of the $4s^2$ and $3d^1$ electrons to form the noble-gas configuration). All of the lanthanides form +3 ions, but +2 and +4 oxidation states also occur in several of them. These latter oxidation states may be expected in some of the lanthanides because they result in either half-filled (Eu^{2+} and Tb^{4+}) or completely filled (Yb^{2+}) *f* orbitals; and as in *d* valence orbitals, there is evidence that either half-filled or completely filled *f* orbitals contribute to the stability of the ion. Not all +2 and +4 lanthanide series ions fit this pattern, however.

The chemistry of Sc, Y, and La is somewhat analogous to that of aluminum in that they are all active metals that will react with H^+ to form H_2 gas, form compounds in which they are in the +3 state only, and form insoluble oxides. Unlike Al_2O_3, however, Sc_2O_3, Y_2O_3, and La_2O_3 are basic (not amphoteric) oxides.

TABLE 21.3

Electronic Configurations and Properties of Group IIIB Elements

NAME	SYMBOL	ELECTRONIC CONFIGURATION	IONIC RADIUS (M^{3+}) (Å)	$\mathscr{E}^0$ ($M^{3+} + 3e^- = M$) (V)
Scandium	$_{21}Sc$	$[Ar]4s^23d^1$	0.81	2.08
Yttrium	$_{29}Y$	$[Kr]5s^24d^1$	—	2.37
Lanthanum	$_{57}La$	$[Xe]6s^25d^1$	1.061	2.52
Cerium	$_{58}Ce$	$[Xe]6s^24f^2$	1.034	2.48
Praseodymium	$_{59}Pr$	$[Xe]6s^24f^3$	1.013	2.47
Neodymium	$_{60}Nd$	$[Xe]6s^24f^4$	0.995	2.44
Promethium	$_{61}Pm$	$[Xe]6s^24f^5$	0.979	2.42
Samarium	$_{62}Sm$	$[Xe]6s^24f^6$	0.964	2.41
Europium	$_{63}Eu$	$[Xe]6s^24f^7$	0.950	2.41
Gadolinium	$_{64}Gd$	$[Xe]6s^25d^14f^7$	0.938	2.40
Terbium	$_{65}Tb$	$[Xe]6s^24f^9$	0.923	2.39
Dysprosium	$_{66}Dy$	$[Xe]6s^24f^{10}$	0.908	2.35
Holmium	$_{67}Ho$	$[Xe]6s^24f^{11}$	0.894	2.32
Erbium	$_{68}Er$	$[Xe]6s^24f^{12}$	0.881	2.30
Thulium	$_{69}Tm$	$[Xe]6s^24f^{13}$	0.869	2.28
Ytterbium	$_{70}Yb$	$[Xe]6s^24f^{14}$	0.858	2.27
Lutetium	$_{71}Lu$	$[Xe]6s^25d^14f^{14}$	0.848	2.25

Since the difference in electronic configuration of the lanthanides is "buried" deep within the atom in 4*f* orbitals, one might expect considerable similarity in the chemistry of these elements. The extent of this similarity can be seen in Table 21.3, which shows how impressively similar their standard oxidation potentials are. The regularity of the small gradations in $\mathscr{E}^0$ is also notable.

Because of the similarities in the lanthanides, they are difficult to separate by purely chemical means. Pure preparations of the elements or their compounds were hard to obtain before 1940. At that time it was shown that a clear separation could be made by making use of a difference in the extent to which the ions of the lanthanide elements adsorb on cation-exchange materials. Interest in the separation of the lanthanides was stimulated by the intense efforts made in nuclear fission research during World War II. It was known as early as 1939 that some of the lanthanides were among the fission products of uranium.

Cerium

The chemistry of only one of the individual lanthanides is mentioned in this book, and that one (cerium) because of its fairly common use in analytical chemistry. The detailed chemistry of the other lanthanides will be left for a later course.

The principal oxidation states of cerium are +4 (ceric ion) and +3 (cerous ion). Ce(IV) is a strong oxidizing agent that reduces to Ce(III) in acid solution:

$$Ce^{4+} = Ce^{3+} + e^-$$

with a potential that closely matches that of the MnO_4^--Mn^{2+} couple. Hence, ceric ion can be used as an oxidizing agent in almost all the analytical applications that MnO_4^- is, and it has the advantages of being significantly more stable and of having only one reduction product. Since the ceric ion is yellow and the cerous ion is colorless, the end point of a titration can be detected by the first appearance of yellow color in the solution being titrated. Special indicators may also be used to increase the sensitivity of detection of the end point.

Ceric ion is always used in highly acidic solution (usually sulfuric acid). As might be expected from its +4 charge, it hydrolyzes extensively to insoluble ceric hydroxide in neutral or even slightly acidic solutions. Hydrochloric acid, however, cannot be used to acidify ceric solutions because the ceric ion readily oxidizes the chloride ion in hot solution:

$$2\,Cl^- + 2\,Ce^{4+} = 2\,Ce^{3+} + Cl_2(g)$$

21.2 Titanium Group

Members of Group IVB, the titanium family, are titanium (Ti), zirconium (Zr), and hafnium (Hf). Some of their properties are listed in Table 21.4. The maximum oxidation state for this family is seen to be +4 (noble-gas configuration). Compounds in which the metals are also +2 and +3 are observed in compounds

TABLE 21.4

Properties of Group IVB Elements

	Ti	Zr	Hf
Atomic number	22	40	72
Electronic configuration	$[Ar]4s^2 3d^2$	$[Kr]5s^2 4d^2$	$[Xe]6s^2 5d^2$
Atomic radius (Å)	1.32	1.45	1.44
Density (g/cm^3)	4.5	6.4	12.1
Melting point (°K)	1950	2125	2495
Ionization energy (I) (kjoules/mole)	658	661	669
ΔH_f^0 (MO_2) (kjoules/mole)	−912	−1080	−1140
ΔG_f^0 (MO_2) (kjoules/mole)	−854	−1020	—

of the two lighter elements, Ti and Zr, but absent in Hf. We will see similar behavior in the other transition families; that is, the lower oxidation states tend to become less prevalent in the heavier elements of the family.

Note that the opposite generalization can be made about the elements of a representative group. In Group IVA, for example, the +2 state is important only with the heavy elements, Sn and Pb.

Note from Table 21.4 the high melting points of the Group IVB elements. The high melting point of titanium, combined with its resistance to corrosion (because of an oxide surface layer) and its relatively low density, makes it useful as a structural material in jet engines and as an alloying material in the production of some very hard and tough steels.

Note also in Table 21.4 the very high enthalpies of formation of TiO_2, ZrO_2, and HfO_2. At high temperature, all three of the metals react readily with oxygen, nitrogen, and carbon. This complicates their preparation. Titanium metal is produced commercially by the conversion of the dioxide to the tetrachloride, followed by reduction with magnesium:

$$TiCl_4(l) + 2\ Mg(s) = Ti(s) + 2\ MgCl_2(s)$$

All three metals form surface coatings of oxides that protect the metal from further oxidation. Hence, while the pure metals, without their oxide coatings, are quite reactive, the coated metals are quite inert.

The most important oxidation state of all three metals is the +4, and the most important compounds are probably the +4 oxides. The mineral rutile (TiO_2) is one of the main sources of titanium. TiO_2 is widely used as a paint pigment and also finds use in the cosmetic industry.

Semiprecious gemstones are made from TiO_2. Crystalline TiO_2 has an exceptionally high index of refraction of light—even higher than diamond—and, therefore, has more sparkle than diamond. Unfortunately, the stones are not very hard and, consequently, scratch easily.

While very insoluble in water, TiO_2 does dissolve slightly in strong bases and in strong acids. Hence, it has amphoteric properties. When fused with a base, the reaction is

$$TiO_2(s) + 2\ NaOH(s) = Na_2TiO_3(s) + H_2O$$

Treating TiO_2 with a strong acid does not produce Ti^{4+} ion as one might at first expect. It is generally true that quadrivalent ions do not exist in water solution to any appreciable extent because of the almost complete hydrolysis of the highly charged ion. In the case of Ti^{4+}, species such as $Ti(OH)_2^{2+}$ result. Dehydrated, $Ti(OH)_2^{2+}$ becomes the titanyl ion, TiO^{2+}. Titanyl sulfate, $TiO(SO_4)$, is an example of the type of salt produced when TiO_2 is treated with a strong acid.

Titanium in the +2 state is formed by reduction of Ti(IV) compounds with titanium metal:

$$TiO_2(s) + Ti(s) = 2\,TiO(s)$$

and

$$TiCl_4(l) + Ti(s) = 2\,TiCl_2(s)$$

Ti^{2+} decomposes water; hence, it has no aqueous-solution chemistry.

Ti_2O_3 can be formed by reduction of TiO_2 with H_2. It is a stable basic oxide, insoluble in water. The Ti^{3+} ion is stable in water, although it is such a strong reducing agent that it needs to be protected from the oxygen of the air. As would be expected from its high charge and small size, Ti^{3+} hydrolyzes appreciably.

The best-known compounds of Zr and Hf are their +4 oxides, ZrO_2 and HfO_2. Zirconium dioxide (melting point 3100° K) finds use as a refractory material in the lining of high-temperature furnaces. As might be expected from the larger size of the metal atoms, ZrO_2 and HfO_2 are more basic than TiO_2 and do not have the amphoteric properties of TiO_2.

The chemical properties of zirconium and hafnium are remarkably similar. This is not surprising since the chemistry of the elements is determined principally by electronic configuration and size. As can be seen in Table 21.4, the electronic structures of Zr and Hf are similar and their sizes almost identical. The reason for the similarity in size has to do with the lanthanide elements intervening between Zr and Hf. As shown in Table 21.4, there is a contraction in size of the atoms (or ions) in the lanthanide series as one moves across the series. This decrease, known as the *lanthanide contraction* is a result of increased nuclear charge within the same period of elements. This decrease very nearly compensates for the increase that would be expected in going from a period 5 element (Zr) to a period 6 element (Hf). The result is that Hf and Zr atoms are almost identical in size. A similar size-relationship applies to Nb and Ta, Mo and W, and, indeed, to some extent, to all the corresponding pairs of transition elements in the fifth and sixth periods.

21.3 Vanadium Group

Table 21.5 lists some of the properties of the vanadium family, group VB. The electronic configuration suggests that the maximum oxidation state is +5 (noble-gas configuration). This is the principal oxidation state for the heavier elements, Nb and Ta. The lighter member of the group, V, has important compounds in the +2, +3, +4, and +5 states.

TABLE 21.5

Properties of Group VB Elements

	V	Nb	Ta
Atomic number	23	41	73
Electronic configuration	$[Ar]4s^2 3d^3$	$[Kr]5s^1 4d^4$	$[Xe]6s^2 5d^3$
Atomic radius (Å)	1.22	1.34	1.34
Density (g/cm^3)	5.96	8.57	16.6
Melting point (°K)	2190	2770	3270
Ionization energy (I) (kjoules/mole)	650	665	761
ΔH_f^0 (M_2O_5) (kjoules/mole)	−1560	−1940	−2090
ΔG_f^0 (M_2O_5) (kjoules/mole)	−1440	—	−1970

Also to be especially noted from Table 21.5 are the very high melting points (even higher than the IVB elements) and the high negative ΔG_f^0 values of the oxides, which show the metals to have high reactivity toward oxygen. The enthalpies of reaction, as shown by the ΔH_f^0 data, are seen to be highly exothermic. Like the IVB elements, the VB elements react at high temperatures with nitrogen and carbon as well as oxygen. Also, like the IVB elements, the full reactivity of the pure VB elements is not generally evident because of the existence of protective oxide coatings.

Also evident in Table 21.5 is the identical (to three significant figures) atomic radii of Nb and Ta. As explained earlier, this is the result of the contraction in size that occurs with the intervening lanthanide elements. As a result of both similar electronic configuration and size, the chemistry of Nb and Ta would be expected to be—and in fact is—very similar, which, as we saw, was the analogous case in Group IVB with Zr and Hf.

Vanadium

The most important element of Group VB is vanadium, and probably the most important compound of vanadium is the pentoxide, V_2O_5. It is formed by direct combination of the elements, is a red solid, and finds use as a catalyst, the commercial process for manufacture of sulfuric acid being a well-known example of that use of V_2O_5.

Vanadium pentoxide, like TiO_2 of Group IVB, is amphoteric. It dissolves in acids to form pervanadyl ion, $VO_2^+(aq)$:

$$V_2O_5 + 2\,H^+(aq) = 2\,VO_2^+(aq) + H_2O$$

In basic solution, V_2O_5 dissolves to produce the vanadate ion VO_4^{3-}, analogous to the phosphate and arsenate ions:

$$V_2O_5 + 6\,OH^- = 2\,VO_4^{3-} + 3\,H_2O$$

Acidic solutions of vanadium in the +5 state, $VO_2^+(aq)$, can be reduced with zinc metal to vanadium in the +4 state, VO^{2+}, vanadyl ion. Vanadyl ion,

Normally when an amphoteric oxide is treated with acid, the metal ion results. In the case of V_2O_5, however, one would not expect to get V^{5+} in an aqueous solution. One or more steps of hydrolysis of the highly charged ion would result. The formation of

$VO_2^+(aq)$, which is also written as $V(OH)_4^+$, is shown below:

$$V^{5+} + 4\,H_2O = V(OH)_4^+ + 4\,H^+$$

Dehydration produces pervanadyl salts:

$$V(OH)_4^+ = VO_2^+ + 2\,H_2O$$

in turn, is further reduced (also by zinc) to +3 vanadium, V^{3+}, which in turn can be reduced to V^{2+}. The standard reduction-potential diagram shown below summarizes these changes. The step-by-step reduction in acid of vanadium (V) to vanadium (II) is a colorful series of reactions: V_2O_5 is red, VO^{2+} is a bright blue, V^{3+} is green, and V^{2+} is violet.

$$VO_2^+ \xrightarrow{1.0} VO^{2+} \xrightarrow{0.36} V^{3+} \xrightarrow{-0.25} V^{2+} \xrightarrow{-1.18} V$$

From the diagram it can be seen that vanadium (V) and (IV) are easily reduced, that V^{2+} is a moderately good reducing agent, and that vanadium metal (as stated earlier) is a very good reducing agent.

21.4 Chromium Group

Some of the properties of Group VIB elements—chromium (Cr), molybdenum (Mo), and tungsten (W)—are listed in Table 21.6. As can be seen from the electronic-configuration information in the table, the maximum oxidation state for this group is +6 (noble-gas configuration), and this is the principal oxidation state for the heavier elements of the group, Mo and W. As we saw in Group VB, the first element of a transition group is usually quite different from the others, and Cr, in addition to important compounds in the +6 state, forms +2 and +3 ions.

The very high melting and boiling points of the Group VIB elements should be noted. Tungsten has the highest melting and boiling points of any of the metallic elements. Tungsten is also one of the most dense elements, with a density comparable to that of gold. Only osmium ($d = 22.7$ g/cm^3), platinum ($d = 21.45$ g/cm^3), and rhenium ($d = 21.40$ g/cm^3) are more dense. The Group VIB elements are hard metals and corrosion-resistant. The latter property is because of protective oxide coatings.

TABLE 21.6

Properties of Group VIB Elements

	Cr	Mo	W
Atomic number	24	42	74
Electronic configuration	[Ar]$4s^1 3d^5$	[Kr]$5s^1 4d^5$	[Xe]$6s^2 5d^4$
Atomic radius (Å)	1.17	1.29	1.30
Density (g/cm^3)	6.92	10.2	19.3
Melting point (°K)	2176	2890	3650
Boiling point (°K)	2900	4900	5800
Ionization energy (I) (kjoules/mole)	652	686	770
ΔH_f^0 (MO_3) (kjoules/mole)	−577	−753	−841
ΔG_f^0 (MO_3) (kjoules/mole)	—	−678	−761

Chromium

Chromium is not an abundant element (0.03% of the earth's crust), but it is still approximately fifty times more abundant than the other two members of Group VIB. It is a hard, silvery-white metal that is quite resistant to corrosion. It takes a high polish, which it retains because of the transparency of the self-forming protective oxide coating. It finds much use, especially in the automobile industry, as a plating material for both decorative and protective purposes. Large quantities of chromium are used in alloy steels. High-chrome steels (up to 30% Cr), such as stainless steel, are well-known for their corrosion resistance.

Table 21.7 shows typical chromium compounds and ions representing the common oxidation states of the element.

We have used the oxides of chromium previously as examples of the effect of oxidation number on acid-base properties. Chromium (VI) oxide is typically acidic, Cr(III) oxide is amphoteric, and Cr(II) oxide is typically basic. Their reactions with acid and base are

$$CrO + 2\,H^+ = Cr^{2+} + H_2O$$
$$Cr_2O_3 + 6\,H^+ = 2\,Cr^{3+} + 3\,H_2O$$
$$Cr_2O_3 + 3\,H_2O + 2\,OH^- = 2\,Cr(OH)_4^-$$
$$CrO_3 + 2\,OH^- = CrO_4^{2-} + H_2O$$

Chromium metal, minus its usual oxide coating, is a moderately strong reducing agent:

$$Cr = Cr^{3+} + 3\,e^-; \qquad \mathscr{E}^0 = 0.74\text{ V}$$

It reacts, therefore, with H^+ to produce H_2 gas under standard state conditions:

$$2\,Cr + 6\,H^+ = 2\,Cr^{3+} + 3\,H_2$$

Note that it is Cr(III) ion, not Cr(II) ion, that results from this reaction. Aqueous solutions of Cr(II) ion are unstable in the presence of air. Cr(II) ion is one of the strongest reducing agents that can exist in water solution:

$$Cr^{2+} = Cr^{3+} + e^-; \qquad \mathscr{E}^0 = 0.41\text{ V}$$

TABLE 21.7

Representative Compounds and Ions of Chromium

OXIDATION STATE	
+6	CrO_3, CrO_4^{2-}, $Cr_2O_7^{2-}$
+3	Cr_2O_3, Cr^{3+}, $Cr(OH)_4^-$
+2	CrO, Cr^{2+}
0	Cr

Solutions of chromous ion are oxidized rapidly and almost completely to Cr^{3+} by the oxygen of the air.

Cr^{2+} is a beautiful blue ion. It can be prepared by the reduction of Cr^{3+} with zinc.

Chromium (III) ion, like Al^{3+}, is an "acidic" cation; that is, it hydrolyzes moderately to produce an acidic solution:

$$Cr(H_2O)_6^{3+} = Cr(H_2O)_5OH^{2+} + H^+; \qquad K = 1.3 \times 10^{-4}$$

or (if you prefer)

$$Cr^{3+} + H_2O = CrOH^{2+} + H^+; \qquad K = 1.3 \times 10^{-4}$$

Its strength as an acid is seen to be about an order of magnitude greater than that of acetic acid.

An outstanding characteristic of Cr(III) ion is its ability to form complex ions. Examples are $Cr(H_2O)_6^{3+}$, $Cr(NH_3)_6^{3+}$, and CrF_6^{3-}. It is typical of chromium complex ions that they form slowly and dissociate slowly.

When base is added to Cr^{3+} ion, the hydroxide first precipitates, but since it is amphoteric, it will dissolve in an excess of hydroxide ion to form the complex chromite ion:

$$Cr^{3+} + 3\,OH^- = Cr(OH)_3(s)$$
$$Cr(OH)_3(s) + OH^- = Cr(OH)_4^-$$

Chromite ion is readily oxidized to chromate ion by hydrogen peroxide in basic solution:

$$2\,Cr(OH)_4^- + 3\,HO_2^- = 2\,CrO_4^{2-} + OH^- + 5\,H_2O \qquad [21.1]$$

CrO_4^{2-} is a bright-yellow ion. When CrO_4^{2-} is acidified, a condensation takes place, and the orange-colored dichromate ion, $Cr_2O_7^{2-}$, is formed:

The name chromium comes from the Greek word for color, *chroma*. As mentioned earlier, Cr^{2+} is a bright blue, Cr^{3+} is a blue-gray-purple, $Cr(OH)_4^-$ is green, CrO_4^{2-} is yellow, and $Cr_2O_7^{2-}$ is orange. The complex ions such as the ammine and halo complexes are also highly colored.

$$\underset{\text{yellow}}{2\,CrO_4^{2-}} + 2\,H^+ = \underset{\text{orange}}{Cr_2O_7^{2-}} + H_2O; \qquad K = 4.2 \times 10^{14}$$

In spite of the large equilibrium constant for the reaction, addition of reasonable quantities of base to $Cr_2O_7^{2-}$ solution reverses the reaction above, and the orange-colored $Cr_2O_7^{2-}$ gives way to the yellow of CrO_4^{2-}. All the reactions involving chromium are quite colorful.

Dichromate ion in acid solution is a strong oxidizing agent:

$$Cr_2O_7^{2-} + 14\,H^+ + 6\,e^- = 2\,Cr^{3+} + 7\,H_2O; \qquad \mathscr{E}^0 = 1.33\text{ V}$$

It is interesting that in basic solution, H_2O_2 oxidizes Cr(III) to Cr(VI) (equation [21.1]), but in acidic solution, H_2O_2 reduces Cr(VI) to Cr(III) (equation [21.2]).

Two examples of its oxidizing ability are given below:

$$Cr_2O_7^{2-} + 6\,Fe^{2+} + 14\,H^+ = 2\,Cr^{3+} + 6\,Fe^{3+} + 7\,H_2O$$
$$Cr_2O_7^{2-} + 3\,H_2O_2 + 8\,H^+ = 3\,O_2(g) + 2\,Cr^{3+} + 7\,H_2O \qquad [21.2]$$

21.5 Manganese Group

Group VIIB, the manganese family, consists of the elements manganese (Mn), technetium (Tc), and rhenium (Re). Manganese is by far the most abundant and most commercially important element of the group. Technetium is a man-made element. All its isotopes are radioactive and of such short half-lives that technetium is not found in nature. Rhenium is quite rare. It was not discovered until 1925. Some of the properties of Group VIIB are tabulated in Table 21.8.

Manganese

Representative oxides and ions of the principal oxidation states of manganese are listed below:

OXIDATION STATE	REPRESENTATIVE OXIDES AND IONS
+7	Mn_2O_7, MnO_4^-
+6	MnO_4^{2-}
+4	MnO_2
+3	Mn_2O_3
+2	MnO, Mn^{2+}
0	Mn

The principal manganese-containing mineral is pyrolusite, MnO_2. The element is obtained by reduction with carbon:

$$MnO_2 + 2\,C = Mn + 2\,CO$$
$$MnO_2 + 2\,CO = Mn + 2\,CO_2$$

The principal use of manganese is as an additive in steel. Manganese steels are hard and tough. Manganese is also a fairly strong reducing agent:

$$Mn = Mn^{2+} + 2\,e^-; \qquad \mathscr{E}^0 = 1.18\text{ V}$$

TABLE 21.8

Properties of Group VIIB Elements

	Mn	Tc	Re
Atomic number	25	43	75
Electronic configuration	$[Ar]4s^23d^5$	$[Kr]5s^14d^6$	$[Xe]6s^25d^5$
Atomic radius (Å)	1.17	—	1.28
Density (g/cm^3)	7.2	11.5	21.40
Melting point (°K)	1517	(2300)	3453
Ionization energy (I) (kjoules/mole)	717	702	761
ΔH_f^0 (MO_2) (kjoules/mole)	−519	—	−350
ΔG_f^0 (MO_2) (kjoules/mole)	−464	—	—

It reacts with H^+ to produce H_2 gas and Mn^{2+} ions (standard state conditions):

$$Mn(s) + 2\,H^+ = H_2(g) + Mn^{2+}$$

Note that it is Mn^{2+}, and not Mn^{3+}, that forms in the reaction of H^+ with Mn(*s*). Manganese does exist in the +3 oxidation state, but for reasons evident in the reduction-potential information given at right, the aqueous solution chemistry of +3 manganese is, for practical purposes, nonexistent. Mn(III) oxide, $Mn_2O_3(s)$, however, is a stable basic oxide.

$$Mn^{3+} + e^- = Mn^{2+}; \qquad \mathscr{E}_1^0 = 1.51\ V$$
$$Mn^{3+} + 2\,H_2O = MnO_2(s) + 4\,H^+ + e^-; \qquad \mathscr{E}_2^0 = -0.95\ V$$
$$2\,Mn^{3+} + 2\,H_2O = 2\,Mn^{2+} + MnO_2(s) + 4\,H^+; \qquad \mathscr{E}^0 = 0.56\ V$$

$$\ln K = \frac{n\mathscr{F}\mathscr{E}^0}{RT} = \frac{(1\ eq)(9.65 \times 10^4\ coulombs/eq)(0.56\ V)}{(8.31\ joules/^\circ K)(298^\circ K)}$$

$$K = 3 \times 10^9$$

Treated with a basic solution, Mn^{2+} forms a gelatinous precipitate of $Mn(OH)_2(s)$, which can be dehydrated to form MnO. Manganous oxide is an entirely basic oxide. Most manganese (II) compounds have a pink color, which is not very common among chemicals. The color is so pale, however, that dilute solutions of manganous ion appear practically colorless.

Manganese dioxide, MnO_2, is the only common compound of Mn in the +4 state. It is a dark-brown solid. In acid solution, MnO_2 is a good oxidizing agent:

$$MnO_2 + 4\,H^+ + 2\,e^- = Mn^{2+} + 2\,H_2O; \qquad \mathscr{E}^0 = +1.23\ V$$

Addition of concentrated hydrochloric acid to MnO_2 is a convenient method for preparing chlorine gas:

$$MnO_2(s) + 2\,Cl^- + 4\,H^+ = Mn^{2+} + Cl_2(g) + 2\,H_2O$$

MnO_4^{2-} can also be produced by reduction of MnO_4^- in alkaline solution; for example,

$$2\,MnO_4^- + SO_3^{2-} + 2\,OH^- = 2\,MnO_4^{2-} + SO_4^{2-} + H_2O$$

When MnO_2 is fused with $KNO_3(s)$ and KOH(*s*), the NO_3^- oxidizes the +4 manganese to the +6 compound K_2MnO_4, potassium manganate. This salt, bright green in color, is stable in water solution as long as the solution is basic. In acid solution, however, manganate ion disproportionates:

$$3\,MnO_4^{2-} + 4\,H^+ = 2\,MnO_4^- + MnO_2 + 2\,H_2O; \qquad \mathscr{E}^0 = 1.70\ V$$

The most important compound of manganese is found in the +7 oxidation state, permanganate ion, MnO_4^-. The dark-purple solution of MnO_4^- is a good oxidizing agent that finds considerable use in analytical chemistry. The reduction product in *acid* solution is always the Mn^{2+} ion:

The +7 oxide of manganese, Mn_2O_7, can be prepared by adding concentrated H_2SO_4 to $KMnO_4$. As might be expected, the +7 oxide is an acidic oxide. It is also a highly explosive liquid.

$$MnO_4^- + 8\,H^+ + 5\,e^- = Mn^{2+} + 4\,H_2O; \qquad \mathscr{E}^0 = 1.51\ V$$

An example of the reduction of MnO_4^- in acid is

$$MnO_4^- + 8\,H^+ + 5\,Fe^{2+} = 5\,Fe^{3+} + Mn^{2+} + 4\,H_2O$$

In neutral or basic solutions, the reduction product of MnO_4^- is MnO_2:

$$2\,MnO_4^- + 3\,SO_3^{2-} + H_2O = 3\,SO_4^{2-} + 2\,MnO_2(s) + 2\,OH^-$$

21.6 Group VIII Elements

The Group VIII elements are best discussed in terms of a triad of the elements consisting of iron (Fe), cobalt (Co), and nickel (Ni); and a second grouping of the six elements ruthenium (Ru), osmium (Os), rhodium (Rh), iridium (Ir), palladium (Pd), and platinum (Pt), known collectively as the platinum metals.

Some of the properties of the iron-triad elements are listed in Table 21.9. It should be apparent after only a brief examination of this table why we discuss Fe, Co, and Ni as a group. The close similarity in both physical and chemical properties of the elements of this triad is very striking.

Iron, cobalt, and nickel are hard, high-melting metals of moderate reactivity. Their chemistry is principally that of the +2 and +3 oxidation states. We have seen this preference for the lower oxidation states in the lighter members of other transition groups that we have discussed. Higher oxidation states are known in the iron-triad elements, but they are generally unstable, and we leave further discussion of them for a later course.

Iron

Iron, constituting 4.7% of the earth's crust, is second only to aluminum in its prevalence among the metals.

The principal ore of iron is the mineral hematite, Fe_2O_3. Iron is obtained from the ore on a massive scale by reduction in continuously operating blast furnaces. A mixture of the iron ore, limestone, and coke is fed into the top of the furnace, and molten iron and a lava-like slag are extracted at the bottom. The heat for the furnace comes from the reaction of coke with the oxygen from

There is a possibility that the center of the earth may be iron, or mostly iron. Earthquake waves produce indirect evidence that the core of the earth is liquid with a density corresponding to that of liquid iron under high pressure.

TABLE 21.9

Properties of the Iron-Triad Elements

	Fe	Co	Ni
Atomic number	26	27	28
Electronic configuration	$[Ar]4s^23d^6$	$[Ar]4s^23d^7$	$[Ar]4s^23d^8$
Atomic radius (Å)	1.16	1.16	1.15
Density (g/cm^3)	7.86	8.9	8.90
Melting point (°K)	1812	1768	1728
Ionization energy (I) (kjoules/mole)	759	757	736
ΔH_f^0 (MO) (kjoules/mole)	−267	−239	−240
ΔG_f^0 (MO) (kjoules/mole)	−244	−213	−216
$\mathcal{E}^0$ ($M^{2+} + 2\,e^- = M$) (V)	−0.440	−0.277	−0.250

Coke is the solid material (mostly carbon) left over from the destructive distillation of coal. When coal is heated in the absence of air, a wide variety of volatile and valuable chemicals are formed. The end product after all the volatile material is removed is coke.

Fe_3O_4 is another example of a mixed oxide (see the discussion of Pb_3O_4 in Chapter 18). One-third of the iron is in the +2 state; two-thirds is in the +3 state. The formula can be written $FeO \cdot Fe_2O_3$. Fe_3O_4 occurs in nature as the mineral magnetite, which gets its name from its ferromagnetic properties.

The oxides of iron are rather easily converted from one into another in spite of being solids. The reason is apparent when one knows the crystal structure of the iron oxides. The structures of all three are related in that the framework of each consists of oxide ions in a cubic close-packed lattice. The iron ions occupy the interstitial holes. Thus, the oxidation number of the iron atom, and therefore the composition of the oxide, can change without seriously disrupting the oxide framework.

Iron from which practically all of the carbon has been removed is called wrought iron. It is relatively soft and malleable, and finds use in manufacture of soft iron wire and nails.

Cast iron is a hard, brittle substance produced from pig iron by partial removal of

a hot blast of air that is introduced through the bottom of the furnace:

$$2\,C(s) + O_2 = 2\,CO(g); \qquad \Delta H^0 = -221 \text{ kjoules}$$

Despite the long history of blast-furnace technology, the chemical reactions in the blast furnace are still not well understood. It is generally agreed that CO (not C) is the principal reducing agent, and that the ore is probably reduced in steps. At the top of the furnace, where the temperature is only 250° C, the first step appears to be a reduction of Fe_2O_3 to Fe_3O_4:

$$CO(g) + 3\,Fe_2O_3(s) = 2\,Fe_3O_4(s) + CO_2(g)$$

As the Fe_3O_4 settles deeper into the furnace where the temperature is higher, it, in turn, is reduced to FeO:

$$Fe_3O_4(s) + CO(g) = 3\,FeO(s) + CO_2(g)$$

Finally, at the bottom of the furnace, where it is the hottest, the complete reduction to elemental iron occurs:

$$FeO(s) + CO(g) = Fe(l) + CO_2(g)$$

The overall reaction for the reduction of Fe_2O_3 can be written as

$$Fe_2O_3(s) + 3\,CO(g) = 2\,Fe(l) + 3\,CO_2(g)$$

Other reactions that take place in the blast furnace are reduction of CO_2 to CO by coke, formation of CaO from heating limestone, and formation of slag by reaction of the basic oxide, CaO, with the acidic oxides in the ore:

$$C(s) + CO_2(g) = 2\,CO(g)$$
$$CaCO_3 = CaO + CO_2$$
$$CaO + SiO_2 = CaSiO_3$$

The last equation is an idealized, highly simplified reaction for the formation of the slag. The molten glasslike slag floats on top of the liquid iron and can, therefore, be tapped off at a higher level in the furnace than can the iron. The slag floating as it does on the top of the iron protects the iron from oxidation.

The iron produced in the blast furnace (pig iron) contains approximately 4% carbon (as iron carbides), smaller amounts of silicon, phosphorus, and manganese, and a trace of sulfur. Pig iron is changed into steel by burning out the impurities to leave a controlled amount of carbon. In the United States this is usually done in an open-hearth process where carbon is removed by oxidation with air and iron oxide, the latter usually added in the form of rusted scrap metal. Limestone is added to provide CaO, which combines with the acidic oxides of the impurities (P_4O_{10}, SiO_2, SO_2) to form a slag. Metals such as Cr, Mn, V, Ni, W, Mo, and Co are then added in controlled amounts to produce steels of the desired properties (hardness, toughness, springiness, corrosion resistance, impact resistance, workability, and so on).

carbon. Cast iron, which is used in such items as stoves and steam radiators, contains over 2% carbon plus various amounts of other impurities.

Iron is a moderately good reducing agent as shown by its oxidation potential:

$$Fe = Fe^{2+} + 2\,e^-; \qquad \mathscr{E}^0 = 0.44\ V$$

It therefore reacts with H^+ to produce H_2 gas under standard state conditions:

$$Fe(s) + 2\ H^+ = Fe^{2+} + H_2(g)$$

In aqueous solution, ferrous ion is pale green. This is apparently the color of the hydrated ion, however, since anhydrous salts of Fe(II) are usually colorless.

Ferrous ion can be easily oxidized to ferric ion by oxygen (standard state conditions):

	$\mathscr{E}^0$ (V)
$4\,Fe^{2+} = 4\,Fe^{3+} + 4\,e^-$	−0.771
$O_2 + 4\,H^+ + 4\,e^- = 2\,H_2O$	1.229
$4\,Fe^{2+} + O_2 + 4\,H^+ = 4\,Fe^{3+} + 2\,H_2O$	0.46

When treated with concentrated nitric acid, iron goes into a "passive" state and no longer reduces H^+. The passivity is probably caused by the formation of a protective oxide film. When scratched or subjected to a mechanical shock (as by a sharp blow with a hammer), its reactivity is restored. Other metals that exhibit similar behavior are Cr, Mo, Co, and Ni.

The concentration of O_2 in air is also sufficient to bring about the oxidation of Fe^{2+} to Fe^{3+}. Because of this, aqueous solutions of ferrous ions are almost always contaminated with some ferric ions.

Aqueous solutions of Fe^{3+} range in color from light yellow to reddish-brown, depending on the concentration. The color is believed to be from hydrolysis products of Fe^{3+}, since addition of H^+ (from a noncomplexing acid such as perchloric or nitric) destroys the yellow color. Like other small, trivalent ions, Fe^{3+} hydrolyzes appreciably:

Pure, solid $Fe(NO_3)_3$ has a pale-violet color. The yellow color usually associated with this and other solid iron salts is from a coating of hydrolysis products of Fe^{3+}. One can readily wash away the coating of a crystal of $Fe(NO_3)_3$ to reveal its true color. Acidic solutions of $Fe(NO_3)_3$ also have a pale-violet color.

$$Fe^{3+} + H_2O = FeOH^{2+} + H^+; \qquad K = 9 \times 10^{-4}$$

or

$$Fe(H_2O)_6^{3+} = Fe(H_2O)_5OH^{2+} + H^+; \qquad K = 9 \times 10^{-4}$$

The extent of hydrolysis is seen to be comparable to that of Cr^{3+} and about an order of magnitude greater than that of Al^{3+}. The second step in the hydrolysis of Fe^{3+} is of about equal importance to the first:

$$FeOH^{2+} + HOH = Fe(OH)_2^+ + H^+; \qquad K = 5.5 \times 10^{-4}$$

or

$$Fe(H_2O)_5OH^{2+} = Fe(H_2O)_4(OH)_2^+ + H^+; \qquad K = 5.5 \times 10^{-4}$$

The third step in the hydrolysis of Fe^{3+} results in the formation of $Fe(OH)_3(s)$, a red-brown precipitate.

EXAMPLE 21.1 Calculate the lowest pH at which $Fe(OH)_3$ will precipitate in a 0.10 *M* Fe^{3+} solution.

SOLUTION

$$Fe(OH)_3(s) = Fe^{3+} + 3\ OH^-; \qquad K_{sp} = 1 \times 10^{-36}$$

At equilibrium,

$$[Fe^{3+}] = 0.10$$
$$[OH^-] = x$$
$$(0.10)(x)^3 = 1 \times 10^{-36}$$
$$x^3 = 1 \times 10^{-35}$$
$$x = 2 \times 10^{-12} = [OH^-]$$
$$pOH = -\log(2 \times 10^{-12}) = 11.7$$
$$pH = 14.0 - pOH = 2.3$$

Hence, $Fe(OH)_3$ starts to form in quite acidic solution.

Cobalt

Cobalt and nickel are present in the earth's crust in much smaller quantities than iron (Co, 0.002%; Ni, 0.008%). Their principal ores are sulfides and arsenides.

Their metallurgy involves roasting to form the oxide, followed by reduction with hydrogen or carbon. Both metals are employed in the manufacture of special steels.

As can be seen from Table 21.9, cobalt is a weaker reducing agent than iron, but it is still capable of reducing H^+:

$$Co(s) + 2\ H^+ = Co^{2+} + H_2(g)$$

Note that it is Co(II) ion that forms. Co(III) ion does not exist in water solution to any measurable extent. As can be seen from its reduction potential,

$$Co^{3+} + e^- = Co^{2+}; \qquad \mathscr{E}^0 = 1.82\ V$$

it is a very strong oxidizing agent, fully capable of oxidizing water. As a result, there is no aqueous chemistry involving Co^{3+} ion. Complex ions of Co(III), however, are stable, two common examples being $Co(NH_3)_6^{3+}$ and $Co(CN)_6^{3-}$. Addition of a base to Co^{2+} solutions precipitates $Co(OH)_2(s)$, which can be converted to CoO by heating.

CoO can be oxidized to Co_3O_4, analogous to Fe_3O_4. The cobalt analog of Fe_2O_3, however, has not been reported.

Nickel

Nickel is a reducing agent of about the same strength as cobalt; it therefore reduces H^+ to $H_2(g)$ and is itself oxidized to Ni^{2+}. Like several other metals, however, its reactivity is masked to some extent by the formation of an oxide coating.

The aqueous solution chemistry of nickel is confined entirely to the +2 oxidation state. Ni^{2+} ion has a bright-green color in aqueous solution. It forms an insoluble hydroxide, a typically basic oxide, and a series of complex ions. The bright-blue ammine complex, $Ni(NH_3)_4^{2+}$ is an example of a stable complex ion of nickel (II). When aqueous ammonia is slowly added to Ni^{2+} solution, the light-green hydroxide first precipitates and then dissolves to form the bright-blue solution of tetraammine nickel (II) ion:

In basic solution, Ni is quite inert and is used in several laboratory applications. Nickel crucibles, for example, are often used for fusion of basic substances.

$$NH_3 + H_2O = NH_4^+ + OH^-$$

$$\underset{\text{green}}{Ni^{2+}} + 2\,OH^- = \underset{\text{green}}{Ni(OH)_2(s)}$$

$$\underset{\text{green}}{Ni(OH)_2(s)} + 4\,NH_3(aq) = \underset{\text{blue}}{Ni(NH_3)_4^{2+}} + 2\,OH^-$$

Further addition of NH_3 converts the blue-colored $Ni(NH_3)_4^{2+}$ to the violet-colored $Ni(NH_3)_6^{2+}$.

Large quantities of elemental nickel are used in coinage and in the manufacture of stainless steel (8% Ni, 18% Cr, and most of the remainder Fe). Finely divided nickel has the interesting property of absorbing large quantities of H_2 gas, which enters the metallic lattice in the form of H atoms. The resulting product is an excellent catalyst for hydrogenation reactions in organic chemistry.

Nickel in the +3 and +4 oxidation states is found in the solid oxides Ni_2O_3 and NiO_2. Both are strong oxidizing agents. The former is a component of the Ni-Fe battery known as the Edison cell, whose reaction is

$$Ni_2O_3(s) + Fe + 3\,H_2O \underset{\text{charge}}{\overset{\text{discharge}}{\rightleftharpoons}} 2\,Ni(OH)_2 + Fe(OH)_2$$

Platinum Metals

The six elements (ruthenium, osmium, rhodium, iridium, palladium, and platinum) that lie just below the iron triad are commonly referred to as the platinum metals. The most significant characteristic that they have in common is their general inertness. Platinum, in particular, is widely used in jewelry and in many laboratory applications because of its inertness. The only common substance that attacks platinum is aqua regia (a mixture of concentrated hydrochloric and nitric acids). As mentioned in Chapter 20, it is the combination of the oxidizing power of the nitrate ion in acid solution with the complexing power of chloride ion that makes the oxidation and dissolution of platinum possible:

Ruthenium and osmium however, do form oxides (RuO_4 and OsO_4) fairly readily. Note the +8 oxidation state of the metal in these compounds.

$$Pt(s) + 6\,Cl^- + 4\,NO_3^- + 8\,H^+ = PtCl_6^{2-} + 4\,NO_2(g) + 4\,H_2O$$

Neither HNO_3 nor HCl by itself reacts with platinum.

Table 21.10 lists some of the physical properties of the platinum metals. They are seen to be high-melting, very dense metals.

TABLE 21.10

Properties of the Platinum Metals

	Ru	Rh	Pd	Os	Ir	Pt
Atomic number	44	45	46	76	77	78
Electronic configuration	$5s^1 4d^7$	$5s^1 4d^8$	$4d^{10}$	$6s^2 5d^6$	$5d^9$	$6s^1 5d^9$
Ionization energy (I) (kjoules/mole)	711	720	803	837	837	877
Atomic radius (Å)	1.24	1.25	1.25	1.26	1.26	1.29
Density (g/cm^3)	12.4	12.4	12.0	22.7	22.6	21.5
Melting point (°K)	2670	2239	1823	2970	2727	2043

21.7 Group IB Elements

Group IB consists of copper (Cu), silver (Ag), and gold (Au). Because of their scarcity and relative resistance to corrosion, they have been used as coinage metals and jewelry since ancient times. They are all highly malleable and ductile, relatively soft for transition metals, and excellent conductors of heat and electricity. Other properties of this group are given in Table 21.11.

The IB elements are seen (Table 21.11) to have completed *d*-level orbitals, which suggests that their chemistry is determined principally (although not entirely) by the lone valence *s* electron each has. In this respect, Group IB shows some resemblance to Group IA. However, comparing Table 18.2 for the IA elements with Table 21.11, one sees immediately that the properties of the two groups are quite different. Group IB elements are seen to have very much higher densities, melting points, and ionization energies, and very much smaller atomic radii than their counterparts in Group IA. As reducing agents, Group IB elements are very poor compared with the Group IA elements. You will recall that IA metals will reduce water to produce H_2 gas. The IB metals, on the other hand, not only do not react with water, but are not even capable of reducing H^+

TABLE 21.11

Properties of Group IB Elements

	Cu	Ag	Au
Atomic number	29	47	79
Electronic configuration	$4s^1 3d^{10}$	$5s^1 4d^{10}$	$6s^1 5d^{10}$
Ionization energy (I) (kjoules/mole)	745	728	891
Atomic radius (Å)	1.17	1.34	1.34
Density (g/cm^3)	8.92	10.5	19.3
Melting point (°K)	1356	1234	1336
ΔH_f^0 (M_2O) (kjoules/mole)	−167	−31	—
ΔG_f^0 (M_2O) (kjoules/mole)	−146	−11	—

under standard state conditions. These differences in both physical and chemical properties between the IA and IB groups are explainable to a large extent by the smaller atomic radii of the IB elements. The significant decrease in size in going from K (Group IA) to Cu (Group IB), both in period 4, can be accounted for by the larger nuclear charge of Cu. Ten elements intervene between K and Cu in period 4, ten between Rb and Ag in period 5, and twenty-four between Cs and Au in period 6 without a change in the principal quantum number.

The electronic configurations given in Table 21.11 suggest that +1 (pseudo-noble-gas structure) is a likely oxidation state for the IB elements. All of them do exist in the +1 state; however, the +2 state, as well, is important in copper, and the +3 state is the more important state for gold.

Copper

The principal ores of copper are sulfide minerals. Their metallurgy is rather simple. They are first roasted in the presence of air to produce a mixture of sulfide and oxide, and then smelted. The reactions, using Cu_2S as an example, are

$$2\,Cu_2S(s) + 3\,O_2(g) \overset{\text{heat}}{=} 2\,Cu_2O(s) + 2\,SO_2(g)$$

$$Cu_2S(s) + 2\,Cu_2O(s) \overset{\text{heat}}{=} 6\,Cu(l) + SO_2(g)$$

The product is refined electrolytically as was described in Chapter 11.

There are both Cu(I) and Cu(II) compounds. The former are all either insoluble compounds (such as Cu_2S, Cu_2O, CuCl, CuBr) or complex ions [such as $Cu(CN)_2^-$, $Cu(NH_3)_2^+$, $CuCl_2^-$], since Cu^+ ion disproportionates almost completely in aqueous solution, as can be seen from the following half-reaction potentials:

	$\mathscr{E}^0$ (V)
$Cu^+ = Cu^{2+} + e^-$	−0.153
$Cu^+ + e^- = Cu(s)$	+0.521
$2\,Cu^+ = Cu^{2+} + Cu(s)$	+0.368

Important copper-containing minerals are Cu_2S (chalcocite), $CuFeS_2$ (chalcopyrite), CuS (covellite), Cu_2O (cuprite), $CuCO_3 \cdot Cu(OH)_2$ (malachite), and CuO (tenorite).

Sulfide ores of copper are generally concentrated by "flotation," a process in which air is bubbled through a suspension of the finely ground ore (copper sulfide minerals and waste rock). Air bubbles selectively attach themselves to the sulfide minerals and float them to the surface, where they are skimmed off. The silicate waste rock sinks to the bottom of the tank.

To promote the flotability of the sulfide minerals, a heteropolar chemical called a "collector" is added to the suspension in the flotation tanks. The collector molecule has a polar end, which is attracted to the sulfide mineral, and a nonpolar, hydrocarbon part, which will attach to the air bubble.

The mechanism involved is similar to that by which soap acts as a cleaning agent. In a soap ion, $C_{17}H_{35}COO^-$ for example, the long chain hydrocarbon end of the ion dissolves in oil-like particles, and the polar end dissolves in water. When the water is drained away, the dirt particles are carried with it.

EXAMPLE 21.2 Calculate the concentration of Cu^+ that exists in equilibrium with Cu metal and 0.10 *M* Cu^{2+} ion solution.

SOLUTION

$$2\,Cu^+ = Cu^{2+} + Cu(s); \qquad \mathscr{E}^0 = 0.368\text{ V}$$

$$\ln K = \frac{n\mathscr{F}\mathscr{E}^0}{RT}$$

$$\ln K = \frac{(2\text{ eq})(9.65 \times 10^4\text{ coulombs/eq})(0.368\text{ V})}{(8.31\text{ joules/}^\circ\text{ K})(298^\circ\text{ K})}$$

$$\ln K = 28.7$$

$$K = 2.9 \times 10^{12}$$

$$K = \frac{[Cu^{2+}]}{[Cu^+]^2} = \frac{0.10}{[Cu^+]^2} = 2.9 \times 10^{12}$$

$$[Cu^+] = 1.9 \times 10^{-7}\ \frac{\text{moles}}{\text{liter}}$$

This is such a low concentration that for most purposes, there is no Cu^+ ion chemistry in aqueous solution.

Two examples of reactions involving Cu(I) compounds are given below:

$$Cu_2O(s) + 2\,H^+ = Cu(s) + Cu^{2+} + H_2O$$
$$2\,Cu^{2+} + 4\,I^- = 2\,CuI(s) + I_2(aq)$$

Note that Cu^+ ion does not appear in the overall reaction in either of these equations. The first reaction might be thought of as taking place in two steps: (a) formation of Cu^+ ion followed by (b) disproportionation of Cu^+ ion. The second equation above is one that finds application in the analysis of copper. In this reaction, the I^- reduces Cu(II) to Cu(I), but only because the Cu(I) compound (cuprous iodide, in this case) is insoluble in water.

In the analysis of copper, the copper, in solution as Cu^{2+} ion, is treated with excess I^- ion. $CuI(s)$ and I_2 (present as the complex ion I_3^- because of the excess I^- present) result. The I_3^- is then reduced with thiosulfate solution of known concentration. The equations involved are

$$2\,Cu^{2+} + 5\,I^- = 2\,CuI(s) + I_3^-$$
$$I_3^- + 2\,S_2O_3^{2-} = S_4O_6^{2-} + 3\,I^-$$

Knowing the number of moles of $S_2O_3^{2-}$ used in the titration, one can calculate the number of moles of I_3^- that were produced, which in turn is related through the first equation to the moles of Cu^{2+} present in the sample.

As seen from the standard oxidation potential

$$Cu = Cu^{2+} + 2\,e^-; \qquad \mathscr{E}^0 = -0.34\text{ V}$$

copper will not reduce H^+ under standard state conditions:

$$Cu + 2\,H^+ = \text{no reaction}$$

Copper can, however, be put into solution with nitric acid (because of the strong oxidizing power of NO_3^- in acid solution):

$$Cu + \underset{\text{(conc.)}}{4\,H^+} + 2\,NO_3^- = Cu^{2+} + 2\,NO_2 + 2\,H_2O$$
$$3\,Cu + \underset{\text{(dilute)}}{8\,H^+} + 2\,NO_3^- = 3\,Cu^{2+} + 2\,NO + 4\,H_2O$$

Copper (II) ion in aqueous solution is blue (see the discussion of color of hydrated Cu^{2+} ion in Chapter 20). If OH^- ion is added to Cu^{2+}, light-blue $Cu(OH)_2(s)$ precipitates, which is a typical basic hydroxide and does not, therefore, dissolve appreciably in excess OH^- ion. Addition of $NH_3(aq)$, however, dissolves $Cu(OH)_2(s)$ to form the deep-blue tetraammine copper (II) complex ion:

$$Cu(OH)_2(s) + 4\,NH_3(aq) = Cu(NH_3)_4^{2+} + 2\,OH^-$$

If the basic solution above is carefully neutralized, one can destroy the ammine complex and still have enough OH^- to re-precipitate $Cu(OH)_2(s)$:

$$Cu(NH_3)_4^{2+} + 4\,H^+ = Cu^{2+} + 4\,NH_4^+$$
$$Cu^{2+} + 2\,OH^- = Cu(OH)_2(s)$$

Further addition of acid will then dissolve the hydroxide to form the blue hydrated Cu^{2+} ion. If the acid used is hydrochloric, excess acid gives rise to complex ions of the type $CuCl_3^-$ and $CuCl_4^{2-}$, which are green to yellow in color.

One of the most insoluble compounds of copper is the sulfide CuS. It is a black solid with a K_{sp} of 8×10^{-37}. One of the most commonly encountered Cu(II) compounds is copper sulfate pentahydrate $CuSO_4 \cdot 5H_2O$. The lattice of this substance is such that each Cu^{2+} ion is surrounded by four water molecules. The fifth water molecule is not directly tied to the Cu^{2+} ion. In high concentrations, Cu^{2+} ion is toxic, and $CuSO_4 \cdot 5H_2O$ is commonly used as a fungicide.

Silver

The principal ores of silver contain Ag_2S, AgCl, and elemental Ag. Elemental silver can be extracted by dissolution in Hg. The Ag is then separated from the amalgam by distilling the mercury away from the silver.

The cyanide leaching process is commonly used for extracting Ag from AgCl and Ag_2S. In this process, the silver is put into solution as the dicyanoargentate ion by treatment with NaCN solution and air:

$$2\,Ag_2S(s) + 8\,CN^- + O_2(g) + 2\,H_2O = 4\,Ag(CN)_2^- + 2\,S(s) + 4\,OH^-$$

$$AgCl(s) + 2\,CN^- = Ag(CN)_2^- + Cl^-$$

It is the great stability of the cyano complex ion that accomplishes the feat of dissolving the very insoluble Ag_2S ($K_{sp} = 6 \times 10^{-52}$) and AgCl ($K_{sp} = 1.8 \times 10^{-10}$). The silver is recovered from the leachate by reduction with Zn:

$$Zn(s) + 2\,Ag(CN)_2^- = Zn(CN)_4^{2-} + 2\,Ag(s)$$

Elemental silver can also be extracted in the cyanide leaching process. Air is blown through the suspension for a period of about two weeks. The equation that applies is

$$4\,Ag(s) + 8\,CN^- + 2\,H_2O + O_2(g) = 4\,Ag(CN)_2^- + 4\,OH^-$$

Again, it is the driving force associated with the formation of the stable cyano complex ion that makes the reaction go.

Silver is a relatively soft metal, and for most uses it is alloyed with copper to give it greater hardness. Besides coinage, silver finds use in jewelry and various decorative arts.

The standard oxidation potential for silver

$$Ag = Ag^+ + e^-; \qquad \mathscr{E}^0 = -0.799\ \text{V}$$

is seen to be even lower than that of copper. Accordingly, there is no reaction between elemental Ag and H^+. Silver, like Cu, however, is oxidized by nitrate ion in acid solution (nitric acid):

$$Ag(s) + \underset{\text{(conc.)}}{2\,H^+ + NO_3^-} = Ag^+ + NO_2(g) + H_2O$$

Note that it is Ag(I) ion that results. This is the commonly encountered oxidation state for silver. When Ag^+ solutions are treated with OH^- ion, Ag_2O precipitates:

$$2\,Ag^+ + 2\,OH^- = Ag_2O(s) + H_2O$$

Silver (I) oxide is a typical basic oxide, showing no amphoteric character. Therefore, it does not dissolve in strong base. It will dissolve in $NH_3(aq)$, however, because of the formation of the ammine complex:

$$Ag_2O(s) + 4\,NH_3(aq) + H_2O = 2\,Ag(NH_3)_2^+ + 2\,OH^-$$

Other insoluble compounds and complex ions of silver were discussed in Chapter 20.

One of the principal uses of silver is in photographic processing. Black and white photographic film is essentially a dispersion of AgBr in gelatin. When exposed to light, the grains of the dispersed solid AgBr become activated by the light and are thereby made more subject to chemical reduction by a mild reducing agent ("developer"). The reduced product is small grains of metallic silver, which appear black because of their finely divided state. The area of the film that has been exposed to the most light appears the darkest on the developed film. The second step in the processing of the film is to remove the excess and unreduced AgBr from the developed film. Since AgBr is a very insoluble substance, some special procedure is required. The chemical sodium thiosulfate, called "hypo" in the photographic industry, provides an answer to the problem. After being developed, the film is placed in a bath of "hypo" where the following industrially important complexation reaction takes place:

$$AgBr(s) + 2\,S_2O_3^{2-} = Ag(S_2O_3)_2^{3-} + Br^-$$

This process is referred to as "fixing." After the "hypo" bath, the film can be washed and brought out into the light.

Gold

Gold is an incredibly ductile and malleable metal. As an example, in ordinary gold leaf the thickness of the metal has been reduced to only 10^{-5} cm, a mere 350 atoms. Gold is a soft metal, and for most purposes, such as jewelry, it must be alloyed with other metals.

Its principal chemical characteristic is its extreme inactivity. Like platinum, it is very resistant to chemical attack. It does not react with oxygen or the halogens, except superficially. The few compounds it does form readily decompose to give back the metal. Unlike Cu and Ag, but like Pt, it is not attacked by nitric acid. Like platinum, however, it is dissolved by *aqua regia:*

$$Au(s) + 6\,H^+ + 3\,NO_3^- + 4\,Cl^- = AuCl_4^- + 3\,NO_2(g) + 3\,H_2O$$

In one step in the assaying of Ag and Au, a small pellet of the mixture of Ag and Au is obtained and weighed to get the sum of the Ag and Au content of the sample. The next step is to treat the Ag + Au pellet with HNO_3, which dissolves the silver. The Au remains and can be weighed. The Ag mass is determined by difference.

The driving force of the reaction is the combination of the strong oxidizing ability of NO_3^- in acid solution and the great stability of the $AuCl_4^-$ complex ion.

In the form of complex ions, gold exists in both +1 and +3 states. Examples are $Au(CN)_2^-$, $Au(CN)_4^-$, and $AuCl_4^-$. There is no aqueous solution chemistry of either Au^+ or Au^{3+} ions, however, since both are sufficiently strong oxidizing agents that they oxidize water. A second reason that we need not expect to find Au^+ ions is that, like the Cu^+ ion, they undergo disproportionation:

$$3\,Au^+ = Au^{3+} + 2\,Au(s); \qquad \mathscr{E}^0 \cong 0.2\text{ V}$$

The Group IIB elements are zinc (Zn), cadmium (Cd), and mercury (Hg). As mentioned previously, these elements are somewhat different from the other transition groups in some of their physical properties. As can be seen in Table 21.12, they are d^{10} elements. With the d level complete, d electrons are not as available to participate in metallic bonding; therefore, those properties associated with bonding, such as freezing points, heats of fusion, and hardness, are significantly lower in the Group IIB elements than in those of the other transition groups.

The principal oxidation state of the Group IIB elements is +2, although mercury also exists in the +1 state. Mercury is quite different from the other two members of the group in other ways. Note (Table 21.12) its very high reduction potential and low melting point, for example.

Zinc

The principal ores of zinc contain the minerals ZnS, $ZnCO_3$, and ZnO. The metallurgy is simple, in principle. The ores are usually roasted to give the oxide and then reduced with coke.

$$2\,ZnS(s) + 3\,O_2(g) = 2\,ZnO(s) + 2\,SO_2(g)$$
$$ZnO(s) + C(s) = Zn(g) + CO(g)$$

The second reaction is carried out at temperatures above the boiling point of Zn; hence, zinc vapor is the product. Upon condensation "zinc dust" forms, which can then be melted and solidified into massive zinc.

Zinc is a moderately reactive reducing agent:

$$Zn = Zn^{2+} + 2\,e^-; \qquad \mathscr{E}^0 = 0.763\ V$$

TABLE 21.12

Properties of Group IIB Elements

	Zn	Cd	Hg
Atomic number	30	48	80
Electronic configuration	$4s^23d^{10}$	$5s^24d^{10}$	$6s^25d^{10}$
Ionization energy (I) (kjoules/mole)	906	866	1004
Atomic radius (Å)	1.25	1.41	1.44
Density (g/cm^3)	7.14	8.65	13.5
Melting point (°K)	693	594	234
Ionic radius (M^{2+}) (Å)	0.74	0.97	1.10
ΔH_f^0 (MO) (kjoules/mole)	−348	−255	−90.8
ΔG_f^0 (MO) (kjoules/mole)	−318	−225	−58.6
$\mathscr{E}^0$ ($M^{2+} + 2\,e^- = M$) (V)	−0.763	−0.403	+0.854

It thus reacts with H^+:

$$Zn(s) + 2\,H^+ = Zn^{2+} + H_2(g)$$

and reduces nitrogen in NO_3^- all the way from $+5$ down to -3:

$$4\,Zn + \underset{\text{(dilute)}}{NO_3^-} + 10\,H^+ = 4\,Zn^{2+} + NH_4^+ + 3\,H_2O$$

When OH^- is added to Zn^{2+} in controlled amounts, $Zn(OH)_2(s)$ precipitates:

$$Zn^{2+} + 2\,OH^- = Zn(OH)_2(s)$$

Addition of excess OH^- causes the $Zn(OH)_2(s)$ to dissolve, showing it to be amphoteric:

$$Zn(OH)_2(s) + 2\,OH^- = Zn(OH)_4^{2-}$$

Zinc metal added to a concentrated solution of strong base reacts to form $H_2(g)$:

The reaction of Zn with OH^- to produce $H_2(g)$ is typical of those metals that form amphoteric hydroxides. Reactions similar to this occur with Al, Cr, Be, Sn, and Pb.

$$Zn(s) + 2\,OH^- + 2\,H_2O = Zn(OH)_4^{2-} + H_2(g)$$

Zinc forms a number of complex ions (Chapter 20). Because of its value in making analytical separations, we mention especially the ammine complex, which is sufficiently stable to put $Zn(OH)_2(s)$ into solution as the tetraammine zinc ion:

$$Zn(OH)_2(s) + 4\,NH_3(aq) = Zn(NH_3)_4^{2+} + 2\,OH^-$$

Another stable and readily formed complex ion of zinc is tetracyanozincate ion. Its use in the metallurgy of silver was referred to earlier. The sulfide of zinc is remarkable for its lack of color. The insoluble sulfides of almost all metals are colored (commonly black) whereas ZnS is white. It finds use as a white-paint pigment, as does ZnO.

Large quantities of elemental zinc are used as a protective coating for iron (galvanized sheet iron and galvanized pipe, for example). The first step in the corrosion of iron is oxidation of iron to the $+2$ state. Zinc, being a better reducing agent than iron, protects the iron by reversing the reaction for formation of Fe^{2+}:

	$\mathscr{E}^0$ (V)
$Zn = Zn^{2+} + 2\,e^-$	0.763
$Fe^{2+} + 2\,e^- = Fe$	-0.440
$Zn + Fe^{2+} = Zn^{2+} + Fe$	$+0.323$

Hence, the zinc coating corrodes, but the structural iron in contact with the zinc is unaffected as long as the zinc lasts.

Cadmium

Cadmium is in many ways similar to zinc. They are found together in nature and separated by distillation as the final step in their metallurgy. As can be seen from the $\mathscr{E}^0$ data in Table 21.12, cadmium will reduce H^+:

$$Cd(s) + 2\,H^+ = Cd^{2+} + H_2$$

Addition of a base to Cd^{2+} produces a precipitate of $Cd(OH)_2$. Dehydration of $Cd(OH)_2$ produces CdO, which is a typical basic oxide.

Among other complex ions, cadmium forms an ammine complex, sufficiently stable to cause $Cd(OH)_2$ to dissolve in $NH_3(aq)$:

$$Cd(OH)_2(s) + 4\,NH_3 = Cd(NH_3)_4^{2+} + 2\,OH^-$$

Cadmium compounds are highly toxic when ingested over long periods of time, and they have proved to be serious health hazards in some industries.

Mercury

The principal source of mercury is the mineral cinnabar, HgS, whose metallurgy is simple in both principle and practice. Roasting the ore produces HgO, which then decomposes to O_2 (g) and $Hg(l)$:

$$2\,HgS(s) + 3\,O_2 = 2\,HgO(s) + 2\,SO_2(g)$$
$$2\,HgO(s) \overset{500^\circ C}{=} 2\,Hg(l) + O_2(g)$$

Mercury is distilled away from the waste rock. The resulting product is quite pure since the only substances that dissolve in Hg are metals, and none of these that are likely to be present as impurities boil at all near the relatively low boiling point (630° K) of Hg.

Many metals are soluble in liquid mercury, especially the soft metals such as copper, silver, gold, and the alkali metals. The resulting alloys are called amalgams.

Mercury and its soluble compounds are toxic. Extreme care should be taken to avoid breathing mercury vapors.

Mercury is a very poor reducing agent, as evidenced by its low oxidation potential:

$$Hg = Hg^{2+} + 2\,e^-; \qquad \mathscr{E}^0 = -0.854\ V$$

Mercury is one of the metals that we are likely to run out of during the twentieth century. The once sizable deposits of HgS that existed in California and Nevada are now well depleted. Spain currently produces most of the world's supply of Hg.

Among the commonly encountered metals, only Au and Pt are less active than mercury. Mercury obviously does not react with H^+, but it is oxidized by nitric acid:

$$Hg(l) + 4\,H^+ + \underset{\text{(conc.)}}{2\,NO_3^-} = Hg^{2+} + 2\,NO_2(g) + 2\,H_2O$$

Other evidence for the formula Hg_2^{2+} for mercurous ion comes from equilibrium studies of the disproportionation reaction of mercurous ion. If mercurous ion is written as Hg_2^{2+}, the reaction and equilibrium expression are

$$Hg_2^{2+} = Hg^{2+} + Hg(l)$$

$$K = \frac{[Hg^{2+}]}{[Hg_2^{2+}]}$$

but if written as Hg^+, we would have

$$2\ Hg^+ = Hg^{2+} + Hg(l)$$

$$K' = \frac{[Hg^{2+}]}{[Hg^+]^2}$$

When the concentrations of mercuric and mercurous ions are measured in various equilibrium solutions, it is found that the ratio of mercuric to mercurous is constant as predicted in the former case, but that mercuric to mercurous squared is not constant. That is, K is a true equilibrium constant, and K' is not. Therefore, $[Hg_2^{2+}]$, not $[Hg^+]$, is the correct formula for mercurous ion. The most direct evidence for the existence of Hg_2^{2+} comes from x-ray crystallographic studies of solid mercurous salts.

Examples of other salts with pronounced covalent character are Al_2Cl_6, $GaCl_3$, $SbCl_3$, and, to a lesser extent, $BeCl_2$ and AgCl.

Mercury shows both +1 and +2 oxidation states in its compounds. The mercurous ion is unusual in that it is a diatomic ion, Hg_2^{2+}. The two mercury atoms are bound together by a covalent bond. There are several pieces of evidence that show that the formula for mercurous ion is Hg_2^{2+} rather than Hg^+. One of these evidences is the lack of paramagnetism in mercurous ion. Hg^+ has an unpaired electron $[(Xe)5d^{10}6s^1]$ and would, therefore, be paramagnetic, whereas Hg_2^{2+} has no unpaired electrons:

$$Hg^{\dot{+}} + \cdot Hg^+ = Hg{:}Hg^{2+}$$

Mercurous ions disproportionate to some extent in aqueous solution; but the extent of the reaction is slight, as shown by the small K for the reaction:

$$Hg_2^{2+} = Hg(l) + Hg^{2+}; \qquad K = 5.9 \times 10^{-3}$$

The above reaction, however, becomes important when attempts are made to precipitate mercury (I) hydroxide by adding OH^- to Hg_2^{2+} ion. Instead of the expected mercury (I) hydroxide (or oxide), disproportionation takes place and mercury (II) oxide and elemental Hg result:

$$Hg_2^{2+} + 2\ OH^- = HgO(s) + H_2O + Hg(l)$$

A similar reaction takes place when mercurous ion is treated with H_2S:

$$Hg_2^{2+} + H_2S(g) = HgS(s) + Hg(l) + 2\ H^+$$

Thus, neither Hg(I) oxide or Hg(I) sulfide exists. Mercurous salts that do exist are the nitrate, sulfate, and perchlorate. The Hg(I) halides are quite insoluble. For example, Hg_2Cl_2, known commonly as calomel, has a K_{sp} of 1×10^{-18}.

Mercuric halides (the fluoride excepted) are noteworthy because of their covalent character. $HgCl_2$, for example, displays all the common characteristics of a principally covalent compound. Its lattice is composed of discrete, linear, covalent molecules. It melts at a temperature of only 280° C, and when melted it does not conduct an electric current. In aqueous solution, $HgCl_2$ behaves as a very weak electrolyte. Note the small equilibrium constants for its ionization reactions:

$$HgCl_2 = HgCl^+ + Cl^-; \qquad K_1 = 3.2 \times 10^{-7}$$
$$HgCl^+ = Hg^{2+} + Cl^-; \qquad K_2 = 1.8 \times 10^{-7}$$

Summary

Approximately 60% of the elements are classified as transition metals. These are the B-group elements, those whose differentiating electron is in either a *d* or an *f* orbital. As a group, the transition metals differ markedly in their physical properties from the representative metals. In general, the former are more dense, higher melting, and harder than the latter.

The chemistry of transition metals is characterized by variable oxidation states, colored compounds, and complex ions, characteristics that are directly related to unfilled *d* or *f* orbitals.

Several groupings of transition elements are noteworthy because of their very close similarity in chemical behavior. The series of lanthanides is an example. With the differentiating electron (4*f*) being deep inside the atom, only very minor differences in chemical properties are discernible among the elements of this series.

Other classifications in which the elements have marked similarities are the first row of transitional elements (the 3*d* series), the iron-triad elements, and the six platinum-group metals. Starting with Group IVB the elements in the 4*d* series and the 5*d* element just below (Zr and Hf, Nb and Ta, Mo and W, . . .) have pronounced similarities. The two elements in each of these pairs are separated in the table by the lanthanides. The contraction in size of the elements that occurs in that series is just enough to make the 4*d* element and its corresponding 5*d* element almost identical in size. With similarity in both electronic configuration and size, similarity in properties is very predictable.

Because of their abundance or industrial importance, or especially because of their interesting chemistry, some of the transition elements are of special interest to us. In your review of this chapter pay particular attention to the elements Cr, Mn, Fe, Co, Ni, Cu, Ag, Zn, Cd, and Hg, and their compounds.

Exercises

21-1.
Complete the following table:

NAME	FORMULA	NAME	FORMULA
Manganate ion		Vanadate ion	
	$Cr_2O_7^{2-}$		VO_2^+
Cupric chloride		Hexachloro-platinate (IV) ion	
	Ce^{4+}		$Cr(NH_3)_6^{3+}$

21-2.
The chemistry of the transition-metal ions in solution is colorful, and this color plays an important part in identification of the ions. Give the color of the following ions: V^{2+}, V^{3+}, Cr^{2+}, Cr^{3+}, $Cr(OH)_4^-$, CrO_4^{2-}, $Cr_2O_7^{2-}$, Mn^{2+}, MnO_4^{2-}, MnO_4^-, Fe^{2+}, Fe^{3+}, Ni^{2+}, $Ni(NH_3)_4^{2+}$, Cu^{2+}, $Cu(NH_3)_4^{2+}$.

21-3.
Complete the following equations, which represent reactions of Group IIIB (including the lanthanides), IVB, and VB elements.

(a) $La + H^+ =$
(b) $La^{3+} + OH^- =$
(c) $Ce^{4+} + Cl^- =$
(d) $TiCl_4 + Mg =$
(e) $Ti^{3+} + Ce^{4+} =$
(f) $V_2O_5 + H^+ =$
(g) $V_2O_5 + OH^- =$
(h) $V^{3+} + Zn =$
(i) $TiO_2 + NaOH \overset{\text{fusion}}{=}$
(j) $V^{2+} + Fe^{3+} =$

21-4.
Complete the following equations, which represent reactions of the Group VIB elements.

(a) $Cr + H^+ =$
(b) $Cr^{3+} + Zn =$
(c) $Cr_2O_7^{2-} + H_2SO_3 =$ (acid solution)
(d) $Cr_2O_7^{2-} + H_2O_2 =$ (acid solution)
(e) $Cr(OH)_4^- + HO_2^- =$ (basic solution)
(f) $CrO_4^{2-} + H^+ =$
(g) $Cr^{3+} + OH^- =$
(h) $Cr^{2+} + O_2 =$ (acid solution)
(i) $Cr(OH)_3 + OH^- =$

21-5.
Complete the following equations, which represent reactions of the Group VIIB elements.

(a) $Mn + H^+ =$
(b) $MnO_4^- + H_2SO_3 =$ (acid solution)
(c) $MnO_4^- + SO_3^{2-} =$ (neutral solution)
(d) $MnO_2 + Br^- =$ (acid solution)
(e) $MnO_4^{2-} =$ (acid solution)

21-6.
Complete the following equations, which represent reactions of the Group VIII elements.

(a) $Fe_2O_3 + CO \overset{\text{heat}}{=}$
(b) $Ni(OH)_2 + NH_3 =$
(c) $Pt + H^+ + NO_3^- + Cl^- =$
(d) $FeO + H^+ + O_2 =$
(e) $Co^{3+} + H_2O =$ (acid solution)
(f) $Fe^{3+} + NH_3(aq) =$
(g) $CoS + H^+ =$

21-7.
Complete the following equations, which represent reactions of Group IB and IIB elements.

(a) $Cu^{2+} + I^- =$
(b) $Cu^{2+} + NH_3 =$
(c) $Ag^+ + I^- =$
(d) $Au + H^+ + NO_3^- + Cl^- =$
(e) $Hg_2^{2+} + Cl^- =$
(f) $Hg + H^+ + NO_3^- =$
(g) $Zn + OH^- =$
(h) $Zn^{2+} + NH_3 =$
(i) $Zn + Cu^{2+} =$
(j) $Ag_2O + NH_3 =$
(k) $Cd(OH)_2 + NH_3 =$
(l) $HgO \overset{\text{heat}}{=}$

21-8.
Write chemical reactions to illustrate how you could distinguish between the following:

(a) Cr^{3+} and Fe^{3+}
(b) Mn^{2+} and Cu^{2+}
(c) Ni^{2+} and Ag^+
(d) Ce^{4+} and $Cr_2O_7^{2-}$
(e) Zn^{2+} and Hg^{2+}

21-9.
Write the chemical reactions that would occur as a result of each of the following processes:

(a) Hot concentrated hydrochloric acid is mixed with solid manganese dioxide.
(b) A brown precipitate forms in a ferric nitrate solution.
(c) Silver (I) sulfide is dissolved in hot, concentrated nitric acid.
(d) A potassium chromate solution is first acidified, then reduced with hydrogen peroxide.
(e) Excess potassium iodide is added to a solution containing cupric nitrate and silver nitrate.

21-10.
Write chemical reactions to illustrate how you would make the following conversions:

(a) $Cr_2O_7^{2-} \rightarrow Cr^{2+}$
(b) $TiO_2 \rightarrow Ti^{3+}$
(c) $Hg \rightarrow Hg_2Cl_2$
(d) $MnO_2 \rightarrow MnO_4^{2-}$
(e) $Zn \rightarrow Zn(NH_3)_4^{2+}$
(f) $Cr^{3+} \rightarrow CrO_4^{2-}$
(g) $Co^{2+} \rightarrow Co(NH_3)_6^{3+}$

21-11.
100 grams of iron nails are added to 1.00 liter of 1.00 *M* ferric nitrate solution, and the solution is stirred until equilibrium is attained for the reaction

$$Fe + 2\,Fe^{3+} = 3\,Fe^{2+}$$

Use $\mathscr{E}^0$ values to calculate K for the reaction, and then calculate the concentration of Fe^{3+} in the final solution.

22

Descriptive Chemistry Review

No new theory and only a minimal number of new facts will be presented in this chapter, yet it could be one of the most valuable sections of the text for you. It is a shortcoming of most of us that we tend to compartmentalize information without sufficient regard for an overall picture. This chapter is designed to help you review descriptive chemistry in such a way that it relates the small compartments of knowledge that you have already been exposed to earlier in the text. You will find that we will repeat many reactions over and over, each time in a different context. By this device we hope to help you tie together this large body of information.

If you will carefully work through the reactions in this chapter and recognize the generalizations that are involved, you will find that you have mastered a significant amount of descriptive chemistry. We have adopted the "Solved Exercise" technique for presenting the discussion. We strongly suggest that you attempt to work the exercises before checking the solutions. Merely reading the exercises and then reading the solutions will be of limited value. As you work the exercises, be thinking about principles and generalizations that will help you write correct reactions. Think in terms of categories of reactions: precipitation, acid-base, oxidation-reduction, and complex-ion types. Watch for analogous reactions by members of the same family of elements.

Exercise 22.1

Water is such a common and important compound, you should be well aware of its reactions with the commonly encountered elements and compounds.

Write balanced equations for the following. (If no significant amount of reaction occurs under standard state conditions, write "N.R."—no reaction.)

1. $H_2O + Na(s) =$
2. $H_2O + K(s) =$
3. $H_2O + Ca(s) =$
4. $H_2O + Zn(s) =$
5. $H_2O + Sn(s) =$
6. $H_2O + Cu(s) =$
7. $H_2O + Ag(s) =$
8. $H_2O + Na_2O(s) =$
9. $H_2O + K_2O(s) =$
10. $H_2O + ZnO(s) =$
11. $H_2O + MnO(s) =$
12. $H_2O + CO_2(g) =$
13. $H_2O + SO_2(g) =$
14. $H_2O + SO_3(g) =$
15. $H_2O + N_2O_5(s) =$
16. $H_2O + P_4O_{10}(s) =$
17. $H_2O + SiO_2(s) =$
18. $H_2O + NO_2(g) =$
19. $H_2O + O^{2-} =$
20. $H_2O + S^{2-} =$
21. $H_2O + CO_3^{2-} =$
22. $H_2O + PO_4^{3-} =$
23. $H_2O + Be^{2+} =$
24. $H_2O + Sn^{2+} =$
25. $H_2O + Fe^{3+} =$
26. $H_2O + SiCl_4(l) =$
27. $H_2O + GeCl_4(l) =$
28. $H_2O + SnCl_4(l) =$
29. $H_2O + TiCl_4(l) =$
30. $H_2O + BCl_3(g) =$
31. $H_2O + PCl_3(l) =$
32. $H_2O + PCl_5(s) =$
33. $H_2O + CCl_4(l) =$
34. $H_2O + F_2(g) =$

Solution

If this problem could be attacked only by a memorization of each of 34 separate reactions, it would be a futile exercise. It is fortunate that what appears to be a near-impossible list can be broken into a few sets of similar and understandable reactions. We can separate out reactions in which water is reduced to give H_2, reactions in which water is oxidized to give O_2, reactions in which acidic and basic oxides react with water to give acidic and basic solutions, and reactions in which anions and cations hydrolyze to give stable ions or insoluble precipitates.

1. $2\ H_2O + 2\ Na(s) = 2\ Na^+ + 2\ OH^- + H_2(g)$
2. $2\ H_2O + 2\ K(s) = 2\ K^+ + 2\ OH^- + H_2(g)$
3. $2\ H_2O + Ca(s) = Ca(OH)_2(s) + H_2(g)$

Sodium, potassium, and calcium are very strong reducing agents ($\mathscr{E}^0_{ox} \cong 3$ V) and are thus easily capable of reducing water ($\mathscr{E}^0_{red} = -0.83$ V) to $H_2(g)$. All the alkali metals and alkaline earth metals are capable of reducing H_2O. The reaction of Be and Mg with water, however, is very slow unless the system is heated.

4. $H_2O + Zn(s) =$ N.R.
5. $H_2O + Sn(s) =$ N.R.
6. $H_2O + Cu(s) =$ N.R.
7. $H_2O + Ag(s) =$ N.R.

Zinc, tin, lead, copper, and silver have standard oxidation potentials too low (0.76, 0.14, 0.13, −0.34, −0.80, respectively) to reduce water ($\mathscr{E}^0_{red} = -0.83$); hence, there is no reaction under standard state conditions and room temperature for numbers 4. through 7.

8. $H_2O + Na_2O(s) = 2\ Na^+ + 2\ OH^-$
9. $H_2O + K_2O(s) = 2\ K^+ + 2\ OH^-$
10. $H_2O + ZnO(s) =$ N.R.
11. $H_2O + MnO(s) =$ N.R.
12. $H_2O + CO_2(g) = H_2CO_3$
13. $H_2O + SO_2(g) = H_2SO_3$
14. $H_2O + SO_3(g) = H^+ + HSO_4^-$
15. $H_2O + N_2O_5(s) = 2\ H^+ + 2\ NO_3^-$
16. $6\ H_2O + P_4O_{10}(s) = 4\ H_3PO_4$
17. $H_2O + SiO_2(s) =$ N.R.
18. $H_2O + 2\ NO_2(g) = H^+ + NO_3^- + HNO_2$

The oxides of the alkali metals, being soluble, dissolve in water to produce the hydroxides. You will recall that this is the definition of a soluble basic oxide (Chapter 16). The reaction takes place because of the extensive hydrolysis of oxide ion.

Zinc oxide and manganese (II) oxide are examples of metallic oxides that are insoluble in water; hence, there is no reaction with water. Being basic oxides, however, they will dissolve in acids; for example,

$$ZnO + 2\,H^+ = Zn^{2+} + H_2O$$

Carbon dioxide, sulfur dioxide, sulfur trioxide, nitrogen pentoxide, and phosphorus pentoxide are examples of soluble nonmetal oxides. They are the acid anhydrides of carbonic, sulfurous, sulfuric, nitric, and phosphoric acid, respectively. Carbonic, sulfurous, and phosphoric acids are weak, and are, therefore, written as molecules. Nitric and sulfuric acids, being strong, are written in the ionic form.

Silicon dioxide is an insoluble nonmetal oxide. It will react with a base, but not with water alone.

Nitrogen dioxide dissolves in cold water to form two acids, nitric and nitrous acids.

19. $H_2O + O^{2-} = 2\,OH^-$
20. $H_2O + S^{2-} = HS^- + OH^-$
21. $H_2O + CO_3^{2-} = HCO_3^- + OH^-$
22. $H_2O + PO_4^{3-} = HPO_4^{2-} + OH^-$
23. $H_2O + Be^{2+} = Be(OH)^+ + H^+$
24. $H_2O + Sn^{2+} = Sn(OH)^+ + H^+$
25. $H_2O + Fe^{3+} = Fe(OH)^{2+} + H^+$
26. $2\,H_2O + SiCl_4(l) = SiO_2(s) + 4\,HCl(g)$
27. $2\,H_2O + GeCl_4(l) = GeO_2(s) + 4\,HCl(g)$
28. $2\,H_2O + SnCl_4(l) = SnO_2(s) + 4\,HCl(g)$
29. $2\,H_2O + TiCl_4(l) = TiO_2(s) + 4\,HCl(g)$
30. $6\,H_2O + 2\,BCl_3(g) = 2\,H_3BO_3 + 6\,HCl(g)$
31. $3\,H_2O + PCl_3(l) = H_3PO_3 + 3\,HCl(g)$
32. $4\,H_2O + PCl_5(s) = H_3PO_4 + 5\,HCl(g)$
33. $CCl_4(l) + H_2O =$ N.R.
34. $2\,F_2(g) + 2\,H_2O = O_2(g) + 4\,HF(g)$

Reactions 19. through 22. are examples of the hydrolysis of anions. Oxygen ion, for all practical purposes, hydrolyzes completely ($K_h > 10^{22}$). The others listed are three of the moderately hydrolyzed anions ($K_h = 10^{-2}$–10^{-4}). Only the first, and most significant, steps of the complete hydrolysis of these polyvalent ions are shown.

Reactions 23. through 25. are examples of the hydrolysis of a few common "acidic" cations, $K_h = 10^{-5}$ to 10^{-2}.

Reactions 26. through 29. are examples of quadrivalent metal halides. These hydrolyze almost completely and generally quite vigorously to release $HCl(g)$.

Reactions 30. through 32. are examples of nonmetal halides that hydrolyze (also with vigor) to release $HCl(g)$. Carbon tetrachloride, and the other tetrahalides of carbon, however, are exceptional in that they do not hydrolyze.

Equation 34. is an example of the oxidation of water. In this example, the powerful oxidizing agent F_2 is used.

Exercise 22.2

Which of the following salts will dissolve in water sufficiently to form at least a 0.10 *M* solution?

1. NaCl
2. KI
3. Rb_3PO_4
4. $Ca(NO_3)_2$
5. $Al(NO_3)_3$
6. $BaCl_2$
7. $HgCl_2$
8. $MgSO_4$
9. $CuSO_4$
10. $ZnSO_4$
11. $MnSO_4$
12. $NiSO_4$
13. $CdSO_4$
14. $BeSO_4$
15. $BaSO_4$
16. $RaSO_4$
17. $PbSO_4$
18. AgCl
19. Hg_2Cl_2
20. $PbCl_2$
21. TlCl
22. $BaCO_3$
23. MgO
24. MgF_2
25. HgS
26. Ag_2S
27. CuS
28. Cu_2S
29. PbS
30. Al_2O_3
31. Fe_2O_3
32. PbO
33. NiO

Solution

The first fourteen are soluble. Refer to the rules on solubility in Chapter 17. Numbers 1. through 3. would be expected to be soluble because they are salts of Group IA metals. The next two, 4. and 5., would be predicted to be soluble since all nitrates are soluble. Numbers 6. and 7. are soluble, as are almost all chlorides. The four common exceptions are numbers 14. through 17., AgCl, Hg_2Cl_2, $PbCl_2$, and TlCl. $PbCl_2$ is the most soluble of the four ($K_{sp} = 2 \times 10^{-4}$ at 25° C). It is quite soluble in hot water. Numbers 8. through 14. are some common soluble sulfates. The remaining compounds 15. through 33. are all well-known for their insolubility.

Exercise 22.3

Complete the following acid-base reactions:

1. $OH^- + H^+ =$
2. $Fe(OH)_3(s) + H^+ =$
3. $Mg(OH)_2(s) + H^+ =$
4. $Mg(OH)_2(s) + CH_3COOH =$
5. $BaO(s) + H^+ =$
6. $CaO(s) + H^+ =$
7. $FeO(s) + H^+ =$
8. $SO_2(g) + OH^- =$
9. $SO_3(g) + OH^- =$
10. $CO_2(g) + OH^- =$
11. $N_2O_5(s) + OH^- =$
12. $P_4O_{10}(s) + OH^- =$
13. $NO_2 + OH^- =$
14. $HSO_3^- + OH^- =$
15. $HCO_3^- + OH^- =$
16. $H_2PO_4^- + OH^- =$
17. $HPO_4^{2-} + OH^- =$
18. $Zn(OH)_2(s) + H^+ =$
19. $ZnO(s) + H^+ =$
20. $Zn(OH)_2(s) + OH^- =$
21. $ZnO(s) + OH^- + H_2O =$
22. $Al(OH)_3(s) + OH^- =$
23. $Cr(OH)_3(s) + OH^- =$
24. $Be(OH)_2(s) + OH^- =$
25. $Sn(OH)_2(s) + OH^- =$
26. $Pb(OH)_2(s) + OH^- =$

Solution

1. $OH^- + H^+ = H_2O$
2. $Fe(OH)_3(s) + 3\ H^+ = Fe^{3+} + 3\ H_2O$
3. $Mg(OH)_2(s) + 2\ H^+ = Mg^{2+} + 2\ H_2O$
4. $Mg(OH)_2(s) + 2\ CH_3COOH = Mg^{2+} + 2\ CH_3COO^- + 2\ H_2O$
5. $BaO(s) + 2\ H^+ = Ba^{2+} + 2\ H_2O$
6. $CaO(s) + 2\ H^+ = Ca^{2+} + 2\ H_2O$
7. $FeO(s) + 2\ H^+ = Fe^{2+} + 2\ H_2O$

Reactions 1. through 3. are neutralization reactions of OH^- with a strong acid. Reaction 4. is an example of a neutralization reaction using a weak acid. Being only partially ionized, acetic acid is written in the molecular form. Equations 5. through 7. are typical reactions of basic oxides with acids.

8. $SO_2(g) + 2\ OH^- = SO_3^{2-} + H_2O$
9. $SO_3(g) + 2\ OH^- = SO_4^{2-} + H_2O$
10. $CO_2(g) + 2\ OH^- = CO_3^{2-} + H_2O$
11. $N_2O_5(s) + 2\ OH^- = 2\ NO_3^- + H_2O$
12. $P_4O_{10}(s) + 12\ OH^- = 4\ PO_4^{3-} + 6\ H_2O$
13. $2\ NO_2(g) + 2\ OH^- = NO_2^- + NO_3^- + H_2O$
14. $HSO_3^- + OH^- = SO_3^{2-} + H_2O$
15. $HCO_3^- + OH^- = CO_3^{2-} + H_2O$
16. $H_2PO_4^- + OH^- = HPO_4^{2-} + H_2O$
17. $HPO_4^{2-} + OH^- = PO_4^{3-} + H_2O$

Reactions 8. through 13. are all examples of nonmetal oxides reacting with a base. Sulfur dioxide, sulfur trioxide, and carbon dioxide are anhydrides of diprotic acids; therefore, depending upon the amount of base added, either (1) the acid salt (one mole of base to one mole of acidic oxide) or (2) the normal salt (two moles of base to one mole of the oxide) will form. Nitrogen dioxide is a "mixed" anhydride of nitrous and nitric acids; hence, neutralization of the acidic oxide produces a mixture of nitrite and nitrate ions. Reactions 14. through 17. represent examples of the neutralization of acid salts.

18. $Zn(OH)_2(s) + 2\ H^+ = Zn^{2+} + 2\ H_2O$
19. $ZnO(s) + 2\ H^+ = Zn^{2+} + H_2O$
20. $Zn(OH)_2(s) + 2\ OH^- = Zn(OH)_4^{2-}$
21. $ZnO(s) + 2\ OH^- + H_2O = Zn(OH)_4^{2-}$
22. $Al(OH)_3(s) + OH^- = Al(OH)_4^-$
23. $Cr(OH)_3(s) + OH^- = Cr(OH)_4^-$
24. $Be(OH)_2(s) + 2\ OH^- = Be(OH)_4^{2-}$
25. $Sn(OH)_2(s) + OH^- = Sn(OH)_3^-$
26. $Pb(OH)_2(s) + OH^- = Pb(OH)_3^-$

Reactions 18. through 26. all involve amphoteric hydroxides or oxides. These compounds react with both acids and bases. The list is not meant to be exhaustive, but it does include many of the commonly encountered amphoteric compounds.

Exercise 22.4

Complete and balance the following reactions, in which solids are dissolved by the addition of a strong acid.

1. $AgC_2H_3O_2(s) + H^+ =$
2. $BaSO_4(s) + H^+ =$
3. $Ga(OH)_3(s) + H^+ =$
4. $AgCl(s) + H^+ =$
5. $ZnS(s) + H^+ =$
6. $MgCO_3(s) + H^+ =$
7. $Ag_2CrO_4(s) + H^+ =$

Solution

1. $CH_3COOAg(s) + H^+ = Ag^+ + CH_3COOH$
2. $BaSO_4(s) + H^+ = Ba^{2+} + HSO_4^-$
3. $Ga(OH)_3(s) + 3\ H^+ = Ga^{3+} + 3\ H_2O$
4. $AgCl(s) + H^+ =$ N.R.
5. $ZnS(s) + 2\ H^+ = Zn^{2+} + H_2S(g)$
6. $MgCO_3(s) + 2\ H^+ = Mg^{2+} + H_2O + CO_2(g)$
7. $2\ Ag_2CrO_4(s) + 2\ H^+ = 4\ Ag^+ + Cr_2O_7^{2-} + H_2O$

Equations 1. through 3. take place because of the formation of weak acids. The equilibrium constant for each of them can be shown to be $K = K_{sp}/K_a$, where K_{sp} is the solubility product of the insoluble salt that is treated with H^+, and K_a is the ionization equilibrium constant of the weak acid that is formed in the reaction. Since K_a is in the denominator, the weaker the acid formed, the more complete is the reaction. This thermodynamic consequence is sometimes stated in the form: "A strong acid will displace a weak acid from its salt." The weak acid in reaction 3. is water. Since the ionization of water is so slight, the equilibrium point in reaction 3. lies very far to the right. The solubility of AgCl (reaction 4.) is not affected by a strong acid because the hydrochloric acid that would be formed if a reaction were to take place is also a strong acid.

Using $BaSO_4$ as an example, we can write its reaction with a strong acid in the following two steps:

$BaSO_4(s) = Ba^{2+} + SO_4^{2-};\ K_{sp}$

$H^+ + SO_4^{2-} = HSO_4^-;\ 1/K_a$

$BaSO_4(s) + H^+ = Ba^{2+} + HSO_4^-;\ K = K_{sp}/K_a$

The driving force in reactions 5. and 6. is derived not only from the formation of a weak acid, but also from the production of a gaseous product, which, if allowed to escape, drives the reaction to the right. Equation 7. can be thought of as the sum of the following two steps:

$$2\ Ag_2CrO_4(s) = 4\ Ag^+ + 2\ CrO_4^{2-}$$
$$2\ CrO_4^{2-} + 2\ H^+ = Cr_2O_7^{2-} + H_2O$$

The effect of H^+ on the overall reaction is seen to be the result of the effect of H^+ on the equilibrium of the second equation above.

Exercise 22.5

Sometimes students get so accustomed to working only with chemical formulas that they forget the names of even simple compounds. This exercise is designed to give you a little drill not only on reactions and properties, but also on nomenclature.

Write balanced net ionic reactions for the following reactions:

1. Silver nitrate solution is added to potassium bromide solution.
2. Sodium sulfate solution is added to lead (II) nitrate solution.
3. Potassium carbonate solution is added to calcium chloride solution.
4. Rubidium hydroxide solution is mixed with ferric nitrate solution.
5. Magnesium sulfate solution is mixed with barium hydroxide solution.
6. Hydrogen sulfide gas is bubbled through a ferrous sulfate solution.
7. Carbon dioxide gas is bubbled through a calcium hydroxide solution.
8. White phosphorus is burned in air.
9. Copper metal reacts with concentrated nitric acid to produce a blue solution and a red-brown gas.
10. Hydrogen peroxide solution is added to a potassium permanganate solution that is acidified with sulfuric acid. The purple permanganate ion solution loses its intense color, and a gas is formed.
11. Phosphoric acid is treated with excess barium hydroxide solution.
12. Solid manganese dioxide is added to concentrated hydrochloric acid solution. A pale yellow-green noxious gas is produced.
13. Platinum dissolves in aqua regia. A red-brown gas is produced.
14. Manganous ion is added to an acidified periodate solution. An intense pink-purple color develops.
15. A solution of potassium chromate is acidified with dilute sulfuric acid. The color changes from yellow to orange.
16. Silver chloride dissolves in ammonia solution.
17. Excess thiocyanate ion is added to an iron (III) ion solution. A deep-red color develops.
18. Chlorine gas is bubbled through a sodium hydroxide solution.
19. Chlorine gas is bubbled through a dilute hydroiodic acid solution.
20. A ferrous sulfate solution decolorizes an acidified permanganate ion solution.
21. Dichromate ion in acid solution reacts with iodide ion.
22. Calcium carbonate dissolves in carbonated water.
23. Ammonium chloride (solid) is treated with concentrated sodium hydroxide. The odor of ammonia is evident.
24. Sodium sulfite is treated with hydrochloric acid. The odor of burning sulfur is evident.

Solution

1. $Ag^+ + Br^- = AgBr(s)$
2. $Pb^{2+} + SO_4^{2-} = PbSO_4(s)$
3. $Ca^{2+} + CO_3^{2-} = CaCO_3(s)$
4. $Fe^{3+} + 3\ OH^- = Fe(OH)_3(s)$
5. $Mg^{2+} + SO_4^{2-} + Ba^{2+} + 2\ OH^- = Mg(OH)_2(s) + BaSO_4(s)$
6. $H_2S(g) + Fe^{2+} = FeS(s) + 2\ H^+$
7. $CO_2(g) + 2\ OH^- + Ca^{2+} = CaCO_3(s) + H_2O$
8. $P_4 + 5\ O_2(g) = P_4O_{10}(s)$
9. $Cu + 2\ NO_3^- + 4\ H^+ = Cu^{2+} + 2\ NO_2(g) + 2\ H_2O$

10. $5\ H_2O_2 + 2\ MnO_4^- + 6\ H^+ = 5\ O_2(g) + 2\ Mn^{2+} + 8\ H_2O$
11. $2\ H_3PO_4 + 3\ Ba^{2+} + 6\ OH^- = Ba_3(PO_4)_2(s) + 6\ H_2O$
12. $MnO_2(s) + 4\ H^+ + 2\ Cl^- = Mn^{2+} + Cl_2(g) + 2\ H_2O$
13. $Pt(s) + 4\ NO_3^- + 6\ Cl^- + 8\ H^+ = PtCl_6^{2-} + 4\ NO_2(g) + 4\ H_2O$
14. $2\ Mn^{2+} + 5\ IO_4^- + 3\ H_2O = 2\ MnO_4^- + 5\ IO_3^- + 6\ H^+$
15. $2\ CrO_4^{2-} + 2\ H^+ = Cr_2O_7^{2-} + H_2O$
16. $AgCl + 2\ NH_3(aq) = Ag(NH_3)_2^+ + Cl^-$
17. $6\ CNS^- + Fe^{3+} = Fe(CNS)_6^{3-}$
18. $Cl_2(g) + 2\ OH^- = Cl^- + OCl^- + H_2O$
19. $Cl_2(g) + 2\ I^- = 2\ Cl^- + I_2(s)$
20. $5\ Fe^{2+} + MnO_4^- + 8\ H^+ = 5\ Fe^{3+} + Mn^{2+} + 4\ H_2O$
21. $Cr_2O_7^{2-} + 6\ I^- + 14\ H^+ = 2\ Cr^{3+} + 3\ I_2(s) + 7\ H_2O$
22. $CaCO_3(s) + H_2CO_3 = Ca^{2+} + 2\ HCO_3^-$
23. $NH_4Cl(s) + OH^- = NH_3(g) + Cl^- + H_2O$
24. $Na_2SO_3(s) + 2\ H^+ = SO_2(g) + 2\ Na^+ + H_2O$

Exercise 22.6

Complete and balance the following reactions involving formation of complex ions:

1. $Zn(OH)_2(s) + NH_3(aq) =$
2. $Cu(OH)_2(s) + NH_3(aq) =$
3. $AgCl(s) + NH_3(aq) =$
4. $AgCl(s) + CN^- =$
5. $Fe^{3+} + SCN^- =$
6. $AgBr(s) + S_2O_3^{2-} =$
7. $Al(OH)_3(s) + OH^- =$
8. $ZnO(s) + H_2O + OH^- =$
9. $AgCl(s) + Cl^- =$
10. $Au(s) + NO_3^- + Cl^- + H^+ =$
11. $Cu^{2+} + Cl^- =$

Solution

1. $Zn(OH)_2(s) + 4\ NH_3(aq) = Zn(NH_3)_4^{2+} + 2\ OH^-$
2. $Cu(OH)_2(s) + 4\ NH_3(aq) = Cu(NH_3)_4^{2+} + 2\ OH^-$
3. $AgCl(s) + 2\ NH_3(aq) = Ag(NH_3)_2^+ + Cl^-$
4. $AgCl(s) + 2\ CN^- = Ag(CN)_2^- + Cl^-$
5. $Fe^{3+} + 6\ SCN^- = Fe(SCN)_6^{3-}$
6. $AgBr(s) + 2\ S_2O_3^{2-} = Ag(S_2O_3)_2^{3-} + Br^-$
7. $Al(OH)_3(s) + OH^- = Al(OH)_4^-$
8. $ZnO(s) + H_2O + 2\ OH^- = Zn(OH)_4^{2-}$
9. $AgCl(s) + Cl^- = AgCl_2^-$
10. $Au(s) + 3\ NO_3^- + 4\ Cl^- + 6\ H^+ = AuCl_4^- + 3\ NO_2(g) + 3\ H_2O$
11. $Cu^{2+} + 4\ Cl^- = CuCl_4^{2-}$

The above reactions are representative of the formation of stable ammine, chloro, cyano, thiocyanato, thisulfato, and hydroxo complexes. Both cations [for example, tetraamminezinc (II) ion] and anions [for example, hexathiocyanatoferrate (III) ion] are represented.

Exercise 22.7

Complete and balance the following typical oxidation-reduction reactions. You may need to refer to a table of standard oxidation-reduction potentials to decide in a few cases whether the reaction will go. In the reactions involving oxidation with *concentrated* nitric acid, consider the main reduction product to be $NO_2(g)$. In reaction 8., where the conditions are not specified, write the reaction with NO as the principal reduction product. In reaction 9. the principal reduction product is NH_4^+.

1. $Zn(s) + H^+ =$
2. $Al(s) + H^+ =$
3. $Fe(s) + H^+ =$
4. $Cd(s) + H^+ =$
5. $Cu(s) + H^+ =$
6. $Ag(s) + H^+ =$
7. $Hg(l) + H^+ =$
8. $Cu(s) + H^+ + NO_3^- =$
9. $Zn(s) + H^+ + NO_3^- =$
10. $Zn(s) + Cu^{2+} =$
11. $Cu(s) + Ag^+ =$
12. $Na(s) + H_2O =$
13. $Li(s) + O_2(g) =$
14. $Na(s) + O_2(g) =$
15. $K(s) + O_2(g) =$
16. $NO(g) + O_2(g) =$
17. $\underset{\text{(conc)}}{NO_3^-} + H^+ + Ag(s) =$
18. $\underset{\text{(conc)}}{NO_3^-} + H^+ + P_4(s) =$
19. $\underset{\text{(conc)}}{NO_3^-} + H^+ + Cu(s) =$
20. $\underset{\text{(conc)}}{NO_3^-} + H^+ + As(s) =$
21. $Pt(s) + H^+ + \underset{\text{(conc)}}{NO_3^-} + \underset{\text{(conc)}}{Cl^-} =$
22. $Au(s) + H^+ + \underset{\text{(conc)}}{NO_3^-} + \underset{\text{(conc)}}{Cl^-} =$
23. $Hg(l) + H^+ + \underset{\text{(conc)}}{NO_3^-} + \underset{\text{(conc)}}{Cl^-} =$
24. $P_4(s) + O_2(g) =$
25. $S(s) + O_2(g) =$
26. $SO_2(g) + O_2(g) \overset{\text{catalyst}}{=}$
27. $SO_3^{2-} + S(s) =$
28. $S_2O_3^{2-} + I_2(aq) =$
29. $H_2SO_4(l) + NaBr(s) =$
30. $H_2O_2(aq) + H_2S(aq) =$
31. $H_2O_2(aq) + SO_2(g) =$
32. $MnO_4^- + H_2O_2(aq) + H^+ =$
33. $MnO_4^- + Fe^{2+} + H^+ =$
34. $MnO_2(s) + H^+ + Cl^- =$
35. $Ce^{4+} + Fe^{2+} =$
36. $Cr_2O_7^{2-} + Fe^{2+} + H^+ =$
37. $Cr(OH)_4^- + HO_2^- =$
38. $Cr_2O_7^{2-} + H^+ + H_2O_2(aq) =$
39. $Hg^{2+} + Sn^{2+} + Cl^- =$
40. $Cu^+ =$
41. $Fe_2O_3(s) + CO(g) \overset{\text{heat}}{=}$
42. $Cu_2O(s) + H_2(g) \overset{\text{heat}}{=}$
43. $H_2(g) + F_2(g) =$
44. $H_2(g) + Cl_2(g) =$
45. $Cl_2(g) + H_2O =$
46. $Cl_2(g) + OH^- =$
47. $ClO^- \overset{\text{heat}}{=}$
48. $ClO_3^- \overset{\text{heat}}{=}$
49. $Cl_2(g) + Br^- =$
50. $Br_2(l) + I^- =$
51. $Cu^{2+} + I^- =$
52. $Fe^{3+} + I^- =$
53. $I^- + I_2(s) =$
54. $I_3^- + S_2O_3^{2-} =$
55. $IO_3^- + I^- + H^+ =$
56. $ClO^- + H_2S(g) =$
57. $Mn^{2+} + IO_4^- + H_2O =$
58. $TiCl_4(l) + 2\,Mg(s) =$
59. $VO_2^+ + Zn(s) =$
60. $Sc(s) + H^+ =$

Solution

1. $Zn(s) + 2\ H^+ = Zn^{2+} + H_2(g)$
2. $2\ Al(s) + 6\ H^+ = 2\ Al^{3+} + 3\ H_2(g)$
3. $Fe(s) + 2\ H^+ = Fe^{2+} + H_2(g)$
4. $Cd(s) + 2\ H^+ = Cd^{2+} + H_2(g)$
5. $Cu(s) + H^+ =$ N.R.
6. $Ag(s) + H^+ =$ N.R.
7. $Hg(l) + H^+ =$ N.R.
8. $3\ Cu(s) + 8\ H^+ + 2\ NO_3^- = 3\ Cu^{2+} + 2\ NO + 4\ H_2O$
9. $4\ Zn(s) + 10\ H^+ + NO_3^- = 4\ Zn^{2+} + NH_4^+ + 3\ H_2O$
10. $Zn(s) + Cu^{2+} = Zn^{2+} + Cu(s)$
11. $Cu(s) + 2\ Ag^+ = Cu^{2+} + 2\ Ag(s)$
12. $2\ Na(s) + 2\ H_2O = 2\ Na^+ + 2\ OH^- + H_2(g)$
13. $Li(s) + O_2(g) = Li_2O(s)$
14. $Na(s) + O_2(g) = Na_2O_2(s)$
15. $K(s) + O_2(g) = KO_2(s)$
16. $2\ NO(g) + O_2(g) = 2\ NO_2(g)$
17. $NO_3^- + 2\ H^+ + Ag(s) = NO_2(g) + Ag^+ + H_2O$
18. $20\ NO_3^- + 20\ H^+ + P_4(s) = 20\ NO_2(g) + 4\ H_3PO_4 + 4\ H_2O$
19. $2\ NO_3^- + 4\ H^+ + Cu(s) = 2\ NO_2(g) + Cu^{2+} + 2\ H_2O$
20. $5\ NO_3^- + 5\ H^+ + As(s) = 5\ NO_2(g) + H_3AsO_4 + H_2O$
21. $Pt(s) + 8\ H^+ + 4\ NO_3^- + 6\ Cl^- = 4\ NO_2(g) + PtCl_6^{2-} + 4\ H_2O$
22. $Au(s) + 6\ H^+ + 3\ NO_3^- + 4\ Cl^- = 3\ NO_2(g) + AuCl_4^- + 3\ H_2O$
23. $Hg(l) + 4\ H^+ + 2\ NO_3^- + 4\ Cl^- = 2\ NO_2(g) + HgCl_4^{2-} + 2\ H_2O$
24. $P_4(s) + 5\ O_2(g) = P_4O_{10}(s)$
25. $S(s) + O_2(g) = SO_2(g)$
26. $2\ SO_2(g) + O_2(g) = 2\ SO_3(g)$
27. $SO_3^{2-} + S(s) = S_2O_3^{2-}$
28. $2\ S_2O_3^{2-} + I_2(aq) = S_4O_6^{2-} + 2\ I^-$
29. $3\ H_2SO_4(l) + 2\ NaBr(s) = SO_2(g) + Br_2(g) + 2\ H_2O + 2\ NaHSO_4(s)$
30. $H_2O_2(aq) + H_2S(aq) = S(s) + 2\ H_2O$
31. $H_2O_2(aq) + SO_2(g) = H^+ + HSO_4^-$
32. $2\ MnO_4^- + 5\ H_2O_2 + 6\ H^+ = 5\ O_2(g) + 2\ Mn^{2+} + 8\ H_2O$
33. $MnO_4^- + 5\ Fe^{2+} + 8\ H^+ = Mn^{2+} + 5\ Fe^{3+} + 4\ H_2O$
34. $MnO_2(s) + 4\ H^+ + 2\ Cl^- = Mn^{2+} + Cl_2(g) + 2\ H_2O$
35. $Ce^{4+} + Fe^{2+} = Ce^{3+} + Fe^{3+}$
36. $Cr_2O_7^{2-} + 6\ Fe^{2+} + 14\ H^+ = 2\ Cr^{3+} + 6\ Fe^{3+} + 7\ H_2O$
37. $2\ Cr(OH)_4^- + 3\ HO_2^- = 2\ CrO_4^{2-} + 5\ H_2O + OH^-$
38. $Cr_2O_7^{2-} + 8\ H^+ + 3\ H_2O_2 = 3\ O_2(g) + 2\ Cr^{3+} + 7\ H_2O$
39. $2\ Hg^{2+} + Sn^{2+} + 8\ Cl^- = Hg_2Cl_2(s) + SnCl_6^{2-}$
40. $2\ Cu^+ = Cu^{2+} + Cu(s)$
41. $Fe_2O_3(s) + 3\ CO(g) = 2\ Fe(s) + 3\ CO_2(g)$
42. $Cu_2O(s) + H_2(g) = 2\ Cu(s) + H_2O$
43. $H_2(g) + F_2(g) = 2\ HF(g)$
44. $H_2(g) + Cl_2(g) = 2\ HCl(g)$

Note that in equation 37., peroxide oxygen oxidizes Cr(III) to Cr(VI), whereas in equation 38., peroxide oxygen is oxidized by Cr(VI). There is no discrepancy. Equation 37. is in basic solution and equation 38. in acid solution. Cr(VI) in acid solution is a strong oxidizing agent.

Note that equation 46 can be reversed by increasing the H^+ concentration:

$HOCl + H^+ + Cl^-$
$= Cl_2(g) + H_2O$

If the reactants in equation 45. are heated, the equation is

$3\ Cl_2(s) + 3\ H_2O$
$= ClO_3^- + 5\ Cl^- + 6\ H^+$

This reaction can also be reversed by addition of H^+.
Addition of base will reverse equation 55.

45. $Cl_2(g) + H_2O = HOCl + H^+ + Cl^-$
46. $Cl_2(g) + 2\ OH^- = Cl^- + OCl^- + H_2O$
47. $3\ ClO^- = 2\ Cl^- + ClO_3^-$
48. $4\ ClO_3^- = Cl^- + 3\ ClO_4^-$
49. $Cl_2(g) + 2\ Br^- = 2\ Cl^- + Br_2(l)$
50. $Br_2(l) + 2\ I^- = 2\ Br^- + I_2(s)$
51. $2\ Cu^{2+} + 4\ I^- = 2\ CuI(s) + I_2(s)$
52. $2\ Fe^{3+} + 2\ I^- = 2\ Fe^{2+} + I_2(s)$
53. $I^- + I_2(s) = I_3^-$
54. $I_3^- + 2\ S_2O_3^{2-} = 3\ I^- + S_4O_6^{2-}$
55. $IO_3^- + 5\ I^- + 6\ H^+ = 3\ I_2(s) + 3\ H_2O$
56. $ClO^- + H_2S(g) = Cl^- + S(s) + H_2O$
57. $2\ Mn^{2+} + 5\ IO_4^- + 3\ H_2O = 2\ MnO_4^- + 5\ IO_3^- + 6\ H^+$
58. $TiCl_4(l) + 2\ Mg(s) = Ti(s) + 2\ MgCl_2(s)$
59. $2\ VO_2^+ + 3\ Zn(s) + 8\ H^+ = 3\ Zn^{2+} + 2\ V^{2+} + 4\ H_2O$
60. $2\ Sc(s) + 6\ H^+ = 2\ Sc^{3+} + 3\ H_2$

Exercise 22.8

Write balanced equations for the following groups of reactions, which involve the formation of a gas. If the gas is allowed to escape, the reaction goes to completion.

1. $Ni(s) + H^+ =$
2. $FeS(s) + H^+ =$
3. $NaCN(s) + H^+ =$
4. $K_2SO_3(s) + H^+ =$
5. $Na_2CO_3(s) + H^+ =$
6. $NaF(s) + H_2SO_4(l) =$
7. $NaCl(s) + H_2SO_4(l) =$
8. $NH_4Cl(s) + OH^- =$
9. $Ca(HCO_3)_2(s) \overset{heat}{=}$
10. $CaCO_3(s) \overset{heat}{=}$

Solution

1. $Ni(s) + 2\ H^+ = Ni^{2+} + H_2(g)$
2. $FeS(s) + 2\ H^+ = Fe^{2+} + H_2S(g)$
3. $NaCN(s) + H^+ = Na^+ + HCN(g)$
4. $K_2SO_3(s) + 2\ H^+ = 2\ K^+ + H_2O + SO_2(g)$
5. $Na_2CO_3(s) + 2\ H^+ = 2\ Na^+ + H_2O + CO_2(g)$
6. $NaF(s) + H_2SO_4(l) = NaHSO_4(s) + HF(g)$
7. $NaCl(s) + H_2SO_4(l) = NaHSO_4(s) + HCl(g)$
8. $NH_4Cl(s) + OH^- = Cl^- + H_2O + NH_3(g)$
9. $Ca(HCO_3)_2(s) = CaCO_3(s) + H_2O + CO_2(g)$ (heat required)
10. $CaCO_3(s) = CaO(s) + CO_2(g)$ (heat required)

Exercise 22.9

An excellent review of the many aspects of water-solution chemistry is afforded by the "fill in the square with the predominant species" type of question asked below. In following through this solved exercise, keep in mind possibilities for (a) ionization, (b) hydrolysis, (c) precipitation, (d) oxidation-reduction, (e) neutralization, and (f) complex-ion formation. You may need to look up standard single-electrode potentials or equilibrium constants to be certain of some of your answers, but, for the most part, you will have learned enough descriptive chemistry to make intelligent guesses for all of the squares.

Write the formula of the *most* abundant species containing the *underlined* element when 0.10 mole of the compound in the left-hand column is equilibrated with 1.0 liter of the solution listed in the row at the top of the table.

SUBSTANCE	WATER	1 MOLAR NITRIC ACID	1 MOLAR SODIUM HYDROXIDE	1 MOLAR HYDROIODIC ACID	1 MOLAR AMMONIA
$\underline{Cu}SO_4$					
$\underline{Ag}Cl$					
$K_2\underline{Cr}O_4$					
$K\underline{Mn}O_4$					
$\underline{Al}(OH)_3$					
$\underline{Fe}(OH)_2$					
$\underline{Mg}(OH)_2$					
$\underline{Fe}(NO_3)_3$					
$\underline{Zn}O$					
$\underline{P}_4O_{10}$					
$Na_2\underline{O}$					

An example will help you get started. Consider the first substance, $CuSO_4$. It is soluble in water and, therefore, will be present as Cu^{2+}, which is to be understood to mean hydrated copper (II) ion. Cu^{2+} ion hydrolyzes fairly extensively, and this should be taken into account, but after we make the calculation, we find Cu^{2+} is by far the predominant form.

The cation that is noted for its hydrolysis is Sn(II) ion, and even in this case, the simple divalent ion is the predominant species in water solution:

$$\underset{0.10-x}{Sn^{2+}} + HOH = \underset{x}{SnOH^+} + \underset{x}{H^+}; \quad K = 1 \times 10^{-2}$$

$$\frac{x^2}{0.10 - x} = 1 \times 10^{-2}$$

$$x = 0.03$$

$$[Sn^{2+}] = 0.10 - 0.03 = 0.07$$

$$[SnOH^+] = 0.03$$

In column 2, we need to consider the effect of the strong acid (H^+) and the effect of a moderate oxidizing agent (dilute NO_3^-). Neither changes the form of Cu^{2+}

In column 3, look for a neutralization reaction or, as in this case, the possibility for precipitation of an insoluble hydroxide. Not being amphoteric, the $Cu(OH)_2$ that forms does not react further with the strong base but remains as solid $Cu(OH)_2$.

In the fourth column, consider the effect of the strong acid (H^+) and the good reducing, complexing, and possible precipitating agent (I^-). In our example, iodide ion reduces copper (II) to copper (I) and precipitates it as $CuI(s)$. Another possibility that should be looked at in arriving at the answer in this square is the formation of iodo complexes of copper.

In the fifth column, we need to be concerned with the effect of the OH^- concentration in 1 M NH_3 ($\sim 10^{-3}$ M in OH^-) and the complexing effect of the NH_3 molecule. In our example, the $[OH^-]$ above would have precipitated $Cu(OH)_2(s)$, $K_{sp} = 10^{-19}$), but the ammine complex is sufficiently stable ($K_d = 5 \times 10^{-14}$) that under the conditions given in the problem $Cu(OH)_2(s)$ would convert completely to $Cu(NH_3)_4^{2+}$.

Solution

The solved exercise is shown below:

SUBSTANCE	WATER	1 MOLAR NITRIC ACID	1 MOLAR SODIUM HYDROXIDE	1 MOLAR HYDROIODIC ACID	1 MOLAR AMMONIA
$CuSO_4$	Cu^{2+}	Cu^{2+}	$Cu(OH)_2(s)$	$CuI(s)$	$Cu(NH_3)_4^{2+}$
$AgCl$	$AgCl(s)$	$AgCl(s)$	$AgOH(s)$	$AgI(s)$	$Ag(NH_3)_2^+$
K_2CrO_4	CrO_4^{2-}	$Cr_2O_7^{2-}$	CrO_4^{2-}	Cr^{3+}	CrO_4^{2-}
$KMnO_4$	MnO_4^-	MnO_4^-	MnO_4^-	Mn^{2+}	MnO_4^-
$Al(OH)_3$	$Al(OH)_3(s)$	Al^{3+}	$Al(OH)_4^-$	Al^{3+}	$Al(OH)_3(s)$
$Fe(OH)_2$	$Fe(OH)_2(s)$	Fe^{3+}	$Fe(OH)_2(s)$	Fe^{2+}	$Fe(OH)_2(s)$
$Mg(OH)_2$	$Mg(OH)_2(s)$	Mg^{2+}	$Mg(OH)_2(s)$	Mg^{2+}	$Mg(OH)_2(s)$
$Fe(NO_3)_3$	Fe^{3+}	Fe^{3+}	$Fe(OH)_3(s)$	Fe^{2+}	$Fe(OH)_3(s)$
ZnO	$ZnO(s)$	Zn^{2+}	$Zn(OH)_4^{2-}$	Zn^{2+}	$Zn(NH_3)_4^{2+}$
P_4O_{10}	H_3PO_4	H_3PO_4	PO_4^{3-}	H_3PO_4	HPO_4^{2-}
Na_2O	OH^-	H_2O	OH^-	H_2O	OH^-

A few additional comments may help you understand some of the answers given above: In the presence of acid, CrO_4^{2-} condenses and splits out water to form $Cr_2O_7^{2-}$:

$$2\,CrO_4^{2-} + 2\,H^+ = Cr_2O_7^{2-} + H_2O; \qquad K = 4.2 \times 10^{14}$$

The dichromate ion in acid solution is a strong oxidizing agent that easily oxidizes the reducing agent, iodide ion. The reduction product is Cr^{3+}.

Aluminum hydroxide is amphoteric and dissolves readily in 1 M OH^-. The concentration of OH^- in 1 M NH_3 is almost sufficient, but not quite, to make $Al(OH)_4^-$ the predominant aluminum species in that solution.

If you add the standard oxidation potential of the Fe^{2+}, Fe^{3+} couple (-0.77) to the standard reduction potential of the NO_3^-, NO couple in acid solution ($+0.96$), you see that NO_3^- will easily oxidize Fe^{2+} to Fe^{3+}, even taking into account the nonstandard state conditions in the problem.

A quick and easy approach to determining the predominant species in a weak acid neutralized to a given pH is to calculate the pH at half-neutralization of the acid. Consider the first ionization of H_3PO_4:

$$H_3PO_4 = H^+ + H_2PO_4^-$$

$$\frac{[H^+][H_2PO_4^-]}{[H_3PO_4]} = K_{a(1)}$$

Taking the negative of the common logarithm of both sides, we obtain:

$$-\log\,[H^+] - \log\frac{[H_2PO_4^-]}{[H_3PO_4]} = -\log K_{a(1)}$$

Now defining $pK = -\log K$ (analogous to the definition of $pH = -\log [H^+]$), we can write

$$pH = pK + \log \frac{[H_2PO_4^-]}{[H_3PO_4]}$$

At the point of *half-neutralization* of the first H^+, $[H_2PO_4^-] = [H_3PO_4]$, and, at that point, pH = pK.

Since K_{a1} for H_3PO_4 is 7.5×10^{-3}, its $pK_{a1} = -\log 7.5 \times 10^{-3} = 2.1$. At a pH of 2.1, there are approximately equal concentrations of H_3PO_4 and $H_2PO_4^-$. At any pH lower than 2.1, H_3PO_4 is the predominant species.

We are able to set $[H_2PO_4^-] = [H_3PO_4]$ only because the ionization of H_3PO_4 is small, making any production of $H_2PO_4^-$ from the ionization of H_3PO_4 negligible.

$K_{a2} = 6.8 \times 10^{-8}$, and, therefore, $pK_{a2} = 7.2$. At pH = 7.2 there will be approximately equal concentrations of $H_2PO_4^-$ and HPO_4^{2-}. At pH values between 2.1 and 7.2, $H_2PO_4^-$ will be predominant.

$K_{a3} = 4 \times 10^{-13}$, and, therefore, $pK_{a3} = 12.4$. At pH = 12.4, there will be approximately equal amounts of HPO_4^- and PO_4^{3-}. Between pH = 7.2 and pH = 12.4, HPO_4^{2-} will be predominant, and above pH = 12.4, PO_4^{3-} will predominate.

Keeping the above in mind, it is easy to fill in the squares for P_4O_{10}. H_3PO_4 is by far the predominant form in the 1 *M* strong acids (pH ~ 0), and PO_4^{3-}, the predominant form in the 1 *M* NaOH solution (pH ~ 14). Only the 1 *M* NH_3 solution requires a calculation. Considering the equilibrium

$$NH_3 + H_2O = NH_4^+ + OH^-; \qquad K_b = 1.8 \times 10^{-5}$$

we calculate that the order of magnitude of the $[OH^-] = 10^{-3}$. The pOH is, thus, approximately 3, and the pH is approximately 11. This is in the range in which HPO_4^- is the predominant species.

Exercise 22.10

"Road map" questions are a good way to review descriptive chemistry. In the examples given below, a route is provided for you to carry an element through a series of reactions in which the product of a preceding reaction is the starting material for the next reaction. In some of these examples, one travels full circle and returns to the starting material.

1. (a) Dilute $NH_3(aq)$ is added to a solution of Cu^{2+} until a light-blue precipitate occurs. Write the reaction.
   (b) Concentrated $NH_3(aq)$ is added until the precipitate in (a) dissolves and a dark-blue solution results. Write the reaction.
   (c) Dilute hydrochloric acid is added slowly to the dark-blue solution in (b) until a precipitate appears. Explain and give the reaction.
   (d) Additional hydrochloric acid is added, and the precipitate in (c) dissolves. Write the reaction.
   (e) Concentrated hydrochloric acid is now added to the solution in (d). Write the reaction to explain the change in color.

2. (a) Dilute sodium hydroxide solution is added to a Cr(III) ion solution until a precipitate appears. Write the reaction.
   (b) Concentrated sodium hydroxide solution is now added until the precipitate in (a) disappears. Write the reaction.
   (c) Hydrogen peroxide solution is then added, and the system is heated. A yellow solution results. Write the reaction.
   (d) Concentrated sulfuric acid is added to the solution in (b) until the color of the solution changes to orange. Write the reaction.
   (e) Iron (II) ion is added to the acidified orange solution of (d). Write the reaction.
3. (a) A solution of MnO_4^- and Mn^{2+} is heated. A brown precipitate forms. Write the reaction.
   (b) The brown precipitate is separated by filtration, and hydrochloric acid is added to it. A pale-yellow-green, poisonous gas is produced, and the precipitate disappears to produce a colorless solution. Write the reaction.
   (c) The excess chloride is removed from the solution in (b) by heating it with concentrated sulfuric acid until dense white fumes appear. Write the reaction.
   (d) Potassium periodate is now added, and the solution is heated. A purple solution results. Write the reaction.
   (e) Arsenious acid is added to the purple solution above, and the color disappears. Write the reaction.

Solution

1. (a) $\underset{\text{(blue)}}{Cu^{2+}} + 2\,OH^- = \underset{\text{(blue)}}{Cu(OH)_2(s)}$
   (b) $\underset{\text{(blue)}}{Cu(OH)_2(s)} + 4\,NH_3(aq) = \underset{\text{(dark blue)}}{Cu(NH_3)_4^{2+}} + 2\,OH^-$
   (c) Addition of H^+ decomposes the ammine complex:

$$Cu(NH_3)_4^{2+} + 4\,H^+ = Cu^{2+} + 4\,NH_4^+$$

   There is still sufficient OH^- left in the solution for the hydroxide to exist:

$$Cu^{2+} + 2\,OH^- = Cu(OH)_2(s)$$

   (d) $Cu(OH)_2(s) + 2\,H^+ = Cu^{2+} + 2\,H_2O$
   (e) $\underset{\text{(blue)}}{Cu^{2+}(aq)} + 4\,Cl^- = \underset{\text{(yellow-green)}}{CuCl_4^{2-}}$
2. (a) $\underset{\text{(light gray-purple)}}{Cr^{3+}} + 3\,OH^- = \underset{\text{(light gray-purple)}}{Cr(OH)_3(s)}$
   (b) $\underset{\text{(light gray-purple)}}{Cr(OH)_3(s)} + OH^- = \underset{\text{(green)}}{Cr(OH)_4^-}$

(c) $2\ Cr(OH)_4^- + 3\ H_2O_2 + 2\ OH^- = 2\ CrO_4^{2-} + 8\ H_2O$
(green) (yellow)

(d) $2\ CrO_4^{2-} + 2\ H^+ = Cr_2O_7^{2-} + H_2O$
(yellow) (orange)

(e) $Cr_2O_7^{2-} + 6\ Fe^{2+} + 14\ H^+ = 2\ Cr^{3+} + 6\ Fe^{3+} + 7\ H_2O$

3. (a) $2\ MnO_4^- + 3\ Mn^{2+} + 2\ H_2O = 5\ MnO_2(s) + 4\ H^+$
(purple) (pale pink) (dark brown)

(b) $MnO_2(s) + 4\ H^+ + 2\ Cl^- = Mn^{2+} + Cl_2(g) + 2\ H_2O$
(pale yellow-green)

(c) $Cl^- + H_2SO_4 = HCl(g) + HSO_4^-$

(d) $5\ IO_4^- + 2\ Mn^{2+} + 3\ H_2O = 5\ IO_3^- + 2\ MnO_4^- + 6\ H^+$
(purple)

(e) $5\ H_3AsO_3 + 2\ MnO_4^- + 6\ H^+ = 5\ H_3AsO_4 + 2\ Mn^{2+} + 3\ H_2O$

Exercise 22.11

The table below lists *incorrect* species that have been given by students. In each case, state clearly why the indicated species *cannot* be the correct one. Assume that 0.1 mole quantities of the compounds in the left-hand column are added to one liter of the solutions in the top row.

COMPOUND	1 MOLAR HYDROIODIC ACID	WATER	1 MOLAR AMMONIA	1 MOLAR SODIUM HYDROXIDE	1 MOLAR NITRIC ACID
$KMnO_4$	MnO_2 (1)			MnO_4^{2-} (2)	
$Hg(NO_3)_2$		$Hg(NO_3)_2(s)$ (3)		$HgOH^+$ (4)	
$NiCl_2$			$Ni(OH)_2(s)$ (5)	$Ni(OH)_4^{2+}$ (6)	
$Zn(OH)_2$		Zn^{2+} (7)	$Zn(NH_4)_4^{2+}$ (8)	Zn^{2+} (9)	$Zn(OH)_2(s)$ (10)
$Ag(NH_3)_2^+$		Ag^+ (11)	$AgOH$ (12)		$Ag(NH_3)_2^+$ (13)
$AlCl_3$		$AlCl_3$ (14)	$Al(NH_3)_6^{3+}$ (15)	$Al(OH)_3(s)$ (16)	
K_2CrO_4	$Cr_2O_7^{2-}$ (17)			$Cr(OH)_4^-$ (18)	CrO_4^{2-} (19)
Cu_2O		Cu^+ (20)	$Cu(NH_3)_4^{2+}$ (21)		Cu^+ (22)
$Cu(NO_3)_2$	Cu^{2+} (23)	$Cu(OH)_2(s)$ (24)			
$Fe(OH)_3$		Fe^{3+} (25)	$Fe(NH_3)_6^{3+}$ (26)	$Fe(OH)_4^-$ (27)	$FeOH^{2+}$ (28)
$NaIO_3$	IO_3^- (29)	$HIO_3(aq)$ (30)			
SO_3		$SO_3(aq)$ (31)		$SO_3(aq)$ (32)	
HI		$HI(aq)$ (33)			
N_2O_5		$N_2O_5(s)$ (34)		NO_2^- (35)	$HNO_3(aq)$ (36)
BeO				$Be(OH)_2(s)$ (37)	$BeO(s)$ (38)
Na_2O		O^{2-} (39)			OH^- (40)

Solution

1. In the presence of a reducing agent, MnO_4^- is reduced to Mn^{2+} in *acid* solution. MnO_2 is a common reduction product in neutral, but not acidic, solutions.
2. No oxidation-reduction is involved. Answer should be MnO_4^- (+7 Mn, not +6 as in MnO_4^{2-}).
3. All nitrates are soluble. Answer should be Hg^{2+}.
4. $HgOH^+$ is a plausible first step, but in 1 *M* OH^-, $Hg(OH)_2(s)$ will precipitate.
5. Nickel (II) ion will form an ammine complex, $Ni(NH_3)_4^{2+}$.
6. The student apparently thought that nickel ion would form a hydroxo complex, which it does not, and if it did, it would be an anion, not a cation. The correct answer is $Ni(OH)_2$.
7. $Zn(OH)_2$ is insoluble in water, as is generally true of polyvalent metal hydroxides. $Ba(OH)_2$ is the most soluble of the divalent metal hydroxides, and its solubility is less than 0.2 moles/liter of solution. The K_{sp} of $Zn(OH)_2$ is 10^{-17}.
8. The correct answer is $Zn(NH_3)_4^{2+}$. It is the ammonia molecule (NH_3), not ammonium ion (NH_4^+), that is the ligand in the ammine complexes.
9. Zinc ion would first form the hydroxide and then, being amphoteric, go on to form the hydroxo complex, $Zn(OH)_4^{2-}$.
10. A base in acid solution?! The answer is Zn^{2+}.
11. Diamminesilver (I) ion is stable in water.
12. Silver hydroxide is not a ridiculous answer, but if you make the calculation from the equilibria data, you find that even in solutions as dilute as 1 M NH_3, the ammine complex wins out. The answer is $Ag(NH_3)_2^+$.
13. Hydrogen ion will decompose the complex: $Ag(NH_3)_2^+ + 2\,H^+ = Ag^+ + 2\,NH_4^+$.
14. Aluminum chloride is soluble.
15. Aluminum ion does not form an ammine complex. The answer is $Al(OH)_3(s)$.
16. Aluminum hydroxide, being amphoteric, would convert to $Al(OH)_4^-$.
17. It is true that CrO_4^{2-} in acid solution condenses to $Cr_2O_7^{2-}$, but the $Cr_2O_7^{2-}$ formed would oxidize I^- in acid solution and be reduced to Cr^{3+}.
18. The correct answer is CrO_4^{2-}. The student's answer implies that NaOH reduces K_2CrO_4, which is not true.
19. In acid solution, CrO_4^{2-} condenses to $Cr_2O_7^{2-}$.
20. Cu_2O is insoluble. Cu^+ ion is unstable under standard state conditions. It disproportionates as follows:

$$2\,Cu^+ = Cu(s) + Cu^{2+}$$

21. $NH_3(aq)$ does not oxidize copper (I) to copper (II). The correct formula for the copper (I) ammine complex is $Cu(NH_3)_2^+$. $Cu(NH_3)_4^{2+}$ is the formula for the copper (II) complex.
22. Cu^+ is unstable. It would either disproportionate or be oxidized by the nitric acid to Cu^{2+}.

23. The iodide ion reduces copper (II) ion to CuI(*s*). This is the basis for an analytical method for determining copper. In the reaction, I^- is oxidized to I_2 (or I_3^-), which is then titrated with a standard thiosulfate solution.
24. The correct answer is Cu^{2+}.
25. $Fe(OH)_3$ is very insoluble in water.
26. Fe(III) does not form an ammine complex. Neither does Fe(II).
27. Fe(III) does not form a hydroxo complex. Neither Fe(II) nor Fe(III) hydroxides is amphoteric.
28. A basic ion will not exist in strong acid solution.
29. Iodate ion oxidizes iodide ion to I_2 (or I_3^-) in acid solution. I_2 (or I_3^-) is also the reduction product.
30. The answer is IO_3^-.
31. SO_3 reacts with water to form H_2SO_4 or $H^+ + HSO_4^-$. Under the conditions of this problem, the latter would be produced.
32. Nonmetal oxides react with bases:

$$SO_3 + 2\,OH^- = SO_4^{2-} + H_2O$$

33. Hydroiodic acid is fully ionized. The answer is I^-.
34. $N_2O_5 + H_2O = 2\,H^+ + 2\,NO_3^-$
35. The answer should be NO_3^-: nitrate ion, not nitrite ion.
36. Nitric acid is a strong acid, fully ionized. The answer is NO_3^-.
37. BeO is an amphoteric oxide:

$$BeO + H_2O + 2\,OH^- = Be(OH)_4^{2-}$$

38. Amphoteric oxides react with acids:

$$BeO + 2\,H^+ = Be^{2+} + H_2O$$

39. Oxide ion hydrolyzes almost completely:

$$O^{2-} + H_2O = 2\,OH^-; \qquad K > 10^{22}$$

40. Hydroxide ion in 1 *M* H^+ solution?!

$$H^+ + OH^- = H_2O; \qquad K = 10^{14}$$

Exercise 22.12

Mixtures of substances are conveniently separated by causing the precipitation of one and leaving the others in solution. The precipitate is separated from the solution usually by either filtration or centrifugation.

Write the formula of a chemical that could be used to separate the ions or compounds in each of the following pairs. Write the formulas of the precipitate and the species that remains in solution.

MIXTURE	ADD	PRECIPITATE	SOLUTION
1. Zn^{2+} and Cu^{2+}			
2. Mg^{2+} and Ni^{2+}			
3. Ba^{2+} and Na^{+}			
4. $AgCl(s)$ and $PbCl_2(s)$			
5. Cl^- and NO_3^-			
6. $BeO(s)$ and $SiO_2(s)$			
7. Fe^{3+} and Al^{3+}			
8. Cd^{2+} and Zn^{2+}			
9. Mg^{2+} and Ba^{2+}			
10. Zn^{2+} and Fe^{2+}			

Solution

There is not always just one right answer. For example, equally good answers for number 10. are

	ADD	PRECIPITATE	SOLUTION
10. Zn^{2+} and Fe^{2+}	OH^-	$Fe(OH)_2$	$Zn(OH)_4^{2-}$
	$NH_3(aq)$	$Fe(OH)_2$	$Zn(NH_3)_4^{2+}$

One set of reasonable answers follows:

MIXTURE	ADD	PRECIPITATE	SOLUTION
1. Zn^{2+} and Cu^{2+}	OH^-	$Cu(OH)_2$	$Zn(OH)_4^{2-}$
2. Mg^{2+} and Ni^{2+}	$NH_3(aq)$	$Mg(OH)_2$	$Ni(NH_3)_4^{2+}$
3. Ba^{2+} and Na^{+}	SO_4^{2-}	$BaSO_4$	Na^+
4. $AgCl$ and $PbCl_2$	$NH_3(aq)$	$PbCl_2$	$Ag(NH_3)_2^+$
5. Cl^- and NO_3^-	Ag^+	$AgCl$	NO_3^-
6. BeO and SiO_2	H^+	SiO_2	Be^{2+}
7. Fe^{3+} and Al^{3+}	OH^-	$Fe(OH)_3$	$Al(OH)_4^-$
8. Cd^{2+} and Zn^{2+}	OH^-	$Cd(OH)_2$	$Zn(OH)_4^{2-}$
9. Mg^{2+} and Ba^{2+}	SO_4^{2-}	$BaSO_4$	Mg^{2+}
10. Zn^{2+} and Fe^{2+}	$NH_3(aq)$	$Fe(OH)_2$	$Zn(NH_3)_4^{2+}$

Exercise 22.13

Write the formula of a chemical that could be used to distinguish between the two substances in each of the following pairs. Tell in words what you would see, and then write equations for the reactions. Is the substance

1. $S_2O_3^{2-}$ or SO_4^{2-}?
2. H_2O or $H_2O_2(aq)$?
3. I^- or NO_3^-?
4. NH_4^+ or Na^+?
5. SO_3^{2-} or NO_3^-?
6. $NaCl(s)$ or $MgCl_2(s)$?
7. $NaF(s)$ or $NaI(s)$?
8. Cl^- or I^-?
9. $NaHCO_3(s)$ or $NaHSO_4(s)$?
10. $K_2CO_3(s)$ or $MgCO_3(s)$?
11. $Ag(s)$ or $Pt(s)$?
12. Pb^{2+} or Be^{2+}?
13. OH^- or H^+?
14. $Hg_2Cl_2(s)$ or $HgCl_2(s)$?
15. IO_3^- or I^-?
16. $CCl_4(l)$ or $SiCl_4(l)$?
17. $Zn(s)$ or $Ag(s)$?

Solution

Here are some examples of reasonable answers.

1. $S_2O_3^{2-}$ or SO_4^{2-}: We make use of the reducing ability of the $S_2O_3^{2-}$. Add I_3^- and soluble starch solution. The dark color of the iodine-starch complex would disappear if the $S_2O_3^{2-}$ were present:

$$2\,S_2O_3^{2-} + I_3^- = S_4O_6^{2-} + 3\,I^-$$

No reaction would take place if the ion were SO_4^{2-}.

A second test would be to add acid to each sample. The $S_2O_3^{2-}$ ion would decompose:

$$S_2O_3^{2-} + 2\,H^+ = S(s) + SO_2(g) + H_2O$$

No visible reaction would take place with the SO_4^{2-}.

2. H_2O or H_2O_2: Several simple distinguishing tests could be performed based on the oxidizing capabilities of H_2O_2. For example, iodide ion and starch solution could be added to each sample. There would be no reaction with water. A dark-colored solution caused by the formation of iodine-starch complex would develop if the compound were H_2O_2:

$$H_2O_2 + 3\,I^- + 2\,H^+ = I_3^- + 2\,H_2O$$

A test making use of the reducing properties of H_2O_2 would be to add acidified MnO_4^- solution. MnO_4^- will oxidize H_2O_2. Evidence for the reaction would be the color change and evolution of a gas:

$$\underset{\text{(purple)}}{2\,MnO_4^-} + 5\,H_2O_2 + 6\,H^+ = 2\,Mn^{2+} + 5\,O_2(g) + 8\,H_2O$$

3. I^- or NO_3^-: Add $AgNO_3$ solution. A precipitate forms if the ion is I^-:

$$Ag^+ + I^- = AgI(s)$$

No reaction occurs with NO_3^-.

An alternative test makes use of the reducing ability of I^-. For example, acidify the solution and add MnO_4^- solution by drops. There would be no reaction with NO_3^- (nitrogen is already in its highest oxidation state), but the MnO_4^- would be reduced and I_2 formed from I^-:

$$2\,MnO_4^- + 10\,I^- + 16\,H^+ = 5\,I_2(s) + 2\,Mn^{2+} + 8\,H_2O$$

4. NH_4^+ or Na^+: Add sodium hydroxide solution and warm. If NH_4^+, the odor of NH_3 gas would be evident:

$$NH_4^+ + OH^- = NH_3(g) + H_2O$$

No reaction occurs with Na^+.

5. SO_3^{2-} or NO_3^-: Add an acid. Sulfite ion would decompose, and the odor of burning sulfur would be evident:

$$SO_3^{2-} + 2\,H^+ = H_2O + SO_2(g)$$

No reaction occurs with NO_3^-.

Inasmuch as SO_3^{2-} is capable of reducing such strong oxidizing agents as MnO_4^-, MnO_2, and $Cr_2O_7^{2-}$, another simple test is to add the unknown solution to acidified MnO_4^- or $Cr_2O_7^{2-}$. If SO_3^{2-} is present, the MnO_4^- or $Cr_2O_7^{2-}$ solution is decolorized:

$$Cr_2O_7^{2-} + 3\,SO_3^{2-} + 11\,H^+ = 3\,HSO_4^- + 2\,Cr^{3+} + 4\,H_2O$$

No reaction occurs with NO_3^-.

6. NaCl or $MgCl_2$: Dissolve in water and add sodium hydroxide solution. There would be no reaction with NaCl. A precipitate would form with $MgCl_2$:

$$Mg^{2+} + 2\,OH^- = Mg(OH)_2(s)$$

7. NaF or NaI: Add concentrated sulfuric acid and warm. A colorless gas would form if NaF; a purple vapor would form if NaI:

$$NaF(s) + H^+ + HSO_4^- = Na^+ + HSO_4^- + HF(g)$$

$$2\,NaI(s) + 3\,H^+ + HSO_4^- = \underset{\text{(purple)}}{I_2(g)} + SO_2(g) + 2\,Na^+ + 2\,H_2O$$

Alternatively, Ag^+ could be added to a solution of the substance. An insoluble pale-yellow precipitate forms with I^-, but AgF is soluble:

$$Ag^+ + I^- = AgI(s)$$

8. Cl^- or I^-: Add $Cl_2(aq)$. If Cl^-, nothing happens; if I^-, $I_2(I_3^-)$ forms, which would be detected with a starch indicator:

$$Cl_2 + 2\,I^- = 2\,Cl^- + I_2$$

Another method depends upon the difference in solubility of AgCl and AgI. First, add Ag^+ and form the silver halide. Add 6 *M* $NH_3(aq)$. AgCl is soluble in the ammonia; AgI is not:

$$AgCl(s) + 2\,NH_3 = Ag(NH_3)_2^+ + Cl^-$$

9. $NaHCO_3$ or $NaHSO_4$: Add acid. The carbonate would release $CO_2(g)$:

$$NaHCO_3 + H^+ = CO_2(g) + H_2O + Na^+$$

No reaction would occur if the compound were $NaHSO_4$.

10. K_2CO_3 or $MgCO_3$: Add water. $MgCO_3$ is insoluble. K_2CO_3, being an alkali metal compound, would be soluble.
11. Ag or Pt: Add concentrated nitric acid. Ag would dissolve:

$$Ag(s) + 2\,H^+ + NO_3^- = Ag^+ + NO_2(g) + H_2O$$

Pt would not dissolve in nitric acid. (It would dissolve in aqua regia; the stability of hexachloroplatinate (IV) ion is responsible.)
12. Pb^{2+} or Be^{2+}: Both ions form white insoluble hydroxides that are amphoteric. Neither forms an ammine complex ion. However, $PbSO_4$ is very insoluble, whereas $BeSO_4$, being the sulfate of one of the light members of Group IIA, would be very soluble. Therefore, add SO_4^{2-}:

$$Pb^{2+} + SO_4^{2-} = PbSO_4(s)$$

13. OH^- or H^+: Add Mg^{2+}:

$$Mg^{2+} + 2\,OH^- = Mg(OH)_2(s)$$

No reaction occurs with H^+.
14. $Hg_2Cl_2(s)$ or $HgCl_2(s)$: Add water. $HgCl_2$ is soluble, whereas Hg_2Cl_2 is one of the few insoluble chlorides.
15. IO_3^- or I^-: IO_3^- is a strong oxidizing agent. Add iodide ion (acid solution):

$$IO_3^- + 5\,I^- + 6\,H^+ = 3\,I_2 + 3\,H_2O$$

The presence of I_2 could be made more easily detectable by adding starch solution to the reaction.
16. CCl_4 or $SiCl_4$: Add water. CCl_4 neither reacts nor mixes with water. $SiCl_4$ would hydrolyze vigorously in water to produce HCl gas:

$$SiCl_4 + 2\,H_2O = SiO_2(s) + 4\,HCl(g)$$

17. Zn or Ag: Add dilute hydrochloric acid:

$$Zn(s) + 2\,H^+ = Zn^{2+} + H_2(g)$$
$$Ag(s) + H^+ = \text{N.R.}$$

Exercise 22.14

Circle the correct answer.

1. Weakest acid:	HF	HCl	HBr	HI
2. Strongest acid:	H_3BO_3	H_2CO_3	HNO_3	
3. Weakest acid:	H_3AsO_4	$H_2AsO_4^-$	$HAsO_4^{2-}$	
4. Strongest acid:	$HClO_4$	$HClO_3$	$HClO_2$	HClO
5. Most acidic oxide:	N_2O	NO	NO_2	N_2O_5

6. Most basic oxide:	CrO	Cr_2O_3	CrO_3	
7. Strongest oxidizing agent:	HCl	$HClO$	H_3PO_4	HI
8. Most hydrolyzed:	NO_2^-	NO_3^-	Cl^-	ClO_4^-
9. Strongest reducing agent:	H_3PO_4	HNO_3	HF	H_2SO_3
10. Best dehydrating agent:	$Na_2SO_4 \cdot 10\,H_2O$	P_4O_{10}	$C_6H_{12}O_6$	$NaCl$
11. Strongest oxidizing agent:	O_2	O_3	F_2	MnO_4^-
12. Strongest reducing agent:	NO_3^-	Br^-	I^-	H_5IO_6
13. Most hydrolyzed:	CO_3^{2-}	HSO_4^-	Cl^-	O^{2-}

Solution

1. HF. It is the only hydrohalic acid that is not fully ionized.
2. HNO_3. Other factors being constant, the strength of an oxygen acid generally increases with an increase in electronegativity of the central atom.
3. $HAsO_4^{2-}$. As a general rule, a second ionization is four to five orders of magnitude weaker than a first ionization, and the third ionization is four to five orders of magnitude weaker than a second ionization. It is obviously much easier to take a proton from a neutral molecule (as in a first ionization) than it is to take it from an already negatively charged ion (as in succeeding ionizations).
4. $HClO_4$. The central atom is the same in all four choices. In this circumstance, the acid in which the central atom has the highest oxidation state is the strongest. The rationale for this rule is that the higher the oxidation number, the larger the number of oxygen atoms that are available to help draw the electron cloud away from the hydrogen atom, which allows it to escape as a proton.
5. N_2O_5. The central atom being the same in all four choices, the oxide in which the central atom has the highest oxidation number will be the most acidic.
6. CrO. The metal atom being the same in the three cases, the oxide with the lowest oxidation state is the most basic. CrO is a basic oxide, Cr_2O_3 is amphoteric, and CrO_3 is an acidic oxide. Other series of metallic oxides that range all the way from basic oxides to acidic oxides as the oxidation number increases are MnO, MnO_2, Mn_2O_7; and VO, V_2O_3, VO_2, V_2O_5.
7. HClO. The oxyhalogen acids are, as a group, strong oxidizing agents. H_3PO_4 is noted for its poor oxidizing ability. Cl^- and I^- could not be oxidizing agents (good or poor) since they are already in their lowest oxidation state. Iodide ion finds many applications as a good *reducing* agent.
8. NO_2^-. It is an anion from the *weak* acid HNO_2. NO_3^-, Cl^-, and ClO_4^- would not hydrolyze at all since they are derived from fully ionized acids.
9. H_2SO_3. It can be oxidized to HSO_4^-. HNO_3, HF, and H_3PO_4 are already in their highest oxidation states; hence, they could not be reducing agents.

10. P_4O_{10} is the only possible choice. It is a very effective drying agent. Concentrated H_2SO_4, $CaSO_4$, $CaCl_2$, CaO, and $Mg(ClO_4)_2$ are other effective drying agents.
11. F_2. All listed are strong oxidizing agents, but F_2 heads the list. $\mathscr{E}^0_{red} = 3.06$ in acid solution.
12. I^-. Only I^- and Br^- are capable of acting as reducing agents. The halide ions listed in order of increasing ability as reducing agents are $F^- < Cl^- < Br^- < I^-$.
13. O^{2-}. It is almost completely hydrolyzed. CO_3^{2-}, coming from a rather weak acid, is moderately hydrolyzed. Cl^- and HSO_4^-, being anions of strong acids, do not hydrolyze.

Exercise 22.15

Write equations to describe how one could make the following syntheses using only the specified starting materials. These exercises are not meant to illustrate either commercial or convenient laboratory preparations. Practical syntheses usually begin with commonly available substances, whether elements or compounds, and contain as few steps as possible. These exercises are designed to help you recall as many properties and reactions as possible. Look upon them as learning tools rather than as suggested methods for syntheses.

1. $Na_2S_2O_3$ from Na, S, O_2, and H_2O.
2. $SnCl_2$ from Sn, Cl_2, and H_2O.
3. $H_2O_2(aq)$ from Na, O_2, S, H_2O, and an appropriate catalyst.
4. $K_2Cr_2O_7$ from Na, Cr, S, O_2, and H_2O.
5. $KClO_4$ from K, Cl_2, and H_2O.
6. Na_2CO_3 from Na, C, O_2, and H_2O.
7. $(NH_4)_2SO_4$ from N_2, H_2, S, O_2, H_2O, and appropriate catalysts.
8. KNO_3 from N_2, H_2, O_2, K, and appropriate catalysts.

Solution

1. $Na_2S_2O_3$.

(a) Prepare a solution of sodium hydroxide:

$$2\,Na + 2\,H_2O = 2\,Na^+ + 2\,OH^- + H_2(g)$$

(b) Burn sulfur and collect the SO_2 gas:

$$S + O_2 = SO_2(g)$$

(c) Bubble the SO_2 gas into the sodium hydroxide solution, using an excess of base:

$$SO_2 + 2\,OH^- = SO_3^{2-} + H_2O$$

(d) Dissolve sulfur in the basic solution of sulfite ion:

$$SO_3^{2-} + S(s) = S_2O_3^{2-}$$

(e) Filter out the excess sulfur. The solution now contains only Na^+, $S_2O_3^{2-}$, and some OH^-.

(f) Concentrate the solution by evaporation to the saturation point. Then cool and crystallize the $Na_2S_2O_3$. Collect the crystals by filtration or centrifugation. Dry.

2. $SnCl_2$.

(a) Tin and chlorine gas will combine directly, but the product is the Sn(IV), not the desired Sn(II) chloride:

$$Sn + 2\ Cl_2(g) = SnCl_4(l)$$

(b) $SnCl_4$ hydrolyzes readily and almost completely. It can, therefore, be used to prepare a hydrochloric acid solution:

$$SnCl_4 + 2\ H_2O = SnO_2(s) + 4\ HCl(g)$$

If the gas is not allowed to escape, a solution of hydrochloric acid would result.

(c) Remove the solid SnO_2 from the HCl solution and add Sn:

$$Sn + 2\ H^+ = Sn^{2+} + H_2(g)$$

(d) The solution now contains only Sn^{2+} and Cl^- ions. Concentrate by evaporation, cool, and crystallize the $SnCl_2$. Filter and dry crystals.

3. H_2O_2.

(a) Prepare Na_2O_2 by direct combination of the elements. You will recall that with sodium it is Na_2O_2 not Na_2O that results:

$$2\ Na + O_2 = Na_2O_2(s)$$

Caution: Peroxide in the presence of sodium metal is an explosive mixture that detonates easily.

(b) Prepare a solution of sulfuric acid:

$$S(s) + O_2 = SO_2(g)$$
$$2\ SO_2(g) + O_2 = 2\ SO_3(g) \qquad \text{(catalyst required)}$$
$$SO_3(g) + H_2O = H^+ + HSO_4^-$$

(c) Treat the sodium peroxide from (a) with the sulfuric acid solution from (b):

$$Na_2O_2(s) + 2\ H^+ = H_2O_2(aq) + 2\ Na^+$$

(d) Separate $H_2O_2(aq)$ from sulfuric acid solution by fractional distillation. A mixture of H_2O_2 and H_2O will distill near 100° C at 1 atm pressure and, hence, can be easily separated from sulfuric acid, which boils at a much higher temperature.

4. $K_2Cr_2O_7$.
   (a) Prepare sulfuric acid as in 3(b) and dissolve chromium metal, $\mathscr{E}^0_{ox} = 0.74$, in it:

$$2\,Cr(s) + 6\,H^+ = 2\,Cr^{+3} + 3\,H_2(g)$$

   This reaction is inhibited by a protective coating of Cr_2O_3 on the surface of the chromium. Since Cr_2O_3 is an amphoteric oxide, however, sulfuric acid should remove it:

$$Cr_2O_3(s) + 6\,H^+ = 2\,Cr^{3+} + 3\,H_2O$$

   (b) Prepare potassium hydroxide solution:

$$2\,K(s) + 2\,H_2O = 2\,K^+ + 2\,OH^- + H_2(g)$$

   (c) Add sufficient potassium hydroxide solution (b) to the Cr(III) ion solution (a) to precipitate $Cr(OH)_3(s)$:

$$Cr^{3+} + 3\,OH^- = Cr(OH)_3(s)$$

   (d) Filter and wash $Cr(OH)_3$ precipitate above, and then dissolve in excess base solution:

$$Cr(OH)_3(s) + OH^- = Cr(OH)_4^-$$

   (e) Add Na_2O_2, prepared as in 3(a), and heat:

$$2\,Cr(OH)_4^- + 3\,HO_2^- = 2\,CrO_4^{2-} + OH^- + 5\,H_2O$$

   Add sufficient sulfuric acid prepared in (a) to solution in (e) until color changes from yellow to orange:

$$2\,CrO_4^{2-} + 2\,H^+ = Cr_2O_7^{2-} + H_2O$$

   (g) Concentrate and crystallize.

5. $KClO_4$.
   (a) Prepare potassium hydroxide solution as in 4(b).
   (b) Bubble $Cl_2(g)$ into the basic solution:

$$Cl_2(g) + 2\,OH^- = Cl^- + OCl^-$$

   (c) Heat. The hypochlorite ion will disproportionate to chlorate and chloride ion:

$$3\,OCl^- = ClO_3^- + 2\,Cl^-$$

(d) The solution now contains K^+, Cl^-, ClO_3^-, and smaller amounts of OH^-. The solubility of $KClO_3$ (7 g/100 ml) is so much smaller than that of KCl (35 g/100 ml) and KOH (107 g/ml) that it should be possible to fractionally crystallize the $KClO_3$ from the mixture.

(e) Wash and dry the $KClO_3$, and heat:

$$4\ KClO_3 = 3\ KClO_4 + KCl$$

(f) $KClO_4$ + KCl can be separated on the basis of solubility in water. $KClO_4$, with a solubility of only 1.3 g/100 ml, has an exceptionally low solubility for the salt of a Group IA element. KCl is approximately twenty-seven times more soluble. Put the mixture of salts into solution, and fractionally crystallize the perchlorate. Filter, wash, and dry.

The disproportionation reaction shown in (e) has a competing reaction:

$$2\ KClO_3(s) = 2\ KCl(s) + 3\ O_2(g)$$

Uncatalyzed, this latter reaction is slow compared to the one in (e), and in any event would not interfere except to lower the yield of $KClO_4$. The reaction above, when catalyzed by MnO_2, is often used for a laboratory preparation of oxygen gas.

6. Na_2CO_3.

(a) Prepare a sodium hydroxide solution as in 1(a).

(b) Burn carbon to produce CO_2:

$$C + O_2 = CO_2(g)$$

(c) Bubble CO_2 into the basic solution:

$$CO_2 + 2\ OH^- = CO_3^{2-} + H_2O$$

(d) Add additional base to ensure that any hydrogen carbonate that was present has all been converted to carbonate. Concentrate and crystallize. The solubilities of Na_2CO_3 and NaOH are 7 and 42 g/100 ml, respectively. Hence, fractional crystallization of Na_2CO_3 from the solution would be feasible.

7. $(NH_4)_2SO_4$.

(a) Use the Haber process to produce NH_3 gas:

$$N_2 + 3\ H_2 = 2\ NH_3 \quad \text{(catalyst, high pressure, high temperature)}$$

(b) Prepare a solution of sulfuric acid as in 3(b).

(c) Mix (a) and (b):

$$2\ NH_3(g) + H^+ + HSO_4^- = 2\ NH_4^+ + SO_4^{2-} + H_2O$$

(d) Concentrate and crystallize.

8. KNO_3.
   (a) Prepare a solution of potassium hydroxide as in 4(b).
   (b) Prepare ammonia gas as in 7(a).
   Oxidize ammonia (Ostwald process):

$$4\ NH_3 + 5\ O_2 = 4\ NO + 6\ H_2O \qquad \text{(catalyst and high temperature)}$$

$$2\ NO + O_2 = 2\ NO_2$$

$$3\ NO_2 + H_2O = 2\ H^+ + 2\ NO_3^- + NO(g)$$

   (d) Heat to expel the $NO(g)$ in the last equation. Then add potassium hydroxide solution until nitric acid is neutralized. Concentrate and crystallize.

Exercise 22.16

This last set of "equations" is a collection of errors made by students on examinations. Besides the incorrect balancing in most of the equations, at least one other serious infraction of fundamental chemistry is present in each reaction. Describe the errors. *Remember:* These are all incorrect equations.

1. $Cr_2O_7^{2-} + 22\ H^+ + MnO_4^- = 2\ Cr^{3+} + Mn^{2+} + 11\ H_2O$
2. $H_2S(g) + 2\ I^- = S(s) + I_2(g) + 2\ H^+$
3. $2\ Na(s) + HOH = 2\ Na^+ + O^{2-} + H_2(g)$
4. $Fe(s) + F_2(g) = FeF_2(s)$
5. $MnO_4^- + 4\ H^+ + Fe^{2+} = Fe^{3+} + Mn^{2+} + 4\ OH^-$
6. $MnO(s) + Cl_2(g) = Mn(s) + Cl_2O(g)$
7. $CuO(s) + H_2(g) = Cu(OH)_2(s)$
8. $CrO_3(s) + 2\ OH^- + 3\ H_2O = 2\ Cr(OH)_4^-$
9. $MnO(s) + 2\ OH^- + 2\ H_2O = Mn(OH)_4^{2-}$
10. $2\ Cu(s) + 2\ H^+ = 2\ Cu^+ + H_2(g)$
11. $Al^{3+} + 3\ Ag(s) = Al(s) + 3\ Ag^+$
12. $2\ ZnO(s) + 2\ H^+ + 2\ H_2O = Zn(OH)_4^{2-}$
13. $2\ MnO_4^- + 6\ I^- + 8\ H^+ = 2\ MnO_2(s) + 4\ H_2O + 3\ I_2(aq)$
14. $N_2O_5(s) + 3\ H_2O = 2\ H_3NO_4$
15. $SO_2(g) + H_2O = H_2SO_4(l)$

Solution

1. There are two strong oxidizing agents among the reactants. Both are reduced, but nothing is oxidized! Reduction cannot take place without oxidation.
2. The reactants are both oxidized, but nothing is reduced! Oxidation cannot take place without reduction.
3. The oxide ion does not exist to any appreciable extent in aqueous solution. It hydrolyzes almost completely. The products of reaction 3 are $H_2(g)$, OH^-, and Na^+.

4. Fluorine is the strongest oxidizing agent of all the elements. It will oxidize iron to the +3 oxidation state, FeF_3.
5. A base (OH^-) would not be a product in an acidic (H^+) solution. Water, not OH^-, is the correct product.
6. Chlorine is a strong oxidizing agent. It will not reduce Mn(II).
7. Hydrogen gas is a good reducing agent. It will reduce CuO and many other metal oxides when heated. The reaction in 7. should be

$$CuO(s) + H_2(g) = Cu(s) + H_2O(g)$$

8. The chromium has been reduced from +6 in CrO_3 to +3 in $Cr(OH)_4^-$, but there was no reducing agent present! Nothing was oxidized.
9. MnO is a basic not amphoteric oxide. $Mn(OH)_4^{2-}$ does not form.
10. The standard oxidation potential of Cu is lower than that of $H_2(g)$. No reaction takes place. Furthermore, since copper (I) ion disproportionates to Cu^{2+} and Cu, Cu^+ would not be the correct species even if the reaction did take place.
11. The oxidation potential of aluminum is higher than that of Ag. The reverse reaction will take place.
12. Zinc oxide will react with an acid to form Zn^{2+} and water. The zincate ion will form in strong basic solution.
13. The reduction product of MnO_4^- in acid solution is the Mn(II) ion. MnO_2 is a common reduction product in neutral or basic solution.
14. The formula for nitric acid is HNO_3, or, since it is a strong acid, $H^+ + NO_3^-$.
15. Sulfur dioxide is the anhydride of sulfurous acid, H_2SO_3 or $SO_2(aq)$. Sulfur trioxide is the anhydride of sulfuric acid.

23

Nuclear Chemistry

To this point we have considered chemical reactions involving the extranuclear electrons of atoms; now we consider reactions of the nucleus. While much of the study of the nucleus falls rightfully in the field of physics, there are topics that are of great interest to chemists. These include the transmutation of elements, radioactivity, activation analysis, and isotopic tracer studies.

23.1 Nuclear Size

Rutherford's experiments, described briefly in Chapter 2, in which gold foil was bombarded with α particles (helium nuclei), led him to propose a nuclear model for the atom. Rutherford observed that most particles passed through the foil with no deflection but that a few were deflected through very large angles. Only by postulating that the mass and positive charge of the atom are concentrated in a very small volume while most of the atom's volume is occupied by the electrons could he account for the unexpected behavior of the α particles.

Studies of the scattering patterns from Rutherford's experiments gave a first estimate of the nuclear radius. The estimated value was 10^{-12} to 10^{-13} cm, or about 0.1% the radius of an atom. A better estimate of the size of the nucleus is obtained from neutron bombardment; a large number of nuclear radii have been estimated by this technique. These radii are surprisingly well related by the equation

$$R = R_0 A^{\frac{1}{3}} \qquad [23.1]$$

where $R_0 = 1.33 \times 10^{-13}$ cm (a constant common to all nuclei), R is the nuclear radius, and A is the mass number. Example 23.1 illustrates the use of equation [23.1] for the calculation of the radius and density of a nucleus.

EXAMPLE 23.1

(a) Calculate the radius of an $^{16}_{8}O$ nucleus.

(b) Assume that the nucleus is a sphere with this radius, and calculate its density in g/cm³.

SOLUTION

(a) Equation [23.1] applies with $A = 16.0$:

$$R = (1.33 \times 10^{-13}\ \text{cm})(16.0)^{\frac{1}{3}} = 3.35 \times 10^{-13}\ \text{cm}$$

(b) To get the density, we calculate the mass and volume of the nucleus:

$$m = \frac{16.0\ \text{g/mole}}{6.02 \times 10^{23}\ \text{nuclei/mole}} = 2.66 \times 10^{-23}\ \text{g/nucleus}$$

$$V = \frac{4}{3}\pi R^3 = \frac{4}{3}\pi(3.35 \times 10^{-13}\,\text{cm})^3 = 1.57 \times 10^{-37}\,\text{cm}^3$$

$$d = \frac{m}{V} = \frac{2.66 \times 10^{-23}\,\text{g}}{1.57 \times 10^{-37}\,\text{cm}^3} = 1.69 \times 10^{14}\,\frac{\text{g}}{\text{cm}^3}$$

$$\text{or } 3.05 \times 10^{9}\,\frac{\text{tons}}{\text{in}^3}!$$

From equation [23.1] one sees that the volume of a nucleus is proportional to A, for

$$V \propto R^3 \propto A$$

This suggests that the neutrons and protons pack together in the nucleus somewhat as hard spheres would pack, making the total volume approximately equal to the sum of the volumes of the individual protons and neutrons.

Scattering experiments along with other experiments also provide information about the shapes of nuclei. Some nuclei, such as $^{1}_{1}H$, $^{4}_{2}He$, $^{12}_{6}C$, $^{16}_{8}O$, $^{40}_{20}Ca$, $^{58}_{28}Ni$, and $^{208}_{82}Pb$, are spherical. Nuclei having atomic numbers near those in the preceding list are spherical or nearly so. Most other nuclei are prolate spheroids (slightly elongated), but this departure from the spherical shape toward a football shape is never extreme.

23.2 Nuclear Binding Energy and Stability

The nature of the **nuclear binding energy,** the very large force that holds the protons and neutrons together in a nucleus, is not well understood, but the magnitude is easily calculated. To do this one notes the mass deficiency obtained by subtracting the mass of the nucleus from the sum of the masses of the neutrons and protons that make up the nucleus. Example 23.2 illustrates how this is done.

EXAMPLE 23.2 Calculate the nuclear binding energy in the $^{16}_{8}O$ nucleus.

SOLUTION The $^{16}_{8}O$ atom has a mass of 15.9994 amu. If we substract the mass of the eight electrons that surround the nucleus, we obtain the mass of the nucleus. The mass of the electron is 5.486×10^{-4} amu. Hence,

$$\text{Mass of } ^{16}_{8}\text{O nucleus} = 15.9994 - (8)(5.486 \times 10^{-4}) = 15.9950\text{ amu}$$

The masses of the neutron and proton are 1.008665 amu and 1.007825 amu, respectively. The mass of eight neutrons plus eight protons is 16.1319 amu. Therefore, the $^{16}_{8}O$ nucleus is lighter than the sum of the particles by the difference $16.1319 - 15.9950 = 0.1369$ amu.

This loss in mass can be attributed to the energy released when the eight protons and eight neutrons are fused to form the $^{16}_{8}O$ nucleus. The loss in mass can be converted to energy through the Einstein mass-energy equivalence expression

$$E_b = mc^2 \qquad [23.2]$$

where E_b is the binding energy, m is the loss in mass, and c is the speed of light. Hence,

$$E_b = \left(0.1369\,\frac{\text{amu}}{\text{atom}}\right)\left(1.6606 \times 10^{-24}\,\frac{\text{g}}{\text{amu}}\right)\left(2.9979 \times 10^{10}\,\frac{\text{cm}}{\text{sec}}\right)^2$$

$$= 2.043 \times 10^{-4}\,\frac{\text{ergs}}{\text{atom}}$$

or

$$E_b = \left(2.043 \times 10^{-4}\,\frac{\text{ergs}}{\text{atom}}\right)\left(6.022 \times 10^{23}\,\frac{\text{atom}}{\text{mole}}\right)\left(10^{-7}\,\frac{\text{joule}}{\text{erg}}\right)\left(10^{-3}\,\frac{\text{kjoules}}{\text{joule}}\right)$$

$$= 1.230 \times 10^{10}\,\frac{\text{kjoules}}{\text{mole}}$$

The magnitude of this binding energy is more fully appreciated when it is compared with the 498.3 kjoules/mole required to separate the two oxygen atoms in the O_2 molecule, or the 1313 kjoules/mole required to remove an electron from the oxygen atom (ionization energy).

In the above example, we calculated the total binding energy in the $^{16}_{8}O$ nucleus. A more interesting number is the binding energy per *nucleon*, where a nucleon is a proton or a neutron. The binding energy per nucleon is obtained by dividing the total binding energy by the mass number A. For oxygen there are sixteen nucleons (eight protons plus eight neutrons). Hence, the binding energy is

$$\left(2.043 \times 10^{-4}\,\frac{\text{ergs}}{\text{atom}}\right)\left(\frac{1\text{ atom}}{16\text{ nucleons}}\right) = 1.277 \times 10^{-5}\,\frac{\text{ergs}}{\text{nucleon}}$$

Nuclear energies are usually expressed in the units of million electron volts, or MeV. Using the conversion factor of 6.2414×10^5 MeV/erg gives the binding energy for $^{16}_{8}O$ as 7.970 MeV/nucleon.

An electron volt (eV) is the kinetic energy gained by an electron when accelerated through a potential of one volt.

$1\text{ MeV} = 10^6\text{ eV}$

$1\text{ eV} = 1.6022 \times 10^{-12}\text{ ergs}$

Binding energies per nucleon (E_b/A) have been calculated for all the elements; the results are plotted in Figure 23.1 as a function of mass number. After a very rapid increase among the lightest nuclei, the binding energy reaches a maximum value of approximately 8.5 MeV/nucleon at about $A = 60$, and then decreases slowly with increasing mass.

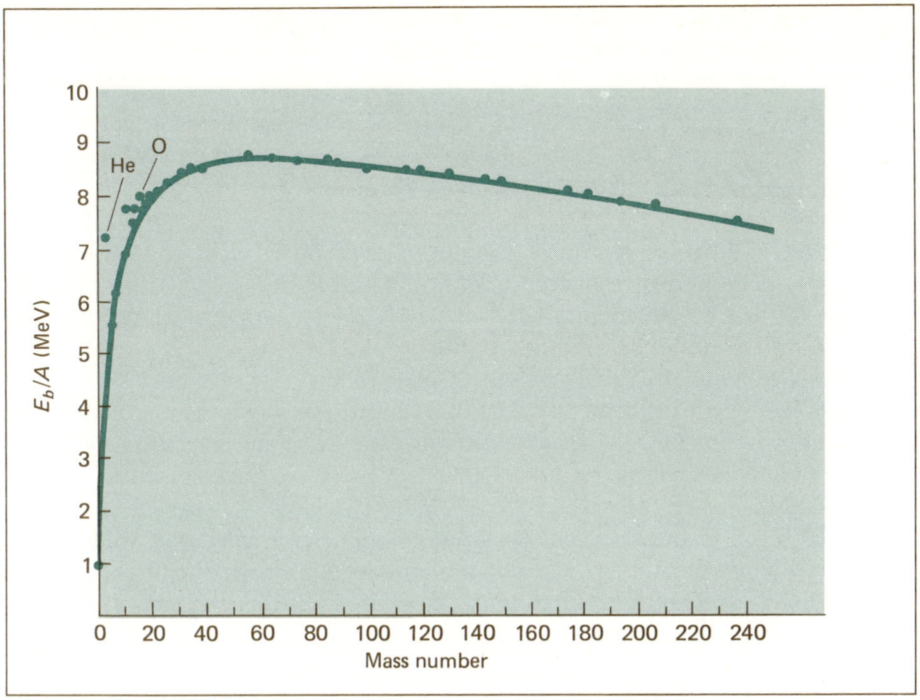

Figure 23.1
Plot of the binding energy per nucleon versus mass number of the nucleus.

A good approximation of the binding energy for a nucleus can be calculated from the mass number by use of the semi-empirical equation

$$E_b = 14.1A - 13A^{\frac{2}{3}} - \frac{0.6Z^2}{A^{\frac{1}{3}}} \qquad [23.3]$$

where E_b is the total binding energy, A is the mass number, and Z is the atomic number or nuclear charge. This equation is interesting because its terms can be related to the major contributions to the binding energy.

The first term on the right of equation [23.3] shows a directly proportional relationship between the binding energy and the number of nucleons. This suggests that the attractive forces are of very short range, extending strongly only to nearest neighbors. If the attractive force from a nucleon were of longer range, that nucleon would be attracted to $A - 1$ other nucleons, and the total binding energy would be proportional to $A(A - 1)$, or approximately to A^2. Hence we see that the possibility of the long-range attraction is ruled out since equation [23.3] fits the observed binding energies.

The second term on the right is a correction for the surface effects. Nucleons on the surface of the nucleus are not completely surrounded by other nucleons and do not, therefore, contribute fully to the binding energy. The number of surface nucleons is proportional to the surface area, and this in turn varies with the square of the radius. The radius (equation [23.1]) is proportional to $A^{1/3}$, making the number of nucleons proportional to $A^{2/3}$. The term is negative since the surface nucleons cause a decrease in binding energy.

The third term corrects for the coulombic repulsion between protons. Coulombic repulsion varies directly with the square of the charge (Z^2) and inversely with the nuclear radius, which is proportional to $A^{1/3}$. This term is also negative because it represents a decrease in binding energy.

23.3 Unstable Nuclei and Radioactivity

Of the almost 2000 known nuclear isotopes (1965 by a recent count) only 284 are stable, that is, not radioactive. The remainder change to other isotopes through **nuclear decay,** the emission of particles or energy, until eventually a stable isotope is obtained and the process stops.

There is evidence that suggests a "shell model" for the nucleus, in that most stable nuclei have even numbers of both protons and neutrons. Table 23.1 shows how important the **neutron/proton ratio** is to **nuclear stability:** there are only eight known stable nuclei that have odd numbers of both protons and neutrons.

In addition to the apparent preference for nuclei having even numbers of both kinds of nucleons, there are certain numbers of protons or neutrons, sometimes called "magic numbers," that are associated with particularly stable and abundant isotopes. These magic numbers are 2, 8, 20, 28, 50, 82, and 126, and suggest an analogy with the noble gases and their filled energy levels.

Natural **radioactivity** may involve several processes. In β^- decay, the nucleus emits a β particle, an electron. The net result of $\boldsymbol{\beta^-}$ **emission** is that a neutron is replaced by a proton. An example is

In this and all other nuclear reactions we consider, the mass number A and the charge or atomic number Z must be conserved.

$$^{14}_{6}C \rightarrow {}^{14}_{7}N + {}^{0}_{-1}\beta$$

Positrons or β^+ particles may also be emitted, with the result that the nucleus gains one neutron and loses one proton. An example of $\boldsymbol{\beta^+}$ **emission** is

A positron has the same mass as the electron. Its charge is the same magnitude as that of the electron but positive.

$$^{22}_{11}Na \rightarrow {}^{22}_{10}Ne + {}^{0}_{1}\beta$$

In ***K*-electron capture,** one of the extranuclear electrons is captured by the nucleus. The result is almost the same as with β^+ emission. The electron enters

TABLE 23.1

Stable Nuclei

MASS NUMBER (A)	NUMBER OF NEUTRONS	NUMBER OF PROTONS	NUMBER OF STABLE NUCLEI
Even	Even	Even	166
	Odd	Odd	8
Odd	Even	Odd	57
	Odd	Even	53

the nucleus, and a proton is replaced by a neutron. For example,

$$^{37}_{18}Ar + {}^{0}_{-1}e \rightarrow {}^{37}_{17}Cl$$

In **α emission,** the nucleus emits a helium nucleus, also known as an α particle. The mass decreases by four mass numbers and the atomic number by two:

$$^{238}_{92}U \rightarrow {}^{234}_{90}Th + {}^{4}_{2}\alpha$$

In **spontaneous fission,** the nucleus splits into two large fragments of approximately equal size. Usually, several neutrons are emitted. Finally in γ decay, a high-energy photon is emitted; **γ emission** provides a way for an excited nucleus to rid itself of excess energy.

In many instances, it is possible to predict which nuclei will be radioactive and what type of decay can be expected. We will now see how this is accomplished.

β Decay and *K*-Electron Capture

Figure 23.2 shows a plot of the atomic number Z versus the number of neutrons N for all stable nuclei. These nuclei are seen to fall within a well-defined belt. Most stable nuclei for elements having an atomic number up to 20 are characterized by having equal numbers of neutrons and protons. For the heavier elements, the stable nuclei contain more neutrons than protons, with the ratio increasing with increasing mass number.

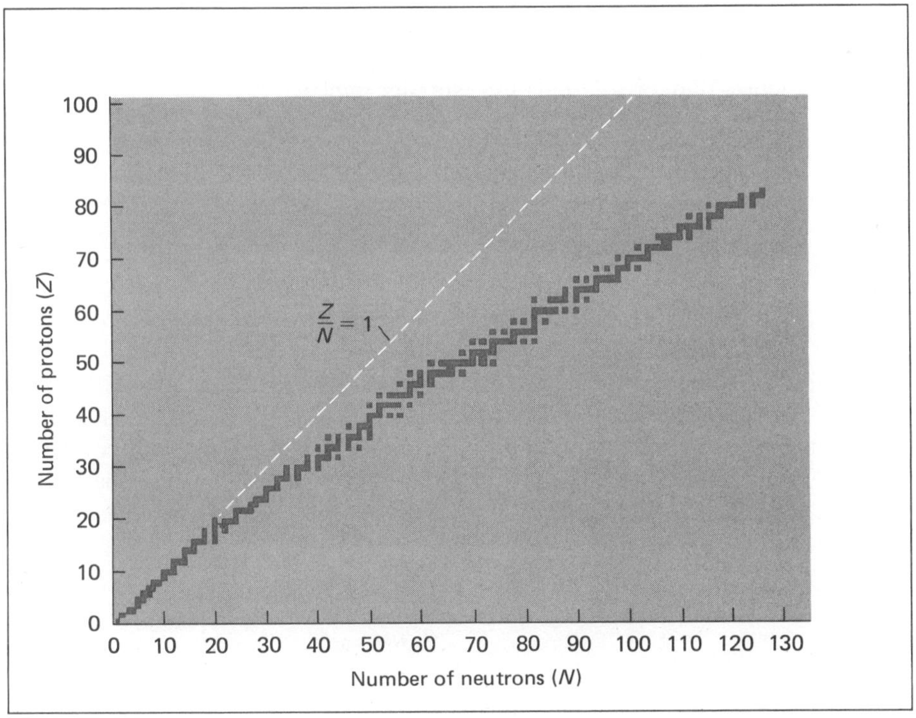

Figure 23.2
A plot of the number of protons as a function of the number of neutrons for the stable nuclei.

It seems probable that the cause of this behavior is the excessive coulombic repulsion in nuclei with a large Z. This repulsion can be diminished by increasing the neutron number, which increases the nuclear size.

Nuclei lying outside this belt of stability are radioactive, and they decay in a manner to form nuclei with a neutron-to-proton ratio closer to the optimum value. For a nucleus with too many neutrons, β^- decay occurs. The result of the β^- decay is that there are one fewer neutron and one more proton in the nucleus, which produces a more stable isotope, one that lies closer to the belt of stability. For nuclei that have a deficiency of neutrons, either β^+ decay or electron capture occurs. Carbon and oxygen isotopes provide examples of these processes:

Positron Emission	Stable	Electron Emission
${}^{10}_{6}C$, ${}^{11}_{6}C$	${}^{12}_{6}C$, ${}^{13}_{6}C$	${}^{14}_{6}C$, ${}^{15}_{6}C$
Example: ${}^{10}_{6}C \rightarrow {}^{10}_{5}B + {}^{0}_{1}\beta$		Example: ${}^{14}_{6}C \rightarrow {}^{14}_{7}N + {}^{0}_{-1}\beta$

Electron Capture	Stable	Electron Emission
${}^{14}_{8}O$, ${}^{15}_{8}O$	${}^{16}_{8}O$, ${}^{17}_{8}O$, ${}^{18}_{8}O$	${}^{19}_{8}O$
Example: ${}^{14}_{8}O + {}^{0}_{-1}e \rightarrow {}^{14}_{7}N$		Example: ${}^{19}_{8}O \rightarrow {}^{19}_{9}F + {}^{0}_{-1}\beta$

α Decay and Spontaneous Fission

Both α decay and **spontaneous fission** represent ways in which large nuclei with high Z values can become stable by losing mass and charge. These processes occur almost without exception among isotopes with mass numbers greater than 200. For the lighter elements in this range, α decay is more common. In general, spontaneous fission becomes more common as the mass increases, and although there are many exceptions, it becomes the preferred route for the largest of the nuclei. Apparently these nuclei are so large that they cannot hold together; hence, they break apart into smaller nuclei.

γ Emission and Energetics of Radioactive Decay

Large amounts of energy are usually released during radioactive decay. This energy can be calculated from a corresponding loss in mass. As an example consider the β^- decay of ${}^{14}C$:

$${}^{14}_{6}C \rightarrow {}^{14}_{7}N + {}^{0}_{-1}\beta$$

To determine the energy released (in this case as kinetic energy with the β particle), we calculate the mass difference between ${}^{14}_{7}N$ and ${}^{14}_{6}C$ and convert this to energy. In the decay the ${}^{14}_{6}C$ atom with six orbital electrons is converted to a ${}^{14}_{7}N$ ion, again with six orbital electrons. This ion plus the β particle is equivalent in mass to the neutral ${}^{14}_{7}N$ atom. The masses are as follows:

${}^{14}_{7}N$	14.003074 amu
${}^{14}_{6}C$	14.003242 amu
Difference	0.000168 amu

This mass difference corresponds to 0.155 MeV of energy.

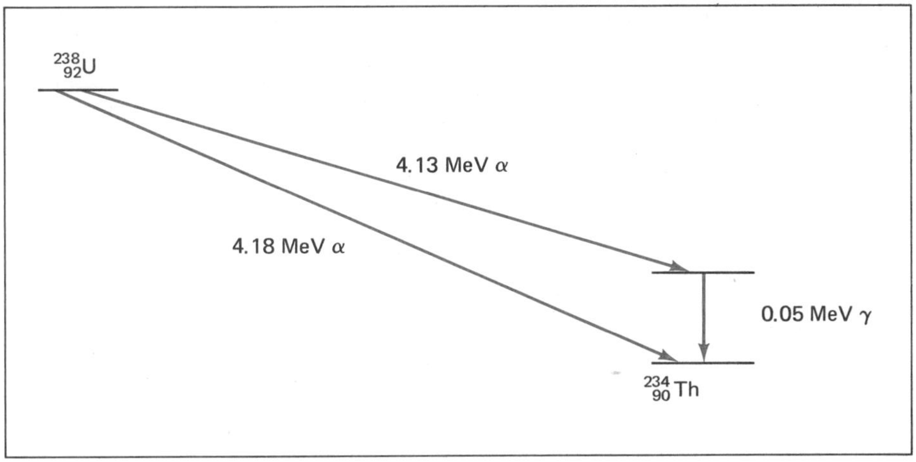

Figure 23.3
The decay of $^{238}_{90}U$ to give alpha particles and a γ-ray.

Frequently, not all the excess energy in the radioactive process is carried away by the α or β particle. Often the new nucleus is produced in an excited state. This nucleus releases the excitation energy by emitting a γ-ray, a photon of very high energy and short wavelength.

As an example of γ decay, α particles with 4.13 MeV or 4.18 MeV of energy are released when $^{238}_{92}U$ decays to $^{234}_{90}Th$. The lower energy α particle results when $^{234}_{90}Th$ is produced in an excited nuclear-energy state, which then decays to the ground state by the emission of a 0.05 MeV γ-ray (see Figure 23.3).

As a further example, consider the decay of $^{60}_{27}Co$ shown in Figure 23.4. The cobalt decays by β^- emission to $^{60}_{28}Ni$ in an excited nuclear state. The $^{60}_{28}Ni$ then loses this excitation energy in two steps by emitting γ-rays with energies of 1.17 and 1.33 MeV, respectively.

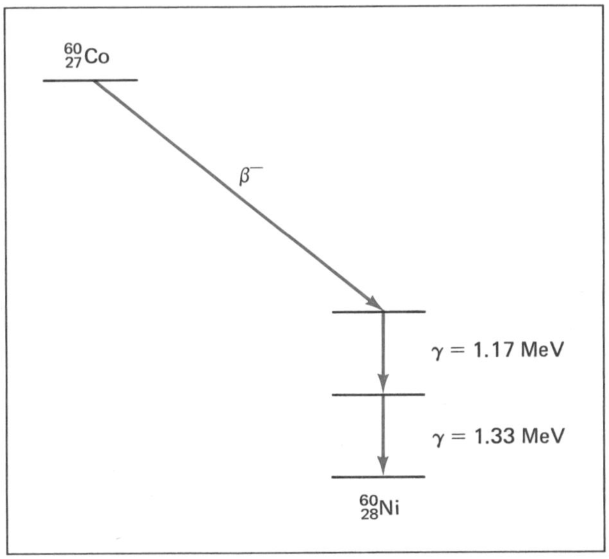

Figure 23.4
The nuclear decay scheme for $^{60}_{27}Co$.

The electronic and nuclear energy-level spacings are usually very different. Compare the 1.13 MeV nuclear energy spacing in $^{60}_{27}Co$ with the 1.02×10^{-5} MeV energy difference between the 1*s* and 2*s* atomic orbitals in hydrogen.

Experimental results such as the above show that a nucleus has discrete quantized energy levels in many ways similar to the electronic energy levels. Gamma-ray spectroscopy, the study of the energies of the γ-rays emitted during radioactive decay, has in many cases enabled the nuclear chemist to construct detailed energy-level patterns for nuclei.

Radioactive Decay Rates

The spontaneous decay of a radioactive nucleus is a first-order process, and the **radioactive decay rate** is expressed mathematically in the same form as for the first-order chemical reaction we studied in Chapter 12. For nuclear reactions, the number of nuclei present rather than the concentration is followed as a function of time. The equation that applies is

$$\ln \frac{N}{N_0} = -\lambda t \qquad [23.4]$$

Equation [23.4] is analogous to equation [12.1] in Chapter 12.

where N is the number of nuclei present at time t, N_0 is the number at $t = 0$, and λ is a constant known as the decay constant. The **half-life,** the time required for one-half of the sample present at any given time to decay, is commonly reported for nuclear reactions. The half-life is easily calculated from the decay constant. When $t = t_{1/2}$, $N/N_0 = 1/2$. Substitution into equation [23.4] gives

The half-life of a radioactive nucleus is independent of the amount of sample present. Unlike the first-order chemical reaction, the half-life of the nuclear process is also independent of temperature and pressure.

$$\ln \tfrac{1}{2} = -\lambda t_{\frac{1}{2}}$$

or

$$t_{\frac{1}{2}} = \frac{\ln 2}{\lambda} \qquad [23.5]$$

The following example illustrates the use of this relationship in calculations of half-life.

EXAMPLE 23.3 The decay constant for the β decay of $^{14}_{6}C$ to $^{14}_{7}N$ is 1.244×10^{-4} years^{-1}. Calculate $t_{1/2}$, the half-life for $^{14}_{6}C$, and calculate the time for 75% of a sample of $^{14}_{6}C$ to decay to $^{14}_{7}N$.

SOLUTION The half-life is calculated from equation [23.5]:

$$t_{\frac{1}{2}} = \frac{\ln 2}{\lambda} = \frac{\ln 2}{1.244 \times 10^{-4}\ \text{years}^{-1}} = 5570\ \text{years}$$

After one half-life (5570 years), 50% of a sample will have decayed. After another 5570 years, one-half of the remainder will decay, for a total of 3/4 or 75% decayed. Hence, it takes two half-lives or 11,140 years for 75% of the $^{14}_{6}C$ sample to decay.

Figure 23.5 is a plot of equation [23.4] for $^{14}_{6}C$ illustrating the exponential decay and the decrease in amount with each half-life.

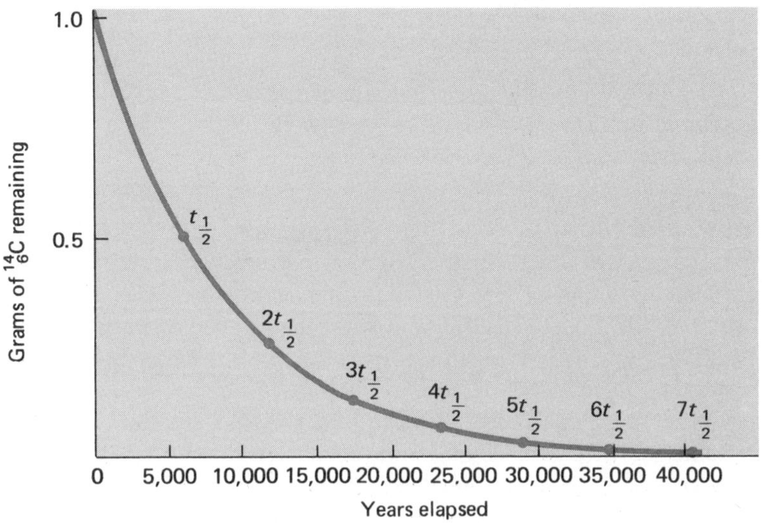

Figure 23.5
The nuclear decay curve for $^{14}_{6}C$. Every 5570 years, the amount of carbon-14 remaining is halved.

The half-life is used as a convenient measure of the stability of a radioactive isotope. There is an extraordinary range in the rates of spontaneous nuclear reactions; for example, the half-life of $^{215}_{86}Rn$ is of the order of 10^{-6} sec while that of $^{209}_{83}Bi$ is of the order of 10^{17} years. This latter number may be better appreciated if it is remembered that estimates of the age of the universe place it at about 5×10^9 years.

23.4 Nuclear Reactions

We turn now from spontaneous **nuclear reactions** to reactions that occur because of the addition of a particle to a nucleus. Through such reactions the alchemist's dream of changing lead into gold becomes a reality. We will consider reactions that provide the means of creating new elements and of creating many radioactive isotopes of known elements.

The product of a spontaneous nuclear reaction can be used as the projectile for causing another reaction to occur. An interesting three-step process beginning with a spontaneous reaction leads to the formation of $^{60}_{27}Co$. This isotope of cobalt is prepared from stable $^{59}_{27}Co$ when the latter is exposed to a beam of neutrons:

$$^{59}_{27}\text{Co} + {}^{1}_{0}\text{n} \rightarrow {}^{60}_{27}\text{Co}$$

The neutrons are conveniently provided by bombarding $^{9}_{4}Be$ with high energy α particles,

$$^{9}_{4}\text{Be} + {}^{4}_{2}\alpha \rightarrow {}^{12}_{6}\text{C} + {}^{1}_{0}\text{n}$$

and the α particles can be obtained from the spontaneous decay of $^{239}_{94}Pu$,

$$^{239}_{94}Pu \rightarrow ^{235}_{92}U + ^{4}_{2}\alpha$$

Hence, the $^{239}_{94}Pu$ spontaneously decays to give α particles; the α particles collide with $^{9}_{4}Be$ nuclei, causing an emission of neutrons; and the neutrons cause $^{59}_{27}Co$ to be transformed to $^{60}_{27}Co$. As we saw earlier, the $^{60}_{27}Co$ is unstable and undergoes subsequent spontaneous reaction.

Sources of Projectiles for Nuclear Reactions

Nuclear reactions are usually performed by bombarding a "target" containing the reacting nuclei with high-energy particles such as the α particles or neutrons described above. While α particles and neutrons may be used from the reactions just described, they and other particles are often obtained as more intense beams from other sources.

A **nuclear reactor** is capable of providing very intense beams of neutrons. The reactor is charged with an easily fissionable material such as $^{235}_{92}U$ or $^{239}_{94}Pu$. Stray neutrons collide with uranium or plutonium nuclei and cause fission to occur in such reactions as

$$^{235}_{92}U + ^{1}_{0}n \rightarrow ^{99}_{40}Zr + ^{134}_{52}Te + 3\ ^{1}_{0}n$$

These additional neutrons cause more fission, and the result is a chain reaction. This chain reaction is controlled by inserting rods of a neutron-absorbing substance such as cadmium or boron into the reactor, with the result that a steady-state condition is reached in which neutrons are produced at the same rate as they are consumed.

In addition to α particles ($^{4}_{2}He$) and neutrons ($^{1}_{0}n$), other such particles as protons ($^{1}_{1}H$), deuterons ($^{2}_{1}H$), and even heavier nuclei such as $^{12}_{6}C$, $^{16}_{8}O$, and $^{40}_{18}Ar$ are used as projectiles in nuclear reactions. To be effective, these particles must be accelerated to high energies so that when they are directed onto a target they can penetrate the coulomb barrier of the target nucleus. These high-energy particles are produced in such devices as the cyclotron and the linear accelerator.

In the **linear accelerator,** shown schematically in Figure 23.6, a beam of charged ions from an ion source (hydrogen for protons, helium for α particles, and so on) is introduced into one end of a long evacuated tube. The tube contains a number of coaxial cylindrical sections. Alternate sections are connected together, and a high-frequency alternating voltage is applied between the two groups of electrodes. An ion passing down the tube is accelerated between adjacent pairs of electrodes if the voltage is in proper phase. Since the ion accelerates in traveling down the tube, the lengths of successive electrodes must increase for the ion to arrive at the next gap at the proper time. If the frequency and the lengths of successive sections are chosen correctly, the ion is accelerated in each gap and emerges from the far end of the tube with energies in the MeV range.

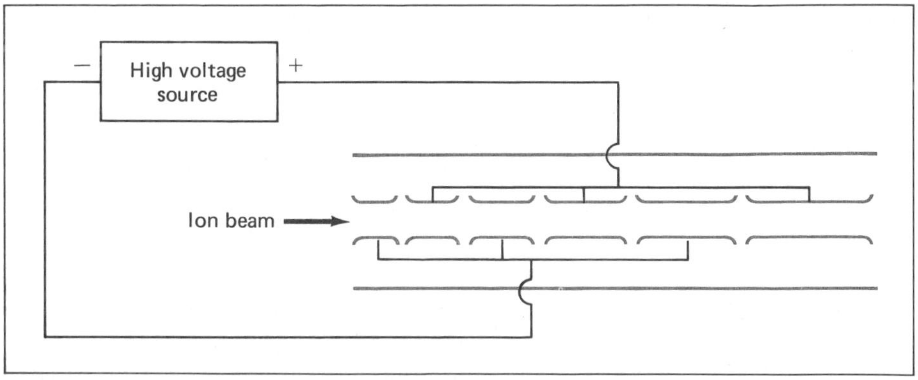

Figure 23.6
Schematic diagram of the accelerating tube of a linear accelerator.

The construction of the ingenious device known as the **cyclotron** is shown schematically in Figure 23.7. Ions are introduced into the cyclotron near the center of the gap between two hollow semicircular electrode boxes called "dees" because of their shape. The dees are enclosed in a vacuum container and are located between the circular pole faces of an electromagnet. A high-frequency alternating voltage is applied between the two dees. A positively charged ion starting from the center is accelerated toward the electrode that has a negative potential at that instant. When the particle reaches the field-free interior of the electrode, it no longer is acted upon by electric forces but is constrained to move in a circular path by the magnetic field. If the frequency of the alternating potential is such that the particle reaches the gap between the electrodes just as the potential reverses itself, the particle will again be accelerated. After acceleration, the particle will move in a larger radius through the next electrode. The time it spends in the electrode stays constant, and it enters the gap at the correct time to be accelerated again. The result is that the ion works its way toward the outer edge of the electrode and constantly gains energy as it does so. Finally, when the ion reaches the periphery of the dee system, it is directed onto the target.

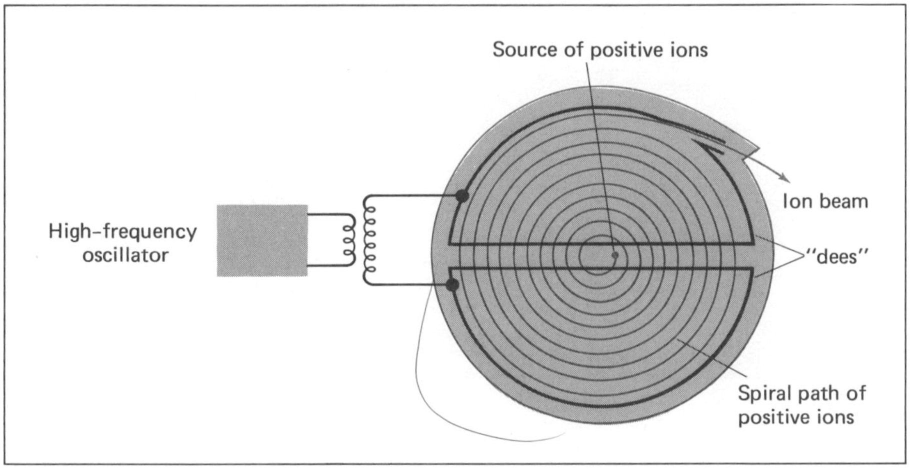

Figure 23.7
Schematic representation of a cyclotron. There are circular magnets placed above and below the cyclotron, but these are not shown.

Synthesis of New Elements

A claim has been made for the synthesis of element 106.

Of the 105 known elements, most of those through number 92 have been found in nature. For some elements, namely Tc ($Z = 43$), Pm ($Z = 61$), At ($Z = 85$), and Fr ($Z = 87$), there are uncertainties; each of these is radioactive with known isotopes having half-lives short enough that little would remain in the estimated age of the earth.

All the elements with atomic numbers above 83 (bismuth) are radioactive. All past uranium ($Z = 92$) have half-lives so short that they are not found in nature. Uranium and thorium have very long half-lives and thus are naturally occurring. As they decay, small quantities of Pa, Ac, Ra, Rn, and Po are produced. Hence, these latter elements, although they have short half-lives, are continually regenerated and are present in uranium ores. An example of a sequence of nuclear reactions by which some of these elements are formed from ^{238}U is as follows:

$$^{238}_{92}U \xrightarrow{t_{\frac{1}{2}} = 4.5 \times 10^9 \text{ years}} {}^{234}_{90}Th + {}^{4}_{2}\alpha$$

$$^{234}_{90}Th \xrightarrow{t_{\frac{1}{2}} = 24 \text{ days}} {}^{234}_{91}Pa + {}^{0}_{-1}\beta$$

$$^{234}_{91}Pa \xrightarrow{t_{\frac{1}{2}} = 1 \text{ minute}} {}^{234}_{92}U + {}^{0}_{-1}\beta$$

$$^{234}_{92}U \xrightarrow{t_{\frac{1}{2}} = 2 \times 10^5 \text{ years}} {}^{230}_{90}Th + {}^{4}_{2}\alpha$$

$$^{230}_{90}Th \xrightarrow{t_{\frac{1}{2}} = 8 \times 10^4 \text{ years}} {}^{226}_{88}Ra + {}^{4}_{2}\alpha$$

$$^{226}_{88}Ra \xrightarrow{t_{\frac{1}{2}} = 1622 \text{ years}} {}^{222}_{86}Rn + {}^{4}_{2}\alpha$$

$$^{222}_{86}Rn \xrightarrow{t_{\frac{1}{2}} = 3.8 \text{ days}} {}^{218}_{84}Po + {}^{4}_{2}\alpha$$

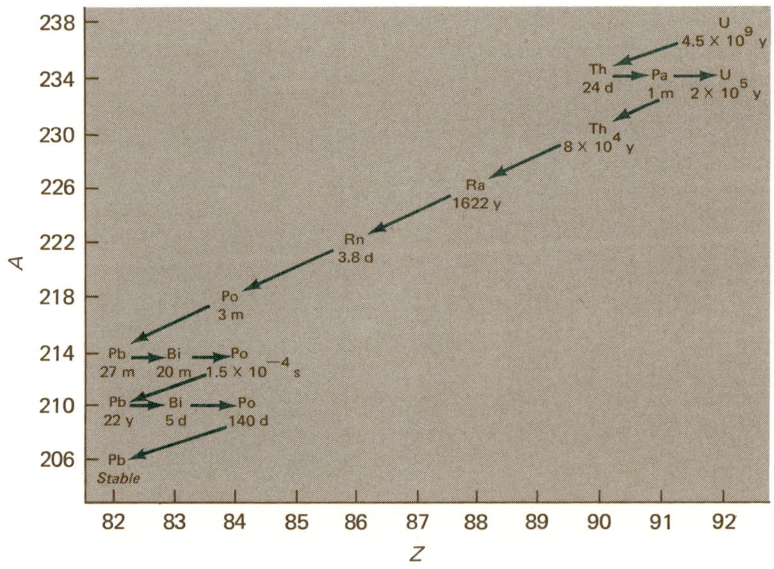

Figure 23.8
The $^{238}_{92}U$ radioactive-decay series. The arrows pointing diagonally to the left correspond to the emission of an α particle by a nucleus; the shorter arrows pointing to the right correspond to nuclear emission of β particles. The figures below the symbols for the elements represent the half-lives of the corresponding isotopes. Abbreviations are as follows: y, years; d, days; h, hours; m, minutes; s, seconds.

The $^{218}_{84}Po$ in turn decays, and the final product is $^{206}_{82}Pb$, a stable isotope. The total sequence is shown in Figure 23.8.

Isotopes of Tc, Pm, At, Fr, and all the elements with $Z > 92$ have been synthesized through appropriate nuclear reactions. Let us now describe the techniques used in the synthesis of some of the more important isotopes and speculate on the possible **synthesis of new elements** with still larger Z values.

A very small amount of francium does exist in nature as the result of a very rare branched decay of ^{227}Ac.

Elements Before Uranium

The first synthetic element to be produced was technetium, element 43. It was first obtained from the bombardment of molybdenum with deuterons. For example,

$$^{97}_{42}Mo + ^{2}_{1}H \rightarrow ^{99}_{43}Tc$$

It can also be made by irradiation of molybdenum with neutrons to form $^{99}_{42}Mo$, which decays to $^{99}_{43}Tc$:

$$^{98}_{42}Mo + ^{1}_{0}n \rightarrow ^{99}_{42}Mo$$

$$^{99}_{42}Mo \xrightarrow{t_{\frac{1}{2}} = 67 \text{ hours}} ^{99}_{43}Tc + ^{0}_{-1}\beta$$

The most common source of Tc, however, is as a fission product in a nuclear reactor. When uranium nuclei break apart, they split into two smaller fragments plus some neutrons. A range of particle sizes is obtained, the distribution or percent yield of which is shown in Figure 23.9. Notice that particles with masses

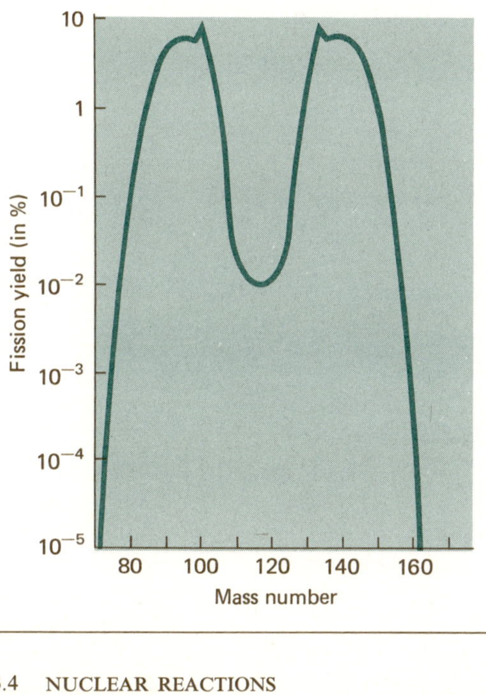

Figure 23.9
Yields of products from slow-neutron-induced fission of ^{235}U.

near either 100 or 133 are most probable. Technetium, with a mass number of 99, is one of those isotopes produced in very high yield. In fact, ^{99}Tc constitutes over 6% of the eventual fission products in a nuclear reactor.

Promethium can be obtained from the neutron irradiation of neodymium followed by β decay. However, as with technetium, the most common source of promethium is as a fission product of uranium, for promethium also lies in the range of high yield (Figure 23.9).

Astatine is difficult to obtain by neutron bombardment since the element immediately preceding it is polonium, which is itself radioactive and available only in very small amounts. Astatine can be obtained, however, by adding an α particle to bismuth:

$$^{209}_{83}\text{Bi} + {}^{4}_{2}\alpha \rightarrow {}^{211}_{85}\text{At} + 2\,{}^{1}_{0}\text{n}$$

Francium is obtained by irradiating $^{232}_{90}$Th with neutrons to produce $^{233}_{90}$Th:

$$^{232}_{90}\text{Th} + {}^{1}_{0}\text{n} \rightarrow {}^{233}_{90}\text{Th}$$

This thorium then decays through a series of steps in which francium appears as a temporary product:

$$^{233}_{90}\text{Th} \xrightarrow{t_{\frac{1}{2}} = 23.3 \text{ minutes}} {}^{233}_{91}\text{Pa} + {}^{0}_{-1}\beta$$

$$^{233}_{91}\text{Pa} \xrightarrow{t_{\frac{1}{2}} = 27.4 \text{ days}} {}^{233}_{92}\text{U} + {}^{0}_{-1}\beta$$

$$^{233}_{92}\text{U} \xrightarrow{t_{\frac{1}{2}} = 1.62 \times 10^{5} \text{ years}} {}^{229}_{90}\text{Th} + {}^{4}_{2}\alpha$$

$$^{229}_{90}\text{Th} \xrightarrow{t_{\frac{1}{2}} = 7340 \text{ years}} {}^{225}_{88}\text{Ra} + {}^{4}_{2}\alpha$$

$$^{225}_{88}\text{Ra} \xrightarrow{t_{\frac{1}{2}} = 14.8 \text{ days}} {}^{225}_{89}\text{Ac} + {}^{0}_{-1}\beta$$

$$^{225}_{89}\text{Ac} \xrightarrow{t_{\frac{1}{2}} = 10.0 \text{ days}} {}^{221}_{87}\text{Fr} + {}^{4}_{2}\alpha$$

Elements past Uranium

Neptunium ($Z = 93$) and plutonium ($Z = 94$) are easily obtained by the neutron irradiation of ^{238}U, followed by beta decay. The reactions are

$$^{238}_{92}\text{U} + {}^{1}_{0}\text{n} \rightarrow {}^{239}_{92}\text{U}$$

$$^{239}_{92}\text{U} \xrightarrow{t_{\frac{1}{2}} = 23.5 \text{ minutes}} {}^{239}_{93}\text{Np} + {}^{0}_{-1}\beta$$

$$^{239}_{93}\text{Np} \xrightarrow{t_{\frac{1}{2}} = 2.33 \text{ days}} {}^{239}_{94}\text{Pu} + {}^{0}_{-1}\beta$$

The half-life of ^{239}Pu is 24,360 years; consequently ^{239}Pu is stable enough that large quantities can be produced by this reaction.

Plutonium is the starting material for the formation of heavier elements. When it is irradiated in a very intense neutron beam, elements up to fermium ($Z = 100$) can be produced, as shown in Figure 23.10. The horizontal arrows represent neutron capture, and the vertical arrows β decay or electron capture. As one proceeds up the chain, however, the process becomes very inefficient. The neutrons are added one at a time, and each step produces fewer and fewer

We will see later that ^{239}Pu is a fissionable material that can be used as a nuclear reactor fuel. It is now the only transuranium element produced and available in large quantities.

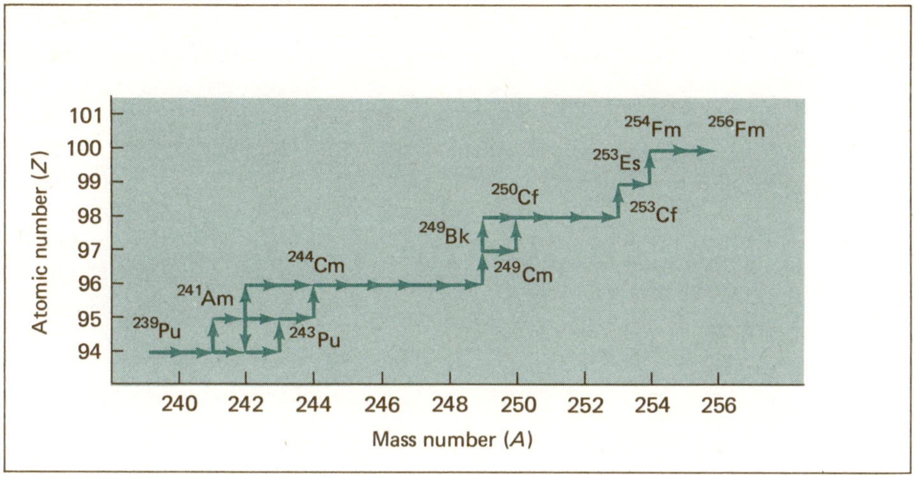

Figure 23.10
Nuclear reactions for production of important heavy nuclides by intense neutron bombardment of plutonium 239. The horizontal arrows result from neutron addition. The vertical arrows from β decay (upward) or electron capture (downward).

nuclei. Other methods that add more mass and charge in one reaction are more effective. For example, curium ($Z = 96$), berkelium ($Z = 97$), and californium ($Z = 98$) are produced by α-particle bombardment.

$$^{239}_{94}\text{Pu} + ^{4}_{2}\alpha \rightarrow ^{242}_{96}\text{Cm} + ^{1}_{0}\text{n}$$
$$^{241}_{95}\text{Am} + ^{4}_{2}\alpha \rightarrow ^{243}_{97}\text{Bk} + 2\ ^{1}_{0}\text{n}$$
$$^{242}_{96}\text{Cm} + ^{4}_{2}\alpha \rightarrow ^{245}_{98}\text{Cf} + ^{1}_{0}\text{n}$$

The elements einsteinium ($Z = 99$) and fermium ($Z = 100$) were first identified from the fallout debris of an atomic-bomb explosion. The bomb explosion produces a very intense neutron flux capable of adding the required neutrons. Larger quantities of these elements are now produced under the controlled conditions of the nuclear reactor.

It has not been possible to produce elements beyond $Z = 100$ by neutron capture since some of the isotopes in the intermediate steps have such short half-lives that insufficient quantities are produced to continue the reaction chain. Mendelevium ($Z = 101$) was first produced by bombarding einsteinium with α particles:

$$^{253}_{99}\text{Es} + ^{4}_{2}\alpha \rightarrow ^{256}_{101}\text{Md} + ^{1}_{0}\text{n}$$

Nobelium ($Z = 102$) and lawrencium ($Z = 103$) were produced by reaction with heavier particles:

$$^{246}_{96}\text{Cm} + ^{12}_{6}\text{C} \rightarrow ^{254}_{102}\text{No} + 4\ ^{1}_{0}\text{n}$$
$$^{252}_{98}\text{Cf} + ^{10}_{5}\text{B} \rightarrow ^{257}_{103}\text{Lw} + 5\ ^{1}_{0}\text{n}$$

These reactions using heavier reacting particles represent a significant advancement in the technology of nuclear reactions.

TABLE 23.2

Half-lives† of the Most Stable Isotopes of Some Actinides

Z	ISOTOPE	HALF-LIFE	Z	ISOTOPE	HALF-LIFE
89	^{227}Ac	22 y	97	^{247}Bk	1380 y
90	^{232}Th	1.4×10^{10} y	98	^{251}Cf	898 y
91	^{231}Pa	3.4×10^{4} y	99	^{254}Es	276 d
92	^{235}U	7.1×10^{8} y	100	^{257}Fm	100.5 d
92	^{238}U	4.5×10^{9} y	101	^{258}Md	55 d
93	^{237}Np	2.2×10^{6} y	102	^{259}No	58 min
94	^{244}Pu	8×10^{7} y	103	^{260}Lw	3 min
95	^{243}Am	7400 y	104	$^{261}[104]$	65 sec
96	^{248}Cm	4×10^{5} y	105	$^{262}[105]$	40 sec

† Abbreviations: y, years; d, days.

Elements 104 and 105 have been synthesized by the reactions

$$^{249}_{98}Cf + ^{12}_{6}C \rightarrow ^{257}[104] + 4\ ^{1}_{0}n$$
$$^{248}_{96}Cm + ^{18}_{8}O \rightarrow ^{261}[104] + 5\ ^{1}_{0}n$$
$$^{249}_{98}Cf + ^{15}_{7}N \rightarrow ^{260}[105] + 4\ ^{1}_{0}n$$

Predictions for the Synthesis of New Elements

Predictions have been made that isotopes in the region of $^{184}[114]$ and perhaps even in the region of $^{360}[164]$ may have reasonably long half-lives.

The half-lives of the heavy elements tend to decrease with increasing atomic number (Table 23.2), thus making synthesis and detection of the elements more difficult. It appears that nuclear forces are not great enough to hold together large numbers of protons and neutrons, and perhaps a limit is being reached. However, there are theoretical considerations that lead to predictions of heavier elements having "magic numbers" with their associated stability. It will be interesting to watch developments in this area.

23.5 Applications of Nuclear Chemistry

Let us now consider some very important applications of nuclear reactions.

Nuclear Energy

It is painful knowledge that the world population is rapidly consuming such fossil fuels as petroleum, natural gas, and coal. With the supplies of those fuels being depleted, it is natural that there is interest in **nuclear energy.** Very large quantities of energy are available from nuclear reactions. The fission of just one gram of ^{235}U produces about as much energy as does the combustion of a ton of coal. Put another way, the fission of ^{235}U releases about 2×10^{10} kjoules/mole while the heat of combustion of coke is about 4×10^{2} kjoules/mole.

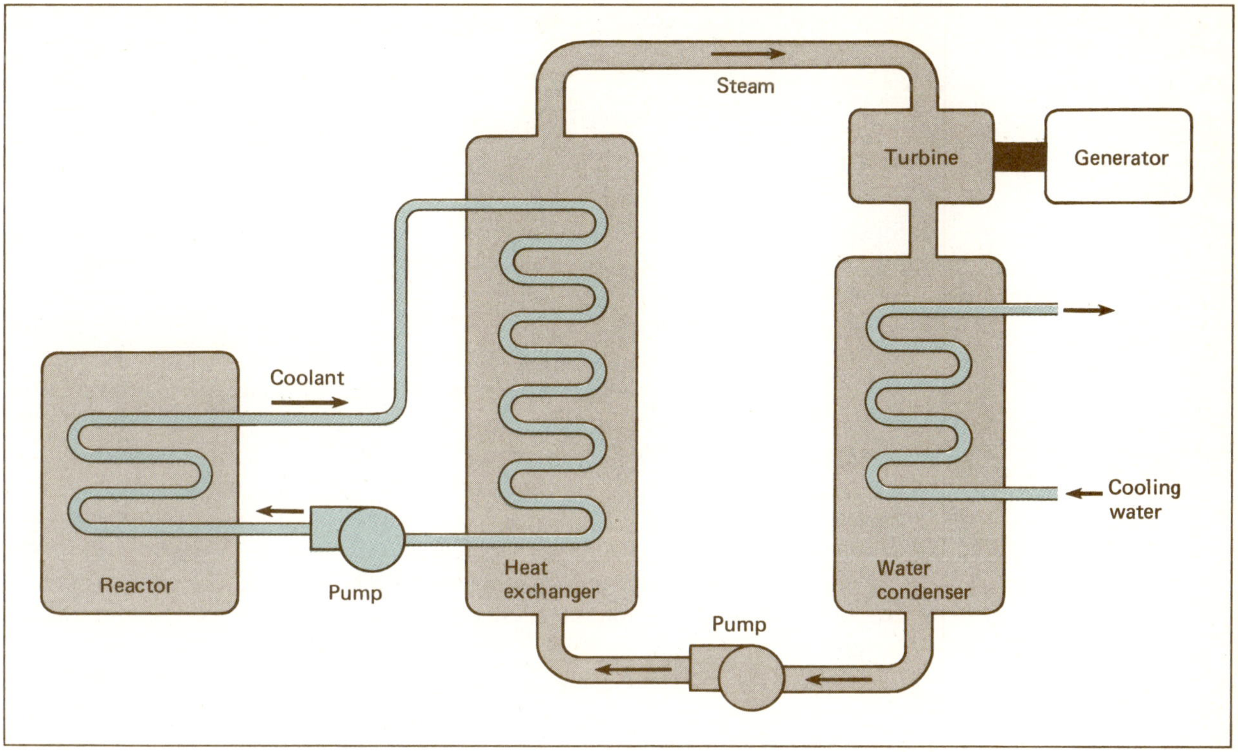

Figure 23.11
Diagram of a nuclear power plant. Heat from the reactor produces steam that drives the electric generator.

Fission

Electrical energy is obtained (Figure 23.11) from the **fission** reaction through a process in which heat from the reaction is used to produce steam, which drives the turbine of an electrical generator. Nuclear power plants are currently being used for the production of electrical power in many parts of the world. There are two very difficult problems associated with the use of nuclear fission as a source of electrical power. First, fission with slow neutrons depends upon the relatively rare isotope ^{235}U, which constitutes less than 1% of naturally occurring uranium. It has been estimated that if the fission of ^{235}U were used as a major energy source, the world supply would last only about 50 years. A second problem is the accumulation of radioactive wastes from the fission process. Some of the products have long half-lives and pose serious disposal or storage problems. The storage of dangerous substances is a difficult problem, and this becomes especially difficult if the storage must be safe for thousands of years.

The first problem can be solved by the development of the breeder reactor, which produces energy through a fission reaction and also produces more fissionable matter than is consumed. The reactor contains ^{238}U in addition to fissionable isotopes. Some of the plentiful neutrons are captured by the ^{238}U

You should refer back to Chapter 7 for a review of the operating properties of a thermodynamic engine. Remember that the maximum efficiency η for the conversion of heat into work is given by

$$\eta = \frac{T_2 - T_1}{T_2}$$

In the power plant, T_2 would be the temperature of the steam entering the turbine and T_1 the temperature at the condenser. The unavoidable discharge of heat into a river or other heat sink (by means of the cooling water) gives rise to thermal-pollution problems that are of concern.

with the following results:

$$^{238}_{92}U + ^{1}_{0}n \rightarrow ^{239}_{92}U$$

$$^{239}_{92}U \xrightarrow{t_{\frac{1}{2}} = 24 \text{ minutes}} ^{239}_{93}Np + ^{0}_{-1}\beta$$

$$^{239}_{93}Np \xrightarrow{t_{\frac{1}{2}} = 2.3 \text{ days}} ^{239}_{94}Pu + ^{0}_{-1}\beta$$

Since the $^{239}_{94}Pu$ is fissionable, there is an increase in the fissionable matter as long as ^{238}U is available; it is estimated that the world's energy needs could be met for thousands of years through the use of the breeder reactor.

The breeder reactor offers much help in the solution to the first problem mentioned above but seriously exacerbates the second. The plutonium produced in large quantity in the breeder reactor is very toxic; the maximum dose tolerated by the human body is less than 10^{-6} g. Since the half-life of ^{239}Pu is about 24,000 years, its containment and storage are critical matters.

Fusion

The problems associated with fission reactors have led to strong interest in the possibility of using controlled **fusion** reactions as an energy source. These reactions occur in an uncontrolled manner in a hydrogen or thermonuclear bomb:

$$^{2}_{1}H + ^{2}_{1}H \rightarrow ^{3}_{2}He + ^{1}_{0}n$$

$$^{2}_{1}H + ^{2}_{1}H \rightarrow ^{3}_{1}H + ^{1}_{1}H$$

$$^{2}_{1}H + ^{3}_{1}H \rightarrow ^{4}_{2}He + ^{1}_{0}n$$

Although the deuterium isotope makes up only 0.02% of naturally occurring hydrogen, this isotope is available in such large quantities in the waters of the oceans that the estimated world's energy needs could be met for billions of years through the use of the fusion reactions. There are neither toxic nor radioactive wastes, but instead an inert and useful gas is produced. The problem faced is to perform the reaction under controlled conditions. To fuse, the two nuclei must collide with enough energy to overcome the coulombic repulsion. A temperature of about $10^{8\circ}$ K is required, leading to enormous problems of containment of the gas. Much work is being done on this problem; from an optimistic point of view, it seems probable that significant progress will be made in the next few decades toward the use of this virtually inexhaustible energy source.

Tracer Studies

Tracer studies, or the use of chemical compounds containing radioactive isotopes at specific positions in the molecule, have proved very useful in following the course of chemical reactions. For example, when Cl^- containing some ^{38}Cl is oxidized by ClO_3^-, the reaction proceeds as follows:

$$2\,Cl^{*-} + 2\,ClO_3^- + 4\,H^+ \rightarrow Cl_2^* + 2\,ClO_2 + 2\,H_2O$$

where the * indicates the ^{38}Cl isotope. The results show clearly that Cl^- is oxidized to Cl_2 and that there is not some oxidation of Cl^- to ClO_2. Such information is most helpful in determining the mechanism of the reaction.

If a radioactive isotope is not available, a stable isotope can sometimes be used as a tracer. In such an instance, mass spectrometry can be used to determine the location of the unusual isotope. A classic example of this approach is in the formation of ethyl acetate. The reaction of ethanol (C_2H_5OH) with acetic acid (CH_3COOH) to form ethyl acetate by the elimination of water could possibly occur by either of the two following mechanisms:

$$CH_3C(=O)\text{—}\boxed{O^*\text{—}H + H}\,O\text{—}C_2H_5 \rightarrow CH_3C(=O)\text{—}O\text{—}C_2H_5 + H_2O^*$$

or

$$CH_3C(=O)\text{—}O^*\text{—}\boxed{H + HO}\text{—}C_2H_5 \rightarrow CH_3C(=O)\text{—}O^*\text{—}C_2H_5 + H_2O$$

When the starred oxygen in the acetic acid was labeled with ^{18}O, the ^{18}O appeared in the water, indicating that the first mechanism is correct.

Radioisotopes in Analytical Chemistry

Radioisotopes find applications in chemical analysis procedures. In **activation analysis,** the sample is irradiated with a flux of neutrons. The neutrons add to stable nuclei in the sample and produce radioactive nuclei. By determining the type, energy, and amount of radioactivity given off by the activated sample, one can determine which elements are present and in what amounts.

The procedure is very sensitive for the analysis of trace amounts of many of the metallic elements. It is possible to detect as little as 10^{-10} grams of copper, sodium, or tungsten by this method; many other elements can be detected with only slightly lower sensitivity.

Isotopic dilution is a technique used to analyze for a compound that can be isolated in a pure state but only with poor yield. A known amount of the substance of interest containing an active isotope is added to the unknown and thoroughly mixed with it. A sample of the pure substance is then isolated from the mixture, and its activity determined. A simple calculation then gives the quantity of the substance in the original material.

Activation analysis has another advantage in that it does not destroy the sample. After a relatively short time period, the short-lived isotopes produced by the activation decay, and since not enough were originally produced to affect the physical appearance of the sample, it is essentially returned to its original condition. The method has been used successfully to determine the authenticity of works of art claimed to have been painted by the old masters. Analysis of the trace metals present in the paint often enables the expert to determine when and where the paint was formulated.

EXAMPLE 23.4 It is possible to determine the amount of the amino acid glycine in a mixture with other amino acids. A 0.500 g portion of a sample of glycine containing ^{14}C with an activity of 25,000 counts per minute per gram is mixed with the unknown. A 0.200 g sample of pure glycine with an activity of 105 counts per minute is isolated from the mixture. Calculate the weight of glycine in the original sample.

SOLUTION First, we will derive an equation that will apply to this and other similar problems. Let W equal the weight of glycine in the original mixture,

w_1, the weight of radioactive glycine added, and w_2, the weight of glycine isolated. Let A be the activity (counts per minute) of the radioactive glycine before mixing, and B, the activity of the final isolated glycine. We can think of A as the activity of the total sample of weight $W + w_1$, and B as the activity of the final isolated glycine (of weight w_2). Therefore, B must be the fractional part of A that w_2 is of $W + w_1$:

$$B = \frac{w_2}{W + w_1}(A)$$

Solving for W gives

$$W = w_2\left(\frac{A}{B}\right) - w_1$$

In our problem

$$w_1 = 0.500 \text{ g}$$

$$A = \left(25{,}000 \frac{\text{cpm}}{\text{g}}\right)(0.500 \text{ g}) = 12{,}500 \text{ cpm}$$

$$w_2 = 0.200 \text{ g}$$

$$B = 105 \text{ cpm}$$

Hence,

$$W = (0.200 \text{ g})\left(\frac{12{,}500 \text{ cpm}}{105 \text{ cpm}}\right) - 0.500 \text{ g}$$

$$= 23.3 \text{ g}$$

The ^{40}K-^{40}Ar dating technique was used to date the age of moon rocks obtained by the Apollo 11 astronauts as 3.5 billion years. This age is considerably greater than expected and tends to discount the theory that the moon was torn from the outer surface of the earth after the earth was formed.

The oldest rocks found on the earth are from ocean-floor basalt that has been pushed up into islands in the mid-Atlantic. Their age, 4.5 billion years, as dated by the ^{238}U-^{206}Pb ratio, gives a lower limit for the age of the earth.

Radiometric Dating

An interesting application of radioisotopes is that of **radiometric dating** to determine the age of objects. For example, $^{40}_{19}K$ is used by geologists to determine the age of rocks. This unstable isotope with a half-life of 1.30×10^9 years is present to a small percent in naturally occurring potassium. It decays by two mechanisms. Eighty-nine percent forms ^{40}Ca by β decay, and 11 percent forms ^{40}Ar by electron capture. The ^{40}Ca content is not useful for dating purposes since it is a common constituent of many rocks. However, ^{40}Ar is a nonreactive gas. If one makes the reasonable assumption that a molten-rock sample contains no ^{40}Ar when it solidifies, then a measurement of the ^{40}Ar trapped in the rocks from ^{40}K plus measurement of the ^{40}K remaining can be used to calculate the time when the rock solidified.

The ^{238}U-^{206}Pb ratio has also been used to date geological samples. Since ^{238}U decays with a half-life of 4.49×10^9 years and ^{206}Pb is the eventual product of the sequence, a measurement of the ^{238}U and of the ^{206}Pb enables one to calculate the age of the rock. (This calculation must take into account the fraction of ^{206}Pb that is in natural lead when no uranium is present.)

One of the most useful dating techniques uses ^{14}C to date the time of death of plants or animals. Reactions in the upper atmosphere with cosmic neutrons maintain a constant supply of $^{14}_{6}C$ by the reaction

$$^{14}_{7}N + ^{1}_{0}n \rightarrow ^{14}_{6}C + ^{1}_{1}H$$

This radioactive carbon, which decays by β emission with a half-life of 5570 years, is assimilated by plants through photosynthesis of CO_2, and from there it is carried into all organisms. As long as the organism is alive, it maintains a stationary level of $^{14}_{6}C$ radioactivity of 15.3 ± 0.1 disintegrations per minute per gram of carbon. When life ceases, the assimilation of $^{14}_{6}C$ stops, and the radioactivity decreases. By measuring the present radioactivity level of the dead object, one can calculate its time of death. Example 23.5 illustrates such a calculation.

EXAMPLE 23.5 Calculate the age of the wood from a coffin of the Ptolemaic period of Egypt. The ^{14}C activity in this sample has been measured as 11.7 disintegrations per minute per gram of carbon.

SOLUTION From equation [23.4],

$$\ln \frac{N}{N_0} = -kt$$

where N_0 is the number of carbon atoms at $t = 0$, and N is the number at time t. However, the number of ^{14}C atoms is proportional to the disintegration rate d (this proportionality is true for a first-order reaction). Hence,

$$\ln \frac{d}{d_0} = -kt$$

or

$$t = \frac{1}{k} \ln \frac{d_0}{d}$$

In this problem, $k = \ln 2/t_{1/2}$, $d_0 = 15.3$ cpm, and $d = 11.7$ cpm. Putting these numbers into the above equation gives

$$t = \frac{5570 \text{ years}}{\ln 2} \ln \frac{15.3 \text{ cpm}}{11.7 \text{ cpm}}$$

$$t = 2160 \text{ years old}$$

The use of ^{14}C dating has been extended to objects nearly 16,000 years old, although the low level of activity in these very old samples makes the measurements difficult. For many objects, alternative methods of dating permit a check on the accuracy of the ^{14}C technique. For example, historical records, tree rings, and dated coins buried with the object have provided data for comparison. The ^{14}C method gives excellent agreement with the known dates.

Summary

The nuclear radius is only about 0.1% of the atomic radius. The nuclear particles are held together by very strong forces of such short range that there is apparently little attraction among nucleons except for nearest neighbors. The binding energy per nucleon reaches a maximum around $A = 60$ (where A is the mass number). Since this binding energy per nucleon is lower at both smaller and larger mass numbers, there is energy released in either the fusion of light nuclei or the fission of heavy nuclei.

Nearly 60% of the 284 stable nuclei have even numbers of both protons and neutrons; only 8 have odd numbers of both. This observation, coupled with the appearance of particularly stable nuclei having "magic" numbers of nucleons, strongly suggests a shell model for the nucleus analogous to that for the electrons of an atom.

Natural radioactivity takes place by β^- decay, positron emission, *K*-electron capture, α decay, spontaneous fission, and γ emission. A nucleus lying outside the stability belt will undergo one of these processes and thus be moved closer to or into the stability belt. The spontaneous decay of a radioactive nucleus is a first-order process, and its rate is independent of temperature and pressure.

High-energy accelerators provide projectiles for causing nuclear reactions. Isotopes that are not known to occur naturally and elements above $Z = 92$ have been prepared by such reactions.

Both fission and fusion reactions offer potential as sources of energy; a fusion reaction, if controlled, would have the great advantage of creating no radioactive waste. Radioisotopes are especially useful to chemistry in tracer studies for following reactions and mechanisms, in analytical techniques, and in dating objects.

Important Terms and Concepts

Nuclear binding energy
Nuclear decay
Nuclear stability
Neutron/proton ratio
Radioactivity
β^- emission
β^+ emission
K-electron capture
α emission
Spontaneous fission
γ emission and energetics of radioactive decay
Radioactive decay rate
Half-life
Nuclear reactions
Nuclear reactor
Linear accelerator
Cyclotron
Synthesis of new elements
Nuclear energy
Fission
Fusion
Tracer studies
Activation analysis
Isotopic dilution
Radiometric dating

Exercises

23-1.
The diameter of the helium atom is 2.65 Å. Calculate the diameter of the $^{4}_{2}He$ nucleus, and calculate the percentage of the total volume of the helium atom occupied by the nucleus.

23-2.
Write equations for the following examples of radioactive decay:

(a) Positron emission by $^{61}_{29}Cu$
(b) β^- emission from $^{32}_{15}P$
(c) α emission from $^{244}_{98}Cf$

23-3.
The successive steps in the artificial decay chain for $^{210}_{86}Rn$ involve α decay, K capture, and β^+ decay. Show the chain, including intermediate isotopes.

23-4.
Complete the following nuclear reactions:

(a) $^{41}_{20}Ca + ^{0}_{-1}e \rightarrow$
(b) $^{36}_{17}Cl \rightarrow ^{0}_{-1}\beta +$
(c) $^{52}_{25}Mn \rightarrow ^{0}_{+1}\beta +$
(d) $^{174}_{72}Hf \rightarrow ^{4}_{2}\alpha +$
(e) $^{237}_{94}Pu \rightarrow ^{96}_{40}Zr + \qquad + 3\,^{1}_{0}n$

23-5.
Predict the likely modes of decay for the following unstable nuclei, and write the decay reactions.

(a) $^{3}_{1}H$ (c) $^{12}_{7}N$ (e) $^{227}_{92}U$
(b) $^{6}_{4}Be$ (d) $^{144}_{55}Cs$ (f) $^{19}_{8}O$

23-6.
Write equations for the following nuclear reactions. In each case, the parentheses give the particle absorbed and the particle emitted.

(a) $^{35}_{17}Cl$ (neutron, proton)
(b) $^{130}_{52}Te$ ($^{2}_{1}H$, 2 neutrons)
(c) $^{237}_{93}Np$ (α, neutron)
(d) $^{238}_{92}U$ ($^{22}_{10}Ne$, 4 neutrons)
(e) $^{82}_{35}Br$ (neutron, γ)
(f) $^{43}_{20}Ca$ (α, proton)

23-7.
How much of a 0.100 g sample of $^{7}_{4}Be$ remains after 30 days if $t_{1/2} = 53$ days?

23-8.
How long will it take for 75% of a sample of $^{233}_{87}Fr$ to disappear? ($t_{1/2} = 22$ minutes)

23-9.
The unstable isotope $^{18}_{9}F$ decays as follows:

$$^{18}_{9}F \rightarrow ^{18}_{8}O + ^{0}_{+1}\beta$$

The rate is such that 10.0% of the original quantity of $^{18}_{9}F$ remains after 369 minutes.

(a) What is the decay constant for $^{18}_{9}F$?
(b) What is the half-life of $^{18}_{9}F$?

23-10.
Radium emits α particles with energies of 4.795 MeV and 4.611 MeV. What is the wavelength of the γ radiation accompanying this decay?

23-11.
In the K-electron capture of $^{51}_{24}Cr$, a γ-ray with an energy of 0.75 MeV is released. If the mass of an atom of the decay product $^{51}_{23}V$ is 50.95953 amu, what is the mass of a $^{51}_{24}Cr$ atom?

23-12.
Given the following atomic masses, calculate the energy consumed or evolved in the decay

$$^{192}_{78}Pt \rightarrow ^{188}_{76}Os + ^{4}_{2}He$$

ISOTOPE	MASS (amu)
$^{192}_{78}Pt$	191.9614
$^{188}_{76}Os$	187.9560
$^{4}_{2}He$	4.00387

23-13.
A sample of wood from an Egyptian mummy case gives a $^{14}_{6}C$ activity of 9.4 counts/minute-g of carbon. How old is the mummy case?

23-14.
The carbon from the center of a giant sequoia tree gives 10.8 $^{14}_{6}C$ counts/minute-gram of carbon. Wood from the outer layer where growth occurs gives 15.3 counts/minute-gram of carbon. How old is the tree?

Problems

23-1.

Given the data below for chromium isotopes, calculate ΔH^0, ΔS^0, and ΔG^0 for the following nuclear reaction at 25° C:

$$3\ {}^{52}_{24}\text{Cr} = {}^{50}_{24}\text{Cr} + 2\ {}^{53}_{24}\text{Cr}$$

You may assume $\Delta H^0 = \Delta E^0$ for this process.

ISOTOPE	PERCENT ABUNDANCE	ATOMIC WEIGHT	NUCLEAR ENTROPY (joules/°K-mole)
${}^{50}_{24}\text{Cr}$	4.31	49.9461	0
${}^{52}_{24}\text{Cr}$	83.76	51.9405	0
${}^{53}_{24}\text{Cr}$	9.55	52.9407	11.3
${}^{54}_{24}\text{Cr}$	2.38	53.9389	0

23-2.

The naturally occuring radioisotope ${}^{232}_{90}\text{Th}$ decays through a series of steps until it ends as stable ${}^{208}_{82}\text{Pb}$. The particles successively emitted in one route are α, β, β, α, α, α, β, α, β, α. Write the decay reaction for each step of the decay chain.

In a sample of uraninite ore taken from the Black Hills of South Dakota, the weight of the lead was 22.8% of the weight of the uranium present. Use the data to estimate the minimum age of the ore sample and, hence, the minimum age for the earth.

23-4.

The antibiotic aureomycin can be produced by fermentation. A 1.00 mg portion of aureomycin containing ${}^{14}\text{C}$ (specific activity = 150 cpm/mg of aureomycin) was added to 1.00 kg of a fermentation broth that was known to contain aureomycin. From the mixture, 0.18 mg of crystalline aureomycin was isolated that had a net activity of 360 counts in 60 min. Calculate the weight of aureomycin originally present in the 1.00 kg of broth.

23-5.

A man weighing 71 kg was injected with 5.09 ml of water containing tritium (the radioactive isotope of hydrogen, ${}^{3}_{1}\text{H}$) which had an activity of 9×10^9 counts/minute. Three hours later, when the tritrated water had equilibrated with the body water of the patient, a 1 ml sample of plasma water was removed. It had an activity of 1.8×10^5 counts/minute. Estimate the weight percent of water in the human body.

24

Introduction to the Chemistry of Carbon Compounds and the Chemistry of Living Systems

The vital-force theory was convincingly disproved in 1828 when Fredrick Wöhler demonstrated that urea, a distinctly organic compound, could be prepared from ammonium cyanate, a compound that is not associated with living systems. The conversion takes place when cyanate is gently heated:

$$NH_4CNO = (NH_2)_2CO$$

The chemistry of carbon compounds is so extensive and so different from the chemistry of the other elements that the compounds of carbon are singled out for special attention in the study of chemistry. The chemistry of the compounds of carbon is called **organic chemistry.** The name was coined in the late eighteenth century when it was thought that a special *life force* was necessary for the synthesis of compounds found in living systems. The name has remained, even though the *vital-force* theory was soon discarded. The chemistry that is concerned with reactions in *living systems* we now call **biochemistry.**

24.1 Tetrahedral Carbon Atom

We have emphasized the structure of molecules several places in the text previously, but nowhere does structure become so important as in the chemistry of the carbon compounds. The rapid development of this extensive field can be traced directly to the organic chemists' early deductions about structure. Their ability to write structural formulas that conveyed the arrangement of atoms in molecules, followed by postulates concerning the distribution of electronic charge in molecules, is the basic reason for the rapid significant advances of organic chemistry.

Later in the chapter, when we discuss biochemical reactions, we see that the *shapes* of molecules play an all-important role in the chemistry of living systems.

The first and probably largest contribution toward the elucidation of the structure of organic compounds was the concept of the tetrahedral arrangement of bonds about the carbon atom. It is now well established that CH_4 is a tetrahedral molecule. We know that the bond angles are the 109°28′ tetrahedral value, and we even account for the structure in terms of bonding theories. Long before the advent of our modern experimental methods and theories, however, the Dutch chemist Jacobus van't Hoff deduced that the carbon atom in the methane molecule lay in the center of a tetrahedron. This significant work (1874) marked the beginning of structural organic chemistry.

The basis for van't Hoff's deduction was the experimental fact that no isomers existed for either of the compounds CH_3Cl (methyl chloride) or CH_2Cl_2 (methylene chloride). These compounds can be considered to be derived from

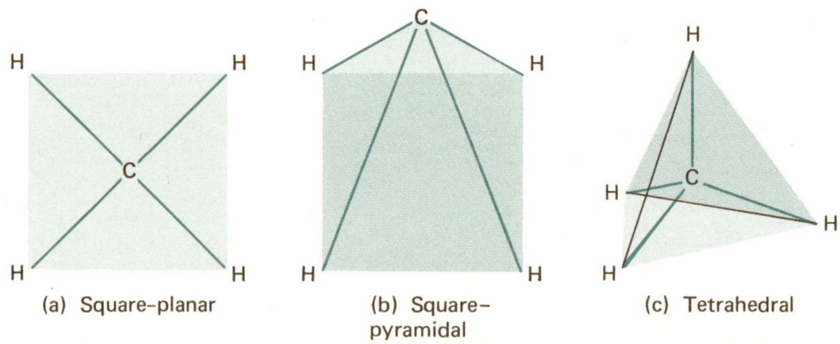

Figure 24.1 *Possible structures considered by van't Hoff for the methane molecule.* The four C-H bonds in each structure would be equivalent.

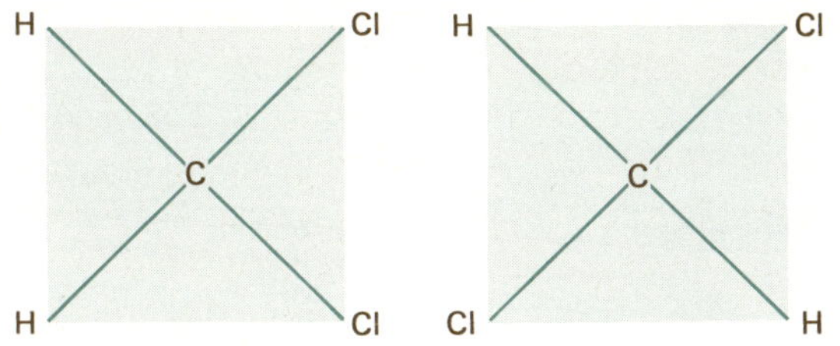

Figure 24.2
Geometrical isomers would be possible if CH_2Cl_2 were a square-planar structure. Failure to find isomers in CH_2Cl_2 showed the square-planar structure to be incorrect.

CH_4 by replacing hydrogen atoms in methane by chlorine atoms. Regardless of the process used to prepare CH_3Cl from CH_4, only one compound of the formula CH_3Cl could be produced. This suggested to van't Hoff that all C-H bonds in CH_4 were equivalent, and hence it made no difference which H atom was replaced by the Cl atom. Any one of three simple geometric structures would account for that observation: (a) a square-planar molecule, (b) a square-pyramidal molecule, or (c) a tetrahedral molecule (see Figure 24.1).

The experimental fact that only one type of CH_2Cl_2 molecule exists immediately eliminated the first two structures, for if the square-planar configuration were true, for example, one should be able to have the *cis* and *trans* geometrical isomers shown in Figure 24.2. Similar reasoning eliminated the (b) structure in Figure 24.1, leaving the tetrahedral arrangement of bonds shown in (c) as the structure that met the requirements that (1) all four C-H bonds are equivalent in CH_4, and (2) that the arrangement of the bonds around the C atom are such that the H-C-H bond angles are also equivalent. When this second condition is met, there can be no "across from" (*trans*) or "next to" (*cis*) positions. Thus we account for the *single* compound with formula CH_2Cl_2.

Refer to Chapter 20 if you need to refresh your memory on the subject of geometrical isomers.

The importance of recognizing the tetrahedral structure of the C atom in CH_4 cannot be overemphasized. This discovery provided the chemist with the important key to the understanding of the structure of a large number of complex organic compounds, and knowing the structures is the first step in devising a method of synthesis.

24.2 Hydrocarbons

We made reference previously to the large number of organic compounds. Over two-million carbon compounds have been identified and described. By way of comparison, all noncarbon compounds amount to fewer than 10 percent of that number. The total carbon compounds that are possible is an astronomical number, as we will shortly see.

Carbon-to-Carbon Bonds

One reason for the profusion of carbon compounds is the ability that carbon has to bond to itself to form stable chain compounds. For example, the compounds ethane, propane, butane, pentane, and hexane have 2, 3, 4, 5, and 6 carbon atoms, respectively, bonded together as shown in Figure 24.3.

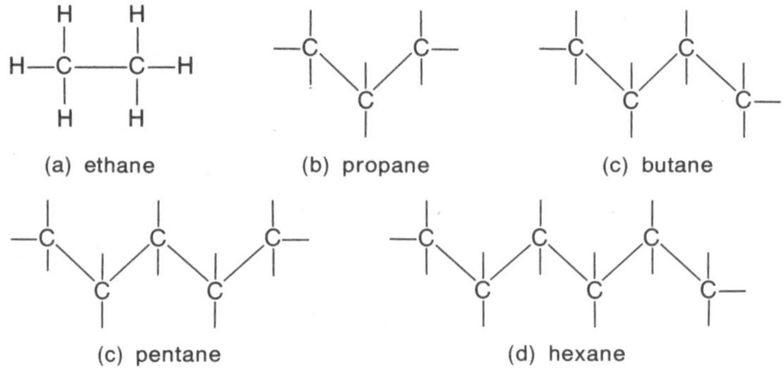

Figure 24.3 *Models illustrating carbon-to-carbon linkages.* The zigzag pattern of the carbon chains is a consequence of the tetrahedral pattern of bonds around the carbon atom. The H atoms have been omitted in (b)–(d) for clarity.

The equivalent of the stable chains of carbon-to-carbon bonds that form the backbone of such organic molecules as those shown in Figure 24.3 is not found in other elements. Several of the other elements will bond two atoms of the same element together in a molecule, such as the oxygens in a peroxide ion, $:\ddot{\underset{..}{O}}-\ddot{\underset{..}{O}}:^{2-}$, for example, or the nitrogens in hydrazine, $\begin{matrix}H\\H\end{matrix}\!\!>\underset{..}{N}-\underset{..}{N}<\!\!\begin{matrix}H\\H\end{matrix}$. Except for C, however, *stable* compounds with more than two atoms of the same element bonded to each other are rare. Silicon hydrides come as near as any class of compounds do to being analogous to the **hydrocarbon** chains. Silicon hydrides of as many as six silicons have been prepared in the laboratory, but unlike the hydrocarbons, the silicon hydrides are exceedingly unstable. Hydrocarbons, on the other hand, occur in natural gas and petroleum with formulas that range from one carbon (CH_4) to chains of as many as 40 and more carbon atoms.

The silicon hydrides should not be confused with the very stable silicones, which are also chain compounds but compounds in which the silicon atoms are not bonded directly to each other but through an oxygen atom. An example of a silicone chain is

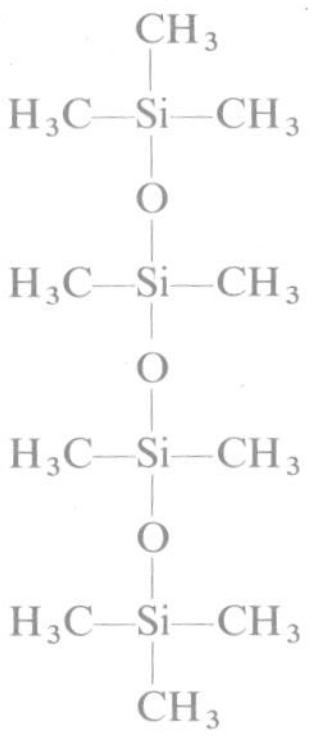

The silicones are useful lubricants. They have the desirable property of undergoing only a very slight change in viscosity when subjected to rather large changes in temperature.

Alkanes

Table 24.1 lists some of the low-molecular-weight, straight-chain, single-bonded hydrocarbons found in natural gas and petroleum, the chief sources of these compounds. Note that the empirical formulas of this class of compounds are

TABLE 24.1

Names and Boiling Points of the Lower-Molecular-Weight, Straight-Chain Alkanes

FORMULA	NAME	BOILING POINT (°C)
CH_4	methane	−162
CH_3CH_3	ethane	−89
$CH_3CH_2CH_3$	propane	−42
$CH_3(CH_2)_2CH_3$	*n*-butane	0
$CH_3(CH_2)_3CH_3$	*n*-pentane	36
$CH_3(CH_2)_4CH_3$	*n*-hexane	69
$CH_3(CH_2)_5CH_3$	*n*-heptane	98
$CH_3(CH_2)_6CH_3$	*n*-octane	126

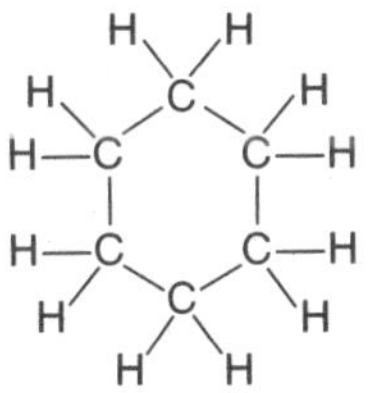

Figure 24.4
Cyclohexane, an example of a closed-ring compound.

Figure 24.5
C-C-C bond angles of 120° would be needed to produce a planar-hexagonal molecule.

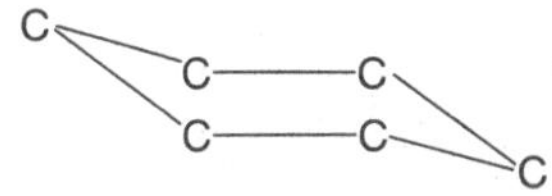

(a) chair

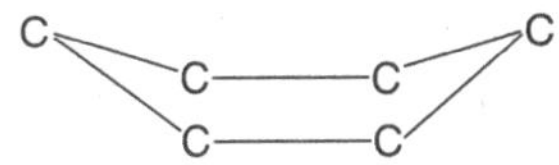

(b) boat

Figure 24.6
Boat *and* chair *conformations of C_6H_{12}.* The cyclohexane molecule is being viewed from its side. Puckering of a molecule can be in either a boat or chair conformation. Puckering is necessary to accommodate 109°28′ tetrahedral C-C-C bond angles.

given by the general formula C_nH_{2n+2}, where n is the number of carbon atoms in the chain. As a group, these are referred to as the **alkanes,** the common "ane" ending denoting single bonds between the carbon atoms. The increase in boiling points with increased molecular weight is a reflection of the increased van der Waals attractions between molecules as the total number of electrons in the molecules increases.

Cyclic Alkanes

In addition to open-chain hydrocarbons, ring compounds, or **cyclic alkanes,** are possible; for example, C_6H_{12} is the formula of cyclohexane (Figure 24.4).

To form a planar molecule, the C-C-C bond angles in a six-membered ring would need to be 120° (see Figure 24.5). Hence, puckering of the ring is necessary to accommodate the 109°28′ tetrahedral angle. The molecule can assume two equally strain-free conformations, as shown in Figure 24.6. Since no one has been able to isolate two isomers of cyclohexane, the two forms (boat and chair) are thought to be readily interconvertible.

Double and Triple Bonds Between Carbon Atoms

In addition to singly bonded carbon-to-carbon compounds, there are series of hydrocarbons containing double and triple carbon-to-carbon bonds. The simplest member of the **double-bonded hydrocarbons** is ethylene, shown in Figure 24.7. Refer to Chapter 14 to review the bonding in ethylene. It is a planar molecule, and there is no free rotation about the double bond. Hence, geometrical isomerism is possible with some of its derivatives. For example, $C_2H_2Cl_2$ can take the structures shown in Figure 24.8. The *cis* and *trans* forms are distinct compounds with different physical and chemical properties.

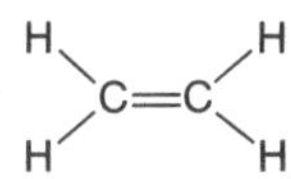

Figure 24.7
Ethylene, an example of a double-bonded carbon compound. The ethylene molecule is planar.

The first member of the **triple-bonded hydrocarbons** is acetylene, a linear molecule, shown in Figure 24.9. Refer to Chapter 14 to review the bonding in this compound. One *sigma* and two *pi* bonds are involved in the carbon-to-carbon linkage.

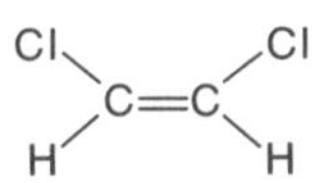

(a) *cis*-dichloroethylene
Boiling point: 60° C
Melting point: −80° C

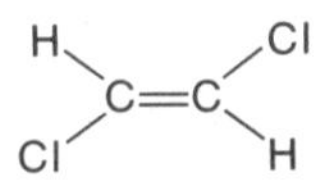

(b) *trans*-dichloroethylene
Boiling point: 48° C
Melting point: −50° C

Figure 24.8
Geometrical isomers of $C_2H_2Cl_2$.

Figure 24.9
Acetylene, an example of a triple-bonded carbon compound.

$H—C\equiv C—H$

Unsaturated liquid vegetable fats are often hydrogenated (addition of hydrogen to unsaturated bonds) to form solid saturated fats. The latter have more-desirable cooking characteristics since they are more stable at high temperatures. Such commercial products as Crisco and Snowdrift are examples of saturated fats produced from partially unsaturated (some double bonds) vegetable oils.

Compounds containing double- and triple-bonded carbon-to-carbon linkages are collectively referred to as *unsaturated compounds*. The most typical reaction of this class of compounds is *addition* to the unsaturated bond. The following are examples of *addition reactions*.

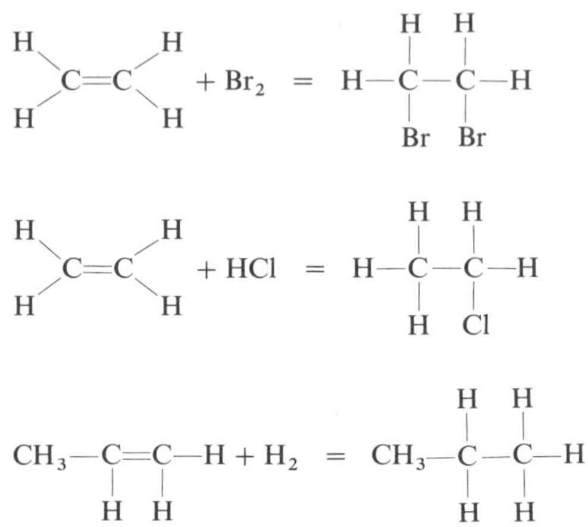

(Correct conditions of temperature, pressure, and catalyst must be observed.)

Nomenclature of organic compounds in its full detail is quite complex. Problems 24-7 to 24-9 at the end of the chapter give the rules as adopted by the International Union of Pure and Applied Chemistry (IUPAC) for naming simple organic compounds. For example, the IUPAC names for the five isomeric hexanes are *n*-hexane, 2-methylpentane, 3-methylpentane, 2,2-dimethylbutane, and 2,3-dimethylbutane. Other names are often used for some very common compounds. For example, 2-methylpropane is commonly referred to as isobutane, and 2-propanol as isopropyl alcohol.

Structural Isomerism

There are several reasons for the profusion of carbon compounds. The stability of carbon-to-carbon bonds, which makes possible chain compounds of many carbon atoms, has already been mentioned. The possibilities for **structural isomerism** in these chains is a second reason for the large number of organic compounds. There is only one structural arrangement each that is possible for CH_4, C_2H_6, and C_3H_8 molecules. Starting with C_4H_{10}, however, several possibilities are available. For example, one can write two compounds with the formula C_4H_{10} (see Figure 24.10). Both are found in nature.

Note that in *n*-butane there are two CH_3 groups (one at each end of the molecule) and two CH_2 groups, while in isobutane there are three CH_3 groups and one CH group. We often write structural formulas by an abbreviated formula of the type $CH_3CH_2CH_2CH_3$ for *n*-butane, and $CH(CH_3)_3$ for isobutane.

There are three possible structures for C_5H_{12} and five for C_6H_{14} (see Figure 24.11). As the number of carbon atoms in the molecule becomes large, the number of possible isomers becomes astronomical. For $C_{40}H_{82}$ there are 62,491,178,805,831 different structures that are possible!

(a) normal butane n-C_4H_{10}

(b) isobutane i-C_4H_{10}

Figure 24.10
No attempt to show correct bond angles has been made in the drawings above. These drawings are intended to show as clearly as possible the number and types of bonds in the isomers.

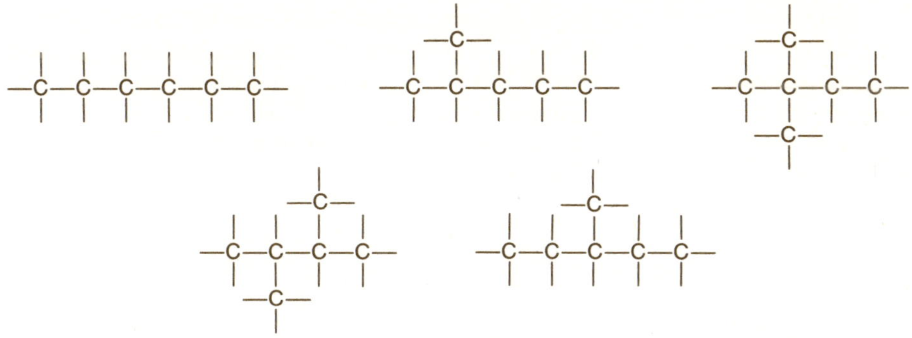

Figure 24.11
The carbon skeletons for the five structural isomers of C_6H_{14}.

24.3 Functional Groups

One way of classifying organic compounds is to consider them as derivatives of hydrocarbons in which one or more hydrogens from the hydrocarbon are replaced by another element or radical. The substituent group may, for example, be a hydrocarbon radical such as the methyl ($\cdot CH_3$) or ethyl ($\cdot CH_2CH_3$) radicals, or it may be a hydroxy ($\cdot OH$), amino ($\cdot NH_2$), or a number of other radicals. The chemistry of the compound is generally well determined by these substituent groups; hence, they are referred to as **functional groups.**

Alcohols

When the functional group is the $\cdot OH$ radical, we have the class of compounds called **alcohols.** If R is allowed to stand for any aliphatic hydrocarbon radical—methyl ($\cdot CH_3$), ethyl ($\cdot CH_2CH_3$), or propyl ($\cdot CH_2CH_2CH_3$), for example—then the general formula of an alcohol is ROH. Table 24.2 lists the names of the monohydroxy alcohols for carbon chains of one to eight carbon atoms. The names of the eight alkyl groups (methyl, ethyl, and so forth) are put in bold face for emphasis; they appear in the names of the other classes of compounds we will consider.

The drug-store variety of "rubbing alcohol" is usually isopropyl alcohol,

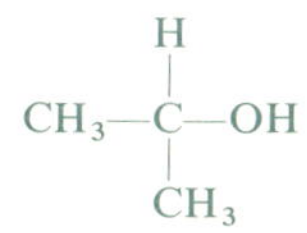

Ethylene glycol, used extensively as an antifreeze for automobile radiators, is a dihydroxy alcohol,

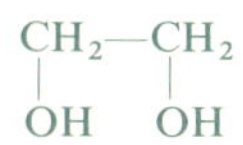

Glycerol, the starting material for nitroglycerin, is a trihydroxy alcohol,

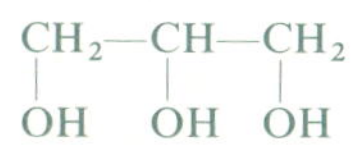

TABLE 24.2

Monohydroxy, Straight-Chain Alcohols Containing One to Eight Carbon Atoms

ROH	COMMON NAME
CH_3OH	**methyl** alcohol
CH_3CH_2OH	**ethyl** alcohol
$CH_3(CH_2)_2OH$	*n*-**propyl** alcohol
$CH_3(CH_2)_3OH$	*n*-**butyl** alcohol
$CH_3(CH_2)_4OH$	*n*-**pentyl** alcohol
$CH_3(CH_2)_5OH$	*n*-**hexyl** alcohol
$CH_3(CH_2)_6OH$	*n*-**heptyl** alcohol
$CH_3(CH_2)_7OH$	*n*-**octyl** alcohol

Methyl alcohol has many industrial applications, including its use as a solvent and starting material for the synthesis of a host of other chemicals. At one time it was obtained from the destructive distillation of wood (heating wood in the absence of air and condensing the resulting vapors). Although its industrial preparation now is from CO and H_2 (under pressure, at an elevated temperature, and in the presence of a catalyst), the name *wood alcohol* is still often used. Methyl alcohol is quite poisonous; when taken internally it sometimes produces permanent blindness.

Ethyl alcohol is obtained from the fermentation of carbohydrates. It is the intoxicant in beers, wines, and hard liquors. Other applications include its use as a solvent in varnishes and lacquers, its use in the synthesis and purification of other organic chemicals, and its use as an antiseptic.

Physical and Chemical Properties of Alcohols

In spite of the OH group in their formulas, alcohols show none of the characteristics of an Arrhenius base. In the extremely small amount of ionization that does take place—less than that of the ionization of water—it is a proton, not a hydroxide ion, that separates from the molecule. Alcohols can be thought of as derivatives of water in which a hydrogen has been replaced by an alkyl radical, and hence alcohols show some of the physical and chemical properties of water. For example, intermolecular hydrogen bonding occurs in alcohols (see Figure 24.12) as it does in water, as evidenced by their boiling points. Methyl alcohol, for example, with a molecular weight of only 32, boils at 65° C. By way of comparison, the boiling point of ethane (molecular weight 30) is −89° C.

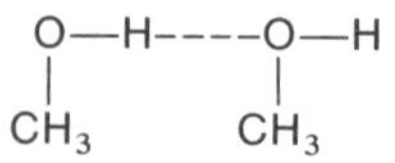

Figure 24.12
Hydrogen bonding in methyl alcohol.

Alcohols react with the very active metals, such as sodium, to release H_2 gas in a manner similar to the reactions of active metals with water:

$$2\,CH_3CH_2OH + 2\,Na(s) = \underset{\text{sodium ethoxide}}{2\,CH_3CH_2ONa} + H_2(g)$$

Note that it is the hydrogen bonded through the oxygen that is the reactive hydrogen. Those hydrogens bonded to the carbon atoms are relatively inert.

Probably the most typical reactions of an alcohol is its reaction with an organic acid. This type of reaction is discussed in the section on esters.

Amines

If the functional group is the NH_2 radical, the compound belongs to a class of compounds called **amines,** whose general formula is RNH_2. Here again, R stands for a hydrocarbon radical. The compound CH_3NH_2 is called methylamine; $CH_3CH_2NH_2$, ethylamine; and so on. The amines are the bases in organic chemistry. They react with acids to form organic salts. Just as alcohols can be considered to be derivatives of water, the amines can be thought of as derivatives of NH_3 (the R radical taking the place of a hydrogen atom). The reactions of amines with acids thus parallel that of NH_3:

$$NH_3 + H^+ = NH_4^+$$

$$CH_3NH_2 + H^+ = \underset{\text{methylammonium ion}}{CH_3NH_3^+}$$

Organic Acids

If the substituent of a hydrocarbon is the carboxyl group (·COOH), the result is an **organic acid.** The structure of the carboxyl group is

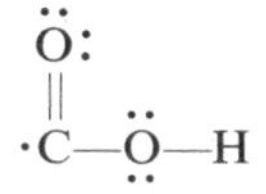

Both oxygens are bonded directly to the carbon, and the hydrogen (as in an alcohol) is bonded to the carbon through an oxygen. This is the acidic hydrogen—the one that ionizes—in the organic acid. The general formula for the organic acids is RCOOH. In this class of compounds, R can be either a hydrogen atom or a hydrocarbon radical. The simplest organic acid is, therefore, H-COOH, formic acid. A list of several common organic acids and their acid dissociation constants are in Table 24.3.

Formic acid, the simplest of the carboxylic acids is found in nature in pine needles, perspiration, and ants.

An early and interesting, but crude, method for preparing formic acid consisted of distillation of red ants.

Acetic acid is the acid of vinegar. It can be prepared by the fermentation of apple juice. An earlier step in the process produces ethyl alcohol, which, if the fermentation is allowed to continue, is oxidized in the next step to acetic acid.

Butyric acid is notable for its unpleasant odor. The odor of rancid butter comes from the formation of butyric acid.

Oxalic acid was included in Table 24.3 as an example of a dicarboxylic acid. A series of such acids exists with carboxyl groups on either end of the molecule and a hydrocarbon chain in between.

The carboxylic acids are weak acids as can be seen in Table 24.3. Some of the strongest are the halogen-substituted acids shown in Table 24.4.

TABLE 24.3

Acid Dissociation Constants for Some Organic Acids

FORMULA	NAME	K_a
HCOOH	formic	1.2×10^{-4}
CH_3COOH	acetic	1.8×10^{-5}
CH_3CH_2COOH	propionic	1.3×10^{-5}
$CH_3CH_2CH_2COOH$	butyric	1.5×10^{-5}
COOH–COOH	oxalic	5.9×10^{-2} 6.4×10^{-5}
COOH–CH_2–COOH	malonic	1.5×10^{-3} 2.0×10^{-6}

TABLE 24.4

Acidities of Halogen-Substituted Carboxylic Acids

ORGANIC ACID	K_a
CH_3COOH	1.8×10^{-5}
$CClH_2COOH$	1.4×10^{-3}
CCl_2HCOOH	5.5×10^{-2}
CCl_3COOH	2.3×10^{-1}
$CH_3CH_2CH_2COOH$	1.5×10^{-5}
$CH_3CH_2CHClCOOH$	1.4×10^{-3}
$CH_3CHClCH_2COOH$	8.9×10^{-5}
$ClCH_2CH_2CH_2COOH$	3.0×10^{-5}
CH_3COOH	1.8×10^{-5}
FCH_2COOH	2.6×10^{-3}
$ClCH_2COOH$	1.4×10^{-3}
$BrCH_2COOH$	1.3×10^{-3}
ICH_2COOH	6.7×10^{-4}

The increased acidity of the chloroacetic acids over acetic acid can be explained by the inductive effect of the electronegative chlorine groups, as illustrated in the formula

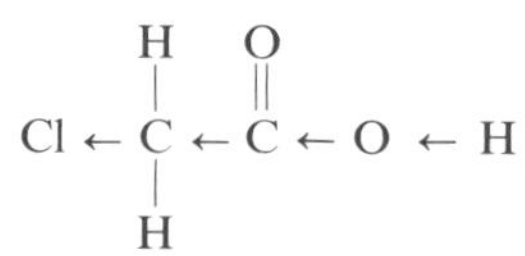

The arrowheads on the bonds show the direction toward which the electron cloud of the molecule is drawn because of the attraction of the chlorine atom for electrons. This allows the acidic hydrogen atom to be more easily removed as a proton. The acidity is seen to increase markedly as additional chlorine atoms are substituted for hydrogens. Trichloroacetic acid with K_a of the order of 10^{-1} approaches the acidity of the HSO_4^- ion.

The second group of acids in Table 24.4 shows the interesting effect of inserting saturated hydrocarbon groups between the chlorines and the carboxyl group. The insulating effect of even one CH_2 group is evident. When two CH_2 groups intervene, the acidity is seen to fall back close to that of the unsubstituted butyric acid.

The third group of acids illustrates how the acidity of the acid varies with the electronegativity of the halogen substituent. As one would expect, the more electronegative the substituent, the greater the inductive effect, and the greater the dissociation of the proton.

Esters

Compounds of the general formula RCOOR, where the two R's can be either the same or different hydrocarbon radicals, are called **esters.** Ethyl acetate,

$$CH_3{-}\overset{\overset{\displaystyle O}{\|}}{C}{-}O{-}C_2H_5$$

is an example. The low-molecular-weight esters such as this one have pleasant, sweet-smelling odors; they are responsible for the fragrant odors of flowers and fruits. Ethyl acetate, for example, is one of the esters that contributes to the odor of apples, isoamyl acetate is known commercially as *banana oil*, and the odor of oranges comes from the ester octyl acetate.

Esters are synthesized in the laboratory from alcohols and organic acids. Ethyl acetate can be made by warming a mixture of ethyl alcohol and acetic acid, with H_2SO_4 added as a catalyst. The reaction is

$$CH_3{-}\overset{\overset{\displaystyle O}{\|}}{C}\boxed{{-}O{-}H + H{-}}O{-}C_2H_5 = CH_3{-}\overset{\overset{\displaystyle O}{\|}}{C}{-}O{-}C_2H_5 + H_2O$$

(esterification)

The process is general between any alcohol and any acid and is referred to as *esterification.* It can be made to go either way. Heating in the presence of a strong base will cause an ester to hydrolyze to an alcohol and water:

$$CH_3COOC_2H_5 + H_2O = C_2H_5OH + CH_3COOH \qquad \text{(hydrolysis)}$$

The acid formed would immediately be neutralized to the organic salt by the NaOH present:

$$CH_3COOH + OH^- = CH_3COO^- + H_2O$$

Animal fat is a naturally occurring ester of importance and interest. This is an ester of the trihydroxy alcohol glycerol, $\underset{\displaystyle OH}{\underset{|}{CH_2}}{-}\underset{\displaystyle OH}{\underset{|}{CH}}{-}\underset{\displaystyle OH}{\underset{|}{CH_2}}$, and long-chain organic acids of which stearic acid, $C_{17}H_{35}COOH$, is an example. Soaps, which are sodium and potassium salts of such long-chain organic acids, are manufactured by hydrolysis of fats. The reaction, using glyceryl tristearate as an example, is shown below:

$$\begin{array}{l} C_{17}H_{35}COOCH_2 \\ \qquad\qquad\quad | \\ C_{17}H_{35}COOCH \\ \qquad\qquad\quad | \\ C_{17}C_{35}COOCH_2 \end{array} + 3\,NaOH = 3\,C_{17}H_{35}COONa + \begin{array}{l} CH_2OH \\ \ | \\ CHOH \\ \ | \\ CH_2OH \end{array}$$

glyceryl tristearate, a fat — sodium stearate, a soap — glycerol

Pioneer women used wood ashes for their source of alkalinity in making soap from animal fat.

Aldehydes and Ketones

Two other common classes of compounds are the **aldehydes** $R{-}\overset{\overset{\displaystyle O}{\|}}{C}{-}H$ and the **ketones** $R{-}\overset{\overset{\displaystyle O}{\|}}{C}{-}R$. Both contain carbonyl ($C{=}O$) groups. Commonly encountered examples of aldehydes are formaldehyde ($HCHO$), which at one time was used as an embalming fluid and which is now a starting material for making an important class of resins, and acetaldehyde (CH_3CHO). The most commonly encountered ketone is acetone (CH_3COCH_3), a much-used organic solvent. Its odor is familiar to most people because of its extensive use in quick-drying glues and fingernail polish.

Ethers

The **ethers** (ROR) are the last classification that we mention. Dimethyl ether (CH_3OCH_3), the simplest of the ethers, and ethyl alcohol (CH_3CH_2OH) are structural isomers. The two have distinctly different physical properties, such as boiling points, for example: $-24.9°$ C for the ether and $+78.5°$ C for the alcohol. Intermolecular hydrogen bonding in the alcohol makes the difference. The two isomers also differ widely in physiological and chemical properties.

Several simple chemical tests could be made to see which of the two compounds should be assigned the formula of an alcohol and which the ether. Probably the simplest and most direct method would be to see which compound reacted with Na. We mentioned earlier that a hydrogen atom connected to a carbon through an oxygen was active enough to be reduced by Na metal to H_2 gas, while those H atoms connected to C are inert in this reaction. Hence, the compound that was an alcohol would give off H_2 when treated with Na; the ether would not react.

Benzene and Its Derivatives

When coal is heated to 1100–1300° C in the absence of air, the products are a distillate of a black viscous substance called *coal tar* and a residue of coke (mostly carbon). The coke is used in the operation of blast furnaces; the coal tar contains several very useful chemicals. One of the most important of these is **benzene,** C_6H_6. We have discussed its bonding in Chapter 14. You may recall that it is a hexagonal, planar molecule with C-C linkages that are intermediate in length and bond energy between a single and double bond. Its most significant feature is the delocalized pi electron system. The structural formula generally used to describe these characteristics is 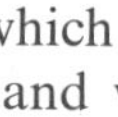.

Two of the many other valuable hydrocarbons that are extracted from coal tar are toluene, $C_6H_5CH_3$, and naphthalene, $C_{10}H_8$. Their structures are shown in Figure 24.13. Each of these is valuable as a starting material for the preparation of dyes, explosives, and medicines. The reactivity of the aromatic ring makes possible a variety of chemical reactions. Probably the most typical is the *substitution* reaction, in which one or more hydrogen atoms are replaced

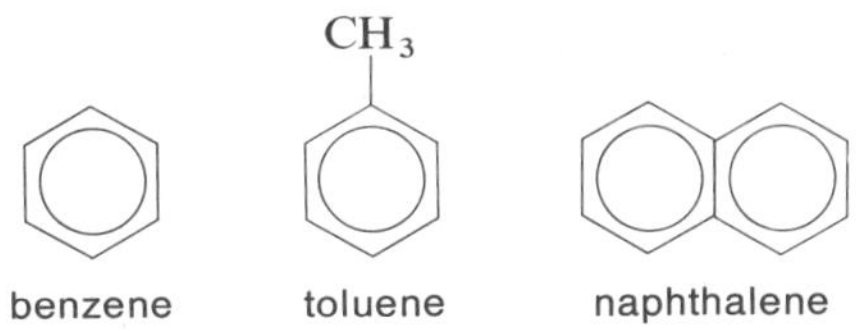

Figure 24.13
Three valuable hydrocarbons found in coal tar.

by another element or group. The reactions shown below with a halogen are examples:

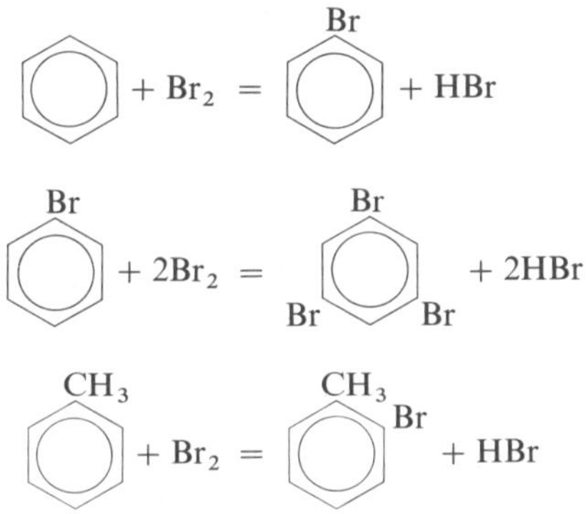

24.4 Organic Chemicals, Unlimited

In this section we briefly describe a wide spectrum of organic molecules, some that are naturally occurring and some that are synthesized only in the laboratory. Our purpose is to illustrate the diversity, the complexity, the value, and the beauty of what has been called by some "the architecture of carbon compounds." It is not expected that you be able to reproduce the formulas and structures of this section, but only that you arrive at a general impression of the vast field of organic structures.

Aspirin

The controlled transformation of one compound to another that has specific desired properties is a great triumph of organic chemistry. Large numbers of products common to developed societies are the direct result of such transformations. We now outline a process by which a readily available starting compound is converted to a substance of wide medicinal use. There is impressive reasoning involved in the development of the conditions under which the desired reactions will occur and also in the recognition of precisely what substance has been produced, separated, and purified at each step in the process.

While the production of aspirin is hardly one of the large industrial processes of the chemical industry, it is impressive to note that about 38 tons of aspirin are consumed in the United States each day.

Shown below is the series of reactions by which benzene, obtained from coal tar, can be converted to aspirin. This is just one example of a wide variety of syntheses by which valuable substances can be made.

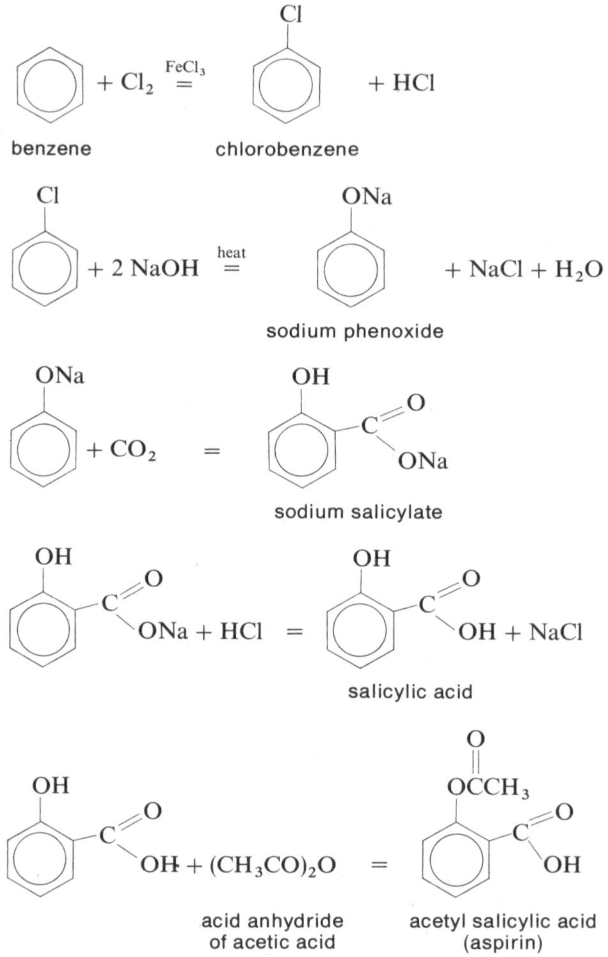

Penicillin

Few developments in the past half-century have had such a profound effect upon the medical care of mankind as has the use of antibiotics. Of these, the penicillins are of great importance. The natural penicillins found in molds, including penicillin G, are now well known as antibiotics.

Penicillin G is a complex molecule having the structure

In the structural formulas for penicillin G and ampicillin, we have followed a common practice of organic chemists and have not shown the carbon atoms at the intersections of the bonds in the rings. Hydrogen atoms are also present where needed to give the C atom its four bonds.

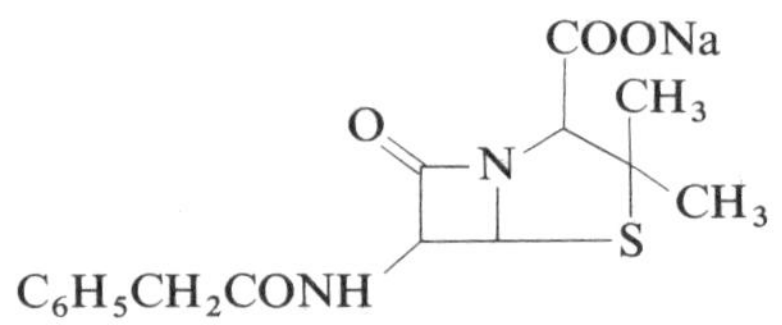

Penicillin G

The isolation and purification of this substance, which is produced in a penicillium (mold) culture, was a complex and laborious process. Following the purification of the natural substance, organic chemists undertook the determination of its structure and then its synthesis. Having accomplished that amazing task, they then undertook the work of synthesizing similar compounds that might have greater antibacterial action. A result of this effort is ampicillin, similar to penicillin G but differing from it by an added NH_2 group and having properties considered to be more valuable in the treatment of infections.

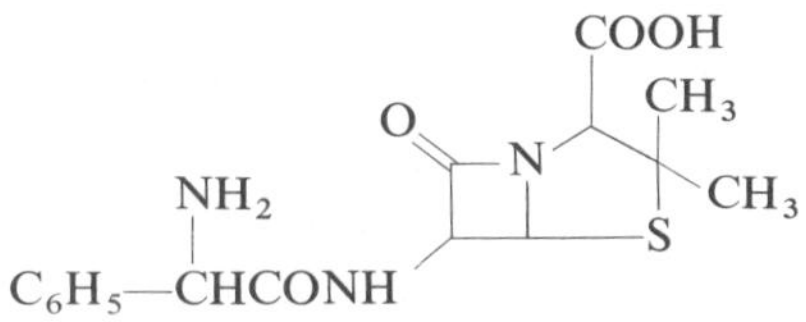

Ampicillin

Such a cycle has been repeated many times: A complex natural product having medicinal value is isolated and identified through a laborious effort of experimental work and deductive reasoning; then synthesis of the compound is undertaken; and finally similar compounds are synthesized and tested for their usefulness.

Sulfanilamide

The sulfa drugs, of which sulfanilamide was one of the earliest, have been used since 1935 to combat several infectious diseases. Prior to the discovery of the antibiotics, the sulfa drugs were the principal chemical treatment for infections.

The mechanism by which the drug is thought to operate is interesting. Instead of destroying bacteria directly, it acts to prevent their multiplication. Sulfanilamide is similar in shape to PABA (para-aminobenzoic acid), an amino acid that is essential for growth of the bacteria. It is thought that ingestion of the sulfa drug by the bacteria is by the same mechanism that it uses to absorb PABA. *Ab*sorption of either chemical is preceded by *ad*sorption of the molecule on the surface of the bacteria. This takes place only when the shape of the surfaces of the molecule and the cell wall of the bacteria are complementary; that is, they fit together in a "lock and key" combination. With the shapes of PABA and sulfanilamide so similar, the bacteria are unable to distinguish between the two. Sulfanilamide adsorbed on the surface blocks the adsorption of the needed PABA and enters the cell. The sulfanilamide, however, does not perform the same physiological function as does PABA, and the cells die out without reproducing. The similarity in shape can be seen from their structural formulas:

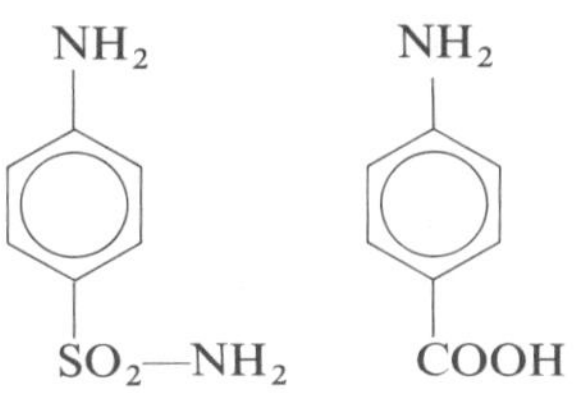

Sulfanilamide PABA

Vitamin A

Vitamin A is a solid alcohol with the structure

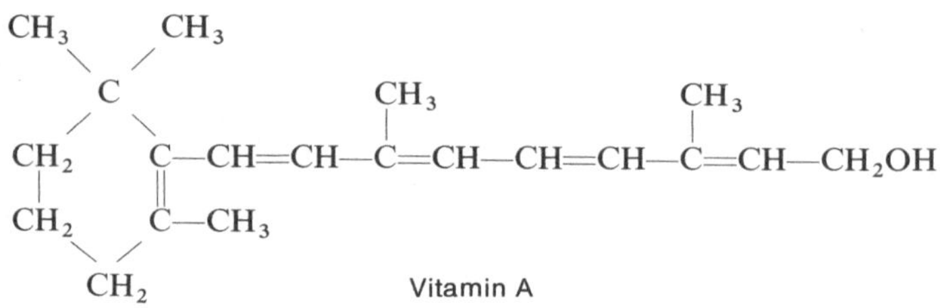

Vitamin A

Alternating single and double bonds in a molecule, such as we have in this compound, generally absorb light in the visible range of the spectrum, resulting in color. Vitamin A is yellow.

When one component of white light is absorbed by an object, the color of the reflected light (the color of the object) is the complementary color of the absorbed light.

The long hydrocarbon chain makes vitamin A insoluble in water but soluble in fat. Hence, it can be retained in the body. This is both good and bad. It is good because it makes daily intake of the vitamin unnecessary. However, since vitamin A in high concentration becomes toxic (as is true of almost anything in high concentration), one should not take massive supplemental amounts of vitamin A. The same is true of vitamin D, which is also fat soluble. On the other hand, the B vitamins and vitamin C are water soluble. The excess quantities of these vitamins are excreted, and daily intake is necessary. The structures of the common vitamins are all well established, and these compounds are synthesized commercially.

Vitamin K

Vitamin K is the blood-clotting vitamin. Animals fed on diets lacking in this vitamin hemorrhage easily because of the inability of their blood to clot. Vitamin K was first isolated from alfalfa; it is also present in spinach. Its structural formula is

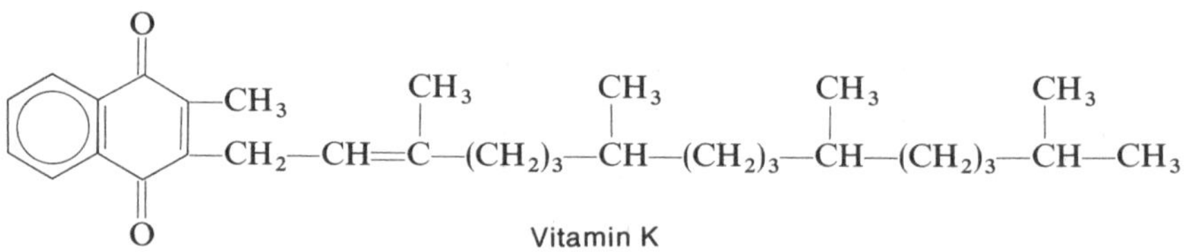

Vitamin K

The effective part of this molecule is apparently not the long hydrocarbon chain, since replacement of the chain with a hydrogen atom produces a molecule that is three times more effective as a blood-clotting agent than the naturally occurring vitamin K.

Quinine

Centuries ago the natives of South America discovered that malaria could be effectively treated by chewing the bark of a certain kind of tree. The early

Jesuit missionaries to South America relayed this information to Europe, and the bark became the standard treatment for malaria for several centuries. In 1827, quinine was isolated from the bark and shown to be the active ingredient. Its structure is

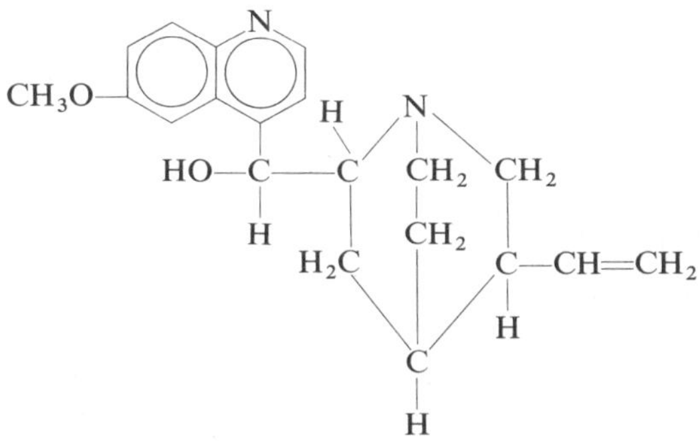

Quinine

Although the trees from which quinine was extracted were discovered in South America, they were in recent years cultivated principally in Java and the East Indies. During World War II, this source of quinine was cut off from the Western world. Organic chemists, however, responded to the challenge and synthesized several antimalarial drugs. The complete synthesis of quinine was accomplished in 1944.

Adrenalin

As an example of a hormone, we consider adrenaline, the first hormone to be obtained in the form of a pure compound. Its structural formula is

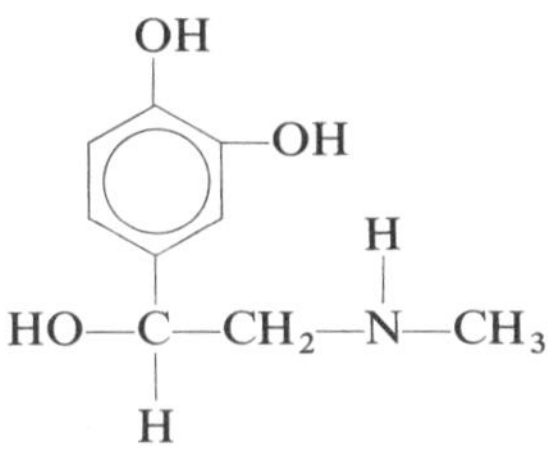

Adrenaline

Adrenaline is secreted by small, ductless glands on each side of the kidneys, the adrenal glands. The secretions are stimulated by the emotions of anger and fear, and it is therefore appropriately known as the "fight or flight" hormone. Adrenaline has several medicinal uses, including its use as a heart stimulant in cases of shock.

TNT

Trinitrotoluene is the explosive in dynamite. It is prepared by the nitration of toluene, a constituent of coal tar, by treatment with nitric acid. Its structure,

as indicated by its name, is

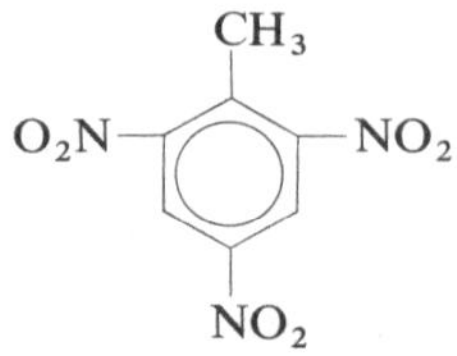

Trinitrotoluene (TNT)

Caffeine

Caffeine, found in coffee and tea, belongs to a class of drugs called stimulants. In small doses (2 to 3 grains), caffeine increases the frequency and depth of respiration and blood circulation. In higher concentrations it acts as a depressant. Theobromine, found in cocoa, has similar physiological behavior. This is not surprising when you compare similarities of the structures of caffeine and theobromine:

The amount of caffeine in an average cup of coffee (or tea) is roughly 2 grains.

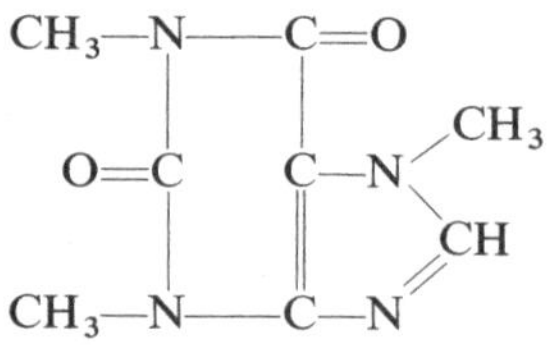

Caffeine

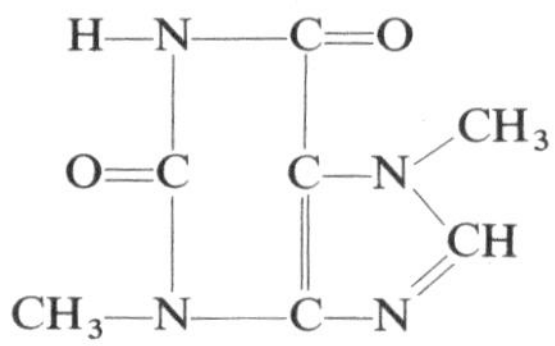

Theobromine

Polycyclic Aromatic Compounds

A large number of hydrocarbons exist that can be thought of as fused benzene rings. Three examples are shown below:

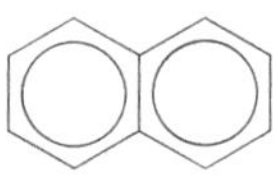

Naphthalene

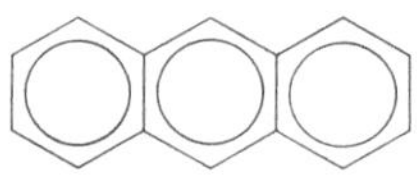

Anthracene

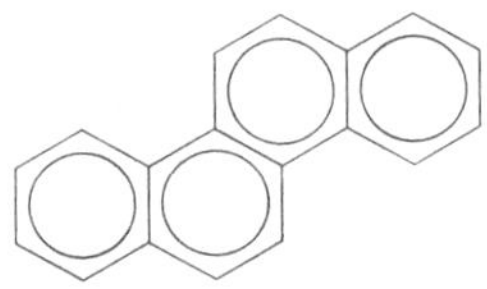

Chrysene

These can be extracted from coal tar. Some of the polycyclic compounds in coal tar (and cigarette tars) are carcinogenic. Two examples are

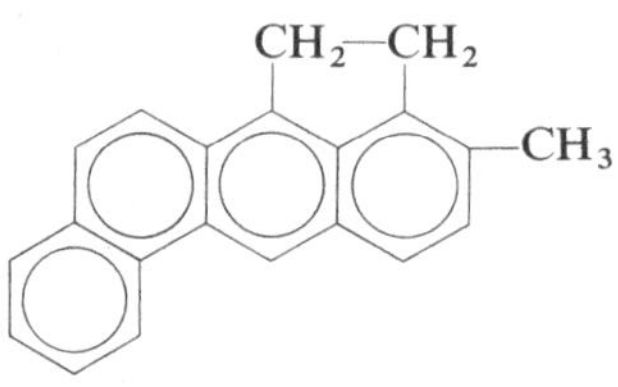

Methylcholanthrene

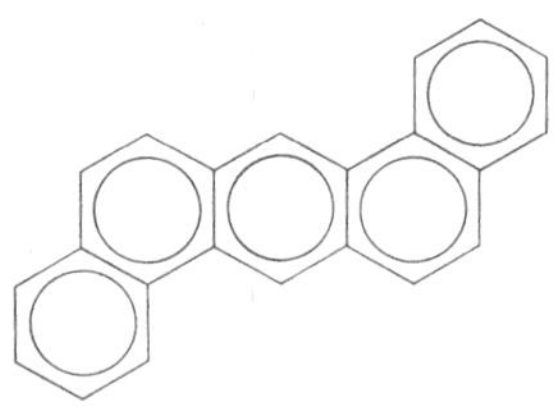

Dibenzanthracene

24.5 Some Compounds and Reactions of Biological Interest

Amino Acids

Amino acids are a group of compounds that are vital to life. They are the "building blocks" of protein molecules, the substance of which such important parts of the body as blood, muscle, brain, and nervous system are composed.

As the name implies, amino acids contain two functional groups. An amino acid is both an amine and an organic acid. The simplest amino acid is glycine:

$$\begin{array}{l} CH_2\text{—}COOH \\ | \\ NH_2 \end{array}$$

The second member of the series is alanine:

$$\begin{array}{l} CH_3CH\text{—}COOH \\ \quad\;\;\; | \\ \quad\;\; NH_2 \end{array}$$

Less vital, but also principally composed of protein, are hair and finger and toe nails.

These and the other amino acids of importance to living systems have the amino substituent on the carbon next to the carboxyl group (the α carbon) and are thus referred to as α-amino acids.

Altogether there are twenty-three α-amino acids that are known to be important constituents of the protein of the human body. Not all of these need to be present in the food we eat, since the body is capable of synthesizing all but nine of them. These nine, known as the *essential amino acids*, must be present in the food we eat. That means that they must be constituents of the protein part of our food.

The isomerism of the α-amino acids is highly interesting. All these molecules, except for the simplest, glycine, have an asymmetric carbon atom, that is, a carbon atom that is bonded to four *different* groups. This is the condition necessary for a type of isomerism called **optical isomerism.** This concept is most easily grasped by using the left and right hands as examples. Each has the same structure (four fingers and one thumb), but it is obvious that they are not identical, because one hand cannot be superimposed on the other. Try it. Lay one hand on top of the other. If the fingers on both hands are pointed in the same direction, the thumbs point in opposite directions to each other. If one hand is rotated till the thumbs line up, the fingers do not. If the palms of the hands are brought together facing each other, then the fingers and thumbs can be aligned in the same way that one hand and its image in a mirror can be matched. The left and right hands are thus mirror images of each other.

In the same way, there can be two alanine molecules each with the same structure; that is, each with a tetrahedral structure and the same four bonds

The general formula for an α-amino acid is

$$\begin{array}{c} H \\ | \\ R\text{—}C\text{—}COOH \\ | \\ NH_2 \end{array}$$

where R represents that part of the molecule that makes one amino acid different from another. A few representative R groups and the names of the corresponding amino acids are given below:

AMINO ACID	R
glycine	H—
alanine	CH_3—
valine	$(CH_3)_2CH$— (drawn as CH_3 and CH_3 joined to CH—)
lysine	$H_2NCH_2(CH_2)_2CH_2$—
aspartic acid	$HO\text{—}\overset{\displaystyle O}{\overset{\|}{C}}\text{—}CH_2$—
phenylalanine	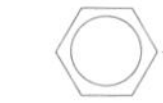—CH_2—
cysteine	HS—CH_2—

Figure 24.14
Optical isomers of glycine. One is the mirror image of the other. The central carbon atom is not shown. The lines outline the tetrahedral structure; they are not bonds.

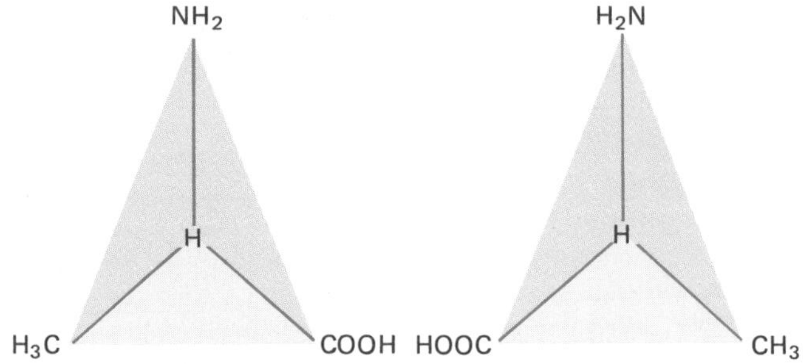

One can postulate (not much else) on why nature prefers the *L* to the *D* configuration. As we will see shortly, the structure of protein is such that either *L*- or *D*-amino acid, but not a mixture, could have been used in its synthesis. It has been suggested that the first living organism started with the *L* configuration by chance, and that all succeeding forms of life inherited the *L* configuration from the original organism.

Whatever the explanation for the phenomenon, we can be grateful that we have only one amino acid configuration natural to this planet. It is easily demonstrated that *D*-amino acids (synthesized in the laboratory) cannot be used by living organisms. Hence, if two types of proteins existed (one constructed from *D*- and one from *L*-amino acids) considerable confusion could result. One could not readily distinguish between them. After all, a steer and its mirror image look very much alike, and yet a beef steak from one would provide the ingredients to form human protein, and the other would be useless.

From what we know now, it is in the realm of the possible that life based on the *D*-amino acid configuration could exist on another planet. If we were to make contact with each other, some interesting situations could result. Although we may look alike, we could not eat each other's foods. We could intermarry but not produce progeny.

(C-H, C-CH_3, C-NH_2, and C-COOH), but with four different groups (H, CH_3, NH_2, and COOH) arranged in such a way that the two molecules are not superimposable but are mirror images, as shown in Figure 24.14.

A particular spatial arrangement of the four groups around the carbon atom of the α-amino acids has been designated *L* (levo), and its mirror image as *D* (dextro). It is a most interesting fact that only the *L*-configuration amino acids are found in nature. *D* and *L* configurations are equally easily synthesized in the laboratory. In fact, unless special techniques are employed, a laboratory synthesis of an α-amino acid always results in an equal mixture of the *L* and *D* forms. The protein in nature, however, both plant and animal, is composed of *L*-amino acids exclusively.

Proteins

Several times in the last section we referred to amino acids as the principal "building blocks" of the **proteins.** Early evidence for this was obtained by Emil Fischer (1852–1919), whose work in the early twentieth century showed the amino acids in protein to be combined in long polypeptide chains. The **peptide linkage** that ties amino acids together in protein is shown below.

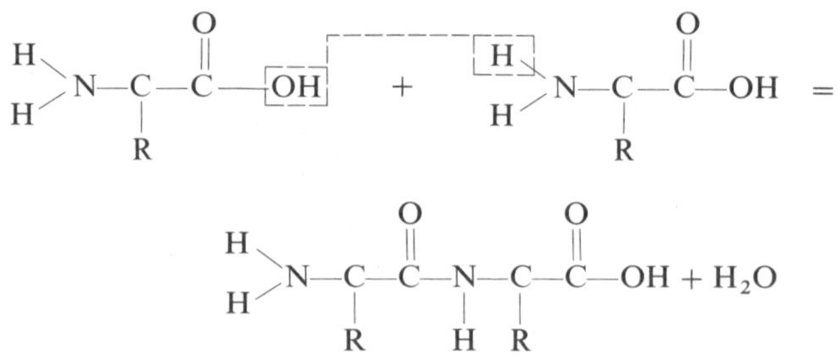

The process shown above for two molecules can be repeated over and over again on both ends of the resulting structure to give a polypeptide chain.

One protein can differ from another in (a) the number of amino acid residues in a chain, (b) the order in which the residues are connected, and (c) the way in which the chains are folded. The protein lysozyme, for example, has 125 amino acid units linked together in one polypeptide chain. The insulin protein molecule, on the other hand, consists of four polypeptide chains, two of which have 21 amino acid residues each and the other two have 30 each. It is readily apparent that the number of possible structures for proteins is very large.

Lysozyme is found in the white of eggs and also—interestingly—in tears. It has the useful property of being able to destroy bacteria.

There is another important feature in the structure of proteins beyond that of the number, kind, and arrangement of the amino acid units. The size and shape of protein molecules indicate that the polypeptide chains are not stretched out but are tightly folded units. It is calculated that the lysozyme molecule, for example, would be 450 Å long if the polypeptide chain were stretched out. As it is, the molecule is globular in shape with a diameter of only about 25 Å. The way in which polypeptide chains are folded has been sorted out only fairly recently. Skillful interpretation of x-ray diffraction measurements show some proteins to be in pleated sheets, but more common is the right-handed helical structure known as the **α helix.** Hydrogen bonding between electronegative elements holds the chain in its coiled structure. For example, the helix for α-keratin (hair, horn, and nails) has 3.6 amino residues per turn. This is just right to provide room for the side chains and to allow all possible hydrogen bonds to form. To fit into a helix, all the amino acid residues must be of the same configuration. Models show that the *L*-configuration is required for a right-handed helix.

Nucleic Acids

As with organic chemistry, the subject of biochemistry is so extensive that we cannot possibly survey the whole field in this one chapter. Recent advances in the subject of **nucleic acids,** however, is of such importance that we must include some, even though limited, discussion of this topic.

Nucleic acids are large polymeric molecules found in and around the nuclei of cells. One type, **DNA** (deoxyribonucleic acid), is believed to contain the genetic code of the cell, that is, the instructions that control all the characteristics of the cell. Another type of nucleic acid, *RNA* (ribonucleic acid), assists with the synthesis of protein. One kind of RNA called **messenger RNA** copies the genetic message contained in the DNA molecule and carries it to the site of protein synthesis. **Transfer RNA** then "reads" the code and transfers each free amino acid to the correct site for synthesis into a protein molecule. The detailed mechanism is fascinating. The discovery of the role of the nucleic acids in (a) containing and preserving the genetic instructions during cell division and (b) using the coded message to synthesize protein represents one of the most exciting scientific events to happen in recent times. We will, however, restrict our discussion in this book to the subject in part (a), preservation of the genetic instructions.

The nucleic acids can be thought of as having been formed from three chemicals: (a) a sugar, (b) phosphoric acid, and (c) a nitrogen-containing ring compound that is referred to as a *base*. The sugar is either ribose or 2-deoxyribose. If the former, the nucleic acid is RNA (ribonucleic acid); if the latter, DNA (deoxyribonucleic acid), although—as we shall see later—this is not the only difference between the two. The backbone of the nucleic acid is formed from the condensation of sugar and phosphoric acid molecules. The process is esterification, the reaction of the acid with the alcoholic groups of the sugar to split out water.

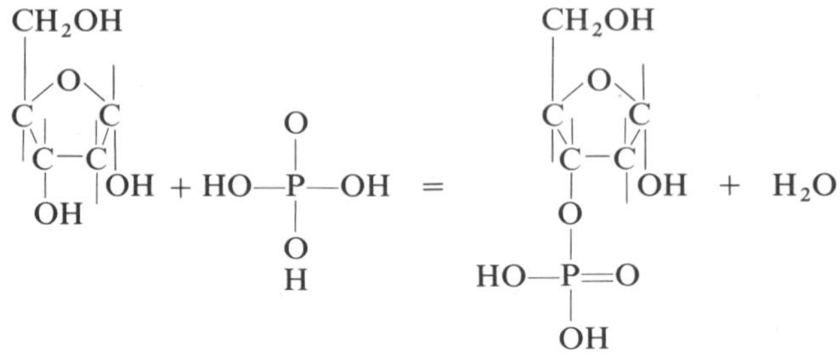

When this process is repeated on both ends of the condensed molecule, the result is a chain of alternating sugar and phosphate residues as shown below:

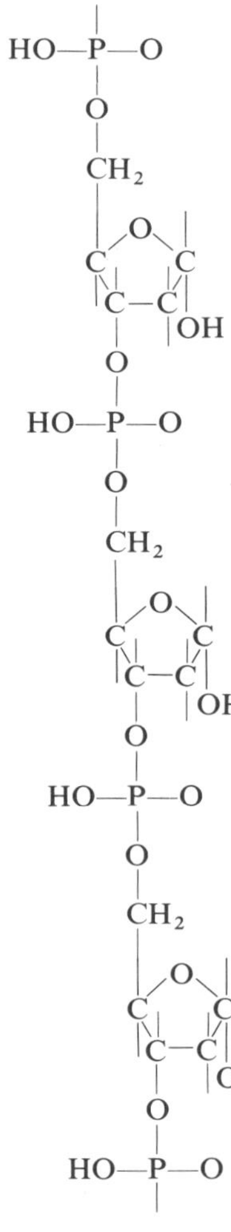

The third constituent, the base (one of five nitrogen-containing ring compounds) is attached to the sugar residue in place of an OH group. A small section of an RNA molecule is shown below. (For simplicity, we have followed the practice of organic chemists and omitted the symbols for carbon atoms in the rings and for hydrogen atoms at the ends of the bonds.)

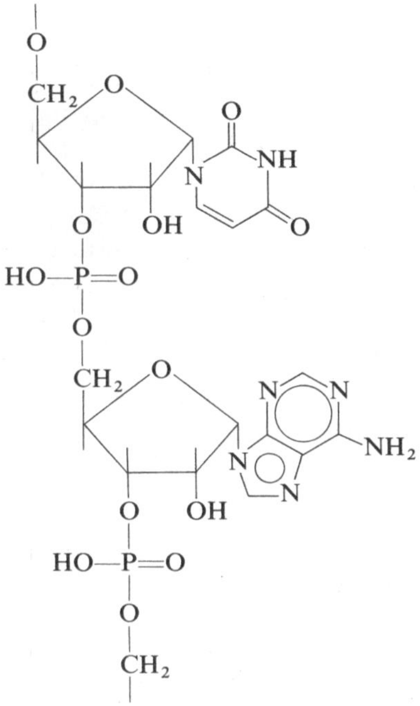

The bases play a most important role. Two pairs of the bases found in DNA and two pairs found in RNA will form multiple H-bonds with each other. The two pairs that are found in DNA are shown below:

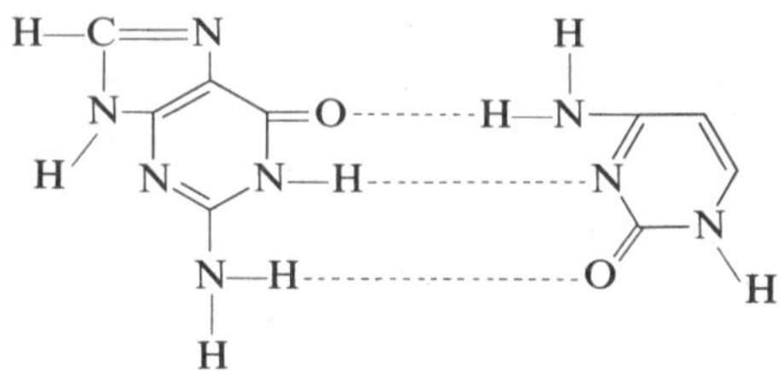

Guanine   Cytosine

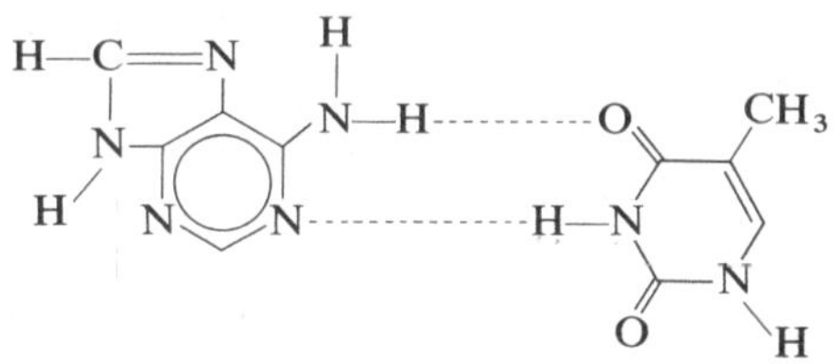

Adenine   Thymine

Both DNA and RNA contain four bases. In DNA they are adenine, guanine, cytosine, and thymine. RNA contains the first three and uracil in place of thymine. Their structures are

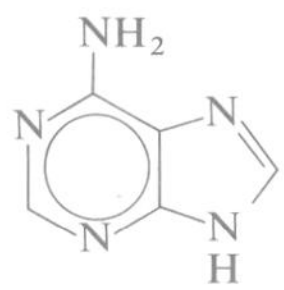

Adenine

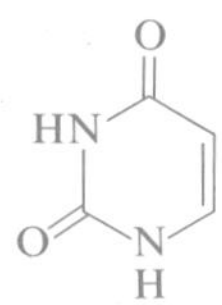

Uracil

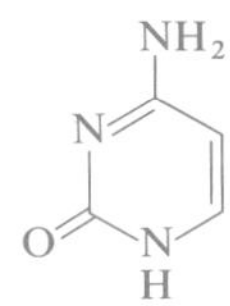

Cytosine

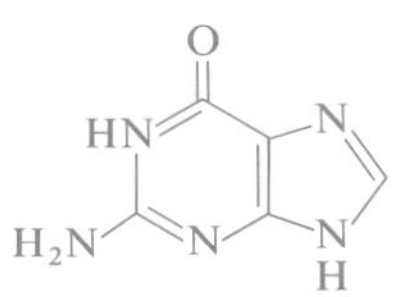

Guanine

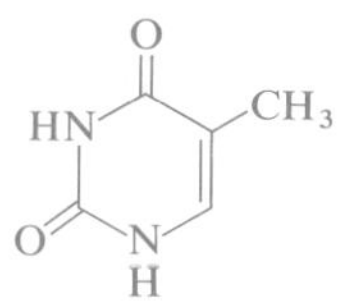

Thymine

Notice that their geometry is just right for the molecules to be tied together in a "lock and key" arrangement with multiple H-bonding. Another pair that have the proper spatial relationships is adenine and uracil. Other combinations do not fit together.

A simplified representation of the base pairs shown above is:

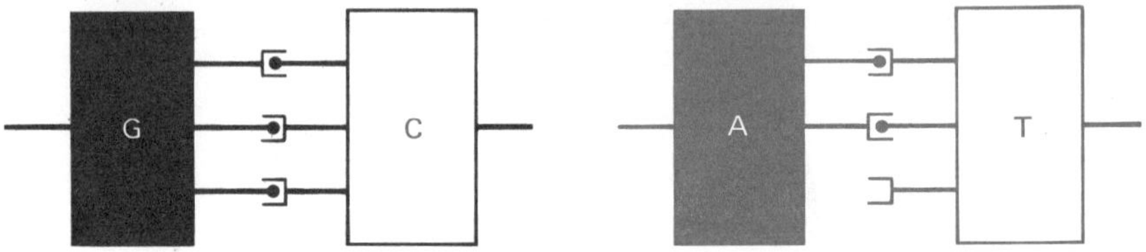

Guanine–Cytosine   Adenine–Thymine

In this diagram, the symbol —⊏ represents an electronegative element (either O or N) that can accept a share in a H atom, and the symbol —• is an oxygen or nitrogen that can donate the H atom to a hydrogen bond. These base pairings, as we will see shortly, play an important role in the structure and replication of DNA.

In 1953 Watson and Crick of Cambridge University postulated the double-helix structure for DNA, which in diagrammatic form would look like this:

Both helixes are right-handed, but headed in opposite directions. Base pairing is responsible for the double-stranded structure. Opposite every guanine molecule of one strand is a cytosine molecule in the other, and opposite each adenine is a thymine. A diagram of the base pairing in a portion of a DNA molecule is shown at the top of page 721.

The Watson-Crick model accounts for not only the x-ray data and base composition, but other physical properties as well, and suggests a method by which genetic information is replicated. The mechanism by which DNA duplicates itself and thus preserves and passes the genetic code on to daughter cells is a story of molecular structure and shape and their controlling effect on biochemical reactions.

The duplication process begins with the strands of the double helix separating. Then by a process of base pairing, each strand serves as a template

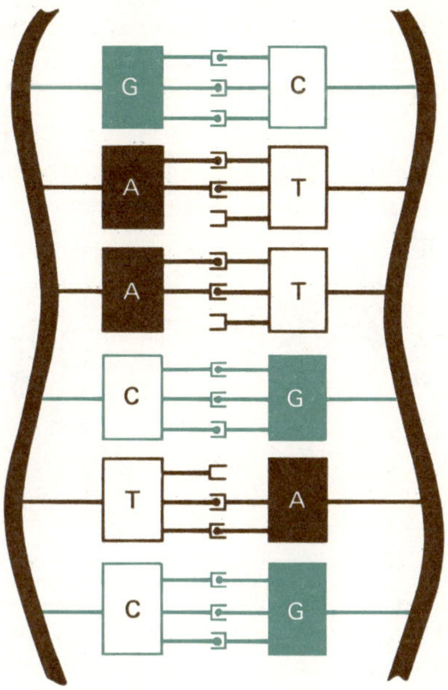

Here the strands are shown before separation. The heavy lines represent the sugar-phosphate backbones of the nucleic acids. See page 718 for the structure.

for the formation of its complementary structure. The separated strands of the portion of the DNA molecule shown above would appear as

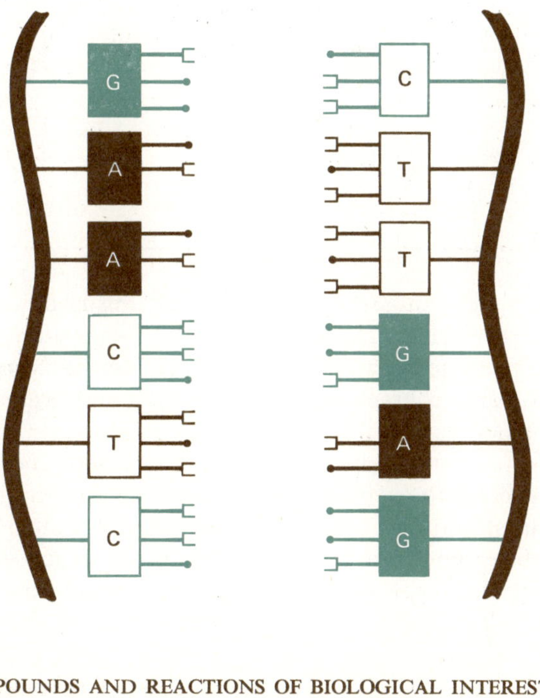

DNA replication. The right side represents one of the original strands of DNA; a daughter molecule is being replicated on the left side. Already, two cytosine molecules are paired with guanine, and a thymine molecule is paired with adenine. An adenine molecule is just coming into position to form an adenine-thymine pair. The process continues until a daughter molecule, identical to the original left-hand strand, is completed.

and the formation of a daughter molecule would take place as shown below:

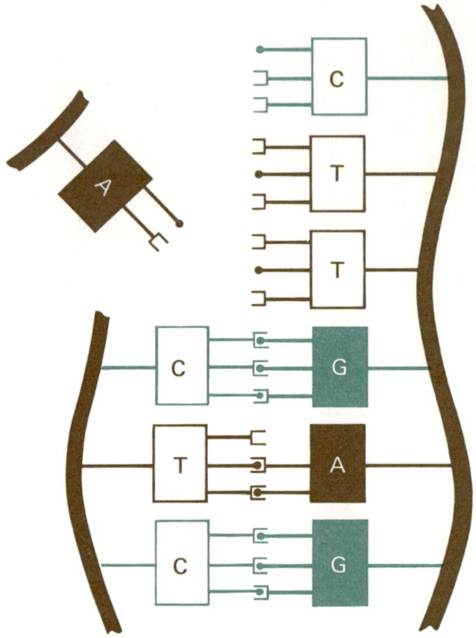

Two daughter molecules, each containing one strand of the original parent, thus result. The process has been described by F. H. Crick as a hand-in-glove arrangement. When the glove is removed, a new glove forms around the hand by attaching itself piece by piece, and a new hand forms inside the original glove by a similar mechanism. Thus the DNA molecule duplicates itself so that when the cell divides, a complete copy of the genetic code will be present in each daughter cell.

Summary

Organic chemistry is the subject that treats the chemistry of carbon-containing compounds. The subject is extensive, interesting, and vital. To date, several million organic compounds have been synthesized (or have been isolated from natural products) and characterized, and an astronomical number of such compounds are possible.

The basic reason for the profusion of carbon compounds is the ability of carbon atoms to bond to each other to produce chains and rings, which can exist in numerous structural-, geometrical-, and stereo-isomers.

Because of the extent of isomerism and the profound effect of structure on the properties of molecules, knowledge of the structure and the use of structural formulas is especially important in organic chemistry. A most important key to the elucidation of the structure of complex organic molecules was van't Hoff's early deductions concerning the "tetrahedral carbon" atom.

Hydrocarbons form an important class of naturally occurring carbon compounds. These are valuable not only as fuel but also as starting material in the synthesis of numerous other compounds, such as plastics, synthetic fibers, and drugs.

Several other important classes of compounds can be thought of as substituted hydrocarbons, in which an H atom has been replaced with any one of several functional groups. Using R to represent a hydrocarbon radical, we have the following general formulas for important classes of compounds:

$R{-}OH$	alcohols
$R{-}NH_2$	amines
$R{-}\overset{\overset{\displaystyle O}{\|}}{C}{-}OH$	organic acids
$R{-}\overset{\overset{\displaystyle O}{\|}}{C}{-}OR'$	esters
$R{-}\overset{\overset{\displaystyle O}{\|}}{C}{-}H$	aldehydes
$R{-}\overset{\overset{\displaystyle O}{\|}}{C}{-}R'$	ketones
$R{-}O{-}R'$	ethers

A very limited number of organic reactions are discussed in this chapter. Of those shown, you should pay particular attention to the following general types:

(1) Esterification—reaction of an alcohol with an organic acid to form an ester.
(2) Hydrolysis of an ester to form an alcohol and an organic acid.
(3) Reaction of an organic acid with an amine to form a peptide linkage in protein formation.
(4) Addition of a halogen or hydrogen halide to a double or triple carbon-to-carbon bond.
(5) Substitution for H on a benzene ring by a halogen.

The chemistry of *living systems* is called biochemistry. It, too, covers an immense field of knowledge, and in this chapter we made no attempt to survey the entire subject.

Amino acids, the "building blocks" of proteins, contain both the amine and the carboxyl functional groups in the same molecule. Amino acids react to form long polypeptide chains that fold and coil into large protein molecules. Elucidation of the important α-helix structure is one of the highlights of modern science.

Advances in the chemistry of the nucleic acids DNA and RNA, and the understanding of the role they play in preserving the genetic code of the cell, are some of the most exciting and far-reaching scientific events in recent times.

Important Terms and Concepts

- Organic chemistry
- Biochemistry
- Tetrahedral carbon atom
- Hydrocarbons
- Alkanes
- Cyclic alkanes
- Double- and triple-bonded hydrocarbons
- Structural isomerism
- Functional groups
- Alcohols
- Amines
- Organic acids
- Esters
- Aldehydes and ketones
- Ethers
- Benzene
- Amino acids
- Optical isomerism
- Proteins
- Peptide linkage
- α helix
- Nucleic acids
- DNA
- Messenger RNA
- Transfer RNA

Problems

24-1.
Vitamin B_{12} has the formula $C_{63}H_{90}N_{14}O_{14}Co$. A typical vitamin capsule contains 1.0 microgram of vitamin B_{12}. How many moles of this vitamin would a person consume in a lifetime, assuming he took one capsule a day for 70 years?

24-2.
Draw structural formulas for five isomers of C_6H_{14}.

24-3.
Write structural formulas for seventeen isomers of C_6H_{12}. (*Hint:* Some are cyclic compounds.)

24-4.
Show all isomers of the following compounds and indicate if they are geometrical isomers or optical isomers.

(a) CH_2Cl_2
(b) $CH_3CH{=}CHCH_3$
(c) $CH_3CHClCH_2CH_3$

24-5.
Write four structural isomers of C_4H_9Cl and determine which one is optically active, that is, which one has an optical isomer. (*Hint:* An asymmetric carbon atom is necessary before a molecule can have an optical isomer. An asymmetric carbon is one that is bonded to four *different* groups.)

24-6.
As an exercise in naming organic compounds, match the structural formula on the left with the name on the right.

FORMULA	NAME
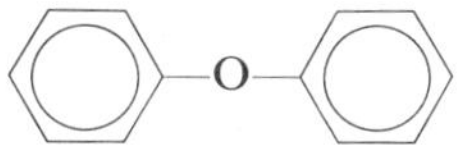	isopropyl alcohol
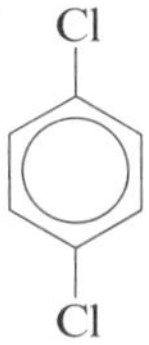	ethyldipropylamine
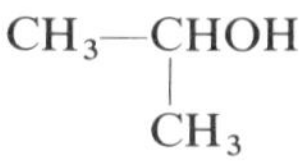	citric acid
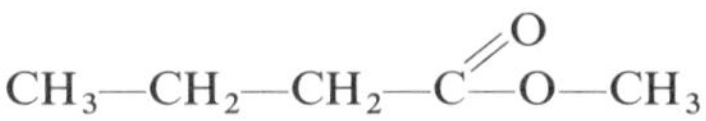	diphenyl ether
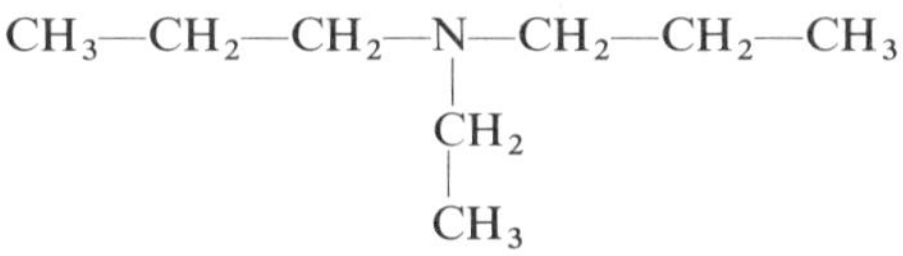	para-dichlorobenzene
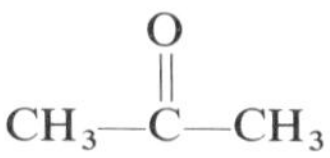	methyl butyrate
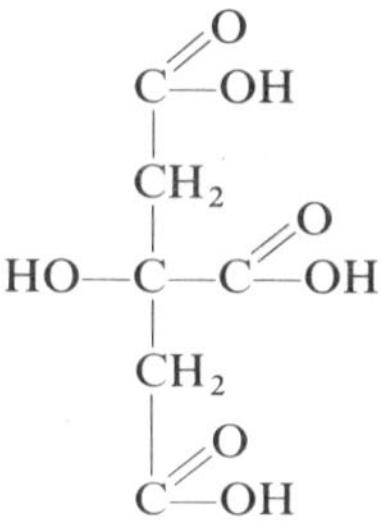	acetone

24-7.
The IUPAC rules for naming alkanes and halogen derivatives of alkanes are as follows:

(1) Select as the parent structure the longest continuous chain, and consider the compound to have been derived from this structure by thc replacement of hydrogen by various alkyl or halogen groups. The names of the parent compounds and the alkyl groups are as follows:

NUMBER OF CARBON ATOMS	PARENT COMPOUND	ALKYL GROUP
1	methane	methyl
2	ethane	ethyl
3	propane	propyl
4	butane	butyl
5	pentane	pentyl
6	hexane	hexyl
7	heptane	heptyl
8	octane	octyl
9	nonane	nonyl
10	decane	decyl

(2) Where necessary, indicate by a number the carbon to which the alkyl group is attached.
(3) In numbering the parent chain, start at the end giving the lowest numbers.
(4) If the alkyl or halo group occurs more than once, as a side chain, indicate this by the prefix di (2), tri (3), tetra (4), penta (5), hexa (6), and so on, to show how many occur, and indicate by various numbers the position of each group.
(5) Name the side groups in alphabetical order or in order of increasing size.

Several examples are as follows:

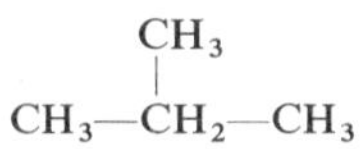
2-methylpropane

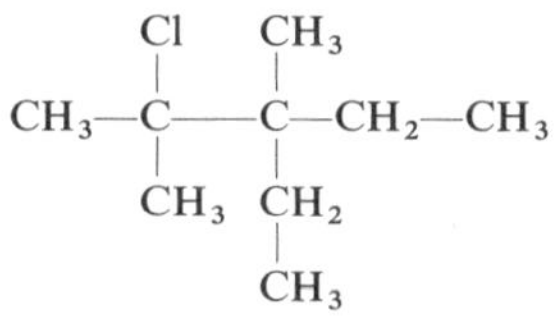
2-chloro-3-ethyl-2,3-dimethylpentane

(a) Write structural formulas for the following compounds:

2,2-dichloropropane
2-methylhexane
3-methyl-4-ethyloctane
methylcyclohexane

(b) Give the IUPAC name for the following compounds:

CCl_4

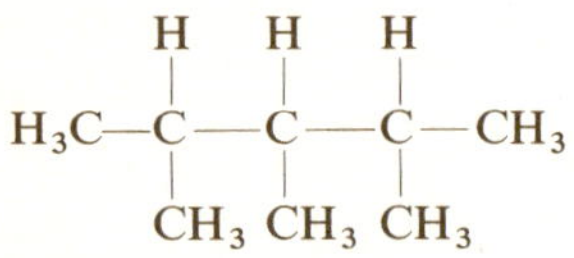

$Br—CH_2—CH_2—Cl$

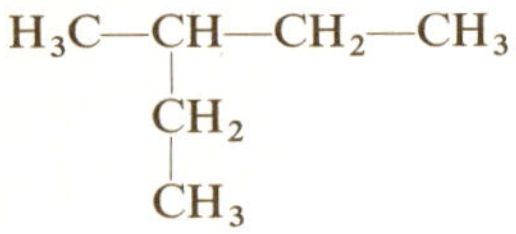

24-8.
The IUPAC rules for naming alkenes are modified from those for naming alkanes (see Problem 24-7) as follows:

(1) Select as the parent structure the longest continuous chain that contains the carbon-carbon double bond and name as with the alkanes, except change the ending of the parent compound from -ane to -ene.
(2) Indicate by a number the position in the chain of the carbon first encountered in the double bond.
(3) Indicate by numbers the positions of other groups.

Examples are as follows:

$CH_3—CH{=}CH—CH_2—CH_3$ 2-pentene

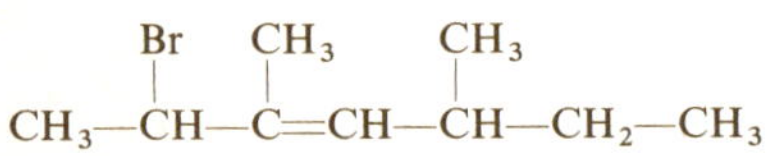

2-bromo-3,5-dimethyl-3-heptene

(a) Write the structural formulas for the following compounds:

3-hexene
1,3-butadiene
2,3-dichloro-2-butene
cyclohexene

(b) Give the IUPAC names for the following four compounds:

$H_3C—CH{=}CH—CH_3$

$CH_3—CH{=}CH—CH_2—CH_2—CH_3$

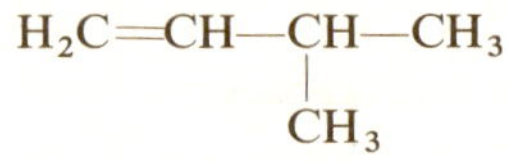

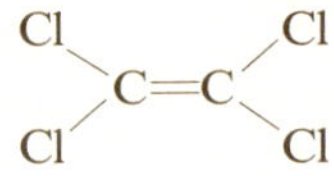

24-9.
The IUPAC rules for naming alcohols are modified from those for naming alkanes (see Problem 24-7) as follows:

(1) Select as the parent structure the longest continuous chain that contains the OH group and name as with the alkanes, except change the ending of the parent compound from -ane to -ol.
(2) Indicate by number the position of the OH group, using the lowest possible number.
(3) Indicate by numbers the positions of other groups.

Examples are:

2-butanol

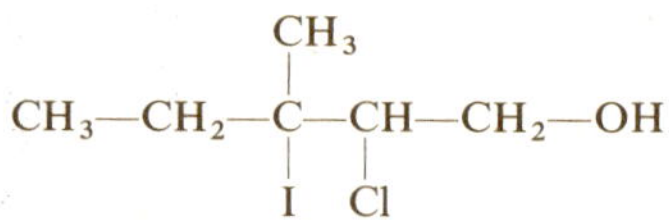

2-chloro-3-iodo-3-methyl-1-pentanol

(a) Give the IUPAC names for the following compounds:

$CH_3—CH_2—OH$

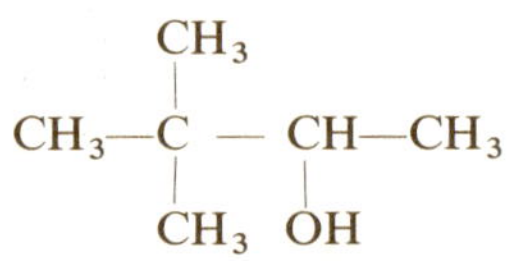

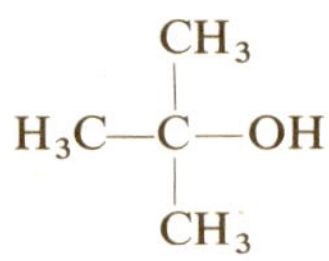

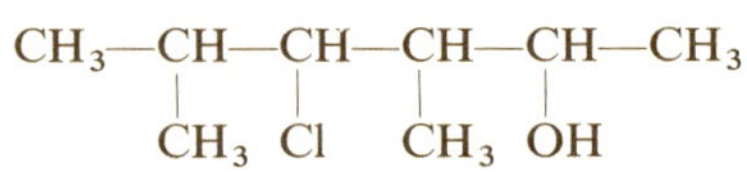

(b) Write structural formulas for the following alcohols:

2-methyl-1-propanol
3-hexanol
2,4-hexadiol
3,3,3-trifluoro-1-propanol

24-10.
Complete the following reactions:

(a) $CH_3—CH_2—CH_2—OH + Na =$

(b) 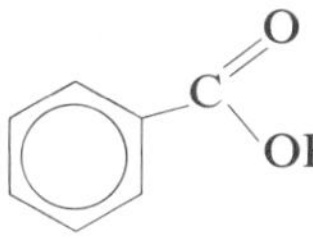 $+ CH_3OH =$

(c) $H_3O^+ + CH_3CH_2NH_2 =$

(d) $CH_3—CH{=}CH—CH_3 + HI =$

(e) $CH_3—CH(NH_2)—COOH +$

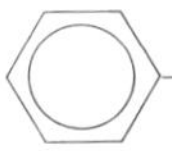 $—CH_2—CH(NH_2)—COOH =$

(f) $HC(=O)OH + NH_3 =$

(g) $CH_2Cl—CH_2—CH{=}CH_2 + Br_2 =$

(h) $HC(=O)OCH_3 + OH^- =$

(i) 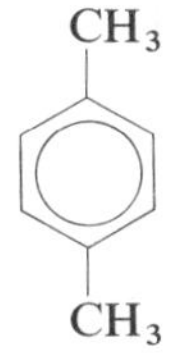 $+ Cl_2 =$

24-11.
For practice in predicting the products of organic reactions, match the following set of reactants on the left with the correct products on the right.

REACTANTS	PRODUCTS
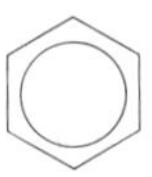 $+ Cl_2 \longrightarrow$	$HC(=O)O^- + CH_3NH_3^+$
$2NH_2CH_2—C(=O)OH \longrightarrow$	$CH_3—CH_2—C(=O)—OCH_3 + H_2O$
$CH_3—CH_2—C(=O)—OH + CH_3OH \longrightarrow$	$CH_3—CH_2—CH_3$
$HC(=O)—OH + CH_3NH_2 \longrightarrow$	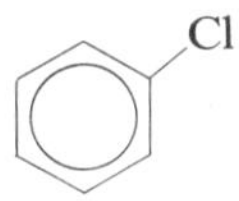 $+ HCl$
$CH_3—CH{=}CH_2 + H_2 \xrightarrow[\text{catalyst}]{\text{Pt}}$	$NH_2—CH_2—C(=O)—NH—CH—C(=O)—OH + H_2O$

24-12.
Markovnikov's rule helps predict the product when an acid (such as HCl, HBr, HI, HOH, and so forth) adds to a double bond. The rule is as follows: "In the addition of an acid to the carbon-carbon double bond of an alkene, the hydrogen of the acid attaches itself to the carbon atom that already holds the greatest number of hydrogens." For example,

$$CH_3—\underset{}{C}(CH_3){=}CH_2 + HBr \rightarrow CH_3—C(CH_3)(Br)—CH_3$$

Use Markovnikov's rule to correctly predict the products of the following reactions, and name the product. See Problems 24-7 to 24-9 for rules for naming organic compounds.

(a) $CH_3—CH{=}CH_2 + HCl \rightarrow$
(b) $CH_3—CH{=}CH—CH_3 + HI \rightarrow$
(c) $CH_3—C(CH_3){=}CH—CH_3 + H_2O \rightarrow$

APPENDIX A

Van der Waals Constants

GAS	a(liter^2-atm/mole^2)	b(liter/mole)
H_2	0.2444	0.02661
He	0.03412	0.02370
N_2	1.390	0.03913
O_2	1.360	0.03183
CO	1.485	0.03985
NO	1.340	0.02789
CO_2	3.592	0.04267
H_2O	5.464	0.03049

Critical-Point Data

GAS	T_c(°K)	P_c(atm)	V_c(cm^3/mole)
He	5.3	2.26	61.6
H_2	33.3	12.8	69.7
N_2	126.1	33.5	90.0
CO	134.0	34.6	90.0
O_2	154.3	49.7	74.4
C_2H_4	282.9	50.9	127.5
CO_2	304.2	72.8	94.2
NH_3	405.6	112.2	72.0
H_2O	647.2	217.7	55.4
Hg	1735.0	1036.0	40.1

APPENDIX B

Thermodynamic Properties at 298.15° K†

Enthalpies of formation (ΔH_f^0), free energies of formation (ΔG_f^0), entropies (S^0), and heat capacities (C_P^0) for selected compounds at 298.15° K.

COMPOUND	STATE	ΔH_f^0 (kjoules/mole)	ΔG_f^0 (kjoules/mole)	S^0 (joules/°K-mole)	C_P^0 (joules/°K-mole)
Aluminum					
Al	solid	0	0	28.32	24.34
Al_2O_3	solid; corundum	−1669.8	−1576.4	50.98	78.99
Al_2Cl_6	solid	−1391	−1274	340	178
Arsenic					
As	solid; gray	0	0	35	25
As_2O_5	solid	−914.6	−1057	105	116.5
As_4O_6	octahedral	−1313.5	−1152.1	214	191
Barium					
Ba	solid	0	0	67	26.4
BaO	solid	−558.1	−528.4	70.3	47.45
$BaCl_2$	solid	−860.06	−810.9	130	75.3
$BaCl_2 \cdot H_2O$	solid	−1165	−1059	170	118
$BaCl_2 \cdot 2\,H_2O$	solid	−1461.7	−1296	203	155.2

† All values except those in parentheses were taken from NBS Circular 500.

COMPOUND	STATE	ΔH_f^0 (kjoules/mole)	ΔG_f^0 (kjoules/mole)	S^0 (joules/°K-mole)	(C_P^0) (joules/°K-mole)
Bromine					
Br_2	liquid	0	0	152	—
Br_2	gas	30.7	3.14	245.3	36.0
HBr	gas	−36.2	−53.22	198.48	29.1
Calcium					
Ca	solid	0	0	41.6	26.3
CaO	solid	−635.5	−604.2	40	42.8
$Ca(OH)_2$	solid	−986.59	−896.76	76.1	84.5
CaF_2	solid	−1215	−1162	68.87	67.03
$CaCl_2$	solid	−795.0	−750.2	114	72.63
$CaSO_4$	solid; anhydrite	−1432.7	−1320.3	107	99.6
Carbon					
C	solid; graphite	0	0	5.6940	8.644
C	solid; diamond	1.896	2.866	2.439	6.063
C	gas	718.384	672.975	157.992	20.838
CO	gas	−110.523	−137.268	197.91	29.14
CO_2	gas	−393.513	−394.383	213.64	37.13
CH_4	methane gas	−74.848	−50.794	186.2	35.72
C_2H_6	ethane gas	−84.667	−32.89	229.5	52.656
C_3H_8	propane gas	(−103.1)	(−23.5)	(271)	—
C_2H_4	ethylene gas	52.283	68.124	219.5	43.56
C_2H_2	acetylene gas	226.75	209.20	200.82	43.928
C_6H_6	benzene; liquid	(49.028)	(124.50)	175	—
HCOOH	formic acid; liquid	−409	−346	129.0	99.04
CH_3OH	methyl alcohol; liquid	−238.64	−166.3	127	81.6
C_2H_5OH	ethyl alcohol; liquid	−277.63	−174.8	161	111.5
HCHO	formaldehyde; gas	−116	−110	218.7	35.4
CH_3COOH	acetic acid; liquid	−487.0	−393	160	123
$H_2C_2O_4$	oxalic acid; solid	−826.8	−697.9	120	110
CCl_4	liquid	−139.5	−68.74	214.4	131.8
$CHCl_3$	liquid	−121	−60.7	203	116
$COCl_2$	gas	−223.0	−210.5	289.2	60.71
CS_2	liquid	87.9	63.6	151.0	75.7
C_2N_2	gas	307.9	296	242.1	56.90
HCN	gas	131	120	201.8	35.9
HCN	liquid	105	121	112.8	70.63
Chlorine					
Cl_2	gas	0	0	222.95	33.9
Cl	gas	121.39	105.40	165.088	21.842
HCl	gas	−92.312	−95.265	186.68	29.1
ClO_2	gas	103	123	249	—

COMPOUND	STATE	ΔH_f^0 (kjoules/mole)	ΔG_f^0 (kjoules/mole)	S^0 (joules/°K-mole)	C_P^0 (joules/°K-mole)
Fluorine					
F_2	gas	0	0	203	31.5
F	gas	134.9	118.0	158.65	22.74
F_2O	gas	23	41	246.7	—
HF	gas	−269	−271	173.5	29.1
Hydrogen					
H_2	gas	0	0	130.59	28.84
H	gas	217.94	203.24	114.61	20.79
H_2O	gas	−241.826	−228.593	188.72	33.58
H_2O	liquid	−285.840	−237.191	69.940	75.295
H_2O_2	gas	−133.2	—	—	—
H_2O_2	liquid	−187.6	—	—	—
Iodine					
I_2	gas	62.241	19.4	260.58	36.9
I_2	solid	0	0	117	54.98
I	gas	106.62	70.149	180.682	20.786
I_2O_5	solid	−177.2	—	—	—
HI	gas	25.9	1.3	206.33	29.2
Magnesium					
Mg	solid	0	0	32.5	23.9
MgO	solid	−601.83	−569.57	27	37.4
$MgCl_2$	solid	−641.83	−592.33	89.5	71.30
Mercury					
Hg	gas	60.84	31.8	174.9	20.79
Hg	liquid	0	0	77.4	27.8
HgO	solid; red	−90.71	−58.534	72.0	45.73
Nitrogen					
N_2	gas	0	0	191.49	29.12
N	gas	358.01	340.90	153.195	20.79
NO	gas	90.374	86.688	210.62	29.86
NO_2	gas	33.85	51.840	240.5	37.9
N_2O	gas	81.55	103.6	220.0	38.71
N_2O_4	gas	9.661	98.286	304.3	79.08
NH_3	gas	−46.19	−16.64	192.5	35.66
N_2H_4	liquid	50.42	—	—	—
HNO_3	liquid	−173.23	−79.914	155.6	109.9
NH_4Cl	solid	−315.4	−203.9	94.6	84.1

COMPOUND	STATE	ΔH_f^0 (kjoules/mole)	ΔG_f^0 (kjoules/mole)	S^0 (joules/°K-mole)	C_P^0 (joules/°K-mole)
Oxygen					
O_2	gas	0	0	205.03	29.36
O	gas	247.52	230.10	160.95	21.91
O_3	gas	142	163.4	238	38.2
Phosphorus					
P	solid; white	0	0	44.4	23.2
P	solid; red	−18	—	—	—
P	solid; black	−43.1	—	—	—
P_4	liquid	2.52	—	—	—
P_4O_{10}	solid	−3013	—	—	—
H_3PO_2	liquid	−599.1	—	—	—
H_3PO_3	liquid	−958.6	—	—	—
Potassium					
K	solid	0	0	63.6	29.2
K_2O	solid	−362	—	—	—
K_2O_2	solid	−494	—	—	—
K_2O_3	solid	−523	—	—	—
K_2O_4	solid	−561	—	—	—
KOH	solid	−425.85	—	—	—
KCl	solid	−435.868	−408.33	82.68	51.51
KBr	solid	−392.2	−379.2	96.44	53.64
KI	solid	−327.7	−322.3	104.4	55.06
Sodium					
Na	solid	0	0	51.0	28.4
Na_2O	solid	−416	−377	72.8	68.2
NaCl	solid	−411.00	−384.03	72.38	49.71
Sulfur					
S	solid; rhombic	0	0	31.9	22.6
S	solid; monoclinic	0.30	0.096	32.6	23.6
SO_2	gas	−296.9	−300.4	248.5	39.8
SO_3	gas	−395.2	−370.4	256.2	50.63
H_2S	gas	−20.15	−33.02	205.6	34.0
H_2SO_4	liquid	−811.32	—	—	—

APPENDIX C

Thermodynamic Properties of Phase Changes[†]

COMPOUND	NORMAL MELTING POINT (°K)	ENTHALPY OF FUSION (joules/mole)	NORMAL BOILING POINT (°K)	ENTHALPY OF VAPORIZATION (joules/mole)
Elements				
He	—	—	4.216	80
Ne	24.57	335	27.07	1770
Ar	83.85	1180	87.29	6519
Kr	115.95	1640	119.93	9029
Xe	161.3	2300	165.1	12,640
Rn	202	2900	211	16,400
H_2	13.96	117	20.39	902.9
F_2	55.20	1560	85.24	6318
Cl_2	172.18	6406	239.11	20,410
Br_2	265.95	10,540	331.4	30,000
I_2	386.8	15,650	456	41,710
O_2	54.39	444.8	90.19	6819.5
S_8 (equilibrium S)	392	1230	717.76	10,500
N_2	63.15	720	77.36	5577
P_4 (white)	317.4	630	553	12,430
Hg	234.29	2296	629.88	59,149
Li	453.70	3024	1604	134,700
Na	370.97	2602	1163	89,040

[†] Normal melting and boiling points, and enthalpies of fusion and vaporization, are tabulated by type of compound and by position in the periodic table of the central atom to emphasize periodic relationships. Data were taken from the NBS Circular 500 and from "Physical Properties of Chemical Compounds, R. R. Dreisbach, ed., *Advances in Chemistry Series*, Vols. 15, 22, and 29, published by the American Chemical Society.

COMPOUND	NORMAL MELTING POINT (°K)	ENTHALPY OF FUSION (joules/mole)	NORMAL BOILING POINT (°K)	ENTHALPY OF VAPORIZATION (joules/mole)
Elements (*cont.*)				
K	336.4	2320	1039	77,530
Rb	312.0	2340	974	69,200
Cs	301.8	2100	958	65,900
Carbon Compounds				
CH_4	90.68	941	111.67	8180
C_2H_6	89.88	2859	184.52	14,720
C_3H_8	85.46	3524	231.08	18,770
$n\text{-}C_4H_{10}$	134.80	4661	272.65	22,390
$n\text{-}C_5H_{12}$	143.43	8393.1	309.22	25,770
$n\text{-}C_6H_{14}$	177.80	13,029	341.89	28,850
$n\text{-}C_7H_{16}$	182.54	14,033	371.58	31,690
$n\text{-}C_8H_{18}$	216.36	20,740	398.82	34,470
C_6H_{12} (cyclohexane)	279.70	2665	353.89	30,080
$C(CH_3)_4$	256.60	3256	282.65	22,750
C_2H_4	103.97	3351	169.45	13,540
C_2H_2	—	—	189.2‡	21,300
C_6H_6	278.683	9832	353.25	30,770
$C_6H_5CH_3$	178.16	6619	383.78	33,460
$o\text{-}C_6H_4(CH_3)_2$	247.97	13,600	417.56	36,820
$m\text{-}C_6H_4(CH_3)_2$	225.28	11,570	412.25	36,440
$p\text{-}C_6H_4(CH_3)_2$	286.41	17,100	411.50	36,070
$C_{10}H_8$ (naphthalene)	353.36	18,050	491.11	42,200
CH_3OH	175.26	3170	337.9	35,270
C_2H_5OH	158.6	5000	351.7	38,580
HCHO	154.9	—	253.9	24,480
CH_3CHO	155	3220	293.3	27,200
HCOOH	281.46	12,680	373.7	22,260
CH_3COOH	289.77	11,700	391.4	24,400
CH_3NH_2	179.70	6134	266.84	25,820
$(CH_3)_2NCHO$	212.67	7895	422.71	38,340
CCl_4	250.3	2510	349.9	30,000
$CHCl_3$	209.7	9200	334.4	29,370
$COCl_2$	145.39	5740	280.71	24,400
CS_2	161.1	4390	319.41	26,780
Oxides				
SO_2	197.70	7401	263.14	24,900
N_2O	182.34	6540	184.68	16,550
NO	109.55	2299	121.42	13,770

‡ Sublimation point.

COMPOUND	NORMAL MELTING POINT (°K)	ENTHALPY OF FUSION (joules/mole)	NORMAL BOILING POINT (°K)	ENTHALPY OF VAPORIZATION (joules/mole)
N_2O_4	261.96	14,650	294.31	38,120
CO	68.10	838	81.67	6042
CO_2	—	—	194.68‡	25,230
Hydrides				
HF	189.39	4577	292.67	6732
HCl	158.97	1992	188.13	16,150
HBr	186.30	2406	206.44	17,620
HI	222.37	2830	237.81	19,770
H_2O	273.150	6010	373.150	40,660
H_2S	187.63	2377	212.85	18,670
H_2Se	207.43	2520	231.9	19,330
H_2Te	222	4200	270.9	23,220
NH_3	195.42	5653	239.76	23,350
PH_3	139.41	1130	185.44	14,600
AsH_3	156.9	2340	210.7	17,490

APPENDIX D

Standard Reduction Potentials at 298.15° K

HALF-REACTION	$\mathscr{E}^0$ (V)	HALF-REACTION	$\mathscr{E}^0$ (V)
$Li^+ + e^- = Li$	−3.045	$Cd^{2+} + 2e^- = Cd$	−0.403
$K^+ + e^- = K$	−2.925	$PbI_2 + 2e^- = Pb + 2I^-$	−0.365
$Ba^{2+} + 2e^- = Ba$	−2.90	$PbSO_4 + 2e^- = Pb + SO_4^{2-}$	−0.356
$Sr^{2+} + 2e^- = Sr$	−2.89	$Co^{2+} + 2e^- = Co$	−0.277
$Ca^{2+} + 2e^- = Ca$	−2.87	$H_3PO_4 + 2H^+ + 2e^- = H_3PO_3 + H_2O$	−0.276
$Na^+ + e^- = Na$	−2.714	$Ni^{2+} + 2e^- = Ni$	−0.250
$Mg^{2+} + 2e^- = Mg$	−2.37	$CuI + e^- = Cu + I^-$	−0.185
$\frac{1}{2}H_2 + e^- = H^-$	−2.25	$AgI + e^- = Ag + I^-$	−0.151
$Be^{2+} + 2e^- = Be$	−1.85	$Sn^{2+} + 2e^- = Sn$	−0.136
$Al^{3+} + 3e^- = Al$	−1.66	$Pb^{2+} + 2e^- = Pb$	−0.126
$Mn^{2+} + 2e^- = Mn$	−1.18	$2H^+ + 2e^- = H_2$	0.000
$V^{2+} + 2e^- = V$	−1.18	$P + 3H^+ + 3e^- = PH_3(g)$	0.06
$TiO^{2+} + 2H^+ + 4e^- = Ti + H_2O$	−0.89	$AgBr + e^- = Ag + Br^-$	0.095
$H_3BO_3 + 3H^+ + 3e^- = B + 3H_2O$	−0.87	$CuCl + e^- = Cu + Cl^-$	0.137
$SiO_2 + 4H^+ + 4e^- = Si + 2H_2O$	−0.86	$S + 2H^+ + 2e^- = H_2S$	0.141
$Zn^{2+} + 2e^- = Zn$	−0.763	$Sn^{4+} + 2e^- = Sn^{2+}$	0.15
$Cr^{3+} + 3e^- = Cr$	−0.74	$Cu^{2+} + e^- = Cu^+$	0.153
$H_3PO_2 + H^+ + e^- = P + 2H_2O$	−0.51	$HSO_4^- + 3H^+ + 2e^- = H_2SO_3 + H_2O$	0.17
$H_3PO_3 + 2H^+ + 2e^- = H_3PO_2 + H_2O$	−0.50	$AgCl + e^- = Ag + Cl^-$	0.222
$Fe^{2+} + 2e^- = Fe$	−0.440	$Cu^{2+} + 2e^- = Cu$	0.337
$Cr^{3+} + e^- = Cr^{2+}$	−0.41	$Fe(CN)_6^{3-} + e^- = Fe(CN)_6^{4-}$	0.36

Source: W. M. Latimer, THE OXIDATION STATES OF THE ELEMENTS AND THEIR POTENTIALS IN AQUEOUS SOLUTIONS, 2nd Ed ©1952, pp. 339–348. Reprinted by permission of Prentice-Hall, Inc., Englewood Cliffs, New Jersey.

HALF-REACTION	$\mathscr{E}^0$ (V)	HALF-REACTION	$\mathscr{E}^0$ (V)
$H_2SO_3 + 4\,H^+ + 4\,e^- = S + 3\,H_2O$	0.45	$Cr_2O_7^{2-} + 14\,H^+ + 6\,e^- = 2\,Cr^{3+} + 7\,H_2O$	1.33
$Cu^+ + e^- = Cu$	0.521	$Cl_2 + 2\,e^- = 2\,Cl^-$	1.3595
$I_2 + 2\,e^- = 2\,I^-$	0.535	$HIO + H^+ + e^- = \frac{1}{2}\,I_2 + H_2O$	1.45
$I_3^- + 2\,e^- = 3\,I^-$	0.536	$PbO_2 + 4\,H^+ + 2\,e^- = Pb^{2+} + 2\,H_2O$	1.455
$Cu^{2+} + Cl^- + e^- = CuCl$	0.538	$Au^{3+} + 3\,e^- = Au$	1.50
$MnO_4^- + e^- = MnO_4^{2-}$	0.564	$MnO_4^- + 8\,H^+ + 5\,e^- = Mn^{2+} + 4\,H_2O$	1.51
$Cu^{2+} + Br^- + e^- = CuBr$	0.640	$BrO_3^- + 6\,H^+ + 5\,e^- = \frac{1}{2}\,Br_2 + 3\,H_2O$	1.52
$O_2 + 2\,H^+ + 2\,e^- = H_2O_2$	0.682	$HBrO + H^+ + e^- = \frac{1}{2}\,Br_2 + H_2O$	1.59
$Fe^{3+} + e^- = Fe^{2+}$	0.771	$H_5IO_6 + H^+ + 2\,e^- = IO_3^- + 3\,H_2O$	1.6
$Hg_2^{2+} + 2\,e^- = 2\,Hg$	0.789	$Ce^{4+} + e^- = Ce^{3+}$	1.61
$Ag^+ + e^- = Ag$	0.799	$HClO + H^+ + e^- = \frac{1}{2}\,Cl_2 + H_2O$	1.63
$Cu^{2+} + I^- + e^- = CuI$	0.86	$HClO_2 + 2\,H^+ + 2\,e^- = HClO + H_2O$	1.64
$2\,Hg^{2+} + 2\,e^- = Hg_2^{2+}$	0.920	$Au^+ + e^- = Au$	1.68
$NO_3^- + 4\,H^+ + 3\,e^- = NO + 2\,H_2O$	0.96	$NiO_2 + 4\,H^+ + 2\,e^- = Ni^{2+} + 2\,H_2O$	1.68
$HNO_2 + H^+ + e^- = NO + H_2O$	1.00	$PbO_2 + HSO_4^- + 3\,H^+ + 2\,e^- = PbSO_4 + 2\,H_2O$	1.685
$Br_2\,(l) + 2\,e^- = 2\,Br^-$	1.065	$MnO_4^- + 4\,H^+ + 3\,e^- = MnO_2 + 2\,H_2O$	1.695
$N_2O_4 + 2\,H^+ + 2\,e^- = 2\,HNO_2$	1.07	$H_2O_2 + 2\,H^+ + 2\,e^- = 2\,H_2O$	1.77
$ClO_4^- + 2\,H^+ + 2\,e^- = ClO_3^- + H_2O$	1.19	$Co^{3+} + e^- = Co^{2+}$	1.82
$IO_3^- + 6\,H^+ + 5\,e^- = \frac{1}{2}\,I_2 + 3\,H_2O$	1.195	$O_3 + 2\,H^+ + 2\,e^- = O_2 + H_2O$	2.07
$ClO_3^- + 3\,H^+ + 2\,e^- = HClO_2 + H_2O$	1.21	$F_2 + 2\,e^- = 2\,F^-$	2.87
$O_2 + 4\,H^+ + 4\,e^- = 2\,H_2O$	1.229	$F_2 + 2\,H^+ + 2\,e^- = 2\,HF(aq)$	3.06
$MnO_2 + 4\,H^+ + 2\,e^- = Mn^{2+} + 2\,H_2O$	1.23		

APPENDIX E

Dissociation Constants at 298.15° K

1. ACID AND BASE CONSTANTS

SUBSTANCE	FORMULA	DISSOCIATION CONSTANT
	ACIDS	
Acetic	CH_3COOH	1.8×10^{-5}
Arsenic	H_3AsO_4	$K_1 = 5.6 \times 10^{-3}$
		$K_2 = 1.7 \times 10^{-7}$
		$K_3 = 3 \times 10^{-12}$
Arsenious	H_3AsO_3	$K_1 = 6 \times 10^{-10}$
		$K_2 = 3 \times 10^{-14}$
Benzoic	C_6H_5COOH	6×10^{-5}
Boric	H_3BO_3	$K_1 = 6.4 \times 10^{-10}$
Carbonic	H_2CO_3	$K_1 = 3 \times 10^{-7}$
		$K_2 = 6 \times 10^{-11}$
Chloracetic	$ClCH_2COOH$	1.5×10^{-3}
Chromic	H_2CrO_4	$K_1 = 1.8 \times 10^{-1}$
		$K_2 = 3.2 \times 10^{-7}$
Formic	$HCOOH$	2×10^{-4}
Hydrocyanic	HCN	6×10^{-10}
Hydrofluoric	HF	7.4×10^{-4}
Hydrogen sulfide	H_2S	$K_1 = 1.1 \times 10^{-7}$
		$K_2 = 1 \times 10^{-13}$
Hypochlorous	$HClO$	1.1×10^{-8}
Iodic	HIO_3	2×10^{-1}
Nitrous	HNO_2	4.5×10^{-4}

1. ACID AND BASE CONSTANTS

SUBSTANCE	FORMULA	DISSOCIATION CONSTANT
ACIDS		
Oxalic	$H_2C_2O_4$	$K_1 = 5.9 \times 10^{-2}$
		$K_2 = 6 \times 10^{-5}$
Phosphoric	H_3PO_4	$K_1 = 7.5 \times 10^{-3}$
		$K_2 = 6.8 \times 10^{-8}$
		$K_3 = 4 \times 10^{-13}$
Phosphorous	H_3PO_3	$K_1 = 5 \times 10^{-2}$
Sulfuric	H_2SO_4	$K_2 = 1.2 \times 10^{-2}$
Sulfurous	H_2SO_3	$K_1 = 1.3 \times 10^{-2}$
		$K_2 = 5 \times 10^{-6}$
BASES		
Ammonia	$NH_3 + HOH = NH_4^+ + OH^-$	1.8×10^{-5}
Aniline	$C_6H_5NH_2$	4×10^{-10}
Hydrazine	H_2NNH_2	3×10^{-6}
Hydroxylamine	$HONH_2$	1×10^{-8}

2. DISSOCIATION CONSTANTS OF COMPLEX IONS

COMPLEX ION	DISSOCIATION CONSTANT
$Al(OH)_4^- = Al(OH)_3(s) + OH^-$	2×10^{-2}
$Sb(OH)_4^- = Sb(OH)_3(s) + OH^-$	10^{-3}
$Cd(NH_3)_4^{2+} = Cd^{2+} + 4\ NH_3$	2×10^{-7}
$Cr(OH)_4^- = Cr(OH)_3(s) + OH^-$	10^2
$Cu(CN)_3^{2-} = Cu^+ + 3\ CN^-$	5×10^{-28}
$Cu(NH_3)_2^+ = Cu^+ + 2\ NH_3$	1×10^{-11}
$Cu(NH_3)_4^{2+} = Cu^{2+} + 4\ NH_3$	5×10^{-14}
$CuBr^+ = Cu^{2+} + Br^-$	0.5
$CuCl^+ = Cu^{2+} + Cl^-$	0.8
$FeBr^{2+} = Fe^{3+} + Br^-$	2.5×10^{-1}
$FeCl^{2+} = Fe^{3+} + Cl^-$	5.0×10^{-2}
$FeOH^{2+} = Fe^{3+} + OH^-$	1.7×10^{-10}
$FeSCN^{2+} = Fe^{3+} + SCN^-$	7.3×10^{-3}
$Fe(SCN)_2^+ = FeSCN^{2+} + SCN^-$	41
$Pb(OH)_3^- = Pb(OH)_2(s) + OH^-$	50
$HgCl_4^{2-} = Hg^{2+} + 4\ Cl^-$	1×10^{-16}

2. DISSOCIATION CONSTANTS OF COMPLEX IONS

COMPLEX ION	DISSOCIATION CONSTANT
$HgCl_4^{2-} = HgCl_2 + 2\ Cl^-$	1×10^{-2}
$HgI_4^{2-} = Hg^{2+} + 4\ I^-$	5×10^{-31}
$HgS_2^{2-} = Hg^{2+} + 2\ S^{2-}$	2×10^{-55}
$Ni(NH_3)_4^{2+} = Ni^{2+} + 4\ NH_3$	5×10^{-8}
$SiF_6^{2-} = SiF_4(g) + 2\ F^-$	2×10^{-15}
$Ag(NH_3)^+ = Ag^+ + NH_3$	4×10^{-4}
$Ag(NH_3)_2^+ = Ag^+ + 2\ NH_3$	6.8×10^{-8}
$AgCl_2^- = Ag^+ + 2\ Cl^-$	6×10^{-6}
$AgCl_3^{2-} = Ag^+ + 3\ Cl^-$	5×10^{-6}
$Sn(OH)_3^- = Sn(OH)_2(s) + OH^-$	2×10^3
$Sn(OH)_6^{2-} = Sn(OH)_4(s) + 2\ OH^-$	5×10^3
$Zn(NH_3)_4^{2+} = Zn^{2+} + 4\ NH_3$	3×10^{-10}
$Zn(OH)_4^{2-} = Zn(OH)_2(s) + 2\ OH^-$	10

APPENDIX F

Solubility Product Constants at 298.15° K

COMPOUND	SOLUBILITY PRODUCT	COMPOUND	SOLUBILITY PRODUCT
Barium:		**Copper:**	
$BaCO_3$	7×10^{-9}	Cupric:	
$BaCrO_4$	3×10^{-10}	$Cu(IO_3)_2$	2×10^{-7}
BaF_2	3×10^{-6}	CuC_2O_4	3×10^{-8}
$Ba(IO_3)_2$	1×10^{-9}	CuS	10^{-42}
$BaC_2O_4 \cdot 2\,H_2O$	2×10^{-8}	Cuprous:	
$BaSO_4$	1×10^{-10}	CuBr	4×10^{-8}
		CuCl	1×10^{-6}
Cadmium:		CuI	4×10^{-12}
CdC_2O_4	1×10^{-8}	CuSCN	4×10^{-14}
CdS	10^{-28}		
		Iron:	
Calcium:		Ferric:	
$CaCO_3$	1×10^{-8}	$Fe(OH)_3$	10^{-36}
CaF_2	3×10^{-11}	Ferrous:	
$Ca(OH)_2$	5×10^{-6}	$Fe(OH)_2$	10^{-15}
$Ca(IO_3)_2 \cdot 6\,H_2O$	6×10^{-7}	FeC_2O_4	2×10^{-7}
$CaC_2O_4 \cdot H_2O$	2×10^{-9}	FeS	10^{-22}
$CaSO_4$	2×10^{-4}		
		Lead:	
Cerium:		$PbBr_2$	7×10^{-5}
$Ce(OH)_3$	10^{-20}	$PbCl_2$	2×10^{-4}
$Ce(IO_3)_3$	3×10^{-10}	PbClF	3×10^{-9}
$Ce_2(C_2O_4)_3 \cdot H_2O$	3×10^{-29}	$PbCrO_4$	8×10^{-14}
		PbF_2	5×10^{-8}
Cobalt:		$Pb(IO_3)_2$	2×10^{-13}
CoS	10^{-27}	PbI	1×10^{-8}

COMPOUND	SOLUBILITY PRODUCT	COMPOUND	SOLUBILITY PRODUCT
PbC_2O_4	3×10^{-11}	**Silver:**	
$PbSO_4$	2×10^{-8}	CH_3COOAg	4×10^{-3}
PbS	10^{-28}	AgBr	5×10^{-13}
		Ag_2CO_3	5×10^{-12}
Magnesium:		AgCl	1.8×10^{-10}
$MgCO_3 \cdot 3\,H_2O$	3×10^{-5}	Ag_2CrO_4	2×10^{-12}
MgF_2	7×10^{-9}	AgOH	10^{-8}
$Mg(OH)_2$	10^{-11}	$AgIO_3$	2×10^{-8}
$MgC_2O_4 \cdot 2\,H_2O$	1×10^{-8}	AgI	1×10^{-16}
		Ag_2SO_4	1.2×10^{-5}
Manganese:		Ag_2S	6×10^{-52}
$Mn(OH)_2$	10^{-14}		
MnS	10^{-16}	**Strontium:**	
Mercury:		$SrCO_3$	2×10^{-9}
Mercuric:		$SrCrO_4$	3×10^{-5}
HgS	9×10^{-52}	SrF_2	3×10^{-9}
Mercurous:		$SrC_2O_4 \cdot H_2O$	1×10^{-9}
Hg_2Br_2	1×10^{-21}	$SrSO_4$	5×10^{-7}
Hg_2Cl_2	2×10^{-18}		
Hg_2I_2	7×10^{-29}	**Zinc:**	
Hg_2SO_4	6×10^{-7}	$ZnCO_3$	3×10^{-8}
		$Zn(OH)_2$	10^{-17}
Nickel:		$ZnC_2O_4 \cdot 2\,H_2O$	3×10^{-9}
$Ni(OH)_2$	6×10^{-18}	ZnS	10^{-25}
NiS	10^{-19}		

APPENDIX G

The Reliability of Numbers

The preferred method of writing a number in order to show clearly its reliability is in *scientific notation.* The decimal point is placed after the first digit, only those digits that are significant in expressing the known accuracy of the number are written to the right of the decimal, and the magnitude is shown by a product of 10 raised to the appropriate power. Thus, 0.00023 is written 2.3×10^{-4}, and 341,000,000 is written 3.41×10^8, if we assume that there are three meaningful figures in this number. The problem of the reliability of the number is evident in the second example just considered; when one sees 341,000,000 one has no way of knowing precisely what is meant. Depending upon the accuracy it might be written as 3×10^8, 3.4×10^8, or with still other numbers of digits. In the following paragraphs we will present in greater detail some techniques for indicating reliability.

You should recognize that scientific notation is designed for convenience; when its use is inconvenient, there is little point in making a fetish of consistency. For example, the volume of solution delivered by a volumetric buret is estimated to a hundredth of a milliliter. A reading might be 23.08 ml. Few chemists would choose to write this as 2.308×10^1 ml. On the other hand, if this volume were expressed in liters, many would then write 2.308×10^{-2} liter instead of 0.02308 liter.

It can be seen from the above discussion that a number should convey two kinds of information: the magnitude of the quantity involved and the accuracy to which this quantity is known. There are some numbers known with unlimited accuracy; for instance, there is exactly 1 atom of sulfur in a single

Portions of this appendix were taken from Ernest H. Swift and Eliot A. Butler, QUANTITATIVE MEASUREMENTS AND CHEMICAL EQUILIBRIA (San Francisco: W.H. Freeman and Company, 1972), pp. 23–25.

molecule of H_2SO_4, and the fraction $\frac{3}{7}$ can be expressed as a decimal fraction having as many meaningful (significant) digits as desired. On the other hand, numbers obtained by measurement in the laboratory are, in general, limited in accuracy by the nature of the experiment and by the measurement techniques used. It is essential that experimenters state their data in such a way that limitations in accuracy are clearly evident.

Significant Figures

The simplest means of indicating the reliability of a number is through the use of significant figures; according to this convention, the digits composing the number are extended to, but not beyond, one digit whose value is uncertain. Thus, if a length measurement is given as 1.02 cm, the implication is that the length is not 1.01 cm and not 1.03 cm; the length is given to three significant figures. The same length could be recorded to the same accuracy (if we ignore the use of scientific notation) as 0.0102 m; there are still three significant figures since the zeros before the 1 only place the decimal point. In the preferred notation, 1.02×10^{-2} m, the number of significant figures is more clearly seen. If the same length were expressed in microns ($10^6\ \mu = 1$ m), the magnitude is given by 10200 μ, but the reliability is incorrectly stated, because the five digits imply that the length is between 10199 and 10201 μ. The proper statement of the length in these units is $1.02 \times 10^4\ \mu$.

Significant Figures in Mathematical Operations

When data are involved in mathematical operations, care must be used that the final value reflects properly the accuracy of the data; accuracy should not be lost, nor should the final value imply an unwarranted increase in the accuracy. Suppose that the following numbers, each properly expressed to its known reliability, are to be added: 1027.1, 1.33249, 0.2, and 7×10^{-3}. First write the numbers in a column for addition with the decimal points aligned:

$$\begin{array}{lr} & 1027.1 \\ & 1.33249 \\ & 0.2 \\ & 0.007 \\ \hline \text{The sum is} & 1028.6 \end{array}$$

Columns with unknown digits are not used. The number of significant figures in the sum is not limited to the least reliable number that was added. Thus 0.2 has only one significant figure, but it contributes to the final sum; 7×10^{-3} also has one significant figure but makes no contribution; 1.33249 has six significant figures, of which only two contribute.

One should notice, however, that each number was used to the same *absolute* reliability. For convenience, assume that the numbers added above are lengths expressed in centimeters. Each number was added to the nearest 0.1 cm, even though the numbers of significant figures were very different.

The situation is different in operations involving multiplication. In the operation

$$\frac{(3.60 \times 10^{-3})(2.9 \times 10^{2})}{0.351}$$

the result is correctly expressed as 3.0, because one number in the product is known only to two significant figures. The reliability of a result in a series of multiplications and divisions cannot possibly exceed that of the least reliable term.

Rounding Off Significant Figures

If one is to maintain significant figures in arithmetical operations, one must round off the number properly. If 27.3 and 19.76 are to be added, the answer is rounded to 47.1.† If the first digit to be dropped is greater than 5, the last digit retained is raised by 1; if the first digit to be dropped is less than 5, the last digit retained is unchanged. If the first digit to be dropped is 5 and is followed by zeros or by no digits, then the last digit retained is made the nearest even number. The examples that follow illustrate these rules. In each case the number is rounded to three significant figures.

1.237	becomes	1.24
24.92	becomes	24.9
29.65	becomes	29.6
29.651	becomes	29.7
4.315	becomes	4.32

† A good practice commonly followed in making calculations is to carry one more significant figure in the arithmetical steps than is justified by the data. The answer is then rounded to the proper number of digits. This avoids the possibility of cumulative errors from rounding several numbers in a calculation.

APPENDIX H

Nomenclature of Inorganic Compounds

During the beginnings of chemistry, names such as water, ammonia, and saltpeter were sufficient to identify the few known chemical compounds. As chemistry grew in complexity, different naming systems were developed. The IUPAC† system is an attempt to bring order into the nomenclature of compounds. It is general and descriptive and the generally preferred system of nomenclature. However, several systems are still in use, and we must learn to apply all of them. We describe here the rules for naming the simple inorganic compounds. When differences occur between the various systems, we will so indicate.

Binary Compounds‡

In binary compounds containing a metal or hydrogen and a nonmetal, the metal is named first and the ending "-ide" is used in place of the ending of the nonmetal. For example:

$CaBr_2$	calcium bromide
NaF	sodium fluoride
KI	potassium iodide
BaO	barium oxide
Mg_3P_2	magnesium phosphide
H_2S	hydrogen sulfide

† This is the acronym for the International Union of Pure and Applied Chemistry.

‡ These are compounds containing only two elements.

Two different procedures are used to distinguish binary compounds with the metal in different oxidation states. In the first system, which is applicable when only two oxidation states occur, the ending of the name of the metal is changed to "-ous" for the lower oxidation state and "-ic" for the higher oxidation state. For example:

Hg_2Cl_2	mercurous chloride
$HgCl_2$	mercuric chloride
SnO	stannous oxide
SnO_2	stannic oxide

Note that the Latin name is used for tin. In this naming system, the Latin name is used for tin (stannum), iron (ferrum), copper (cuprum), silver (argentum), gold (aurum), and lead (plumbum). Thus the two fluorides of lead are plumbous fluoride (PbF_2) and plumbic fluoride (PbF_4).

A better system, and the one endorsed by IUPAC, has a Roman numeral following the name of the metal to designate its oxidation state. Examples are

Hg_2Cl_2	mercury (I) chloride
$HgCl_2$	mercury (II) chloride
MnO	manganese (II) oxide
Mn_2O_3	manganese (III) oxide
MnO_2	manganese (IV) oxide
Mn_2O_7	manganese (VII) oxide

When the binary compound contains two nonmetals, the more electropositive (less electronegative) element is named first, and prefixes are used to show the number of atoms of each element in the molecule. The numerical prefixes are

mono- = 1	hexa- = 6
di- = 2	hepta- = 7
tri- = 3	octa- = 8
tetra- = 4	nona- = 9
penta- = 5	deca- = 10

Examples are

CO	carbon monoxide
CO_2	carbon dioxide
PCl_3	phosphorus trichloride
PCl_5	phosphorus pentachloride
N_2O	dinitrogen monoxide
N_2O_3	dinitrogen trioxide

N_2O_5	dinitrogen pentoxide
P_4O_{10}	tetraphosphorus decaoxide

Notice that the "mono-" prefix is often left off the name of the first element. We do not often say monocarbon dioxide.

Sometimes this last system is used for naming metal compounds. Also, the IUPAC system is sometimes used for compounds containing only nonmetals. For example:

$SnCl_4$	tin (IV) chloride tin tetrachloride stannic chloride
P_2S_5	diphosphorus pentasulfide phosphorus (V) sulfide

In addition, a special class of compounds containing more than two elements are named like binary compounds. These compounds contain a covalently bonded complex ion that replaces one of the elements in the binary compounds. An abbreviated list of these ions follows:

OH^-	hydroxide
CN^-	cyanide
NH_2^-	amide
NH_4^+	ammonium†
CNO^-	cyanate

Also, the peroxide ion O_2^{2-}, although it contains only one element, fits with this grouping. Examples of compounds containing these radicals are

NH_4Cl	ammonium chloride
NaOH	sodium hydroxide
$LiNH_2$	lithium amide
$Fe(CN)_2$	iron (II) cyanide or ferrous cyanide
BaO_2	barium peroxide

Acids

Binary hydrogen compounds of the Group VIA and VIIA elements ionize in water solution to form acids. We name them in the gas phase as given above

† NH_4^+ takes the place of the metal atom; the others replace the nonmetals.

(for example, H_2Se: hydrogen selenide), and we name them in water solution as acids. The acid name is obtained by giving first the hydrogen by its stem "hydro-" and coupling this with the stem of the other element and adding "-ic." For example:

HCl	hydrochloric acid
H_2S	hydrosulfuric acid
H_2Se	hydroselenic acid
HF	hydrofluoric acid

Ternary acids contain hydrogen and usually oxygen. The third element is usually a nonmetal or a metal in a high oxidation state. The common names of these acids are obtained by adding prefixes and suffixes to the name of the third element to distinguish the oxidation state of this element. The prefixes and suffixes listed as a function of oxidation state are as follows:

Increasing oxidation number ↓	
	hypo ______ ous
	______ ous
	______ ic
	per ______ ic

The prefixes are used only if more than two oxidation states are present. For example:

H_2SO_3	sulfurous acid
H_2SO_4	sulfuric acid
HClO	hypochlorous acid
$HClO_2$	chlorous acid
$HClO_3$	chloric acid
$HClO_4$	perchloric acid

Salts

We have already described the naming of binary salts, such as sodium chloride. For more complex salts, we name the metallic element first followed by the root name of the acid from which the salt is considered to have been derived. Where necessary, "-ous" or "-ic" endings or Roman numerals are used to distinguish oxidation states of the metal. Again, the ammonium ion is considered as a metal. The name of the negative ion is obtained from the acid by replacing the "-ous" ending with "-ite" and the "ic" ending with "-ate."† Thus,

† A few complex ions occur so frequently in chemical compounds that you should commit their names to memory. The ions and their names are as follows:

NH_4^+	ammonium	ClO_4^-	perchlorate	SO_4^{2-}	sulfate	$Cr_2O_7^{2-}$	dichromate
OH^-	hydroxide	CN^-	cyanide	CO_3^{2-}	carbonate	MnO_4^-	permanganate
NO_3^-	nitrate	CrO_4^{2-}	chromate	PO_4^{3-}	phosphate		

nitric acid (HNO_3) becomes nitrate (NO_3^-), nitrous acid (HNO_2) becomes nitrite (NO_2^-), hypobromous acid (HBrO) becomes hypobromite (BrO^-), and perchloric acid ($HClO_4$) becomes perchlorate (ClO_4^-). Examples of the use of this system are as follows:

Na_2SO_4	sodium sulfate
$Sn(NO_3)_2$	tin (II) nitrate or stannous nitrate
$KClO_4$	potassium perchlorate
$(NH_4)_2CO_3$	ammonium carbonate
$FePO_4$	iron (III) phosphate

Sometimes the salt contains more than one positive ion. In these cases, both positive elements are named:

$NaHCO_3$	sodium hydrogen carbonate
NaH_2PO_4	sodium dihydrogen phosphate
Na_2HPO_4	disodium hydrogen phosphate
$MgNH_4PO_4$	magnesium ammonium phosphate

Other complications also occur. For example, sulfur can replace oxygen in the anion. A "thio-" prefix is used to indicate this replacement. For example:

SO_4^{2-} sulfate	$S_2O_3^{2-}$ thiosulfate
OCN^- cyanate	SCN^- thiocyanate

Thus Na_2SO_4 is sodium sulfate and $Na_2S_2O_3$ is sodium thiosulfate.

We have, of course, only begun the discussion of nomenclature rules. In Chapter 20 we expand the rules to include complex ions, and in Chapter 24 we give the very beginnings of the nomenclature of organic compounds. With the rules given here and in Chapter 20, you can name a very large number of chemical compounds. Expansion of the rules to include more complicated systems is reserved for more advanced study.

APPENDIX I

SI Units

In order to establish some uniformity in the use of units, the General Conference of Weights and Measures in 1960 adapted a set known as the SI† units. The SI system is based on the seven fundamental units shown in Table I.1. Various derived units can be constructed from these seven. The most common derived units have the special names and symbols shown in Table I.2. Multiples of the units in Tables I.1 and I.2 are formed with the help of the prefixes shown in Table I.3. The special names of some multiples and submultiples are given in Table I.4. The SI equivalents of some common non-SI units are listed in Table I.5.

Perhaps some day in the future SI units will be used exclusively in a textbook such as this. For the present, we will continue to use energy units such as the calorie, electron volt, and erg in addition to the joule; distances such as centimeters and angstroms; volumes in liters; and mass in grams. Conversion factors for changing from one set of units to another are given in Tables I.4 and I.5 and on the inside back cover of the book.

† SI is an abbreviation from the official name in French "le Systéme International d'Unites."

TABLE I.1

Names and Symbols of the SI Base Units

PHYSICAL QUANTITY	NAME OF SI UNIT	SYMBOL FOR SI UNIT
length	meter	m
mass	kilogram	kg
time	second	s
electric current	ampere	A
thermodynamic temperature	kelvin	K
amount of substance	mole	mol

TABLE I.2

SI Derived Units Having Special Names and Symbols

PHYSICAL QUANTITY	NAME OF UNIT	SYMBOL FOR UNIT	DEFINITION OF UNIT
frequency	hertz	Hz	s^{-1}
energy	joule	J	$m^2 \cdot kg \cdot s^{-2}$
force	newton	N	$m \cdot kg \cdot s^{-2} = J \cdot m^{-1}$
power	watt	W	$m^2 \cdot kg \cdot s^{-3} = J \cdot s^{-1}$
pressure	pascal	Pa	$m^{-1} \cdot kg \cdot s^{-2} = N \cdot m^{-2} = J \cdot m^{-3}$
electric charge	coulomb	C	$s \cdot A$
electric potential difference	volt	V	$m^2 \cdot kg \cdot s^{-3} \cdot A^{-1} = J \cdot s^{-1} A^{-1}$
electric resistance	ohm	Ω	$m^2 \cdot kg \cdot s^{-3} \cdot A^{-2} = V \cdot A^{-1}$

TABLE I.3

SI Prefixes

SUBMULTIPLE	PREFIX	SYMBOL	MULTIPLE	PREFIX	SYMBOL
10^{-1}	deci	d	10	deca	da
10^{-2}	centi	c	10^2	hecto	h
10^{-3}	milli	m	10^3	kilo	k
10^{-6}	micro	μ	10^6	mega	M
10^{-9}	nano	n	10^9	giga	G
10^{-12}	pico	p	10^{12}	tera	T
10^{-15}	femto	f			
10^{-18}	atto	a			

TABLE I.4

Decimal Fractions and Multiples of SI Units Having Special Names

PHYSICAL QUANTITY	NAME OF UNIT	SYMBOL FOR UNIT	DEFINITION OF UNIT
length	Ångstrom	Å	10^{-10} m
length	micron	μ	10^{-6} m = μm
force	dyne	dyn	10^{-5} N
pressure	bar	bar	$10^5 \ N \cdot m^{-2}$
energy	erg	erg	10^{-7} J

TABLE I.5

Non-SI Units

PHYSICAL QUANTITY	NAME OF UNIT	SYMBOL FOR UNIT	DEFINITION OF UNIT
length	inch	in	2.54×10^{-2} m
mass	pound (avoirdupois)	lb	0.45359237 kg
force	kilogram-force	kgf	9.80665 N
pressure	atmosphere	atm	$101{,}325\ \mathrm{N \cdot m^{-2}}$
pressure	torr	torr	$(101{,}325/760)\ \mathrm{N \cdot m^{-2}}$
energy	British thermal unit	BTU	1055.056 J
energy	kilowatt-hour	kWh	3.6×10^{6} J
energy	thermochemical calorie	$\mathrm{cal_{th}}$	4.184 J
energy	electron volt	eV	1.60219×10^{-19} J
mass	unified atomic mass unit	amu	$1.6605655 \times 10^{-27}$ kg
electric dipole moment	Debye	D	$3.33564 \times 10^{-30}\ \mathrm{A \cdot m \cdot s}$
charge per molecule	Faraday	F	$9.648456 \times 10^{4}\ \mathrm{C\,mol^{-1}}$

APPENDIX J

Derivation of the Ideal Gas Equation from Kinetic-Molecular Theory

Starting with the two postulates of the kinetic-molecular theory described in Chapter 6—(1) *matter is composed of molecules* and (2) *the molecules are in motion*—and adding a third postulate—(3) *the average kinetic energy of the molecules in a gas is directly proportional to the absolute temperature*—we can derive the ideal gas equation obtained from experimental observation in Chapter 5. We will see that the derivation is based on two assumptions: (1) *the molecules in the gas are point masses* (they occupy no volume), and (2) *there are no attractions or repulsions between the molecules.*

These two assumptions actually define the ideal gas on the molecular level.

The Root-Mean-Square Velocity

Let us consider N ideal gas molecules enclosed in a sphere of volume V. Since the gas is ideal, the molecules have no volume; consequently, V is the free volume in the sphere. As the molecules move in the sphere, they collide with the walls and exert a force F and pressure P. The pressure and force are related by the equation

The shape of the container is of no importance. The derivation is made for a spherical container. The same result would be obtained for a cube or any other shape.

$$P = \frac{F}{A} \qquad [J.1]$$

where A is the surface area of the sphere. Because the gas is ideal, no attractive forces exist between the molecules. Therefore, each molecule behaves independently, and the total force is the sum of the forces exerted by the individual molecules:

Equation [J.2] applies only when the molecules act independently, that is, when there are no attractive or repulsive forces between the molecules.

$$F = \sum F_i \qquad [J.2]$$

The force exerted by the collisions of a moving object is the change in momentum per unit of time. By canceling the units we see that for each molecule

$$F_i = \text{change of momentum/sec} = \frac{\text{(distance traveled/sec)(change of momentum/impact)}}{\text{(distance traveled/impact)}} \qquad [J.3]$$

The force is expressed in this manner because the quantities on the far right can easily be calculated. The first term in the numerator is the velocity of the molecule:

$$\text{distance traveled/sec} = u_i \qquad [J.4]$$

To calculate the other terms in Equation [J.3], consider a molecule moving in a sphere, as in Figure J.1, in such a way that the molecule collides at point A, moves along the diameter of the sphere and collides again at point B, then bounces straight back. The distance traveled per impact, the denominator in Equation [J.3], is the distance $AB = 2r$. The momentum of the molecule just before the collision is mu_i. After the collision, the direction of the motion is reversed and the momentum is $m(-u_i)$. The change in momentum, the second term in the numerator of Equation [J.3], becomes

$$mu_i - m(-u_i) = 2mu_i$$

and the force calculated from Equation [J.3] is

$$F_i = \frac{(2mu_i)(u_i)}{2r} = \frac{mu_i^2}{r} \qquad [J.5]$$

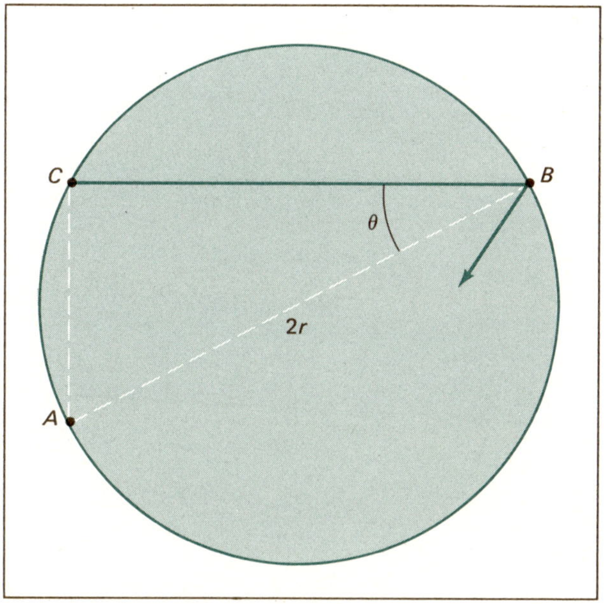

Figure J.1.
A gas molecule collides with the walls of a sphere at point *C*, rebounds, and collides again at *B* after traveling a distance *CB* = 2*r*cos θ. The collision at *B* changes the momentum of the molecule by 2*mu*cos θ.

In general the molecule is not moving along the radius. Instead it may take a path as shown in Figure J.1, in which it collides first at C then at B. In this case,

$$\frac{\text{distance traveled}}{\text{impact}} = CB = 2r\cos\theta \qquad [\text{J.6}]$$

where θ is the angle shown in the triangle in which ACB is a right angle. Only the component of momentum perpendicular to the surface of impact contributes to the force (Figure J.2). The perpendicular component before collision is $mu_i \cos\theta$, while after collision it is $m(-u_i \cos\theta)$. Hence,

$$\frac{\text{change in momentum}}{\text{impact}} = mu_i\cos\theta - m(-u_i\cos\theta) = 2mu_i\cos\theta \qquad [\text{J.7}]$$

Combining Equations [J.4], [J.6], and [J.7] with [J.3] gives the same result as obtained earlier:

$$F_i = \frac{(2mu_i\cos\theta)(u_i)}{(2r\cos\theta)} = \frac{mu_i^2}{r} \qquad [\text{J.8}]$$

The pressure can be related to the velocity of the molecules and size of the sphere by combining Equations [J.8], [J.2], and [J.1]. The ratio m/r is a constant that can be factored out of the summation. The resulting equation is

$$P = \frac{F}{A} = \frac{m}{Ar}\sum u_i^2 \qquad [\text{J.9}]$$

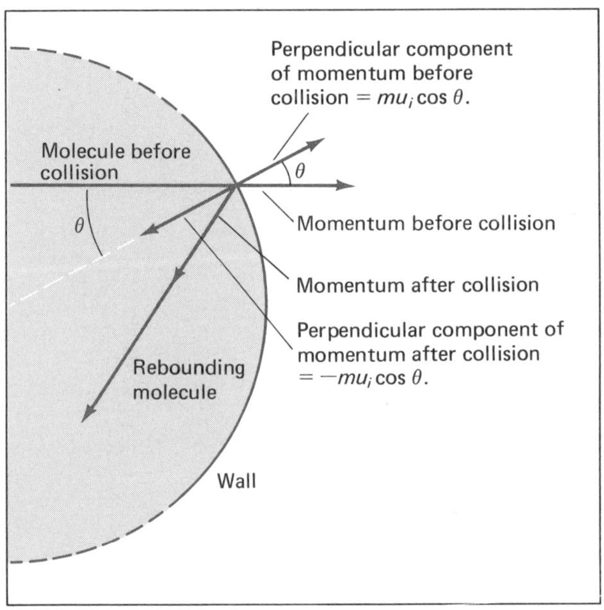

Figure J.2.
The change in the perpendicular component of momentum is $mu_i \cos\theta - (-mu_i\cos\theta) = 2mu_i\cos\theta$.

We now make substitutions for A and $\sum u_i^2$. The surface area of a sphere is given by

$$A = 4\pi r^2 \qquad [J.10]$$

$\sum u_i^2$ is related to $\overline{u^2}$, the average squared velocity, through the relation

$$\overline{u^2} = \frac{\sum u_i^2}{N} \qquad [J.11]$$

Combining [J.9], [J.10], and [J.11] gives

$$P = \frac{Nm\overline{u^2}}{4\pi r^3} \qquad [J.12]$$

For a sphere, the volume is given by

$$V = \frac{4}{3}\pi r^3$$

from which we can write

$$3V = 4\pi r^3 \qquad [J.13]$$

Equation [J.13] applies only if V is the "free volume" available for movement of the gas molecules. This stipulation is obtained when the gas molecules are point masses, that is, when they occupy no volume.

Substituting [J.13] into [J.12] and rearranging gives

$$PV = \frac{1}{3}Nm\overline{u^2} \qquad [J.14]$$

We can express N in terms of moles (n) by the relation

$$n = \frac{N}{N_0} \qquad [J.15]$$

where N_0 is Avogadro's number. Likewise

$$m = \frac{M_w}{N_0} \qquad [J.16]$$

where M_w is the molecular weight. Substitution of [J. 15] and [J. 16] into [J. 14] gives

$$PV = \frac{1}{3}nM_w u^2 \qquad [J.17]$$

The root mean square velocity u_{rms} is defined as the square root of $\overline{u^2}$. That is,

$$u_{rms} = \sqrt{\overline{u^2}} \qquad [J.18]$$

Substituting [J.18] into [J.17] relates u_{rms} to the PV products of the gas:

Equation [J.19] is a mathematical statement of the law of Boyle. At a given temperature, u_{rms} is a constant as are n and M. Hence

$$PV = \text{const}$$

(See Equation [5.2].)

$$PV = \frac{1}{3}nM_w u_{rms}^2 \qquad [J.19]$$

We now relate $\bar{E}_{\text{trans}}$, the average translational kinetic energy, to u^2. For each molecule, the translational kinetic energy is related to its velocity by the equation as given below for the "i"th molecule:

$$(E_{\text{trans}})_i = \frac{1}{2}mu_i^2 \qquad [J.20]$$

The average translational energy is then given by

$$\bar{E}_{\text{trans}} = \frac{\sum(E_{\text{trans}})_i}{N} = \frac{\frac{1}{2}m\sum u_i^2}{N} \qquad [J.21]$$

Combining [J.11] and [J.21] gives

$$\bar{E}_{\text{trans}} = \frac{1}{2}m\overline{u^2} \qquad [J.22]$$

To complete the derivation of the ideal gas equation, we make use of the third postulate of the kinetic-molecular theory to write

$$\bar{E}_{\text{trans}} \propto T \qquad [J.23]$$

or

$$\bar{E}_{\text{trans}} = CT \qquad [J.24]$$

where C is the proportionality constant that relates kinetic energy of translation to the absolute temperature T. Combining Equations [J.24] and [J.22] gives

$$m\overline{u^2} = 2CT \qquad [J.25]$$

Substituting this result into Equation [J.17] and recognizing that $m = M_w/N_0$ (Equation [J.16]) gives

$$PV = \frac{2}{3}nN_0CT \qquad [J.26]$$

Combining the constants 2/3, N_0, and C into one constant R gives the ideal gas equation:

$$PV = nRT \qquad [J.27]$$

APPENDIX K

Logarithms

The logarithm of a number is the power to which a base must be raised in order to give the number. That is,

$$\text{base}^{\text{logarithm}\,M} = M$$

Common logarithms (log) use 10 as the base. Hence,

$$10^{\log M} = M$$

The log of 10 is 1, the log of 1000 is 3, the log of 0.001 is -3, and so on. The table that follows can be used to obtain values of logarithms for numbers that are not even multiples of 10.

Natural logarithms (ln) use 2.71828 . . . as their base. This number, known by the symbol e, results from a mathematical operation and is of great usefulness in calculus. For natural logarithms,

$$e^{\ln M} = M$$

or

$$(2.71828\ldots)^{\ln M} = M$$

A common logarithm (log) from Table K.1 can be converted to the natural logarithm (ln) by use of the factor 2.3026. That is, $\ln A = 2.3026 \log A$. A few numerical examples will illustrate.

A	$\log A$	$\ln A$
23.10	1.3636	3.1398
562.3	2.7500	6.3320
7.2×10^{-5}	-4.14	-9.54

TABLE K.1

Four-Place Common Logarithms

											Proportional Parts								
N	0	1	2	3	4	5	6	7	8	9	1	2	3	4	5	6	7	8	9
10	0000	0043	0086	0128	0170	0212	0253	0294	0334	0374	4	8	12	17	21	25	29	33	37
11	0414	0453	0492	0531	0569	0607	0645	0682	0719	0755	4	8	11	15	19	23	26	30	34
12	0792	0828	0864	0899	0934	0969	1004	1038	1072	1106	3	7	10	14	17	21	24	28	31
13	1139	1173	1206	1239	1271	1303	1335	1367	1399	1430	3	6	10	13	16	19	23	26	29
14	1461	1492	1523	1553	1584	1614	1644	1673	1703	1732	3	6	9	12	15	18	21	24	27
15	1761	1790	1818	1847	1875	1903	1931	1959	1987	2014	3	6	8	11	14	17	20	22	25
16	2041	2068	2095	2122	2148	2175	2201	2227	2253	2279	3	5	8	11	13	16	18	21	24
17	2304	2330	2355	2380	2405	2430	2455	2480	2504	2529	2	5	7	10	12	15	17	20	22
18	2553	2577	2601	2625	2648	2672	2695	2718	2742	2765	2	5	7	9	12	14	16	19	21
19	2788	2810	2833	2856	2878	2900	2923	2945	2967	2989	2	4	7	9	11	13	16	18	20
20	3010	3032	3054	3075	3096	3118	3139	3160	3181	3201	2	4	6	8	11	13	15	17	19
21	3222	3243	3263	3284	3304	3324	3345	3365	3385	3404	2	4	6	8	10	12	14	16	18
22	3424	3444	3464	3483	3502	3522	3541	3560	3579	3598	2	4	6	8	10	12	14	15	17
23	3617	3636	3655	3674	3692	3711	3729	3747	3766	3784	2	4	6	7	9	11	13	15	17
24	3802	3820	3838	3856	3874	3892	3909	3927	3945	3962	2	4	5	7	9	11	12	14	16
25	3979	3997	4014	4031	4048	4065	4082	4099	4116	4133	2	3	5	7	9	10	12	14	15
26	4150	4166	4183	4200	4216	4232	4249	4265	4281	4298	2	3	5	7	8	10	11	13	15
27	4314	4330	4346	4362	4378	4393	4409	4425	4440	4456	2	3	5	6	8	9	11	13	14
28	4472	4487	4502	4518	4533	4548	4564	4579	4594	4609	2	3	5	6	8	9	11	12	14
29	4624	4639	4654	4669	4683	4698	4713	4728	4742	4757	1	3	4	6	7	9	10	12	13
30	4771	4786	4800	4814	4829	4843	4857	4871	4886	4900	1	3	4	6	7	9	10	11	13
31	4914	4928	4942	4955	4969	4983	4997	5011	5024	5038	1	3	4	6	7	8	10	11	12
32	5051	5065	5079	5092	5105	5119	5132	5145	5159	5172	1	3	4	5	7	8	9	11	12
33	5185	5198	5211	5224	5237	5250	5263	5276	5289	5302	1	3	4	5	6	8	9	10	12
34	5315	5328	5340	5353	5366	5378	5391	5403	5416	5428	1	3	4	5	6	8	9	10	11
N	0	1	2	3	4	5	6	7	8	9	1	2	3	4	5	6	7	8	9

N	0	1	2	3	4	5	6	7	8	9	Proportional Parts 1	2	3	4	5	6	7	8	9
35	5441	5453	5465	5478	5490	5502	5514	5527	5539	5551	1	2	4	5	6	7	9	10	11
36	5563	5575	5587	5599	5611	5623	5635	5647	5658	5670	1	2	4	5	6	7	8	10	11
37	5682	5694	5705	5717	5729	5740	5752	5763	5775	5786	1	2	3	5	6	7	8	9	10
38	5798	5809	5821	5832	5843	5855	5866	5877	5888	5899	1	2	3	5	6	7	8	9	10
39	5911	5922	5933	5944	5955	5966	5977	5988	5999	6010	1	2	3	4	5	7	8	9	10
40	6021	6031	6042	6053	6064	6075	6085	6096	6107	6117	1	2	3	4	5	6	8	9	10
41	6128	6138	6149	6160	6170	6180	6191	6201	6212	6222	1	2	3	4	5	6	7	8	9
42	6232	6243	6253	6263	6274	6284	6294	6304	6314	6325	1	2	3	4	5	6	7	8	9
43	6335	6345	6355	6365	6375	6385	6395	6405	6415	6425	1	2	3	4	5	6	7	8	9
44	6435	6444	6454	6464	6474	6484	6493	6503	6513	6522	1	2	3	4	5	6	7	8	9
45	6532	6542	6551	6561	6571	6580	6590	6599	6609	6618	1	2	3	4	5	6	7	8	9
46	6628	6637	6646	6656	6665	6675	6684	6693	6702	6712	1	2	3	4	5	6	7	7	8
47	6721	6730	6739	6749	6758	6767	6776	6785	6794	6803	1	2	3	4	5	5	6	7	8
48	6812	6821	6830	6839	6848	6857	6866	6875	6884	6893	1	2	3	4	4	5	6	7	8
49	6902	6911	6920	6928	6937	6946	6955	6964	6972	6981	1	2	3	4	4	5	6	7	8
50	6990	6998	7007	7016	7024	7033	7042	7050	7059	7067	1	2	3	3	4	5	6	7	8
51	7076	7084	7093	7101	7110	7118	7126	7135	7143	7152	1	2	3	3	4	5	6	7	8
52	7160	7168	7177	7185	7193	7202	7210	7218	7226	7235	1	2	2	3	4	5	6	7	7
53	7243	7251	7259	7267	7275	7284	7292	7300	7308	7316	1	2	2	3	4	5	6	6	7
54	7324	7332	7340	7348	7356	7364	7372	7380	7388	7396	1	2	2	3	4	5	6	6	7
55	7404	7412	7419	7427	7435	7443	7451	7459	7466	7474	1	2	2	3	4	5	5	6	7
56	7482	7490	7497	7505	7513	7520	7528	7536	7543	7551	1	2	2	3	4	5	5	6	7
57	7559	7566	7574	7582	7589	7597	7604	7612	7619	7627	1	2	2	3	4	5	5	6	7
58	7634	7642	7649	7657	7664	7672	7679	7686	7694	7701	1	1	2	3	4	4	5	6	7
59	7709	7716	7723	7731	7738	7745	7752	7760	7767	7774	1	1	2	3	4	4	5	6	7
60	7782	7789	7796	7803	7810	7818	7825	7832	7839	7846	1	1	2	3	4	4	5	6	6
61	7853	7860	7868	7875	7882	7889	7896	7903	7910	7917	1	1	2	3	4	4	5	6	6
62	7924	7931	7938	7945	7952	7959	7966	7973	7980	7987	1	1	2	3	3	4	5	6	6
63	7993	8000	8007	8014	8021	8028	8035	8041	8048	8055	1	1	2	3	3	4	5	5	6
64	8062	8069	8075	8082	8089	8096	8102	8109	8116	8122	1	1	2	3	3	4	5	5	6
65	8129	8136	8142	8149	8156	8162	8169	8176	8182	8189	1	1	2	3	3	4	5	5	6
66	8195	8202	8209	8215	8222	8228	8235	8241	8248	8254	1	1	2	3	3	4	5	5	6
67	8261	8267	8274	8280	8287	8293	8299	8306	8312	8319	1	1	2	3	3	4	5	5	6
68	8325	8331	8338	8344	8351	8357	8363	8370	8376	8382	1	1	2	3	3	4	4	5	6
69	8388	8395	8401	8407	8414	8420	8426	8432	8439	8445	1	1	2	2	3	4	4	5	6
70	8451	8457	8463	8470	8476	8482	8488	8494	8500	8506	1	1	2	2	3	4	4	5	6
71	8513	8519	8525	8531	8537	8543	8549	8555	8561	8567	1	1	2	2	3	4	4	5	5
72	8573	8579	8585	8591	8597	8603	8609	8615	8621	8627	1	1	2	2	3	4	4	5	5
73	8633	8639	8645	8651	8657	8663	8669	8675	8681	8686	1	1	2	2	3	4	4	5	5
74	8692	8698	8704	8710	8716	8722	8727	8733	8739	8745	1	1	2	2	3	4	4	5	5
N	0	1	2	3	4	5	6	7	8	9	1	2	3	4	5	6	7	8	9

N	0	1	2	3	4	5	6	7	8	9	Proportional Parts 1	2	3	4	5	6	7	8	9
75	8751	8756	8762	8768	8774	8779	8785	8791	8797	8802	1	1	2	2	3	3	4	5	5
76	8808	8814	8820	8825	8831	8837	8842	8848	8854	8859	1	1	2	2	3	3	4	5	5
77	8865	8871	8876	8882	8887	8893	8899	8904	8910	8915	1	1	2	2	3	3	4	4	5
78	8921	8927	8932	8938	8943	8949	8954	8960	8965	8971	1	1	2	2	3	3	4	4	5
79	8976	8982	8987	8993	8998	9004	9009	9015	9020	9025	1	1	2	2	3	3	4	4	5
80	9031	9036	9042	9047	9053	9058	9063	9069	9074	9079	1	1	2	2	3	3	4	4	5
81	9085	9090	9096	9101	9106	9112	9117	9122	9128	9133	1	1	2	2	3	3	4	4	5
82	9138	9143	9149	9154	9159	9165	9170	9175	9180	9186	1	1	2	2	3	3	4	4	5
83	9191	9196	9201	9206	9212	9217	9222	9227	9232	9238	1	1	2	2	3	3	4	4	5
84	9243	9248	9253	9258	9263	9269	9274	9279	9284	9289	1	1	2	2	3	3	4	4	5
85	9294	9299	9304	9309	9315	9320	9325	9330	9335	9340	1	1	2	2	3	3	4	4	5
86	9345	9350	9355	9360	9365	9370	9375	9380	9385	9390	1	1	2	2	3	3	4	4	5
87	9395	9400	9405	9410	9415	9420	9425	9430	9435	9440	0	1	1	2	2	3	3	4	4
88	9445	9450	9455	9460	9465	9469	9474	9479	9484	9489	0	1	1	2	2	3	3	4	4
89	9494	9499	9504	9509	9513	9518	9523	9528	9533	9538	0	1	1	2	2	3	3	4	4
90	9542	9547	9552	9557	9562	9566	9571	9576	9581	9586	0	1	1	2	2	3	3	4	4
91	9590	9595	9600	9605	9609	9614	9619	9624	9628	9633	0	1	1	2	2	3	3	4	4
92	9638	9643	9647	9652	9657	9661	9666	9671	9675	9680	0	1	1	2	2	3	3	4	4
93	9685	9689	9694	9699	9703	9708	9713	9717	9722	9727	0	1	1	2	2	3	3	4	4
94	9731	9736	9741	9745	9750	9754	9759	9763	9768	9773	0	1	1	2	2	3	3	4	4
95	9777	9782	9786	9791	9795	9800	9805	9809	9814	9818	0	1	1	2	2	3	3	4	4
96	9823	9827	9832	9836	9841	9845	9850	9854	9859	9863	0	1	1	2	2	3	3	4	4
97	9868	9872	9877	9881	9886	9890	9894	9899	9903	9908	0	1	1	2	2	3	3	4	4
98	9912	9917	9921	9926	9930	9934	9939	9943	9948	9952	0	1	1	2	2	3	3	4	4
99	9956	9961	9965	9969	9974	9978	9983	9987	9991	9996	0	1	1	2	2	3	3	3	4
N	0	1	2	3	4	5	6	7	8	9	1	2	3	4	5	6	7	8	9

Answers

Chapter 2

EXERCISES

1. (a) 7 (b) 6 (c) 1 (d) 4 (e) 3 (f) 6 (g) 2 (h) 3 (i) 3,4,6 (j) 6 (k) 1

2. 2.6 g/cm^3

3. (a) $\sim 490^\circ$ K, $\sim 575^\circ$ K (b) dark solid (c) At_2

4. (a) $1s^22s^22p^63s^1$ (b) $1s^22s^22p^63s^23p^4$ (c) $1s^22s^22p^5$ (d) $1s^22s^22p^63s^23p^6$ (e) $1s^22s^22p^63s^23p^1$

5. (a) $1s^22s^22p^63s^23p^64s^2$ (b) $[Xe]4f^{14}5d^{10}6s^26p^1$ (c) $[Ar]3d^{10}4s^24p^5$ (d) $[Kr]4d^25s^2$ (e) $[Kr]4d^{10}5s^25p^6$

6. (a) $[Xe]6s^2$ (b) $[Xe]4f^{14}5d^46s^2$ (c) $[Xe]4f^46s^2$ (d) $[Xe]4f^{14}5d^{10}6s^26p^2$ (e) $[Xe]4f^{14}5d^{10}6s^26p^5$

7. (a), (b), (e), (g), and (h) are ionic; (c), (d), and (f) are covalent

8.

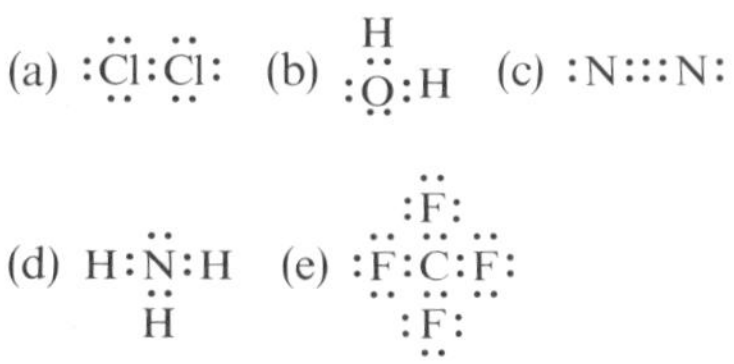

9.

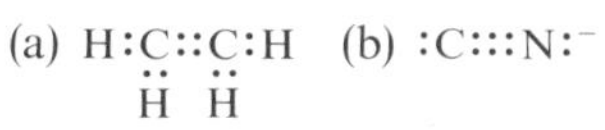

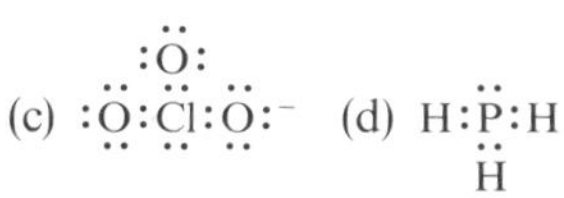

10.

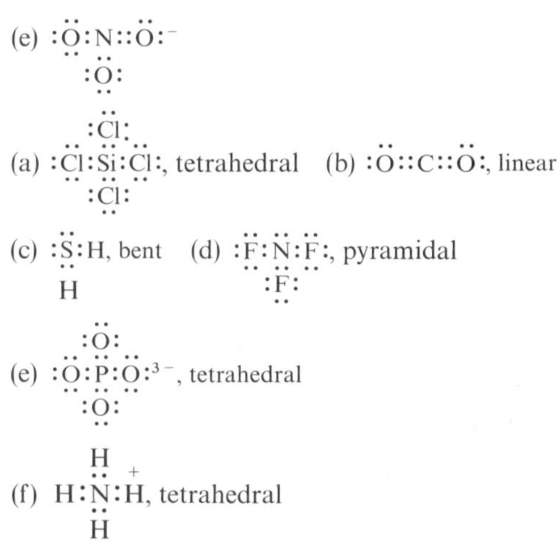

11. (a) Li_2S (b) Al_2O_3 (c) $CaSO_4$ (d) $Sr(ClO_4)_2$ (e) $Mg(NO_3)_2$ (f) $(NH_4)_2CO_3$

Chapter 3

EXERCISES

1. 0.99 garf

2. 27.0

3. 10.0 g B

4. 20.18 amu/atom

5. (a) 5.02×10^{23} atoms (b) 3.76×10^{23} atoms (c) 2.11×10^{23} atoms

6. (a) 0.585 mole (b) 3.52×10^{23} molecules
   (c) 7.75×10^{24} atoms
7. (a) 3.33×10^{-3} mole (b) 9.63×10^{22} atoms
8. 4.00×10^3 g
9. (a) 43.38% Na, 11.33% C, 45.29% O
   (b) 29.44% Ca, 23.55% S, 47.01% O
   (c) 18.53% Na, 25.84% S, 51.58% O, 4.06% H
10. (a) 3.75 moles (b) 0.937 mole
11. (a) 5.32×10^{-23} g (b) 1.40×10^{-21} g
12. (a) 0.3320 mole Ba, 0.6640 mole Cl
    (b) 1.999×10^{23} molecules
13. 6.71×10^4 g/mole
14. CH_2O
15. C_2H_5
16. SnF_4
17. (a) MnO_2 (b) MnO_2 (c) Na_3PO_4 (d) C_2H_6O
18. BaO_2
19. Ta_2O_4
20. (a) 0.50 mole (b) 6.0 g (c) 6.02×10^{23} atoms
21. 17.3 g
22. (a) 27 g (b) 1.5×10^{23} molecules
23. (a) 1.07 g (b) 2.19×10^{22} molecules
24. 13.5 g
25. 49.0 g
26. 15.8 g C_4H_{10}, 66.3 g O_2
27. (a) 7.41 *M* (b) 0.149 mole fraction
28. (a) 0.694 *M* (b) 0.617 *m*
29. (a) 2.172 *m* (b) 0.9624 mole fraction
30. d = 0.937 g/ml
31. 4.63 g
32. 47 ml
33. 86.0 ml
34. 375 ml
35. 1.68 l
36. (a) $m = 3.70$ (b) $M = 2.79$
37. 46 g

PROBLEMS

1. 0.35 g
2. 83%
3. (a) 4.98 moles (b) 399 g (c) 327 g
   (d) 6.14×10^{24} molecules (e) 110 l
4. (a) 50.0 g/mole (b) 140 g
5. (a) 76.64 g $C_7H_6O_3$, 33.32 g $C_2H_4O_2$
   (b) 308.6 tablets
6. (a) 5.0183×10^{24} atoms/mole
   (b) 8.39975 amu/atom (c) 0.19
7. 1.60×10^4 g/mole
8. (a) 6.0×10^4 light years (b) 5.4×10^8 millenia

Chapter 4

EXERCISES

1. (a) 0.437 *M* (b) 437 ml
2. 0.27 g
3. 15.0 ml
4. 385 ml
5. 0.0765 *M*
6. 0.572 *M*
7. 0.06173 *M*
11. 1266 ml
12. 0.1071 *M*
13. 55.0 g
14. 3.67 *M*
16. (a) 1.3×10^{-3} *M* (b) 1.3×10^{-3} *M* (c) 0.10 *M*
17. (a) 8.2×10^{-3} *M* (b) 1.5×10^{-3} *M*
18. 1.9×10^{-3} *M*
19. 1.8×10^{-5} *M*
20. 1.8×10^{-5} *M*
21. (1) pH = 3.0 (2) pH = 11.0 (3) pH = 1.60
    (4) pH = 10.17
22. (1) $[H^+] = [OH^-] = 1.0 \times 10^{-7}$
    (2) $[H^+] = 3.2 \times 10^{-4}$, $[OH^-] = 3.1 \times 10^{-11}$
    (3) $[H^+] = 5.6 \times 10^{-10}$, $[OH^-] = 1.8 \times 10^{-5}$
    (4) $[H^+] = 1.0 \times 10^{-1}$, $[OH^-] = 1.0 \times 10^{-13}$
23. (a) 2.89 (b) 1.00
24. (a) 11.11 (b) 13.00
25. (a) 1.3×10^{-5} *M* (b) 1.8×10^{-9} *M*
    (c) 1.8×10^{-9} *M*
26. (a) 7.9×10^{-5} *M* (b) 2.0×10^{-10} *M*
    (c) 2.2×10^{-6} *M*
27. 5×10^{-52}

PROBLEMS

1. 74.30%
2. no precipitate

Chapter 5

EXERCISES

1. 64.8° C
2. −273° C
3. 390° K
4. 9.85 l
5. 2.33×10^{-5} g
6. 1.79 g/l
7. 4.55 g

8. (b)
9. 12.2° K
10. 0.430 atm
11. 3×10^7 molecules
12. 2.3×10^{7}° K
13. 2.69×10^{22} molecules
14. (a) 0.89 moles (b) 2.0 g/l
    (c) 5.4×10^{23} molecules (d) 27 g
15. 4.53 l
16. 300° K, 1.02 moles
17. 0.6 atm
18. (a) 38.8 l (b) 1.27 atm
19. (a) 12.7 atm (b) 586° K
20. 24.6 g/mole
21. (a) CH_2 (b) C_6H_{12}
22. $C_2F_2H_4$
23. 400 torr
24. 0.467 atm
25. 0.560 atm
26. 19%
27. 834 torr
28. 0.610 atm H_2, 0.0440 atm N_2, 0.0280 atm CO_2, 0.0770 atm CH_4
29. 34.1 g
30. (a) 22.9 mole percent (b) 4.98 g
31. (a) 2.50 atm (b) 2.00 atm
32. 118 l
33. 271 ml

PROBLEMS

1. 45.6 cm^3
2. (a) 5×10^4 g (b) 1.1×10^5 g
3. 0.044 moles
4. (a) 0.082061 l-atm/°K-mole
   (b) $P = 22.415 - 0.151\ P$
5. (a) 17.038 g/mole (b) 14.014 g/mole
6. (a) 8.74 atm (b) 8.63 atm
7. (a) 0.403 l (b) 0.404 l
8. $d = PM_W/(RT + BP)$
9. 10.9%
10. Alcoholics Anonymous

Chapter 6

EXERCISES

1. (a) 63 mph (b) 5.29×10^4 cm/sec
2. (a) 1.2×10^7 cm/sec (b) 1.92×10^5 cm/sec
   (c) 1.45×10^4 cm/sec
3. (a) 329 atm (b) 2410° K (c) 1.2×10^4 cal
4. directly proportional to u_{rms}
5. 9850° K
6. 5.17×10^2 m/sec, 4.84×10^2 m/sec, 1.93×10^3 m/sec
7. (a) 6240 joules (b) 1.77×10^5 cm/sec
   (c) 0.0974 g/l
8. $u_{rms}(H_2$ at 27° C$) = 1.99\ u_{rms}(O_2$ at 927° C$)$
9. (a) He (b) same for each (c) same (d) O_2
   (e) He
10. 0.831 g/l
11. (a) F (b) F (c) T (d) F (e) F (f) T (g) T
    (h) T (i) T (j) F (k) T (l) F (m) F (n) T
    (o) F (p) F (q) T

PROBLEMS

1. (b) 63.9 g/mole
2. (a) $u_{rms} = 2.50 \times 10^5$ cm/sec, $\bar{u} = 2.30 \times 10^5$ cm/sec
   $\alpha = 2.04 \times 10^5$ cm/sec (b) 1500° K
3. (a) 3780 cal/mole
   (b) $He < O_2 < CO_2 < BF_3 < CH_4$

Chapter 7

EXERCISES

1. −1.040 cal
2. -4.95×10^3 joules
3. (a) 617 joules (b) 1.30 kjoules
4. $Q = 6.24 \times 10^3$ joules, $W = 0$
5. (a) 1 joule/mole (b) 9832 joule/mole
6. (a) −1.01 kjoules (b) 2.51 kjoules
   (c) 1.50 kjoules
7. (a) -2.27×10^3 joules (b) 5.69×10^3 joules
   (c) 3.42×10^3 joules (d) 5.69×10^3 joules
8. 4.16×10^3 joules
9. −8 kjoules
10. −35.19 kjoules
11. −92.3 kjoules
12. 1.90 kjoules
13. 53.3 kjoules
14. −49 kjoules
15. (a) −2220.6 kjoules (b) −726.5 kjoules
    (c) −871.6 kjoules
16. −33,670 joules
17. 284 kjoules (b) 285 kjoules
18. −657.6 kjoules
19. (a) 107.23 joules/°K (b) 118.78 joules/°K
    (c) 2.92 joules/°K

20. (a) −163.17 joules/°K (b) 19.82 joules/°K
(c) −118.78 joules/°K
21. −197,162 joules/mole
22.

	$\Delta H°$ (kjoules)	$\Delta G°$ (kjoules)
(a)	44.014	8.598
(b)	−196.6	−140.0
(c)	−136.950	−101.01
(d)	−88.09	−14.3
(e)	−275	−346
(f)	−1299.62	−1235.16
(g)	33.7	34.6

23. (a) −688.1 kjoules (b) −597 kjoules
(c) −307 joules/°K
24. (a) 10.0 kjoules (b) −55.0 kjoules
25. (a) −80.0 kjoules (b) −50.0 kjoules
(c) −0.100 kjoules/°K (d) 5.0 joules/°K-mole
(e) spontaneous (f) exothermic
26. (a) −332.30 joules/°K (b) 126.79 joules/°K-mole
(c) spontaneous (d) exothermic (e) product
(f) $\Delta S > 0$ (universe)
27. (a) 326,200 joules (b) 163,100 joules
(c) endothermic (d) not spontaneous
28. $\Delta H° = -118.3$ kjoules, $\Delta G° = -268.5$ kjoules
29. (a) No, $\Delta G° = 73.2$ kjoules
(b) Yes, $\Delta H° = 112.5$ kjoules (c) Less order
(d) Less order, $\Delta S° = 131.7$ joules/°K

2. 1 joule/°K
3. (a) −20.61 joules/°K-mole
(b) surroundings: 21.41 joules/°K-mole, universe: 0.80 joules/°K-mole

Chapter 8

EXERCISES

1. (a) $\Delta S > 0, \Delta G = 0$ (b) $\Delta S < 0, \Delta G < 0$
(c) $\Delta S > 0, \Delta G = 0$
2. (a) red (b) 674° K
3. 2.00 atm
4. 0.32 torr
5. 540 torr
6. 0.142 atm
7. 146 torr
8. 0.5957
9. (a) 1.16×10^6 torr/molal
(b) 6.46×10^7 torr/mole fraction
10. 8.87×10^{-3} g
11. 9.6×10^7 g
12. 83 torr
13. 0.616 moles/kg
14. −5.80° C
15. 256 g/mole, S_8

PROBLEMS

1. (a)

	Path I			Path II		
	INITIAL	INTERMEDIATE	FINAL	INITIAL	INTERMEDIATE	FINAL
P	1.00 atm	1.00 atm	0.250 atm	1.00 atm	0.250 atm	0.250 atm
V	25.0 l	80.0 l	80.0 l	25.0 l	25.0 l	80.0 l
T	305° K	974° K	244° K	305° K	76.1° K	244° K

(b)

	Path I			Path II		
kjoules	(1)	(2)	Total	(1)	(2)	Total
W	−5.56	0	−5.56	0	−1.39	−1.39
Q	13.90	−9.10	4.80	−2.86	3.49	0.63
ΔE	8.34	−9.10	−0.76	−2.86	2.10	−0.76

16. 202 g/mole, $C_{16}H_{10}$
17. Se_2S_6
18. 198 g/mole
19. Se_8
20. (a) $-7.56°$ C-kg/mole (b) 10,500 joules/mole
21. (a) 23.64 torr (b) 100.141° C (c) $-0.512°$ C
22. 42.5%
23. 2%
24. 23.070 torr

PROBLEMS

1. 22 g
2. (b) solid I
3. 100 g/mole
6. 4.1×10^{-21}° C

Chapter 9

EXERCISES

1. 7.7×10^{21}
2. (a) 2.40×10^{33} (b) 1.95×10^{140} (c) 2.75×10^{38}
4. (a) 128.12 kjoules, 29.0 kjoules, 0.332 kjoules/° K
   (b) 1.3×10^{-2} atm
5. 6.77×10^{13} atm
6. 8.3×10^{-4} atm
7. 4.61×10^{-2} atm
8. 52.3 kjoules; 0.140 kjoules/° K (b) 8.4×10^{-2} atm
9. (a) 0.09 (b) 60 kjoules (c) 2.1×10^5 joules, no
10. (a) 1.65×10^{12}, 1.71×10^8
    (b) -1.04×10^5 joules, yes
11. (a) 6.76×10^4 joules (b) 2.49×10^3 joules
    (c) 130 joules/°K
12. 52.7 atm
13. (a) left (b) right (c) left (d) left (e) no effect
    (f) left
14. (a) right (b) left (c) right (d) left (e) left
    (f) right (g) no effect

PROBLEMS

1. (a) 0.108 (b) 0.767 atm
2. (a) 1.50×10^{-23} atm (b) 165 kjoules
3. (a) (1) left (2) no effect (3) left (4) left
   (b) 119.847 kjoules, no
   (c) 172.467 kjoules, shift right
   (d) 3.174×10^{-11} atm (e) P_{co} will decrease
   (f) 66400 joules
4. I (a) no effect (b) decrease II (a) increase
   (b) decrease III (a) increase (b) decrease
   IV (a) decrease (b) decrease V (a) decrease
   (b) increase VI (a) decrease (b) increase

Chapter 10

EXERCISES

1. (a) $\Delta G° = 12.5$ kcal, $\Delta S° = 33.4$ cal/°K
   (b) 0.084 atm
2. (a) 1.74×10^{57} (b) 2.40×10^{-29} atm
   (c) increase T and increase P
3. (a) 80.96 kjoules (b) 6.55×10^{-15}
   (c) 3.74×10^{-7} moles/l
4. 2.25×10^{-3} atm
5. (a) 0.14 atm (b) 0.087 atm
6. 6.4×10^{-22}
7. (a) 0.442 mole H_2 and CO_2,
   0.558 mole H_2O and CO (b) same as for (a)
   (c) 0.88 mole H_2 and CO_2, 1.12 moles H_2O and CO
   (d) 1.72 moles H_2, 2.72 moles CO_2,
   2.28 moles H_2O, 3.28 moles CO
8. $0.10\ T^2$
9. $7.0 \times 10^2/T^2$
10. $[COCl_2] = 0.85$ mole/l, $[Cl_2] = 0.15$ mole/l,
    $[CO] = 1.15$ mole/l
11. 1.61 atm
12. 4.61×10^{-2} atm
13. (a) 0.387 atm (b) 7.88 g
14. (a) 1.4×10^{19} atm (b) oxidation will not occur
15. (a) 6.15×10^{-72} (b) 2.48×10^{-36} atm
16. 2.1×10^{-3}
17. $[H^+] = [F^-] = 5.7 \times 10^{-4}$, $[HF] = 4.3 \times 10^{-4}$
18. $[H^+] = 1.6 \times 10^{-7}$, $[G^-] = 1.0 \times 10^{-7}$
19. 4.8
20. 4×10^{-3}
21. 1×10^{-8}
22. 9.6×10^{-8}
23. 1.3×10^{-5}
24. (a) 3.0×10^{-3} (b) 3.3×10^{-12}
    (c) pH = 11.5, pOH = 2.5
25. 9.2
26. 4.0×10^{-4}
27. 0.71 moles/l
28. 1.1×10^{-9}
29. 8.3
30. 1.5×10^{-9}
31. 6×10^{-4}
32. 11
33. 3.0
34. $[H^+] = 1.1 \times 10^{-3}$, $[F^-] = 0.10$, $[HF] = 0.15$

35. (a) 3.4 (b) 4.7 (c) 8.4
36. 1×10^{-3}
37. 2×10^{-3} moles/l
38. 2.25×10^{-16}
39. 3.2×10^{-11}
40. 1×10^{-10} mole/l
41. (a) 0.075 g (b) 0.0011 g
42. 1.0×10^{-16}
43. 1×10^{-4} g
44. (a) 2.8×10^{-2} (b) 1.8×10^{-9}
45. 10^{-6}

PROBLEMS

1. 0.33 atm
2. 3.2
3. (a) -1.89×10^5 joules (b) 11.3
4. (a) $K = a^2_{C_6H_{12}O_6}/a_{C_{12}H_{22}O_{11}}a_{H_2O}$ (b) $a_{H_2O} = 0.9356$
   (c) $[C_6H_{12}O_6] = 3.45$, $[C_{12}H_{22}O_{11}] = 1.27$
5. 6×10^5 ions
6. (a) 9.2 (b) 11.0
7. 9.2
8. 8.1×10^{-8}
9.

	ml acid added	0	10	50	90	99	100	101	110	150
(a)	pH (strong base)	13.0	12.9	12.5	11.7	10.7	7.0	3.3	2.3	1.7
(b)	pH (weak base)	11.0	10.0	9.0	8.0	7.0	5.1	3.3	2.3	1.7

10. Any mixture of NH_3 and NH_4^+ with $[NH_3]/[NH_4^+] = 1.76$
11. 85 ml
12. 0.34 g
13. (a) 3.7 (b) 3.5 (c) 3.9 (d) 1.0
14. Choose $[HPO_4^{2-}]/[H_2PO_4^-]$; ratio = 0.68
15. $[H^+] = 0.0080$, $[HSO_4^-] = 0.0020$, $[SO_4^{2-}] = 0.0030$
16. 2.02
17. 1×10^{-13}
18. 7×10^{-9}
19. $[S^{2-}] = 1 \times 10^{-12}$, $[HS^-] = 3.5 \times 10^{-4}$
20. $[H_3AsO_4] = 8 \times 10^{-2}$,
    $[H_2AsO_4^-] = [H^+] = 2.1 \times 10^{-2}$
    $[HAsO_4^{2-}] = 1.7 \times 10^{-7}$, $[AsO_4^{3-}] = 2 \times 10^{-17}$,
    $[H^+] = 2.1 \times 10^{-2}$
21. 1.9
22. $[Na^+] = 0.040$, $[S^{2-}] = 0.003$,
    $[HS^-] = [OH^-] = 0.017$
23. (a) The hydrolysis reaction dominates
    (b) pH = 10.1
24. (a) 11.6 (b) 9.7
25. (a) 4.6 (b) 11.0
26. 8.7
27. $[Ag^+] = 7.0 \times 10^{-7}$, $[CN^-] = 2.2 \times 10^{-8}$,
    $[Br^-] = 6.8 \times 10^{-7}$
28. 10 g $PbCl_2$, 3.4×10^{-7} g AgCl
29. 14% $AgNO_2$
30. 9.3×10^{-4}
31. Both CuS and FeS precipitate
32. (a) 1.1×10^{-4} mole/l (b) 2.7×10^{-2} mole/l
33. 4×10^{-6} mole/l
34. (a) 1.8×10^{-4} (b) 1.7×10^{-7}
35. $[Cu^{2+}] = 2.0 \times 10^{-7}$, $[CuCl^+] = 1.2 \times 10^{-4}$,
    $[CuCl_2] = 5.0 \times 10^{-3}$
    $[CuCl_3^-] = 1.5 \times 10^{-2}$, $[CuCl_4^{2-}] = 8.0 \times 10^{-2}$,
    $[Cl^-] = 1.0$
36. (a) 18 mole % (b) 29 mole %

Chapter 11

EXERCISES

1. (a) −0.02 V (b) 0.74 V (c) 1.09 V
2. (a) 0.91 V (b) −0.01 V (c) −1.04 V
3. (a) $Cu + Cu^{2+} + 2\,Cl^- = 2\,CuCl$ (b) 0.164
4. 0.215 V
5. 4.0
6. 5.04
7. 0.46
8. (a) $Cu^+ + e^- = Cu$, $Cu^+ = Cu^{2+} + e^-$
   (b) 1.67×10^6 (c) 1.34×10^{-4}
9. (a) 8.92×10^{40} (b) 6.15×10^{-21}
10. 1.67×10^{-8}
11. 1.08×10^{-6}
12. 6.49×10^{-14}
13. Pb, Cr, Zn
14. Cl_2 and MnO_4^- are examples
15. (a) Cu^{2+}, MnO_4^-, O_2, and Fe^{3+} are examples
    (b) Co^{2+}, Cu^{2+}, I_2, and H_2 are examples
    (c) Cl_2, MnO_4^-, Au^{3+}, H_2O_2, O_3, and F_2 are examples
    (d) H_5IO_6, Ce^{4+}, and F_2 are examples
    (e) H_2S, H_2, Pb, and Sn are examples
    (f) Br^-, Ag, Fe^{2+}, I^-, Cu, and Cd are examples
    (g) Ag, Fe^{2+}, I^-, and Cu are examples
    (h) Ag, Hg, Fe^{2+}, and I^- are examples
16. (a) no reaction
    (b) $Ca(s) + 2\,H_2O = Ca(OH)_2(s) + H_2(g)$
    (c) no reaction

(d) $Hg^{2+} + Cu(s) = Hg(l) + Cu^{2+}$
(e) $HOCl + 2\,Fe^{2+} + H^+ = Cl^- + 2\,Fe^{3+} + H_2O$
(f) $H_2O_2 + H_2SO_3 = H^+ + HSO_4^- + H_2O$

17. (a) $Cl_2 + 2\,Br^- = Br_2 + 2\,Cl^-$
(b) $2\,I^- + 2\,Fe^{3+} = I_2 + 2\,Fe^{2+}$
(c) $Cr_2O_7^{2-} + 6\,Fe^{2+} + 14\,H^+ =$
$2\,Cr^{3+} + 6\,Fe^{3+} + 7\,H_2O$
(d) $H_2(g) + 2\,Ag^+ = 2\,Ag(s) + 2\,H^+$

18. (a) H_2 and O_2 (b) Cu and Br_2
(c) (H_2 and O_2) and (Zn and O_2) about equally likely
(d) H_2 and O_2

19. (a) 1.12×10^3 sec (b) 0.368 g Cu

20. 388 g Pb

21. 4

PROBLEMS

1. (a) +5 (b) 1.50 V (c) 2.1×10^{-4} mole/l

2. (a) 0.459 V (b) (i) 0.500 V (ii) 0.518 V
(iii) 0.629 V (iv) 0.739 V (v) 0.781 V

3. 1.0×10^{-12}

4. 4.6 l H_2O/day

5. 9.5×10^8 amperes

Chapter 12

EXERCISES

1. Rate $= k[O_2][NO]^2$

2. $R_2 = 16\,R_1$

3. (a) Rate $= k[B]^2$ (b) 125 liters/mole-sec
(c) 80 moles/liter-sec

4. (a) Rate $= k[A]^2[B]$ (b) 0.12 liters²/moles²-sec
(c) 1.5×10^{-2} moles/liter-sec

5. (a) Rate $= k[A]^2[C]$ (b) 4 liters²/moles²-min
(c) 60 moles/liter-min

6. (a) 41 minutes
(b) 0.060 moles/liter, 2.3×10^{-12} moles/liter

7. 4.55 hours

8. (a) 9.9×10^{-4} year (b) 2300 years

9. (a) 0.12 hour^{-1} (b) 19 hours

10. Rate $= k_1[A]^2$

11. (a) Rate $= k[A][B]^2$ (b) $k = k_2k_1/k_{-1}$
(c) 2.3 moles/liter

12. (a) $\text{Rate}_r = k_r[C_2H_4]/[C_2H_2]^2$
(b) $\text{Rate}_r = k_r[O_2]^{1/3}[H_2O]^2/[H_2]$
(c) $\text{Rate}_r = k_r[NO_2]^2[O_2]^{1/2}$

13. 13.9

14. (a) Rate $= k[A]^2[B]$ (c) exothermic

PROBLEMS

1. (a) $k = 0.056$ min^{-1}
(b) 1.8×10^{-3} moles/liter-min

2. 2640 sec

3. (a) first order (b) 63 min

4. Rate $= k[H_2][I_2]$, in which $k = k_2k_1/k_{-1}$

5. Rate $= k[HNO_2][NH_4^+]$, in which
$k = k_1k_2k_3/k_{-1}k_{-2}$

6. (a) 1.00×10^5 joules (b) 9.0×10^{-4} sec^{-1}

7. 7.7×10^5 liters²/mole²-sec

Chapter 13

EXERCISES

1.

λ(nm)	ε(ergs)	ν(sec^{-1})
200	9.95×10^{-12}	1.50×10^{15}
500	3.98×10^{-12}	6.00×10^{14}
1000	1.99×10^{-12}	3.00×10^{14}

2. 32.39 cm

3. 340 nm (ultra-violet)

4. (a) −3267.61 kjoules (b) 3.636×10^{-8} g

5. 0.51 Å³

6. (a) 2.6×10^{-24} nm (b) 1.4×10^{-28} nm
(c) 4.0×10^{-2} nm

7. 9.9×10^7 cm/sec

8. 6.5×10^{-13} cm

9. 2.70 cm/sec

10.

n	l	m	type orbital
2	1	1	p
3	2	2	d
3	2	2	d
4	3	3	f

11. (a), (b), and (f) are not possible; (c) and (e) are possible

12.

l	m
7	0, ±1, ±2, ±3
2	0, ±1, ±2
1	0, ±1
0	0

13. $n = 4; l = 3; m = 0, \pm 1, \pm 2, \pm 3; s = \pm\frac{1}{2}$

14. (a) 5 (b) 9 (c) $n = 7; l = 4;$
$m = 0, \pm 1, \pm 2, \pm 3, \pm 4; s = \pm\frac{1}{2}$
15. $1s^2$: $1, 0, 0, \pm\frac{1}{2}$
$2s^2$: $2, 0, 0, \pm\frac{1}{2}$
$2p^6$: $2, 1, 0, \pm\frac{1}{2}$
$2, 1, 1, \pm\frac{1}{2}$
$2, 1, -1, \pm\frac{1}{2}$
16. 82.09 kjoules/mole
17. (a) 1.634×10^{-11} ergs (b) 1.29×10^{-8} ergs

PROBLEMS

1. (a) 7.3 Å (b) 3.7×10^{-3} Å (c) 1.8×10^{-5} Å
2. 1.5×10^4 volts
3. (a) 1.09×10^8 cm/sec
(b) $d_1 = 1.058$ Å, $d_2 = 4.22$ Å, $d_3 = 9.50$ Å
4. (b) 656.2 nm; 486.1 nm; 434.0 nm; 410.1 nm
5. (a) 1025 nm, 2.924×10^{15} sec^{-1}
(b) 656.2 nm, 4.566×10^{14} sec^{-1}
(c) 1875 nm, 1.599×10^{14} sec^{-1}
6. (a) $1s^3, 2s^3, 2p^4$ (b) 19
7. 7.1×10^{-14} erg
8. 241.3 nm
10. (a) 6.2×10^{-2} cm (b) 7.04 *m* (c) 2.3×10^{23} *m*
(d) 2.5×10^{42} kg-m/sec^2

Chapter 14

EXERCISES

1. (a) Ar_2^+ (b) Ar_2^- (c) Ar_2^+ are the more stable
2. (a), (b), (c), (d), and (f) are paramagnetic;
(e) is not
3. (a) Cl_2 (b) S_2 (c) P_2 (d) Si_2^-
(e) P_2 are the more stable
4. $F_2^+ > F_2 > F_2^-$
5. (a) 1 (b) $\frac{1}{2}$ (c) $\frac{1}{2}$ (d) $\frac{1}{2}$ (e) 0
6. (a) 2π and $\frac{1}{2}\sigma$ (b) CN^+ (c) CN^-
7. (a) one σ and $1\frac{1}{2}\pi$ (b) NO (c) NO
8. stability: PS > S_2 > ClS
9. (a) and (c) are paramagnetic; (b) is not
12. (a) *sp* (b) sp^2 (c) sp^3
13. (a) *sp* (b) sp^3
15. (a) sp^2 (b) sp^3 (c) sp^3
16. sp^3, pyramidal
17. linear
18. (a) sp^2, triangular (b) sp^3, bent
(c) sp^3d, triangular
19. (a) sp^3d, linear (b) sp^3d^2, square planar
20. (a) sp^2, bent (b) sp^3, tetrahedral (c) *sp*, linear
(d) sp^2, triangular
21. (a) sp^3, tetrahedral (b) sp^3d, trigonal bipyramid with a pair of electrons (instead of an atom) at one equatorial corner (c) sp^3d^2, square planar
(d) same as (b)
22. (a) tetrahedral (b) see example 14.12 (page 409)
(c) triangular (d) triangular
23. (a) sp^3, tetrahedral (b) sp^3d, trigonal bipyramidal
(c) sp^3d^2, octahedral (d) sp^3d^2, octahedral
24. (a), (b), (d), (e), and (f)
25. (a) and (e)

PROBLEMS

2. (a) 89 kcal, 154 kcal (b) $Cl_2^{2+} > Cl_2^+ > Cl_2$
4. (a) one σ bond (b) 278 nm
7. PCl_4^+, sp^3, tetrahedral; PCl_6^-, sp^3d^2, octahedral;
PCl_5, sp_3d, trigonal bipyramidal
8. (a) sp^3d^3 (b) pentagonal bipyramidal

Chapter 15

EXERCISES

1. 26%
2. 2.56 g/cm^3
3. (a) 4.05 Å (b) 2.86 Å
4. 45.6 cm^3
5. Body centered cubic lattice
6. 0.58 cm^3/mole
7 0.155
8. (a) zinc blende (b) rock salt (c) CsCl (d) CsCl
10. (a) 6 (b) 3
11. (a) 0.28 (b) 0.53
12. AB_3
13. (a) $NaWO_3$ (b) 7.37 g/cm^3
14. Al_2ZnO_4
15. (a) 4 Zn ions and 4 S ions (b) 5.41 Å
16. (a) $r_{Na^+} = 0.940$ Å, $r_{Cl^-} = 1.88$ Å
(b) 30.1%
17. 65.0 Å^3
18. (a) 6.72° (b) 13.6°
19. 3.08 Å
20. −3924 kjoules/mole

PROBLEMS

1. (a) decrease
2. 6.06×10^{23}
3. 0.12
4. 6.022×10^{23}
5. 4.32 g/cm^3

6. (a) 2.565 Å (b) 2.221 Å
7. (a) 2.82 Å (b) 2.82 Å (c) 4.00 Å
8. HF, −10 kcal; HCl, 47 kcal
9. 60¢/box

Chapter 16

EXERCISES

1. (a) Br (b) Ar (c) Cl (d) K
2. (a) Ga (b) Sr (c) Cs (d) Similar sizes
3. (a) Na (b) F^- (c) Ca (d) K^+
4. (a) I^- (b) I^- (c) Se^{2-} (d) Sr^{2+}
5. (a) Na (b) Mg (c) Ne (d) F
6. (a) Ar (b) Cl^- (c) I (d) Ba
7. (a) Li^+ (b) Li^+ (c) He^+ (d) F
8. (a) Sr^{2+} (b) Na^+ (c) F^-
9. For example:
   (a) Cl^-, K^+, Ca^{2+} (b) Br^-, Kr, Rb^+
   (c) Xe, Cs^+, Ba^{2+} (d) Cl^-, Ar, Ca^{2+}
11. (a) SO_2 (b) SiO_2 (c) Al_2O_3 (d) As_2O_3

PROBLEMS

2. (a) 45 cm^3/mole (b) 75 $Å^3$/atom
3. 1.278 Å
4. 1.519 Å
5. (a) 3.145 Å (b) 1.813 Å, 1.332 Å

Chapter 17

PROBLEMS

1. 3.5×10^{-3} g
2. (a) 1.3×10^{-5} mole/liter (b) 8×10^{-5} mole/liter
   (c) 4×10^{-2} mole/liter
3. (a) 1.0×10^{-10} (b) 2×10^{-12} (c) 1.8×10^{-18}
4. 6.7×10^{-2} g
5. 8×10^{-26} mole/liter
6. 16%
7. (a) 1.0×10^{-4} moles/liter
   (b) 1.3×10^{-5} moles/liter
   (c) 1.0×10^{-4} moles/liter
8. 0.081 g
9. 2×10^{-2}%
10. $[Cl^-] = 0.040$, $[Br^-] = 1.1 \times 10^{-4}$,
    $[I^-] = 2.2 \times 10^{-8}$
11. 0.223 V
12. 94 mg AgI, 14 mg AgCl
13. 0.7 g AgCl/100 ml, 0.05 g AgBr/100 ml
14. 4.0×10^{-5} moles/liter

Chapter 18

PROBLEMS

4. (a) 7 (b) 5.1 (c) 1.8
5. 7.11×10^{-7} molecules
6. 5×10^{-3} M

Chapter 19

PROBLEMS

3. $\Delta G^\circ = -130.60$ kcal
   $\Delta H^\circ = -56.46$ kcal
4. 18
5. (a) 1.62 (b) 4.01 (c) 10.09 (d) 12.6 (e) 6.9

Chapter 20

NO ANSWERS GIVEN

Chapter 21

EXERCISES

11. $K = 8.56 \times 10^{40}$, 6.28×10^{-21} M Fe^{3+}

Chapter 23

EXERCISES

1. 4.22×10^{-13} cm, 4.0×10^{-13}%
5. (a) β (b) β^+ or K capture (c) β^+ or K capture
   (d) β (e) α (f) β
7. 0.068 g
8. 44 min
9. (a) 0.00624 min^{-1} (b) 111 min
10. 0.0677 Å
11. 50.95978 amu
12. −1.43 Mev/atom
13. 3910 years
14. 2800 years

PROBLEMS

1. $\Delta H^\circ = 5.4 \times 10^{11}$ joules, $\Delta S^\circ = 22.6$ joules/°K,
   $\Delta G^\circ = 8.1 \times 10^{11}$ joules
3. 1.52×10^9 years
4. 8.71%
5. 70%

Chapter 24

PROBLEMS

1. 1.93×10^{-5} moles

6. diphenyl ether, para-dichlorobenzene, isopropyl alcohol, methyl butyrate, ethyl dipropylamine, acetone, citric acid

7. (a) $CH_3{-}CCl_2{-}CH_3$, $CH_3{-}CH(CH_3){-}CH_2{-}CH_2{-}CH_2{-}CH_3$, $CH_3{-}CH_2{-}CH(CH_3){-}CH(C_2H_5){-}CH_2{-}CH_2{-}CH_2{-}CH_3$,

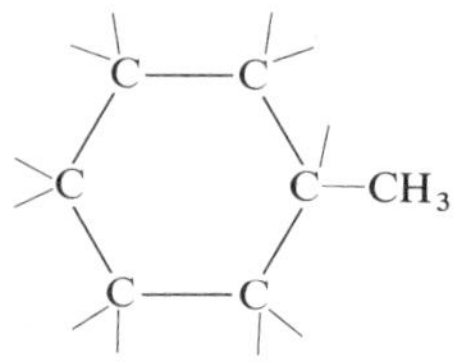

(b) tetrachloromethane, 2,3,4-trimethylpentane, 1-bromo-2-chloroethane, 3-methylpentane

8. (a) $CH_3{-}CH_2{-}CH{=}CH{-}CH_2{-}CH_3$, $CH_2{=}CH{-}CH{=}CH_2$, $CH_3{-}CCl{=}CCl{-}CH_3$,

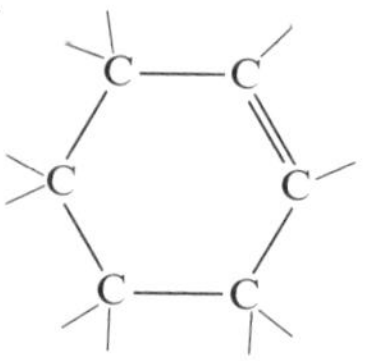

(b) 2-butene, 2-hexene, 3-methyl-1-butene, 1,1,2,2-tetrachloroethene

9. (a) ethanol, 3,3-dimethyl-2-butanol, 2-methyl-2-propanol, 3,5-dimethyl-4-chloro-2-hexanol

(b) $CH_3{-}CH(CH_3){-}CH_2{-}OH$, $CH_3{-}CH_2{-}CH_2{-}CH(OH){-}CH_2{-}CH_3$, $CH_3{-}CH_2{-}CH(OH){-}CH_2{-}CH(OH){-}CH_3$, $CF_3{-}CH_2{-}CH_2{-}OH$

10. (a) $CH_3{-}CH_2{-}CH_2O^- + Na^+ + H_2(g)$

(b) 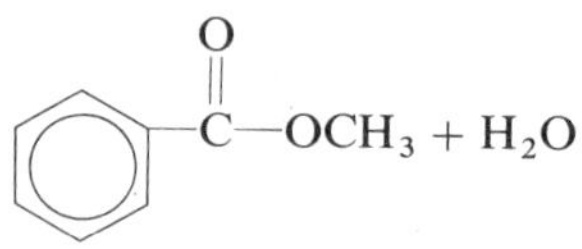

(c) $CH_3{-}CH_2{-}NH_3^+ + H_2O$

(d) $CH_3{-}CH_2{-}CHI{-}CH_3$

(e) 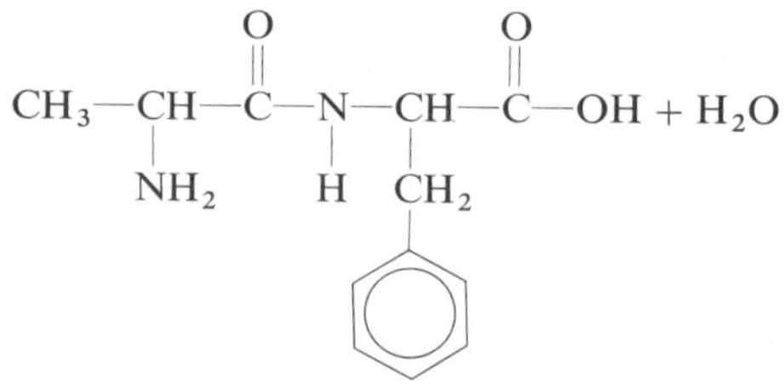

(f) $H{-}C(=O){-}NH_2 + H_2O$

(g) $CH_2Cl{-}CH_2{-}CHBr{-}CH_2Br$

(h) $HC(=O){-}O^- + CH_3OH$

(i) 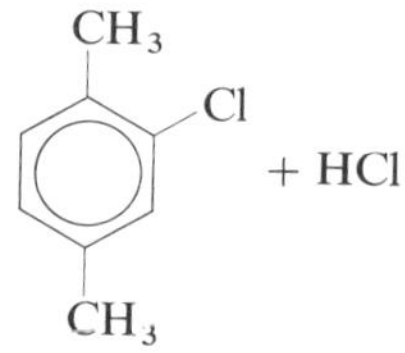

11. 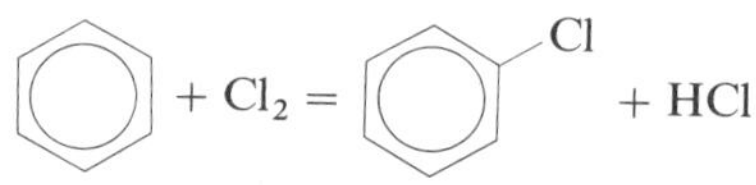

$$2\left[NH_2CH{-}C(=O){-}OH\right] = NH_2{-}CH{-}C(=O){-}NH{-}CH{-}C(=O){-}OH + H_2O$$

$$CH_3{-}CH_2{-}C(=O){-}OH + CH_3OH = CH_3{-}CH_2{-}C(=O){-}OCH_3 + H_2O$$

$$HC(=O){-}OH + CH_3NH_2 = HC(=O){-}N(H)(CH_3) + H_2O$$

$$CH_3{-}CH{=}CH_2 + H_2 \xrightarrow[\text{catalyst}]{\text{Pt}} CH_3{-}CH_2{-}CH_3$$

12. 2-chloropropane, 2-iodobutane, 2-methyl-2-butanol

Index

A

B

D

E

F

G

H

I

J

K

L

M

N

O

P

Q

R

S

T

U

V

W

X

Y

Z

A 0
B 1
C 2
D 3
E 4
F 5
G 6
H 7
I 8
J 9